Property of
FAMILY OF FAIT
LIBRARY
W9-AOO-374

thomson.com
changing the way the world learns℠

SOCIOLOGY

A Global Perspective

Third Edition

Joan Ferrante
Northern Kentucky University

Wadsworth Publishing Company

I(T)P® An International Thomson Publishing Company

Belmont, CA • Albany, NY • Bonn • Boston • Cincinnati • Detroit • Johannesburg • London • Madrid
Melbourne • Mexico City • New York • Paris • Singapore • Tokyo • Toronto • Washington

Publisher: Eve Howad
Developmental Editor: Alan Venable
Assistant Editor: Jennifer Burke
Editorial Assistant: Barbara Yien
Marketing Manager: Chaun Hightower
Project Editor: Jerilyn Emori
Print Buyer: Karen Hunt
Permissions Editor: Robert Kauser
Production Coordinator: Ruth Cottrell
Interior and Cover Designer: Jeanne Calabrese
Copy Editor: Laura Larson
Photo Researcher: Stephen Forsling
Illustrators: Donna Macaluso, Karen Minot, and Dan Swanson
Cover Photographs: © PhotoDisc Inc. 1997
Compositor: New England Typographic Service
Printer: World Color Book Services/Taunton

Printed in the United States of America
1 2 3 4 5 6 7 8 9 10

For more information, contact Wadsworth Publishing Company, 10 Davis Drive, Belmont, CA 94002, or electronically at http://www.thomson.com/wadsworth.html

International Thomson Publishing Europe
Berkshire House 168-173
High Holborn
London, WC1V 7AA, England

Thomas Nelson Australia
102 Dodds Street
South Melbourne 3205
Victoria, Australia

Nelson Canada
1120 Birchmount Road
Scarborough, Ontario
Canada M1K 5G4

International Thomson Publishing GmbH
Königswinterer Strasse 418
53227 Bonn, Germany

International Thomson Editores
Campos Eliseos 385, Piso 7
Col. Polanco
11560 México D.F. México

International Thomson Publishing Asia
221 Henderson Road
#05-10 Henderson Building
Singapore 0315

International Thomson Publishing Japan
Hirakawacho Kyowa Building, 3F
2-2-1 Hirakawacho
Chiyoda-ku, Tokyo 102, Japan

International Thomson Publishing
Southern Africa
Building 18, Constantia Park
240 Old Pretoria Road
Halfway House, 1685 South Africa

Library of Congress Cataloging-in-Publication Data
Ferrante-Wallace, Joan.
Sociology : a global perspective / Joan Ferrante.—3rd ed.
p. cm.
Includes bibliographical references (p. 509) and index.
ISBN 0-534-52551-2
1. Sociology. 2. Social history—Cross-cultural studies.
I. Title.
HM51.F47 1997
301—dc21 97-26146

To my mother,
Annalee Taylor Ferrante

and in memory of my father,
Phillip S. Ferrante
(March 1, 1926–July 8, 1984)

Brief Contents

Contents

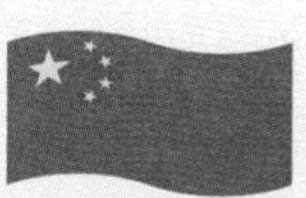

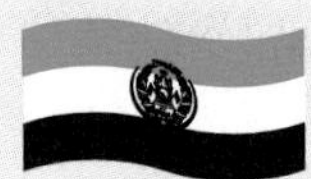

14 Religion
With Emphasis on the Islamic State of Afghanistan 435

15 Social Change
With Emphasis on the Internet 467

Preface

My approach to teaching and writing is driven by a need to show that sociology's concepts and theories constitute more than a list of terms and descriptions to be memorized. My goal is to present the "vocabulary" of the discipline in such a way that readers come to see that sociology offers a powerful tool for thinking about any issue. Ideally, if students grasp the conceptual power of the sociological perspective, they will not think to ask "Why do we have to learn all these terms?" or to say "I memorize the definitions to terms but really don't know that they mean."

The third edition of *Sociology: A Global Perspective* incorporates information about life in countries other than the United States, and it introduces sociology in a coordinated and integrated way such that the final outcome is not simply an encyclopedia-style overview of the field. I have kept these features in the third edition because my teaching experiences have taught me that students are fascinated with comparative examples. Such examples give them hope that "things can be otherwise." In addition, comparative examples lend support to the sociological perspective that encourages us to look below the surface and see the world in which we live in a new light.

My experiences have also taught me that students grasp and remember concepts and theories more easily when they are applied toward answering a central question. Thus the key to my integrative approach lies in defining a central question or theme for each chapter. For example, the concepts and theories covered in Chapter 6 — Social Interaction and the Social Construction or Reality (with emphasis on Zaire, now the Democratic Republic of the Congo) — are framed in such a way that they address a central question: In light of the fact that one of the oldest cases of HIV-infected blood is an unidentified sample taken in 1959 and stored in a Zairean blood bank, how it is that AIDS has grown from a few cases to become a problem in the United States and around the globe?

For the third edition I have worked to ease concerns that this approach is sometimes overwhelming and compromises coverage of the United States. In addition, I have made a number of major changes that I believe strengthen the book's global, comparative, integrative features as well as keeping it current.

New to This Edition

- To ease the concern that the book emphasizes life in other countries to the neglect of the United States, each chapter now begins with a standard heading "Why Focus On . . ." This new feature explains why the focus country is a good choice for broadening readers' understanding of a particular sociological topic. For example, Chapter 2 (Theoretical Perspectives) focuses on Mexico—specifically on the approximately 2,200 U.S. manufacturing plants in Mexico. The three theoretical perspectives are presented as important frameworks for analyzing this fact and for tempering hasty and oversimplistic judgments. In essence, the "why focus" explanations that begin each chapter identify a central issue, question, or theme to which relevant concepts and theories are applied.

  Throughout the chapter readers are reminded through examples, maps, boxes, charts, and tables that studying the focus country (1) offers insights into the workings of and important aspects of U.S. society; (2) illustrates that the lives of people in the United States are intertwined closely with those of people in other countries; (3) enables us to clarify the issues we face as a country; and/or (4) shows the extent to which decisions made by leaders in the United States affect its people and the lives of people in other countries.

- Boxes within chapters have a clear purpose and pedagogical focus. There are one or two boxes in each chapter titled "U.S. in Perspective;" they place some aspect of American life in the context of a global situation, compare it directly or indirectly with life in another country, or address an issue of special concern. For example, Chapter 3 includes a U.S. in Perspective box that considers where students study abroad. Specifically we learn the top ten host countries to U.S. students and the top ten countries that send students to the United States. Statistics show that 45,000 students from Japan studied in the United States in 1994–95 compared to 2,310 U.S. students who studied in Japan. This information leads us to ask several questions, including "Why is this the case? Does it matter that Japan sends approximately 20 times more students to the United States than the United States sends to Japan? What are the consequences of this arrangement?

• Each chapter includes several maps that were created to help readers visualize an important domestic issues such as inequality, interconnections between the United States and other countries, and/or significant issues within the focus country. For example, the opening map for Chapter 7 — Social Organizations with Emphasis on the Multinational Corporation in India— show locations of Union Carbide plants within and outside the United States. Union Carbide is significant to India because in 1989 it agreed to pay $470 to its government as compensation to the victims of the world's worst chemical accident. Many claims have yet to be settled. As with all the maps in the book, this map vividly illustrates the significance of an important issue (i.e., multinational corporations) and helps to frame critical thinking questions.

• Chapter 1 (Sociological Imagination) now considers and gives contemporary applications to three classic works — *Suicide* by Emile Durkheim, *An Invitation to Sociology* by Peter Berger, and *The Souls of Black Folk* by W. E. B. Du Bois.

• Chapter 4 (Culture) gives more emphasis to the dynamic nature of culture in light of ever-increasing interaction on a global scale. Simply consider that as the 33,000 exchange students from South Korea and the 35,000 U.S. troops stationed in South Korea go about their lives, they are exposed to versions of the other's cultures. In Chapter 4 we ask, Is the Korean student still "Korean" after spending four or more years in the United States? Is the U.S. soldier still "American" after spending time in South Korea? And what do we mean by "U.S. culture" or "Korean culture"?

• Chapter 10 incorporates some of the most cutting-edge work related to the idea of race — in particular, that race is a social construction with real consequences. Chapter 10 gives new emphasis to the U.S. government's obsession with maintaining a racial classification system in which everyone fits into clear-cut racial categories.

• The focus country for Chapter 14 (Religion) has changed from Lebanon to Afghanistan. While Lebanon is an important country, its fate is intertwined with that of Israel, the West Bank, and Gaza, which are covered in Chapter 5. The Islamic Republic of Afghanistan is a strategically located Central Asian country that has been at war since 1979, the year the former Soviet Union invaded. The United States, in midst of the Cold War, supported Islamic Afghan factions known as the *mujahidin*, who resisted the invasion in the name of Islam. The county has remained at war even though the Soviets troops left in 1989. An emphasis on various religious (including civil religious) ideologies justifying U.S. and Soviet involvement in Afghanistan and the various Islamic-inspired responses to that involvement challenges us to rethink popular conceptions of Islam.

• Instead of focusing on the post-Cold-War era, Chapter 15 (Social Change) focuses on the Internet (a U.S. invention and U.S.-dominated technology) and its role in promoting and supporting global interdependence. Some people argue that the consequences of this technology are as great as those of the printing press.

• Each chapter ends with a "Summary and Implications" section that highlights important points and suggests wider applications.

• An Internet assignment at the end of each chapter encourages students to apply sociological concepts and theories.

Although this text cannot be all and things to all people, I do believe that I have written a book that introduces sociology in a meaningful and engaging way. We have been fortunate to have received much positive feedback from instructors and students, and this has helped to shape and improve this edition. The third edition has been carefully crafted and designed to highlight the sociological perspective. Still, it is difficult to imagine what this book is like simply from glancing at its title, table of contents, and preface. Ultimately you must decide whether a global, comparative, and integrative approach to sociology, framing life in the United States, meets your needs.

Ancillary Materials

The *Study Guide*, which I also wrote, contains study questions, concept applications, applied research questions, Internet sites, and movie recommendations.

The web site is a wonderful interactive learning tool, which includes on-line quizzes and chapter-by-chapter links to related Internet sites. Visitors to the site are encouraged to participate in on-going discussions on current hot topics in sociology and to take advantage of the additional sociology and Internet-related resources in Virtual Society, the Wadsworth Sociology Resource Center (http://sociology.wadsworth.com).

Acknowledgments

The third edition builds on the efforts of those who helped me with the first and second editions. Four people stand out as particularly influential: Sheryl Fullerton (the editor who signed this book), Serina Beauparlant (the editor who saw the first and second editions through to completion), Maggie Murray (the developmental editor for the first edition), and John Bergez (the developmental editor for the second edition).

For the third edition, I was fortunate to work with Alan Venable as my developmental editor. Alan helped me to define a revision strategy. According to my count, Alan sent me more than 110 e-mail messages (many of which were several pages long), asking me to respond to questions about the writing, photo ideas, and map suggestions. As one measure of my satisfaction, I have elected to save my e-mail from Alan to paper to remind me of the intellectual exchange and collaboration that went into creating the maps and selecting the photos. In addition, I must credit Alan for writing the intellectually stimulating descriptions that accompany each map and photo.

I give special thanks to my editor, Eve Howard, who gave me fresh insights about how to revise *Sociology: A Global Perspective* to better highlight its global, comparative, and integrative pedagogy. Eve encouraged me to give special attention to chapter introductions, boxes, tables, maps, and conclusions because these elements function to enhance and highlight the core coverage of the discipline. Of course, a revision plan depends on thoughtful, constructive, and thorough reviewer critiques. In this regard I wish to extent my deepest appreciation to those who have reviewed this edition:

R. Carlos Cavazos, Texas State Technical College, Harlingen

Paul Ciccantell, Kansas State University

Kevin Early, Oakland University

Kimberly Folse, Southwest Texas State University

Jack Harkins, College of DuPage

Gary R. Lemons, Scottsdale Community College

M. Cathey Maze, Franklin University

Robert Perry, Johnson County Community College

Richard Sweeney, Modesto Junior College

Leslie Wang, The University of Toledo

I also was fortunate to work with a team of professionals whose names are listed in the most unassuming manner on the copyright page of the text. In particular, I have worked directly with Ruth Cottrell (production editor), Stephen Forsling (photo researcher), Laura Larson (copy editor), and Barbara Yien (editorial assistant). Unless you have access to the "inside stuff" as I do, their professional titles alone cannot begin to convey the high-level organizational, conceptual, and detail skills they must possess to turn writing into a book.

For the third edition, I have had the privilege of working with Lindsay Hixson (NKU, Class of 1997) and Angela Vaughn (NKU, Class of 1998), my research assistants on this book. They worked behind the scenes helping me gather hundreds of articles, documents, and statistics required to revise and update this book. Angela and Lindsay ran errands, compiled the bibliography and glossary, and searched for material sometimes with only the vaguest of clues. In addition to being dependable, persistent, thorough, and independent workers, they were just plain fun to work with.

I thank Angela for introducing me to Rubi Simon, who became my typist. As any author knows, the typist can make or break the project. Rubi is the quickest and most accurate typist with whom I have ever worked. Rubi is a great typist because she doesn't just type the words, she thinks about what she is typing and is able to correct mistakes as she goes.

I am grateful to Kim Bo-Kyung and Kevin Kirby for their thoughtful reading and critique of Chapter 4, "Culture: With emphasis on South Korea." I am also grateful to Arture Castillo for taking time to research the prices of various products in Mexico.

I also appreciate the help of librarians at the University of Cincinnati, the Public Library of Cincinnati, and especially Northern Kentucky University. The NKU library faculty handled my requests for information in the most professional manner. For this I thank Nancy Campbell, Allen Ellis, Mary Ellen Elsbernd, Rebecca Kelm, Laura Sullivan, Sharon Taylor, Emily Werrel, and Theresa Wesley. Phil Yannerella and Judy Brueggen in government documents deserve special thanks because they spent considerable time helping me to find the materials I needed.

As always I wish to express my appreciation to my mother, Annalee Taylor Ferrante, who files, clips newspaper articles, and cooks gourmet meals for me (an author and her husband have to eat) when I am busy meeting deadlines (which is almost always). Thanks also go to my father-in-law, Walter D. Wallace, who sends me newspaper clippings several times a week, many of which give me ideas for examples. Thanks also go out to Renee McCafferty and her mother, Jilda Carroll, who also work at clipping newspaper articles from *The New York Times* and the *Los Angeles Times*.

In closing, I acknowledge, as I have done in all editions of this and other books, the tremendous influence of Dr. Horatio C Wood IV, M.D. on my philosophy of education. As time passes my feelings of warmth and gratitude toward him only deepen. Finally, I express my love for the most important person behind this book, my husband and colleague, Robert K. Wallace. I dedicate this book to the memory of my father, Phillip S. Ferrante, and to my mother, Annalee Taylor Ferrante.

SOCIOLOGY

A Global Perspective Third Edition

1 The Sociological Imagination

Opening ceremonies of the 1984 Olympics, Los Angeles. ©David Burnett/ Contract Press Images

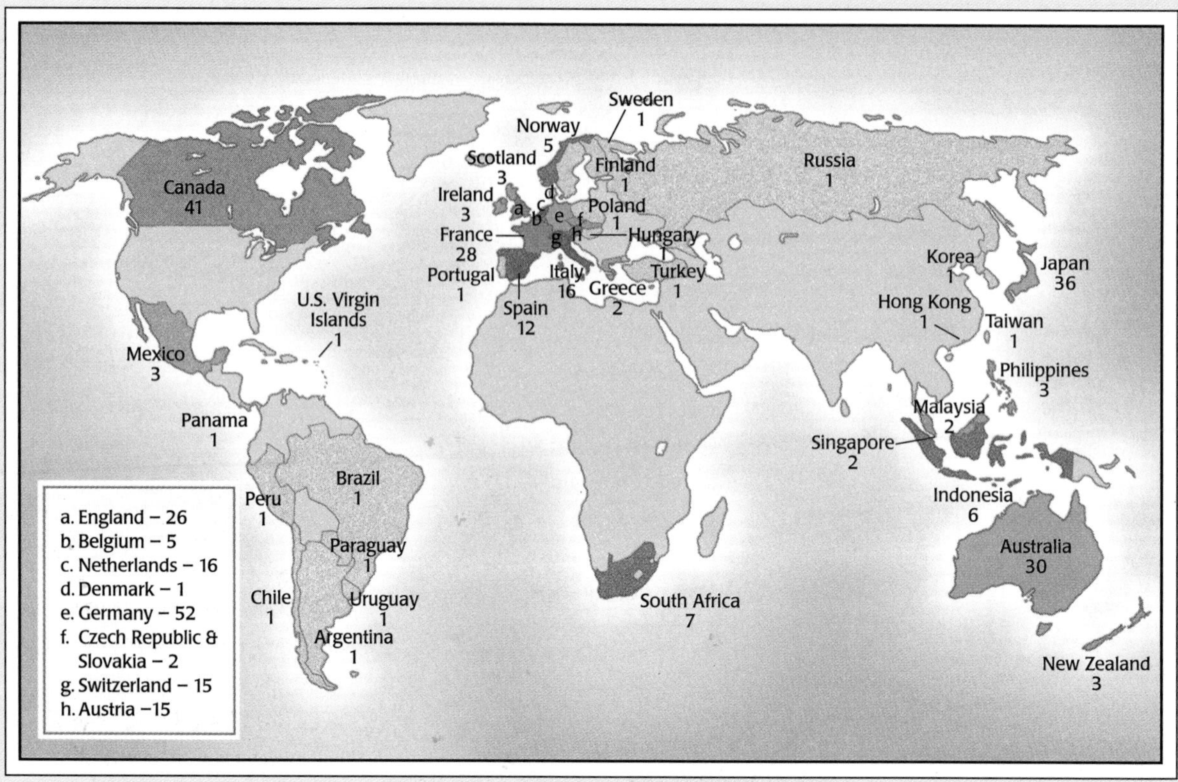

Sources: Michael Erlewine, ed., *All American Music Guide: The Best CDs, Albums, and Tapes* (San Francisco: Miller Freeman, 1994), and Quest Interactive Media 1996 Tour Schedule.

Who Went Where?

This is where popular U.S. musicians performed abroad between November 1, 1996, and January 15, 1997. In those two-and-a-half months, at least 58 major U.S. popular musical groups toured at least 183 cities in 43 countries around the world.

The map says something about how important U.S. culture is in the minds of young people throughout the world today. Even if you do not plan to be a rock star some day, the map may also suggest that your future way of living—your career, your political and economic choices, even your choices in marriage and family—are likely to take a global leap.

Of course, it is not a one-way street. More music from abroad than ever before is for sale at your local CD outlet. In every sense—social, educational, economic, political—cultures are crossing international boundaries.

These are some reasons that this book takes a global approach to the study of human interaction. By seeing societies in a global context, you will understand better the nature and future of your own society.

Why Focus on the United States from a Global Perspective?

In this book, a global perspective is used to introduce sociological concepts and theories. We apply those concepts and theories to a wide range of critical issues and events affecting the United States that cannot be separated from a larger global context. These issues include unemployment and underemployment, the trade deficit, cultural misunderstandings, immigration, the information explosion, AIDS, crime, legal and illegal drug use, inequality, race and ethnic relations, gender inequality, and family "breakdown."

We consider the United States in the context of a global perspective for reasons besides understanding domestic issues. Those reasons have to do with its role in the world. After the Cold War ended, the United States assumed the status of the only military superpower in the world. This status is indicated by its military presence in 140 countries and every region of the world (U.S. Department of Defense 1996). Other evidence can be found in the amount of money it spends on defense. The $284.4 billion the United States spends each year represents 37.9 percent of the approximately $750 billion all governments spend on defense (U.S. Central Intelligence Agency 1995).

In addition, the United States has the most powerful and diverse economy in the world. Approximately 2,000 of the top 8,000 corporations in the world are headquartered in the United States (Center for International Financial Analysis and Research 1992). The significance of this fact is evident when we consider the evolving global economy is "spearheaded by a few hundred corporate giants," many of which have revenues larger than most countries in the world (Barnet and Cavanagh 1994:14). As one measure of the power of the United States, consider that 4.6 percent of the world's population live within its borders, yet its people purchase 18 percent of the world's exports (U.S. Central Intelligence Agency 1995). The United States is also the largest exporter of goods and services, and it is especially known for its entertainment-related exports: films, television programs, videos, radio programming, and recorded music. In this regard, Australian writer Peter Carey (1995) maintains:

> Americans have no understanding of the power that their culture has on everybody else; so even when there are people who are your political enemies, they are people who have taken on and internalized your popular music, your sitcoms, and there's a part in their heart which loves this country and its culture. I don't know whether it's Mary Tyler Moore or Bruce Springsteen or what it is, but these parts of this country are out there in the most unexpected places. (pp. 124–125)

The U.S. Department of Commerce (1993) estimates that films produced by United States–based studios are shown in more than 100 countries and that U.S. television programs have audiences in more than 90 countries. As the executive summary of a Department of Commerce (1993) special report, *Globalization of the Mass Media*, states, "the products of the U.S. electronic mass media industry are vehicles through which ideas, images, and information are dispersed across the United States and throughout the world. As such these mass media can be a powerful agent for political and social change."

On August 30, 1995, the White House designated Miami International Airport [MIA], the second busiest international gateway to the United States, as a federal "reinvention lab." This status allows Miami International to create new systems of passenger and cargo processing shortening the amount of time it takes them to pass through Custom Service checkpoints while simultaneously strengthening law enforcement and safety at the borders. A reinvention lab team[1] will have the flexibility to implement and test a variety of proposals aimed at

- *developing a system to pre-select "high-risk" travelers for special screening;*
- *working with the airline industry to improve quality and quantity of data on passengers;*
- *eliminating unnecessary work processes, forms, duplicative data collection.*

After 18–24 months of testing, Miami International's successes may be implemented in other U.S. ports of entry. (White House Press Release 1995)

[1]The team will include representatives from the Customs Service, Immigration and Naturalization Service, the Department of Agriculture, Animal and Plant Health Inspection Service, the State Department's Passport Agency, the U.S. Fish and Wildlife Service, U.S. and foreign airlines, the Airline Management Council of South Florida, the Dade County Aviation Department, and airport-airline employee representatives.

The White House press release on "reinvention lab status" speaks to a contradiction behind global interdependence, an important historical force that is pushing us into the twenty-first century. In the most general sense, **global interdependence** is a state in which the lives of people around the world are intertwined closely and in which each country's problems—unemployment, underemployment, drug abuse, pollution, inequality, disease — are part of a larger global situation. Yet what we see are simultaneous efforts to protect borders and enforce boundaries while also opening borders and erasing boundaries. In the case of Miami International Airport, the reinvention lab team wishes to weed out drug traffickers, terrorists, smugglers, illegals, and anyone else they define as representing a threat to the United States from those who are entering for purposes defined as legitimate.

This contradiction goes beyond the general area of trade, travel, and tourism. It seems that global interdependence is breaking down boundaries in other areas of life as well. For example, consider the roles medical technology, the Internet, and the production process have played in erasing boundaries. In medical technology, the large numbers of soldiers injured by machine gun shrapnel and bombshells during World War II motivated doctors to create a system of collecting and preserving blood plasma. That technological breakthrough marked the beginning of the movement of blood across national boundaries. Today the United States exports blood to sixty-five countries and is the world's largest exporter of blood and blood products. It also imports blood and blood products from twenty-nine countries. This means that people from different places of the world "interact" with one another through the exchange of blood without ever meeting. At the same time, the movement of blood and blood products across international boundaries is accompanied by efforts to make the blood supply more safe by screening out people who are believed to be high-risk donors.

The Internet is another technology that erases boundaries. The facts of the Internet are these: millions of computers across the world are connected to one another. The Internet is not run by a central agency, office, group, or person. Tim Berners-Lee, the inventor of the World Wide Web, predicts that in the next several years virtually everything that has been written, produced, filmed, photographed, recorded, painted, or otherwise created will be accessible via the Web. The Internet has the potential to put individuals at the center of information in that they can access information quickly and directly, bypassing hierarchies that have traditionally managed and granted access to information.[2] As the

Global interdependence A state in which the lives of people around the world are intertwined closely and each country's problems are part of a larger global situation.

[2]As one example, consider that we rely on the media for most of our information. Under the current arrangements, reporters go to the source of the news, decide what is important, and then relay it to their viewers and readers for a price. With the Internet, users can go to the source of the news and read the press releases organizations put out to the media, thus bypassing the hierarchy that brings us "filtered" news.

The growth of international airports and the relatively cheap cost of rapid international travel are but two signs of how globally oriented our lives and society are becoming. Consider also the many ways we may now communicate with or affect the lives of people halfway around the globe while never leaving home.

©Kevin R. Morris/Tony Stone Images

"boundaries" controlling access to information melt away, a simultaneous movement has arisen among some segments of the population to set boundaries concerning the kind of information available via the Internet.

A final example of growing global interdependence concerns the production process, which increasingly draws labor and raw materials from around the world. We can no longer say with certainty that a product is "made in the U.S.A." or produced by a U.S. company. "One of the ironies of our time is that an American who buys a Ford automobile or an RCA television is likely to get less American workmanship than if he bought a Honda or Matsushita TV" (Reich 1988:78). In the area of product ownership, it may come as no surprise to learn of the popularity of American music around the world, yet it may be startling to learn that Sony and five other foreign multinationals own virtually the entire U.S. recording industry (Hilburn and Philips 1992; *Hoover's Guide to Media Companies* 1996). As the geographic origin of workmanship and corporate ownership become more difficult to determine, some segments, such as the Buy America Foundation, are interested in determining which products are "made in the U.S.A." and/or made by U.S. corporations (Lohr 1995).

Troubles, Issues, and Opportunities: The Sociological Imagination

The individual experience of global interdependence is played out in a virtually endless number of ways, as the following statements from student response papers illustrate. Note that each speaks in some way to dissolving boundaries: (1) corporations expanding their operations across boundaries, requiring their employers to have the language skills to cross boundaries; (2) Internet users overwhelmed by the amount of information they can access with the touch of a few keys; (3) a foreign exchange student visiting the United States and comparing it with the images of it she acquired from the movies.

> The company I work for — EG&G Technologies — has several companies throughout the world. Job listings are always posted for work in many of these foreign countries. In some cases the jobs require not only technical knowledge in a particular field but language skills and knowledge about how business operates (in other words, the business customs) in those countries.

> At my job, I answer many phone inquiries from foreigners. I studied Spanish and thought I knew

enough to get by. I had so many problems communicating with Spanish-speaking people over the phone. I couldn't understand a word they were saying. This was the first time I came into any contact with global interdependence. I realized that I didn't know the language as well as I thought I did.

The construction company my father worked for is based in Ohio. The owners decided there were more job opportunities in Russia. They closed the construction company to go there.

With the Internet you can find and access information in minutes that in the world of paper would have taken weeks. For example, one site gives me access to at least 100 foreign newspapers. If I want to talk to someone I can go to a CyberFriends site and I can select a "friend" from a list that gives background information and E-mail addresses for thousands of people around the world.

An exchange student from Bulgaria lived with us for about six months. She had seen more American movies and videos than I had (and I love movies). I think she was very disappointed in the "real, everyday" experience of the United States as she expected life here to be glamorous and perfect. We joked about how when she returned to Bulgaria that she would whisper in her friends' ears during movies: "I have been there and it is nothing like the movies." (See Table 1.1.)

These comments represent the personal experiences and struggles associated with a major historical force, global interdependence. The discipline of **sociology**, which as we will learn is the systematic study of social interaction, offers us a perspective known as the sociological imagination to help us connect such personal experiences to larger historical forces. In his now-classic book *The Sociological Imagination,* sociologist C. Wright Mills (1959) explains that the key element of the sociological imagination is an ability to make a distinction between troubles and issues. Mills defines **troubles** as personal problems and difficulties that can be explained in terms of individual characteristics such as motivation level, mood, personality, or ability. The resolution of a trouble, if it can be resolved, lies in changing an individual's character or immediate relationships. Mills states that when only one man or woman is unemployed in a city of 100,000, that is his or her personal trouble, and for its relief we properly look to that person's character, skills, and immediate opportunities (that is, "She is lazy," "He has a bad attitude," "She has always been a moody person," "He didn't try very hard in school," "She had the opportunity but didn't take it.").

An **issue**, on the other hand, is a matter that can only be explained by factors outside an individual's control and immediate environment. "But when in a nation of 50 million employees, 15 million men [and women] are unemployed, that is an issue, and we may not hope to find its solution within the range of opportunities open to any one individual. The very structure of opportunities has collapsed" (Mills 1959:9). Unemployment in this situation cannot be explained simply in terms of personal shortcomings, character flaws, or missed opportunities because not enough opportunities are available to accommodate everyone who needs a job. In this case, the **structure of opportunities** — the chances available in a society to achieve a valued goal—have collapsed or no longer exist.

Mills argues that people often feel "trapped" by issues because, even though they respond to their personal situations, it is as if they remain spectators feeling little control over the larger forces affecting their lives. When a war breaks out, for example, an ROTC student becomes a rocket launch technician, a teenager buys and wears a sweatshirt that displays his or her support (or lack of support) for the war, an automobile owner thinks about the role of oil in the war while pumping gas, and a child whose parents are in the military has to live without them for a while — or forever. "The personal problem of war, when it occurs, may be how to survive it or how to die in it with honor; how to make money out of it; how to climb into the higher safety of the military apparatus; or how to contribute to the war's termination" (Mills 1959:9). The issues of war concern its causes: for example, a threat to the resources (such as oil) on which a country depends (see Table 1.2), a threat to a country's major trading partners, or a threat to a country's credibility or image in the world if it does not act. Such threats are significant because they affect the structure of opportunities. In the case of a threat to

Sociology The systematic study of social interaction.

Troubles Personal problems and difficulties that can be explained in terms of individual characteristics such as motivation level, mood, personality, or ability.

Issue A matter that can only be explained by factors outside an individual's control and immediate environment.

Structure of opportunities The chances available in a society to achieve a valued goal.

Table 1.1 The Fifty Top-Grossing U.S. Films

These are the fifty top U.S. films of 1995 ranked in terms of gross worldwide incomes. U.S. films are an important factor in how people in other countries view life in the United States. Based on this list, what impressions do you think they get of us? What impressions might be accurate? Which might be just a little off the mark?

Title/Distributor	Domestic	Foreign	World
1. *The Lion King* (BV)	$298.9	$341.4	$640.3
2. *Forrest Gump* (Par)	298.1	182.8	480.9
3. *True Lies* (Fox/UIP)	146.2	208.1	354.3
4. *The Flintstones* (U)	130.5	211.0	341.5
5. *Mrs. Doubtfire* (Fox)	107.4	202.6	310.0
6. *Schindler's List* (U)	91.1	209.0	300.1
7. *Speed* (Fox)	121.2	161.6	282.8
8. *Four Weddings & a Funeral* (Gramercy/various)	52.7	190.2	242.9
9. *The Mask* (NLC/various)	118.6	93.7	212.3
10. *Clear & Present Danger* (Par)	121.7	66.5	188.2
11. *Maverick* (WB)	101.6	79.0	180.6
12. *Philadelphia* (Sony)	76.9	111.0	177.9
13. *The Client* (WB)	92.1	51.8	143.9
14. *Interview with the Vampire* (WB)	100.0	40.7	140.7
15. *The Specialist* (WB)	55.8	83.8	139.6
16. *The Santa Clause* (BV)	134.5	4.5	139.0
17. *The Pelican Brief* (WB)	48.8	87.2	136.0
18. *Naked Gun 33$\frac{1}{3}$* (Par)	51.1	71.3	122.4
19. *Wolf* (Sony)	65.0	41.1	106.1
20. *When a Man Loves a Woman* (BV)	50.0	54.5	104.5
21. *Beverly Hills Cop III* (Par)	42.6	59.6	102.2
22. *Pulp Fiction* (Mrmx/various)	62.4	33.7	96.1
23. *Cool Runnings* (BV)	7.7	86.0	93.7
24. *The Crow* (Mrmx/Summit)	50.6	43.0	93.6
25. *Ace Ventura* (WB/various)	72.2	16.1	88.3
26. *A Perfect World* (WB)	2.8	77.0	79.8
27. *Sister Act 2* (BV)	21.7	57.8	79.5
28. *Natural Born Killers* (WB)	50.2	24.9	75.1
29. *Stargate* (MGM)	68.2	6.3	74.5
30. *On Deadly Ground* (WB)	38.6	33.6	72.2
31. *Timecop* (U)	44.4	26.5	70.9
32. *Star Trek Generations* (Par)	70.4	—	70.4
33. *Beethoven's 2nd* (U)	26.3	41.1	67.4
34. *I Love Trouble* (BV)	31.1	27.0	61.9
35. *In the Name of the Father* (U)	25.0	36.3	61.3
36. *The Three Musketeers* (BV)	5.9	55.3	61.2
37. *Grumpy Old Men* (WB)	57.9	3.2	61.1
38. *The River Wild* (U)	45.2	14.1	59.3
39. *Free Willy* (WB)		59.2	59.2
40. *Dumb and Dumber* (NLC)	59.1		59.1
41. *The Little Rascals* (U)	51.9	7.0	58.9
42. *Tombstone* (BV/Summit)	39.6	16.7	56.3
43. *Wyatt Earp* (WB)	25.0	30.9	55.9
44. *The Piano* (Mrmx/CB2000)	24.7	30.5	55.2
45. *Demolition Man* (WB)	1.6	52.9	54.5
46. *Angels in the Outfield* (BV)	50.2	3.8	54.0
47. *Intersection* (Par)	20.9	32.5	53.4
48. *Blown Away* (MGM)	30.1	22.6	52.7
49. *Major League II* (WB/various)	30.6	21.3	51.9
50. *D2: The Mighty Ducks* (BV)	45.6	3.7	49.3

Table 1.2 1995 U.S. Reliance on Foreign Supplies of Critical Nonfuel Mineral Materials

U.S. dependence on oil makes the headlines, but the United States also relies on other nonfuel mineral materials. This chart shows the extent to which the United States relied on foreign countries for critical materials and the major uses for which those materials are employed.

Commodity	Percentage	Major Sources (1991–94)	Major Uses
Arsenic	100	China, Chili, Mexico	Wood preservatives, glass manufacturing, agricultural chemicals
Columbium (niobium)	100	Brazil, Canada, Germany	Steelmaking, superalloys
Graphite (natural)	100	Mexico, Canada, China, Madagascar, Brazil	Refractories, brake linings, packings
Manganese	100	South Africa, Gabon, France, Brazil	Steelmaking
Mica, sheet (natural)	100	India, Brazil, Finland, China	Electronic and electrical equipment
Strontium (celestite)	100	Mexico, Germany	Television picture tubes, ferrite magnets, pyrotechnics
Thallium	100	Belgium, Canada, United Kingdom	Rat poisoning
Yttrium	100	China, United Kingdom, Hong Kong, Japan, France	—
Bauxite and alumina	99	Australia, Jamaica, Guinea, Brazil	Aluminum production, abrasives, refractories
Gemstones	98	Israel, India, Belgium, United Kingdom	Jewelry, carvings, gem and mineral collections
Fluorspar	92	China, South Africa, Mexico	Hydrofluoric acid production, steelmaking
Tungsten	87	China, Germany, Bolivia, Peru	Machinery, lamps and lighting
Tin	84	Brazil, Bolivia, Indonesia, China	Cans and containers, electrical, transportation
Cobalt	82	Zambia, Norway, Canada, Zaire, Finland	Aerospace alloys, catalysts, paint driers, magnetic alloys
Tantalum	80	Australia, Germany, Canada, Thailand	Electronic components
Chromium	78	South Africa, Turkey, Zimbabwe, Russia, Finland	Ferroalloys, chemicals, refractories
Potash	74	Canada, Belarus, Germany, Israel, Russia	Fertilizer
Barite	65	China, India, Mexico	Oil and gas well drilling fluids
Iodine	62	Japan, Chile	Animal feed supplements, catalysts, inks, disinfectants
Nickel	61	Canada, Norway, Australia, Dominican Republic	Stainless steel, other alloys
Antimony	60	China, Mexico, South Africa, Hong Kong	Flame retardants, batteries
Stone (dimension)	57	Italy, Spain, India, Canada	Construction
Peat	55	Canada	Horticulture/agriculture
Magnesium compounds	50	China, Canada, Mexico, Greece, Austria	Refractories, agriculture, chemicals

Source: U.S. Department of the Interior, Bureau of Mines (1994).

oil, the opportunity to drive a vehicle at a certain price per gallon is affected.

Likewise, when the CEOs of corporations based in the United States decide that if they are to compete in a global economy, they must downsize their operation or move some of their daily operations to a foreign country or to a lower-wage area of the United States, those affected react in many ways. The laid-off employees may file for unemployment, return to college, open a business, take on third-shift work, choose early retirement, or move to another city to find work. The personal problems connected with this kind of economic

Citizens of the United States are not much accustomed to thinking that events in "far-off" corners of the world are likely to involve them. Yet issues centering thousands of miles beyond U.S. borders can quickly express themselves as personal troubles of our own. A sudden call to duty, for example, may mean at least a disrupted career and a break in one's family life.

©Alexandra Boulat/Sipa Press

restructuring may include finding a way to live on a reduced budget, fighting drowsiness while driving home from third-shift work, or trying to stay cheerful while hunting for a job. Such responses do not change the issue of global interdependence, which is driven largely by an economic system that measures success in terms of profit. Profit is achieved by lowering production costs, hiring employees who will work for lower wages, introducing labor-saving technologies (computerization), and moving productions out of high-wage zones inside or outside the country. Such strategies affect employment opportunities and leave workers vulnerable to unemployment.

Mills asks, "Is it any wonder that ordinary people feel they cannot cope with the larger worlds with which they are so suddenly confronted?" (pp. 4–5). He believes that to gain some sense of control over their lives, people need "a quality of mind that will help them to use information" in a way that they can think about "what is going on in the world and of what may be happening within themselves" (p. 5). Mills equates this quality of mind with the sociological imagination. Those who possess the sociological imagination can view their inner life and human career in terms of opportunity structures and the larger historical forces that affect those structures. The opportunity structure may be related to almost any valued goal—the opportunity to attend a high school prom, the opportunity to become an effective reader, the opportunity to work, the opportunity to marry, the opportunity to live a long life. The sociological imagination empowers us to see how opportunities, whatever they may be, are shaped by the way in which behavior is patterned or organized in society (that is, structured). In the United States, the opportunity to attend a high school prom, for example, is constrained for many students by the fact that they (or school officials) define a date as a social occasion planned in advance between two people. Ideally one partner (the male) is taller than the other (the female). Some school officials and students have recognized how this definition limits students' opportunities to participate and have taken steps to widen the opportunity structure by encouraging and accepting other dating arrangements such as group dates whose members may involve any combination of males and females.

Mills maintains that the payoff for those who possess this quality of mind is that they will understand the forces shaping their lives and thus be able to respond in ways that benefit their lives as well as those of others. They will "come to know that every individual lives, from one generation to the next, in some society; that [people] live out a biography, and that [they live] it out within some historical sequence" (p. 6). Yet, if only by living in a society, people shape that society, however minutely, even as the society shapes them.[3]

In summary, sociology offers a perspective that permits people to define and analyze the structure of opportunities and the historical forces affecting that structure. It does not promise that if people understand this structure, then their personal troubles will be solved easily. The sociological imagination, however, does decrease our chances of responding in inappro-

[3]Mills (1963) maintains that knowing one's place in the larger scheme of things is "In many ways . . . a terrible lesson; in many ways, a magnificent one" (p. 5). This knowledge can be a terrible lesson because we learn that human agony, hatred, selfishness, sadness, tragedy, and hopelessness have no limits. It can be a magnificent lesson because we learn that human dignity, pleasure, love, self-sacrifice, and effort also are unlimited. We will "come to know that the limits of 'human nature' are frighteningly broad" (p. 6).

priate or misguided ways. The ability to see the structure of opportunities becomes most evident to people when they can grasp two interrelated concepts: social relativity and the transformative powers of history.

Social Relativity as a Key to Analyzing Opportunity Structures

Social relativity is the view that ideas, beliefs, and behavior vary according to time and place. People who have lived in more than one culture can teach us much about social relativity. Consider the following excerpts from diaries kept by college students in my classes who were asked to interview foreign students about their adjustment to life in the United States:

> I asked François (from France) what was the biggest adjustment he had to make coming to America. He answered in this way. When he came to America he took a job. The first day on the job he expected to take a lunch break around 12 o'clock. What seemed like several hours passed before his boss told him to take a half-hour lunch break. Lunch is not taken so lightly in France. François was accustomed to a larger meal and to taking as long as two hours to eat.

> I asked Irma if she had any problems readjusting to her native culture when she visited Indonesia. She said that she did have problems. Irma mentioned that in Indonesia everyone she knows drops in on a relative or friend without advance notice. In the United States, on the other hand, she had to learn to call people first. When she was home over Christmas break, she began dialing the phone number of a relative to let the person know that she and her mother were coming over. Her mother asked her why she was behaving so strangely. Irma replied that she wanted to check to see if this was a convenient time. Her mother thought Irma was going crazy. In Indonesia calling before a visit is equivalent to asking someone if they will go out of their way to make special preparations. Indonesians view such a request as rude.

> I asked Hannah what were some of the hardest things for her to get used to since coming to the United States. She explained that in Jordan a girl would never introduce a boy as a boyfriend. If that were to happen, the girl would be disowned not only by the family but by the whole society. In general, the nature of family ties is very different in the two countries. In Jordan people do not always do what they would like. They have a deep respect for family. This respect is so strong that family comes first and individual achievement second. Hannah described how marriage partners are chosen. The first and most important consideration is not the marriage partner per se, but the person's family. If someone was from a bad family, you would not marry the person for two reasons. First, the prevailing assumption is that if the family is bad, so is an individual raised in that family. Second, you would not disgrace the family by marrying a person of questionable character.

These accounts show that ideas about eating, visiting relatives, and dating vary across regions and cultures and that these ideas affect opportunities — in these cases, structured opportunities to relax in the middle of the day, to visit friends and relatives, and to select potential mates. Furthermore, they help us recognize and understand that many of our ideas and behaviors do not originate from any one individual but are products of the environment into which we were born.[4] The concept of social relativity helps us to see that things "'could be otherwise.' That is, social life — the way school, work, families, daily experience are organized — may feel permanent and given, but the arrangements are socially constructed, have changed over time [and place], and can be changed" (Thorne 1993:xxxi).

Transformative Powers of History and Opportunity Structures

According to the concept of the **transformative powers of history**, the most significant historical events have dramatic consequences on people's opportunities. This concept, like social relativity, sees events in the context of time and place. To understand the transformative power of an event, a person must have some

Social relativity The view that ideas, beliefs, and behavior vary according to time and place.

Transformative powers of history The concept that most significant historical events have dramatic consequences on people's opportunities and that sees events in the context of time and place.

[4]Two other examples from student journals follow:

> Alex, from Brazil, observed that the Americans he has encountered seem more individualistic than Brazilians. He thinks this comes from how people are raised. For example, when U.S. parents buy a toy, it is usually for a particular child—that child does not have to share. In Brazil, when parents buy a toy, it belongs to all their children to share.

> Monika, a Swedish-born student, told me that in her country, love is a very strong word. Swedes use the word more selectively than do Americans. People in the U.S. might say, "I love the outdoors." Based on her experiences, if a Swedish person said this, people would literally think something was wrong with the person.

idea of the way people thought or behaved before and after that event.

Consider the tremendous transformative power of the Industrial Revolution of the nineteenth century. Before the Industrial Revolution, most people's lives varied little from one generation to the next. Thus, parents could assume that their children's daily lives would be much like their own and that their children would face the same challenges and problems that they faced. Consequently, societies embarked on long-term projects that extended to subsequent generations. For example, in the Middle Ages people built cathedrals that took several lifetimes to complete, "presumably never doubting that such edifices would be used and appreciated by their great-grandchildren when construction was complete" (Ornstein and Ehrlich 1989:55).[5]

During the Industrial Revolution, however, the pace of change increased to such a degree that parents could no longer assume that their children's lives would be like their own. "Imagine the reactions of an American today if asked to contribute to a building that would take 150 years to finish: 'We don't want to tie up our capital in something with no return for one hundred and fifty years!' 'Won't a new design and construction process make this one obsolete long before it's finished?'" (Ornstein and Ehrlich 1989:55).

The sociological imagination permits those who adopt it to see how biography and history intersect—to understand that even some of the most personal experiences are shaped by time and place and by the transformative powers of history. This type of imagination is characteristic of all the great sociologists, including three of the most influential: Emile Durkheim, Max Weber, and Karl Marx. All had a tremendous impact on the discipline of sociology; all spent much of their professional careers attempting to understand the nature and consequences of one event: the Industrial Revolution.

The Industrial Revolution and the Emergence of Sociology

Between 1492 and 1800, an interdependent world began to emerge. Europeans learned of, conquered, or colonized much of North America, South America, Asia, and Africa and set the tone of international relations for centuries to come. During this time, colonists forced local populations to cultivate and harvest crops and to extract minerals and other raw materials for export to the colonists' home countries. When the Europeans' labor needs could not be met by indigenous populations, they imported slaves from Africa or indentured workers from Asia. In fact, an estimated 11.7 million Africans survived their journey to the "New World"[6] between the mid-fifteenth century and 1870 (Chaliand and Rageau 1995; Holloway 1996; Conrad 1996).

In light of these events, one can argue that the world has been interdependent for at least 500 years. The scale of social and economic interdependence changed dramatically, however, with the Industrial Revolution, which gained momentum in England in about 1850 and soon spread to other European countries and the United States. The Industrial Revolution drew people from even the most remote parts of the world into a process that produced unprecedented quantities of material goods, primarily for the benefit of the colonizing countries.

Between 1880 and 1914, European annexation and colonization expanded to meet Europe's growing demand for raw materials and labor. This period in history is known as the Age of Imperialism. The Age of Imperialism represents the most rapid colonial expansion in history, in which rival European powers (for example, the United Kingdom, France, Germany, Belgium, Portugal, the Netherlands, Italy) competed to secure colonies and spheres of influence in Asia, the Pacific, and especially Africa. By 1914, all of Africa had been divided into European colonies. In that year, 84 percent of the world's land area had been affected by colonization, and an estimated 500 million people lived as members of European colonies (*Random House Encyclopedia* 1990; see Figure 1.1).[7]

[5]This is a "romantic" example of the transformative powers of the Industrial Revolution appropriate for this point in the chapter. As the material that follows will show, the Industrial Revolution was a disruptive, even violent, event that affected people all over the world.

[6]The European name for the Americas.

[7]These figures do not include those areas of the world imperialist powers divided according to "spheres of influence" such as China. Compared with the European countries, the United States and Japan were latecomers to colonialism and thus are labeled as "minor" powers. During the Age of Imperialism, the U.S. government declared Oklahoma and other territories that were home to Native Americans open to white settlement, entered its last major military conflict with Native Americans resisting resettlement, annexed the independent republic of Hawaii, purchased the Virgin Islands from Denmark, and established military governments in the Dominican Republic and Haiti. In addition, the United States fought and won the Spanish-American War and acquired Puerto Rico, Guam, and the Philippines from Spain. The United States also gained temporary control over Cuba, turning it into a protectorate. Japan annexed Korea, Taiwan, Karafuto (the southern half of the former Soviet island Sakhalin) and the southern half of the Pacific Islands.

One fundamental feature of the Industrial Revolution was **mechanization**, the addition of external sources of power such as oil or steam to hand tools and modes of transportation. This innovation turned the spinning wheel into a spinning machine, the hand loom into a power loom, and the blacksmith's hammer into a power hammer. It replaced wind-powered sailboats with steamships and horse-drawn carriages with trains. On a social level, it changed the nature of work and the ways in which people interacted with one another.

The Nature of Work

Before the mechanization brought on by the Industrial Revolution, goods were produced and distributed at a human pace, as illustrated by the effort required to bake bread or to make glass:

- Bakeries produced bread almost entirely by manual labor, the hardest operation being that of preparing the dough, "usually carried on in one dark corner of a cellar, by a man, stripped naked down to the waist, and painfully engaged in extricating his fingers from a gluey mass into which he furiously plunges alternately his clenched fists."
- In glassmaking, everything was done by hand, "the gatherers taking the metal from the furnace at the end of an iron rod, the blower shaping the body of the bottle with his breath, while the maker who finished the bottle off . . . tooled the neck with a light spring-handled pair of tongs. Each bottle was individually made no matter what household, shop, or tavern it was destined for." (Zuboff 1988:37–38)

Workers paid a price for the reduction of the physical requirements necessary to produce goods, because skills and knowledge were intertwined with the physical effort.

> Before mechanization, knowledge was inscribed in the laboring body—in hands, fingertips, wrists, feet, nose, eyes, ears, skin, muscles, shoulders, arms, and legs—as surely as it was inscribed in the brain. It was knowledge filled with intimate detail of materials and ambiance—the color and consistency of metal as it was thrust into a blazing fire, the smooth finish of the clay as it gave up its moisture, the supple feel of the leather as it was beaten and stretched, the strength and delicacy of glass as it was filled with human breath. (Zuboff 1988:40)

Mechanization The addition of external sources of power such as oil or steam to hand tools and modes of transportation.

Lewis Hine's remarkable photograph "The Steamfitter" exemplifies many people's concern about the molding of human beings to machines brought about by the Industrial Revolution.

International Museum of Photography at George Eastman House

Industrialization transformed individual workshops into factories, craftspeople into machine operators, and hand production into machine production. Products previously designed as unique entities and assembled by a few people were now standardized and assembled by many workers, each performing only one function in the overall production process. This division and standardization of labor meant that no one person could say, "I made this; this is a unique product of my labor." The artisans yielded power over the production process to the factory owner because their skills were rendered obsolete by the machines. Now people with little or no skill could do the artisan's work, and at a faster pace.

The Nature of Interaction

Between 1820 and 1860, a series of developments—the railroad, the steamship, gas lighting, running water, central heating, stoves, iceboxes, the telegraph, and mass-circulation newspapers—transformed the ways in which people lived their daily lives. Although all of these developments were important, the railroad and the steamship were perhaps the most crucial. These new modes of travel connected people to one another in reliable, efficient, less time-consuming ways. These inventions caused people to believe they had "annihilated" time and space (Gordon 1989). They permitted people to travel day and night; in rain, snow, or sleet; across smooth and rough terrain.

Before the steam engine train was introduced, a trip by coach between Nashville, Tennessee, and Washington, D.C., took one month. Before the advent of the steamboat, a 150-mile trip on a canal took thirty-two hours (Gordon 1989). An overseas trip from Savannah, Georgia, to Liverpool, England, took months. Each year following the discovery of steam power, transportation became faster and more reliable. In 1819, the year of the first ocean crossing by a steam-driven vessel, the trip took one month. By 1881, the time needed to make that trip had been reduced to seven days (*The New Columbia Encyclopedia* 1975).

Railroads and steamships increased personal mobility as well as the freight traffic between previously remote areas. These modes of transportation facilitated an unprecedented degree of economic interdependence, competition, and upheaval. Now people in one area could be priced out of a livelihood if people in another area could provide goods and materials at a lower price (Gordon 1989).

Industrialization did more than change the nature of work and interactions; it affected virtually every aspect of daily life. "Within a few decades a social order which had existed for centuries vanished, and a new one, familiar in its outline to us in the late twentieth century, appeared" (Lengermann 1974:28). The changes triggered by the Industrial Revolution, and especially by mechanization, are incalculable; they are still taking place.

Sociological Perspectives on Industrialization

Sociology emerged as an effort to understand the dramatic and almost incalculable effects of the Industrial Revolution on human life across the globe. Thus, the perspective sociology offers for understanding human affairs can be traced to those who first tried to make sense of this dramatic event: Karl Marx, Emile Durkheim,[8] and Max Weber.

Karl Marx
German Information Center

Karl Marx (1818–1883)

Karl Marx was born in Germany but spent much of his professional life in London, working and writing in collaboration with Friedrich Engels. Two of Marx and Engels's most influential treatises are *The Communist Manifesto* and *Das Kapital*. *The Communist Manifesto*, a pamphlet issued in 1848, outlines the principles of Communism, among other things. It begins, "A specter is haunting Europe—the spectre of communism," and includes these lines: "The workers have nothing to lose but their chains; they have a whole world to gain. Workers of all countries unite." *Das Kapital*, a massive multivolume work published in 1867, 1885, and 1894, is critical of the capitalist system and predicts its

[8]In 1888, Durkheim began an opening lecture for a course in sociology with the following words: "Charged with the task of teaching a science born only yesterday, one can as yet claim but a small number of principles to be definitely established, it would be rash on my part not to be awed by the difficulties of my task" (p. 43).

defeat by a more humane and more cooperative economic system, socialism.

Marx's vision of the way to build a society as outlined in these treatises has profoundly influenced economic, social, and political life around the world, although Marx would not necessarily have approved of the effect. At his funeral, Engels ([1883] 1993) spoke of Marx

> as the best hated and most calumniated[9] man of his time. Governments, both absolutist and republican, deported him from their territories. Bourgeois, whether conservative or ultra-democratic, vied with one another in heaping slanders upon him. All this he brushed aside as though it were a cobweb, ignoring it, answering only when extreme necessity compelled him. And he died beloved, revered and mourned by millions of revolutionary fellow workers—from the mines of Siberia to California, in all parts of Europe and America—and I make bold to say that, though he may have had many opponents, he had hardly one personal enemy.
>
> His name will endure through the ages, and so also will his work.

According to Marx, the sociologist's task is to analyze and explain conflict, the major force that drives social change. The character of conflict is shaped directly and profoundly by the means of production, the resources (land, tools, equipment, factories, transportation, and labor) essential to the production and distribution of goods and services. Marx viewed every historical period as characterized by a system of production that gave rise to specific types of confrontation between an exploiting class and an exploited class. For Marx, class conflict was the vehicle that propelled people from one historical epoch to another. Over time, free people and slaves, nobles and commoners, barons and serfs, guildmasters and trade workers have confronted one another.

The appearance of machines as a means of production was accompanied by the rise of two distinct classes: the **bourgeoisie**, the owners of the means of production, and the **proletariat**, those who must sell their labor to the bourgeoisie. Marx expressed profound moral outrage over the plight of the proletariat, who at the time of his writings were unable to afford the products of their labor and whose living conditions were deplorable. Marx devoted his life to documenting and understanding the causes and consequences of this inequality, which he connected to a fatal flaw in the organization of production (Lengermann 1974).

Karl Marx believed that an economic system—capitalism—ultimately caused the explosion of technological innovation and the enormous, unprecedented increase in the amount of goods and services produced during the Industrial Revolution. In a capitalist system, profit is the most important measure of success. To maximize profit, the successful entrepreneur reinvests profits to expand consumer markets and obtain technologies that allow products and services to be produced in the most cost-effective way.

The capitalist system is a vehicle of change in that it requires the instruments of production to be revolutionized constantly. Marx believed that capitalism was the first economic system capable of maximizing the immense productive potential of human labor and ingenuity. He also felt, however, that capitalism ignored too many human needs and that too many people could not afford to buy the products of their labor. Marx stated that capitalism already had unleashed "wonders far surpassing Egyptian pyramids, Roman aqueducts, and Gothic cathedrals—[and] expeditions that put in the shade all former Exoduses of nations and crusades" ([1881] 1965:531). He believed that if this economic system were in the right hands—those of socially conscious people motivated not by a desire for profit or self-interest but by an interest in the greatest benefit to society—public wealth would be more than abundant and would be distributed according to need.

Instead, according to Marx, capitalism survived and flourished by sucking the blood of living labor. The drive for profit (which Marx maintained is derived from the labor of those directly involved in the production process) is a "boundless thirst—[a] werewolflike hunger—[that] takes no account of the health and the length of life of the worker unless society forces it to do so" (Marx 1987:142). The thirst for profit "chases the bourgeoisie over the whole surface of the globe" ([1881] 1965:531).

Bourgeoisie The owners of the means of production.

Proletariat Those who must sell their labor to the bourgeoisie.

[9]*Calumniated* means to be a victim of unjust and false accusations intended to damage one's reputation. In an interview with the *Chicago Tribune* in 1879, Marx said if he took the time to deny untrue and unjust statements made and written about him that he "would require a score of secretaries."

Emile Durkheim (1858–1918)

The Frenchman Emile Durkheim observed that as society became industrialized, the nature of the ties that had bound people to one another changed in profound ways. He believed that the sociologist's task is to analyze and explain **solidarity**, the ties that bind people to one another, and the mechanisms through which solidarity is achieved. Just as Marx defined the means of production as a central concern to sociologists, Durkheim regarded solidarity as the essential concern. His preoccupation with this theme is evident throughout his writings.[10] By way of introduction, we explore how this theme drove Durkheim's analysis of suicide.

In his book *Suicide,* Durkheim argued that it is futile to study the immediate circumstances that lead people to kill themselves because an infinite number of such circumstances exists. For example, one person kills herself in the midst of newly acquired wealth, while another kills herself in the lap of poverty. One kills himself because he is unhappy in his marriage and feels trapped, while another kills himself because he has just ended in divorce a marriage that made him unhappy. In one case a soldier ends his or her life after having been punished for a crime he or she did not commit; in another a criminal whose crimes have remained secret and thus unpunished commits suicide. Because almost any personal circumstance can serve as a pretext for suicide, Durkheim concludes that there is no situation that could not serve as an occasion for someone's suicide.

Durkheim also reasoned that there was no central emotional quality common to all suicides. At best one could argue that all suicides are related to disappointments or sorrows. Yet it is impossible to say how great the disappointment or sorrow must be to have such a tragic consequence. We can point to cases in which people resist horrible misfortune while others kill themselves over seemingly slight troubles. Moreover, we can cite examples in which people renounce life at times when it is most comfortable or at times of great achievement.

Given these conceptual difficulties, Durkheim offered a definition of suicide that goes beyond its popular meaning (that is, the act of intentionally killing oneself) and that takes the spotlight off the victim and points it outward and toward the character of the ties that bind (or fail to bind) people to others in the society. He defined *suicide* as the severing of relationships. To make his case, Durkheim argued that every group has a greater or lesser propensity for suicide. The suicide rates for various age, sex, and race groups in the United States, for example, show that some categories of people have a higher suicide rate than do other categories (see Table 1.3). From a sociological point of view, the differences in the rates of suicide can be explained not by pointing to the victim's immediate circumstances but by examining the nature of the social ties that bind or fail to bind someone in that social category to others. For example, everyone who finds themselves in the position of being suddenly unemployed must adjust to life without a job (even if for only a day). That adjustment may entail finding a way to live on a reduced budget, trying to stay cheerful while hunting for a job, or feeling uncomfortable around friends who have a job. According to Durkheim, it is inevitable a certain number of people in this position will succumb to the social pressures associated with being among the unemployed and choose to sever the relationships from which such pressures emanate.

Emile Durkheim

Corbis-Bettmann

[10]We will visit Durkheim's theme of solidarity in Chapter 6 (Social Interaction), Chapter 8 (Deviance), Chapter 13 (Education), and Chapter 14 (Religion).

Solidarity The ties that bind people to one another.

Table 1.3 1994 U.S. Death Rates and Rates per 100,000, Firearm Suicides

If Durkheim were alive today, how might he explain the differences in rates of suicides by firearms?

	Total		Male		Female	
Age Group	**Deaths**	**Rate**	**Deaths**	**Rate**	**Deaths**	**Rate**
00–04	0	0.00	0	0.00	0	0.00
05–09	1	0.01	1	0.01	0	0.00
10–14	187	1.00	139	1.45	48	0.52
15–19	1,377	7.82	1,204	13.32	173	2.02
20–24	1,967	10.72	1,784	19.14	183	2.03
25–29	1,750	9.11	1,531	15.89	219	2.29
30–34	1,782	8.04	1,506	13.63	276	2.48
35–39	1,733	7.90	1,424	13.05	309	2.80
40–44	1,531	7.78	1,275	13.11	256	2.57
45–49	1,351	8.10	1,124	13.74	227	2.67
50–54	1,092	8.28	894	13.95	198	2.92
55–59	902	8.25	794	15.15	108	1.90
60–64	895	8.88	776	16.37	119	2.23
65–69	958	9.60	841	18.68	117	2.14
70–74	1,052	12.04	950	25.07	102	2.06
75–79	991	15.8	920	34.64	71	1.81
80–84	693	15.94	649	41.87	44	1.57
85+	497	14.13	469	47.88	28	1.10
Unknown	6		6			
Total	18,765	7.21	16,287	12.82	2,478	1.86

Source: U.S. Center for Disease Control (1994).

Durkheim identified four problematic relationships: (1) egoistic, (2) altruistic, (3) anomic, and (4) fatalistic. **Egoistic** describes a state in which the ties attaching the individual to others in the society are weak. When individuals are detached from others, they encounter less resistance to suicide. The circumstances of the chronically ill are often characterized by excessive individuation if friends, family, and other acquaintances avoid interacting with the ill person out of fear of upsetting the patient or themselves. **Altruistic**, the opposite of egoistic, describes a state such that individuals have no life of their own and, in fact, strive to blend in with the group to have a sense of being. When such people take their life, it is for the group that they love more than themselves. The classic example is a member of a military unit: the first quality of a soldier is a sense of selflessness. The soldier must be trained to set little value on the self and to sacrifice him- or herself upon being ordered to do so.

Anomic describes a state brought on by dramatic changes in economic circumstances — a recession, a depression, or economic boom.[11] In all cases, something like a declassification occurs that suddenly casts individuals into a lower or higher status than before. When people are cast into a lower status, they must reduce their requirements, restrain their needs, and practice self-control. When individuals are cast into a higher status, they must adjust to increased prosperity, which unleashes aspirations and expands desires such that there are no limits on them. A thirst to acquire goods and services arise that cannot be satisfied.

[11]Durkheim also wrote about conjugal anomic suicide, which occurs when people face divorce.

Egoistic A state in which the ties attaching the individual to others in the society are weak.

Altruistic A state such that individuals have no life of their own and strive to blend in with the group to have a sense of being.

Anomic A state brought on by dramatic changes in economic circumstances.

Fatalistic, the opposite of anomic, describes a state in which there is no hope of change and thus oppressive discipline against which there is no chance of appeal or release. Under such conditions, individuals see their futures as hopelessly blocked. Durkheim (1951) asked, "Do not the suicides of slaves, said to be frequent under certain conditions, belong to this type?" (p. 276).

Max Weber (1864–1920)

The German-born scholar Max Weber has had a monumental influence not only on sociology but also on political science, history, philosophy, economics, and anthropology. Although Weber recognized the significance of economic and material conditions in shaping history, he did not believe, as did Marx, that these conditions were the all-important forces in history. Weber regarded Marx's arguments as one-sided in that they neglected the interplay of economic forces with religious, psychological, social, political, and military pressures (Miller 1963).

According to Weber, the sociologist's task is to analyze and explain the course and the consequences of **social actions** — actions that people take in response to others—with emphasis on the meanings that the involved parties attach to their behavior.

> Certainly Weber's preoccupation with the forces that move people to action was influenced by the fact that he saw the thrust behind sociological curiosity as residing in the endless variety of societies. Everywhere one looks one sees variety. Everywhere one finds [people] behaving differently. . . . Scrutiny of the facts can show endless ways of dealing with the problems of survival, and an infinite wealth of ideas. The sociological problem is to make sense of this variety. (Lengermann 1974:96)

In view of this variety, Weber suggested that sociologists focus on the broad reasons that people pursue goals, whatever those goals may be (for example, to make a profit, to earn a college degree, or to be recognized by others). He believed that social actions could be classified as belonging to one of four important types: (1) **traditional** (a goal is pursued because it was pursued in the past), (2) **affectional** (a goal is pursued in response to an emotion such as revenge, love, or loyalty), (3) **value-rational** (a goal is pursued because it is valued, and it is pursued with no thought of foreseeable consequences and often without consideration of the appropriateness of the means chosen to achieve it), and (4) **instrumental** (a goal is pursued after it has been evaluated in relation to other goals and after thorough consideration of the various means to achieve it) (Abercrombie, Hill, and Turner 1988; Coser 1977; Freund 1968).

Max Weber
AKG Photo/London

An example will help clarify the distinctions among the four types of action. Consider, for example,

Fatalistic A state in which there is no hope of change and thus oppressive discipline against which there is no chance of appeal or release.

Social actions Actions that people take in response to others.

Traditional The type of goal being pursued because it was pursued in the past.

Affectional The type of goal being pursued in response to an emotion such as revenge, love, or loyalty.

Value-rational The type of goal being pursued because it is valued, with no thought of foreseeable consequences and often no consideration of the appropriateness of the means chosen to achieve it.

Instrumental The type of goal being pursued after evaluation in relation to other goals and after thorough consideration of the various means to achieve it.

the goal of earning a college degree. If individuals pursue a college degree because everyone in their family for the past five generations is college educated, the action can be classified as traditional. If they pursue a degree for the love or pure pleasure of learning, the action is affectional. The pursuit of a college education is value-rational if the person decides to attend college because potential employers value and demand a diploma. Sometimes the people who approach college in this way view it only as a ticket to apply for jobs; they lose sight of all but this narrow reason for earning a diploma. As a result, they may simply go through the motions in college and do only what they absolutely must do. They may take the easiest courses, have others write their papers, and skip classes while relying on others to take notes for them.

Instrumental action is more complex. Suppose the individual considers other goals before settling on the goal of earning a college degree. He or she might consider whether to travel and see the world, enlist in the military, or work a few years. Once a goal is chosen, that person might consider the various means to obtain it, including taking out a loan, living at home, enlisting in the army to obtain money for college, getting a job and attending school at night, and so on. The individual also may consider the best approach to learning in the context of an increasingly interdependent world. Such a strategy might include taking classes with the most challenging professors; enrolling in mathematics, science, and foreign language classes; and participating in campus activities. All of these behaviors illustrate instrumental action, a well-thought-out, careful approach to defining and achieving goals. With instrumental action, all angles and possibilities are considered.

Weber maintained that in the presence of industrialization, behavior was less likely to be guided by tradition or emotion and was more likely to be value-rational. (He believed that instrumental action was rare.) Weber was particularly concerned about the value-rational approach; he believed it could lead to disenchantment, a great spiritual void accompanied by a crisis of meaning. Disenchantment occurs when people focus so uncritically on the ways they go about achieving a valued goal that they lose sight of that goal.

How the Discipline of Sociology Evolves

Even today, the ideas of Marx, Durkheim, and Weber continue to influence the discipline of sociology because they have survived the test of time. Over many decades, people from a variety of backgrounds have found the ideas of these three scholars useful for thinking about a wide array of situations. (Ideas lose their usefulness if they cannot explain the situations and events that people consider important.)

Harriet Martineau's Society in America *was an important contribution to sociology, but only recently have her name and ideas been included in sociology textbooks.*
Archive Photos

Yet, despite the importance of Marx, Durkheim, and Weber, we must keep in mind that some voices and experiences have been left out of the record of sociology's history; still others were left out initially but were "discovered" later. Harriet Martineau (1802–1836) is one example. This Englishwoman began writing in 1825,[12] but only in recent years has her name been included in some sociology textbooks. From a sociologi-

[12]In her day, Harriet Martineau was a popular author. One measure of her popularity is that her first book sold more copies in a month than John Stuart Mill's *Principles of Economics* sold in four years. Martineau's success with the general public can be attributed in part to her style: she developed a plot, a setting, and a cast of characters and used them to illustrate economic principles at work in a community (Fletcher 1974).

cal viewpoint, Martineau's most important book is *Society in America.*[13] Sociologists can learn much from the way she conducted her research on the United States. Martineau made it a point to see the country in all its diversity, and she believed it was important to hear "the casual conversation of all kinds of people" ([1837] 1968:54):

> I visited almost every kind of institution. The prisons of Auburn, Philadelphia, and Nashville: the insane and other hospitals of almost every considerable place: the literary and scientific institutions, the factories of the north, the plantations of the south, the farms of the west. I lived in houses that might be called palaces, in log-houses, and in a farm house. I traveled much in wagons, as well as stages; also on horseback, and in some of the best and worst of steamboats. I saw weddings, and christenings; the gatherings of the richer at watering places, and of the humbler at country festivals. I was present at orations, at land sales, and in the slave market. I was in frequent attendance on the Supreme Court and the Senate; and witnessed some of the proceedings of state legislatures.
>
> I traveled among several tribes of Indians; and spent months in the southern States, with Negroes. (pp. 52–53)

Perhaps most useful are the methods that Martineau chose to make sense of all this information. First, she wanted to communicate her observations without expressing her judgments of the United States. Second, she gave a focus to her observations by asking the reader to compare the actual workings of the society with the principles on which the country was founded, thus testing the state of affairs against an ideal standard. Third, with this focus in mind, Martineau asked her reader "to judge for themselves, better than I can for them . . . how far the people of the United States lived up to, or fell below, their own theory" (pp. 48, 50).

Another voice that initially was ignored and later was "discovered" as important to sociology is W. E. B. Du Bois (1868–1963), who began "The After-Thought" to his book *The Souls of Black Folk* ([1903] 1996) with these words: "Hear my cry, O God the Reader; vouchsafe that this my book fall not still-born into the world-wilderness. Let there spring, Gentle One, from out of its leaves vigor of thought and thoughtful deed to reap the harvest wonderful." These words indicate that

W. E. B. Du Bois

Corbis-Bettmann

Du Bois was not confident that his ideas would be heard or taken seriously.

Two sentences in the "Forethought" to *The Souls of Black Folk* ([1903] 1996) summarize Du Bois' preoccupation and conceptual contribution to sociology: "Herein lie buried many things which if read with patience may show the strange meaning of being black here in the dawning of the Twentieth Century. The meaning is not without interest to you, Gentle Reader; for the problem of the Twentieth Century is the problem of the color-line." The strange meaning of being black in America includes a double consciousness that Du Bois defined as "this sense of always looking at one's self through the eyes of others, of measuring one's soul by the tape of a world that looks on in amused contempt and pity." The double consciousness includes a sense of two-ness: "an American, a Negro; two souls, two thoughts, two unreconciled strivings; two warring ideals in one dark body, whose dogged strength alone keeps it from being torn asunder."

Du Bois' preoccupation with the "strange meaning of being black" was no doubt affected by the fact that his father was a Haitian of French and African descent and his mother was an American of Dutch and African descent (Lewis 1993). Historically in the United States, a "black" person is a black even when their parents are

[13]Martineau receives the most credit for introducing sociology to England; she translated Auguste Comte's six-volume work *Positive Philosophy* into English. Many people consider Comte to be the father of sociology (Terry 1983; Webb 1960).

of different "races."[14] To accept this idea, we act as if white and blacks do not marry and/or produce offspring, and we act as if one parent, the "black" one, contributes a disproportionate amount of genetic material, so large a genetic contribution that it negates the genetic contribution of the other parent. In *The Philadelphia Negro: A Social Study,* Du Bois ([1899] 1996) wrote about popular ideas of race and reality. Ironically, almost 100 years after this book was published, the flaws of the U.S. system of racial classification, which rests on the assumption that people can be classified into clear-cut racial categories, are just now becoming a standard topic in many sociology texts. Yet Du Bois documented that blacks and whites married and paired off despite laws prohibiting marriage and that they did have children. He reminded us that race amalgamation took place "largely under the institution of slavery and for the most part, though not wholly, outside the bonds of legal marriage" ([1899] 1996:359). It is this kind of data, which Du Bois painstakingly collected, that is used today to discredit the idea of race as a valid way of categorizing humanity and to remind people that "multiracial" people is not a recent phenomenon.

Ignoring some people's ideas not only weakens the discipline; it also does a disservice to those who have been ignored. In *The Mismeasure of Man,* Stephen Jay Gould writes about the agony of being denied the opportunity to participate: "We pass through this world but once. Few tragedies can be more extensive than the stunting of life, few injustices deeper than the denial of an opportunity to strive or even to hope, by a limit imposed from without, but falsely identified as lying within" (1981: 28–29).

Gould's statements can be clarified by the case of Josh Gibson, an African-American baseball player whose talents can be equated to those of Babe Ruth. Legend has it that Gibson hit a thousand home runs while in the Negro Leagues, as they were known then —leagues for players of color who were not permitted to play in "organized baseball." (Baseball was integrated with the arrival of Jackie Robinson in 1947.) Spectators said that Gibson's home runs were "quick smashing blows that flew off the bat and rushed out of the stadium" (Charyn 1978:41). In 1943, Gibson suffered a series of nervous breakdowns and was institutionalized at St. Elizabeth's Hospital in Washington, D.C., leaving to play baseball on the weekends. He began to hear voices; his teammates found him "sitting alone engaged in a conversation with [Yankee star] Joe DiMaggio," a man he was never permitted to play with or against: "'C'mon, Joe, talk to me, why don't you talk to me? . . . Hey, Joe DiMaggio, it's me, you know me. Why don't you answer me? Huh, Joe? Huh? Why not?'" (p. 41). The case of Josh Gibson reminds us that people's physical characteristics (race, gender, age) cannot be the criteria for judging the worth of their ideas and contributions.

Today the discipline of sociology owes much to Marx, Durkheim, Weber, Martineau, and Du Bois. Although they wrote in the nineteenth and early twentieth centuries, their observations are still relevant today. In fact, many insights into the character of contemporary society can be gained by reading their writings, because those who witness and adjust to a significant event are intensely familiar with its consequences in daily life. Because most of us living today know only an industrialized life, we lack the insights that come from living in transitional times. Consequently, the words of witnesses can be revealing, as the following anecdote suggests.

Recently a scientist was interviewed on the radio. He maintained that scientists were close to understanding the mechanisms that govern the aging process and that people might soon be able to live to be 150 years old. If the aging mechanisms in fact are controlled, the first people to witness the change will have to make the greatest adjustment. In contrast, people born after this discovery is made will know only a life in which they can expect to live 150 years. If these postdiscovery humans are curious, they may wish to understand how living to age 150 shapes their lives. To learn this, they will have to look to those who recorded life before the change and who described their adjustments to the so-called advancement.

So it is with industrialization: to understand how it shapes human life, we can look to some of the early sociologists, who laid the foundations of the discipline of sociology. Because the Industrial Revolution (and all that it encompassed) was so important in the profes-

[14]As one of my students wrote, "I can't be anything but what my skin color tells people I am. I am black because I look black. It does not matter that my family has a complicated biological heritage. One of my great-great-grandmothers looked white but she was of French and African-American descent. Another great-great-grandmother looked Indian but she was three-fourths Cherokee and one quarter black. My great-grandfather looked white but his sister was so black she looked purple. My coloring is a middle shade of brown, but I have picked up a lot of red tones in my hair from my Indian heritage. My family is a good example of how classifying people according to skin color is ridiculous." In the United States, Tiger Woods is a "black" golfer and Colin Powell is a "black" general. We ignore the fact that Woods is a mixture of Chinese, Native American, Thai, White, and African. We ignore that Powell's ancestry is African, English, Irish, Scottish, Jewish, and Arawak Indian (Gates 1995; Page 1996).

In spite of his great athletic abilities, Josh Gibson ("the black Babe Ruth") could play professional baseball only in the Negro Leagues. Gibson was profoundly troubled by the fact that he never got to play with or against a player like Joe DiMaggio.

UPI/Corbis-Bettmann

sional and personal lives of the early sociologists, the sociological perspective may be regarded as a "necessary tool for analyzing social life and the recurring issues that confront human beings caught up in the intensity and uncertainty of the modern era" (Boden, Giddens, and Molotch 1990:B1). The insights of early sociologists teach us that "The fascination of sociology lies in the fact that its perspective makes us see in a new light the very world in which we have lived all our lives. . . . It can be said that the first wisdom of sociology is this — things are not what they seem" (Berger 1963:21, 23).

What Is Sociology?

In the classic book *Invitation to Sociology,* Peter L. Berger (1963) presents sociology as a form of consciousness or as a perspective that gives us the theories and concepts to look below the surface of popular meanings and interpretations. Sociologists assume that all human activity has several levels of meaning, some of which are hidden from view. Sociologists are most interested in those hidden from view. Berger equates the sociologist with a curious observer walking the neighborhood streets of a large city fascinated with what he or she cannot see taking place behind the building walls. The facades of the buildings offer few clues beyond the architectural tastes of the people who build the structures and who may no longer inhabit them. According to Berger, the wish to penetrate the facade is analogous to the sociological perspective.

Berger offers the following example to illustrate his point. In Western countries, and especially in the United States, we assume that people marry because they are in love. There is the popular belief that love is a violent, irresistible emotion that strikes at random. Upon investigating which people actually marry each other, we find that the so-called lightning shaft of Cupid seems to be guided by considerations of age, height, income, education, race, and so on. Thus, it is not so much the emotion of love by itself that causes us to marry; rather, when certain conditions are met (for example, a person is the right height in relation to another — usually the male is taller), we allow ourselves to "fall in love."

"The sociologist investigating our patterns of 'courtship' and marriage soon discovers a complex web of motives related in many ways to the entire institutional structure within which an individual lives his life — class, career, economic ambition, aspirations of power and prestige" (Berger 1963:35–36). For exam-

Lennon and McCartney sang, "All you need is love," but when people do attach themselves to someone else, many considerations are quietly at work in their minds. Usually, we have checked out numerous social indicators about the other person before we "fall" in love.

©Anne Dowie

ple, people meet potential spouses in the circles within which they move (work, school, church, neighborhoods, and so on). Consideration of class and education are obviously present if someone meets their spouse at a college campus where the tuition is $1,600 per year versus $20,000. "The miracle of love now begins to look somewhat synthetic" (p. 36). This does not mean that sociologists dismiss the emotion of love as irrelevant. Rather, they look beyond the popular meanings and publicly approved interpretations. In summary, the sociological viewpoint requires those who use it to go beyond official, popular, and widely accepted interpretations and look for other levels of reality.

What exactly do sociologists study? Berger argues that the distinctiveness of the sociological perspective lies with its focus on human interaction. Specifically, he points to two key concepts: society and social. Society has some popular meanings that differ from the way sociologists use the term. On the popular level, the word *society* can be used in references to a group of people (as in "Society for the Prevention of Cruelty to Animals") or people endowed with great prestige or privilege (as in "high society"). Sociologists use the term in a more general way. They think of **society** as denoting a large complex of human relationships, or a system of interaction. The word *large* cannot be assigned a number. Sociologists may study a *society* that includes millions of people (say, "U.S. society"), but they may also study much smaller entities (say, a *society* of athletes known as a basketball team). From a sociological point of view, three people studying together for an exam do qualify, as does an exchange student meeting his host family in an airport terminal for the first time. What constitutes a society, then, cannot be decided on numerical grounds. It instead applies to a complex of relationships among people sufficiently distinct so that it can be analyzed in terms of its social characteristics.

Society A large complex of human relationships; a system of interaction.

Social In sociology, a quality of interaction in which the involved parties are affected by each other whether or not they are in each other's physical presence.

The second concept distinguishing sociology as a discipline is **social.** In popular speech, *social* can refer to the informal quality of a certain gathering ("This is a *social* event, not a business meeting"), an altruistic attitude ("He had a strong sense of *social* responsibility"), or something resulting from contact with people ("a *social* disease"). Sociologists use the term *social*, however, to refer to a quality of interaction in which the involved parties are affected by each other whether or not they are in each other's physical presence. Thus, two strangers passing on a street corner oblivious to one another's presence does not constitute a social relationship. If they acknowledge one another or go out of their way to avoid acknowledging each other, then what transpires between them is social if only in the sense they respond (or act as if they are not responding) to each other with gestures they did not invent or speak to one another in a language they did not create.

Society consists of a complex of such **social interactions**—everyday events in which at least two people communicate[15] and respond to affect one another's behavior and thinking. In the process, the parties involved define, interpret, and attach meaning to the encounter. The preoccupation with social interactions is the cornerstone of the discipline of sociology. Definitions of sociology posted on the Internet by various sociology departments reflect this preoccupation:

> Sociology is the study of the many groups that humans form: families, communities, and states, as well as a variety of social, religious, political, business, and other organizations that exist in society. Sociologists study the behavior and interaction of these groups, their origin and growth, and the influences of group activities on individual members. (Pennsylvania State University 1997)

> Sociology is the systematic study of social relations. The discipline is concerned with the everyday life of people and the forms of collective action common to different social groups. The focus of sociology's concern is not only on observing the everyday patterns of human interaction, but also on understanding the meaning such interaction has for those who engage in it. (Bates College 1997)

> Sociology is the study of life in society. Society may be as small as your family and as inclusive as Canadian or North American society. Sociologists attempt to understand social interaction and the principles by which society is organized. (University of Waterloo 1997)

> Sociologists study the interplay of social interaction, human experience, and more durable institutions. Sociology deals with the full range of social organization — from small groups to towns, communities, and nations — and all realms of social interaction, from economy to polity to culture. Its subject matter can be a small circle of friends, large organizations, governments, or even cross-cultural influences and the global system of capitalism. (Princeton University 1997)

The Importance of a Global Perspective

Marx, Durkheim, Weber, Martineau, and Dubois were wide-ranging and comparative in their outlook; they did not limit their observations to a single academic discipline, a single period in history, or a single society. They were particularly interested in the transformative powers of history, and they located the issues they studied according to time and place. All five lived at a time when Europe was colonizing much of Asia and Africa; when Europeans were migrating to the United States, Canada, South Africa, Australia, New Zealand, and South America; and when slaves and indentured servants were moving to areas to fill demands for cheap labor. Sociologist Patricia M. Lengermann believes that European expansion and movement of peoples had significant consequences for the discipline of sociology. She writes:

> Explorers, traders, missionaries, administrators, and anthropologists recorded and reported more or less accurately the details of life in the multitudes of new social groupings which they encountered. Westerners were deluged with the flood of ethnographic information. Never had man more evidence of the variety of answers which his species could produce in response to the problems of living. This knowledge was built into the foundations of sociology—indeed, one impulse behind the emergence of the field must surely have been Western man's need to interpret this evidence of cultural variation. (1974:37)

In part because of the specialization of knowledge into distinct disciplines, many sociologists abandoned the historical, comparative, and global vision of the classic sociologists. Today the great majority of sociolo-

[15]The content of that communication may include messages of conflict, hostility, sexual attraction, friendship, loyalty, or the intent to sell or purchase goods or service.

Social interactions Everyday events in which at least two people communicate and respond to affect one another's behavior and thinking.

gists in the United States focus disproportionately on contemporary life in the United States, providing limited discussion about life in other countries and about the role of the United States in the world. Sociologists, however, are not unique in their neglect of cultural and historical forces; such neglect is a shortcoming of most American educators. At all levels, education in the United States is structured in such a way that foreign countries, especially non-Western countries, are disregarded entirely. In addition, the United States is one of the few countries in the world in which students are not required to learn a second language.

This book represents a return to a vision of sociology that reaches beyond the borders of the United States and places U.S. domestic concerns in the context of an interconnected world. The most important goal of this text is to cultivate the sociological imagination by (1) introducing readers to essential sociological concepts, perspectives, and topics and (2) demonstrating the value of the sociological perspective for understanding the individual situation and for comprehending a wide array of contemporary issues, especially in the context of the world.

Key Concepts

Use this outline to organize your review of key chapter ideas.

Global interdependence
Sociological imagination
- **Troubles**
- **Issues**
- **Structure of opportunities**
- **Social relativity**
- **Transformative powers of history**

Bourgeoisie
Proletariat
Solidarity
Suicide
- **Egoistic**
- **Altruistic**
- **Anomic**
- **Fatalistic**

Social action
- **Traditional**
- **Affectional**
- **Value-rational**
  - **Disenchantment**
- **Instrumental**

Double consciousness
Sociology
- **Society**
- **Social**

internet assignment

The following is a list of international, national, and regional sociological associations. Although each posts different information, you will frequently find information on student competitions, association activities and meetings, new publications of interest to sociologists, membership details, and links to sociological resources on the Internet.

- International Sociological Association
  http://www.ucm.es/OTROS/isa/
- American Sociological Association
  http://www.asanet.org/
- Mid South Sociological Association
  http://www.uakron.edu/hefe/mssapage.html
- North Central Sociological Association
  http://miavx1.muohio.edu/~ajjipson/ncsa/htmlx
- Pacific Sociology Association
  http://www.csus.edu/psa/psa.html
- Society for Applied Sociology
  http://www.indiana.edu/~appsoc/
- Society for the Study of Symbolic Interaction (SSSI): Papers of Interest
  http://sun.soci.niu.edu/~sssi/
- Southern Sociological Association
  http://www.MsState.Edu/Org/SSS/sss.html

Browse through several of the sites listed here, following any links that look interesting. Do any of these sites provide information or links that make sociology appear attractive as a field of study? Explain.

BANCO MEXICANO SOMEX
PRIME

Theoretical Perspectives

With Emphasis on U.S. Manufacturing Operations in Mexico

The Functionalist Perspective

The Conflict Perspective

The Symbolic Interactionist Perspective

Financial District, Reforma Avenue, Mexico City. ©Larry Reider/Sipa Press

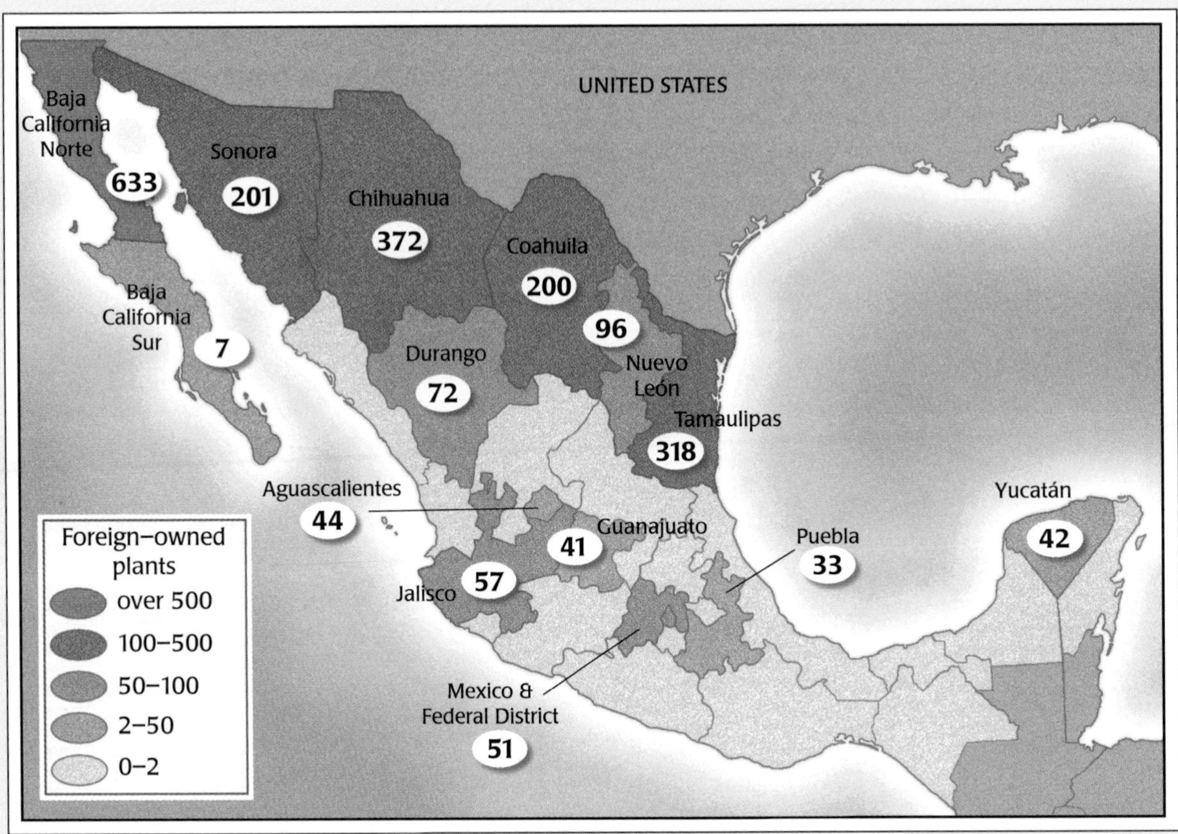

Source: Secretariat of Commerce and Industrial Development, Mexico; *Twin Plant News* (1996c).

The Number of Foreign-Owned Assembly Plants in Mexico

More than 2,200 foreign-owned manufacturing plants operate in Mexico. Most are located along the Mexican side of the U.S.–Mexico border; the rest are located in the interior states of Mexico. Is the presence of so many U.S.-owned factories in Mexico just below the U.S. border good or bad for the people of Mexico? Is it good or bad for the people of the United States? Three basic sociological theories—functional, conflict, and interactionist—provide three ways to examine the issues.

Consider the following facts:

- *Approximately 2,200 foreign-owned manufacturing plants exist in Mexico, with an estimated 710,268 employees. The great majority (over 90 percent) are U.S. owned and are located along the northern border of Mexico.*
- *General Motors is the largest private employer in Mexico, with approximately 65,000 Mexican workers (*Twin Plant News *1996c).*
- *Approximately 3,000 freight trucks cross just the San Diego–Tijuana border each day to deliver equipment, components, and raw materials for assembly and to pick up assembled products (clothing, computers, automobiles) for return to the United States (Krause 1996; Darling and Rotella 1992).*

Why Focus on Mexico?

In this chapter, we pay special attention to Mexico because, although the United States shares a 1,952-mile border with Mexico (which millions have crossed each year to shop and visit vacation spots), most Americans* know little about what binds the two countries together. In truth, few people know about the large number of U.S. corporations with manufacturing operations on the Mexican side of the U.S.–Mexico border or understand their economic significance. In addition, most of us are unaware that Mexico is the United States' third-largest trading partner (after Japan and Canada) or that the United States is Mexico's largest trading partner. In 1995, Mexico purchased 73 percent of its imports from the United States (U.S. Department of State 1996).

Lacking this knowledge, many Americans interpret the presence of U.S. corporations in Mexico as an economic threat to the United States. They argue that Mexico, a major source of cheap labor, lures companies out of the United States, takes away jobs, and drives down U.S. workers' wages. Based on this interpretation, the natural impulse is to revoke trade agreements such as the North American Free Trade Agreement (see "What Is NAFTA?") or to pass laws to stop the flow of Mexicans into the United States, to prevent U.S. companies from investing in Mexico, and to limit the flow of Mexican-assembled products into the United States.

In light of this narrow, yet very common, interpretation of events, the three major sociological perspectives offer us quite different ways to assess the relationship between Mexico and the United States and to put what we hear, read, or experience about Mexico into a broader context. In addition, the three perspectives offer us some constructive ways to think about not just this relationship but also a larger global trend: the transfer of labor-intensive manufacturing or assembly operations out of Western European countries, the United States, and Japan into low-wage, labor-abundant countries such as Mexico, China, Malaysia, Singapore, the Philippines, Taiwan, South Korea, Haiti, Brazil, Dominican Republic, Bangladesh, Honduras, and Colombia. In this chapter, we use U.S. manufacturing operations in Mexico as a vehicle for thinking about this global trend.

*In the broadest sense, the word *Americans* applies to people living in North, South, and Central America, and all of the islands in the surrounding waters. Yet *American* is almost always used in reference to the people, culture, government, history, economy, or other activities of the United States of America. Some critics argue that using *Americans* as synonymous with residents of the United States is an example of "American" arrogance. However, there is no succinct word such as Mexican, Canadian, Brazilian, Central American, South American, or North American that can be applied easily (*United Statesian* is not an accepted word and it is awkward). For this reason I use the word *American* as synonymous with "pertaining to the United States."

In the most general sense, a **theory** is a framework that can be used to comprehend and explain events. Theories in every science serve to organize and explain events that can be observed through the senses (sight, taste, touch, hearing, and smell). A **sociological theory** is a set of principles and definitions that tell how societies operate and how people relate to one another.

Three theories dominate the discipline of sociology: the functionalist, the conflict, and the symbolic interactionist perspectives. This chapter outlines the basic assumptions and definitions of each of these theories, showing how each one offers a distinct framework that can be used to interpret any event.

Each perspective offers a central question to help guide our thinking about a particular event and a vocabulary for answering the central question. Keep in mind that one theoretical perspective is not necessarily better than the others. Each simply gives a different angle from which to analyze an event. We turn first to an overview of the functionalist perspective. The central questions functionalists ask are, Why does a particular arrangement exist? and What are the consequences of this arrangement for society? Following this overview, we use the functionalist perspective to explain the facts listed earlier.

The Functionalist Perspective

Functionalists focus on questions of order and stability in society. They define society as a system of interrelated, interdependent parts. The parts are connected so closely that each one affects all the others as well as the system as a whole. Functionalists consider a **function** to be the contribution of a part to the larger system and its effect on other parts in the system.

To illustrate this complex concept, early functionalists used biological analogies. For example, the human body is made up of parts that include bones, cartilage, ligaments, muscles, a brain, a spinal cord, nerves, hormones, blood, blood vessels, a heart, a spleen, kidneys, lungs, and chemicals, all of which work together in impressive harmony. Each part functions in a unique way to maintain the entire body, but it cannot be separated from other parts, which it affects and which in turn help it function. Consider eyelids: when they blink, they work in conjunction with tear fluid, tear ducts, the nasal cavity, and the brain to keep the corneas (the transparent coat over the eyes) from drying out and clouding over. Furthermore some scientists speculate that blinking functions in some way to activate the brain, which controls the body and performs thought processes (Rose 1988).

Functionalists see society, like the human body, as made up of parts, such as food-growing techniques, sports, medicine, bodily adornments, funeral rites, greetings, religious rituals, laws, language, modes of transportation, appliances, tools, and beliefs. Like the various body parts, each of the social parts functions to maintain a larger system. For example, one of the many functions of the U.S. educational system is to transmit knowledge and skills from one generation to the next. Thus, for instance, young people learn about previous generations' solutions for meeting various environmental challenges and are not forced to start from scratch. Consider, too, sports teams, whether they be Little League, grade school, high school, college, city, Olympic, or professional teams. Sports teams function to unify people who are often extremely different from one another economically, culturally, linguistically, politically, religiously, and so on. Loyalty to a sports team transcends individual differences and helps bind people to the school, a company, a city, or a country associated with it.

In the most controversial form of this perspective, functionalists argue that all parts of society, even those that seem not to serve a purpose, contribute to the system's stability. Functionalists maintain that a part would cease to exist if it did not serve some function. Thus, for example, such phenomena as poverty, illegal immigration, and the transfer of labor from the United States into Mexico contribute to the stability of the social system.

Herbert Gans (1972) argues this point in his classic analysis of the functions of poverty. To his own question, "Why does poverty exist?" he answered that poverty performs at least fifteen functions, several of which are described here:

- The poor have no choice but to do the unskilled, dangerous, temporary, dead-end, undignified, menial work of society at low pay. Hospitals, ho-

Theory A framework that can be used to comprehend and explain events.

Sociological theory A set of principles and definitions that tell how societies operate and how people relate to one another.

Function The contribution of a part to the larger system and its effect on other parts in the system.

What Is NAFTA?

Many Americans have fiercely opposed NAFTA because they believe that companies will take advantage of the agreement to export U.S. jobs to Mexico.

©Louis DeMatteis

The North American Free Trade Agreement (NAFTA) is a 2,000-page document written with the purpose of eventually eliminating all trade barriers between the United States, Canada, and Mexico. The details of the agreement have been worked out and approved by the U.S. Congress. The agreement went into effect January 1, 1994. NAFTA sets guidelines for the gradual elimination of tariffs and duties on goods that are traded between the three countries. In its original form, it calls for the immediate elimination of many tariffs while others would be reduced over a period of up to fifteen years. For example, half of U.S. farm goods exported to Mexico will immediately become duty-free. Throughout North America, all other tariffs on agricultural goods will be phased out over the next fifteen years. Within five years of implementation of the agreement, 65 percent of U.S. goods will gain duty-free status (White and Maier 1992).

In general, the goals of the agreement are to eliminate or decrease trade barriers in order to increase trade between the three countries. Designers of the pact believe that this increase will force the three economies to specialize in whatever they do best and, therefore, become more efficient. The increased efficiency, they theorize, will lead to lower consumer prices, the creation of new jobs, and an increased standard of living for the people of all three countries.

However, one should not conclude that NAFTA will bring about changes without causing problems to some segments of the three countries' economies. After all, the North American Free Trade Agreement has created the largest market in the world: 390 million consumers with a total GNP of $6 trillion. It will unite three countries with very diverse cultures and levels of economic development. The United States has a population of 265.6 million and a per capita GNP of $25,850. Mexico, on the other hand, has a population of 95.7 million and a per capita GNP of $7,900. Canada's population is only about a third of Mexico's, at 28.8 million with a per capita GNP of $22,760 (U.S. Central Intelligence Agency 1995).

NAFTA has the potential to "leave virtually no aspect of the economy untouched" (Risen 1992:A1). The significance and strength of the agreement's effects will vary depending on where the individual, corporation, or agency is positioned within the economy of its particular country. The effects the agreement will have on labor and the environment seem to have drawn the most attention. However, there is also a great deal of debate over the effects it will have on industries such as agriculture, automobile, finance, energy, and transportation.

Although some economic analysts predicted that the *maquila* industry would disappear under NAFTA (Lee and Kraul 1993; Essential Organization 1996), NAFTA has had a positive economic impact on most *maquila* operations because "export quotas on originating textile goods and wearing apparel have been eliminated; U.S. Custom duties on originating goods have been decreased or eliminated altogether; and *maquiladoras* are now allowed to sell more goods into the Mexican national market" (Pina 1996:39).

Lee, Patrick, and Chris Kraul. 1993. "Uniqueness of *Maquiladora* Could Fade." *Los Angeles Times,* November 19, p. D1.
Pina, Rudy A. 1996. "Verifying Origin: U.S. Customs Visits." *Twin Plant News,* 11(9): 39–40.
Risen, James. 1992. "U.S., Mexico, Canada Agree to Form Huge Common Market." *Los Angeles Times,* August 13, pp. A1+.
U.S. Central Intelligence Agency. 1995. *World Factbook.* http://www.ooci.gov/cia/publications/95fact/index/html.
White, George, and Andrea Maier. 1992. "A Closer Look at the Trade Agreement." *Los Angeles Times,* August 13, p. A7.

Source: Kevin Steuart, Northern Kentucky University (August 1996); updated February 1997.

Poor people who eat in soup kitchens often eat food that is donated from restaurants and that would otherwise be thrown away.

©Donna Binder/Impact Visuals

tels, restaurants, factories, and farms draw their employees from a large pool of workers who are forced to work at minimum or below-minimum wages. This hiring policy keeps the costs of their services reasonable and increases profits.

- Affluent persons contract out and pay low wages for many time-consuming activities such as housecleaning, yard work, child care, and grocery shopping. This practice gives them time for other, more "important" activities. This function of poverty was brought to national attention in 1993 when President Clinton's first two nominees for U.S. attorney general (Zoe Baird and Kimba Wood) disclosed that they had employed illegal immigrants to care for their children. Nor is this child care arrangement confined to the most prominent and most affluent Americans. Many middle-class and even lower-class Americans make similar arrangements.
- The poor often volunteer for over-the-counter and prescription drug tests. Most new drugs must be tried on healthy subjects to determine potential side effects (for example, rashes, headaches, vomiting, constipation, drowsiness) and appropriate dosages. Money motivates subjects to volunteer. Because payment is relatively low, however, the tests attract a disproportionate share of poor people as subjects.
- The occupations of some middle-class workers—police officers, psychologists, social workers, border patrol guards, and so on—exist to serve the needs or to monitor the behavior of poor people. For example, about 5,700 immigration agents patrol the U.S.–Mexican border (Pear 1996). Similarly, physicians and grocery store owners are reimbursed for serving poor people. The poor receive food stamps and medical cards, and the providers receive cash.
- Poor people purchase goods and services that otherwise would go unused. Day-old bread, used cars, and secondhand clothes are purchased by or donated to the poor. In the realm of services, the labor of many less competent professionals (teachers, doctors, lawyers), who would not be hired in more affluent areas, is absorbed by low-income communities.

Gans outlines the economic usefulness of poverty to show how a part of the society that everyone agrees is problematic and should be eliminated remains intact: it contributes to the stability of the overall system. Based on this reasoning, the economic system would be strained seriously if poverty were completely eliminated; industries, consumers, and occupational groups that benefit from poverty would be forced to adjust.

Although functionalists emphasize how parts contribute to the stability of the system, they also recognize that the system does not remain static: as one part changes, other parts adjust and change in ways that lessen or eliminate strain.

Critique of Functionalism

As you may have realized by now, the functionalist perspective has a number of shortcomings. First, critics argue that the functional theory is by nature conservative: "merely the orientation of the conservative social scientist who would defend the present order of things" (Merton 1967:91). In other words, when functionalists identify how a problematic part of society such as poverty contributes to the system's stability, by definition they are justifying its existence and legitimating the status quo. Critics argue that so-called system stability is more often than not achieved at a cost to some segment of the society such as those who are poor and powerless. Functionalists reject this criticism, claiming that they are simply illustrating why such controversial practices or "parts" continue to exist despite efforts to change or eliminate them.

Second, critics argue that a part may not be functional when it is first introduced. Often practices and inventions find their usefulness only after they have come into existence. When the automobile was invented, for example, there were no paved roads, so the automobile's use was limited. It was not until some time later that it realized its transportation function:

> The availability of cheap energy and an inexpensive mass [-produced] car soon transformed the American landscape. Suddenly there were roads everywhere, paid for, naturally enough, by a gas tax. Towns that had been too small for the railroads were reached now by roads, and farmers could get to once-unattainable markets. Country stores that sat on old rural crossroads and sold every conceivable kind of merchandise were soon replaced by specialized stores, for people could drive off and shop where they wanted to. Families that had necessarily been close and inwardly focused in part because there was nowhere else to go at night became somewhat weaker as family members got in their cars and took off to do whatever they wanted to do. The car stimulated the expansiveness of the American psyche and the rootlessness of the American people; a generation of Americans felt freer than ever to forsake the region and the habits of their parents and strike out on their own. (Halberstam 1986:78–79)

The example of the automobile complicates the functionalist argument that a part exists because it contributes to order and stability in society. The automobile example also suggests that necessity is not always the mother of invention but rather that the invention often is the mother of necessity. In the case of the automobile, it became a necessity after it was invented.

Third, critics assert that the part of the system that fills a function is not necessarily the only way or the most efficient way to meet the function. Although the automobile functions to connect people with one another, it is not an environmentally sound means of doing so. In fact, the widespread use of and reliance on the automobile hinders the development of more environmentally efficient modes of public transportation. Reliance on automobiles also reduces socializing with others, something that public transportation demands and reinforces.

Finally, because the functionalist perspective assumes that a part's function must support the smooth operation of society, it has difficulty accounting for the origins of social conflict or other forms of instability. This assumption leads functionalists to overlook the fact that inventions may have negative consequences for the system or for certain groups within the society.

To address some of this criticism, sociologist Robert K. Merton (1967) introduced a few concepts that help us think about a part's overall effect on the social system and not just its contribution to stability. These concepts are manifest and latent functions and dysfunctions.

Merton's Concepts

Merton distinguished between two types of functions that contribute to the smooth operation of society: manifest functions and latent functions. **Manifest functions** are the intended, recognized, expected, or predictable consequences of a given part of the social system for the whole. **Latent functions** are the unin-

Manifest functions The intended, recognized, expected, or predictable consequences of a given part of the social system for the whole.

Latent functions The unintended, unrecognized, unanticipated, or unpredicted consequences of a given part of the social system for the whole.

tended, unrecognized, unanticipated, or unpredicted consequences.

To illustrate this distinction, consider the manifest and latent functions associated with annual community-wide celebrations such as fireworks displays on the Fourth of July or during concerts in the park. Often corporate sponsors join with the city to mount such events. Three manifest functions readily come to mind: (1) the celebration functions as a marketing and public relations event for the corporate sponsor or sponsors, (2) the event provides an occasion to plan activities with friends, and (3) the celebration unifies the community through a shared experience.

At the same time, however, several unexpected or latent functions are associated with community celebrations. First, such celebrations often give a visible role to public transportation systems as people take buses or ride trains to avoid traffic jams. Second, such events function to break down barriers across neighborhoods. People who do drive find they have to park some distance from the event, often in neighborhoods that they would not otherwise visit. Consequently, after they park, people have the opportunity to walk through neighborhoods and observe life up close instead of at a distance.

Merton also points out that parts of a social system can also have **dysfunctions**; that is, they can sometimes have disruptive consequences to the system or to some segment in society. Like functions, dysfunctions can be manifest or latent. **Manifest dysfunctions** are the expected or anticipated disruptions to order and stability that a part causes in some segment of the system. Some predictable disruptions that seem to go hand in hand with community wide celebrations are traffic jams, closed streets, piles of garbage, and a shortage of clean public toilets.

Latent dysfunctions are unintended, unanticipated negative consequences. Community-wide celebrations often have some unanticipated negative consequences. Sometimes police departments choose to negotiate contracts with the host city just before the celebration, thereby using the event as a bargaining tool to secure a good contract. (Actually, one might argue that this is a latent function for the police and a latent dysfunction for the city.) Another latent dysfunction of the community-wide celebration is that many people celebrate to the point that they miss class or work the next day to recover.

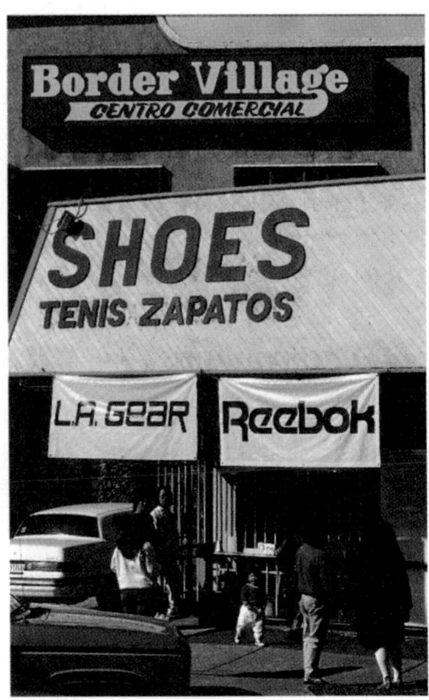

Two distinct cultures and economies are intermingled in the towns on either side of the U.S.–Mexican border.

©Les Stone/Sygma

You can see from just this brief analysis of community-wide celebrations that the concepts of manifest and latent functions and dysfunctions make for a more balanced perspective than does the concept of function alone. The addition of these concepts eliminates many of the criticisms leveled at the functionalist perspective. However, this broader functionalist approach also introduces a new problem: it gives us "no techniques . . . for adding up the pluses and the minuses and coming out with some meaningful overall calculation or quotient or net effect" (Tumin 1964:385). In regard to the community-wide celebrations, we are left with the question of whether the overall impact of this event has had a positive or negative effect on society.

Dysfunctions Disruptive consequences to the system or to some segment in society.

Manifest dysfunctions The expected or anticipated disruptions to order and stability that a part causes in some segment of the system.

Latent dysfunctions Unintended, unanticipated negative consequences.

The Functionalist Perspective on U.S. Manufacturing Operations in Mexico

To see how the functionalist perspective can be applied to a specific event, let us consider how functionalists

The everyday reality of illegal immigration is highlighted by this highway sign in the border city of San Ysidro, California. The sign warns motorists to watch for illegal immigrants and at the same time warns migrants that crossing illegally is prohibited.

would view U.S. manufacturing operations in Mexico, known as *maquiladoras*[1] (pronounced mah-kee-la-doras). In Mexico, there are more than 2,300 *maquiladoras* or foreign-owned manufacturing plants. Although there are Asian-, European-, and Canadian-owned *maquiladoras,* the great majority are U.S. owned and located along the 2,000-mile border that divides Mexico from the United States. Well-known companies with *maquila* operations include Xerox, Johnson & Johnson, GTE, Texas Instruments, Black and Decker, Uniroyal, Singer, Clark Equipment, Ford, General Motors, Chrysler, Honeywell, Kellogg, Foster Grant, and Westinghouse.

Maquiladoras work like this: Foreign companies ship tools, machinery, raw materials, and components duty-free into low-wage, labor-abundant countries; workers use the tools and machinery to process and assemble the raw materials and components into finished or semifinished products, which are exported back to the country of the contracting company. Only the cost of labor to assemble the product is subject to tariff. Mexico, because of its proximity to the United States and because of its abundant, cheap labor force, is the United States' most important partner with regard to assembly operations. Mexican workers perform a variety of labor-intensive tasks, ranging from assembling TV receivers to sorting U.S. retail store coupons. Assembly operations are concentrated, however, in the electronics, automotive, and textile sectors.[2]

In assessing *maquiladoras,* functionalists ask, Why do *maquiladoras* exist, and what consequences do they have for the United States and Mexico? We can apply Merton's concepts to analyze the overall consequences of the *maquiladora* program on the United States and Mexico. The concepts of manifest, latent, function, and dysfunction remind us that in assessing *maquiladoras,* we must look for intended and unintended consequences that lead to both order and disorder.

Manifest Functions To understand the intended, planned function of the *maquiladoras,* we need to understand why they were created in the first place. The *Maquiladora* Program was launched in 1965 after the

[1]*Maquila* is a derivative of the verb *maquilar,* "to do work for another." Originally, *maquila* signified the toll that a farmer paid to a miller for processing grain (Magaziner and Patinkin 1989:319).

[2]A foreign company can take part in the Border Industrialization Program in several ways: it can (a) establish its own plant, (b) subcontract with an existing firm in Mexico, or (c) participate in a shelter plan. A shelter plan is an arrangement with a liaison company that provides facilities (buildings and warehouses), contracts with a work force, and handles paperwork and red tape. The contracting company provides raw materials and components, production equipment, and on-site management.

Bracero Program[3] was terminated. The *Bracero* Program, which began in 1942, allowed Mexicans to work legally in the United States to relieve labor shortages in rural areas and to bolster the American work force during World War II. The *braceros* replaced American workers who were needed to work in defense plants or to serve in the armed forces. After World War II, the *Bracero* Program was extended several times to supplement the American work force during the Korean War and to provide laborers for agricultural and other low-wage work. With the termination of the *Bracero* Program, the Mexican government established the Border Industrialization Program.

One purpose of the program was to encourage foreign investors to locate plants along the border so that employment opportunities could be created for returning *braceros* and so that U.S. corporations could draw from a low-wage labor pool to decrease production costs, allowing them to compete in what was becoming an increasingly global economy. A second purpose was to increase the employment opportunities in border cities that were experiencing a high level of population growth because of a large influx of migrants from the interior of Mexico. (The termination of the *Bracero* Program increased the already high unemployment in the border region.) In addition, the program's creators envisioned that border industrialization would increase economic ties between the border region and the rest of the country, provide foreign exchange, and stimulate technology transfer from the United States to Mexico.

One way that some advocates of the *Maquiladora* Program assess its contribution to stability is to estimate the effect on Mexico and the United States if it were eliminated. They project that under such a scenario, Mexico's gross national product (GNP) would drop by $3.1 billion because *maquilas* are directly responsible for an estimated 710,268 jobs, and for every one *maquila* job, two additional "support" jobs are created (*Twin Plant News* 1996a). The U.S. economy would be affected similarly. In addition, prices for American goods would increase by a projected 36 percent. The higher prices would mean a reduced demand for U.S. products and a corresponding loss of manufacturing, management, marketing, and retail jobs (Cornejo 1988). Proponents argue that this projected loss would occur because *maquilas* function in two important ways to save and create U.S. jobs.

[3]*Bracero* is derived from the Spanish *brazo*, which means "arm." The word refers to manual laborers (those who use their arms).

First, U.S.-based companies offer services[4] and supply raw materials and components for assembly by Mexican *maquila* workers. In 1994, for example, the *maquiladoras* in Baja California, Mexico, alone imported $5.4 billion in goods and services from the United States (California's share was $810 million; Kraul 1996). Second, the low-cost, labor-intensive *maquila* output is sent to sister plants in the United States for final assembly. Thus, the Mexican contribution saves manufacturing jobs and permits U.S. workers to receive higher wages for their work.

This economic arrangement is typified well by the *maquila* plant Delredo, a division of Delco, one of the largest producers of magnets. Delredo is located in Nuevo Laredo, Mexico. Iron components needed to make magnets are purchased from companies in Oklahoma. The magnets are produced at Delredo and then are shipped to a sister plant in Rochester, New York. There they are placed in electric motors that operate power windows and windshield wipers (Jacobson 1988).

Latent Functions On a latent level, the growth of *maquila* plants along the border function to further integrate Mexico with the United States socially, economically, and politically.[5] For example, an estimated 40,000 people from both sides of the border cross the San Diego–Tijuana line every day legally to work, shop, eat, or go to school, making it one of the two busiest land-border crossings in the world (Krause 1996). This integration is symbolized by the bridge and border crossing that connect border cities with their respective sister cities across the Rio Grande. These bridges increase not only the flow of goods but also the exchange of persons, information, and services between the United States and Mexico.

The increased interaction among border peoples is blurring the culture at the border. The culture is neither

[4]Examples of companies located on the U.S. side of the border that supply services are Border Restaurant Supply (designs, furnishes, and installs cafeterias and dining rooms for *maquilas*), Language Plus (language services to the *maquila* industry), and Latin Moves (manages machinery moves between *maquila* plants).

[5]We must remember that, although the *maquila* industry *clearly* has increased the social and economic integration, especially along the border separating the two countries, the United States and Mexico have always been interconnected. This fact is obvious when we simply consider that in 1848 the U.S. government annexed (in the name of manifest destiny) Mexican territory along with many of its inhabitants (who lived in what is now California, Nevada, Texas, Utah, and parts of Arizona, Colorado, New Mexico, and Wyoming). Just because a line was drawn separating the two countries politically does not mean that the social and economic ties between people on either side ceased to exist.

Mexican nor American; it is a blend (Langley 1988). This blurred line of demarcation at the border is functional in that it eases the strains usually caused by differences in language, outlook, and customs that inevitably arise when people from distinct cultures interact. This buffer zone may prove essential for smooth relations between Mexico and the United States as the two countries become more dependent on each other.

Some border residents understand both worlds and act to mediate, facilitate, and smooth exchanges between the two countries. Bilingual consultants who are comfortable with both cultures help U.S. businesses establish assembly operations in Mexico. There are also academic centers that facilitate the blending of cultures. They include the American Center for Mexican Studies at the University of California in Los Angeles, the Center for Frontier Studies at the University of Texas in El Paso, and Mexico's Colegio de la Frontera Norte in Tijuana (Sanders 1987).

The latent functions just described are positive, but you should not take that to mean that no cultural frictions exist at the border. In *Days of Obligation: An Argument with My Mexican Father,* Richard Rodriguez (1992) reminds us that on the U.S. side of the border "are petitions to declare English the official language of the United States [and] the Ku Klux Klan nativists posing as environmentalists, blaming illegal immigration for freeway congestion" (p. 84). Rodriguez's observations suggest that the Mexican–United States economic and social interdependence brings some dysfunctional consequences.

Manifest Dysfunctions Although the *Maquiladora* Program contributes to stable relations between Mexico and the United States, it has several expected (manifest) dysfunctions. Job displacement is one obvious manifest dysfunction: whenever an event alters the way in which significant numbers of people earn their livelihoods, we can anticipate that some adjustments will have to be made. In the United States, for example, many workers have been laid off or fired because their jobs were transferred to Mexico. The disappearance of entry-level manufacturing jobs has left many unskilled workers without jobs and lacking the qualifications to fill newly created jobs. Workers in cities that specialized in handling only routine manufacturing jobs were particularly affected. Flint, Michigan, along with the surrounding Genesee County, a region of 430,000 people, is a notable example.

During the late 1970s, General Motors employed 80,000 people in the Flint area. In a ten-year span 32,000 jobs disappeared permanently. Over the course of the 1990s it is expected that GM will cut down to only 35,000 workers in the Flint area (Jacobson 1988). Rural counties are particularly vulnerable to this form of job displacement because manufacturing jobs in these areas tend to be concentrated in routine manufacturing industries (food, textiles, furniture production, apparel). These industries require less skilled labor (such as garment inspection, repetitive assembly, and simple machine operation) and can be shifted to foreign-based plants (O'Hare 1988).[6]

As you might expect, the U.S. decision to terminate the *Bracero* Program compounded unemployment programs with the border, which the *Maquiladora* Program was expected to alleviate. Although we have little information about the fate of the *braceros,* we know from numbers alone that the Border Industrialization Program did not function to absorb them into the Mexican economy. More than four million permits to work in the United States were issued to Mexican *braceros* between 1951 and 1964 (Garcia 1980), whereas the *maquila* industry has generated only about 700,000 jobs over a thirty-year period.

Latent Dysfunctions Several unexpected or latent dysfunctions are also associated with the growth of *maquilas* along the border. First, because the Border Industrial Program has increased economic interdependence between the United States and Mexico, problems in one country affect the economic life of the other. For example, the 1974–75 and 1981–82 recessions in the United States contributed to *maquila* plant layoffs and closings. When a $100 billion debt crisis hit Mexico in the early 1980s and again in 1995, and when Mexico's stock market crashed in 1987, the peso subsequently was devalued. Each time the president of Mexico announced that in devaluing the peso, he had to make the people poorer, at least for a time (Golden 1995). In all cases, retail businesses on the U.S. side of the border dependent on cross-border shoppers were devastated (Myerson 1995). (It is estimated that under normal economic conditions, 40 to 70 percent of the spending by border Mexicans takes place on the U.S. side.) Although *maquila* owners benefit from a weak

[6]It would be very misleading, however, if we were to attribute job displacement and loss in the United States to just the *Maquiladora* Program. Mexico is one of many locations in the world to which U.S.-based corporations have moved assembly and other operations. Another often overlooked reason for job displacement and loss is the failure of U.S. corporations to invest in the technologies that would allow them to compete in a global economy.

In a world seemingly committed to freer trade among countries, borders become more and more paradoxical. Even as the U.S. Immigration and Naturalization Service puts up more new security fences to block illegal human crossings from Mexico, governments on both sides want to increase *the flow of goods, information, and money.*

©Jeffry D. Scott/Impact Visuals

Mexican economy and a weak peso, the adjustments that the Mexican government imposes on its people—forced austerity, currency devaluation, curtailment of social programs, decline of real wages, privatization of government-owned enterprises — make life difficult[7] for the masses, increasing the flow of illegal immigrants to the United States and sowing the seeds for political upheaval that affects both the Mexican and American economies.

Another latent dysfunction is the transfer of significant numbers of white-collar jobs from the United States into Mexico. When the *maquila* industry started in 1965, most of the white-collar work that was necessary for *maquila* operations was done by workers in the United States. Since then, however, many Mexican citizens have participated in high school equivalency programs offered by *maquilas* or have graduated from one of Mexico's public universities or technical schools whose missions are to prepare young people for skilled work (Uchitelle 1993). Engineers, technicians, administrators, computer operators, accountants, personnel officers, and lower and middle managers now can be drawn from the Mexican population. Because salaries for Mexican industrial relations managers average $14,000 a year, compared with $30,000 or more earned by their U.S. counterparts, this alternative is attractive for *maquila* owners (McAllen Economic Development Corporation 1995; see Table 2.1). Although this situation is functional for Mexico, it is a latent dysfunction for the white-collar segment of the labor force in the United States.

The populations of the border cities, where most *maquila* plants are located, have increased rapidly because of migrations of persons seeking employment from the interior of Mexico. This rapid and unregulated population growth has resulted in another latent dysfunction: the establishment of large human settlements characterized by substandard or nonexistent housing, without proper sanitation, transportation, water, electricity, or social services on both sides of the border. More than 200,000 people in Texas and New Mexico live in such settlements, known as *colonias* (NAFTA 1993; see "Frequently Asked Questions About *Colonias* in the United States" on page 42).

[7]During the 1995 economic crisis in Mexico, the value of the peso was devalued 55 percent against the dollar. In personal terms, that meant that a worker earning 50 pesos a day before the devaluation earned the equivalent of $15.40. After the devaluation, that same worker earned 70 pesos per day but earned only the equivalent of $10 (Levi 1995).

Table 2.1 Professional Average Hourly Minimum Wages in Reynosa (all rates in Mexican pesos)

Pesos		Dollars	Selected U.S. Counterpart ($ U.S.)
2.35	General worker	0.32	10.29
6.90	File clerk	0.93	7.15
15.45	Industrial engineer	2.09	17.00
22.07	Accounting supervisor	2.08	13.33
2.83	Seamstress (factory)	0.38	—
7.08	Truck driver (freight)	0.96	18.88
49.98	Industrial relations manager	6.75	—
6.71	Warehouse clerk	0.91	8.15
10.08	Nurse (licensed)	1.36	—
14.28	Training specialist	1.93	10.08
3.22	Journeyman metal caster	0.44	—
2.91	Pressure casting machine operator	0.39	16.12
2.90	Tool and die maker	0.39	17.67
21.89	Maintenance superintendent	2.96	—
2.84	Journeyman operator of plastic molding machine	0.28	—
30.01	Manufacturing engineering supervisor	4.06	—
3.21	Offset press operator	0.43	—
121.00	Materials control manager	16.35	—
3.05	Journeyman electronics, technician, repair	0.41	—
6.83	Receptionist (bilingual)	0.92	8.43
15.87	Buyer (bilingual)	2.14	15.95
3.84	Stenographer typist (Spanish)	0.52	7.65
6.23	Watchman	0.84	9.23

Rates are subject to daily Mexican peso exchange rates and changes in Mexican minimum wage (7.4 pesos = $1.00).

Source: McAllen Economic Development Corporation (1996).

Problems such as severe health-threatening pollution, traffic congestion, flooding, and disease affect community life on both sides of the border (Herzog 1985). Hence, these problems cannot be solved by one side alone. Problems do not stop at the border, as illustrated by the automobile and truck exhaust generated by traffic slowdowns and delays on the many bridges and other border crossings between Mexico and the United States. It is estimated that vehicles waiting to cross the Cordova Bridge (Bridge of the Americas), which joins El Paso, Texas, and Ciudad Juarez, emit 1,280 tons of carbon monoxide, hydrocarbons, and oxides of nitrogen annually (Roderick and Villalobos 1992). Obviously, emissions from commercial vehicles affect both sides of the border.

Table 2.2 summarizes the manifest and latent functions and dysfunctions associated with the *maquiladora* industry. These concepts help us answer these questions: Why do *maquiladoras* exist, and what consequences do *maquilas* have on the United States and Mexico? The strength of the functionalist perspective is that it gives us a balanced overview of a part's contribution—negative, positive, intended, and unintended. The weakness of this perspective is that it still leaves us wondering whether the overall impact of *maquilas* contributes to stability or instability in the United States and Mexico. (As a case in point, see "U.S. Employers' Preference for Female Labor in the *Maquiladoras:* Latent Function or Dysfunction?" on pages 44–45.) On the other hand, the functionalist perspective does help us see that some segments of each country benefit more than others from this labor transfer arrangement. At the same time, it leaves us believing that the negative consequences are simply the costs for

U.S. In Perspective

Frequently Asked Questions About Colonias *in the United States*

WHAT ARE *COLONIAS*?

Colonias are housing clusters with substandard infrastructures occupied predominantly by Hispanic-American citizens. *Colonias* are notable by the absence of one or more of the following: paved streets, numbered street addresses, sidewalks, storm drainage, sewers, electricity, potable water, or telephone services. The quality of housing varies from brick ranch-style homes to packing sheds.

WHERE ARE THEY LOCATED?

Colonias are found in Texas, New Mexico, Arizona, and California, along the border with Mexico.

WHO LIVES IN *COLONIAS*?

Almost all of the residents of colonias are of Hispanic origin. There are currently approximately 1 to 1.5 million residents. Eighty-five percent are U.S. citizens. Some are fifth-generation American citizens.

WHAT IS THE FAMILY SIZE?

Colonias average 3.6 persons per household, compared with 2.7 persons in urban areas and 2.6 persons in rural areas. The percentage of households headed by single females is lower in the *colonias* (7.5 percent) than in rural areas overall (21.4 percent), urban areas (22.1 percent), and the state of Texas (21.1 percent). Also, 36.6 percent of the population are age seventeen or younger. In urban areas, 27.8 percent are in that age group.

WHAT IS THE AVERAGE ANNUAL INCOME?

According to the 1990 census, the median income for residents of *colonias* was $16,608. In comparison, the median income for residents of the state of Texas was $26,000, and the average for residents in rural areas was $21,000. Approximately 43 percent of all *colonia* residents live in poverty, whereas only 18 percent of all residents of Texas are below the poverty level.

HOW DID *COLONIAS* BECOME ESTABLISHED?

The lots in the *colonias* are typically sold to low-income migrant farm workers. Thousands of low-income border families are not able to find affordable rental units or obtain home financing from traditional sources; thus, they must rely on financing provided by *colonia* developers. Generally, the property does not transfer until the final payment is made.

The lots are small, many only 60 by 100 feet, and sell for $3,000 to $12,000. Lots in *colonia* areas within the corporate limits of border cities cost more: lots in El Paso and Laredo can cost $10,000 to $20,000.

WHAT ARE THE LIVING CONDITIONS IN THE *COLONIAS*, AND ARE THERE HEALTH RISKS?

In Texas, 26 percent of *colonia* residents live in crowded conditions. Overall, 8 percent of the state's residents live in crowded conditions. "Crowded" is defined as more than one person per room. In *colonias*, 8.9 percent of residents live in housing units with seven or more persons. In urban areas, 1.7 percent live in similar situations; in rural areas, 1.9 percent live in units with seven or more people.

Only 35 percent of the homes in *colonias* have public sewage disposal systems (81 percent of the homes in Texas have public sewage). *Colonias* also have a high percentage of homes with septic tanks, cesspools, and no sewage removal system.

Source: U.S. Department of Housing and Urban Development (1996).

overall order and stability. As we will learn in the next section, such a conclusion would receive no support from a conflict theorist.

The Conflict Perspective

In contrast to functionalists, who emphasize order and stability, conflict theorists focus on conflict as an inevitable fact of social life and as the most important agent for social change. Conflict can take many forms besides outright physical confrontation, including subtle manipulation, disagreement, dominance, tension, hostility, and direct competition. The conflict theorists emphasize the role of competition in producing conflict. Dominant and subordinate groups in society compete for scarce and valued resources (access to material wealth, education, health care, well-paying jobs, and so on). Those who gain control of these resources strive to protect their own interests against the resistance of others.

Conflict theorists draw their inspirations from Karl Marx, who focused on class conflict. Marx maintained that there are two major social classes and that class

This maquila *worker is paid about $5 per day to sew pants for export to the United States.*

©Louis DeMatteis

Table 2.2 Latent and Manifest Functions and Dysfunctions of the *Maquiladora* Program for Mexico and the United States

Manifest Functions

- Creates employment opportunities for returning *braceros* and border residents
- Creates a low-wage labor pool from which U.S. corporations can draw
- Increases economic ties between border region and the rest of Mexico
- Provides a major source of foreign exchange for Mexico
- Saves and creates jobs in the United States

Latent Functions

- Integrates Mexico and the United States socially and politically
- Increases exchange of persons, information, and services between the United States and Mexico
- Transfers some white-collar jobs from the United States into Mexico (a plus for Mexico)

Manifest Dysfunctions

- A mismatch arose between skills needed for jobs lost and jobs created in the United States.
- Many workers lost jobs in the United States.
- Returning *braceros* were not hired.

Latent Dysfunctions

- Environmental problems on one side of the border affect the other.
- Transfer of some white-collar jobs from the United States into Mexico (a minus for the United States).
- Rapid growth of *maquila* cities contributes to problems on both sides of the border.

membership is determined by relationship to the means of production. The more powerful class is the **bourgeoisie**, or those who own the means of production (land, machinery, buildings, tools) and purchase labor. The bourgeoisie, motivated by the desire for profit, need constantly to expand markets for their products. In addition, they search for ways to make production more efficient and less dependent on human labor (using machines, robots, and automation, for example), and they strive to find the cheapest labor and raw materials. These needs spread "the bourgeoisie

Bourgeoisie The owners of the means of production (land, machinery, buildings, tools) who purchase labor.

U.S. Employers' Preference for Female Labor: Latent Function or Dysfunction?

The . . . debate over the North American Free Trade Agreement (NAFTA) . . . focused public attention on the Mexican *maquiladora* industry. Although it has never been demonstrated that the movement of U.S. firms to Mexico and elsewhere is the main cause of rising unemployment in the United States, many Americans see Mexican workers as direct competitors with U.S. labor for manufacturing jobs. The knowledge that most *maquiladora* workers are women has heightened the controversy, for many Americans believe that the women hired by U.S.-based firms have less need for wages than the U.S. workers, many of whom are men, whose jobs are being eliminated. The *maquiladoras'* employment practices are also a focus of concern for many Mexican citizens, who lament that U.S. companies' preference for female labor is disrupting Mexican society.

Women have always been the preferred work force for the unskilled, low-wage jobs that characterize the *maquiladoras.* Although most managers and skilled technicians are men, most of the workers who assemble clothing, electronic components, and other products are women. When the *maquiladoras* were first established, most assemblers were young, single, childless women who worked for a few years before dropping out of the labor force to marry and raise families. Over time, as more *maquiladoras* came to the border and the demand for female labor began to exceed the supply of young, unmarried women willing to take assembly jobs, firms changed their recruitment practices to encompass more married women and more women with children. Regardless of their age or family arrangements, most women assemblers make essential economic contributions to households in which men are often unemployed or underemployed.

The large-scale incorporation of women into the manufacturing work force was a latent function of the *Maquiladora* Program. When the Border Industrialization Program (BIP) was established in the mid-1960s, it was intended to alleviate the unemployment and underemployment troubling the border region. Because Mexican culture defines wage work as a male activity and views women in terms of their roles as wives and mothers, it was assumed that men were the primary victims of unemployment and that men would be the key beneficiaries of the new jobs created within the *maquiladoras.* The founders of the BIP did not anticipate that the U.S. firms establishing *maquiladoras* would focus their recruitment efforts on women.

Their hiring practices in Mexico were not unique; corporations investing in various developing countries employ women for assembly jobs. Although *maquiladora* managers attribute their preference for female labor to women's manual dexterity, patience, and tolerance for monotonous work, feminist critics counter that these companies hire women because they are cheaper to employ than men, and because their gender-role socialization makes them more docile, and thus less likely to organize unions, than men. As a result of these employment practices, the *Maquiladora* Program has not reduced male unemployment but instead has drawn a previously nonemployed sector of the population into the labor force.

Was this unintended consequence a latent function or dysfunction? Has the employment of women been beneficial or detrimental to Mexican women, their families, and the society generally? Critics of the program emphasize its negative consequences. Some argue that the *maquiladoras* have harmed women by subjecting them to exploitative, menial, low-wage jobs that impair their physical and mental health. Others claim that women's *maquiladora* employment has damaged the Mexican family. In Mexico, conventional gender roles assign women to the sphere of home and family and men to the world of waged employment, leading to a clear-cut gender division of labor within and outside the household. In

over the whole surface of the globe. It must nestle everywhere, settle everywhere, establish connections everywhere" (Marx [1888] 1961:531). According to Marx, in less than a hundred years of existence, the bourgeoisie "has created more massive and more colossal productive forces than have all preceding generations together" (p. 531).

Proletariat A less powerful class composed of workers who sell their labor to the bourgeoisie.

The less powerful class, the **proletariat**, consists of the workers who own nothing of the production process except their labor. The proletariat's labor is a commodity no different from machines and raw materials. Mechanization combined with the specialization of labor has left the worker with no skills; the worker is an "appendage of the machine, and it is only the most simple, most monotonous, and most easily acquired

encouraging women to take jobs outside the home, critics claim, the *Maquiladora* Program has disrupted family equilibrium by giving women new statuses that weaken their adherence to traditional feminine roles, and by giving women resources that enable them to challenge men's traditional roles as sole family wage-earners.

Critics claim that male alcoholism, divorce, spousal abuse, and other indicators of family disruption are on the rise and that these dislocations are a direct result of women's employment in the *maquiladoras.* Finally, critics argue that the *maquiladoras'* preference for female labor has contributed to the border region's unemployment problems. In their view, rather than providing jobs for those who most need them – men with families to support – the *maquiladoras* have employed women who previously were supported by husbands or fathers. Once women join the economically active population by taking jobs in *maquilas,* they develop a lifestyle requiring continual waged employment. If, as is often the case, their factory lays them off or they quit their jobs because they cannot tolerate the drudgery and monotony of assembly work, they will look for another, different kind of job. When displaced *maquiladora* workers take jobs that could have been filled by men, they contribute to male unemployment by displacing men from the labor force.

Advocates of the *Maquiladora* Program, by contrast, emphasize its positive consequences: in their view, the *maquiladoras* provide jobs for women who must earn a wage to support themselves and their families. Many Mexican women are single household heads supporting their children without assistance from male partners. Those who live in stable partnerships often have little economic security, because many Mexican men are unemployed, and others earn a wage that is insufficient to support their households. Throughout the last decade, Mexico has suffered from a severe economic crisis marked by spiraling inflation and cutbacks on government services, which have made it difficult for households to subsist on a single wage-earner's income. In most households, the earnings of all adult members, whether male or female, have been essential to the family's economic survival. Without the employment opportunities provided by the *maquiladoras,* many border region families might have faced economic destitution. Not only has the *Maquiladora* Program benefited many families by providing women with needed income, but it has offered women jobs that increase their status in society. According to this view, wage work gives women resources that increase their personal autonomy and enhance their bargaining power relative to male partners or fathers. Women who earn an income tend to have more input into household decisions than nonemployed women who are completely dependent on male members to support them. Not only do they have some say-so about how their wages will be spent, but they have the economic wherewithal to leave abusive marriages when necessary to protect themselves and their children. Their employment in the *maquiladoras* thus provides the economic foundation for their liberation from repressive family relationships.

This essay shows how whether one views a set of social facts as functional or dysfunctional can depend on one's own values and theoretical perspectives. What are some of the values and perspectives of those who find positive results from the *Maquiladora* Program? What are some values and perspectives of those who stress its negative results?

Source: Susan Tiano, University of New Mexico (1994).

knack that is required of him" (Marx [1988] 1961:532). As a result, workers produce goods that have no individual character and no sentimental value to either the worker or the consumer.

Conflict exists between the two classes because those who own the means of production exploit workers by "stealing" the value of their labor. They do so by paying workers only a fraction of the profits they make from the workers' labor and by pushing them to increase output. Increased output without a commensurate pay raise reduces wages to an even smaller fraction of the profit.

The exploitation of the proletariat by the bourgeoisie is disguised by a **facade of legitimacy**—an explanation that members in dominant groups give to

Facade of legitimacy An explanation that members in dominant groups give to justify their actions.

justify their actions — or by a justifying ideology. Conflict theorists define **ideologies** as fundamental ideas that support the interests of dominant groups. The notion that poor people are poor because they are lazy rather than because they are paid low wages is an example of an ideology. Likewise, conflict theorists would argue that the idea that corporations "have to hold down wages like it (is) some kind of natural law" (Kernaghan 1996:A16) is an ideology.

On closer analysis, however, ideologies are at best half-truths, based on "misleading arguments, incomplete analyses, unsupported assertions, and implausible premises. . . . [A]ll ideologies foster illusions and cast a veil over clear thinking . . . that enable class divisions in society to persist" (Carver 1987:89–90). For example, many employers in the garment industry justify low wages by pointing to their need to compete on a global scale. Yet labor rights crusader Charles Kernaghan (1996) asks, "Why do we have to accept the system the way it is?" (p. A16). He points to the multinational corporations such as the Walt Disney Co., which earns an estimated $1 billion licensing the use of its name and animated characters to clothing and other manufacturers. Because people are willing to pay a premium for Disney clothes and products, why could Disney not make a reasonable living wage for workers a part of its licensing agreement (Bearak 1996)?

The capitalists' exploitation of the proletariat is justified by the argument that members of the proletariat are free to take their labor elsewhere if they do not like the arrangement. However, this is not the case. "The Capitalist, if he cannot agree with the Labourer, can afford to wait, and live upon his capital. The workman cannot. He has but wages to live upon, and must therefore take work when, where, and at what terms he can get it. The workman has no fair start. He is fearfully handicapped by hunger" (Engels [1881] 1996).

On the most basic level, employers have considerably more leverage over workers than vice versa: if workers make too many demands, are unreliable, or do not produce — or if business is slow — employers can fire or lay off their workers. Workers have no comparable leverage against unreliable and overdemanding employers. Furthermore, many workers must take what employers offer, because hundreds of other workers may be waiting to fill the jobs if they refuse. For the most part, workers are an "incoherent mass scattered over the whole country, and broken up by their mutual competition" (Marx [1888] 1961:533).

Ideologies Fundamental ideas that support the interests of dominant groups.

Marx's ideas inspired most conflict theorists, and the fundamental theory has many variations. Yet, despite these variations, most conflict theorists ask this basic question: Who benefits from a particular pattern or social arrangement and at whose expense? In answering this question, conflict theorists strive to identify practices that the dominant groups have established, consciously or unconsciously, to promote and protect their interests. Exposing these practices helps explain why there is unequal access to valued and scarce resources.

Most conflict theorists also examine the facade of legitimacy that supports existing practices. They observe how exploitive practices are justified logically by those in power. The most common methods of justification are (1) blaming the victims by proposing that character flaws impede their chances of success and (2) emphasizing that the less successful benefit from the system established by the powerful. Consider the argument a Denver woman gives *MacNeil/Lehrer Newshour* correspondent Tom Bearden (1993) for hiring an illegal immigrant to care for her children:

MR. BEARDEN: Does the employer of the undocumented worker have too much power over that person? There are some that believe that people who hire undocumented aliens gain an unfair power over them, it gives them influence over them because they're, in a sense, collaborating in something that's against the law. Do you agree with that, or have any thoughts about that?

DENVER WOMAN: I guess I would disagree with that. The one thing that you get in undocumented child care or the biggest thing that you probably get, my woman from Mexico was available to me twenty-four hours a day. I mean, her cost of living in Mexico and quality of life in Mexico compared to what she got in my household were two extremes. When we hired her, she said, "I'll be available all hours of the day, I'll clean the house, I'll cook." They do everything. And if you hire someone from here in the States, all they're going to do is take care of your children. So not only do you have a differentiation in price, you have a differentiation in services in your household. I have to admit that was, at that point, with a newborn infant, wonderful to have someone who was so available. . . .

MR. BEARDEN: And it's not like indentured servitude?

DENVER WOMAN: That crossed my mind, and after she had been here for six months or so, we went to a schedule where she finished at 6 or 7 o'clock at night. And I don't think I ever really took advantage of her. Once a week I'd have her get up with the baby, so I didn't. . . . [S]he was available to me, but I don't feel like I really took advantage of her, other than the fact that I paid her less and she was certainly more available. But she got paid more here than she would have gotten paid if she'd stayed where she was. (p. 8)

Conflict theorists would take issue with the logic that this Denver woman uses to justify hiring an undocumented worker at a low salary. When it comes right down to it, the Denver woman is protecting and promoting her interests (having someone available all hours of the day to cook, clean, and provide child care) at the expense of the Mexican worker.

As one final example of how capitalists use a facade of legitimacy to justify exploitive practices, let us look at how they explain the fate of jobless workers. More than twelve million American factory workers have lost their jobs because of plant relocations and shutdowns. Many of the companies from which factory workers have been laid off, such as General Electric and General Motors, have established job reentry programs. An article by Peter Kilborn in the *New York Times* highlighted the plight of 1,200 union workers laid off after General Electric moved refrigerator production from Cicero, Illinois, to Decatur, Alabama. There, nonunion workers are paid $9.50 per hour, $4 less than the company paid the Cicero employees. In Kilborn's article, Robert Jones, then the U.S. assistant secretary of labor for employment and training, described the laid-off workers as being "as dysfunctional as you can get." Others who were interviewed described the workers as "not able to meet entry-level requirements of other jobs" and as "relatively old and unskilled." Many were described as functioning at fifth- or sixth-grade levels, and some Hispanics were said to "have never had to communicate in English" (Kilborn 1990:A12).

At face value, these comments suggest that companies like General Electric, in conjunction with the U.S. Department of Labor, are concerned enough to be spending $4 billion in efforts to retrain laid-off employees who, employers believe, lack the intelligence, motivation, skills, and so on, to take advantage of the potential employment opportunities available to them (U.S. Department of Labor 1996). Conflict theorists, however, would point out that the fate of these jobless employees represents the tragic outcome of a production process requiring so little from its workers (other than repetitive manual labor) that the workers have acquired no skills and have nothing marketable to show for decades of employment. From a conflict perspective, the laid-off workers are not to blame for their fate. Those who used them as if they were machines to do mindless, repetitive work are to blame.

Critique of Conflict Theory

Like the functionalist perspective, conflict theory has its shortcomings. A major criticism is that it overemphasizes the tensions and divisions between dominant and subordinate groups and underemphasizes the stability and order that exist within societies. It tends to assume that those who own the means of production are all-powerful and impose their will on workers who have nothing to offer except their labor. The theory also assumes that the owners exploit the natural resources and cheap labor of poor countries at will and without resistance. This is a somewhat simplistic view of the employer-employee relationship and of relationships between corporations and host countries. It also tends to ignore the real contributions of industrialization in improving people's standard of living.

In an August 1, 1995, letter to the editor of the *Los Angeles Times*, Raymond Goldberg (1995), president of New Age Intimates, a company that has manufactured brassieres in the Dominican Republic, Jamaica, Haiti, and Colombia for twenty-five years, points out that his company pays sewers between $0.75 and $1.25 (a competitive wage, comparable to other sewing factories in the area) and provides medical care and breakfast. Given that the real unemployment rate in these countries is 50 percent, these jobs are desperately needed and eagerly sought after.

Moreover, some owners of production, such as the CEO of Sequins International in Queens, New York, believe that educated, healthy workers who are not limited to mindless repetitive work tasks will work more effectively, exhibit higher commitment, and produce higher-quality products, allowing the employer to both charge more and cut costs and to compete against "500,000 women and children with needle and thread sequins" in China, India, and in other low-wage regions of the world (Sexton 1995). Likewise, many companies such as Xerox Corporation, Johnson & Johnson, and IBM deserve attention for their favorable employee policies such as those at the companies named each year by *Working Mother* magazine in its article "100 Best Companies for Working Mothers." Noteworthy companies are also named in the *100 Best Companies to Work for in America* (Levering and Moskowitz 1993).

Finally, conflict theorists tend to neglect situations in which consumers, citizen groups, or workers use economic incentives to modify or control the way capitalists pursue profit. For example, the Environmental Defense Fund negotiated successfully with McDonald's to ban polystyrene packaging at its 8,500 restaurants across the United States. The Earth Island Institute Dolphin Project used lawsuits, court injunctions, and letter-writing campaigns to persuade Starkist to change its fishing techniques so as to make oceans safer for dolphins (Koenenn 1992).

The Conflict Perspective on the Maquiladora *Program*

From a conflict perspective, *maquilas* represent the pursuit of profit. The means of production (machinery, raw materials, tools, and other components) are owned by capitalists in the United States and other foreign countries. The Mexican workers own only their labor and must sell it to the owners of production at a low price. They cannot demand higher wages and better working conditions because the employers can easily replace them from a large pool of Mexican workers. If labor problems emerge, foreign companies can move their operations to the interior of Mexico (or to another country) where the labor is less expensive and there is less competition among employers for workers (Patten 1996).

As for the facade of legitimacy, some maintain that the *Maquiladoras* Program benefits both the United States and Mexico. Mexico gains because its growing workforce gets jobs, and the United States gains because its industries become competitive in the world market. The loss of some American jobs is more than compensated for, considering the alternatives of bankruptcy and widespread unemployment.

However, from a conflict point of view, the facade of legitimacy masks the real purpose of the *maquila* industry, which is to increase profits by exploiting the most vulnerable and least expensive labor. United States businesspeople describe this labor arrangement as being of mutual benefit. But, in reality, they do not want Mexico to prosper because their success is tied to a fragile Mexican economy: the weaker the peso, the lower the wages that company owners have to pay workers (Parra, Edmundo, Fernandez, and Osmond 1996). In fact, the editors of *Twin Plant News*, the leading magazine covering the *maquila* industry, argue that "the *maquila* industry got its start in the mid-1960s, but it wasn't until the peso devaluation of 1988 that it exploded into the giant industry it is today" (*Twin Plant News* 1996a:37). The point is that as long as Mexican labor is a bargain in the world labor market, companies in the United States and other countries will continue to locate assembly plant operations there.

The importance of the Mexican plant to General Motors' overall operations was made plain in a conversation between Bill Moyers and a GM plant manager, presented in the CBS documentary "One River, One Country: The U.S.–Mexican Border." An excerpt of their conversation follows. The plant manager's evasiveness on the question of wages suggests that the owners of production are fully aware that they exploit Mexican workers.

GM MANAGER: We are constructing the rear body wiring harness that goes into all GM autos for assembly in the United States.

MOYERS: These are the electrical wires that run to backup lights and turn signals. . . .

GM MANAGER (interrupts): . . . backup lights and turn signals and to license plate lamps.

MOYERS: How important is this plant to General Motors?

GM MANAGER: If one of these wiring harnesses does not get built, an automobile goes across the assembly line without a wiring harness in it.

MOYERS: If the wiring doesn't get done, then no car gets built?

GM MANAGER: That's right; we are a single-source plant, as all our plants are.

MOYERS: And what about their wage? What do these young people get paid?

GM MANAGER: Well, their wages obviously are on the Mexican wage scale, and we follow those standards as a *maquila* industry.

MOYERS: So what are those?

GM MANAGER (pause): The numbers are ever-changing. The minimum wage has just undergone a new adjustment. It will be a 25 percent increase [to] 1,640 pesos a day for the minimum wage.

MOYERS: So, that's about — under the new devaluation—about $3?

GM MANAGER: I believe that's right.

MOYERS: Is that per hour or per day?

GM MANAGER: Per day. (Moyers 1986)

The Moyers interview shows why Mexico is an attractive place for investment. Mexico's low-cost labor pool is virtually guaranteed for decades to come. Nearly one million new jobs must be created each year in Mexico's job market just to absorb those who reach working age (64 percent of Mexico's population is under twenty-three years of age.) Yet, despite low-cost labor along the border, there is a movement among outside investors to find even cheaper labor pools within Mexico's near and deep interiors.

Compared with the employer, the Mexican worker gains very little (see Table 2.3). *Maquila* jobs are characterized by insecurity, lack of advancement, and exceedingly low wages. Most of the work is mind-numbing and outrageously repetitive. In electronics plants, for example, workers peer "all day through a microscope, bonding hair-thin gold wires to a silicon chip destined to end up inside a pocket calculator" (Ehrenreich and Fuentes 1985:373). In the *maquilas* where manufacturers' coupons are sorted, workers sort up to 1,300 coupons each hour. Tens of thousands of times per day, workers drag coupons across an electronic bar code scanner, which flashes a numeric code that identifies the slot where the worker is to file the coupon (Glionna 1992).

It is important to recognize that conflict theorists focus on exploitive conditions, of which plenty prevail. However, not all assembly plants organize the production process and work environment in the manner described here. In *Exports and Local Development: Mexico's New Maquiladoras,* Patricia Wilson (1992) describes three different environments:

> In a hot, stifling, poorly lighted Quonset hut on the Mexican border where shoes are assembled for well-known U.S. department stores, the noxious fumes overwhelmed me as I watched the women and men work. Using sewing machines, thread, and leather from the United States, rows of women stitched tops to soles. Some men attached heels to shoes with hot glue in rapid succession, while others deftly removed excess rubber from the heel by turning the shoe around the sharp blade of a trimmer with their hands, rapidly, mechanically, under the gaze of the Virgin's image pasted on the machine to remind them that they are human.
>
> In another Mexican city I visited a well-lighted, air-conditioned electronics factory with row after row of women inserting scores of tiny colored pieces into circuit boards in just the right order. At break time they doffed the colored robes that indicate their seniority to reveal attractive dresses and high heels. "We come to get out of the house, to socialize, to meet men, not just to support our families," one young worker told me. On the other side of the factory floor was a new automated high density double-sided insertion machine, attended by one man.
>
> In an auto parts assembly plant I saw young men working in teams aided by machines to stamp, bend, weld, and paint materials. Each team was responsible for an entire subassembly. Not far away in a clearing on the shop floor a soundproof room housed a blackboard where line workers are encouraged to meet with their team managers to discuss production problems and suggest alternative solutions. A handwritten diagnosis was still on the board from the last occupants. (pp. 1–2)

In addition to focusing on those who exploit Mexican labor, conflict theorists focus on the *maquilas'* effect on the surrounding environment. The Mexican equivalent of the U.S. Environmental Protection Agency found that approximately a thousand U.S.-owned *maquila* plants generate hazardous waste. Only about one-third comply with Mexican laws requiring them to file reports on how hazardous waste is handled. Only one in five could document that they dispose of hazardous wastes properly[8] (Suro 1991).

Generally speaking, in the context of Mexico's overall employment problems, the *maquilas* do little to alleviate unemployment. In 1995, about 700,000 persons worked at *maquila* plants, which have been in existence since 1965; to stave off unemployment, though, Mexico would have to generate about a million new jobs a year.

Conflict theorists would argue that U.S. and other foreign employers have the upper hand, not only because Mexico needs a seemingly endless number of jobs but also because the country faces intense pressures to generate foreign currency to pay back an estimated $128 billion in loans (U.S. Central Intelligence Agency 1995).[9] Lending institutions in countries like the United States make profits from the high interest rates that they charge. *Maquila* owners benefit from the conditions of the loans, which stipulate that Mexico must promote export manufacturing as a means of generating foreign currency for loan repayments.

[8]An agreement between Mexico and the United States, known as Annex III, specifies that hazardous waste and materials (HWM) produced at *maquila* plants must be returned to the country of origin. An application must be filed forty-five working days before every HWM shipment. The application specifies, among other things, a detailed description of the HWM, the route and final destination, and emergency measures to be taken in case of accidental spill (Partida and Ochoa 1990).

[9]Foreign companies cannot pay workers in dollars. They must exchange dollars for pesos to ensure that the dollars go to the government to pay off debts rather than to the workers.

Table 2.3 The Costs of Everyday Items for Mexican vs. U.S. Workers

According to the conflict theorist Karl Marx, workers exchange their labor for a wage that goes to purchase what they need to survive. A comparison of the labor time Mexican and American workers pay for a number of everyday items is shown here. Prices of various items in Mexico and metropolitan Los Angeles are based on recent random samplings and adjusted for different packaging sizes and weights. The time worked to earn these goods is based on a forty-hour work week and 1995 U.S. Labor Department statistics for the average wage of manufacturing workers in the United States and Mexico ($11.44 and $1.51 per hour, respectively).

Item	Mexican Price ($ U.S.)	Time to Earn (Hours)	U.S. Price ($ U.S.)	Time to Earn (Hours)
Gasoline (gal)	1.33	0.88	1.42	0.12
Blender–Osterizer eight-speed	56.76	37.61	34.99	3.06
Microwave–Sanyo	247.3	163.8		
Batteries–C 2-pack	2.30	1.50	2.33	0.20
Toothbrush–Oral B	0.95	0.63	2.00	0.17
Razors–Gillette 2-pack	1.23	0.82	7.13	0.62
Corn flakes (500 g)	1.31	0.87	2.89	0.25
Peanut butter–Smucker's 12 oz.	1.81	1.20	2.98	0.25
Beer–Corona 6 pack	0.68	0.45	6.39	0.56
Ketchup–Del Monte (567 g)	0.81	0.54	3.98	0.35
Instant coffee–Nescafe (200 g)	2.68	1.78	5.20	0.45

Sources: Castillo (1996), *Los Angeles Times* (1996), and Sears (1996).

In the meantime, American companies with assembly plants in Mexico gain another measure of control over workers and communities in the United States. In the face of a threatened relocation, employees take pay cuts, work harder, and complain less. Absenteeism declines and productivity rises. Many local, state, and federal governments give land, special tax breaks, and wage concessions to keep operations in their region or to entice companies to locate new operations in their areas (Lekachman 1985).

In response to all of this, conflict theorists ask, Who benefits from the transfer of labor-intensive manufacturing or assembly operations from the United States into low-wage, labor-abundant countries such as Mexico? The answer is the owners of production, or the capitalists: "But if you stretch it a little, we're all a part of it—anyone who has anything to do with Ford, GM, GE, or if you own anything electronic that's not Asian. Anyone in Wisconsin or Michigan working for these firms or using their products does so at the expense of Latin America" (Weisman 1986:133).

Unlike the functionalist perspective, which is unclear about the overall effect that an event or arrangement has on society, the conflict perspective zeros in on its exploitive consequences. Thus, for conflict theorists the overall effect of the *maquiladoras* is clear: *maquiladoras* benefit a small segment—the capitalists. The clear losers are the workers. It is not an arrangement that benefits workers or strengthens their role in the production process. Figure 2.1 invites you to make your own comparison between the functionalist and conflict perspectives.

We turn now to the third theoretical perspective: symbolic interaction. It is distinct from the functionalist and conflict perspectives, which focus on the way social systems are organized. Instead, symbolic interactionists focus on how people experience and understand the social world.

The Symbolic Interactionist Perspective

In contrast to functionalists, who ask how parts contribute to order and stability, and to conflict theorists, who ask who benefits from a particular social arrangement, symbolic interactionists ask, How do people de-

Figure 2.1 Border Activity

This map and its data show the tremendous two-way traffic across the border wall as attempts on the part of the United States to control that traffic. How would a functionalist interpret this activity? What would a conflict theorist make of this activity?

This map and its data show the tremendous two-way traffic across the border as well as attempts on the part of the U.S. to control that traffic. How would a functionalist interpret this activity? What would a conflict theorist make of this activity?

Source: Case 1996
Darling and Rotella 1992
Holland 1996
Hampshire 1995
Pear 1996

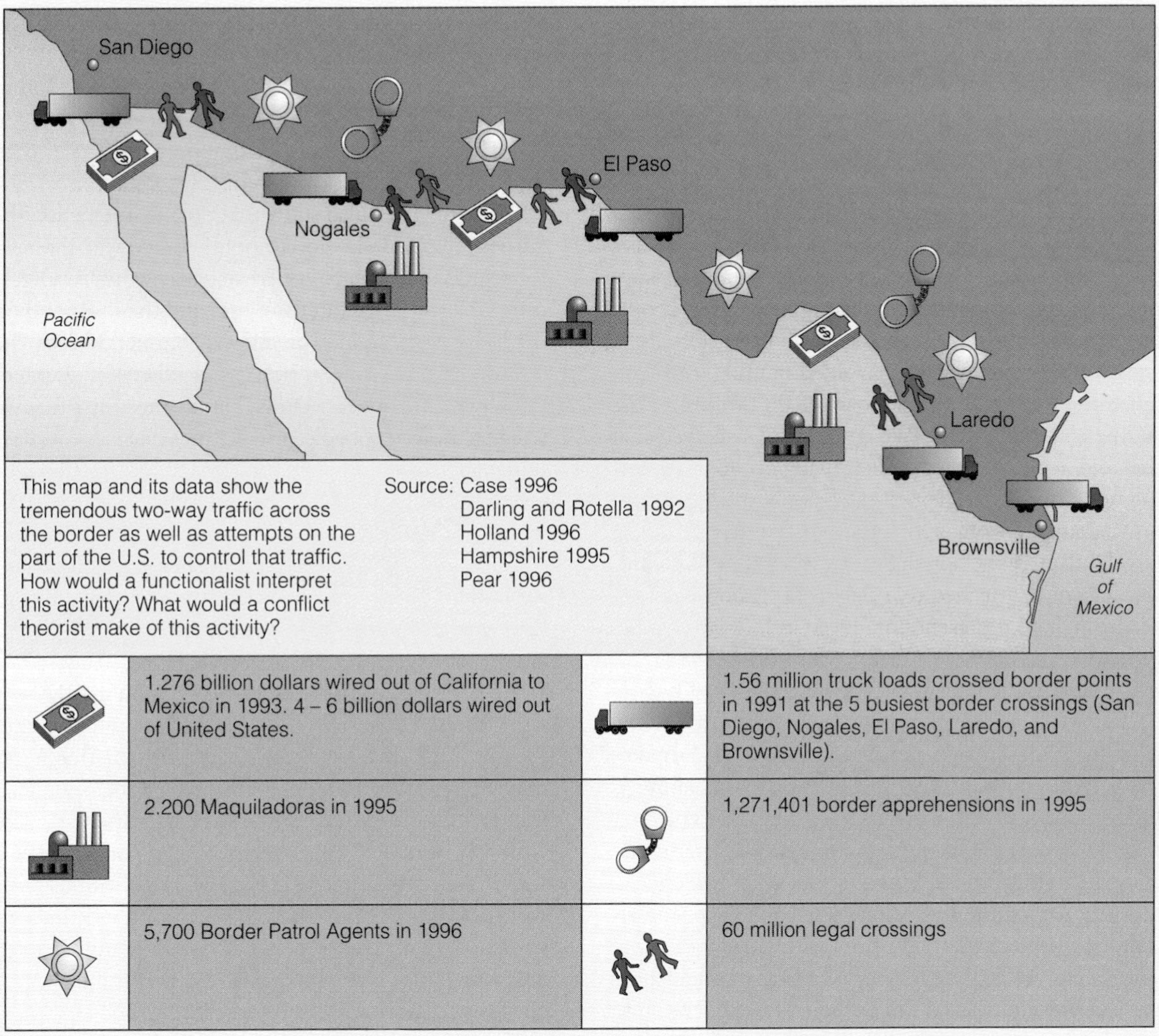

	1.276 billion dollars wired out of California to Mexico in 1993. 4 – 6 billion dollars wired out of United States.		1.56 million truck loads crossed border points in 1991 at the 5 busiest border crossings (San Diego, Nogales, El Paso, Laredo, and Brownsville).
	2.200 Maquiladoras in 1995		1,271,401 border apprehensions in 1995
	5,700 Border Patrol Agents in 1996		60 million legal crossings

fine reality? In particular, they focus on how people make sense of the world, on how they experience and define what they and others are doing, and on how they influence and are influenced by one another. These theorists argue that something very important is overlooked if an analysis does not consider these issues.

Symbolic interactionists have drawn much of their inspiration from American sociologist George Herbert Mead (1863–1931). Mead was concerned with how the self develops, how people attach meanings to their own and other people's actions, how people learn these meanings, and how meanings evolve. Consequently, he focused on people and their relationships with one another. He maintained that we learn meanings from others, that we organize our lives around those meanings, and that meanings are subject to change (Mead 1934).

According to symbolic interactionists, symbols play a central role in **social interaction**, everyday events in which two people communicate, interpret, and respond to each others' words and actions. A **symbol** is any kind of physical phenomenon—a word, object, color, sound, feeling, odor, movement, taste—to which people assign a meaning or value (White 1949). However, the meaning or value is not evident from the physical phenomenon alone. This is a deceptively simple idea that suggests that people decide what something means.

Consider the various meanings assigned to a suntan. In the United States, a tan has at various times represented quite different ideas about social class, youthfulness, and health. The shift in meanings supports the symbolic interactionists' premise that people assign meanings. Around the turn of the century, wealthy persons purposely avoided tanning to distinguish themselves from members of the working class (farmers and laborers). Pale complexions showed that they did not have to make their living outdoors, laboring under the sun. Then, as the basis of the U.S. economy changed from agriculture to manufacturing, a large portion of the population moved indoors to work. The meaning attached to a pale complexion changed accordingly to represent unrelieved indoor labor; a tan came to mean abundant leisure time (Tuleja 1987).

The presence or absence of a suntan also has reflected ideas about health and youthfulness. Many Americans describe a tan as something that makes them look good and feel better. But this meaning is likely to change with increasing reports about the connection between exposure to the sun and premature aging and skin cancer. In the face of such evidence, a tanned complexion symbolizes the skin's desperate attempt to protect the body from ultraviolet radiation. These changes in the meaning of a suntan underscore the fact that a physical form becomes a symbol because people agree on its meaning. Likewise, they demonstrate that meanings of symbols change as conditions change.

Symbolic interactionists maintain that people must share a symbol system if they are to communicate with one another. Without some degree of mutual understanding, encounters with others would be ambiguous and confusing. The importance of shared symbols frequently is overlooked unless a misunderstanding occurs. Several TV situation comedies — including *Third Rock from the Sun, Mork and Mindy, Perfect Strangers,* and *Beverly Hillbillies* — have depended on such misunderstandings by featuring characters who do not share the same symbol system as do other characters.

These comedies show us that problems arise when involved parties place different interpretations on the same event. They also show that during interaction the parties involved do not respond directly to the surroundings and to each other's actions, words, and gestures. Instead, they interpret first and then respond on the basis of those interpretations (Blumer 1962). The interpretation-response process is taken for granted. Usually we are not conscious that the meanings that we assign to objects, people, and settings make encounters understandable and shape our reactions. To make us aware of this fact, something must happen that challenges our interpretations.

To understand what distinguishes the symbolic interactionist perspective from the functionalist and conflict views, consider how the various theorists might answer this question: Why did the United States send 200,000 troops to the Persian Gulf region in 1990–91 after President Saddam Hussein sent Iraqi troops into Kuwait, and why have more than 24,000 troops remained? From a conflict perspective, the American military buildup and continued presence in the Persian Gulf can be analyzed in terms of (1) a conflict over a scarce and valued resource (oil); (2) the minority or working-class background of U.S. troops, especially as compared with the upper-class background of most government leaders who decide whether military presence is necessary; (3) Arab resistance to the U.S. presence in the Middle East; and (4) the slow official reaction to the wide range of debilitating health problems that as many as 20,000 veterans experienced, which some attribute to chemical weapons exposure or to "side-effects" resulting from the experimental drugs and vaccines they took to protect themselves.

A functionalist view would analyze the buildup in terms of its unifying function. That is, at the time it took place (August 1990 to January 1991), the military buildup functioned to divert the American people's attention from divisive domestic issues (for example, a savings and loan crisis, the huge budget deficit, relatively high levels of unemployment, and an impending recession). In addition, the buildup and continued mil-

Social interaction Everyday events in which two people communicate, interpret, and respond to each others' words and actions.

Symbol Any kind of physical phenomenon to which people assign a meaning or value.

itary presence functioned to protect the U.S. national interests related to oil. A functionalist would acknowledge the class differences between military personnel and government leaders but would emphasize that the military functions to offer working-class people, the unemployed or the poor, and those who want to serve their country a chance to earn money to go to college, learn a skill, gain on-the-job experience, or fulfill a moral obligation.

Symbolic interactionists, in contrast, would focus on how the involved parties interpret the situation and construe each other's words and actions. Thus, the symbolic interactionists would focus on interactions between American service women and business women and their Middle Eastern hosts, specifically on situations that one party adjusted its interpretations and actions to facilitate smooth interaction in situations in which interpretations and actions clashed. For example, symbolic interactionists would be interested in how American women in Saudi Arabia interpreted and adjusted to Saudi restrictions on dress or driving.

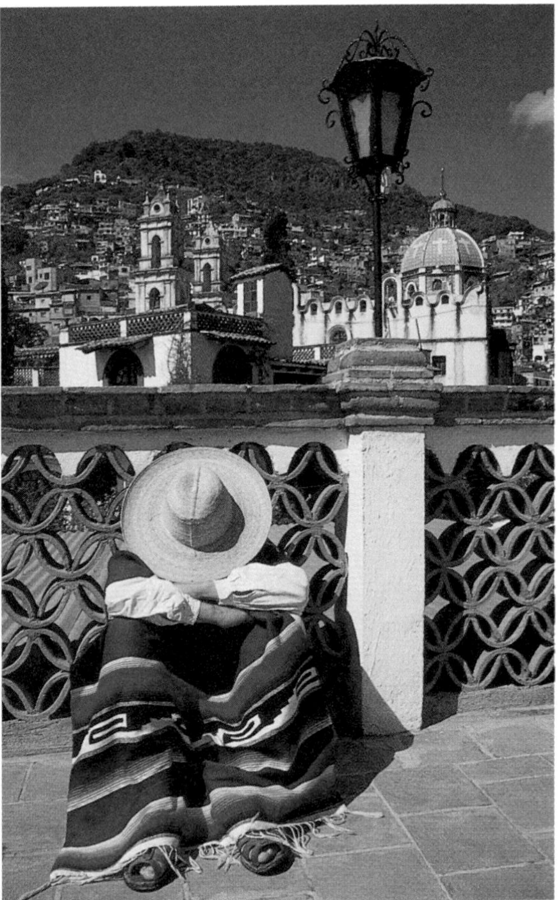

The tourism industry often promotes images of a country and its people that do not correspond with reality or that focus on a small slice of reality. Such images foster and maintain stereotypes, even in the minds of those who never visit the country.

©Rick Strange/The Picture Cube

Critique of Symbolic Interaction

Symbolic interactionists inquire into factors that influence how we interpret what we say and do, especially those factors that promote the same interpretations from significant numbers of people. Related topics of interest include origins of symbolic meaning, the way in which meanings persist, and the circumstances under which people question, challenge, criticize, or reconstruct meanings. Although symbolic interactionists are interested in these topics, they have established no systematic frameworks for predicting what symbolic meanings will be generated, for determining how meanings persist, or for understanding how meanings change.

Because of these shortcomings, the symbolic interactionist perspective does not give precise guidelines about where to focus one's attention. For example, whose interpretation should we focus on when we analyze an event? The cast of characters involved in these events is virtually endless. Even if we were able to consider every interpretation, we would still be left with the questions, What really happened? and Whose interpretation best captures the reality of the situation?

The Symbolic Interactionist Perspective on the Maquiladora *Program*

As we have seen, symbolic interactionists explore meanings that people assign to words, objects, actions, and human characteristics. Such a broad focus allows considerable flexibility when it comes to applying the symbolic interactionist perspective to foreign manufacturing operations in Mexico. We can examine the different meanings assigned to *maquila* operations by people living on opposite sides of the U.S.–Mexico border, policy makers in the United States and Mexico, American and Mexican businesspeople, union leaders in both countries, or displaced workers. We can also analyze why U.S. corporations initially were slow to move assembly operations to Mexico. When the value-added system (in which only the labor costs associated with assembling the product are subject to tariff) first became available in the mid-1960s, Hong Kong, not Mexico, grew to be the most important foreign assembly partner of the United States. Long-standing images associated with Mexican workers and Asian workers may help explain why this happened.

Americans tend to stereotype Asians as hardworking, intelligent, and obedient (Yim 1989). In sharp contrast, Mexicans are often stereotyped as unambitious and lazy and as people who value leisure more than work (Noll 1992). The unflattering image that many people hold is that of the "lazy peon asleep in the sun, his sombrero tipped forward over his eyes" (Dodge 1988:48). Many Americans also tend to associate Mexico with the afternoon siesta and believe that the country closes shop for a couple of hours after a big noontime meal. In view of these contrasting conceptions, it is understandable that a label bearing "Assembled in Mexico" evokes a much different meaning for consumers and businesspeople than a label bearing "Assembled in Hong Kong."

These images of the siesta and the sombrero, although prevalent, are at odds with the fact that the Mexican people through sheer work ethic and sacrifice have pulled out of at least three economic crises over the past twelve years. Further, many who equate the afternoon siesta with lack of ambition and unwillingness to work probably do not know that for many Mexicans the workday begins at 5:30 A.M. and ends at 7:00 or 8:00 P.M. When we consider that Mexico is now the United States' most important foreign assembly partner and that Japan is moving rapidly to capitalize on the strengths of the Mexican labor force, we might expect such negative images to fade. Yet, the results of a survey of 2,800 Japanese youths by the Mexican government suggest that these conceptions are still strong and are shared by people other than Americans. When Japanese youths were asked, "What comes to mind when you hear the word *Mexico*?" some 30 percent answered "sombrero," 20 percent said "dirt," 10 percent said "crime," and 30 percent said "desert" (Pearce 1987).

Although the origins of these images of lazy Mexicans are not clear, they are rooted in part in American views toward poverty and its causes and are perpetuated by the mass media. Much poverty exists among the Mexican masses, and Americans typically define the poor as a drain on society. They tend to attribute poverty to inferior traits of the people themselves—laziness, lack of discipline, lack of skill, and so on. For some reason, many Americans find it difficult to envision people who work hard and yet remain poor.

In spite of the negative images of Mexico, this country has emerged over the past decade as a world leader in international assembly operations. To see Mexico as a viable production site, owners of American companies had to redefine Mexican attributes in such a way that the Mexican labor force, especially now that Asian countries were becoming industrial powers in their own right. Together, Mexican communities trying to attract American investors, corporations offering shelter plans, and U.S. border communities that might benefit from *maquila* operations advertised a new image of Mexico. What attributes did they emphasize to offset the negative stereotypes of Mexicans? Let us take a look at some of the strategies they used (and continue to use) to market (or define) Mexico.

First, advertisers list the names of U.S. companies with established plants in Mexico to assure potential clients that reputable companies are succeeding there. Second, they emphasize the proximity of *maquila* cities to the United States to remind potential clients that the border is "close to home" and that transportation and communication costs are cheaper than in Asia. Third, they describe the size and cost of the Mexican labor force relative to that of other countries. Finally, they show pictures of happy, middle-class-type workers to counter images of exploited foreign workers and to suggest that the workers' standard of living benefits from *maquila* employment. Here are some examples of advertisements from *Twin Plant News* (1990a, 1990b, 1996b) that illustrate these strategies:

- "What Industrial City Is 3 Times the Size of Dallas and 45 Times Closer to You Than Taiwan?" (The advertisement promotes Monterrey, Nuevo Leon.)
- "Shopping the Interior [of Mexico]? Picture This . . . ABUNDANT LABOR . . . Worker Housing for 180,000 is in close proximity to the park."
- "Yucatan is only 90 minutes from the USA. . . . Yucatan has an abundant and efficient labor pool, easy to train and with very low turnover."

The symbolic interactionist perspective adds yet another dimension to the way we approach social events. It asks us to consider how the involved parties interpret an event. In this sense, it enhances the functionalist and conflict perspectives, which focus on the origins (societal need versus profit) and consequences (stability and disruption versus exploitation). Each perspective offers a unique set of questions and concepts to answer those questions and alerts us to avoid making simple statements or generalizations about events.

Summary and Implications

We began this chapter with some facts that describe ways in which Mexico and the United States are interconnected. Taken alone, the facts tell us little about the relationship between the two countries. At first glance, many Americans see these facts as proof that Mexico threatens their economic well-being. The three theoretical perspectives give us a strategy for thinking about these facts and tempering hasty and oversimplistic reactions. The strategy is reflected in the questions and vocabulary of each perspective (see Table 2.4).

Now that we have had a chance to see how each perspective guides analysis of a particular event through its central question and key terms, we can see that no single sociological perspective, of course, can give us a complete picture of social events. Each of the three basic perspectives offers only one way of looking at this phenomenon. Because no one perspective can capture all aspects of a situation, we can know more of a given situation if we apply more viewpoints. All three perspectives are useful in that each makes a distinct contribution to understanding.

With regard to *maquilas,* functionalists emphasize how the arrangements between the United States and Mexico create a pattern that contributes to the overall economic stability and the well-being of both countries. Conflict theorists examine how these arrangements benefit the owners of production and exploit workers in both Mexico and the United States. Symbolic interactionists consider the meanings assigned by various groups to *maquila* industries and explore how these meanings affect relationships between Mexicans and Americans. Significantly, despite the differences among these theories, none can be used to support a claim that Mexico is a drain on the U.S. economy. In fact, all three support the notion that, for better or worse, the two countries are dependent on each other.

Very few sociologists adhere to only one perspective and maintain that it should be adopted, to the neglect of the other two perspectives. Ideally, the three perspectives should not be viewed as clashing or incompatible. In fact, as we have seen, they overlap considerably. The conflict perspective's focus on exploitation overlaps the functionalist perspective's focus on manifest and latent dysfunctions. Both perspectives recognize the loss of U.S. jobs, the low wages paid to Mexican workers, and the vulnerability of communities that rely on assembly jobs. But their emphasis differs. Conflict theorists make exploitation the focus. Given their emphasis on order and stability, functionalists view the negative consequences as the price some segments in each society pay to support economic prosperity in both countries. Yet, despite these differences, the functionalists' understanding of the benefits

Table 2.4 Overview of the Three Theoretical Perspectives

	Functionalist Perspective	Conflict Perspective	Symbolic Interactionist Perspective
Focus	order and stability	conflict over scarce and valued resources	shared meaning
Vision of Society	system of interrelated parts	dominant and subordinate groups in conflict over scarce and valued resources	interaction is dependent on shared symbols
Key Terms	function, dysfunction, manifest, and latent	means of production, facade of legitimacy	symbols
Central Question	How does a part contribute to overall stability of a society?	Who benefits from a particular pattern or social arrangement, and at whose expense?	How are symbolic meanings generated?
Major Criticisms	defends existing social arrangements; offers no technique to establish a part's "net effect"	exaggerates tension and divisions in society	no systematic framework for predicting which symbolic meanings will be generated or for how meanings persist or change

Different theoretical perspectives highlight different observations on social life. The sign above this top seamstress at the Nova/Link maquiladora *in Matamoros, Mexico, means that she meets or exceeds 100 percent of her production goal. How would each theoretical perspective tend to evaluate that detail?*

of exploitive practices complements the conflict theorists' need to define and eliminate such practices. Understanding the function of these practices helps explain why they persist. Knowing why they persist helps in defining policies to eliminate them.

Symbolic interactionists can benefit from the insights of functionalists and conflict theorists as they attempt to understand the various meanings of *maquilas* for different segments of the world population. The functionalist and conflict perspectives can be used to explain the origins of the various symbolic meanings assigned to *maquilas*. Among those segments of society that benefit or profit, *maquilas* are more likely to evoke negative images among exploited segments. It is interesting, however, that the people who benefit from *maquilas* are often unaware that they do so and thus define Mexico as a burden on the American economy. Similarly, persons whom conflict theorists would define as exploited are often unaware that they are exploited and may define the *maquilas* in positive terms. Both symbolic interaction and conflict theory (specifically the concept facade of legitimacy) could be used to explain this incongruence.

With regard to U.S. assembly operations in Mexico, the larger lesson of this chapter is that we must avoid hasty and oversimplistic reactions to the headline news, sound bites, and so-called facts we hear about issues such as illegal immigration and NAFTA. Consider these recent headlines:

> "U.S. Strengthens Patrols Along Mexican Border"
>
> "A Losing Battle Against Illegal Immigration"
>
> "After Peso Collapse, an Unemployed Population Looks North"
>
> "'Free Trade' Proves Costly to U.S. Jobs: A Somber Look at NAFTA"

Headlines such as these should raise questions rather than solidify opinions. For example, with regard to illegal immigration, three sets of questions should immediately come to mind:

1. How does illegal immigration contribute to order and stability in Mexico and the United States? What are the manifest and latent functions and dysfunctions? (functionalist)
2. Who benefits from the existence of illegal immigrants, and at whose expense? (conflict)
3. Does everyone in the United States and Mexico see illegal immigrants in the same way? (symbolic interaction)

Knowing the answer to these questions will not necessarily make decisions about how to respond easier but the answers will give you more information on which to base a constructive response. Although it may be depressing to learn that knowing answers to these questions may lead to agonizing choices about how to respond, keep in mind the choices should be agonizing any time other people's lives are affected by your response.

Key Concepts

Use this outline to organize your review of the key chapter ideas.

Sociological theory
Functionalist perspective
Function
Latent function
Manifest function
Dysfunction
Latent dysfunction
Manifest dysfunction
Conflict perspective
Facade of Legitimacy
Ideology
Symbolic interactionist
Social interaction
Symbols

internet assignment

For more on the ambiguous line known as the U.S.–Mexico border; border activity and issues; and the social, economic, political, cultural, and historical ties between the United States and Mexico, browse the three Web sites listed here. As you browse, imagine that you are a conflict theorist, a functionalist, or a symbolic interactionist. Report on three to five examples of information that would attract your attention. Give at least one example for each perspective.

BorderLines: http://lib.nmsu.edu/subject/bord/bordline

BorderLines is a monthly publication of the U.S.–Mexico Project at the Interhemispheric Resource Center, an Albuquerque-based nonprofit private research and policy institute. This on-line journal focuses on issues related to the ever-changing relationship between the United States and Mexico. Examples of article titles include "Free Trade, Drug Trade" (explores the link between free trade and drug trade along the border), "Cross-Border Links and the Rise of Citizen Diplomacy" (examines the rise of border networks and coalitions), "Workers Succeed in Cross-Border Bid for Justice" (covers the story of *maquila* workers who won a sexual harassment suit filed against a U.S. employer), and "Cross-Border Indigenous Nation: A History" (describes the history of four Native American peoples separated by the border).

LatinoLink News: http://www.latinolink/com

LatinoLink offers news, analysis, commentary, and photo essays that explore the joys and challenges of people who call the United States home but who have roots in the countries of the Spanish-speaking Americas. Many interesting documents focusing on the meaning of the border and border activity.

North American Free Trade (NAFTA) Border Home Page:

http://www.itaiep.doc.gov/border/nafta.htm

The Office of the Assistant Secretary for International Economic Policy created the NAFTA Border Home Page, a collection of Web resources that lead to almost every kind of information about the United States and Mexico.

JR

Research Methods in the Context of the Information Explosion

With Emphasis on Japan

The Information Explosion | **The Scientific Method**

A bullet train with Mount Fuji in the background. ©Tony Stone Images

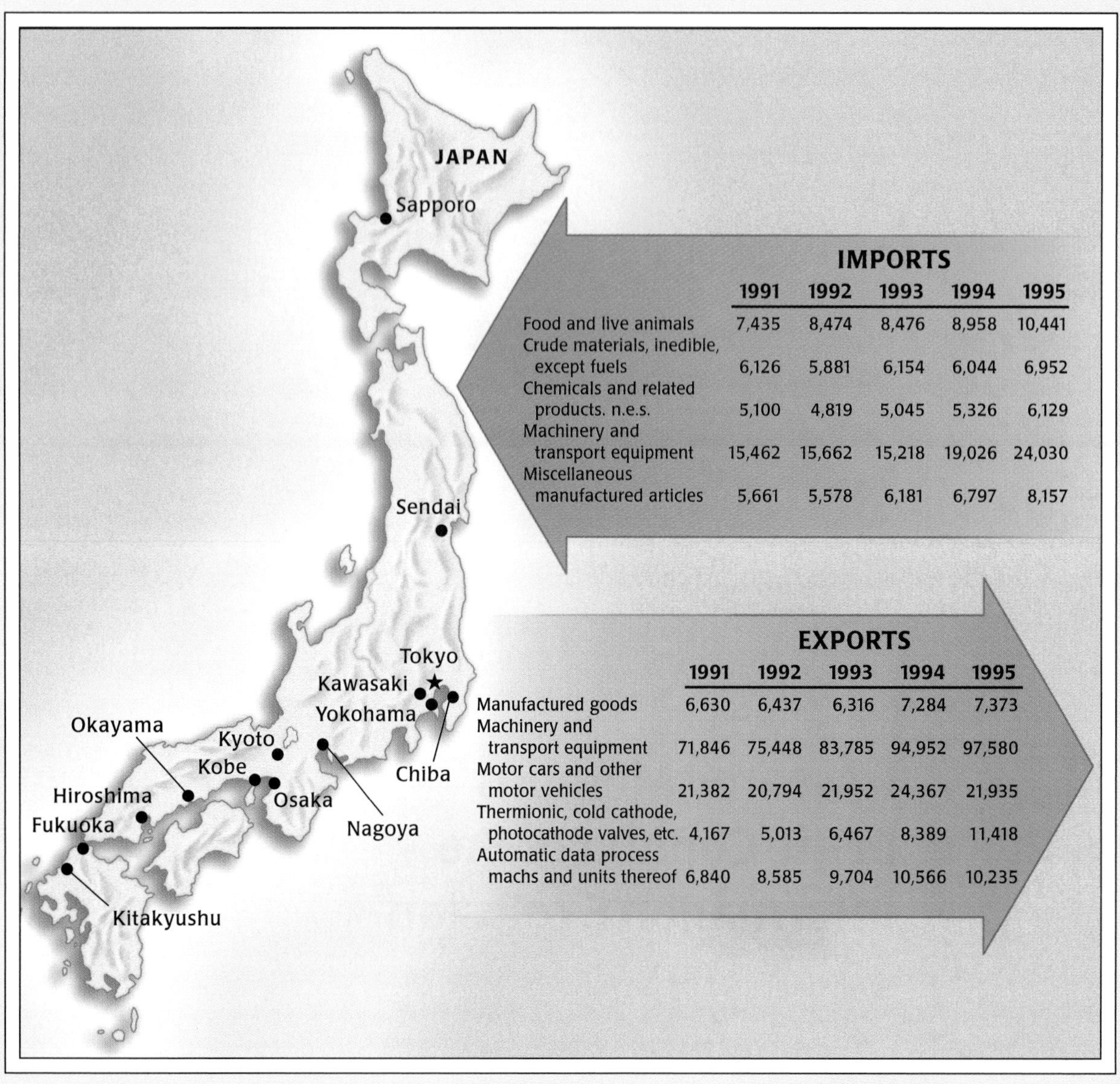

IMPORTS

	1991	1992	1993	1994	1995
Food and live animals	7,435	8,474	8,476	8,958	10,441
Crude materials, inedible, except fuels	6,126	5,881	6,154	6,044	6,952
Chemicals and related products. n.e.s.	5,100	4,819	5,045	5,326	6,129
Machinery and transport equipment	15,462	15,662	15,218	19,026	24,030
Miscellaneous manufactured articles	5,661	5,578	6,181	6,797	8,157

EXPORTS

	1991	1992	1993	1994	1995
Manufactured goods	6,630	6,437	6,316	7,284	7,373
Machinery and transport equipment	71,846	75,448	83,785	94,952	97,580
Motor cars and other motor vehicles	21,382	20,794	21,952	24,367	21,935
Thermionic, cold cathode, photocathode valves, etc.	4,167	5,013	6,467	8,389	11,418
Automatic data process machs and units thereof	6,840	8,585	9,704	10,566	10,235

Source: U.S. Department of State (1996).

The U.S. Trade Deficit with Japan The United States posts a trade deficit with almost every country in the world, but its largest deficit is with Japan. Many informed Americans worry about this. What factors might explain the deficit? Does the deficit show that U.S. society is somehow weaker than Japanese society?

How can we begin to study and make comparisons between two very different societies? This chapter explores how sociologists study social phenomena scientifically. The goal is to gather reliable and valid data and analyze the data according to the rules of the scientific method.

Consider that in the United States, as of June 1997, the amount of printed material about Japan included:

- ***7,653 books in print,***
- ***30,744 magazine articles (since 1960),***
- ***873,000 holdings in the United States Library of Congress,***
- ***12,339 articles listed in* Social Science Research Index, *and***
- ***30,938 articles listed in* Periodical Abstracts.**

A significant number of these articles and books present a holistic image of Japanese society, portraying it as a group-oriented society that places great emphasis on consensus and social harmony. Often this image is juxtaposed against an equally holistic image of the United States portrayed as an individual-oriented society that places high value on competition, personal freedom, and independence. Certainly each society possesses distinct qualities, but is it accurate to present the two as polar opposites? A knowledge of research methods can help us evaluate this practice.

Why Focus on Japan?

Most Americans know that of the 179 countries with which the United States trades, the largest trade deficit is with Japan and that U.S. economic policy toward Japan is aimed at opening the Japanese market to U.S. goods and services. In fact, the trade imbalance with Japan is the single largest contributor to the record trade deficits that the United States incurred in the 1980s and 1990s. Although considerable debate prevails over what this large deficit means, it has increased public awareness about the interdependence between the United States and Japan.

Politicians, the media, and many social critics use the trade deficit as a measure of the overall health of the U.S. economy, Japanese aggressiveness, the patriotism of U.S. consumers, the competitiveness of U.S. corporations, and the overall quality of U.S. workmanship. The attention given to the trade deficit has caused Americans at almost every level to question why they are not selling as many "U.S.-made" products abroad as they are purchasing from foreign countries (particularly Japan) and to question what quality in the Japanese has made them so "successful" and what shortcomings in themselves have contributed to their "failure."

In search of answers to these questions, U.S. politicians, journalists, and social critics have pointed to such things as the special bond between Japanese mothers and their children, the Japanese work ethic (which is said to be so extreme for men that Japanese children grow up "fatherless"), the value that the Japanese place on the group, unfair trading practices, and the general unwillingness of Japanese consumers to purchase U.S.-manufactured products. The stated or unstated implication is that the opposite qualities exist in the United States. A basic understanding of research methods enables us not only to evaluate the accuracy of the things we read, hear, and view about Japan but also to evaluate comparisons made between the United States and Japan.

Japanese children visit the Hiroshima Peace Memorial Museum. Might a study of the clothing worn by Japanese children or other groups in Japan lead to insights about changes in Japan's relations with the United States and the world as part of its rise to economic power? What questions might such a study ask?

©Jodi Cobb/National Geographic Society

Research is a fact-gathering and fact-explaining enterprise governed by strict rules (Hagan 1989). **Research methods** are the various techniques that sociologists and other investigators use to formulate meaningful research questions and to collect, analyze, and interpret facts in ways that allow other researchers to check the results. We need to possess a working knowledge of research methods even if we do not plan to become sociologists or to do research of our own. One important reason is connected with a relatively new global phenomenon — the **information explosion.** This dramatic term describes an unprecedented increase in the volume of data due to the development of the computer and telecommunications. Notice that we distinguish between data and information. **Data** consist of printed, visual, and spoken materials. Data become **information** after someone reads, listens to, or views data.

Research Fact gathering and explaining enterprise governed by strict rules.

Research methods The various techniques used to formulate meaningful research questions and to collect, analyze, and interpret facts in ways that allow other researchers to check the results.

Information explosion A term describing an unprecedented increase in the volume of data due to the development of the computer and telecommunications.

Data Printed, visual, and spoken materials.

Information Data that have been read, viewed, or heard.

In addition to coping with large quantities of data, we also have to consider the data's quality. Most of the data that we hear, read, and see have been created by others. Therefore, we can never be sure that data are accurate. In "Too Much of a Good Thing? Dilemmas of an Information Society," Donald Michael (1984) argues that we cannot assume that more data will lessen uncertainty and increase feelings of control and security. In fact, the opposite may be true: more data can overwhelm us to the point that we conclude we cannot believe anything we hear, read, or see. We need not accept Michael's gloomy assessment, however, if we possess a working knowledge of social research methods. Such knowledge gives us the skills to identify and create high-quality data. In addition, this working knowledge enables us to evaluate the accuracy of the things we read, hear, and view and to cope with the consequences of the information explosion.

The Information Explosion

At least two technological innovations are responsible for the information explosion: computers and telecommunications. Both technologies help people create, store, retrieve, and distribute large quantities of data at mind-boggling speeds. Comparing the size and capabilities of the first computers with those of the present suggests why the volume of data has increased so rapidly over the past fifty years. The computers of the 1940s weighed five tons, stood 8 feet tall and 51 feet long, and contained 17,468 vacuum tubes and 5,000 miles of wiring (Joseph 1982). They performed simple calculations in a few seconds, but they tended to overheat and break down. They also used so much power that the lights in nearby towns often failed when the machines were turned on. And, because of their size and cost, computers were used only by the U.S. Defense Department and the Census Bureau. Today, in contrast, a single silicon chip a quarter-inch thick can process millions of bits of information in a second. The chip has reduced computer size and cost and has made possible the widespread use of the personal computer.

Similarly, telecommunications have increased our ability to send data quickly across space. Although the telephone, radio, and television have existed in some form for up to a hundred years, methods of rapidly transmitting clear signals have changed considerably. Fiber optic cables have replaced wire cables as the means of transmitting images, voices, and data. In 1923, the cable connecting Britain and the United States contained 80,000 miles of iron and steel wire (enough to circle Earth three times) and four million pounds of copper. It could transmit the equivalent of 1,200 letters of the alphabet per minute across the ocean. In contrast, when the capabilities of fiber optics are exploited fully, a single fiber the diameter of a human hair can carry the entire telephone voice traffic of the United States and can transmit the contents of the Library of Congress anywhere in the world in a few seconds (Lucky 1985). This latter example is no small feat if we stop to consider that the Library of Congress houses 100 million items on 532 miles of shelves (Thomas 1992).

One software tool that has helped to increase the amount of data available at our fingertips is **hypertext**. Before hypertext, reading was a linear matter. That is, a book, pamphlet, chapter, column, advertisement, or the like, had a clear beginning and end. Most of us know no other way to read. The linear format, however, does not match the way we think, because the mind operates by association: "with one item in its grasp, it snaps instantly to the next by association" (Deemer 1994).

Sometimes the associations lead us along unexpected paths. Hypertext is the opposite of the "start-and-end-here" approach in that readers are free to pick and choose among keywords (that is, important ideas) and follow links that enable them to learn more about those keywords. Readers can choose to wander off along tangential links, some of which may not be even remotely related to the topic at hand.

Likewise, such innovations as the personal computer and fiber optics have dramatically changed the methods of gathering and distributing information, putting data creation technologies into the hands of the general public and increasing the speed by which data are entered, edited, duplicated, stored, retrieved, and distributed. Sociologist Orrin Klapp writes about the information explosion in *Overload and Boredom: Essays on the Quality of Life in the Information Society.* He uses a vivid metaphor to describe the dilemma of sorting through and keeping up with the massive amounts of data generated. Klapp envisions a person "seated at a table fitting [together] pieces of a gigantic jigsaw puzzle. From a funnel overhead, pieces are pouring onto the table faster than one can fit them. Most of these pieces do not match up. Indeed, they do not all belong to the same puzzle" (Klapp 1986:110). The pieces falling from overhead represent research accumulating at a pace that interferes with people's ability to organize it into a comprehensible pattern. Klapp uses this analogy to show that the speed with which data are produced and distributed overwhelms the brain's capacity to organize and evaluate it.

When we want information on any subject, we must not only select from a large quantity of data but also often sift through distorted, exaggerated presentations. Klapp gives some reasons for these added burdens. New technologies permit large numbers of magazines, newspapers, radio stations, and television channels to exist. As a result, message senders must compete for our attention. Reporters, producers, and others in the media often devise ways to entice us to read and listen. Common strategies include eye-catching headlines, misleading titles, and shocking stories (murders, car accidents, plane crashes). Some eye-catching book titles and news headlines related to Japan, for example, include "Collision Course: Confrontation Looms as Japan Persists with Trade Barriers," "In Japan, Even Toddlers Feel the Pressure to Excel," "Still in Diapers but Cramming for those Competitive Exams," and "The Coming War with Japan." Too often such exaggerated headlines lure people into reading and listening to material that turns out to be trivial, repetitive, contradictory, and ultimately

Hypertext A computer-based, nonlinear text system.

uninformative in any broader sense. The titles and headlines might catch our attention, but they usually mask rather than reveal a more complex reality.

Klapp also cites **dearth of feedback** as a factor in creating poor-quality data. What does he mean by "dearth of feedback"? Much of the data that are televised and published are not subjected to honest, constructive feedback because there are too many messages and not enough critical readers and listeners to evaluate the data before the material is released or picked up by the popular media. Without feedback, the creators cannot correct their mistakes; thus, the data they produce are diminished in quality. Klapp believes that data overload, coupled with distortion, exaggeration, and triviality, is as problematic as a lack of data.

Although it is easy to become pessimistic about the human ability to organize, evaluate, comprehend, and trust or question the growing quantity of data, the situation has another, brighter side. For one thing, the information explosion also increases the chances that good and useful ideas will receive some exposure. For another, the variety means that there is something for everyone. And, even though the data are not well organized, researchers still can draw from and organize the data in new and unexpected ways.

The information explosion does not negate the need to be informed; it simply increases the need to be able to create, identify, and synthesize data and turn data into useful and worthwhile information. Decisions still must be made, actions still must be taken, and policies still must be formed. No constructive decision, action, or policy can be based on haphazard, misleading, or inadequate data. That situation would be equivalent to a physician's decision to perform heart surgery based only on the intuition that such action will solve the patient's health problems, without ordering medical tests, reviewing the patient's history, or interviewing the patient beforehand.

The larger point is that we live in a society in which people need to be more than computer literate (able to operate a computer and use it to input, access, and print out data). People also need to be **research-methods literate**; that is, they must know how to collect data that are worth putting into the computer and how to interpret the data that come out of it. Unless computer-literate people also have research-methods literacy, they merely possess the skills to enter data. We turn now to basic techniques and strategies that sociologists (and all other researchers) use to evaluate and gather reliable data. We start with the guiding principle: the scientific method.

Dearth of feedback A situation in which not enough critical readers and listeners evaluate material before it is used by the popular media.

Research-methods literate The ability to know how to collect data worth entering into a computer and then how to interpret the resulting data.

Scientific method An approach to data collection in which knowledge is gained through observation and its truth confirmed through verification.

Objectivity A state in which personal, subjective views do not influence one's opinions or behavior.

The Scientific Method

Sociologists are guided by the scientific method when they investigate human behavior; in this sense they are scientists. The **scientific method** is an approach to data collection guided by two assumptions: (1) knowledge about the world is acquired through observation, and (2) the truth of the knowledge is confirmed by verification — by others making the same observations. Researchers collect data that they and others can see, hear, taste, touch, and smell. They must report the process by which they make their observations and present conclusions so that interested parties can duplicate that process. If observations cannot be duplicated, or if upon duplication the results differ substantially from those of the original study, the study is considered suspect. Findings endure as long as they can withstand continued reexamination and duplication by the scientific community. "Duplication is the heart of good research" (Dye 1995:D5). No finding can be taken seriously unless other researchers repeat the process and obtain the same results. When researchers know that others are critiquing and checking their work, the result is reinforcement of careful, thoughtful, honest, and conscientious behavior. Moreover, this "checking" encourages researchers to maintain **objectivity**—that is, not to let personal and subjective views about the topic influence the outcome of the research.

Because of continued reexamination and revision, research is both a process and a dialogue. It is a process because findings and conclusions never are considered final. It is a dialogue because a critical conversation between researchers and readers leads to more questions and additional research. This description of the scientific method is an ideal one, because it outlines how researchers and reviewers should behave. In practice, though, questionable acts on both sides do occur sometimes. Some research may be dismissed as unimportant (often even before it is read) and as unworthy of examination simply because the topic is controversial, or departs from mainstream thinking or because the results are reported by someone from a group that is considered "inferior." Moreover, the scientific method works on the assumption that researchers are honest — that they do not manipulate data to support personal, economic, and political agendas. However, the extent to which researchers actually are honest is unknown.

In one survey sponsored by the American Association for the Advancement of Science (AAAS), 25 percent of the 1,500 people surveyed reported that in the past ten years they had witnessed some faking, falsifying, or plagiarizing of data (Marsa 1992). In another survey of 2,000 doctoral candidates and 2,000 faculty[1] from 99 of the largest chemistry, civil engineering, microbiology, and sociology graduate schools in the United States, Acadia Institute researchers found that "22 percent of faculty report instances of their colleagues overlooking sloppy use of data, and 15 percent know of cases where data that would contradict an investigator's own previous research have not been presented" (Swazey, Anderson, and Lewis 1993:545).

Ideally, research is a carefully planned, multistep, fact-gathering and fact-explaining enterprise (Rossi 1988) that involves a number of interdependent steps:

1. Defining the topic for investigation
2. Reviewing the literature
3. Identifying core concepts
4. Choosing a research design, forming hypotheses, and collecting data
5. Analyzing the data
6. Drawing conclusions

Researchers do not always follow the steps in sequence, however. Sometimes they do not define the topic (step 1) until they have familiarized themselves with the literature (step 2). Sometimes an opportunity arises to gather information about a group (step 4), and a project is defined to fit the opportunity (step 1). Although the six steps need not be followed in sequence, all need to be completed to ensure the quality of the project.

[1]Two thousand six hundred people returned the survey, for a response rate of 65 percent.

In the sections that follow, we will examine each stage individually, making reference to a variety of research projects investigating the U.S.-Japan trade deficit (see Figure 3.1), comparing people living in the United States with people living in Japan on some attribute, and documenting some supposedly unique qualities about Japan's society.

Step 1: Defining the Topic for Investigation

The first step of a research project is choosing a topic. It would be impossible to compile a comprehensive list of the topics that sociologists study, because almost any subject involving humans is open to investigation. Sociology is distinguished from other disciplines not by the topics it covers, but by the perspectives it uses to study topics.

Good researchers explain to their readers *why* their chosen topic is significant. Explanation is vital because it clarifies the purpose and significance of the project, as well as the motivation for doing the work. If you do not know why you are conducting a study, it is unlikely to generate much personal or public interest.

Researchers choose their topics for a number of reasons. Personal interest is a common and often underestimated motive. It is perhaps the most significant reason that someone picks a specific topic to study. This is especially true if we consider how a researcher eventually chooses one topic from a virtually infinite set of possibilities. Consider the reasons that researcher Anne Allison (1991) studied *obentōs,* the lunch boxes that Japanese mothers make up for their children when they are in nursery school. An *obentō* is a "small box packaged with a five or six course miniaturized meal whose pieces and parts are artistically arranged, perfectly cut, and neatly arranged" (Allison 1991:196). Allison became interested in the *obentōs* while she was living in Japan and taking her child to a Japanese nursery school. Her son's nursery school teacher talked with Allison "daily about the progress he was making finishing his *obentōs*" (p. 200). As Allison tells it:

> The intensity of these talks struck me at the time as curious. We had just settled in Japan and David, a highly verbal child, was attending a foreign school in a foreign language he had not yet mastered; he was the only non-Japanese child in the school. Many of his behaviors during this time were disruptive: for example, he went up and down the line of children during morning exercises hitting each child on the head. Hamada-sensei [the teacher], however, chose to discuss the *obentōs*. I thought surely David's survival in and adjustment to this environment de-

Figure 3.1 United States Trade Deficit with Top Twenty Countries, 1994 (general imports in millions of dollars)

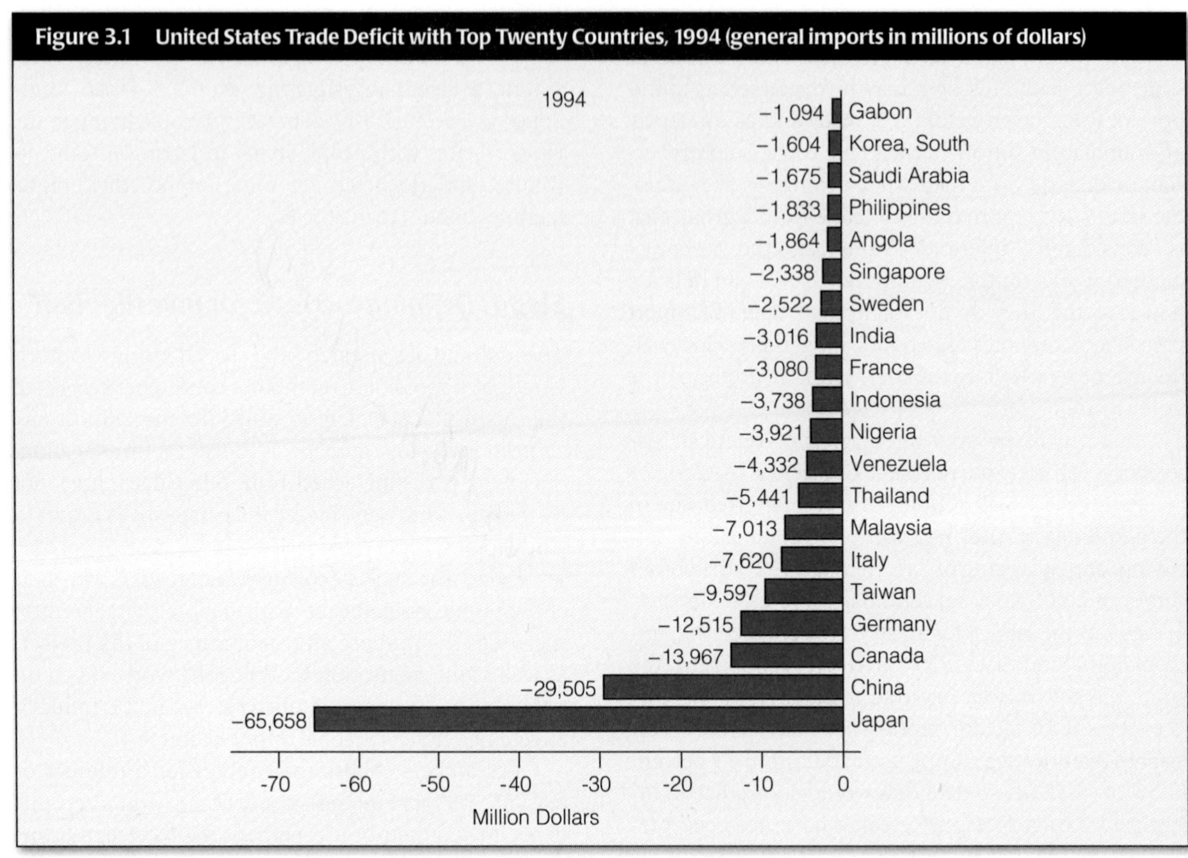

Source: U.S. Department of Commerce (1996), Table 13.

> pended much more on other factors, such as learning Japanese. Yet it was the *obentō* that was discussed with such recall of detail ("David ate all his peas today, but not a single carrot until I asked him to do so three times") and seriousness that I assumed her attention was being misplaced. (pp. 200–201)

On a personal level, Allison undertook this study to learn more about her son's new environment so she could help him adapt to it. Realizing that the *obentō* was somehow significant to his success in a Japanese preschool, Allison picked it as a research topic. The choice of a research topic, however, usually has some further significance that goes beyond personal interest. Allison's research on *obentōs*, for example, examines why Japanese mothers are expected to make such elaborate lunches for their children and why Japanese teachers pay so much attention to how well children eat lunch. In addition, Allison's research offers readers some insights into the Japanese system of preschool education and allows us to see the educational significance of eating lunch — a seemingly routine activity. According to Allison's research, the elaborately prepared lunch is a sign of a Japanese woman's commitment as a mother, which inspires her child to be similarly committed as a student.

Step 2: Reviewing the Literature

All good researchers take existing research into account. They read what knowledgeable authorities have written on the chosen topic, if only to avoid repeating what has already been done. More important, reading the relevant literature can generate insights that the researcher may not have considered. Even if a researcher believes that he or she has revolutionary ideas, the researcher must consider the works of past thinkers and show how the new research verifies, advances, and corrects what has been done in the past.

One example of research that "corrected" existing literature is Roger Goodman's (1993) work on *kikokushijo* — Japanese children who return to Japan after living overseas with their parents who work for

The preschool obentōs *that inspired Anne Allison's study are a good example of how sociologists' curiosity may be aroused by almost any fact of social life.*

©Ulrike Welsch

Japanese corporations based in foreign countries. Goodman found that literature on this subject indicated that *kikokushijo,* especially those who have returned from the United States or Europe, suffer from nonadaptation disease, "all kinds of mental and physical problems on their return to Japan owing to the very nature of Japanese society," which is characterized as "homogeneous, exclusivist, conformist, and harmonious" in contrast to the individualist, heterogeneous, independent, and argumentative West (p. 3). According to the literature, the adaptation problems are so great that special schools known as *ukeireko* have been established to offer "relief education" and to "re-japanize" returnees or "peel off their foreignness." However, Goodman found nothing in the literature investigating the process by which these special schools accomplish these goals.

Goodman's review of the literature shaped his research in that he used it as a contrast against his own experiences. After spending one year observing Japanese junior high schools, he spent a second year observing a *ukeireko.* Goodman found this special school to be very different from that described in the literature.

> The significance of the *ukeireko* depended not so much on the treatment accorded to the *kikokushijo* within it as on the fact that such a school existed at all. The *ukeireko* proved to be more interested in the education of future leaders of society than in "decontaminating" *kikokushijo* of their overseas experience. The school modelled itself more on a British public school than on a refugee camp. The teachers sought to instil a combination of "Japaneseness" and "internationalness" in their education programmes that seemed to undermine the widely held image of Japan's innate exclusivity, and the problems that such exclusivity caused for *kikokushijo.* (p. 4)

"Where Students Study Abroad" gives another perspective on foreign travel.

Step 3: Identifying and Defining Core Concepts

After deciding on a topic and reading the relevant literature (not necessarily in that order), researchers typically state their core concepts. **Concepts** are powerful thinking and communication tools that enable us to

Concepts Thinking and communication tools that are used to give and receive complex information efficiently and to frame and explain observations.

U.S. In Perspective

Where Students Study Abroad

Mike Mansfield, U.S. ambassador to Japan (1977–1988), called the relationship between the two countries "the most important bilateral relationship in the world—bar none" (Ryūzō 1991:274). The map and tables show where students studied abroad and which countries sent the greatest number of students to the United States around 1994–95. What might the former ambassador say in response to these numbers?

What is your own comment on the numbers, in view of how much people in the United States know or will need to know about Japan and other societies around the world?

Why are the patterns as they are? What should or could be done on the U.S. side to change them?

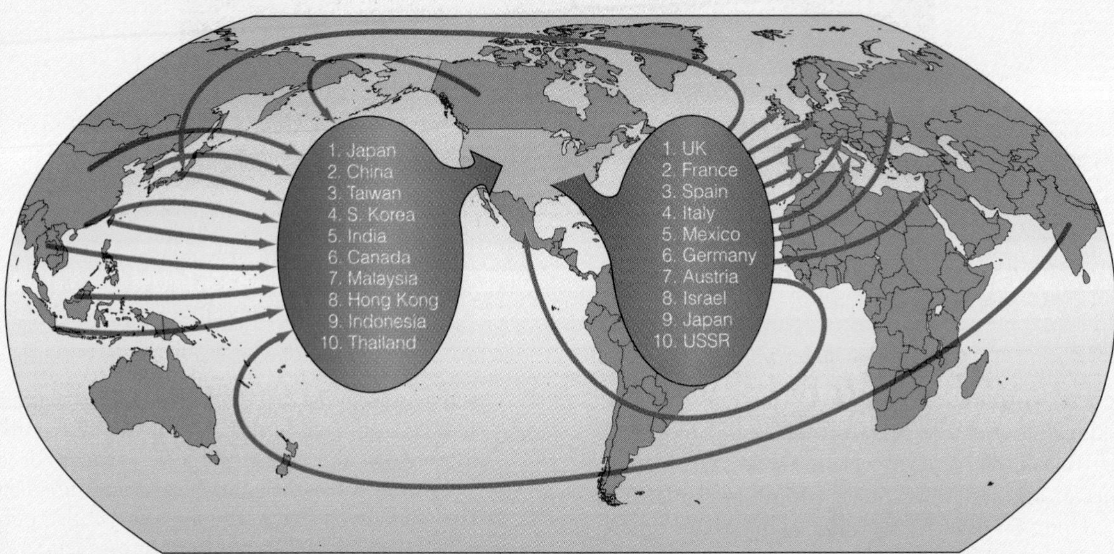

Leading Ten Host Countries of the U.S. Study Abroad Students

	%	*n*
United Kingdom	27.0	29,700
France	12.8	14,080
Spain	10.4	11,440
Italy	8.4	9,240
Mexico	5.0	5,500
Germany	4.7	5,170
Austria	3.0	3,300
Israel	2.6	2,860
Japan	2.1	2,310
USSR	1.9	2,090

110,000 (estimated) U.S. students studied abroad in 1994–95

The Ten Top Countries That Send Students to the United States

		% of all foreign
Japan	45,276	9.9
China	39,403	8.6
Taiwan	36,407	8.0
Rep. of Korea	33,599	7.4
India	33,537	7.3
Canada	22,747	5.0
Malaysia	13,617	3.0
Hong Kong	12,935	2.8
Indonesia	11,872	2.6
Thailand	10,889	2.3
Total	456,000	

Sources: Institute of International Education (1993/94, 1994/95)

give and receive complex information efficiently and to frame and explain observations. Concepts prompt researchers to focus their attention on specific features. The mention of a concept triggers in the minds of others who are familiar with its meaning a definition and a range of important associations. Consider the concept suicide. Among sociologists who hear the word, *suicide* brings to mind the French sociologist Emile Durkheim's (1858–1918) classic definition — "the severing of relationships" — and it focuses attention on the structure of relationships and the kinds of relationships that promote or discourage suicide. Durkheim maintained that those who find themselves in any of the following four situations are likely to have higher suicide rates than others who have more secure and stable attachments to a group: (1) excessively isolated from the group (the chronically ill or elderly), (2) excessively attached to the group to the point that the self cannot be separated from the group (recruits undergoing basic training), (3) suddenly thrown out of a group (the suddenly unemployed, lottery winners, widows and widowers), and (4) hopelessly locked into a group (a prisoner on death row, those without resources to escape a confining environment such as an impoverished rural community) (Durkheim 1951).[2]

This concept of suicide was one that guided David Lester, Yutaka Molohashi, and Bijow Yang (1992) in their study of the effect of four socioeconomic variables — the unemployment rate, the annual percentage change in the gross national product, female participation in the labor force, and the divorce rate per 1,000— on the rate of suicide in Japan and the United States. Among other things, the researchers tested Durkheim's idea that an increase in the rate of divorce would lead to a corresponding increase in the suicide rate because divorce separates those involved from a family unit and reduces regulatory forces on people's lives. The point is that a clear statement of core concepts enables researchers to focus their investigation.[3]

Step 4: Choosing a Research Design and Collecting the Data

Once researchers have clarified core concepts, they decide on a **research design**, a plan for gathering data on the topic they have chosen. A research design specifies the population to be studied and the **method of data collection**, the procedures used to gather relevant data. One research design in itself is not better than another. The population and the data-gathering procedures that are chosen for one study may not be appropriate for another study. Thus, researchers choose a design to fit the circumstances of each study (Smith 1991).

The Population to Be Studied Researchers must decide who or what they are going to study. The most common "thing" social scientists study is individuals, but they may also decide to study traces, documents, territories, households, small groups, or individuals (Rossi 1988).

Traces **Traces** are materials or other evidence that yield information about human activity, such as the items that people throw away, the number of lights on in a house,[4] or changes in water pressure.[5] One example of research that examines traces is the Garbage Project, a University of Arizona program directed by William L. Rathje (Rathje and Murphy 1992). Since the 1970s, Rathje and his team of researchers have been collecting garbage from landfills and selected neighborhood garbage cans. Among other things, Rathje has found that the hazardous waste generated by households across the United States equals the hazardous waste generated by commercial establishments.

Research design A plan for gathering data that specifies the population and method of data collection.

Method of data collection The procedures used to gather relevant data.

Traces Materials or other evidence that yield information about human activity.

[2]Keep in mind that the four categories are not clear-cut. Prisoners on death row are in a situation that isolates them from family and friends. At the same time, prisoners are hopelessly confined to living in a highly structured environment with other inmates.

[3]Lester and his colleagues found that in the United States, the higher the divorce rate, the higher the suicide rate. By contrast, in Japan, the higher the divorce rate, the lower the suicide rate. The researchers note that the "rates of divorce are very different in the two countries, ranging from .73 per 1,000 persons per year to 1.51 in Japan as compared to a range of 2.1 to 5.3 in the U.S.A. Thus, divorce rates may have a different social meaning as a social indicator in Japan than in the U.S.A." (Lester, Motohashi, and Yang 1992:317).

[4]Sociologists Robert and Helen Lynd ([1929] 1956) observed the times at which lights were turned on during winter mornings to determine whether people from working-class households started their days earlier than middle-class people.

[5]Researchers have studied reductions in municipal water pressure to obtain estimates of the number of people watching commercials shown at breaks in prime-time programs. The assumption is that when commercials are interesting, fewer people go to the bathroom than when commercials are uninteresting (Rossi 1988).

Documents **Documents** are written or printed materials, such as magazines, books, calendars, graffiti, birth certificates, and traffic tickets. For her research on *obentōs,* Anne Allison (1991) examined a variety of documents. She did a content analysis of *obentō* magazines and cookbooks and *obentō* guidelines that are issued biweekly by Japanese nursery schools. From examining the documents, Allison realized the cultural importance of the *obentō.* She found that the magazines, cookbooks, and guidelines were filled with recipes, hints, pictures, and ideas about how to prepare food so that children will eat it. One common suggestion was that the mother should design the presentation of food items and the bag in which the food is carried. In other words, the mother should not let someone else prepare the *obentō.*

Territories **Territories** are settings that have borders or that are set aside for particular activities. Examples include countries, states, counties, cities, streets, neighborhoods, classrooms, and buildings. William H. Whyte's *City: Rediscovering the Center* (1988) is an example of research that examines a territory. For more than sixteen years, Whyte visited and observed New York City and other cities across the United States and around the world. He was interested in identifying the conditions under which people use urban spaces. Whyte found that some city spaces simply are not designed to be used: "Steps too steep, doors too tough to open, ledges you cannot sit on because they are too high or too low, or have spikes on them" (p. 1). Often city streets are simply not very interesting places. Among other things, Whyte also found that Tokyo's streets are consistently more interesting than most city streets in the United States. This is because the Japanese "do not use zoning to enforce a rigid separation of uses. They encourage a mixture, not only side by side, but upwards. In the buildings you will see showrooms, shops, pachinko parlors, offices, all mixed together and with glass-walled restaurants rising one on top of the other — three, four, and five stores up" (p. 89).[6]

Households **Households** include all related and unrelated persons who share the same dwelling. Economist Raymond A. Jussaume, Jr., and sociologist Yoshiharu Yamada surveyed households in Seattle and in Kobe, Japan, and found some important differences between the two. The very composition of Kobe and Seattle households is different: 3 percent of Kobe households have no children or elderly residents, in contrast to 15.7 percent of Seattle households. Also, although virtually every Kobe and Seattle household has at least one telephone, only 2.1 percent of Kobe households have unlisted telephone numbers, compared with approximately 20 percent of Seattle households (Jussaume and Yamada 1990).

Small Groups **Small groups** are defined as two to about twenty people who interact with one another in meaningful ways (Shotola 1992). Examples include father-child pairs, doctor-patient pairs, families, sports teams, circles of friends, and committees. An example of research conducted at this level of analysis is the work of sociologist Masako Ishii-Kuntz (1991). She studied father-child relationships in three countries (Japan, Germany, and the United States) to evaluate the popular belief that Japanese fathers are largely absent from the home and uninvolved in their children's lives. The results of Ishii-Kuntz's study are described later in this chapter.

Because of time constraints alone, researchers cannot study entire **populations** — the total number of individuals, traces, documents, territories, households, or groups that could be studied. Instead, they study a **sample**, or a portion of the cases from a larger population.

Samples Ideally, a sample should be a **random sample**, with every case in the population having an equal chance of being selected. The classic, if inefficient, way

Documents Written or printed materials used in research.

Territories Settings that have borders or that are set aside for particular activities.

Households All related and unrelated persons who share the same dwelling.

Small groups Two to about twenty people who interact with one another in meaningful ways.

Populations The total number of individuals, traces, documents, territories, households, or groups that could be studied.

Sample A portion of the cases from a larger population.

Random sample A sample in which every case in the population has an equal chance of being selected.

[6]Pachinko is a type of pinball game. Pachinko parlors are typically located near train stations and commercial districts (Koji 1983).

A systematic look at specific social units such as the family reveals differences between the United States and Japan. For example, households without children or elderly residents are much more common in Seattle than in Kobe.

©Dave Bartruff/Artistry International

of selecting a random sample is to assign every case a number, place the cards or slips of paper on which the numbers are written into a container, thoroughly mix the cards, and pull out one card at a time until the desired sample size is achieved. However, rather than follow this tedious system, most researchers use computer programs to generate their samples. If every case has an equal chance of becoming part of the sample, then theoretically the sample should be a **representative sample** — that is, one with the same distribution of characteristics (such as age, gender, and ethnic composition) as the population from which it is selected. Thus, for example, if 56.4 percent of the population from which a sample is drawn is at least thirty years old, then approximately 56.4 percent of a representative sample should be that age. Sometimes just by chance researchers draw samples that do not represent the population with regard to specified characteristics. In that case researchers may draw additional cases from the categories that are underrepresented in the sample. In theory, if the sample is representative, then whatever is true for the sample is also true for the larger population.

Obtaining a random sample is not as easy as it might appear. For one thing, researchers must begin with a **sampling frame** — a complete list of every case in the population — and each member of the population must have an equal chance of being selected. Securing such a complete list can be difficult. Campus and city telephone directories are easy to acquire, but lists of, say, U.S. citizens, of adopted children in the United States, of U.S.-owned companies in Japan or of Japanese-owned companies in the United States are not so easily obtained. Almost all lists omit some people (perhaps persons with unlisted numbers, members too new to be listed, or between-semester transfer students) and include some people who no longer belong (such as persons who have moved, died, or dropped out). What is important is that the researcher consider the extent to which the list is incomplete and update it before drawing a sample. Even if the list is complete, the researcher also must think of the cost and time required to take random samples and consider the problems of inducing all sampled persons to participate.

Researchers sometimes select samples to study people that they know are not representative of the larger population because they are easily accessible. For example, researchers often sample from high

Representative sample A research sample with the same distribution of characteristics as the population from which it is selected.

Sampling frame A complete list of every case in the population.

school and college students because they are a captive audience. Other researchers may choose nonrepresentative samples for other important reasons: (1) little is known about them, (2) they have special characteristics, or (3) their experiences clarify important social issues. (See "The Innovative Research of Erving Goffman.")

Methods of Data Collection In addition to identifying who or what is to be studied, the design also must include a plan for collecting information. Researchers can choose from a variety of data-gathering methods including self-administered questionnaires, interviews, observations, and secondary sources.

Self-Administered Questionnaire A **self-administered questionnaire** is a set of questions given (or mailed) to respondents, who read the instructions and fill in the answers themselves. This method of data collection is probably most common. The questionnaires found in magazines or books, displayed on tables or racks in service-oriented establishments (hospitals, garages, restaurants, groceries, physicians' offices), and mailed to households are all self-administered questionnaires. This method of data collection has a number of advantages. No interviewers are needed to ask respondents questions; the questionnaires can be given to large numbers of people at one time; and respondents are not influenced by an interviewer's facial expressions or body language, so they feel more free to give unpopular or controversial responses.

Researchers Paul Tuss, Jules Zimmer, and Hsiu-zu Ho (1995) used a questionnaire to study the beliefs that fourth-grade achievers[7] and underachievers in the United States, Japan, and China held about the reasons for successful and unsuccessful mathematics performance. Two of the questions were as follows:

1. People use different reasons to explain why they have done things well or poorly. Think of the last test you did poorly on. Why do you think you did so poorly? Write the reasons on the lines below.
2. Now think of the last math test you did well on. Why do you think you did so well? Write the reasons on the lines below. (p. 414)

Self-administered questionnaires pose some disadvantages. Respondents can misunderstand and ignore some questions. Often questionnaires are mailed, set out on a table or counter, or published in a magazine or newspaper. This leaves researchers wondering whether the people who choose to fill out a questionnaire have different opinions than those who don't. The results of a questionnaire depend not only on respondents' decisions to fill out and return it but also on the quality of the survey questions asked and a host of other considerations.

Interviews In comparison with questionnaires, **interviews** are more personal. They are face-to-face sessions or telephone conversations between an interviewer and a respondent in which the interviewer asks questions and records respondents' answers. As respondents give answers, interviewers must avoid pauses, expressions of surprise, or body language that reflect value judgments. Refraining from such conduct helps respondents feel comfortable and encourages them to give honest answers.

Interviews can be structured or unstructured, or some combination of the two. In a **structured interview**, the wording and sequence of questions are set in advance and cannot be altered during the course of the interview. In one kind of structured interview, respondents choose answers from a response list that the interviewer reads to them. In another kind of structured interview, respondents are free to answer the questions as they see fit, although the interviewer may ask them to clarify or explain answers in more detail.

In contrast, an **unstructured interview** is flexible and open-ended. The question-answer sequence is spontaneous and resembles a conversation in that the questions are not worded in advance and are not asked in a set order. The interviewer allows respondents to take the conversation in directions they define as crucial. The interviewer's role is to give focus to the inter-

[7]Achievers are those who scored in the expected range for someone in that grade.

Self-administered questionnaire A set of questions given to respondents who read the instructions and fill in the answers themselves.

Interviews Face-to-face or telephone conversations between an interviewer and a respondent in which the interviewer asks questions and records respondents' answers.

Structured interview An interview in which the wording and sequence of questions are set in advance and cannot be changed during the interview.

Unstructured interview An interview in which the question-answer sequence is spontaneous, open-ended, and flexible.

The Innovative Research of Erving Goffman (1922)

Erving Goffman made the study of talk, conversation, and interaction his life's work. There is no doubt that Goffman's social "methods for accomplishing this were, to say the least, singular" (Drew and Wootton 1988:6). His writings span thirty years, beginning in 1953 with the completion of his Ph.D. dissertation (*Communication Conduct in an Island Community*) and ending in 1983 with the publication of his presidential address to the American Sociological Association (*The Interaction Order*). Some sociologists believe that Goffman was "the greatest sociologist of the latter half of the twentieth century" (Collins 1988:41). Although other sociologists take issue with such a claim, "no one would question the claim that Erving Goffman was one of the leading sociological writers of the post–[World War II] period" (Giddens 1988:250). Few sociologists writing in the latter half of this century have made the impression that Goffman has on professionals/academics outside the discipline of sociology (Giddens 1988), including anthropology, linguistics, folklore, communication, political science, psychiatry, ethnology (Winkin 1989).

One of the many qualities that makes Goffman unique among sociologists is the sources from which he drew to form, support, and illustrate his ideas and insights about the largely taken-for-granted world of everyday interaction. These sources included newspaper clippings, comic strips, scenes from popular films, cartoons, personal observations and anecdotes, "Dear Abby" columns, etiquette manuals such as those by Emily Post, popular fiction, serious novels, autobiographies, the writings of "respectable" researchers, "informal memoirs written by colorful people"

Erving Goffman

Courtesy American Sociological Association

(Goffman 1959:xi), and contemporary and ancient theatrical writings. In the essay "Radio Talk: A Study of the Ways of Our Errors," Goffman (1981) tells readers that he draws on the following sources: "eight of the LP records and three of the books produced by Kermit Schafer from his recordings (Jubilee Records) of radio bleepers. . .; twenty hours of taped programs from two local stations in Philadelphia and one in the San Francisco Bay area; a brief period of observation and interviewing of a classical DJ at work; and informal note-taking from broadcasts over a three-year period" (p. 197).

Goffman believes that any group of persons – subsistence farmers, prisoners, pilots, patients, disk jockeys, people with disabilities, the socially rejected – "develop a life of their own that becomes meaningful, reasonable, and normal once you get close to it, and that a good way to learn about any of these worlds is to submit oneself in the company of the members to the daily round of petty contingencies to which they are subject" (Goffman 1961:x). This was not idle talk on the part of Goffman. For his doctoral dissertation, Goffman spent one year studying a Shetland Island community of 300 people with a subsistence farming economy. For his book *Asylums: Essays on the Situation of Mental Patients and Other Inmates,* Goffman (1961) assumed the role of an assistant to the athletic director in a 7,000-bed federal mental hospital for one year to learn about "the social world of the hospital inmate, as this world is subjectively experienced by him (p. ix).

Goffman avoided one of the most common methods of research – gathering and generating statistical evidence. He believed that creating statistics interfered with his goals of capturing and conveying to his readers the "tissue and fabric" of the interaction order. Goffman acknowledged that, in addition to the absence of statistical data in his writings, his approach to studying social encounters is biased in the direction of one category of participants over other participants. Goffman described this bias most openly in the preface to *Asylums:*

> *The world view of a group functions to sustain its members and expectedly provides them with a self-justifying definition of their own situation and a prejudiced view of nonmembers, in this case, doctors, nurses, attendants, and relatives. To describe the patient's situation faithfully is necessarily to present a partisan view. (For this last bias I partly excuse myself by arguing that the imbalance is at least on the right side of the scale, since almost all professional literature on mental patients is written from the point of view of the psychiatrist, and he, socially speaking, is on the other side). (Goffman 1961:x)*

view, ask for further explanation or clarification, and probe and follow up interesting ideas expressed by respondents. The interviewer appraises the meaning of answers to questions and uses what was learned to ask follow-up questions. Talk show hosts often use an unstructured format to interview their guests. However, sociologists have much different goals than talk show hosts. For one thing, sociologists do not formulate questions with the goal of entertaining an audience. In addition, sociologists strive to ask questions in a neutral way, and no audience reaction is possible to influence how respondents answer the questions.

Observation As the term implies, **observation** involves watching, listening to, and recording behavior and conversations as they happen. This technique sounds easy, but observation is more than seeing and listening. The challenge of observation lies in knowing what to look for while still remaining open to other considerations; success results from identifying what is worth observing. "It is a crucial choice, often determining the success or failure of months of work, often differentiating the brilliant observer from the . . . plodder" (Gregg 1989:53). Good observation techniques must be developed through practice to learn to know what is worth observing, be alert to unusual features, to take detailed notes, and make associations between observed behavior.

If observers come from a culture different from the one they are observing, they must be careful not to misinterpret or misrepresent what is happening. Imagine for a moment how an uninformed, naive observer might describe a sumo wrestling match: "One big, fat guy tries to ground another big, fat guy or force him out of the ring in a match that can last as little as three seconds" (Schonberg 1981:B9). Actually, for those who understand it, sumo wrestling is "a sport rich with tradition, pageantry, and elegance and filled with action, excitement, and heroes dedicated to an almost impossible standard of excellence down to the last detail" (Thayer 1983:271).

Observation A research technique involving watching, listening to, and recording behavior and conversations as they happen.

Nonparticipant observation A research technique involving detached watching and listening in which the researcher does not interact with the study participants.

Participant observation A research technique in which researchers interact directly with study participants.

Observational techniques are especially useful for studying behavior as it occurs, learning things that cannot be surveyed easily, and acquiring the viewpoint of the persons under observation. Observation can take two forms: participant and nonparticipant. **Nonparticipant observation** is detached watching and listening: the researcher only observes and does not interact or become involved in the daily life of those being studied. A good example of nonparticipant observation is Catherine C. Lewis's (1988) research on Japanese first-grade classrooms. Lewis observed fifteen Japanese classrooms each for a full day, "from morning greetings through the children's departure from school" (p. 161). She observed:

> [I]t was common for two [student] monitors to stand at the front of the class, ask the class to be quiet, announce what subject was about to be studied, and ask the children to rise and greet the teacher. In some classrooms, the monitors picked which group showed the best order, cautioned groups to improve their posture, granted check marks to groups that were quiet and ready to begin the class period, or chose which groups would receive lunch first. Often, the monitors assembled and quieted the class when the teacher was not yet present. Typically, monitors rotated each day, and all children became monitors in turn. (pp. 162–163)

In contrast, researchers engage in **participant observation** when they join a group and assume the role of a group member, interact directly with those whom they are studying, assume a position critical to the outcome of the study, or live in a community under study. Anne Allison's research on *obentōs* described earlier in this chapter is an excellent example of participant observation. Her research revolved around conversations she had with other mothers, daily conversations with the teacher about her son's eating habits, Mothers' Association Meetings, and other school-related events.

In both participant and nonparticipant observation, researchers must decide whether to hide or to announce their identity and purpose. One of the primary reasons for choosing concealment is to avoid the Hawthorne effect,[8] a phenomenon whereby research

[8]The term "Hawthorne effect" originates from a series of studies of workers' productivity conducted in the 1920s and 1930s, which involved female employees of the Hawthorne, Illinois, plant of Western Electric. Researchers found that no matter how they varied working conditions — bright versus dim lighting, long versus short breaks, frequent versus no breaks, piece rate pay versus fixed salary—workers' productivity increased. One explanation for these findings is that workers were responding positively to the fact that they had been singled out for study (Roethlisberger and Dickson 1939).

subjects alter their behavior when they learn they are being observed. If researchers announce their identity and purpose, they must give participants time to adjust to their presence. Usually, if researchers are present for a long enough time, the subjects eventually will display natural, uninhibited behaviors.

Secondary Sources or Archival Data Another strategy for gathering data is to use **secondary sources** or **archival data**. These are data that have been collected by other researchers for some other purpose. Government researchers, for example, collect and publish data on many areas of life including births, deaths, marriages, divorces, crime, education, travel, and trade. "Every researcher who uses an existing data set or who does a literature review in which published research findings are taken out of their original context and applied to a different issue or question is involved in 'archival research'" (Horan 1995:423). For their research on changing norms and expectations related to the care of the elderly in Japan, Naohiro Ogawa and Robert D. Retherford (1993) used data from the National Survey on Family Planning, a biannual random survey conducted by the Population Problems Research Council of the Mainchi Newspapers since 1950. These surveys included a question asking women, "Are you planning to depend on your children in your old age (including adopted children, if any)?"[9]

Another kind of secondary data source consists of materials that people have written, recorded, or created for reasons other than research (Singleton, Straits, and Straits 1993). Examples include television commercials and other advertisements, letters, diaries, home videos, poems, photographs, artwork, graffiti, movies, and song lyrics. Researcher Subir Sengupta (1995) used secondary sources for his research on portrayals of women in television commercials aired in the United States and Japan. He videotaped eighteen hours of television shows aired in Japan and fifteen hours of television shows aired in the United States. All shows selected were aimed at adult audiences. Sengupta then focused on commercials with adult male and adult female characters, removing commercials showing "characters who looked like foreigners in their respective countries" (p. 322). This left him with 507 Japanese television commercials with 367 male and 480 female characters and 227 commercials from the U.S. television market with 310 male and 200 female characters.

[9]The researchers found that in 1950, 65 percent of women expected to depend on their children in old age, and that percentage declined to 18 percent in 1990.

Identifying Variables and Specifying Hypotheses

As researchers acquire a conceptual focus, identify a population, and determine a method of data collection, they also identify the variables they want to study. A **variable** is any trait or characteristic that can change under different conditions or that consists of more than one category. The variable "sex," for example, is generally divided into two categories: male and female. The variable "marital status" is often separated into six categories: single, living together, married, separated, divorced, and widowed.

Researchers strive to find associations between variables in order to explain and/or predict behavior. The behavior to be explained or predicted is the **dependent variable**. The variable that explains or predicts the dependent variable is the **independent variable**. Thus, a change in the independent variable brings about a change in the dependent variable. A **hypothesis**, or trial explanation put forward as the focus of research, predicts how independent and dependent variables are related. This trial idea specifies what outcomes will occur as the independent variable varies.

Consider the hypotheses that Sengupta (1995) advanced in his research on the influence of culture on portrayals of women in television commercials aired in the United States and Japan. He hypothesized that characters' dress and roles would be related to the country in which the advertisement appeared. Two of Subir's hypotheses are as follows:

1. Women in Japanese television commercials dress less seductively (i.e., more demurely) than women in U.S. television commercials.
2. Women in Japanese television commercials are less likely to be portrayed in working roles than women in U.S. television commercials.

Secondary sources or **archival data** Data that have been collected by other researchers for some other purpose.

Variable Any trait or characteristic that can change under different conditions or that consists of more than one category.

Dependent variable The behavior to be explained or predicted.

Independent variable The variable that explains or predicts the dependent variable.

Hypothesis A trial explanation put forward as the focus of research that predicts how independent and dependent variables are related and what outcome will occur.

In arriving at these hypotheses, Sabir reasoned that because Japanese society is more traditional in its outlook toward women than American society and because Japan is a more male-dominated society than the United States, these differences will be reflected in the portrayal of women in television commercials. Consequently, he believed that the independent variable (country in which an advertisement is aired) is related to the dependent variables (women's dress and roles).

One major reason researchers collect data is to test hypotheses. For the findings to matter, other researchers must be able to replicate the study. For this reason, researchers need to give clear and precise definitions and instructions about how to observe and measure the variables being studied.

Operational Definitions In the language of research, such definitions and accompanying instructions are called **operational definitions**. An analogy can be drawn between an operational definition and a recipe. Just as anyone with basic cooking skills can follow a recipe to achieve a desired end, anyone with basic research skills should be able to replicate a study if he or she knows the operational definitions (Katzer, Cook, and Crouch 1991). For example, one operational definition of education is the number of years of formal schooling a person has completed.

If the operational definitions are not clear or do not indicate accurately the behaviors they were designed to represent, they are of questionable value. Good operational definitions are reliable and valid. **Reliability** is the extent to which the operational definition gives consistent results. For example, the question "How many magazines do you read each month?" may not yield reliable answers because respondents may forget some magazines. Thus, if you asked the question at two different times, the respondent likely would give two different answers. One way to increase the reliability of this question is to ask respondents to list the magazines that they have read in the past week. The act of listing forces respondents to think harder about the question, and shortening the amount of time to the past week makes it easier to remember.

Consider the operational definition that Sengupta used to classify the characters in television commercials aired in the United States and Japan. Characters were classified according to type of dress (demure or seductive) and according to role (working or nonworking). If in a working role, they were further classified as being in a high-level business, entertainment, middle-level business, nonprofessional white-collar, blue-collar, or other role. If in a nonworking role, characters were further classified as being in a family, recreational, or decorative role.

Although we can assume that Sengupta gave coders precise instructions about how to classify characters, we might also suspect that coders sometimes disagreed about or had trouble deciding whether a character was in a work or nonwork role, foreign- or native-born, or dressed in a demure or seductive fashion. In such situations, the reliability of the codes would be at issue.

Validity is the degree to which an operational definition measures what it claims to measure. Professors give tests to measure students' knowledge of a particular subject as covered in class lectures, discussions, reading assignments, and other projects. Students may question the validity of this measure if the questions on a test reflect only the material covered in lectures. In such instances, students may argue that the test does not measure what it claims to measure. As a second example, many critics claim that the trade balance as currently figured is not a valid measure of the economic relationship between countries (see pages 78–79).

Steps 5 and 6: Analyzing the Data and Drawing Conclusions

When researchers get to the stage of analyzing collected data, they search for common themes, meaningful patterns, and/or links. In the study of fourth-grade underachievers and achievers in Japan, China, and the United States described earlier in this chapter, Tuss et al. (1995) identified the following patterns: (1) in each country there were no significant differences between underachievers and achievers in how they explained their success or failure on math tests; (2) there were, however, significant differences among countries in how students *explained* success and failure. Specifically, in comparison with their Japanese, but especially Chinese, counterparts, U.S. fourth graders perceived a pronounced lack of control over unsuccessful outcomes. American students who were under- and over-

Operational definitions Clear, precise definitions and instructions about how to observe and measure variables.

Reliability The extent to which the operational definition gives consistent results.

Validity The degree to which an operational definition measures what it claims to measure.

An Operational Measure of Interdependence?

A critical step in the research process is the operationalization of concepts: precise, specific instructions for observing and recording valid, reliable evidence of core concepts and variables.

Could international telephone calls be used as an operational definition of how interconnected people are between two countries and how interdependent nations are? The map shows total phone calls placed between the United States and eleven other countries.

Because the total number of calls will be influenced greatly by the population size of each country, we might want to use a per capita measure instead: total number of calls divided by total population of the country. The table shows the resulting per capita numbers. Do these results make sense to you, based on your knowledge about the countries and their relationships with the United States? What cautions would you suggest about accepting these numbers as an operational definition of interdependence?

Collecting and comparing such numbers over the years is easy. Changes might reflect greater or lesser interdependence between countries. What else might changes in per capita calls suggest about the relationship between two countries?

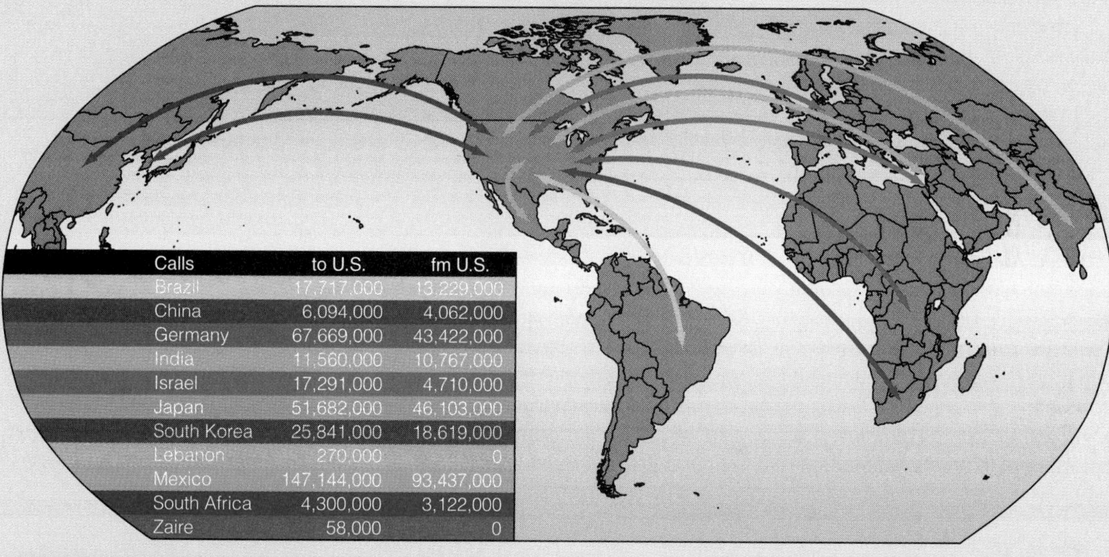

Calls	to U.S.	fm U.S.
Brazil	17,717,000	13,229,000
China	6,094,000	4,062,000
Germany	67,669,000	43,422,000
India	11,560,000	10,767,000
Israel	17,291,000	4,710,000
Japan	51,682,000	46,103,000
South Korea	25,841,000	18,619,000
Lebanon	270,000	0
Mexico	147,144,000	93,437,000
South Africa	4,300,000	3,122,000
Zaire	58,000	0

Average Annual Telephone Messages and Population (by Country)

Country	Total Messages	Population Size	Per Capita Messages
Brazil	30,946,000	153,322,000	0.20
China	10,156,000	1,133,682,501	0.01
Germany	111,091,000	79,753,227	1.39
India	22,327,000	846,302,688	0.03
Israel	22,001,000	4,037,620	5.45
Japan	97,785,000	123,611,167	0.79
South Korea	44,460,000	43,410,899	1.02
Lebanon	270,000	2,126,325	0.13
Mexico	240,581,000	81,249,645	2.96
South Africa	7,422,000	26,288,390	0.28
Zaire	58,000	29,671,407	0.002
TOTAL	587,097,000	2,523,455,869	0.23

Source: U.S. Federal Communication Commission (1994)

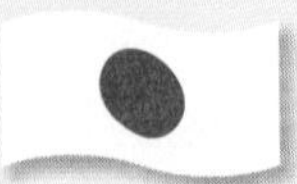

U.S. In Perspective

Is the Trade Balance a Valid Measure of the Economic Relationship Between Japan and the United States?

Consider the trade balance between two countries. It is calculated by adding up the dollar value of the goods and services that each country exports to the other and then subtracting the smaller amount from the larger. On the basis of this formula, the United States has sustained a trade deficit with Japan for approximately a decade and a half.

Some critics argue that this formula is not an accurate or valid measure of the amount of trade between the United States and Japan. It does not consider the dollar value of goods and services produced by American and Japanese companies located in the other's country. A more valid measure would be to add up the goods and services that each country exports to the other, plus what each country's firms produce and sell within the other country (Robinson 1985).

How does this worker in a Honda factory in Marysville, Ohio, illustrate the difficulty of calculating the trade balance between two countries in a global economy?

©Andy Snow/Saba

In 1985, Richard D. Robinson, a professor of international management, figured the U.S. trade deficit with Japan on the basis of two formulas. According to the formula that counts annual American exports to Japan ($25.6 billion) and Japanese exports to the United States ($56.8 billion), the trade deficit was $31.2 billion for the United States. According to the second formula, which includes sales by Japanese-owned businesses in the United States ($12.6 billion) and sales by American-owned businesses

achievers tended to attribute a poor performance to lack of ability or to difficult tests rather than to lack of effort.

In presenting their findings, researchers may make use of graphs, frequency tables, photos, statistical data, and so on. The choice of presentation depends on what results are significant and how they might be best shown. Researchers William T. Bailey and Wade C. Mackey used percentages to summarize the data that they collected in their 1989 nonparticipant observation study of 18,272 child-adult interactions in public settings in Japan and the United States. These researchers constructed tables that show the percentages of child-adult interactions in Japan and the United States involving women only, men only, and men and women. The tables show that, contrary to the popular image of Japanese men as people who work long hours and spend little or no time with their children, Japanese men are observed as frequently as American men interacting with their children in public settings (see Figure 3.3).

In the final stage of the research process, sociologists comment on the **generalizability** of findings, the extent to which the findings can be applied to the larger population from which the sample is drawn. The sample used and the response rate are both important when it comes to generalizability, as is clear from the following example.

In 1991, the U.S. Centers for Disease Control (CDC) abandoned its plans to conduct a nationwide survey of households to determine the prevalence of

Generalizability The extent to which findings can be applied to the larger population from which the sample is drawn.

in Japan ($43.9 billion), the trade surplus was $100 million for the United States, significantly different from the deficit figure of $31.2 billion.[1]

According to the second formula, then, the United States had a slight trade surplus with Japan. Even this formula, however, is not free of problems. For one thing, it does not take into account that countries export goods and services *indirectly* to one another. For example, 70 percent of Japanese males use Schick razors. Schick, with headquarters in Connecticut, exports razors to Japan via Hong Kong (Totten 1990). Similarly, Japan exports cars to the United States from Canada, Mexico, and various countries in Asia.

On the basis of this information, we can argue that the best operational definition is a third formula: for each country, add up the dollar value of goods and services it exports to the other country, *plus* what that country's firms produce and sell within the other country, *plus* what that country exports indirectly to the other country. Then figure the difference between the two totals. Although this third formula is not accepted as a more valid measure of trade, it is virtually impossible to find complete and accurate figures on indirect exports. In addition, anyone who tries to calculate indirect exports faces the challenge of determining the national identity or origin of the exported products. For example, should the approximately 164,000 vehicles produced in Japan and exported to the United States as Chryslers, Dodges, and Chevrolets be considered Japanese exports (Sanger 1992)? Similarly, should the 50,000 Nissan Quest minivans produced at Ford's Avon, Ohio, production plant be considered Japanese exports? As you can see, the issue of trade balance is not simple or clear-cut.

Similarly, when the trade deficit with Japan is used as an operational definition of the openness of the Japanese market, Japanese aggressiveness, or the patriotism of American consumers, we must question whether it is a valid measure of such behaviors. For example, we know that the trade deficit as currently calculated cannot be used as a measure of the openness of Japanese markets to U.S. goods because it does not consider (1) the dollar value of goods and services produced by American and Japanese companies located in the other's country or (2) the dollar value of the indirect exports that each country sends to the other. Some examples of more valid measures of market openness are (1) the number of trade barriers each country puts up against the other, (2) the number of suggestions each country offers the other for openings its markets, and (3) per capita purchases of imports (see Figure 3.2).

[1]To evaluate this difference ($31.2 billion versus $100 million), it is important to understand the difference between a million and a billion. "For example, knowing that it takes only about eleven and a half days for a million seconds to tick away, whereas almost thirty-two years are required for a billion seconds to pass, gives one a better grasp of the relative magnitude of these two common numbers" (Paulos 1988:10).

HIV infection after a pilot study involving 1,724 households in Pittsburgh and Allegheny County, Pennsylvania, showed that they could not secure a high enough response rate to be confident in its estimate. Although 85 percent agreed to a take a blood test, CDC researchers wondered whether the 15 percent who refused to participate might have a higher risk of HIV infection than those who agreed. Researchers contacted those 15 percent a second time and convinced about half to participate. They found that the "reluctant participants" were twice as likely as the original participants to have used intravenous drugs and to have reported male-to-male sex. But were these 7.5 percent representative of the other 7.5 percent who did not participate at all? The fact that the CDC researchers knew nothing about this later group caused them to decide to abandon the study rather than risk making inaccurate generalizations about rates of HIV infection among the general population (Hilts 1991).

If a sample is randomly selected, if almost all subjects agree to participate, and if the response rate for every question is high, we can say that it is representative of the population and that the findings theoretically are generalizable to that population. If a sample is chosen for some other reason — perhaps because it is especially accessible or interesting — then the findings cannot be generalized to the larger population. Keep in mind that even though one goal of drawing conclusions is to make generalizations about the larger population, generalizations are not statements of certainty that apply to everyone. Consider the comparative research on father-child interaction in three cultures conducted by sociologist Ishii-Kuntz (1992). In this case, for example, knowing a respondent's culture did not mean that the researchers could predict the frequency fathers were home for dinner in the evenings to eat with their children. The study data show that Japanese fathers are significantly less likely than U.S. and German

Figure 3.2 Alternative Measures of the Relative Openness of U.S. and Japanese Markets

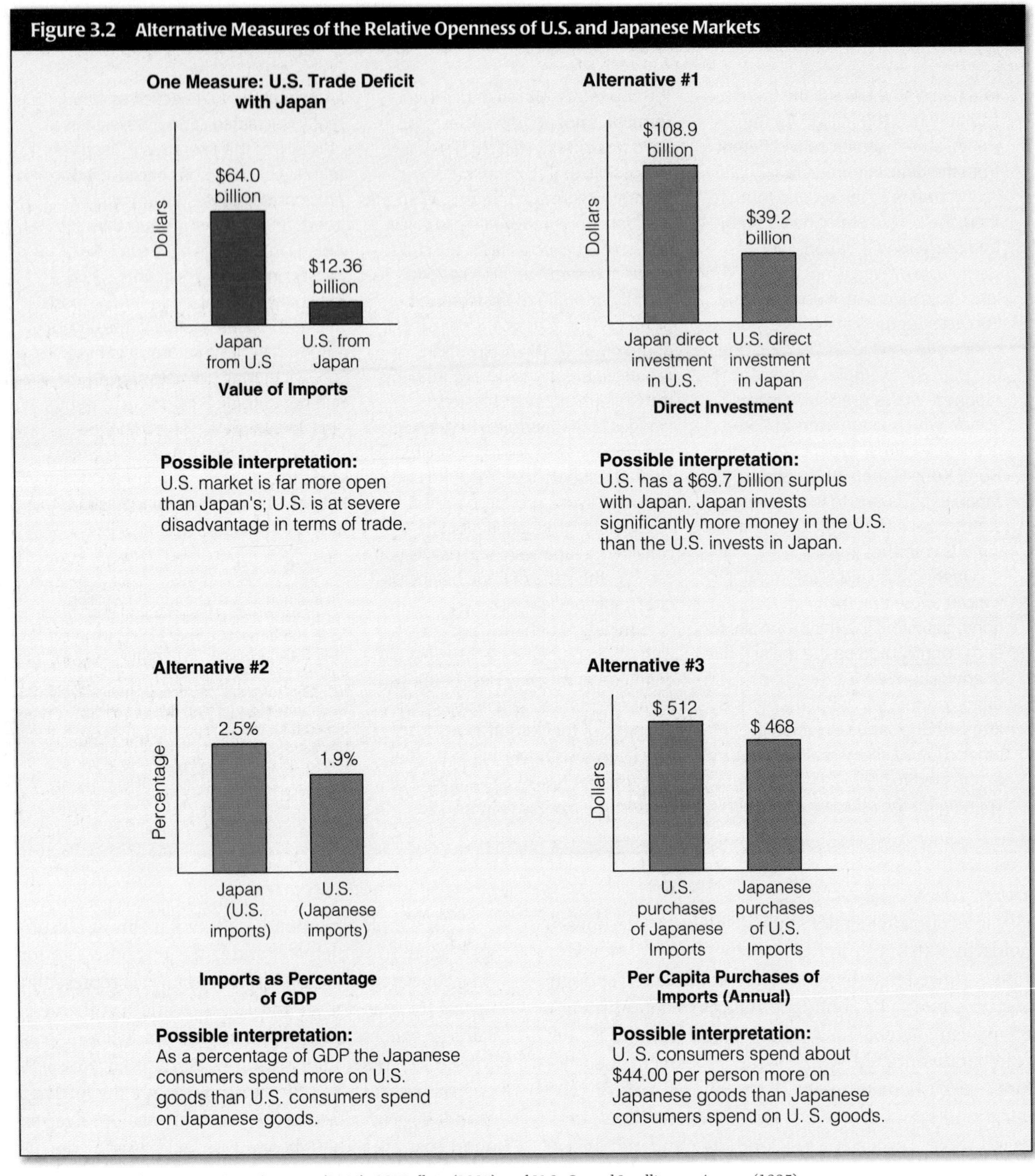

Sources: Adapted from Sanger (1992), Totten (1990), *SOI Bulletin* (1995), and U.S. Central Intelligence Agency (1995).

fathers to eat dinner with their children every day. Approximately 50 percent of Japanese fathers ate dinner every day with their children compared with 70 percent of U.S. fathers and 60 percent of German fathers. Clearly, this does not mean that all *individual* Japanese fathers behave in this way but that as a group they eat dinner with their children less frequently than U.S. and German fathers do.

Because the generalizations do not apply to everyone, it is virtually impossible to claim that one independent variable (in this case, nationalities) causes a dependent variable (in this case, the frequency that fa-

Figure 3.3 Comparisons of Men's Presence with Children After and During Work Hours in the United States and Japan

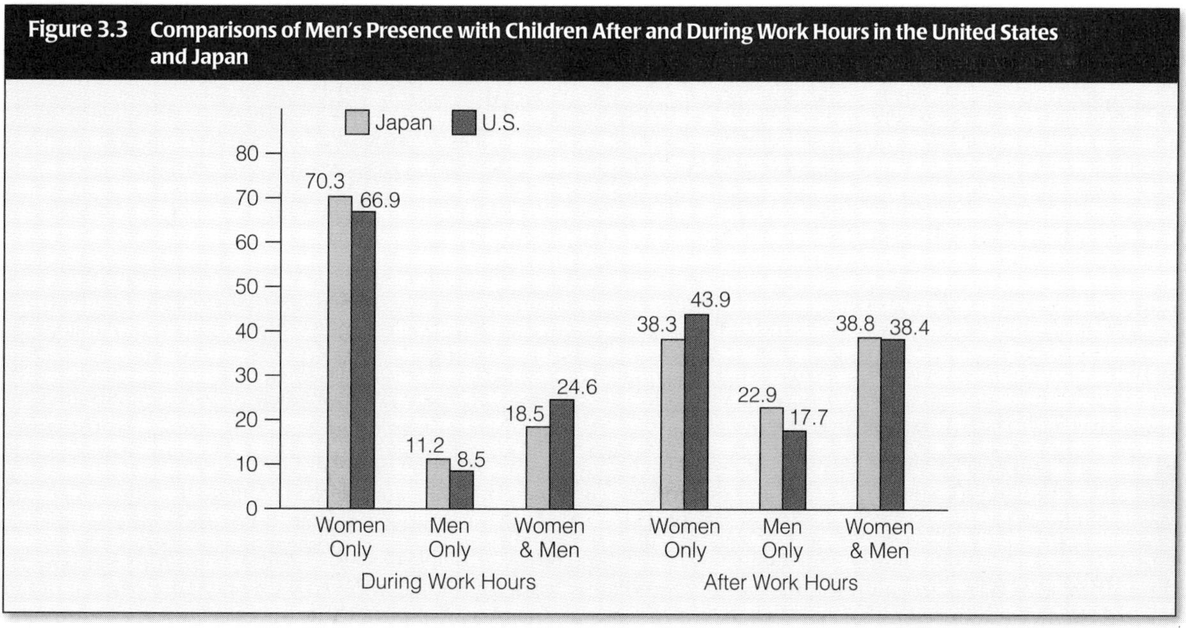

Source: "Observations of Japanese Men and Children in Public Places: A Comparative Study," by W. T. Bailey and W. C. Mackey. P. 733 in *Psychological Reports*, 65. Copyright © 1989 by *Psychological Reports*. Reproduced by permission of the authors and publisher.

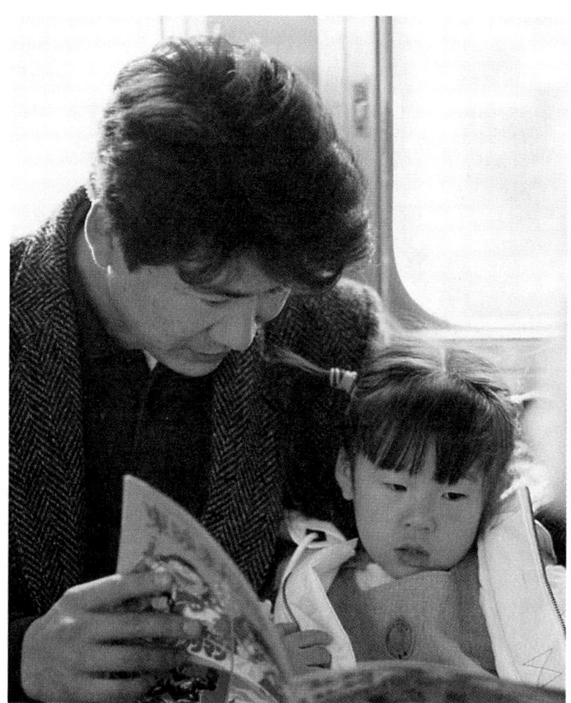

Contrary to a common stereotype, the results of Bailey and Mackey's comparative study of parent-child interactions indicated that Japanese fathers spend about as much time with their children as U.S. fathers do.

©Sonia Katchian/Photo Shuttle: Japan

thers are home for dinner). Consequently, instead of claiming cause, researchers search for independent variables that make significant contributions toward explaining the dependent variable.

At least three conditions must be met before a researcher can claim that an independent variable contributes significantly toward explaining a dependent variable. The first condition is that the independent variable must precede the dependent variable in time. Time sequence can be established easily when the independent variable is a predetermined factor such as sex or birth date. These kinds of factors are fixed before a person is capable of any kind of behavior. Usually, however, time order cannot be established so easily.

A second condition that must be met before a researcher can claim that an independent variable contributes toward explaining a dependent variable is that the two variables must be correlated. The strength of this contribution is often represented by a **correlation coefficient**, a mathematical representation of the extent to which a change in one variable is associated

Correlation coefficient A mathematical representation of the extent to which a change in one variable is associated with a change in another.

with a change in another (Cameron 1963). Correlation coefficients range in value from –1.0 to +1.0, with 0.00 representing no association between variables and 1.0 representing a perfect association. If, when the value of one variable (for example, number of fire trucks at the scene) increases, there is a corresponding increase in the other variable (for example, dollar amount of fire damage), the correlation coefficient is a *positive* number. If, on the other hand, when one variable increases (number of fire trucks at the scene), the other decreases (dollar amount of fire damage), the correlation coefficient is a *negative* number.

Establishing a correlation is a necessary step but not in itself sufficient to prove causation. A correlation shows only that the variables are related; it does not mean that one variable causes the other. For one thing, a correlation can be spurious. A **spurious correlation** is one that is coincidental or accidental; in reality, some third variable is related to both the independent and the dependent variables. The presence of the third variable makes it seem that those two variables are related.

If a researcher is to claim that an independent variable helps explain a dependent variable, there must be no evidence that another variable is responsible for a spurious correlation between the independent and the dependent variables. To check this possibility, sociologists identify **control variables**, variables suspected of causing spurious correlation. Researchers determine whether a control variable is responsible by holding it constant and reexamining the relationship between the independent and the dependent variable.

A good example of a significant relationship between two variables that disappear when we control for a third variable is the strong correlation between the number of fire trucks at the scene and the amount of damage done in dollars (the more fire trucks, the greater the dollar amount of fire damage). However, common sense tell us that this is a spurious correlation. A third variable — the size of the fire — is responsible for both the number of fire trucks sent and the amount of damage done. Although the number of fire trucks called to the scene does help us predict the amount of damage in dollars, that variable is not the variable of cause.

A second example of a possible spurious correlation comes from Sengupta's research on portrayals of women in television commercials aired in Japan and the United States. He found that, contrary to expectations, "the type of dress worn (demure/seductive) by women in the advertisements was not found to be significantly related to the country in which the advertisements appeared" (p. 327). Eighty percent of women in U.S. commercials were judged by coders as demure in dress; 84.4 percent of women in commercials aired in Japan were judged as demure in dress. On the other hand, Sengupta found support for the hypothesis that role portrayals are related to the country in which the commercial aired. "In fact, 30.5 percent of women in advertisements from the United States were depicted in working roles, compared with only 16.5 percent in the Japanese advertisements" (p. 326). Sengupta cautions, however, that the observed relationship between country and roles may be found spurious if we control for type of commercial. It seems that the two countries vary in the distribution of commercials across product categories. Notice in Table 3.1 that Japanese television airs more "food, snack, and soda" commercials than U.S. television does and that U.S. television airs significantly more car commercials. One might argue that

Spurious correlation A correlation that is coincidental or accidental because some third variable is related to both the independent and dependent variables.

Control variables Variables suspected of causing spurious correlation.

Are Japanese employment patterns more stable because Japanese workers are by character more "loyal" to their companies than American workers are to theirs? Or is it more because Japanese society penalizes workers who change jobs?

Table 3.1 Distribution of U.S. and Japanese Television Commercials across Product Categories

Products	U.S. Ads *n*	(%)	Japanese Ads *n*	(%)
Food, snacks, soda	28	(12.3)	148	(29.2)
Personal and beauty care products	25	(11.0)	93	(18.3)
Cars and accessories	34	(15.0)	20	(3.9)
Restaurants and retail outlets	13	(5.7)	11	(2.2)
Drugs and medicines	13	(5.7)	36	(7.1)
Household appliances/furnishings	13	(5.7)	51	(0.1)
Institutional/public service	25	(11.0)	34	(6.7)
Alcoholic beverages	17	(7.5)	17	(3.4)
Pet food and related products	0	(0.0)	4	(0.8)
Household cleaning agents	2	(0.9)	27	(5.3)
Clothing	3	(1.3)	6	(1.2)
Finance and real estate	4	(1.8)	7	(1.4)
Others	50	(22.0)	53	(10.5)
Total	227	(100.0)	507	(100.0)

Note: Percentages may not total 100 percent because of rounding up.

Source: "The Influence of Culture on Portrayals of Women in Television Commercials: A Comparison Between the United States and Japan," by Subir Sengupta. P. 323 in *International Journal of Advertising*, vol. 14. Copyright ©1995 Advertising Association. Reprinted by permission of Blackwell Publishers.

the difference in roles is due to the kind of commercials aired, not cultural differences regarding the role of women. If we controlled for type of commercial, we might see no relationship between country and portrayals of female characters.[10]

Thinking about the possibility of a spurious relationship is especially important when the independent variable is an **ascribed characteristic**: any physical trait that is biological in origin and/or cannot be changed but to which people assign overwhelming significance, such as hair texture and color, eye shape, and skin color. Other examples are age and country of birth. Such findings announcing an ascribed characteristic as a "cause" can be used to stereotype and stigmatize some groups of people. For example, many Americans believe that Japanese workers are more loyal than Americans, that Japanese students are more studious, that Japanese criminals are more contrite, that the Japanese are less litigious than Americans, and that Japanese people in general are more group oriented. Actually, however, such observed differences are beyond the control of individual Japanese and can be explained as due to some other factors:

> The "loyalty" of white-collar workers to their company, in contrast to the constant movement of employees in other countries, is one clear example. [In Japan] the major corporations tacitly agree never to hire someone who has left another firm. Japanese children are studious in large part because admission to the University of Tokyo, which is based on examination scores, is essentially their only hope for having an influential place in society.
>
> Japanese are "nonlitigious," not just because of their alleged love of consensus but also because of the acute shortage of lawyers. The Ministry of Justice controls the Legal Training and Research Institute, where future lawyers and judges must train, and it admits only 2 percent of those who apply. (Of the 23,855 who took the entrance examination in 1985, 486 were admitted.)
>
> Most criminals arrested by the police confess partly out of a sense of remorse but also because they know what a trial would mean: in 99 percent of criminal trials, the verdict is guilty. (Fallows 1989:28)

[10]For example, women in food, snack, and soda commercials might be portrayed in nonworking roles simply because people eat and prepare these items in nonwork environments. The fact that Japanese television airs more of these kinds of ads would explain the differences in how women are portrayed in the two countries.

Ascribed characteristic Any physical trait that is biological in origin and/or cannot be changed but to which people assign overwhelming significance

Summary and Implications

In this chapter, we have identified the technological forces (computer, telecommunications, the Internet, hypertext) that allow people to produce large quantities of data, and we have considered the quality control problems — dearth of feedback and exaggerated, distorted headlines — that are associated with the information explosion. As a result of these issues we have to think critically about the research we encounter.

Despite such problems, a basic knowledge of research methods helps us sort through data in a way that allows us to distinguish media hype from balanced research. To illustrate the power of research methods to help us evaluate "facts," we focused on Japan; specifically, we focused on the trade deficit between Japan and the United States, the meanings people attach to the deficit, and generalizations made about Japanese society (and, by implication, U.S. society) to explain the deficit. We also reviewed research that investigated some of the traits that many people assume are somehow connected with Japan's economic success. Consider some of these highlights from research studies:

- In Japan, an elaborately prepared lunch is a sign of a Japanese woman's commitment as a mother, which inspires her child to be similarly committed as a student. This observation helps us understand the Japanese teacher's attention to eating habits. (By implication, U.S. mothers are more concerned with children's social adjustment than with eating habits.)
- *Kikokushijo* are not "refugee camps" for returning Japanese students who have become individualistic, independent, and argumentative while abroad and must relearn the value of harmony, conformity, and group mindedness.
- Only 2.1 percent of Kobe households have unlisted numbers compared with 20 percent of Seattle households.
- In Japan, student monitors play an important role in maintaining classroom order and discipline. (By implication, in the United States, teachers bear almost total responsibility for maintaining order and discipline.)
- The percentage of women who expect to depend on their children in old age has declined dramatically.
- The trade deficit is not a valid measure of economic exchange between the United States and Japan.
- American fourth graders tend to blame lack of ability and difficult tests as reasons for poor performance on math tests, whereas their Japanese and Chinese counterparts tend to blame lack of effort.
- Japanese fathers are less likely than their U.S. and German counterparts to eat dinner every day with their children.
- Japanese men are observed as frequently as American men interacting with their children in public settings.

Taken together, these findings suggest some interesting differences between Japan and the United States with regard to certain features: *obentōs, kikokushijo,* unlisted phone numbers, attribution for lack of success on math tests, methods of keeping classroom order. For the most part, however, these differences are not absolute. For example, there are some qualities that people living in the United States and Japan share, such as the amount of time fathers appear and interact with their children in public. In addition, when significant differences between the two countries with regard to some attributes are found, some people in both societies are "exceptions to the rule." In other words, although most residents of Japan do not have unlisted numbers, a small percentage do. Likewise, although a significant percentage of U.S. residents have unlisted phone numbers, most choose to make their phone numbers public. Likewise, some U.S. fourth graders in the United States do blame lack of effort for poor scores on math tests just as some fourth graders in Japan and China do blame difficult tests and lack of ability for poor performances.

Although some of the findings cited here suggest that some characteristics, such as the *kikokushijo* or *obentōs,* appear to be uniquely Japanese, we must be careful about such a claim. The world has thousands of cultures, and anyone who claims some quality is unique to one place implies that they are an expert on all the world's culture. The Internet or World Wide Web (WWW) may help in identifying unique characteristics. For example, Darryl Macer (1995), founder of the Eubios Ethics Institute, a Japan-based organization with the goal of promoting an integrated and cross-cultural approach to bioethics and building an international network of interested parties, used the WWW for this purpose when he asked readers of the on-line *Eubios Journal of Asian and International Bioethics* for comments on the following observation:

> On 27 October the University Animal Research Center held its annual memorial service for the research animals. In the past year 13,000 animals were used, and about 110 persons came to the shrine for experimental animals outside the animal research center. It was my first time to participate, in the 20-minute service, during which time the people place a white chrysanthemum on the shrine while saying a short prayer. It is among trees, with birds flying overhead—overall a very interesting experience. It may serve for the relief of guilt of using animals, and recognizing their contribution to research. Earlier in the month the annual memorial service for the families who gave dead relatives bodies for research, education, and autopsies was held. In the past year 214 bodies were given; included were about a dozen or so fetuses. This is a part of Japan that appears to be unique, I wonder if anyone can tell us of parallels. (Macer 1995)

Because everyone in Japan who participated in the research studies covered in this chapter did not respond in the same way suggests that we must reassess "the notion of Japan as a cohesive and tightly knit society based on a single value system" (Mouer and Sugimoto 1990:108). Likewise, we cannot assume that the United States is simply a society of individuals who only value self-expression and personal independence. The larger implication is that we cannot explain away differences between the two countries by labeling one society as group oriented and the other as individualistic.

When researchers report cross-national, comparative findings, readers (and even the researchers) tend to treat the people in the country or society as one unit when, in actuality, the people within each society possess a wide range of characteristics. For example, in his research on the *kikokushijo,* Goodman (1993) noted that the experiences of Japanese children living abroad varies depending on

> where the children went, how long they were overseas, what type of education they received there, and how much contact they had with the local communities. There are also, of course, differences depending on their age, their gender, and their own individual personalities. Yet all such children, when they return to Japan, are classified as *kikokushijo.* They all tend to have the same qualities ascribed to them as a result of their overseas experience. (p. 2)

Goodman maintains that most researchers who study *kikokushijo* do not question this category as a valid way of classifying all returnees. Instead, they proceed to focus on and document the problems returnees are supposed to have while overlooking evidence that suggests healthy adjustments. In a similar vein, many researchers study the trade deficit without considering whether it is a valid measure of an economic relationship between two countries. The information in this chapter suggests that if we plan to conduct research aimed at understanding the economic relationship between the United States and Japan, then we cannot use the trade deficit as an operational definition of that relationship.

The fact that quality and conscientious research yields complex findings may leave some people feeling frustrated about how to approach problems believed to be related to the trade deficit or to cultural differences. On the other hand, the complex (rather than absolute) findings suggest that problems between groups are not insurmountable because differences are not as great as previously assumed.

Key Concepts

Use this outline to organize your review of the key chapter ideas.

Information explosion
- **Information**
- **Data**
- **Hypertext**
- **Dearth of feedback**

Research

Scientific method
- **Objectivity**

Concepts

Research design
- **Population**
  - **Traces**

- **Documents**
- **Households**
- **Small groups**

Sample

- **Random sample**
- **Representative sample**
- **Sampling frame**

Method of data collection

- **Self-administered questionnaire**
- **Interviews**
  - **Structured interviews**
  - **Unstructured interviews**
- **Observation**
  - **Participant observation**
  - **Nonparticipant observation**
  - **Hawthorne effect**
- **Secondary sources or archival data**

Variables

- **Dependent variables**
- **Independent variables**
- **Control variables**

Operational definitions

- **Reliability**
- **Validity**

Generalizability

Correlations

- **Positive correlation**
- **Negative correlation**
- **Spurious correlation**

internet assignment

The following is a list of Web sites that publish sociological research papers:

- Society for the Study of Symbolic Interaction (SSSI): Papers of Interest
  http://sun.soci.niu.edu/~sssi/papers/papers.html
- Current Research in Social Psychology
  http://www.uiowa.edu/~grpproc/crisp/crisp.html
- Electronic Journal of Sociology
  http://olympus.lang.arts.ualberta.ca:8010/
- Sociological Research Online
  http://www.soc.surrey.ac.uk/socresonline

Browse these sites and look for an article in which some part of an author's research represents a particularly good illustration of one of the steps in the research process.

Culture

With Emphasis on South Korea

Demilitarized zone between North and South Korea. ©Nathan Benn/Woodfin Camp & Associates

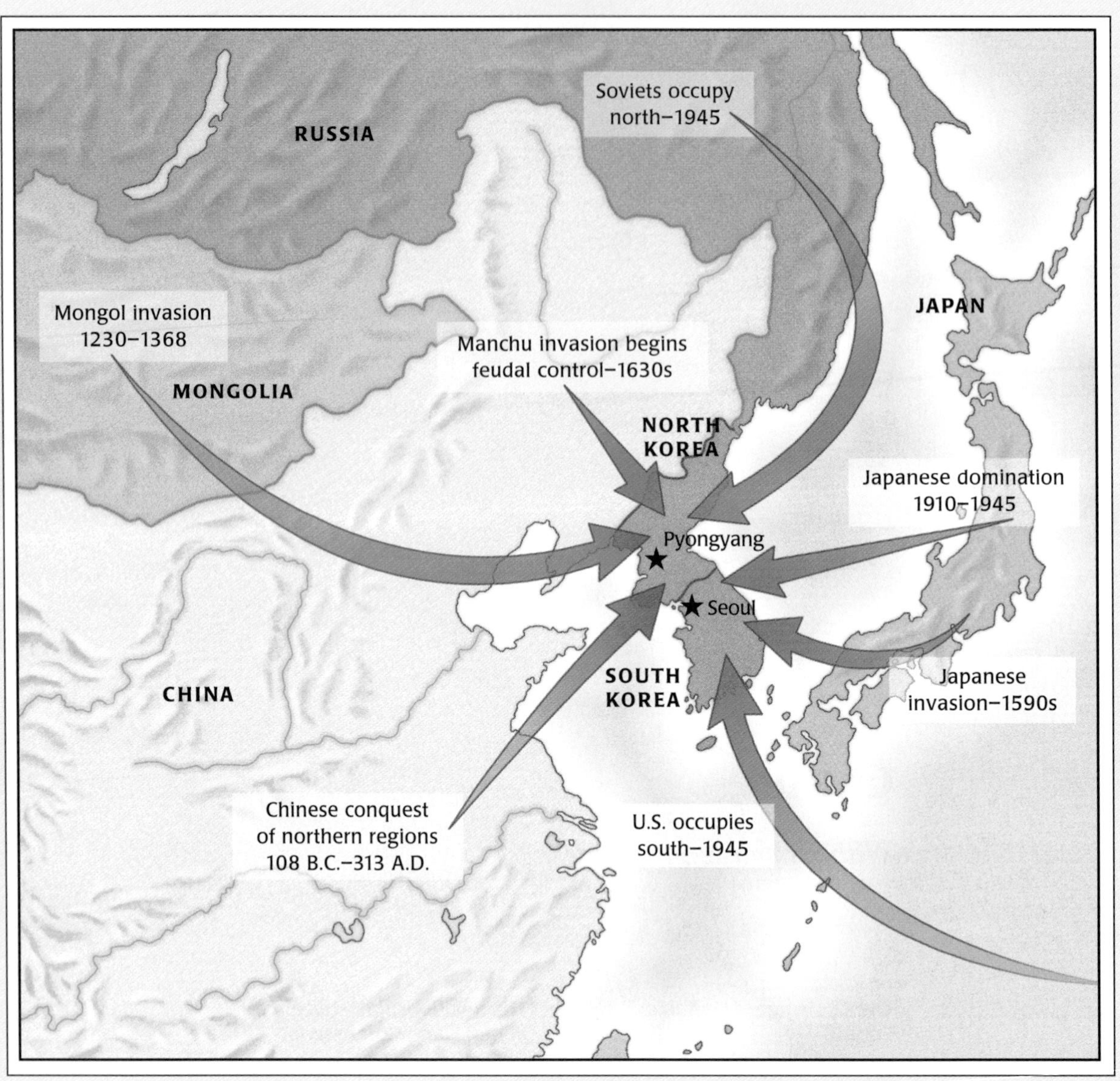

How Does Culture Survive?
At various times in their 2,000-year history, the Korean people have been subject to invasion and control by outsiders. Invasions bring new political and economic forces, religions, languages, and customs. How has Korean culture persisted in the face of this? Does an experience of invasion in the past have anything to do with South Korea's economic success today or the success of Korean emigrants to the United States and other countries? Is the same history related to the quality of "sadness" that has been noted in Korean culture? This chapter provides some tools for understanding why cultures persist and why they change.

They came to Seoul not for pleasure but for their future. They were nervous on arrival, for by and large they were small Midwesterners, owners of modest companies, and many of them had not traveled very much. Now not only were they in a distant and strange land but they had arrived there vulnerable, almost beholden, needing to make a deal, and they might have to give away part of their company in the process. They came because they feared they could no longer compete at home, and they had been told by their most important customers, the giant American assembly companies, to get their costs down. Korea was to them, like it or not, their best hope. Korea, they had been told, was the new Japan and for them a way of holding off the Japanese challenge.

What had once been a trickle of them had become by the spring of 1986 a torrent. In Seoul they met their new partners, men with whom they could not communicate at all. William Vaughn, a Chrysler representative in Korea, had watched them come and witnessed their desperation, knowing they were there to survive, for if they could not work something out, they were convinced, they would soon be out of business. Vaughn had an unusual job; he represented Chrysler in its continuing negotiations to complete the massive deal with Samsung, the vast South Korean manufacturing combine, but in addition to that he was a kind of matchmaker between the American parts manufacturers used by Chrysler and the Koreans, trying to find Korean companies who could succeed in this alien new world of autos. (Halberstam 1986:697)

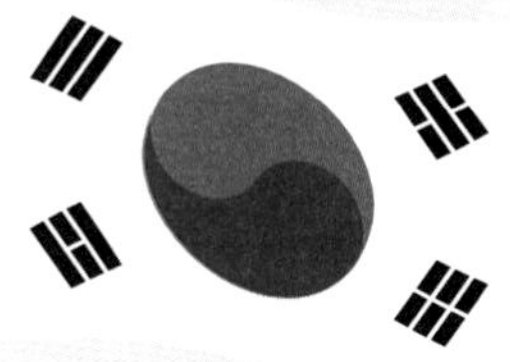

Why Focus on South Korea?

South Korea is one of twelve countries and city-states in the Pacific Rim, a region that is expected to become a formidable economic force in the twenty-first century.[1] Most Americans are unfamiliar with Pacific Rim countries, with the exception of Japan, and are generally unaware of the rapid economic development taking place in that region. In fact, South Korea, along with Taiwan, Hong Kong, and Singapore, are known popularly as "the next Japans" or "the four dragons." Along with the People's Republic of China, they are among the most rapidly developing economies in the world. All four, but especially South Korea, are credited with achieving economic development at an unprecedented speed in a region of the world characterized by relatively few natural resources (Kim 1993). Consider that South Korea arose from the total devastation of the Korean War to become one of the major players in the Asian Economic Community, one of three regional trading blocks expected to dominate the global economy over the next decade (Welch 1991).

Another reason for focusing on South Korea is that thirty-five years after the Korean War, which divided the country into North Korea and South Korea, approximately 35,000 U.S. military personnel are stationed in South Korea. Yet, despite this long involvement in Korean affairs, Americans have interacted with Koreans mainly on the basis of narrowly conceived national security or military interests. Little attention has been paid to how Koreans live or how they view the United States.

Now that South Korea is more than a military concern and is intertwined economically with the United States, Americans who want to do business with Korea are under pressure to inform themselves about Korean life and to learn to interact with Koreans as equal partners or else exclude themselves from one of the dynamic economies in the Asian region (Lie 1995).

[1]The countries and city-states of the Pacific Rim are Japan, South Korea, China, Taiwan, Hong Kong, the Philippines, Vietnam, Thailand, Malaysia, Singapore, and Australia.

During the 1980s and 1990s, tens of thousands of partnerships, joint ventures, and product agreements were formed between U.S. and foreign industries and corporations. As you might imagine, there is considerable debate about the meaning of this trend for the overall economic well-being of the United States. But one implication is clear: the United States is part of a more interconnected and more competitive world economic climate than it was even a decade ago. The new global competition and accompanying concerns about world position have caused many U.S. business leaders, educators, government officials, and social critics to compare the United States with other countries on a host of attributes, including geographic and scientific knowledge, unemployment rates, savings rates, literacy, high school graduation rates, productivity, divorce rates, health costs, military expenditures, and credit card use. The assumption underlying such comparisons is that there is "a connection between the kinds of everyday behaviors a society encourages and its [economic] stability and prosperity" (Fallows 1990:14).

In addition, the new global competition has caused U.S. government, business, and educational leaders to emphasize the need to train citizens, employees, and students to be more sensitive to cultural differences. In fact, an entire service industry has emerged to prepare people to deal with cultural differences. It includes consultants who offer cross-cultural and diversity training programs and publishers of intercultural materials (manuals, guidebooks, videos, and so on). Consultants and publishers aim to prepare people to cope in the face of unfamiliar, unpredictable habits of thought and patterns of daily behavior (Grove and Franklin 1990).

During the past two decades, tens of thousands of partnerships, joint ventures, and product agreements were formed between U.S. and Korean businesses.

The Challenge of Defining Culture

Consider the entry for *culture* in the *Cambridge International Dictionary of English* (1995), which presents core definitions of words and the most common usage of the word among English speakers. The entry shows the "vocabulary of culture" to which people learning about it are typically exposed and reads as follows:

> **culture** WAY OF LIFE *n* the way of life, esp. general customs and beliefs of a particular group of people at a particular time • *youth/working-class/Russian/Roman/mass culture* • *She's studying modern Japanese language and culture.* • *The cultures of Britain and Nigeria are very different.* • *Thatcher's enterprise culture* (= way of thinking and behaving) *of the 1980s brought many changes.* • *There's a* **culture gap** (= difference in ways of thinking and behaving) *between many teenagers and their parents.* • *It was a real* **culture shock** *to find herself in London after living on a small island* (= She felt alone and was confused by the completely different way of life there).

This entry shows that we use the word *culture* in conjunction with specific places (Russia, Rome, Nigeria, Britain, small islands) and categories of people (the masses, teenagers, parents, Russians, Japanese). We also use the word in ways that emphasize differences ("The cultures of X and Y are very different"; "There is a cultural gap between X and Y"; "It is culture shock to come from X and live in Y"). Our use of the word suggests that we think of culture as having clear boundaries, as an explanation for behavior, and as a blueprint for living that people follow in mechanical ways. Our uses also suggest that we think of interaction between people of different cultures as problematic.

In light of the seemingly clear way that we use the word, we may be surprised to learn that the real challenge and the sources of endless debate among people who study culture include the following:

- *Describing a culture.* That is, is it possible to find words to define something so vast as the way of life of a people?

- *Determining who belongs to a group designated as a culture.* Does a person who "looks Korean" and who has lived in the United States most of his or her life belong to Korean or American culture?
- *Identifying the distinguishing characteristics that set one culture apart from others.* For example, is eating rice for breakfast a behavior that makes someone Korean? Is an ability to speak Korean a behavior that makes someone Korean? Are ethnic Koreans who speak English not Korean?

Cynthia K. Mahmood and Sharon Armstrong (1992) faced these nagging questions when they traveled to Eastermar, a village in the Netherlands province of Friesland, to study the Frisian people's reactions to a book published about their culture. They found that the Frisian people were unable to agree on a single "truth" about them as described in the book. At the same time, Frisian villagers could not come up with a list of features that would apply to all Frisians and that would distinguish them from other people living in Eastermar. Yet the Frisians were "convinced of their singularity," and the villagers reacted emotionally to the suggestion that perhaps they did not constitute a culture.

The Frisian situation captures the conceptual challenges associated with the idea of culture: the paradox of recognizing a culture but being unable to define its **boundaries**, the characteristics determining where a culture begins and leaves off or the qualities marking some people off from others as a unified and distinctive group. This chapter offers a framework for thinking about culture that considers both its elusive nature and its importance in shaping human life. The framework includes eight essential principles defining the nature of culture.

A few words of caution are in order first. Although this chapter focuses on South Korea[2] and the United States, we can apply the concepts we discuss here to understand *any* culture and frame other cross-cultural comparisons. As you read about culture, remember that South Korea is referred to broadly as a country possessing an Eastern or Asian culture and the United States is regarded as a country possessing a Western culture. Therefore, many of the patterns described here are not necessarily unique to Korea or the United States but are shared with other Eastern or Western societies. At the same time, do not overestimate the similarity among countries that share a broad cultural tradition.

Koreans have been part of the United States in small numbers for about 150 years, even though their presence has received little attention until recently. This photo of a Korean farmer and his son was taken around 1920.

Courtesy USC Korean Heritage Library

Do not assume, for example, that South Korea (or any other Pacific Rim country or city-state) is just like Japan. As we will see, much of Korean identity is tied up with being "not Japanese" (Fallows 1988). To assume that South Korea is like Japan is equivalent to assuming that the United States is just like a Western European country, such as Germany or England. As we know, the United States is a country that celebrates its independence from European influence.

[2]Republic of Korea is the official name for South Korea. In this chapter, the terms *South Korea* and *Korea* are used interchangeably. Any reference to North Korea specifically is noted as such.

Boundaries The qualities marking some people off from others as a unified and distinctive group.

Material and Nonmaterial Components

Principle 1: Culture consists of material and nonmaterial components. In sociological terms, material culture consists of objects or physical substances. Nonmaterial culture consists of those elements that cannot be directly observed or easily described. At this point we are not concerned about identifying which people share material and nonmaterial culture. Instead, we focus on the kinds of objects, ideas, and behavior people *can* share.

Material Culture

Material culture consists of all the physical objects people have borrowed, discovered, or invented and to which they have attached meaning. Material culture includes natural resources such as plants, trees, and minerals or ores, as well as items that people have converted from natural resources into other forms for a purpose. Examples of the latter include cars and trucks to transport people, animals, and goods; microwave ovens to cook and heat food; computers to make calculations; video cameras equipped with devices that selectively soften facial features to make people look younger than they are; indoor plumbing to bring together "in one room the toilet from the outhouse or closet, the washbowl from the bedchamber, and the tub from the kitchen" (Nasaw 1991:10); radios to entertain, inform, and provide background sound; and so on.

The significant feature of material culture is that people attach meanings to each item, including the purpose for which it is designed, the value placed on it, and the fact that some people are unhappy without it or unhappy about it and direct their energies toward acquiring or eradicating it (Kluckhohn 1949). In thinking about material culture, it is important to learn not only the most obvious and practical uses for which an object is designed but also the meanings assigned to that object by the people who use it (Rohner 1984). As an example, consider the radio, a device for receiving and then broadcasting sound messages that travel through the air in the form of electromagnetic waves. For many people, the radio takes on meaning beyond the obvious purposes mentioned here. People use it to fill the void that can accompany boring tasks, daily routines, or loneliness; to sustain or create a mood (for example, upbeat, romantic, relaxed); and to provide a social lubricant in that people can talk with one another about what they have heard on the air. The importance of the radio in some people's lives is evident in statements like these: "To me, when the radio is off, the house is empty"; "I listen to the radio from the time I get up until I go to bed"; "Radio puts me in a better mood"; and "It makes driving easier" (Mendelsohn 1964: 242–243). Learning the meaning that people assign to objects in the material culture helps us grasp the significance of those objects in people's lives.

Material culture All the physical objects people have borrowed, discovered, or invented and to which they have attached meaning.

Nonmaterial culture Intangible creations or things that we cannot identify directly through the senses.

Beliefs Conceptions that people accept as true, concerning how the world operates and where the individual fits in relationship to others.

Nonmaterial Culture

Nonmaterial culture consists of intangible creations or things that we cannot identify directly through the senses. Three of the most important of these creations are beliefs, values, and norms.

Beliefs The first component of nonmaterial culture is **beliefs**, conceptions that people accept as true, concerning how the world operates and where the individual fits in relationship to others. Beliefs can be rooted in blind faith, experience, tradition, or the scientific method. Whatever their accuracy or origins, they can exert powerful influences on actions and can be used to justify almost any kind of behavior, ranging from the most generous to the most violent. Here are some examples of beliefs that people can share:

- Interruptions or imbalances in the flow of *qi* (pronounced chee) — the vital energy that flows through the body—cause illness.
- Very small organisms called germs cause disease.
- Continuous conversation, rather than silence, validates a relationship.
- Talent derives primarily from genetic propensities to be good at something.
- Talent is essentially a product of hard work, practice, and persistence.
- After death the human spirit returns to earth in a different form.

Values

The second component of nonmaterial culture is **values**, general, shared conceptions of what is good, right, appropriate, worthwhile, and important with regard to conduct, appearance, and states of being. Whereas beliefs are conceptions about how the world and people in it operate, values are conceptions about how the world *should* operate and how people *should* behave.

Perhaps the most significant study on values was made by social psychologist Milton Rokeach (1973). Rokeach identified thirty-six values that people everywhere share to differing degrees, including the values of freedom, happiness, true friendship, broadmindedness, cleanliness, obedience, and national security. He suggests that societies are distinguished from one another not on the basis of which values are present in one society and not in another but rather according to which values are more pervasive and dominant. Americans, for example, place considerable value on the individual as an individual; they stress personal achievement and unique style (free choice). In contrast, Koreans value the individual in relationship to the group (particularly the family); they stress self-discipline and respect toward those who are older.

In sports, for example, the value placed by Americans on the individual is evidenced in the fact that they single out the most valuable player of a game, a season, a league, or a tournament. Furthermore, when Americans view an outstanding athletic feat, they tend to give more credit to the individual's talent or desire to win than to disciplined practice. In addition, American athletes work to find the style that is right for them and are willing to change this style if it does not bring success. Koreans, on the other hand, do spotlight individual achievement but, in doing so, place considerable value on discipline, particularly on form (that is, adhering to time-tested and efficient methods of accomplishing goals). From the Korean point of view, athletic achievement does not occur simply because a person wants to excel or because he or she possesses raw talent. Athletic competence develops over time, after the individual masters and appreciates the steps that combine to produce the intended result. Compared with the American system, the Korean system minimizes individual achievement because the achiever owes success to the mastery of technique.

Values, we should note, transcend any particular situation. For example, the American emphasis on individual achievement and unique style — and the Korean emphasis on the group, form, and discipline — are not confined to one single area of life, such as sports. As we will see later in this chapter, these cultural values permeate many areas of life, including studying (see "Group Study, Cheating, and the Korean Foreign Student Experience").

Dodgers pitcher Chan Ho Park bows to the umpire before throwing the first pitch. What values and beliefs might inspire this behavior?

Norms

The third component of nonmaterial culture is **norms**, written and unwritten rules that specify behaviors appropriate and inappropriate to a particular social situa-

Values General, shared conceptions of what is good, right, appropriate, worthwhile, and important with regard to conduct, appearance, and states of being.

Norms Written and unwritten rules that specify behaviors appropriate and inappropriate to a particular social situation.

U.S. In Perspective

Group Study, Cheating, and the Korean Foreign Student Experience

Even though it is very interesting to live in a foreign country, you cannot help but be frustrated almost every day. You always feel that you are missing something in the culture—whatever you do and wherever you go. At its worst, you may even hurt other people's feelings unintentionally. Of course, yours can get hurt too. To avoid those kinds of occasions as a foreigner living in America, you try to open your eyes and ears as wide as possible to collect any information about the culture, hoping you can use it sometime in the future. When the information seems to be hard to collect, you become even more alert. And even though you have accurate information, there are always cases when you use it at the wrong time and at the wrong place. There are even some cross-cultural nightmares that would never have taken place if anybody had taken the trouble to pass on to you only one sentence beforehand. In some of these nightmares, you may have to live with a mark of shame on your back for what you have done even though you did not intend to do it. The worst part of it is that you may not know that the mark is on your back. On a nice evening after dinner, long after I had graduated from an American university, I learned that the mark many unwitting Korean students were living with was "Master of Cheating."

When I was a graduate student in the late 1980s, with few exceptions the Korean students would associate almost exclusively with Korean students. Students who were married gathered with other families who lived in the same apartment complexes. Those who were not married shared apartments. Because most of us went to the engineering school and majored in a small number of different subjects, many of us took the same classes at the same time, or a little earlier, or a little later. Therefore, we were in a good position to help each other with our studies. We could naturally pool all the old homework assignments and exams. Of course, everyone took good care of the Old Exam Folder, as it was the most important reference. This folder was off-limits to other classmates who were not Koreans. However, any intent to keep it a secret was not mentioned outright by any of us. There were many foreign students at our school, and all of them worked desperately to get good grades. Seeing them studying with their own people in their own languages, we who had to compete with them just assumed that they were taking care of themselves just as we were taking care of ourselves. For Koreans, "racial identity" is a familiar concept, so we did not give this business a second thought. Anyhow, that kind of group study helped me at several points during my student years.

Some years later I married an American who worked for a while as a teaching assistant at the same school. At one point, he and I happened to be sipping a cup of coffee after dinner talking about our graduate student years. We started talking about my old days as a foreign student and Korean students in general. Then, suddenly, I could not believe what I was hearing from him. He said that all his American friends who were teaching assistants had a saying: "Beware of the Korean network; the Korean students are masters of cheating." This was very strange to hear. We were such a close-knit community (as we say in Korean, we knew the count of each other's silverware), so we naturally would have heard if some of us were known to be cheaters! What could this mean?

The answer was very short: our homework solutions, computer programming assignments and written reports were very much the same, and we prepared for our tests with copies of old exams. I said with a shaking voice that students in Korea often prepare for tests with copies of old exams. But that is not cheating, is it? Well, maybe not precisely. But to American students it is not considered very upright, ethically. Let me try to explain.

American universities do not give a fixed course sequence to students in a given major. The university just gives lists of what courses the students need to take to graduate. Accordingly, every student plans her or his own classes independently.

tion. Examples of written norms are rules that appear in college student handbooks, on the backs of lottery tickets, on signs in restaurants ("No Smoking Section"), and on garage doors of automobile repair centers ("Honk Horn to Open"). Unwritten norms exist for virtually every kind of situation: wash your hands before preparing food; do not hold hands with a friend of the same sex in public; leave a 15 percent tip for waiters and waitresses; remove your shoes before entering the house. One unwritten rule followed by a majority of

Very often you find a very mixed class: sophomores sitting next to juniors and seniors, or seniors sitting next to master's students. There is not much class-year unity to be seen. Another difference from Korea is that at many U.S. schools students often have their own part-time jobs. At my school, many American students showed up on campus right before their classes began, then immediately after their classes ended they rushed out to home or work. Because of this, too, they did not have any strong bond with people who majored in the same subjects and did not know (and did not care) which classmate would graduate in what year. There was no notion at all of *sunpae* or *hupae.*[1] It was very natural that they could not think of getting precious old exams from people who took the class in earlier years. Sometimes they organized a study group on their own or at the recommendation of an instructor, but most of the time Americans did homework all alone and prepared tests without the benefit of any old exams.

To most Americans, studying for long hours alone at their desks is part of the college experience. Occasionally, turning in a few incomplete or incorrect homework problems in a physics class, say, is no great shame. Instructors often give students some credit for their partial solutions, and indeed, they may respect the student's efforts even if it did not result in perfection. The American graduate assistants I knew in computer science, for example, considered difficult 1,000-line programming assignments part of a rite-of-passage for majors. When they saw students in their own classes taking the "easy route" by consulting the work of others, they were filled with disdain. Whereas we Koreans saw our goals as merely to learn the subject matter well and achieve a good grade, for Americans, individual solitary effort for its own sake seemed part of the formula.

Of course, all this applies to Asian foreign students in general, who make up large parts of science and engineering graduate programs in the United States. There was even a group of German students who studied together closely at my school. But as I now understand it, the problem was not that we studied together. The problem was that we made one shared project out of succeeding in school, literally sharing old exams and homework, and so the individual effort, not to mention the individual creativity, was simply not there for each person. The harsh reaction of the teaching assistants I have mentioned is not universal here, of course. However, it would be a good idea for students who are planning to study here to ask around about their university's customs or policies in this matter.

It is distressing to think about how many habits we bring from Korea that are judged as failings here. What is more troubling is that these judgments are often hidden from us and that we lead our personal and professional lives in this new country completely unaware. Of course, there is a lesson from the reciprocal of this scenario also. The *American* habits that trouble us deeply should always be placed in context. How much will I really understand deeply while I live here? Maybe that is why I try to be kind to Americans. On occasions when they make me upset, I try to calm myself down first and try to tell them how I view the situation. This kind of direct talk is usually acceptable. In many cases, I end up learning to interpret the occasion in a way I never imagined before. In fact, sometimes I talk and talk about my culture to American listeners until they get sick and tired of hearing me, because I hope that through all my talk they find the false negatives about Korea fading away. I also hope that through my listening I will come to know better the real America, as the America in my mind is certainly misrepresented. However, the greatest hope I have is that when Americans remember my "bad behavior" at some occasions long past, they will be able to judge me anew, more clearly and fairly, after all these years.

[1]Translator's note: In Korea, a *sunpae* is a student at your school who began his or her studies one or more years before you did. A *hupae* is a student at your school who began after you. There is a strict respect hierarchy here. And even decades after graduation, recognizing that a stranger is in fact a *sunpae* or *hupae* establishes an instant bond.

Source: Kim Bo-Kyung, Northern Kentucky University (1997). This essay is adapted from *"Hankuk Yuhaksaeng: 'Khunning Tosa'ranun Numyengi Woeyn Malinka?"* to appear in a collection of essays published in Korea by Hanul Press, 1997.

women in the United States is to shave or otherwise remove facial and body hair.

Consider the folkways that govern how a meal typically is eaten at Korean and American dinner tables. In Korea, diners do not pass items to one another, except to small children. Instead, they reach and stretch across one another and use their chopsticks to lift small portions from serving bowls to individual rice bowls or directly to their mouths. The Korean norms of table etiquette — reaching across instead of passing, having no clear place settings, and using the same utensils to eat and to serve—deemphasize the individual and reinforce the greater importance of the group.

Americans follow different dining folkways. They have individual place settings, marked clearly by place mats or blocked off by eating utensils. It is considered impolite to reach across another person's space and to use personal utensils to take food from the serving bowls. Diners pass items around the table and use special serving utensils. The fact that Americans have clearly marked eating spaces, do not trespass into other diners' spaces, and use separate utensils to take food reinforces values about the importance of the individual.

Koreans and Americans even have different folkways about how they should use resources such as notebook paper and the electricity needed to keep refrigerators cold. Koreans open the refrigerator door only as wide as necessary to remove an item, blocking the opening to minimize the amount of cold air that escapes. Americans open the refrigerator door wide and often leave it open while they decide what they want or until they move the desired item to a stove or countertop.

As for notebook paper, until about fifteen years ago Koreans filled every possible space on a sheet of paper before throwing it away. Sometimes they used a ruler to draw extra lines between those already on the paper to double the writing space. Americans will sometimes throw away a sheet of paper with only one line of writing because they do not like what they wrote. Another more visible folkway related to conservation can be observed when cars stop at red traffic lights in the city; drivers turn off their headlights. When the light turns green, they turn them on again (Kim and Kirby 1996).

Some norms are considered more important than others—and so the penalties for violation are more severe. Depending on the importance of a norm, punishment can range from a frown to death. In this regard we can distinguish between folkways and mores.

Folkways are norms that apply to the mundane aspects or details of daily life: when and what to eat, how to greet someone, how long the workday should be, how many times a day caregivers should change babies' diapers. Waitresses, waiters, and members of occupational groups that serve the public are expected to

Folkways Norms that apply to the mundane aspects or details of daily life.

The Korean table has no clear place settings, a practice that reflects values about the relationships of the individual to the group. In addition, Korean diners use the same utensils for serving and eating.

©David Bartruff/Artistry International

adhere to certain norms: "Oblige customers even if they treat you rudely, refrain from eating garlic and spicy foods, keep toothbrush and toothpaste on hand, avoid nail biting, wear only clear nail polish and keep fingers out of mouths, ears and noses" (Murphy 1994:A1, A14). As sociologist William Graham Sumner (1907) noted, "Folkways give us discipline and support of routine and habit"; if we were forced constantly to make decisions about these details, "the burden would be unbearable" (p. 92). Generally, we go about everyday life without asking "Why?" until something reminds us or forces us to see that other ways are possible.

Books like *The Travelers' Guide to Asian Customs and Manners* (Chambers 1988) list folkways that foreign travelers should follow when visiting a particular country. When in South Korea, this book suggests observing the following norms or folkways:

- When you meet a person older than yourself, be the first to offer greetings.
- Men should not initiate a handshake with a Korean woman; they should wait for her to offer her hand. Western women should initiate handshakes with Korean men.
- The closing paragraph of a letter usually includes good wishes for the recipient's health and business success. Always mention that you expect a reply;

U.S. In Perspective

Culture and the Pacific Rim

Consider the position of the United States in relation to twelve countries and city-states of the Pacific Rim. What aspects of material and nonmaterial culture are you already aware of being shared between the United States and these countries? What aspects of culture would you expect to be increasingly shared in the twenty-first century? Trade is naturally a major force in the exchange of culture. What other forces might promote cultural exchange across the rim? What forces might hold it back?

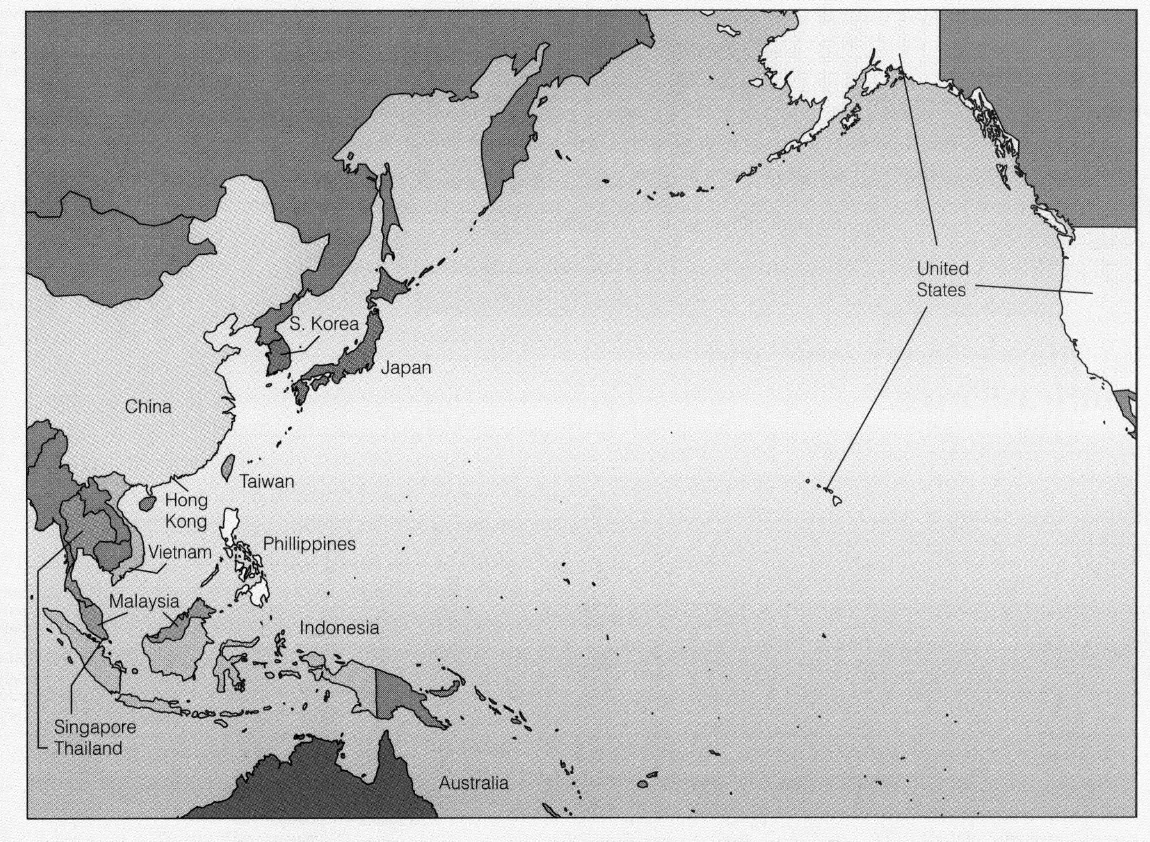

in general, Koreans don't like to write letters.[3] (p. 275)

Mores are norms that people define as essential to the well-being of a group. People who violate mores are usually punished severely—they are ostracized, institutionalized in prisons or mental hospitals, sentenced to physical punishment, condemned to die. In contrast to folkways, people consider mores to be unchangeable, regarding them as "the only way" and "the truth." In a 1994 case that received international attention, Singapore officials found Michael Faye, an eighteen-year-old American, guilty of violating their official mores, which place social order and citizens' general well-being ahead of individual rights. Faye was found

Mores Norms that people define as essential to the well-being of a group.

[3]The interesting point about these suggestions is that many apply to behavior and/or practices that are common in the United States, yet, they are presented as if they represent real differences between the two cultures.

guilty of vandalism because, over a ten-day period, he had damaged cars with eggs and spray paint. Singapore officials defined this behavior as "a calculated course of criminal conduct" and sentenced him to caning and sixty days in jail. President Bill Clinton appealed the sentence; a Clinton administration official called the penalty excessive "for a youthful, nonviolent offender who pleaded guilty to reparable crimes against private property" (Wallace 1994:A5).

One way to explain differences in behavior is to point to differences in Korean and American values and norms. For example, one could argue that Koreans value conservation and Americans value consumption and that they have devised standards of appropriate behavior that reflect these values. Yet, it is not very satisfying simply to say that values and norms guide behavior. We must investigate the geographic and historical circumstances that gave rise to specific norms and values.

The Role of Geographic and Historical Forces

Principle 2: Geographic and historical forces shape the character of culture. Sociologists operate under the assumption that culture is "a buffer between [people] and [their] habitat" (Herskovits 1948:630). That is, material and nonmaterial aspects of culture represent the solutions that people of a society have worked out over time to meet their distinctive historical and geographic challenges and circumstances:

> All mankind shares a unique ability to adapt to circumstances and resolve the problems of survival. It was this talent which carried successive generations of people into the many niches of environmental opportunity that the world has to offer—from forest, to grassland, desert, seashore, and icecap. And in each case, people developed ways of life appropriate to the particular habitats and circumstances they encountered. (Reader 1988:7)

Part of the reason that Koreans and Americans use refrigerators and paper differently has to do with the amount of natural resources in each country. Korea has no oil, only moderate supplies of coal, and depleted forests. Relative to Korea, the United States possesses abundant supplies of oil, wood, and coal. Although Koreans can import these resources, they face pressures unknown to most people in the United States, even as Americans come to realize that resources are dwindling. Because Koreans depend on other nations for most resources, they are vulnerable to any world event that might disrupt the flow of resources into their country. This vulnerability reinforces the need to use resources sparingly and not to take them for granted. The relative lack of natural resources has affected the energy costs in Korea and may ultimately explain conservation-oriented behavior.[4]

In sum, conservation- and consumption-oriented values are rooted in circumstances of shortage and abundance. To understand this connection, recall a time when your electricity or water was turned off. Think about the inconvenience you experienced after a few minutes and how it increased after a few hours. The idea that one must conserve available resources takes root. People take care to minimize the number of times they open the refrigerator door.

Imagine how a permanent resource shortage or the dependence on other countries for resources can affect people's lives. Consider how a long-term resource shortage affected Californians when their state experienced a six-year drought (1987–1992). In some water districts, Californians cut their use of water by 25 to 48 percent. In Contra Costa County in the San Francisco Bay area, for example, customers cut water consumption below the 280 gallons of water per household, per day recommended under the voluntary rationing plan to an average of 165 gallons (Ingram 1992).

In contrast, you can imagine how the abundance of resources breaks down conservation-oriented behaviors. After spending some time in the United States, many Koreans stop double-lining their paper because there are few incentives to do so. Supplies of paper are abundant, and no one else is conserving this way. Likewise, as Korea has improved its economic status, the frugal use of notebook paper has seemed to have disappeared (Kim and Kirby 1996).

For the most part, people do not question the origin of the values they follow and the norms to which they conform "any more than a baby analyzes the atmosphere before it begins to breathe it" (Sumner 1907:76). Nor are they aware of alternatives. This is because many values and norms that people believe in and adhere to were established before they were born. Thus, people behave as they do simply because they know of no other way. And, because these behaviors seem so natural, we lose sight of the fact that culture (in this case, conservation and excessive consumption) is learned.

[4]For example, even wealthy Koreans rarely air-condition their homes but use ordinary room and window fans (Kim and Kirby 1996).

The Transmission of Culture

Principle 3: Culture is learned. Humans are born "with two endowments, or, more properly stated, with one and into one" (Lidz 1976:5). Specifically, we are born with a genetic endowment and into a culture. Our parents transmit via their genes a biological heritage at once common to all humans but uniquely individual. The genetic heritage that we share with all humans gives us a capacity for language development, an upright stance, four movable fingers and an opposable thumb on each hand, and other characteristics. If these traits seem too obvious to mention, consider that they allow humans to speak innumerable languages, perform countless movements, and devise and use many inventions and objects. In fact, "most people are shaped to the form of their culture, because of the enormous malleability of their genetic endowment" (Benedict 1976:14).

Regardless of their physical traits (for example, eye shape and color, hair texture and color, skin color), babies are destined to learn the ways of the culture into which they are born and raised. The point is that our genes endow us with our human and physical characteristics, not our cultural characteristics. We cannot assume that someone comes from a particular culture simply because he or she looks like a person whom we expect to come from that culture. This fact becomes obvious to Korean-American youth who participate in cultural immersion programs that involve study in Korea. "Many say they have never felt so American as when they are slurping noodles in Korea. Even their slurps have an American accent" (Kristof 1995:47).

An excerpt from a letter written by a first-generation Taiwanese-American mother to her daughter shows that even parents have a hard time accepting this idea when their children are raised in a country and culture different from their own:

> To you, Taiwan is just a fun but humid place that you visited one summer, and your grandparents are just fuzzy voices over a telephone. To me, there is a lifetime that sits thousands of miles away, tucked inside the navy blue suit you see on me now. And to me, Taiwan is still ohm. HOME. Yes, our white stucco house with the orange door is home, too, but part of my blood still flows toward Taiwan. Can you understand that? Maybe that's why sometimes I expect you to understand Chinese culture without having experienced any of it firsthand. I think that you have the same blood, and that it pulses to the same beat. You ARE Chinese still, and I know that you have some interest and even some pride in it, but there's so much you don't know. It is your right to know. It is your right to know your family's experiences, even if you don't care about them. Maybe someday, you will care. (Yeh 1991:2)

The Role of Language

The development of language illustrates the relationship between genetic and cultural heritages. Human genetic endowment gives us a brain that is flexible enough to allow us to learn the language(s) that we hear spoken by the people around us. As children learn words and the meanings of words, they learn about their culture. They also acquire a tool that enables them to think about the world — to interpret their experiences, establish and maintain relationships, and convey information. Anyone who speaks only one language might not realize this property of language until he or she learns another language. To become fluent in another language is not merely a matter of reading and conversing in that language but of actually being able to think in that language.[5] Similarly, when young children learn the language of their culture, they acquire a thinking tool.

The following characteristics of language show the relationship between learning the meaning of words and learning the ways of a culture:

- *Language conveys important messages above and beyond the actual meaning of words.* Words have two levels of meaning—denotative and connotative. **Denotation** is literal definition; **connotation** is the set of associations that a word evokes. The connotation of a word is as important, and sometimes more important, to understanding meaning as is the literal definition. Idioms help us see the distinction between denotation and

[5]In the 1920s, linguist Edward Sapir wrote a highly influential book, *Language: An Introduction to the Study of Speech.* Sapir (1949), who is responsible for alerting us to the social issues and structure of language, believed, "No two languages are ever sufficiently similar to be considered as representing the same social reality. The worlds in which different societies live are distinct worlds, not merely the same world with different labels attached" (p. 162). These assumptions underlie the linguistic relativity hypothesis advanced by Sapir: languages are so different that it is nearly impossible to make translations in which words produce approximately the same effects in the speaker of language X as in the speaker of language Y. Today, most social scientists reject this position and argue that, although considerable work may be required, it is possible for people to discover one.

Denotation A literal definition.

Connotation The set of associations that a word evokes.

connotation. An **idiom** is a group of words that when taken together have a meaning different from the internal meaning of each word understood on its own. Examples of idioms include "to bump into," "on the edge of my seat," "to be eating out of someone's hand," and "to be in hot water" (The Comenius Group 1996). In a literal sense, "to bump into" means "to collide with or hit something" but when used in the sentence "Guess who I bumped into the other day?" the phrase connotes "to meet unexpectedly" (The Comenius Group 1996).

- *Words refer to more than things; they also describe relationships.* The word *adoption,* for example, as in the adoption of a child, refers to more than the child or the taking in and raising of that child. It also implies the presence of biological and adoptive parents and evokes assumptions about the relationship among the child, the biological parents, and the adoptive parents. The norms that guide these relationships are important features of the word. In Korea, for example, most people think of adoption as "a system whereby a sonless couple may receive a son from one of the husband's brothers or male cousins" (Peterson 1977:28). An adopted son cannot come from the wife's family or the husband's sister. In contrast, in the United States adoption usually involves a situation in which a child's biological parents release him or her to responsible adults who agree to raise the child as their own. The adoptive parents usually have no connection with the biological parents.

- *Words mirror cultural values.* Language embodies values considered important to the culture. For instance, in Korean society age is an exceedingly important measure of status: the older a person is, the more status, or recognition, he or she has in the society. Korean language acknowledges the importance of age by its use of special age-based hierarchical titles for everyone. In fact, it is nearly impossible to carry on a conversation, even among siblings, without taking age into consideration. Every word referring to one's brother or sister acknowledges his or her age in relation to the speaker. Even twins are not equal because one twin was born first. Furthermore, norms that guide Korean forms of address do not allow the speaker to refer to elder brothers or sisters by first name. A boy addresses his elder brother as *hyung* and his elder sister as *muna;* a girl addresses her elder brother as *oppa* and her elder sister as *unni.* Regardless of gender, however, people always address their younger siblings by their first name (Kim and Kirby 1996).

Idiom A group of words that when taken together have a meaning different from the internal meaning of each word understood on its own.

An American cultural norm is the idea that when a child is adopted, it typically leaves a biological family to join an unrelated family. Is there anything particularly "natural" about this? Adoption practices are quite different in Korean culture.

Among Koreans, the importance of the group over the individual is reflected in rules governing the writing and speaking of one's name. Koreans tend to identify themselves by stating the family name first and then the given name. In effect, the family is more important than the individual. Likewise, a letter is addressed to the country, the province, the city, the street, the house number, and, finally the recipient.

- *Common expressions embody the preoccupations of the culture.* Frequently used phrases and words serve as indicators of cultural preoccupations — stresses, strains, and values. For instance, Americans use the word *my* to express "ownership" of persons or things over which they do not have exclusive rights: my mother, my school, my country. The use of *my* reflects the American preoccupation with the needs of the individual over those of the group. In contrast, Koreans express possession as shared: our mother, our school, our country. The use of the plural possessive reflects the Korean preoccupation with the group's needs over the individual's interests.

Another example of cultural preoccupations is the Korean response to a full moon—"Isn't that sad?"—or to singing birds — "They are weeping!" These comments reflect patterns of response rooted in centuries of invasions and wars. In part because of its geographic location, Korea has had a continuous history of invasions—by the Japanese, the Chinese, and the Russians—that caused severe hardship, substantial loss of life, and widespread devastation. Generations of warfare and occupation by foreigners have created in the Korean people a sadness that is reflected in their responses to many natural phenomena.

These four characteristics of language demonstrate that children do more than learn words and meanings. They also acquire a perspective that reflects what is important to the culture. Nevertheless, learning about and acquiring a cultural perspective does not make people cultural replicas of one another.

The Importance of Individual Experiences

Principle 4: People are products of cultural experiences but are not cultural replicas of one another. Why are people not cultural replicas? A baby enters the world and, by exposure to an already established set of human relationships, is introduced to many versions of the culture. Virtually every event the child experiences — being born, nursing, being cleaned, being talked to, weaning, toilet training, talking, playing, and so on — involves people. The people involved in the child's life at any one time include various combinations of father, mother, grandparents, brothers, sisters, playmates, other adult relatives, neighbors, babysitters, and others (Wallace 1952). All of these people expose the child to their own versions of culture, which they have acquired in the same way and which they pass on to the child in modified terms. The following excerpt from the essay "Faculty Brat: A Memoir," by Emily Fox Gordon (1994), captures this selective and interpretive dynamic:

> My father's background was Jewish, my mother's Presbyterian. Both of them were agnostic rationalists, and I grew up hearing almost nothing of belief or doctrine. My mother preserved the aesthetic parts of her Christian heritage. We spent two weeks before Christmas, my mother, sister, brother and I at the kitchen table mixing food coloring into vanilla icing in small glass dishes — pale green, pink, a shade I called chocolate blue. We used toothpicks to paint striped frosting trousers on the rudimentary legs of gingerbread men, buttoned up their blurred pastel waistcoats with silvery sugar balls. We also collected pine cones and sprayed them, over newspaper, with silver and gold (the wonderful toxic reek of those cans, which were also preternaturally cold to the touch!); we saved the tops and bottoms of tin cans and used metal shears to cut them into stars and spirals for the Christmas tree. We made Santas, gluing triangles of cotton on the chins of walnuts and red felt hats on their foreheads. . .
>
> We children learned nothing of Judaism, except a vague understanding that the pickles and corned beef sandwiches my father loved, and the demonstrative relatives from New York and Philadelphia we occasionally visited, were things from the Jewish side of the universe. . . .
>
> I am the only one of my siblings to marry a Jew, and from her birth my daughter has always been Jewish to me. My brother and sister consider themselves and their children to be unaffiliated, but they celebrate Christmas and Easter. I'm not sure what I call myself, but now, having a child, I find I cannot celebrate the Christian holidays, even though the memory of some carols—"It Came upon a Midnight Clear," and "Lo, How a Rose Ere Blooming"—brings tears to my eyes when I find myself humming them in December. I know more about Judaism now, and I have a great abstract respect for it, but my mother's holiday Christianity, its sweetness, the memories of food and music and the surfaces of familiar things embellished and glittering, is like a beloved country from which I have exiled myself. (pp. 6–8)

This excerpt shows that individuals are products and carriers of cultural experiences and that they pass on those experiences selectively with varying degrees of clarity and confusion. The people to whom they transmit these experiences then repeat the process. Because individuals perceive, interpret, select, manipulate, revive, and create culture, they cannot be viewed as passive agents who absorb one version of culture.

Culture as a Tool for Problems of Living

Principle 5: Culture is the tool that enables the individual to adjust to the problems of living. Although our biological heritage is flexible, it presents all of us with a number of challenges. As noted previously, we are dependent on others for a relatively long time. We also feel emotions; we experience organic drives such as hunger, thirst, and sexual desire; and we age and eventually die. These biological inevitabilities have given rise to functional requisites, arrangements, or "formulas" necessary for the survival of all societies. There are formulas for caring for children; satisfying the need for food, drink, and sex; channeling and displaying emotions; segmenting the stages and activities of the life cycle; and eventually departing this world. In this section, we will focus on the differing cultural formulas for dealing with two biological events—hunger and the social emotions.

Cultural Formulas for Hunger All people become hungry, but the factors that stimulate and satisfy appetite vary considerably across cultures. One indicator of a culture's influence is that people define only a portion of the potential food available to them as edible. Culture determines not only what is defined as edible but also who prepares the food, how the food is served and eaten, what the relationship is among those eating together, how many meals are eaten in a day, and when during the day meals are eaten. For example, dogs and snakes are among the foods defined by many Korean and other Asian peoples as edible, but they are not defined as such by most Americans who cannot understand why someone would eat dog meat. This reaction should not be surprising when we consider that more than one-third of U.S. households include at least one dog. In addition, the United States has more than 10,000 pet shops, 19,000 dog food vendors, 11,000 grooming shops, 7,000 kennels, and 300 pet cemeteries, as well as 200 products for dogs, ranging from feeding dishes to raincoats and sunglasses (Rosenfeld 1987).

Americans may wonder how some Koreans can eat dogs. (They may also wonder how some Americans can eat squid or snakes.) Koreans and others may wonder how an American can stand to let a dog lick her face. Much of the world would be dismayed at the resources that Americans "waste" on pet dogs and cats. Why feed dogs when people starve?

©Ann Cecil/Photo 20-20

Among other factors, these differences in attitudes toward dogs are rooted in historical and environmental factors. Whereas the United States uses an abundance of fertile, flat land for grazing cattle, many Asian countries such as Korea with limited space use available land to grow crops, not to graze cattle. The few existing cattle are important to the agricultural system as a source of labor—to pull plows. Even today, cattle are more efficient than tractors in tilling steep inclines. The agricultural importance of cattle, combined with the lack of land to support a cattle industry, discourages the widespread practice of eating beef and encourages the consumption of dogs and snakes as alternative food sources.[6]

Another interesting cultural difference is that rice is the staple of the Korean diet, whereas corn is the staple of the U.S. diet. Very little in the American diet is not affected by corn, though few Americans realize this. In addition to being a vegetable, corn (in one form or another) is found in soft drinks, canned foods, candy, condensed milk, baby food, jams, instant coffee, instant potatoes, and soup, among other things (see "The Importance of Corn" on page 106).

Corn is a "gift" from Native Americans to all people who settled in the United States. The Native Americans recognized the significance of corn to life by referring to it as "our mother," "our life," or "she who sustains us" (Visser 1988). It is also worth noting that among the many colors of corn, only the yellow and the white varieties are defined as edible by the dominant U.S. culture; the more exotic colors (blue, green, orange, black, and red) are considered fit only for decoration at Thanksgiving time (although blue corn chips and other blue corn products have become available in gourmet food stores and trendy restaurants). One might specu-

[6]Even in the United States, cattle are relatively new as a major source of meat. The widespread use of beef as food began only after the westward expansion of the 1870s and the development of the tractor. Before then, the population was largely confined to the densely wooded eastern states, where pigs rather than cows thrived. Thus, pork was the major source of meat. The settlement of the West opened an abundance of rich flatland for grazing and food production (Tuleja 1987).

late that this arbitrary preference for some colors of corn over others is tied to the early American immigrants' rejection of the "exotic" elements of Native American cultures.

Koreans have no such lack of awareness of rice as a staple. They recognize and appreciate the significance of rice in their lives and eat it at all meals—breakfast, lunch, and dinner. In fact, in the Korean language *rice* can be a word for food.[7] Most Koreans are aware of the many uses of rice and the by-products of rice plants: to feed livestock; to make soap, margarine, beer, wine, cosmetics, paper, and laundry starch; to warm houses; to provide inexpensive fuel for steam engines; to make bricks, plaster, hats, sandals, and raincoats; and to use as packing material to prevent items from breaking in shipping.

Cultural Formulas for Social Emotions

Culture also influences the expression of emotion, just as it influences people's responses to food needs. **Social emotions** are internal bodily sensations that we experience in relationships with other people. Grief, love, guilt, jealousy, and embarrassment are a few examples of social emotions. Grief, for instance, is felt at the loss of a relationship; love reflects the strong attachment that one person feels for another person; jealousy can arise from fear of losing the affection of another (Gordon 1981). People do not simply express social emotions directly, however. They also interpret, evaluate, and modify their internal bodily sensations upon considering "feeling rules" (Hochschild 1976, 1979).

Feeling rules are norms that specify appropriate ways to express the internal sensations. They define sensations that one should feel toward another person. In the dominant culture of the United States, for example, same-sex friends are supposed to like each other but not feel romantic love toward each other. The process by which we come to learn feeling rules is complex; it evolves through interactions with others.

In her novel *Rubyfruit Jungle,* Rita Mae Brown (1988) describes a situation in which feeling rules are articulated. The central character, Molly, who is about seven years old at the time, wonders whether girls can marry each other. She approaches Leota, a girlfriend whom she likes very much, about this possibility:

> "Leota, you thought about getting married?"
> "Yeah, I'll get married and have six children and wear an apron like my mother, only my husband will be handsome."
> "Who you gonna marry?"
> "I don't know yet."
> "Why don't you marry me? I'm not handsome, but I'm pretty."
> "Girls can't get married."
> "Says who?"
> "It's a rule."
> "It's a dumb rule. Anyway, you like me better than anybody, don't you? I like you better than anybody."
> "I like you best, but I still think girls can't get married." (p. 49)

In another scene in the same novel, Molly walks in on her father, Carl, while he is comforting his friend, Ep, whose wife has just died. In this passage Molly reflects on the feeling rules that apply to men:

> I was planning to hotfoot it out on the porch and watch the stars but I never made it because Ep and Carl were in the living room and Carl was holding Ep. He had both arms around him and every now and then he'd smooth down Ep's hair or put his cheek next to his head. Ep was crying just like Leroy. I couldn't make out what they were saying to each other. A couple of times I could hear Carl telling Ep he had to hang on, that's all anybody can do is hang on. I was afraid they were going to get up and see me so I hurried back to my room. I'd never seen men hold each other. I thought the only things they were allowed to do was shake hands or fight. But if Carl was holding Ep maybe it wasn't against the rules. Since I wasn't sure, I thought I'd keep it to myself and never tell. I was glad they could touch each other. Maybe all men did that after everyone went to bed so no one would know the toughness was for show. Or maybe they only did it when someone died. I wasn't sure at all and it bothered me. (p. 28)

These examples show that people learn norms that specify how, when, where, and to whom to display emotions.

Feeling rules apply to male-female relationships as well as to same-sex friends. Different norms, for example, govern the body language that men and women use to show affection toward each other. A Korean husband and wife almost never touch, hug, or kiss in pub-

Social emotions Internal bodily sensations that we experience in relationships with other people.

Feeling rules Norms that specify appropriate ways to express the internal sensations.

[7]In Korean, a more precise word for food is *umsik*.

U.S. In Perspective

The Importance of Corn

Corn is believed to have originated in the area of the world known today as Mexico and Central America. Today it is grown all over the world. The United States is the largest corn producer in the world. It produces half of the world's supply, exporting 40 percent of all that it grows. China is the second largest producer of corn. Other countries that grow significant amounts of corn include Egypt, Thailand, Indonesia, Russia, Iran, and many African countries (Perry 1993). Corn is more than a source of nourishment. There is also the world of "hidden corn" in which corn is the equivalent of "industrial gold" (Shapiro 1992).

> *The driving wheel of the supermarket is not always visible: it is not the business of a driving wheel to be ostentatious. But it is there—everywhere. It is American corn, or maize. You cannot buy anything at all in a North American supermarket which has been untouched by corn, with the occasional and single exception of fresh fish—and even that has almost certainly been delivered to the store in cartons or wrappings which are partially created out of corn. So is milk: American livestock and poultry are fed and fattened on corn and cornstalks. Frozen meat and fish has a light corn starch coating on it to prevent excessive drying. The brown and golden colouring which constitutes the visual appeal of many soft drinks and puddings comes from corn. All canned foods are bathed in liquid containing corn. Every carton, every wrapping, every plastic container depends on corn products—indeed all modern paper and cardboard, with the exception of newspaper and tissues, is coated in corn.*
>
> *One primary product of the maize plant is corn oil, which is not only a cooking fat but is important in margarine (butter, remember, is also corn). Corn oil is an essential ingredient in soap, in insecticides (all vegetables and fruits in a supermarket have been treated with insecticides), and of course in such factory-made products as mayonnaise and salad dressings. The taste-bud sensitizer, monosodium glutanate or MSG, is commonly made of corn protein.*
>
> *Corn syrup—viscous, cheap, not too sweet—is the very basis of candy, ketchup, and commercial ice cream. It is used in processed meats, condensed milk, soft drinks, many modern beers, gin, and vodka. It even goes into the purple marks stamped on meat and other foods. Corn syrup provides body where "body" is lacking, in sauces and soups, for instance (the trade says it adds "mouth-feel"). It prevents crystallization and discolouring; it makes foods hold their shape, prevents ingredients from separating, and stabilizes moisture content. It is extremely useful when long shelf-life is the goal.*
>
> *Corn starch is to be found in baby foods, jams, pickles, vinegar, yeast. It serves as a carrier for the bubbling agents in baking powder; is mixed in with table salt, sugar (especially icing sugar), and many instant coffees in order to promote easy pouring. It is essential in anything dehydrated, such as milk (already corn, of course) or instant potato flakes. Corn starch is white, odourless, tasteless, and easily moulded. It is the invisible coating and the active neutral carrier for the active ingredients in thousands of products, from headache tablets, toothpastes, and cosmetics to detergents, dog food, match heads, and charcoal briquettes.*
>
> *All textiles, all leathers are covered in corn. Corn is used when making things stick (adhesives contain corn)—and also whenever it is necessary that things should not stick: candy is dusted or coated with corn, all kinds of metal and plastic moulds use corn. North Americans eat only one-tenth of the corn their countries produce, but that tenth amounts to one and a third kilograms (3 lb.) of corn—in milk, poultry, cheese, meat, butter, and the rest—per person per day.*
>
> *The supermarket does not by any means represent all the uses of corn in our culture. If you live in North America—and even very possibly if you do not—the house you live in and the furniture in it, the car you drive, even the road you drive on, all depend for their very existence on corn. Modern corn production "grew up" with the industrial and technological revolutions, and the makers of those revolutions were often North American. They turned their problem-solving attention to the most readily*

lic; instead, they express affection inwardly. Indeed, Koreans tend to value the concealment of such emotions. When they see American couples express love overtly by touching, caressing, and kissing in public, they regard that behavior as a sign of insecurity. Korean couples have no need to express love overtly; they know that they love each other (Park 1979).

Sociologist Choong Soon Kim (1989) found that these feeling rules applied even to Korean couples separated from one another for thirty years or more. Choong observed the "Family Reunion" program that took place in South Korea in June 1983. The program was a television campaign designed to reunite relatives living throughout South Korea who had been separated from one another as a result of the Korean War (Jun and Dayan 1986). Kim noted that no husband and wife kissed when reunited, and most couples did not hug. No Korean onlookers commented that they found this behavior unusual. In the United States, by contrast, if reunited couples did not display affection, onlookers would wonder about the quality of the relationship.

As these examples suggest, people learn norms that specify how, when, where, and to whom to display emotions. Another example of the role that culture

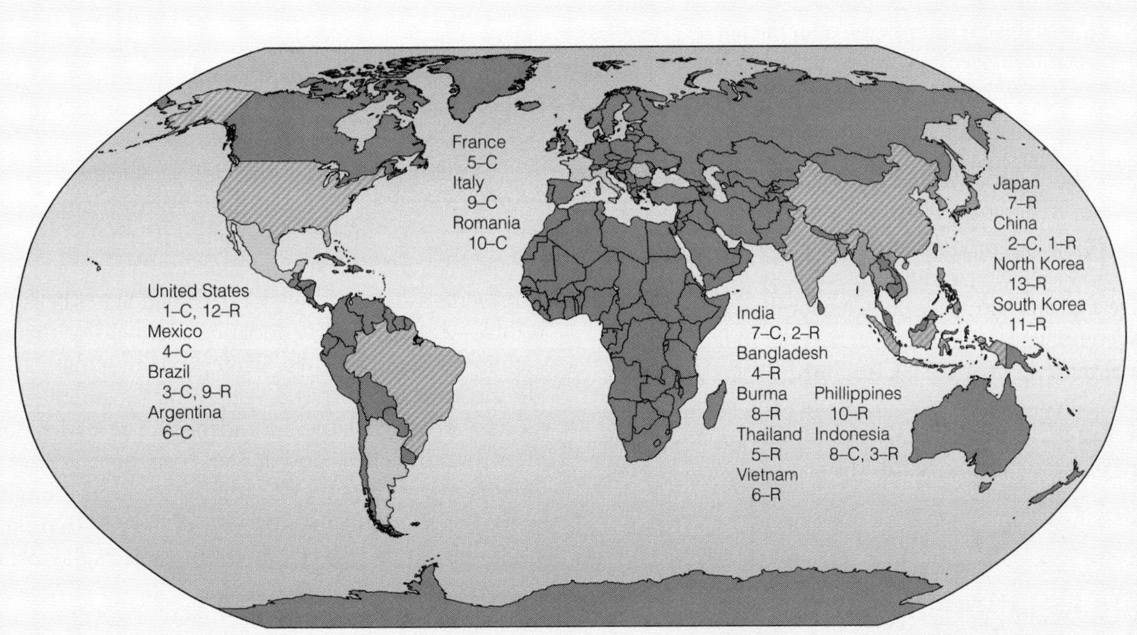

From its origins in the Americas, maize-corn has spread to be a major grain around the globe. The United States produces more than 440 billion pounds of corn each year, as much as the rest of the world put together. From its origins in Asia, rice has also traveled the globe. Both staples permeate the cultures that rely on them.

Numbers indicate rank amongst corn (C)- and rice (R)-producing countries, respectively.

Source: U.S. Department of Agriculture (1995–96), Tables 34 and 43.

available raw materials and made whatever they wanted to make—antibiotics or deep-drilling oil-well mud or ceramic spark plug installators or embalming fluids—out of the material at hand. And that material was the hardy and obliging fruit of the grass which the Indians called maïs.

In English, the word corn *denotes the staple grain of a country. Wheat is "corn" to the people of a country where wheaten bread is the staple. Oats is "corn" to people who eat oats; rye is "corn" if the staple is rye. When Europeans arrived in America they saw that, for the Indians, maize was the basic food, so the English-speaking newcomers called it "Indian corn."*

We continue in North America to recognize the primacy of maize in our culture by calling it "corn."

Source: Pp. 22–23 in *Much Depends on Dinner,* by Margaret Visser. Copyright ©1988 by Grove Press, New York. Reprinted with permission.

plays in channeling expressions of emotion has to do with laughter. Laughter is *not* something that happens only when people are amused. Instead, laughter can be an expression of emotions such as anxiety, sadness, nervousness, happiness, or despair.

No matter what emotion laughter releases or defuses, however, it occurs when something is awry in a given situation or when behavior in a specific circumstance differs from what is expected. Because culture provides the guidelines for what is expected in a specific set of circumstances, this observation about laughter implies that the situations that make someone laugh in one culture may not be funny in another. In fact, communication specialists generally agree that jokes do not translate well. The following anecdote in which a Japanese translator decides not to translate a foreign speaker's joke because the audience will not understand the cultural context that makes the joke funny represents one way of handling this dilemma:

> I began my speech with a joke that took me about two minutes to tell. Then my interpreter translated my story, and about thirty seconds later the Japanese audience laughed loudly. I continued with my talk, which seemed well received but at the end, just to

> make sure, I asked the interpreter, "How did you translate my joke so quickly?" The interpreter replied: "Oh, I didn't translate your story at all. I didn't understand it. I simply said, "Our foreign speaker has just told a joke, so would you all please laugh." (Moran 1987:74)

To this point, we have discussed a number of principles about culture. We have emphasized that culture is a tool that people learn and draw from to meet the challenges of living. The discussion may have led you to believe that the nonmaterial components of culture have a greater influence on behavior than do the material components, but that is debatable. In fact, the material and nonmaterial components of culture are interrelated: the nonmaterial shapes the material, and the material — particularly the introduction of some new object or technological advance—shapes the nonmaterial.

The Relationship Between Material and Nonmaterial Culture

Principle 6: It is difficult to separate the effects of nonmaterial and material cultures on behavior. To see how nonmaterial components of culture shape the material, consider the flags of the United States and South Korea. (The South Korean flag was the flag of Korea before the country was divided at the thirty-eighth parallel into North and South Korea.) A flag can be designed in any number of ways, so why did Korean and American designers settle on these particular designs? One could argue that designers drew on symbols that reflected important themes in each country's history.

The South Korean flag has a circle in the middle that is divided into two comma-shaped halves, the red half for yin (feminine) and the blue half for yang (masculine). The yin/yang design represents contributing, yet opposing, forces in the universe and depicts balance and harmony as a solution to contradictory or conflicting forces. The four trigrams (☰) around the circle represent heaven (☰) opposite earth (☷) and water (☵) opposite fire (☲). The message conveyed in the design is that humans have the "responsibility to balance these forces for optimum social harmony and human progress" (Reid 1988:171). The use of the yin/yang symbol and the trigrams, taken from the *I Ching,* or *Book of Changes,* an ancient Chinese book (1122 B.C.), acknowledges the value of the past and China's influence on Korean culture. The flag's design also reflects a belief in the interdependence between forces rather than the dominance of one force over another. Finally, it represents an important norm by which Koreans are guided: people have the responsibility to balance opposing or contradictory forces to ensure human progress (Yoo 1987).

Similarly, the flag of the United States reflects its nonmaterial culture. Each of the thirteen alternating red and white stripes represents one of the thirteen original colonies. Each of the fifty stars on the blue field represents a state. Although the United States began as a British colony, the flag does not acknowledge that influence (except in the red, white, and blue colors). Nor do the stars and stripes overlap. This separateness reflects the value placed on independence or freedom from a strong central power. The absence of any symbol that represents Britain or Europe also reflects the value placed on shedding the influence of the past to make a clean start. It matches the norm that rejects the role of the past in defining a person.

We can see another influence of nonmaterial culture on material culture in the different burial customs of Koreans and Americans. Most Americans from a Christian background have the deceased placed in a wooden coffin and then in a concrete-lined metal vault for burial. The double enclosure keeps the dead body separate from the earth as it decays. This method ensures that the deceased will not become "confused" with the earth; it reflects the belief that humans and their environment are independent. In contrast, the Korean practice of burying the dead in wooden coffins

The text describes how South Korea's flag reflects a belief in the interdependence among natural forces. How would you interpret the cultural significance of some other flag (besides the U.S. flag) with which you are familiar?

that decay corresponds with the belief that humans and their environment are interdependent and that decay is part of that interdependence.

Just as nonmaterial components of culture shape flags and coffins, material components shape the nonmaterial. For example, the microwave oven is an important material influence on American values and norms, specifically values that emphasize the individual over the group and norms that govern when the members of a family should eat.

Approximately 70 percent of American households have microwave ovens, which have eliminated one of the incentives for families to eat together. Before the microwave was available, it was more efficient for one member of the household to cook for everyone because of the time — whether calculated in person time or in oven time — required to cook a meal. Now that meals take only minutes to prepare, it is no longer considered inefficient for family members to cook and eat separately. The microwave also enhances independence in that a person is not tied to a particular meal schedule or even family group:

> The old dining room table required each individual to give up some personal autonomy and bow to the dictates of the group and the social system. If we stop eating together, we shall save time for ourselves and achieve mealtime self-sufficiency. . . . The communal meal is our primary ritual for encouraging the family to gather together every day. If it is lost to us, we shall have to invent new ways to be a family. It is worth considering whether the shared joy that food can provide is worth giving up. (Visser 1989:42)

In Korean society, the writing system of *hangul,* introduced by King Sejong in 1446, has had far-reaching effects on the culture. This writing system was developed over a twenty-year period with the explicit purpose of encouraging literacy. What makes *hangul* unique is that anyone who speaks Korean can learn to read in a matter of days, because the simple and consistent phonetic rules make it "one of the clearest and most logical systems of writing ever devised" (Iyer 1988:48).

The efficiency of *hangul* is evident when it is contrasted with the English alphabet, whose phonetic rules are inconsistent. Although each of the twenty-six letters theoretically is associated with a distinct sound, some letters symbolize more than one sound, especially when combined with other letters. Combinations of vowels can be quite problematic. Consider the vowel sounds in the following words: *break* versus *freak, sew* versus *few, food* versus *good, paid* versus *said,* and *shoe* versus *foe* ("Letters" 1988).

Hangul has had an immense impact on Korean society. First, it freed Koreans from the writing and reading system that was previously in use — a system based on the complex ideograms of the Chinese system. The relative simplicity of *hangul* is responsible for the nearly 100 percent literacy rate in Korea. Second, *hangul* supports the strong Korean sense of and desire for national unity. Because virtually everybody can read, Koreans have a common base of knowledge. Such shared knowledge encourages unity by breaking down barriers due to differences in literacy levels. Finally, the widespread literacy has contributed to South Korea's rapid rise as an economic force in the world. Higher literacy rates lead to greater employability, lower error rates at work, and fewer industrial accidents — all of which give Korea an advantage over countries with lower rates of literacy. Many people in the United States are learning only now that a literate workforce, even in entry-level positions, is essential to the country's ability to compete in the worldwide marketplace.

The examples discussed in this subsection show that inventions often have consequences that go beyond their intended purpose. The microwave oven not only changed the time and effort it takes to cook a meal; it also changed how family members relate to one another. Likewise, the purpose of a writing system is to record ideas, but the design of the Korean writing system promoted widespread literacy and equality. These examples not only show how material components affect values and norms but also demonstrate that the introduction of a new invention can have profound consequences on the way in which people in a society relate to one another.

Most people tend to think that their material and nonmaterial culture is self-created. They underestimate the extent to which ideas, materials, products, and other inventions are connected in some way to foreign sources or borrowed outright from those sources.

Cultural Diffusion

Principle 7: People borrow ideas, materials, products, and other inventions from other societies. Most people tend to underestimate, ignore, or distort the extent to which familiar ideas, materials, products, and other inventions are connected in some way to outside sources or are borrowed outright from those sources (Liu 1994). The process by which an idea, an invention, or some other cultural item is borrowed from a foreign source is

U.S. In Perspective

Opportunities for Cultural Diffusion Between Americans and South Koreans

Which of these statistics represent the greatest opportunities for cultural diffusion between Americans and South Koreans? What other opportunities for diffusion or exchange might there be? Can the question be answered simply in terms of numbers of people heading one way or the other? In what other ways might the cultures diffuse?

South Korean students in U.S. Colleges, 1994	35,599
U.S. Students in South Korea, 1994	399
South Koreans in U.S. for temporary employment, 1994	7,413
Korean-born residents of U.S., 1990	663,000
Airline passengers between U.S. and South Korea, 1993	2,021,285
U.S. military personnel in South Korea, 1993	35,786
People employed in 1993 by South Korean affiliates in the U.S.	58,800
Phone calls between U.S. and South Korea, 1991	56,641,000
Immigrant visas issued to Korean orphans adopted in U.S., 1987–95	26,126
U.S. tourists to Korea, 1994	383,063

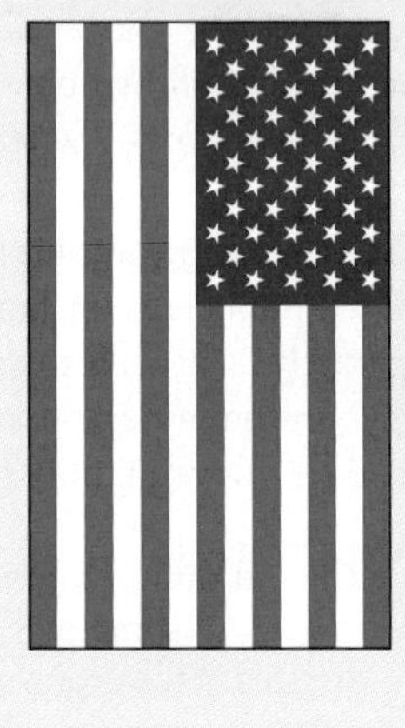

Sources: Korean Overseas Information Service (1995–96), U.S. Department of State (1996), *Chronicle of Higher Education* (1996), UNESCO (1995), U.S. Department of Transportation (1993), U.S. Department of Justice (1995), U.S. Department of Defense (1994), U.S. Federal Communications Commission (1993–94), and U.S. Department of Commerce (1995).

called **diffusion**. The term *borrow* is used in the broadest sense; it can mean to usurp, pirate, steal, imitate, plagiarize, purchase, or copy. The opportunity to borrow occurs whenever two people who have different cultural traits make contact, whether face to face, by phone or fax, or via the Internet.

Basketball, a U.S. invention, has been borrowed by people in seventy-five countries, including South Korea, where twenty-one clubs are registered with the *Federation Internationale de Basketball*. Baseball, another U.S. invention, has been borrowed by people in more than ninety countries, including Korea, which won the twenty-seventh world baseball championships in 1982 after it upset the United States and Japan (An 1997; *World Monitor* 1992, 1993; see Figure 4.1).

Instances of cultural diffusion are endless and can be found by skimming the newspaper headlines. Here are some examples:

> "School for Japan's Executives in the United States"
>
> "In China, Beauty Is a Big Western Nose"
>
> "Thai Publisher Plans to Expand Empire in U.S."
>
> "Invasion of the Discounters: American-Style Bargain Shopping Comes to the United Kingdom"
>
> "Global Goliath: Coke Conquers the World"
>
> "Japan's Favorite Import from America: English"

Diffusion The process by which an idea, an invention, or some other cultural item is borrowed from a foreign source.

Figure 4.1 Number of Baseball Players Around the World Who Belong to the International Baseball Association

What aspects of American baseball culture might appeal to most Koreans? What aspects of it might not be so appealing?

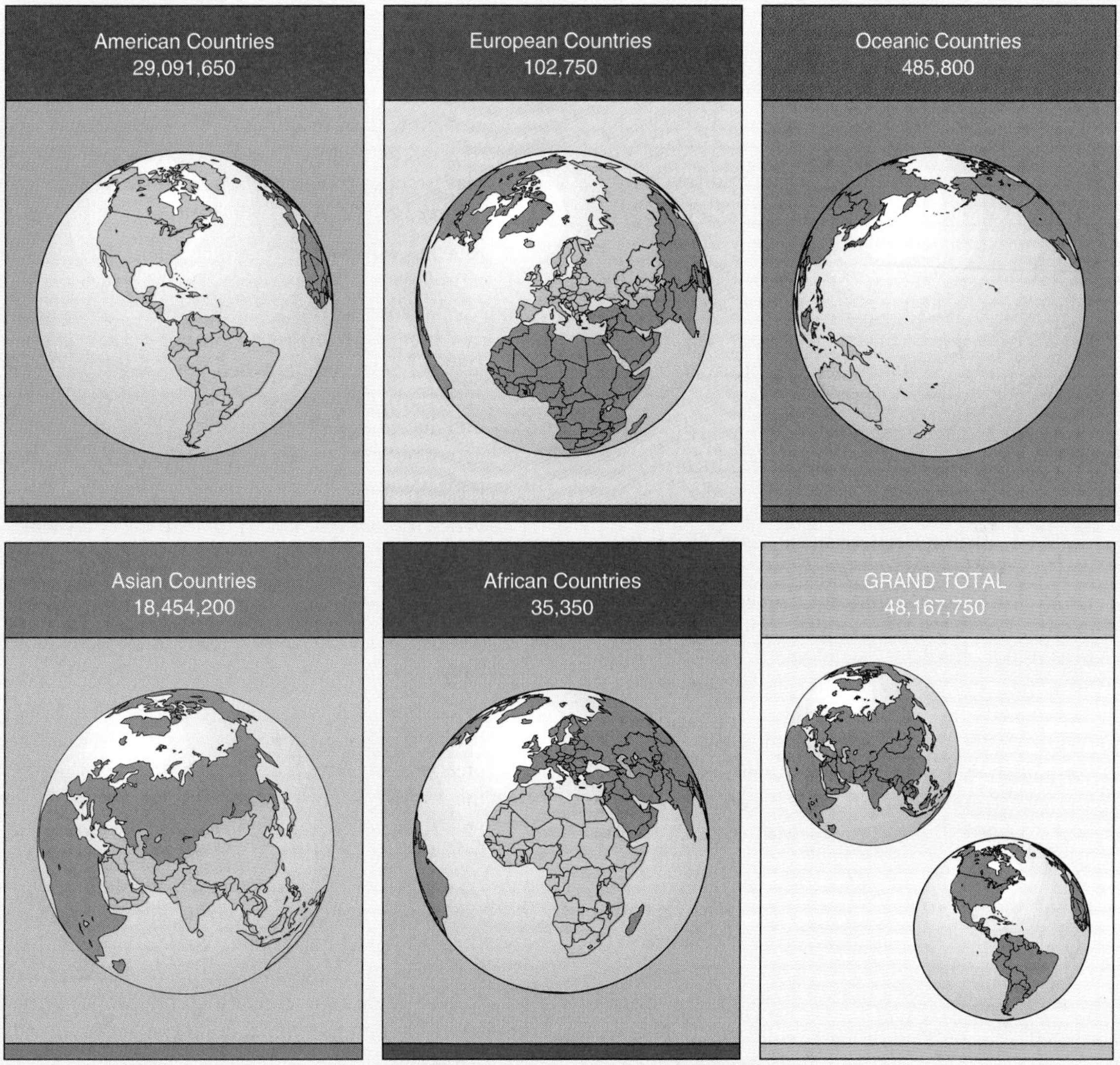

Source: International Baseball Association (1993).

"Skilled Asians Leaving U.S. for High-Tech Jobs at Home"

Opportunities for cultural diffusion between the United States and South Korea exist as a result of the forty-five-year presence of U.S. troops, international trade, and the arrival and return of many Koreans who have attended colleges and universities in the United States. More than 33,000 South Koreans enrolled in colleges and universities in the United States in 1994–95 (Institute of International Education 1996; see "Opportunities for Cultural Diffusion Between Americans and South Koreans").

For example, the presence of the American military in South Korea since 1950 has encouraged diffusion in a number of ways. First, the media service, American Forces in Korea Network (AFKN), exposes Koreans to U.S. television and music and (by extension) to its dominant values and norms. Second, many jobs for U.S. bases are filled by Koreans, and a service industry (bars, gift shops, brothels, tailor shops, laundries) exists to meet the soldiers' needs.[8] Third, U.S. soldiers, in collaboration with Korean black marketeers, leak PX goods into Korean society. At one point, 60 percent of the available PX materials flowed into Korean society. This situation stimulated a desire for these products among the Korean people (Bok 1987).

People of one society do not borrow ideas, materials, or inventions indiscriminately from another society. Borrowing is almost always selective: ideas and inventions are accepted or rejected depending on how useful they are to the receiving culture. They must be useful enough to make their acceptance worth the trouble. Selective borrowing has an important implication for anyone trying to sell products to people who live in foreign markets: if each market has its own tastes, "the producer who doesn't tailor is the producer who will fail" (Magaziner and Patinkin 1989:36). Many Americans have been slow to grasp this concept as they try to convince foreigners to purchase their products:

> American firms rarely customize. "They just send us products made for Americans and say '"Why don't you Koreans buy them?"' [explains Kim, a Korean engineer working for Samsung]. Sometimes American exports are almost unusable. Although Korean households have 220-volt electricity, U.S. firms have been known to try selling refrigerators built for 110 volts, with only rudimentary converters. "Even the chocolate," says Kim. "My kids like chocolate very much, but they don't like American chocolate. Too sweet. You want to sell chocolate here, you have to know our taste." (Magaziner and Patinkin 1989:37–38)

Even if people in one society accept a foreign idea or invention, they are still selective about which features of the item they adopt. Even the simplest invention is really a complex of elements, including various associations and ideas of how it should be used. Thus, not surprisingly, people borrow the most concrete and most tangible elements and then develop new associations and shape the item to serve new ends (Linton 1936). For example, the Japanese borrowed the game of baseball from Americans but modified it considerably to fit important Japanese cultural values. In a Public Broadcasting Service *Frontline* documentary, "American Game, Japanese Rules," reporters asked U.S. athletes playing baseball in Japan about the differences and the similarities in how the game is played in the two countries. One American athlete said, "Well, they play nine innings. That's about the only thing they have in common [with us in the game]. After we put on the uniform I'm not sure what we're doing out there" (*Frontline* 1988:2).

During this documentary, American athletes described a number of striking differences:

> I think that the biggest difference is, they play for ties. You know, that's unheard of back home. You'll play all day and all night to break a tie. But over here, you play for ties and you have a time limit on a game that is three hours and twenty minutes.
>
> A tie is a wonderful thing in Japan. In the U.S., being a professional athlete when I played it, a tie is like kissing your sister. I mean, I'd just as soon not have that happen, I'd just as soon not play the game. In Japan they're ecstatic when you have a tie because everybody came out, everybody had the big fight, they had the big confrontation. They all fought together but nobody lost—no face was lost. So that is a perfect day.
>
> When they win the championship they only win by a couple of games. [Say] they're up by ten games [at some point in the season], by the end of the year maybe they'll be up by two. That looks OK; it doesn't look like one team blew out the other team and embarrassed them.
>
> What they feel is that, for instance, if a foreigner comes over and he's just hitting a lot of home runs, a great deal of home runs, the umpire—I wouldn't say the umpire — the league and people themselves would expect the umpire to expand the strike zone to balance things out. They feel that this foreigner has to be very strong and if the pitchers would keep throwing the ball over the plate and he's going to hit a home run every time up and that's not fair. So what they do is they'll start calling pitches this far outside a strike, that far inside a strike. And their philosophy is they're being fair because they're balancing it out. (*Frontline* 1988:2–3, 8)

A particularly significant global trend is the diffusion of technology from already developed countries to the newly industrializing countries such as Korea and other Pacific Rim countries. From the mid-1960s to early 1970s, these countries moved to produce goods such as calculators, computers, cars, microwave ovens, televisions, radios, shoes, and clothing for already developed markets in the United States, Western Europe, and Canada. After they learned how to make these products, they perfected them by making them more functional, more efficient, higher in quality, and lower in price.

[8]Companies such as Hyundai and Daelin got their start by filling contracts for the U.S. military.

Two international business consultants, Ira Magaziner and Mark Patinkin, visited Korea in 1977 and witnessed the beginnings of the borrowing and perfecting process there. They describe this process in *The Silent War:*

> The factory floors were bare concrete, and people were hand-wheeling parts to and from the production line. I moved on to Samsung's research lab, which reminded me of a dilapidated high-school science classroom. But the work going on there intrigued me. They'd gathered color televisions from every major company in the world — RCA, GE, Hitachi—and were using them to design a model of their own. They were working on refrigerators and other appliances as well. The chief engineer was young, well-trained, a recent graduate of an American university. I asked him about Samsung's color-television strategy, telling him I presumed the company planned to buy parts from overseas, only doing assembly in Korea. Not at all, he said. They were going to make everything themselves — even the color picture tube. They'd already picked the best foreign models, he said, and signed agreements for technical assistance. (Magaziner and Patinkin 1989: 23–24)

Microwave oven production is another case in point. Since 1979, the conglomerate Samsung has been the world's leading producer of this product, invented forty years earlier in the United States. Samsung workers assemble about 80,000 ovens per week; about one in every three microwave ovens sold in the United States is made in Korea (Magaziner and Patinkin 1989). Although Koreans borrowed U.S. technology, they improved on the production process. Cost-efficient Korean production is a major reason that 70 percent of U.S. households own at least one microwave oven.

One of the reasons that people are selective with regard to the items they borrow from another culture is that they evaluate those items in terms of the standards of their home culture.

The Home Culture as the Standard

Principle 8: The home culture is usually the standard that people use to make judgments about the material and nonmaterial cultures of another society. Most people come to learn and accept the ways of their culture as natural. When they encounter other cultures they can experience mental and physical strain. Sociologists use the term **culture shock** to describe the strain that people from one culture experience when they must reorient themselves to the ways of a new culture. In particular, they must adjust to a new language and the idea that the behaviors and responses they learned in their home culture and have come to take for granted do not apply in the foreign setting. The intensity of culture shock depends on several factors: (1) the extent to which the home and foreign cultures are different, (2) the level of preparation or knowledge about the new culture, and (3) the circumstances (vacation, job transfer, or war) surrounding the encounter. Some cases of culture shock are so intense and so unsettling that people become ill. Among the symptoms are "obsessive concern with cleanliness, depression, compulsive eating and drinking, excessive sleeping, irritability, lack of self-confidence, fits of weeping, nausea" (Lamb 1987:270).

In his book *Communication Styles in Two Different Cultures,* Myung-Seok Park, a professor of communication, describes some examples of the kinds of stress encountered when someone enters a foreign society. These experiences are typical of the adjustments that Koreans must make when they come to the United States. Taken as separate incidents, the two encounters are not especially stressful. It is the cumulative effect of a series of such encounters that causes culture shock.

> When I studied at the University of Hawaii, my academic advisor was an old, retired English professor, Dr. Elizabeth Carr. All of the participants were struck by her enthusiasm, deep devotion and her unfailing health. So at the end of the fall semester I said to her, "I would like to extend my sincere thanks to you for the enormous help and enlightening guidance you gave us in spite of your great age." Suddenly she put on a serious look, and I saw a portion of her mouth twisting. I had an inkling that she seemed unhappy about the way I expressed my thanks to her. Understandably enough, I was not a little embarrassed. A few hours later she told me that my remark "in spite of your great age" had reminded her suddenly that she was very old. I felt as if I had committed a big crime. I restrained myself from commenting on age any more. (Park 1979:66)
>
> One Saturday afternoon I was drinking in an American drinking establishment with some of my new American friends. What surprised me at that moment was that whenever the bar girl brought some

Culture shock The strain that people from one culture experience when they must reorient themselves to the ways of a new culture.

bottles of beer, she made change and took away a certain amount out of the money in front of each drinker. It was only I who did not put money on the table. (We Korean people pay the total amount for drinks when we leave.) When the money placed in front of each person ran out, the girl proceeded to take another person's money. What was still more surprising to me was that each one filled his own glass and drank without passing the glass to his friend and without asking him to drink. I was somewhat bewildered because I had never poured my own glass before. . . . I realized that there was something cold and unfriendly in the American way of drinking. (pp. 38–39)

Some accounts from a Japanese physician visiting the United States for the first time provide additional illustrations of culture shock:

> Another thing that made me nervous was the custom whereby the American host will ask a guest, before the meal, whether he would prefer a strong or a soft drink. Then, if the guest asks for liquor, he will ask him whether, for example, he prefers scotch or bourbon. When the guest has made this decision, he next has to give instructions as to how much he wishes to drink, and how he wants it served. With the main meal, fortunately, one has only to eat what one is served, but once it is over one has to choose whether to take coffee or tea, and — in even greater detail — whether one wants it with sugar, and milk, and so on. I soon realized that this was only the American's way of showing politeness to his guest, but in my own mind I had a strong feeling that I couldn't care less. What a lot of trivial choices they were obliging one to make — I sometimes felt — almost as though they were doing it to reassure themselves of their own freedom. My perplexity, of course, undoubtedly came from my unfamiliarity with American social customs, and I would perhaps have done better to accept it as it stood, as an American custom. (Doi 1986:12)

> The "please help yourself" that Americans use so often had a rather unpleasant ring in my ears before I became used to English conversation. The meaning, of course, is simply "please take what you want without hesitation," but literally translated it has somehow the flavor of "nobody else will help you," and I could not see how it came to be an expression of good will. The Japanese sensibility would demand that, in entertaining, a host should show sensitivity in detecting what was required and should himself "help" his guests. To leave a guest unfamiliar with the house to "help himself" would seem excessively lacking in consideration. This increased still further my feeling that Americans were a people who did not show the same consideration and sensitivity towards others as the Japanese. As a result, my early days in America, which would have been lonely at any rate, so far from home, were made lonelier still. (p. 13)

Reentry shock Culture shock in reverse; experienced upon returning home after living in another culture.

Americans in Korea or Japan would be equally confused and bewildered by the Korean habits of pouring and drinking from each other's glasses, by their constant references to age (especially mature age), and by Japanese norms about hospitality.

Do not assume that culture shock is limited to experiences with foreign cultures. People can experience **reentry shock**, or culture shock in reverse, upon returning home after living in another culture (Koehler 1986). In fact, some researchers have discovered that many people find it surprisingly difficult to readjust to the return after spending a significant amount of time elsewhere. As in the experience of culture shock, they are in a situation in which differences jump to the forefront.

As with culture shock, the intensity of reentry shock depends on an array of factors, including the length of time lived in the host culture and the extent of the returnee's immersion in the everyday lives of people in that culture. Symptoms of reentry shock are essentially the mirror image of those associated with culture shock. They include panic attacks ("I thought I was going crazy"), glorification of the host culture, nostalgia for the foreign ways, panic, a sense of isolation or estrangement, and a feeling of being misunderstood by people in the home culture. These comments by Americans returning from abroad illustrate these reactions:

> "People pushed and shoved you in New York subways; they treated you as if you simply don't exist. I hated everyone and everything I saw here and had to tell myself over and over again: 'Whoa, this is your country; it is what you are part of.'" (Werkman 1986:5)

> "America was a smorgasbord. But within two weeks, I had indigestion. Then things began to make me angry. Why did Americans have such big gas-guzzling cars? Why were all the commercials telling me I had to buy this product in order to be liked? Material possessions and dressing for success were not top priorities in the highlands. And American TV? I missed the BBC." (Sobie 1986:96)

Although many people expect to have problems in adjusting to a stay in a foreign culture and even prepare for such problems, most do not expect trouble upon returning home. Because reentry shock is unexpected, many people become anxious and confused and feel guilty about having problems with readjustment ("How could I possibly think the American way was anything but the biggest and the best?"). In addition, they are apprehensive about how their family, friends, and other acquaintances might react to their critical views of the home culture; for example, they are afraid others might view them as unpatriotic.

The experience of reentry shock points to the transforming effect of an encounter with another culture (Sobie 1986). The fact that the returnees go through reentry shock means that they have experienced up close another way of life and have been exposed to new norms, values, and beliefs. Consequently, when they come home, they see things in a new light.

One reason people experience cultural shock upon visiting a foreign culture or returning home after becoming immersed in a foreign culture is that they hold the viewpoint of **ethnocentrism**. That is, they use one culture as the standard for judging the worth of foreign ways. From this viewpoint, "one's group is the center of everything, and all others are scaled and rated with reference to it" (Sumner 1907:13). Thus, other cultures are seen as "strange" or, worse, as "inferior."

Ethnocentrism

There are different levels and consequences of ethnocentrism. The most harmless is simply defining foreign ways as peculiar, as did some American visitors who attended the 1988 Summer Olympic Games in Seoul. Learning that some Koreans eat dog meat, some people made jokes about it. People speculated about the consequences of asking for a doggy bag, and they made puns about dog-oriented dishes — Great Danish, fettuccine Alfido, and Greyhound as the favorite fast food (Henry 1988).[9]

The most extreme and most destructive form of ethnocentrism is **cultural genocide**, in which the people of one society define the culture of another society not only as offensive but as so intolerable that they attempt to destroy it. There is overwhelming evidence, for example, that the Japanese tried to exterminate Korean culture between 1910 and 1945. After the Japanese annexed Korea in 1910, Japanese became the official language, *hangul* was banned, Koreans were given Japanese names, Korean children were taught by Japanese teachers, Korean literature and history were abandoned, ancient temples — important symbols of Korean heritage — were razed and bulldozed, and the Korean flag could not be flown. The Japanese brutally suppressed all resistance on the part of the Korean people. When Koreans tried to declare their right to self-determination in March 1919, thousands of people were injured or killed in clashes with the Japanese military.[10] Unfortunately, American history is filled with instances of this type of ethnocentrism, as when the U.S. Bureau of Indian Affairs forced Native Americans to attend boarding schools (see "Forced Culture Change of Native Americans by Bureau of Indian Affairs Schools").

Sociologist Everett Hughes (1984) identifies yet another type of ethnocentrism:

> One can think so exclusively in terms of his own social world that he simply has no set of concepts for comparing one social world with another. He can believe so deeply in the ways and the ideas of his own world that he has no point of reference for discussing those of other peoples, times, and places. Or he can be so engrossed in his own world that he lacks curiosity about any other; others simply do not concern him. (p. 474)

Ethnocentrism A viewpoint that uses one culture as the standard for judging the worth of foreign ways.

Cultural genocide A form of ethnocentrism in which the people of one society define the culture of another society not only as offensive but as so intolerable that they attempt to destroy it.

[9]After Seoul was awarded the 1988 Summer Olympic Games, the South Korean government took steps to rid the city of snake and dog shops. Several forces caused this move, including Korean sensitivity to foreign opinion and the power of global mass media. The history of foreign involvement in Korean affairs has made the Koreans self-conscious about what people from other countries think about Korean culture. Thus, to avoid ridicule and embarrassment, the South Korean government moved to close the dog and snake shops so they would not be present for the cameras.

[10]The Japanese maintained that their purpose in destroying Korean culture was to help modernize Korea. Even today, many Japanese argue that under their occupation, Korean agricultural and industrial productivity improved and that highways, railroads, ports, communication systems, and industrial plants were built. The Koreans' earnings and output, however, went to support Japanese society and left most Koreans impoverished.

As Japan became more deeply involved in World War II, Koreans were forced to serve in the military or to fill the factory positions of Japanese workers fighting in the war. During this thirty-five-year period, the Koreans were subservient to the Japanese in every way. In essence, the Japanese attempted to "Japanize" the Koreans by taking away their language, ideas, and material culture.

Forced Culture Change of Native Americans by Bureau of Indian Affairs Schools

The Eastern band of Cherokee Indians live on a federal reservation in the mountains of North Carolina and today numbers more than 8,000 people. In the fall of 1990, the North Carolina Cherokees took direct control of their school system. For almost a century, however, dating back to 1892, Cherokee schools, like schools for most other Native Americans, were operated directly by the Bureau of Indian Affairs (BIA) and financed by federal taxes.

This Cherokee school was one of the special schools established by the U.S. Bureau of Indian Affairs to "Americanize" the native people of the United States. To further this goal, the school required pupils to speak English and to wear uniforms.

Courtesy Museum of the Cherokee Indian

For most of the twentieth century, formal education was a tool to wipe out Indian culture and transform Indians into "red-skinned whites." The process of acculturation was accomplished over the objections of parents and tribal leaders and often at the expense of the physical and emotional well-being of Indian students. For most Native Americans, the most oppressive phase of BIA education occurred during the first three decades of the twentieth century. Even today, however, there are lingering effects.

The hallmark of BIA education as it evolved around the turn of the century was the boarding school. In 1920 the BIA superintendent on the Cherokee reservation wrote about the importance of "taking the most promising young men and women away from reservation influences" (Eastern Cherokee Agency File 1920). The "bad influences" of the reservation were not crime or other social problems but parents who spoke Cherokee or wore moccasins.

All over the United States, Indian students were shuffled around to maximize the distance between themselves and their families. Thus, although a boarding school existed on the Cherokee reservation, many of the students were non-Cherokee Indians from distant reservations; in

James F. Larson and Nancy Rivenburgh, professors of communication, offer a vivid example of this type of ethnocentrism. They list a number of comments made by television commentators during the opening ceremonies of the 1988 Summer Olympic Games held in Seoul, Korea. The comments are characterized by simple, sweeping generalizations that provide little insight into the people who live in the countries mentioned:

- Oman is probably the "hottest nation in the world."
- Mexico is "one of the most highly emotional countries in the world."
- Americans are "cool cats," "superstars" from "the most famous of all Olympic nations."
- Ireland, the "home of the leprechaun and four-leaf clover."
- Japan once occupied Korea, but the countries are now "making their peace" (Larson and Rivenburgh 1991:85–86).

Since the end of the Korean War, millions of Americans — military personnel, technicians, social workers, and educators — have spent time in South Korea as advisers but not as learners. For this reason,

turn, many Cherokees, even as young as age six, were shipped to Chilocco Boarding School in Oklahoma, to Haskell in Kansas, to Carlisle in Pennsylvania, or to Hampton in Virginia. Even where day schools were available for elementary school–age students, the situation often changed with junior high or high school where boarding schools became the norm.

In 1918 the Cherokee BIA superintendent commented on one small boy whose whole family "have been known to stand him [the superintendent] off with drawn knives" (Annual Report of the Commissioner of Indian Affairs 1918). For many Native Americans, whether Navajo, Cheyenne, Cherokee, or others, the months of August and September were spent trying to hide their children from the truant officers. Captured children were taken crying off to the out-of-state boarding schools. If they were lucky, they got to return home for a visit the following summer. In many cases, however, it was years before children saw their parents again. It was not uncommon for a weeping six-year-old to be taken away and return to his parents only at age sixteen when he could legally drop out of school.

To make things even worse, the boarding school was often an alien environment where only English could be spoken and uniforms had to be worn. The schools stressed vocational subjects: boys learned to farm—ironic for Cherokee boys who were removed from their family farms—and girls were trained to be servants. Girls at the Cherokee Boarding Schools were taught to do laundry and to cook: "the demand for them as house servants in Asheville and other neighboring towns could not be supplied" (Young 1894:172).

Children caught speaking their native language had their mouths washed out with soap or were beaten if they were repeat offenders. At the Cherokee Boarding School, children who attempted to run away were chained to their dormitory beds at night.

Perhaps the best description of the BIA boarding school era came from the anthropologist John Collier who, as commissioner of Indian affairs under Franklin Roosevelt's New Deal, attempted to repeal many of the oppressive policies. In 1941, Collier, writing in a magazine called *Indians at Work,* reflected on a night in the 1920s when he was allowed to sleep over at the boys dormitory at the Cherokee Boarding School. Collier had gotten lost hiking in the mountains and asked to stay over the night until he could get transportation off the reservation: "The little boy inmates could not or would not talk English to each other, and they dared not talk Cherokee. . . . This was an Oliver Twist place but with every light and shade of imagination disbarred. . . . Horror itself gave up the fight" (Collier 1941:3).

REFERENCES

Collier, John. 1941, September 1–8. Editorial. *Indians at Work.* Commissioner of Indian Affairs. 1918. *Annual Report of the Commissioner of Indian Affairs, Interior Department Report.* Vol. 2. Indian Affairs and Territories, 65th Congress, 3rd Session, House Document 1455.

Eastern Cherokee Agency File. 1892–1958. Eastern Cherokee Agency correspondence. Federal Archives and Records Center, East Point, GA.

Young, Virginia. 1894. "A Sketch of the Cherokee People on the Indian Reservation of North Carolina." *Woman's Progress,* pp. 171–172.

Source: Sharlotte Neely, Northern Kentucky University (1991).

"surprisingly few Americans have come away from their Korean experience with much understanding of the country, its people, language, or culture" (Hurst 1984:1).

A conversation between author Simon Winchester (1988) and the Korean owner of a bar frequented by American service personnel shows that from the Korean viewpoint, many American military people display the kind of ethnocentrism that Everett Hughes describes:

> Just then two burly and unshaven airmen walked past. One wore a patch on his jacket that said "Munitions Storage—We tell you where to stick it!" The other had a T-shirt with the words: "Kill 'Em All: Let God Sort the Bastards Out." Both were sporting newly stitched shoulder patches showing what appeared to be a small plane—the fuselage looking remarkably phallic—beneath the rubric "One Hundred Successful Missions to A-Town." Mr. Kwong shook his head with distaste.
>
> "That's what I can't take. Don't they ever learn? We need to be respected here, and they're not respecting us. They treat us like we're some backward Third World country, and you know we're not. We're proud, we've got good reason to be. But this. . . ." He gestured with despair.

> I said I hadn't found anything very offensive about the two passing airmen. "Maybe not, maybe I react too much," he said. "I've worked for twenty-five years trying to bring the two communities together. I organize them to go out to meet families. I try to persuade them to learn a bit of Korean, to eat some of the food, to understand why they're here. But they don't want to know. And it's the way some of them treat our women, and our men too. Some of them just have no respect for us. The way they see it, they're top of the pile, and everyone else is nothing. It makes me mad." (p. 144)

These examples may lead you to conclude that ethnocentric thinking is an American characteristic, but unfortunately it exists in every society. In Korea, for example, a dissident poet recently released from prison displayed an ethnocentric perspective when he argued that "the Korean race" is destined to take over the world. As another example, the ultimate feel-good movie for Koreans is one in which North and South Koreans team up to defeat Japan (Kim and Kirby 1996).

Another relatively unknown type of ethnocentrism is **reverse ethnocentrism**, in which the home culture is seen as inferior to a foreign culture. People who engage in this kind of thinking often regard other cultures as utopias. For example, the former Soviet Union is labeled as the model of equality, Japanese culture as the model of human connectedness, the United States as the model of self-actualization, India as the model of otherworldliness, Israel as the model of pioneering spirit, Nigeria as the model of family values, or Native Americans as the original environmentalists (Hannerz 1992).

People who engage in reverse ethnocentrism not only idealize other cultures but also reject any information suggesting that things might be otherwise. For example, some people in the United States view the various peoples encompassed by the term *Asian American* as model minorities — as people who work hard, do well in school, and have or cause few problems. Consequently, if someone such as Bong Hwan Kim, director of the Korean Youth Center in Los Angeles, tells them that he works "with juvenile delinquents in Koreatown, they get a blank look on their face and the conversation ends right there" (Schoenberger 1992:A24). The popular image of Koreans as an orderly, conforming people clashes with this fact. A position that runs counter to ethnocentrism is cultural relativism.

Reverse ethnocentrism A type of ethnocentrism in which the home culture is seen as inferior to a foreign culture.

Cultural relativism The perspective that a foreign culture should not be judged by the standards of a home culture and that a behavior or way of thinking must be examined in its cultural context.

Cultural Relativism

A perspective that runs counter to ethnocentrism is cultural relativism. **Cultural relativism** means two things: (1) that a foreign culture should not be judged by the standards of a home culture and (2) that a behavior or way of thinking must be examined in its cultural context — in terms of that society's values, norms, beliefs, environmental challenges, and history. Cultural relativism is a perspective that aims to understand, not condone or discredit, foreign behavior and thinking.

For example, the Korean practice of defining infants as one year of age at birth cannot be understood if it is evaluated according to dominant U.S. values, which emphasize the future over the past. These values support the idea that the birth of a baby represents a fresh start with unlimited possibilities no matter what the social class, ethnicity, or educational level of the baby's parents. From this point of view, it makes sense to define age at birth as zero. The Korean practice, however, must be considered in light of Korean values. Defining age in this way corresponds with Korean values about the importance of the past and beliefs about the relationship of the individual to the group. The message is that generations are interdependent and that past events (even before conception) are important to a person's life. The Korean way of defining age ties together past, present, and future generations.

Similarly, whereas most Americans cannot understand why some Koreans eat dog meat, most Koreans are equally appalled that Americans often let dogs live in their homes, allow them to lick their faces, and spend so much money on them when there are many poor and homeless people in the United States. When we consider the historical and environmental challenges surrounding the Korean decision to eat dog, the practice might not seem so unreasonable.

Cultural relativism, however, does not entail an "anything goes" position. Such a position allows every cultural trait — even some of the most harmful and violent ones (for example, infanticide, human sacrifice, foot binding, death threats against those holding controversial ideas, witch hunts, scientific experimentation on naive human subjects)—to escape judgment or criticism. Unfortunately, uninformed critics often interpret cultural relativism as a search to justify any behavior and as support for one or more of the following attitudes: "Whatever they do is fine"; "That's just the way that culture is"; "It's none of my business what others do"; or "Everything is relative. What's good in one culture is bad in another, and vice versa."

Anthropologist Clifford Geertz (1995) maintains that if one is serious about addressing a problematic cultural trait, the first step is to ask the persons affected (supporters, critics, victims, beneficiaries) about the practice. Investigators should not assume that they know the truth of the situation and proceed as if they have the solution. As a rule of thumb one should avoid making judgments and recommendations that affect another culture when these are based on superficial knowledge. With regard to education, for example, Iranian education critic, reformer, and scholar Samad Behrangi maintains that

> unless we have seen the school's environment and surrounding community, unless we have lived among the people, unless we have been friends with the people, we have not heard their voice and have not known their desires, it is not even proper to have sympathy for the environment, or impose unrelated educational policies, or even to write stories or textbooks for them. (Behrangi 1994:28)

Understanding a cultural trait or taking a position of cultural relativism does not mean that one accepts the trait unconditionally, nor does it mean that one has no values of one's own. As Geertz (1995) argues, the challenge of taking this position is finding "a way to keep one's values and identity while living with other values — values you can neither destroy or approve" (Berreby 1995:47). In Geertz's words:

> I hold democratic values, but I have to recognize that a lot of other people don't hold them. So it doesn't help much to say, "This is the truth." That doesn't mean I don't believe anything. . . . You can't assert yourself in the world as if nobody else was there. Because this isn't [just] a clash of ideas. There are people attached to those ideas. If you want to live without violence, you have to realize that other people are as real as you are. (p. 47)

Cultural relativism is especially useful when people must make a decision supporting one cultural trait over another or must contemplate taking action to change, modify, or even eliminate a cultural trait that "others" share. In all cases, cultural relativism facilitates decision making and/or negotiation by making the cultural elements in question intelligible to everyone involved. Certainly a decision is more appropriately made by people who understand the traits than by those who do not.

Studying and learning about other cultures teaches us that morality is both relative and universal (Redfield 1962). Morality is relative in that norms, values, and beliefs about rightness and wrongness vary across time and place. It is also universal, however, in that every culture has conceptions of morality. Although mores do exist that can make virtually any idea or behavior seem right or wrong, "some mores have a harder time making some things right than others" (Redfield 1962:451).

Subcultures

Principle 8: In every society there are groups that possess distinctive traits that set them apart from the main culture. Groups that share in some parts of the dominant culture but have their own distinctive values, norms, language, or material culture are called **subcultures**. A subculture that conspicuously challenges, rejects, or clashes with the central norms and values of the dominant culture is referred to as a **counterculture**. All of the cultural principles discussed thus far apply to subcultures (which include countercultures).

Often we think we can identify subcultures on the basis of physical traits, ethnicity, religious background, geographic region, age, gender, socioeconomic or occupational status, dress, or behavior defined as deviant by society. However, determining which people constitute a subculture is a complex task that requires careful thought; it must go beyond simply including everyone who shares a particular trait. For example, using broad ethnic categories as a criterion for identifying the various subcultures within the United States makes little sense. Realistically, a biologically and culturally intermixed population numbering hundreds of millions cannot be divided neatly into white, African-American, Hispanic, Native American, and Asian subcultures (Clifton 1989).

Subcultures Groups that share in some parts of the dominant culture but have their own distinctive values, norms, language, or material culture.

Counterculture A subculture that conspicuously challenges, rejects, or clashes with the central norms and values of the dominant culture.

The broad categorization of "Native American," for example, ignores the fact that the early residents of North America "practiced a multiplicity of customs and lifestyles, held an enormous variety of values and beliefs, spoke numerous languages mutually unintelligible to the many speakers, and did not conceive of themselves as a single people — if they knew about each other at all" (Berkhofer 1978:3). The point is that the presence or absence of a single trait, especially a physical trait, cannot be the only criterion for classifying someone as part of a subculture. Sociologists determine whether a group of people constitutes a subculture by learning whether they share a language, values, norms, or a territory and whether they interact with one another more than with people outside the group.

One characteristic central to all subcultures is that their members are isolated in some way from other people in the society. This isolation may be voluntary, or it may result from an accident of geography, or it may be imposed consciously or unconsciously by a dominant group. Or it may be a combination of all three. Whatever the reason, subcultures are cut off in some way. The cutoff may be total or it may be limited to selected segments of life such as work, school, recreation and leisure, dating and marriage, friendships, religion, medical care, or housing.

Some subcultures within the United States are isolated in different ways or to different degrees; their members integrate into mainstream culture when possible but voluntarily or involuntarily accept a segregated role in other areas of life. In general, African Americans who work or attend school primarily with whites are often excluded from personal and social relationships with white colleagues. This exclusion forces them to form their own fraternities, study groups, support groups, and other organizations.

The number and variety of affiliations between members of a subculture and people outside their group are a rough indicator of the subculture's relationship to mainstream culture. In general, the greater the number and the more diverse the affiliations with "outsiders," the more likely it is that the subculture shares a significant portion of the mainstream culture. Some subcultures are **institutionally complete** (Breton 1967); their members do not interact with anyone outside their subculture to shop for food, attend school, receive medical care, or find companionship because the subculture provides for these needs. In the United States, retirement communities are a typical example of a setting in which residents' needs to affiliate with the mainstream culture are minimized. In South Korea, where the total number of non-Koreans is less than 1 percent of the population, members of the largest minority group — composed of 50,000 Chinese — often live in institutionally complete ethnic communities.

Despite the ethnic similarity among the people, Korea has been divided since 1945 into two institutionally complete societies: North Korea and South Korea. The people of the two Koreas have no relationship with one another (including no correspondence by mail, phone, or travel) and technically remain at war with one another. The line that separates the two people at the thirty-eighth parallel is the most heavily militarized region of the world; each side is prepared to stop the other from invading. Both sides recognize their common language, ethnicity, history, and culture, and both believe in unification. Yet, despite these similari-

Institutionally complete Subcultures whose members do not interact with anyone outside their subculture to shop for food, attend school, receive medical care, or find companionship because the subculture provides for these needs.

For more than half a century, the demilitarized zone has divided Korea into what are now a booming capitalist economy in the south and a halting socialist economy in the north. If the two countries were to reunite tomorrow, how would the lack of exchange between the two societies for so long make it difficult to reunite?

U.S. In Perspective

Korean Immigrants to the United States May Not Be from Korea

Korean immigrants to the United States are not always from Korea. Significantly large Korean communities exist outside Korea in Siberia, Canada, Japan, and even Brazil. In addition, Koreans intermarry with people from other ethnic groups. The two case studies presented here profile two "Koreans" living in the United States. As you read each case, consider the following question: To which culture does each person belong?

SOVIETSKIE KOREITSKY

Her speech is not marked by the familiar flowing cadences of a Korean accent. It is not the flawless urbane English of many American-born Koreans. Instead, Ksenia's speech is punctuated with the sharpness of a Soviet accent. Ksenia Choi's exotic beauty and coloring belie her mixed heritage. She is a fourth-generation *Sovietskie Koreitsky,* Soviet Korean, born in Central Asia to a Korean father and a Russian mother.

Now after eleven years in New York City, she fondly describes her life in Tashkent, of the former Soviet Union. *"Koreitsky zhivut koroso*—Koreans live well." Indeed, her family owned a home and even a car, not unusual among the *Sovietskie Koreitsky,* considering that urbanization and industrialization have led to the emergence of a group of professionals and local elites from the collective farms. *"Koveseitsev uvazhaiu za trudoliubie*—Koreans are respected for their hard work," Ksenia states firmly.

Ksenia is one of the more than half million Koreans located in the former Soviet Union. The Korean minority is found mostly in concentrated settlements in the east—the result of an early-twentieth-century exodus fleeing the Japanese occupation of Korea. The Korean refugees fought in Siberia; their heroics are commemorated by sacred monuments in remote regions of Siberia, extolling the contributions of unknown and long-forgotten Koreans in the Soviet Union. In Khabaraovsk, a street is even named after a *Sovietskie Koreitsky,* Kim Yu Chen Street.

The former Soviet Union comprises over 270 million people from dynamically varied national groups. It rivals the United States in its ethnic diversity: there are over 104 officially recognized nationalities, and 130 languages are spoken within the territory. But here, unlike America, there is no celebration of the glorious mosaic. The declared goal of the former Soviet Union was to create a culture that is national in form but socialist in content and a society that fuses all nationalities.

Ksenia explains a little bashfully, "My Korean is not so good. I can speak some and read and write some, but I never really learned too much Korean. There is not too much Korean taught in schools because there is not a demand for it. My family thought that learning Russian would be most useful, so we went to Russian schools." Her experience is reflective of the cultural trend of Koreans in Russia, where a new generation is being assimilated with such speed that there exists the risk of the eventual loss of Korean identity.

FROM SEOUL TO BRASILIA TO NEW YORK CITY

Born in Brazil, Seung Min Lee is a student at Columbia University. His dorm room is an eclectic jumble of American, Brazilian, and Korean items. But despite his name and complexion, the overriding theme of his room seems, oddly enough, to be American, with broad strokes of Brazilian influence indicated in the Portuguese novels that sit on his shelf. His Korean heritage is hinted at only minimally—the statuette of the Virgin Mary on his desk bears a Korean inscription.

"My parents moved to Brazil because they have an open-border policy. Open border countries like Canada, Australia, and Brazil have attracted a great deal of Korean immigrants in recent decades. My dad already had a job lined up there. It's not like in America where you come first and find a job later. My mother's uncle owned a small shipping company, so off they went and now here I am."

Seung Min admits that his Korean is not up to par with his Portuguese and his English. "Yeah, sure, there is a Korean population in Brazil. But it's only like forty some odd thousand. That's nothing compared to the numbers that live in New York City alone—close to 300,000. Down there, it's so much easier to assimilate. I learned some Korean from my parents, but almost everyone else around us spoke Portuguese. I even spoke to my Korean friends in Portuguese. We don't have the strength and the inertia to create a separate Korean community. It's not like here in New York City."

Source: From "The Invisible Nation of Korean Emigrants" by Jennifer Lee. Pp. 39–40 in *Korean Culture.*

ties, neither side has been able to compromise on economic or government structure, the elective process, or the appointive process for government offices.

In view of the unification of Germany, we might speculate that the two Koreas might also reunite soon. South Korean officials predict, however, that it may take twenty years before they eventually reunite. One important difference is that West Germany, in comparison with South Korea, was far more wealthy and thus better able to absorb the staggering economic costs ($250 billion) of reunification. Recently, South Korea's assistant minister for reunification commented: "After seeing what happened in Germany, especially in the economic area, it has changed attitudes toward unification in Korea. The biggest lesson we learned is that we need to prepare steadily, and we need to have realistic expectations. We cannot just be ruled by sentiment" (Protzman 1991; Sterngold 1991:C1).

Often there is a clear association between institutional completeness and language differences. Persons who cannot speak the language of the dominant culture are very likely to live in institutionally complete ethnic communities (for example, Little Italy, Chinatown, Koreatown, Mexican barrios). Of the 750,000 Koreans living in the United States, approximately 300,000 live in southern California. A large portion of these 300,000 live in an institutionally complete Koreatown west of downtown Los Angeles. Still, the Korean experience in the United States cannot be described according to a few generalizations (see "Korean Immigrants to the United States May Not Be from Korea").

Whether an immigrant chooses to belong to a community of like persons depends on a variety of factors: circumstances of migration (forced or voluntary), age, gender, status of mother country relative to host country, knowledge of the host country's language, level of skills, education, and social class. Korean immigrants who cannot obtain bank loans to start businesses may become members of *kyes*,[11] informal financing systems of twenty or so would-be entrepreneurs to which each member promises to contribute several hundred dollars a month for twenty months. Each month one member of the *kye* receives $10,000 and continues to pay into the system until all twenty members have received $10,000 each (Reinhold 1989; Sanchez 1987). The presence of *kyes* in some Korean-American communities means that those Koreans do not have to rely on banks. In this sense *kyes* contribute to institutional completeness.

[11]*Kyes* have been part of Korean society for hundreds of years. They are not something that Korean immigrants invented to ease their transition to the United States.

Summary and Implications

We began this chapter by identifying three challenges sociologists face when they study culture: (1) describing a culture, (2) determining who belongs to a designated culture, and (3) identifying characteristics that set one culture apart from another. Sociologist Immanuel Wallerstern (1990) explains the challenges of describing a culture and identifying its members and its distinguishing characteristics:

> [I]t is surely true that people in different parts of the world, or different epochs, or in different religious or linguistic communities do indeed behave differently from each other, and in certain ways that can be specified and fairly easily observed. For example, anyone who travels from Norway to Spain will note that the hour at which restaurants are most crowded for the "evening meal" is quite different in the two countries. Anyone who travels from France to the U.S. will observe that the frequency with which foreign strangers are invited to homes is quite different. The length of women's skirts in Brazil and Iran is surely strikingly different . . . on the one hand, differences are obvious . . . and yet the degree to which groups are in fact uniform in their behavior is distressingly difficult to maintain. (p. 34)

Several factors account for these challenges. First, although people are products of cultural experiences, people from the same culture are not replicas of one another. This is because people possess the ability to interpret, select, manipulate, revive, and create culture. Thus, they are not passive agents who absorb or transmit to others one version of a culture. Second, people borrow ideas, materials, products, and other inventions from those in other societies. Culture diffusion occurs through many kinds of relationships including everyday minglings (the routine talking, looking, and listening people do as they live their lives). For example as the 33,000 exchange students from South Korea and the 35,000 U.S. troops stationed in Korea go about their lives, they are exposed to the others' culture. A second kind of relationship is the marketplace, the arena in which transactions between sellers and buyers occur. The processes by which goods and services are exchanged between people from two cultures (producing, advertising, shipping, selling) also involves interaction between the parties. Moreover, the goods and services exchanged have the potential of transforming each culture.

Cultural diffusion also occurs when "outsiders" make organized, deliberate efforts to transform, reform, or replace some cultural characteristic. The work of Wycliff Bible Translators represents one example. The organization's aim is to translate the Bible into all the languages of the world. The existence of the Bible transforms those who read it *and* the language into which it is translated transforms the Bible in subtle, and even dramatic, ways.

The facts of cultural diffusion and individual autonomy may leave you wondering whether there is such a thing as Korean or American culture. In spite of these facts, Ulf Hannerz (1993) maintains that *culture* "is still the most useful keyword we have to summarize that peculiar capacity of human beings for creating and maintaining their own lives together" (p. 109). He suggests that the problem lies not so much in the idea of culture itself but in our narrow use of the concept. As we discussed, we use "culture" (1) in conjunction with specific places and categories of people; (2) in ways that emphasize differences (cultural gaps); (3) as if cultures have clear, identifiable boundaries; and (4) in ways that suggest people from one culture are replicas of one another.

The facts of cultural diffusion and individual autonomy demand that we expand our conception of culture. An expanded view prepares us to see overlap, influences, and exchanges between cultures. Thus, culture becomes more than a barrier to interactions and a source of differences. Expanding our view of culture to think of it as a force that shapes individuality (see principle 4), a force that borrows (albeit selectively) from other cultures, and a force that people manipulate, revive, and re-create prepares us to approach people as more than simply representatives of a particular culture and to treat cultures as parts of an intercultural tapestry.

Hannerz (1992) offers the following metaphor that applies to this expanded vision of a culture:

> When you see a river from afar, it may look like a blue (or green, or brown) line across a landscape; something of awesome permanence. But at the same time, "you cannot step into the same river twice," for it is always moving, and only in this way does it achieve its durability. The same way with culture—even as you perceive structure, it is entirely dependent on ongoing process. (p. 4)

Key Concepts

Use this outline to organize your review of the key chapter ideas.

Culture
- **Material culture**
- **Nonmaterial culture**
  - **Beliefs**
  - **Values**
  - **Norms**
    - **Folkways**
    - **Mores**

Social emotion

Diffusion

Culture shock

Reentry shock

Ethnocentrism
- **Cultural genocide**
- **Reverse ethnocentrism**

Cultural relativism

Subculture
- **Institutionally complete**

internet assignment

Visit the Koreatown, Los Angeles, California, Web site at http://www.koma.org/koreatown.html. Browse through the documents posted on this site, but pay particular attention to the "1.5 Generation" link. As you read, identify information posted on this site that clarifies three concepts covered in this chapter.

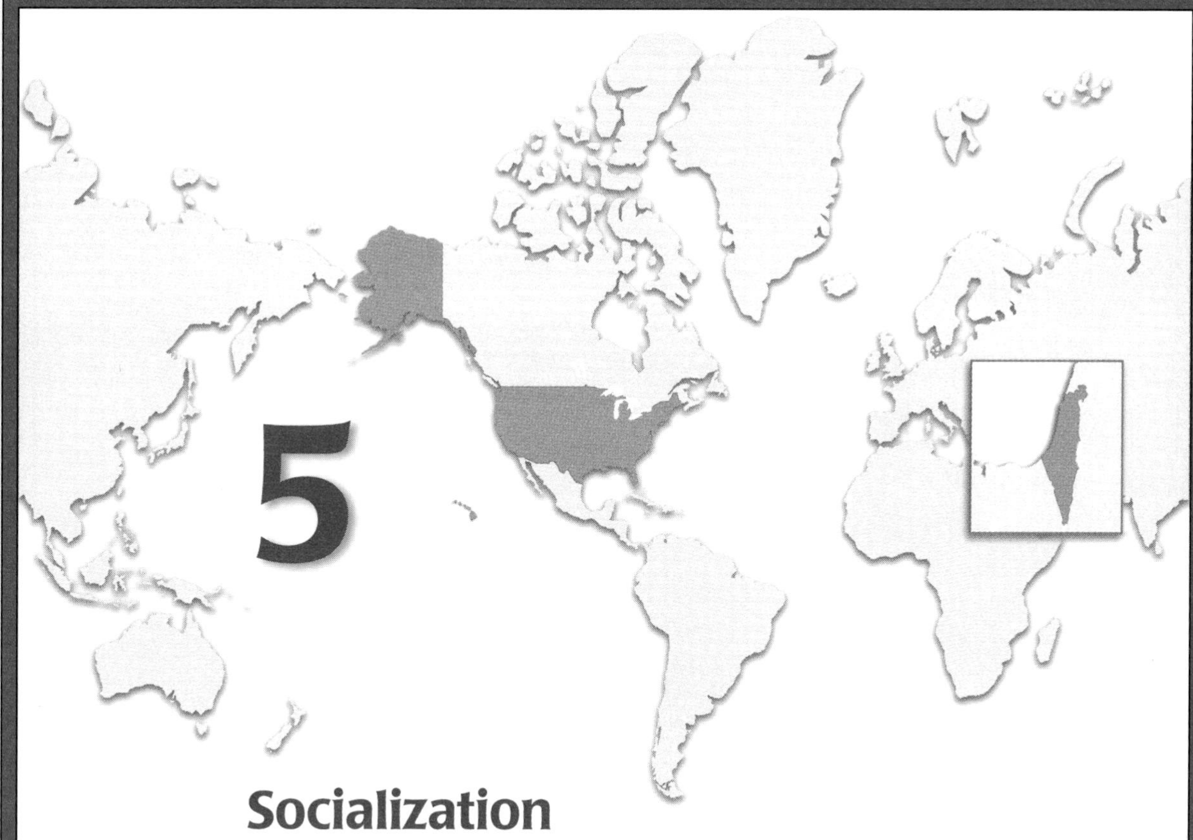

5

Socialization

With Emphasis on Israel, the West Bank, and Gaza

Coming to Terms: The Palestinians and Israelis

Nature and Nurture

The Importance of Social Contact

Individual and Collective Memory

The Role of Groups

Symbolic Interactionism and Self-Development

Cognitive Development

Resocialization

A view of Jerusalem, with the Wailing Wall in the foreground. ©James Stanfield/National Geographic Society

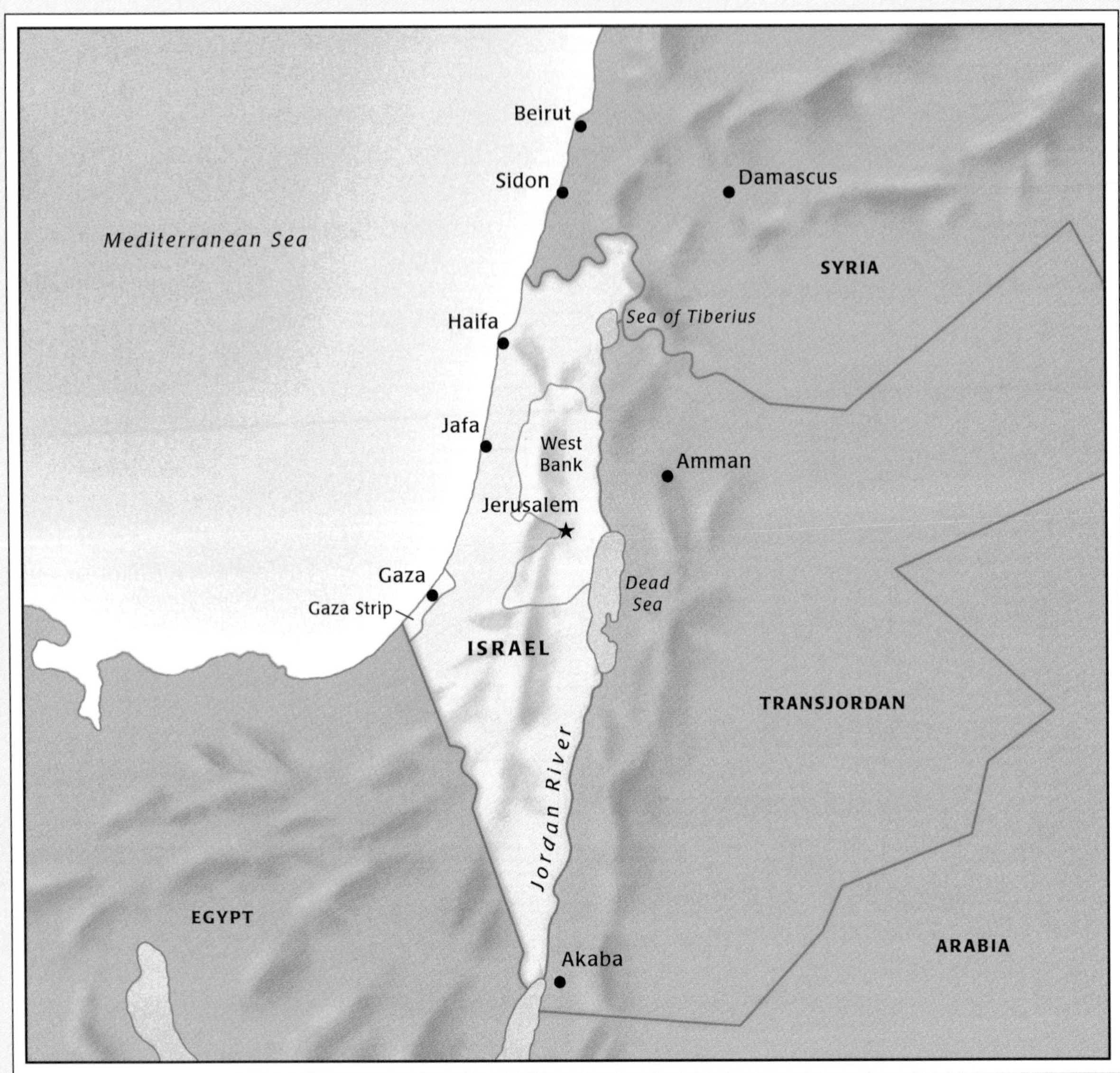

Source: Chaliand and Rageau (1995).

Palestine and Israel

What causes a person to say that he or she is a member of the Jones family or the Keshawi family? What causes him or her to identify with a specific family, class, religion, history, or nation, even when that group may be distant in space or time? This chapter explores the importance of socialization in creating and maintaining a sense of who one is and the groups to which one does or does not "belong."

Let me say to you, the Palestinians, we are destined to live together on the same soil in the same land. We the soldiers who have returned from battles tainted with blood; we who have seen our relatives and friends killed before our eyes; we who have attended their funerals and cannot look into the eyes of their parents; we who have come from a land where parents bury their children; we who have fought against you, the Palestinians, we say to you today in a loud and clear voice: Enough of blood and tears.

Enough! We have no desire for revenge; we have—we harbor no hatred toward you. We, like you, are people—people who want to build a home, to plant a tree, to love, live side by side with you in dignity, in affinity, as human beings, as free men.

We are today giving peace a chance and saying to you—and saying again to you—enough. Let us pray that a day will come when we all say farewell to the arms.

We wish to open a new chapter in the sad book of our lives together, a chapter of mutual recognition, of good neighborliness, of mutual respect, of understanding. We hope to embark on a new era in the history of the Middle East. Prime Minister Yitzhak Rabin (1993:A7) Middle East Peace Signing Ceremony, Washington, D.C.

Why Focus on Israel, the West Bank, and Gaza?

Our emphasis in this chapter is on the fierce century-long conflict between Jews and Arabs in Israel and the West Bank and Gaza. This conflict is one of an estimated sixty internal conflicts going on in the world between people who share a territory but who differ from one another in ethnicity, race, language, or religion. Most of these conflicts have long histories, which means that the conflict has been passed on from one generation to the next. We concentrate on Israel and the West Bank and Gaza because several times in the past (most recently, September 1993 and September 1996), the United States has been a major sponsor of meetings between Israeli and Palestinian representatives. According to President Bill Clinton (1996), "The United States has often played a pivotal role in bringing Arabs and Israelis together to work out their differences in peace. It is our responsibility to do whatever we can to protect the peace process and to help move it forward." Achieving peace will not be easy, however, because everyone involved has been affected in some way by the conflict. For most, the conflict has been a part of their lives since they were born. In this chapter, we draw on socialization concepts and theories to help us understand how the conflict has been "passed down" from one generation to the next.

Keep in mind that socialization is just one factor of many that help us understand why the Israeli-Palestinian conflict has lasted at least a century. Although we concentrate on these two groups (Palestinians and Israelis), we must keep in mind that the issues raised here are relevant to the strategies people everywhere use to teach newcomers how to participate in the society in which they are born. The newcomers, however, do not become carbon copies of their teachers. They learn about the environment they inherit and then come to terms with it in unique ways.

At birth the human cerebral cortex (the seat of complex thought) is not sufficiently developed to permit a sophisticated awareness of self and others or reflection on the rules of social life, which we call norms. These capacities develop, however, as children mature biologically and as they interact with others.

Most two-year-olds are biologically ready to show concern for what adults regard as rules of life. They are bothered when rules are violated: paint peeling from a table, broken toys, small holes in clothing, and persons in distress raise troubling questions. From a young child's point of view, when something is "broken," then somebody somewhere has done something very wrong (Kagan 1988a, 1988b, 1989).[1]

To show this kind of concern with standards, however, two-year-olds must first be exposed to information that leads them to expect behavior, people, and objects to be a certain way (Kagan 1989). They develop these expectations in the course of their social relations with adults. Children go to adults with their questions and needs. Adults respond in different ways: they may offer explanations, express concern, try to help, show no concern, or pay no attention. Through many such simple exchanges, children learn how to think about objects and people. They learn about the social group to which they primarily belong and about other groups to which they do not belong. In addition, through these exchanges children are acquiring basic skills such as the ability to talk, walk upright, and reason. This learning from others about the social world is part of a complex lifelong process called **socialization.**

Socialization begins immediately after birth and continues throughout life. It is a process by which newcomers develop their human capacities and acquire a unique personality and identity. This process also enables **internalization**, in which people take as their own and accept as binding the norms, values, beliefs, and language needed to participate in the larger community. Socialization is also the process by which culture is passed on from generation to generation.

An Israeli soldier accompanies a preschool outing in Galilee near the border of Lebanon in 1995. He is there because terrorist attacks were common in that area at the time. How might socialization in the United States be different if soldiers routinely had to accompany school outings? How might it be different if military service were required of every U.S. citizen?

©Annie Griffiths Belt/Aurora

In this chapter, we explore the socialization process and its significance for both societies and individuals. Keep in mind that socialization is just one factor of many that help us understand the relevant strategies people everywhere use to teach newcomers how to participate in the society into which they are born. The newcomers, however, do not become carbon copies of their teachers. They learn about the environment they inherit and then come to terms with it in unique ways.

Coming to Terms: The Palestinians and Israelis

The following excerpt describes the vicious cycle of Palestinian protest against Israeli occupation of the West Bank and Gaza and Israeli attempts to suppress that protest:

Socialization A process by which people develop their human capacities and acquire a unique personality and identity and by which culture is passed on from generation to generation.

Internalization The process in which people take as their own and accept as binding the norms, values, beliefs, and language needed to participate in the larger community.

[1]Kagan (1989) reports the results of a study in which fourteen- and nineteen-month-old children were given a set of twenty-two toys to play with. Of the twenty-two toys, ten were unflawed, ten were flawed (for example, a boat with holes in the bottom, a doll with black streaks on its face), and two were meaningless wooden forms. No fourteen-month-old child showed concern with the flawed toys, but 60 percent of the nineteen-month-old children showed clear concern with the flawed toys: they brought the toys to their mothers and said things like "broke," "bad," or "yucky." According to Kagan, this "moral sense is one of the most profound accomplishments. We should view it just as we view singing and speaking and walking. It is a maturational milestone that will be acquired by every child as long as they live in a world of human beings" (Kagan 1988a:62).

> One of my friends recently served in the [occupied] territories. He kept on talking about his inner conflict. During the day, he said, you join in everything — beatings, shootings, driving [Palestinian] women out of houses about to be blown up — and you have no problem with any of it. You're ready to maim the first stone thrower you come across without a qualm. But later, back at home, you feel miserable thinking about it, and toss about in bed unable to sleep. But while it's happening you feel nothing. There you stand as a soldier with your moral claims, surrounded by hundreds spitting at you, calling you names, throwing bottles and stones, calling your mother a whore. And as you walk though a village, and out of every window somebody is shouting, you begin to be afraid. An incredible fear takes hold of you, and then comes the moment when you can't stand it any longer and you hit out at the next person you see. For a fleeting moment you simply have to feel you're still stronger and able to defend yourself or you'll go crazy. Most of us, he told me, then just hit out, without looking, and you see these terrible explosions of rage. In the evening or on weekends they sit around and talk about what's happened. Feeling desperate, they must keep talking. He himself, he said, had talked about it so often, and almost everyone he knows feels tormented. (Sichrovsky 1991:61–62)

The situation that this Israeli soldier found himself in is connected to a century-long dispute between Palestinian Arabs and Israeli Jews over the land between the Jordan River and the Mediterranean Sea. Both sides call this land "home." The Palestinians, descendants of Canaanites, Muslims, and Christians, had lived on this land for more than 2,000 years. The Hebrews (Israelites), exiled from Egypt, arrived in the area around 1200 B.C. and established a kingdom (Eretz Israel) with Jerusalem as its capital. The Romans conquered this land around 70 B.C. They treated the Jews harshly and suppressed all expression of Jewish culture, including religion and language. The Jews actively resisted and rebelled against Roman rule. The Romans responded by expelling most of the Jews. Only a very small number of determined Jews remained through successive occupations by Persians, Arabs, European Crusaders, Turks, and, ultimately, the British. The dispersed Jews, victims of discrimination abroad, never forgot their "home" in Palestine.

In the late nineteenth century, a growing climate of anti-Semitism arose throughout Europe and Russia (see "The Jewish Migration Out of Europe"). In response, Theodor Herzl founded the modern Zionist movement.[2] Herzl, a Jewish émigré living in England, believed that the only way to combat European anti-Semitism was to establish a Jewish state. He formulated a plan to return dispersed Jews to Palestine, which he thought belonged rightfully to Jews everywhere. Young European and Russian Jews followed Herzl's lead and emigrated to Palestine to buy land and build settlements.

Shortly after the Jewish return movement began, World War I came to an end. The British, who had defeated the Turks, were given control over Palestine by the Allies. Although the British set limits on the number of returning Jews, the Nazi Holocaust during World War II increased the flow of Jewish refugees. It gave a decisive push and a desperate urgency to the Jewish return movement. Haunted by the Holocaust, which had claimed more than six million lives (approximately one-third of all European Jews), the Jews were determined to establish their own state and their own army to protect themselves from future aggression and to reduce their dependency on other countries for food, shelter, jobs, and passports when they assumed refugee status.[3]

The steady stream of Jewish colonists notwithstanding, Palestine was then home to approximately 1.2 million Palestinian Arabs (Smooha 1980).[4] From the time that Jews began to return, Arabs and Jews fought local battles over the land. Members of each group burned crops, destroyed trees, stole animals, sabotaged irrigation systems, and destroyed agricultural equipment belonging to the other. As the British prepared to withdraw following World War II, they asked the United Nations to act as an outside mediator in the growing dispute. On November 29, 1947, the United Nations voted to partition Palestine into two independent states, one Jewish and the other Arab. Under this plan, 400,000 Palestinians were living in the Jewish half, and an unknown number of Jews were living in the Palestinian side (Rubinstein 1991).

[2]Zionism is the plan and the movement to establish a homeland in Palestine for Jews scattered across the globe. After the establishment of Israel, "Zionism was widened to include material and moral support for Israel" (Patai 1971:1262).

[3]One of the last messages received by the outside world from the Warsaw ghetto as it was being crushed by the Nazis was "The world is silent. The world *knows* (it is inconceivable that it should not) and stays silent. God's vicar in the Vatican is silent; there is silence in London and Washington; the American Jews are silent. This silence is astonishing and horrifying" (Steiner 1967:160).

[4]In his biography of Theodor Herzl, Ernst Pawel (1989) writes that Herzl never acknowledged the presence of large Arab villages, contrary to the findings of Leo Motzkin, the man he commissioned to survey the land. Herzl formulated his plan on the basis of two misconceptions: first, that the land was unpopulated; second, that any people who lived there would welcome the modernization the Jews would bring.

Haunted by the Holocaust and widespread persecution throughout Europe, Jewish refugees fled to the territory they called Israel and the Palestinian people called Palestine.

©Robert Capa/Magnum

With the defeat of Arab armies in 1948, the country of Palestine ceased to exist, and about 1 million Palestinians became refugees.

©UPI/Corbis-Bettmann

The Palestinians could not tolerate this arrangement. From their point of view it was inconceivable that representatives from other countries could vote to divide and give away their land. As the British gradually withdrew troops and military equipment, the struggle between Palestinians and Israelis over the now-evacuated territory escalated.

On May 14, 1948, Jewish leaders declared Israel an independent state. The next day, Palestinian and Arab armies from Egypt, Syria, Jordan, Lebanon, and Iraq attacked from all sides. When the Jews defeated the Arab armies, the country of Palestine ceased to exist, and about one million Palestinian Arabs fled or were driven out to refugee camps controlled by Jordan, Syria, Egypt, and the United Nations. Approximately 160,000 Palestinians remained (Elon 1993). A second large Palestinian displacement occurred after Israel defeated armies from Jordan, Syria, Iraq, and Egypt in the Six-Day War in 1967. Approximately 700,000 Palestinian Arabs fled to neighboring Arab countries when Israeli troops took control of the West Bank of the Jordan River, East Jerusalem, Gaza, and Golan Heights seized from Syria (see "Palestinian Migration Out of Palestine"). Arab governments responded by expelling their Jewish citizens, many of whom found asylum in Israel. Today, approximately 2 million Palestinian refugees and their descendants live in Jordan, Syria, and Lebanon (Broder and Kempster 1993).

Currently, one million Palestinians, descendants of those who remained in 1948, live in Israel (Elon 1993). The West Bank, Gaza, and East Jerusalem are home to about 2.6 million Palestinians and 275,000 Israelis who have built settlements there (Population Reference Bureau 1996; U.S. Central Intelligence Agency 1995; see Figure 5.1).

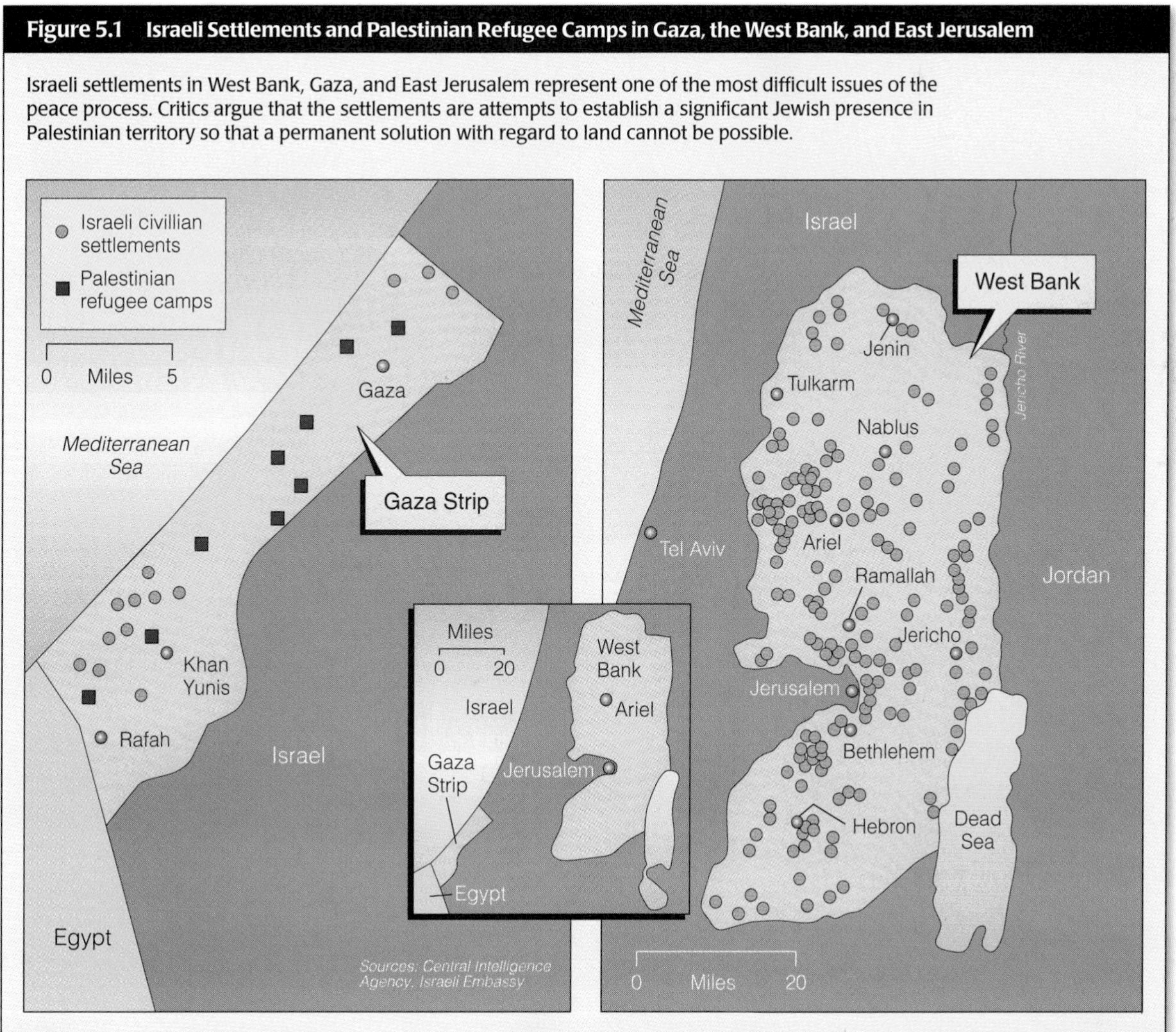

Figure 5.1 Israeli Settlements and Palestinian Refugee Camps in Gaza, the West Bank, and East Jerusalem

Israeli settlements in West Bank, Gaza, and East Jerusalem represent one of the most difficult issues of the peace process. Critics argue that the settlements are attempts to establish a significant Jewish presence in Palestinian territory so that a permanent solution with regard to land cannot be possible.

U.S. In Perspective

The Jewish Migration Out of Europe

This chapter focuses on the creation of the state of Israel and the accompanying "return" of the Jews from Europe to Israel. However, another "great migration" preceded the creation of Israel. Between 1880 and 1914, European *pogroms* (massacres and persecutions instigated by the government) and severe economic crises pushed Jews from Europe to many countries, but especially the United States.

Considering the moment in history and cultural factors, what reasons would you suggest for the fact that large numbers of Jews chose the United States and Canada as their destination over other parts of the world between 1880 and 1939? What factors related to socialization might have shaped Jewish migration patterns?

Percentage of Jewish Immigration to Various Parts of the World, 1880–1939

North America	73%
Palestine	12%
Latin America	10%
Rest of the world	5%

Source: The "Grand Migration (1880–1914)." Maps by Catherine Petit. P. 59 in *The Penguin Atlas of Diasporas* by Gerard Chaliand and Jean-Pierre Rageau. Copyright ©1995 by Gerard Chaliand and Jean-Pierre Rageau. Reprinted by permission of Penguin Books.

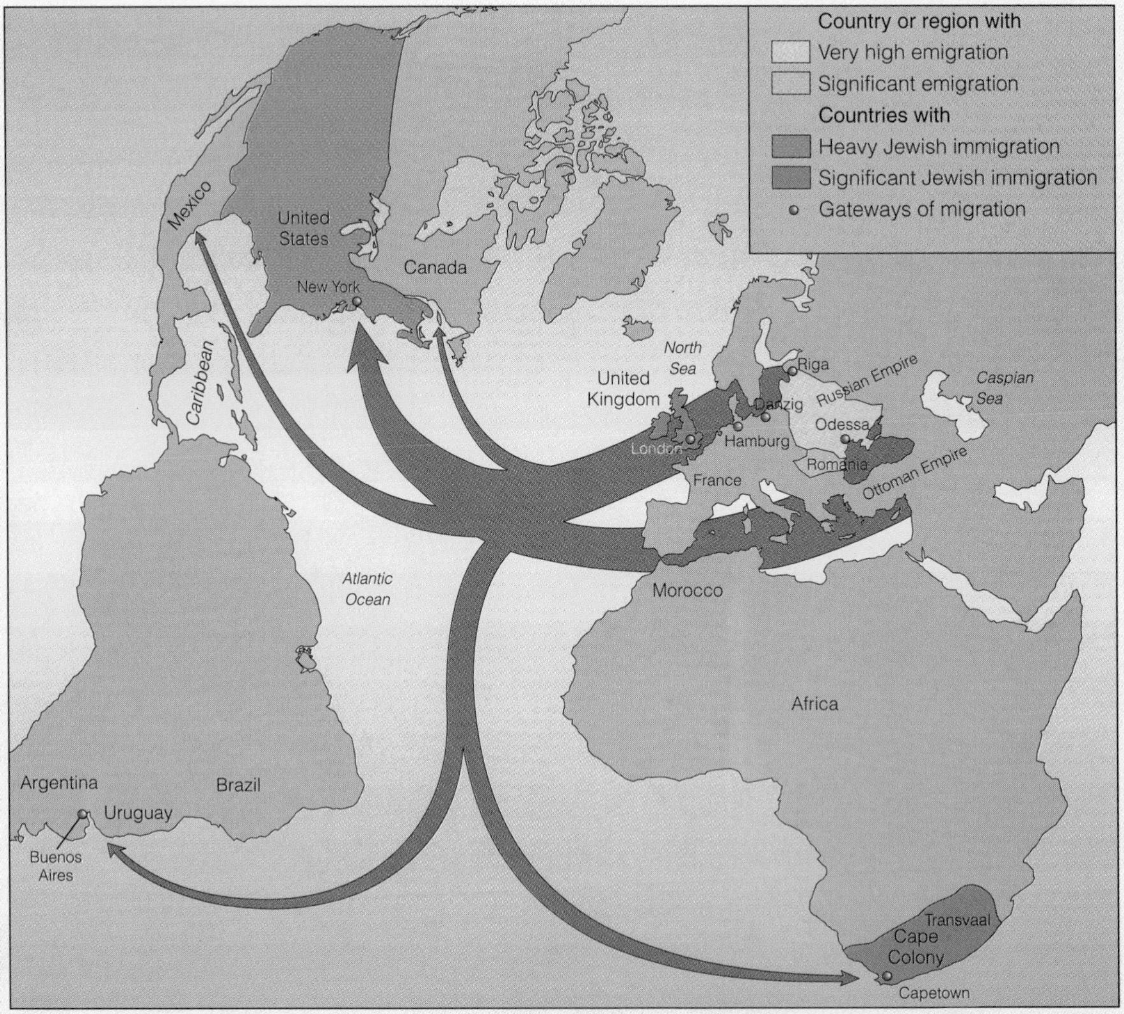

U.S. In Perspective

Palestinian Migration Out of Israel

The creation of the state of Israel and the subsequent 1948–49 Arab-Israeli War resulted in the Palestinian diaspora. While approximately 100,000 Palestinians emigrated to the United States, several million others emigrated elsewhere. The table shows the diaspora as of 1988.

Considering the moment in history and cultural factors, what response might you suggest for the fact that about 100,000 Palestinians came to the United States? Is the number larger or smaller than you might have guessed? What factors related to socialization may have shaped Palestinian migration patterns?

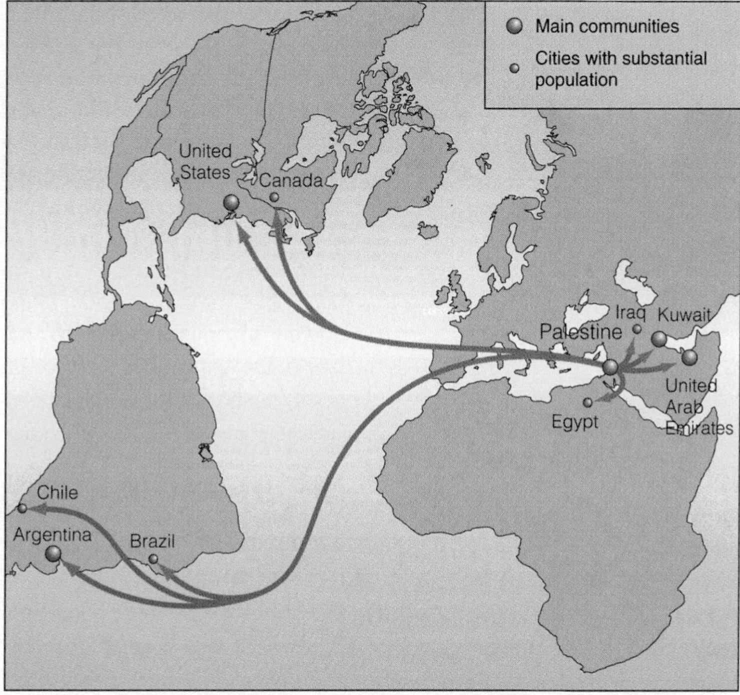

Source: "The Palestinians in the Modern World." Map by Catherine Petit. P. 179 in *The Penguin Atlas of Diasporas* by Gerard Chaliand and Jean-Pierre Rageau. Copyright ©1995 by Gerard Chaliand and Jean-Pierre Rageau. Reprinted by permission of Penguin Books.

Palestinians in Historic Palestine and Diaspora, 1988 Estimates	
Historic Palestine and adjacent countries	4,300,000
Gulf countries (Kuwait, Saudi Arabia, United Arab Emirates, Qatar)	600,000
Iraq	25,000
Libya	20,000
United States	100,000
Other countries	100,000

Source: Adapted from Chaliand and Rageau (1995).

Israeli Prime Minister Benjamin Netanyahu and Palestinian President Yasir Arafat, lifelong enemies, met in Washington, DC, in 1996 at the urging of the Clinton administration to begin talks about one of the most difficult issues of the peace process—the status of Jerusalem.

©Sygma

The conflict between Jews and Arabs has lasted more than a century, and its meaning has passed from generation to generation. Only the form of the conflict has changed. Before 1948, both sides engaged in terrorist activities as well as land and property destruction. Since Israel declared statehood in 1948, Israelis, Palestinians, and Arabs have fought six major wars (1948, 1956, 1967, 1968–71, 1973, and 1982); hardly a day has passed without acts of retaliation by both sides. For almost two decades the Palestinians lived under military occupation, waiting impatiently for a resolution to the "Palestinian question." Then, beginning in 1987, a new generation launched the *intifada*,[5] and fought the occupation with knives, Molotov cocktails, flags, rocks, graffiti, strikes, boycotts, and barricades. The Israelis responded with gunfire, deportations, imprisonments, curfews, and school closures.

This long sequence of historical events in conjunction with the end of the Cold War, pushed Arabs and Israelis to participate in peace talks[6] jointly sponsored by Moscow and Washington. On September 13, 1993, the prime minister of Israel, Yitzhak Rabin, and the chairman of the Palestinian Liberation Organization (PLO), Yasir Arafat, after several years of negotiations (including eighteen months of secret negotiations held in and mediated by leaders in Norway) met on the U.S. White House lawn to sign a peace accord. For the first time, each side acknowledged the other's right to exist and took the first steps toward a permanent settlement of their conflict. However, many unresolved issues persist. To complicate matters, various factions on each side have tried to influence the course of the settlement and the outcome of subsequent talks to match their own vision of the future. Factions on both sides initiated violent confrontations between Palestinians and Israelis and against those pushing the peace process forward with the hope that such actions will put an end to a possible peace settlement and future talks.

In studying this conflict, sociologists ask these questions: (1) how do members of a new generation learn about and come to terms with the environment they have inherited, and (2) how is conflict passed down from one generation to another? For sociologists, part of the answer lies with socialization, a process that involves nature and nurture.

Nature and Nurture

No discussion of socialization can ignore the importance of nature and nurture to physical, intellectual, social, and personality development. **Nature** is the term for human genetic makeup or biological inheritance. **Nurture** refers to the environment or the interaction experiences that make up every individual's life. Some scientists debate the relative importance of genes and environment, arguing that one is substantially more important than the other to all phases of human development. But most consider such a debate futile, be-

Nature Human genetic makeup or biological inheritance.

Nurture The environment or the interaction experiences that make up every individual's life.

[5]Translated literally, the word *intifada* means a "tremor, shudder or shiver." It is derived from the Arabic *nafada*, which means "to shake, to shake off, shake out, dust off, to shake off one's laziness, to have reached the end of, be finished with, to rid oneself of something, to refuse to have anything to do with . . . someone" (Friedman 1989:375). The *intifada* was waged most visibly by the "children of the stones" — young Palestinians who throw rocks and bottles at Israeli settlers and soldiers. Although Palestinian youth play a critical and very confrontational role in the *intifada*, a complex structure of committees and leadership directs the uprising. The *intifada* is supported through boycotts, business closures, and strikes.

[6]This cooperation between Moscow and Washington would not have been possible before the Cold War ended, when foreign policy in the Middle East was dominated by U.S. and Soviet efforts to contain the spread of the other's economic and political systems.

cause it is impossible to separate the influence of the two factors or to say that one is more important. Both are essential to socialization. Trying to distinguish the separate contributions of nature and nurture is analogous to examining a tape player and a cassette tape separately to determine what is recorded on the tape rather than studying how the two work together to produce the sound (Ornstein and Thompson 1984).

The development of the human brain illustrates rather dramatically the inseparable qualities of genes and environment. As part of our human genetic makeup, we possess a cerebral cortex[7] — the thinking part of the brain — which allows us to organize, remember, communicate, understand, and create. The cortex is made up of at least 100 billion neurons or nerve cells (Montgomery 1989); the fibers of each cell form thousands of synapses, which make connections with other cells. The number of interconnections is nearly infinite; it would take 32 million years, counting one synapse per second, to establish the number (Hellerstein 1988; Montgomery 1989).

Perhaps the most outstanding feature of the human brain is its flexibility. Scientists believe that the brain may be "set up" to learn any of the more than 6,000 to 9,000 known human languages. In the first months of life, babies are able to babble the sounds needed to speak all of these languages, but this enormous potential is reduced by the language (or languages) that the baby hears and eventually learns. Evidence suggests that the brain's language flexibility begins to diminish when the child reaches one year of age (Ornstein and Thompson 1984; Restak 1988). The larger implication is that genetic makeup provides essential raw materials but that these materials can be shaped by the environment in many different ways.

The human genetic makeup is flexible enough to enable a person to learn the values, beliefs, norms, behavior, and language of any culture. A multitude of experiences must combine with genetic makeup to create a Palestinian who desires a homeland, believes that he or she should not "pay" for the Holocaust, and protests actively against the occupation. Likewise, nature and nurture combine to create an Israeli who values a homeland, believes the Jews have a legitimate right to land they were forced to leave 2,000 years ago, and fights to preserve the boundaries of his or her country. These ideas, however, are learned in interaction with others. If there is no contact with others, a person cannot ever become a normally functioning human being, let alone learn to become part of society.

The Importance of Social Contact

Cases of children raised in extreme isolation or in restrictive and unstimulating environments show the importance of social contact (nurture) to normal development. Some of the earliest and most systematic work in this area was done by sociologist Kingsley Davis, psychiatrists Anna Freud and Rene Spitz, and sociologist Peter Townsend. Their work shows how neglect and lack of socialization influence emotional, mental, and even physical development.

Cases of Extreme Isolation

In two classic articles, "Extreme Isolation of a Child" and "Final Note on a Case of Extreme Isolation," sociologist Kingsley Davis (1940, 1947) documented and compared the separate yet similar lives of two girls in the United States, Anna and Isabelle. Each girl had received a minimum of human care during the first six years of her life. Both were illegitimate children and for that reason were rejected and forced into seclusion. When authorities discovered the girls, they were living in dark, atticlike rooms, shut off from the rest of the family and from daily activities. Although both girls were six years old when authorities intervened, they exhibited behavior comparable to that of six-month-old children. Anna "had no glimmering of speech, absolutely no ability to walk, no sense of gesture, not the least capacity to feed herself even when food was put in front of her, and no comprehension of cleanliness. She was so apathetic that it was hard to tell whether or not she could hear" (Davis 1947:434).

Like Anna, Isabelle had not developed speech; she communicated with gestures and croaks. Because of a lack of sunshine and a poor diet, she had developed rickets: "Her legs in particular were affected; they 'were so bowed that as she stood erect the soles of her shoes came nearly flat together, and she got about with a skittering gait'" (Davis 1947:436). Isabelle also exhibited extreme fear of and hostility toward strangers.

Anna was placed in a private home for retarded children until she died four years later. At the time of

[7]Ornstein and Thompson (1984) provide an excellent metaphor for visualizing the cortex:

> Form your hands into fists. Each is about the size of one of the brain's hemispheres, and when both fists are joined at the heel of the hand they describe not only the approximate size and shape of the entire brain but also its symmetrical structure. Next, put on a pair of thick gloves, preferably light gray. These represent the cortex (Latin for "bark") — the newest part of the brain and the area whose functioning results in the most characteristically human creations, such as language and art. (pp. 21–22)

her death, she behaved and thought at the level of a two-year-old child. Isabelle, on the other hand, was placed in an intensive and systematic program designed to help her master speech, reading, and other important skills. After two years in the program, Isabelle had achieved a level of thought and behavior normal for someone her age.

On the basis of Anna and Isabelle's case histories, Davis concluded that extreme isolation has a profound and negative effect on mental and physical development. At the same time, Davis concluded, Isabelle's case demonstrates that extreme "isolation up to the age of six, with the failure to acquire any form of speech and hence failure to grasp nearly the whole world of cultural meaning, does not preclude the subsequent acquisition of these" (Davis 1947:437).

In addition, Davis speculated on the question of why Isabelle did so much better than Anna. He offered two possible explanations: (1) Anna may have inherited a less hardy physical and mental constitution than Isabelle's, and (2) Anna's condition may have resulted from not being exposed to the intensive and systematic therapy that Isabelle received. The case history comparisons are inconclusive, though. Because Anna died at age ten, researchers will never know whether she eventually could have achieved a state of normal development.

In drawing these conclusions, Davis overlooked one important factor. Although Isabelle had spent her early childhood years in a dark room, shut off from the rest of her mother's family, she spent most of this time with her deaf-mute mother. Davis and the medical staff who treated Isabelle seemed to equate deafness and muteness with feeble-mindedness (not an unusual association in the 1940s). In addition, they seemed to assume that being in a room with a deaf-mute is equivalent to a state of isolation. A considerable body of evidence, however, now indicates that deaf-mutes have rich symbolic capacities (see Sacks 1989). The fact that Isabelle was able to communicate through gestures and croaks suggests that she had established an important and meaningful bond with another human being. Although the bond was less than ideal, it gave her an advantage over Anna. In view of this possibility, we must question Davis's conclusion that children may be able to overcome the effects of extreme isolation during the first six years of life, provided they have a "good enough" constitution and systematic training. We use the term "good enough" because researchers do not know the exact profile of a person or training regime that would allow someone to overcome such effects.

Children of the Holocaust

Anna Freud and Sophie Dann (1958) studied six German-Jewish children whose parents had been killed in the gas chambers of Nazi Germany. The children were shuttled from one foster home to another for a year before they were sent to the ward for motherless children at the concentration camp at Tereszin. The ward was staffed by malnourished and overworked nurses, themselves concentration camp inmates. After the war, the six children were housed in three different institution-like environments. Eventually, they were sent to a country cottage where they received intensive social and emotional care.

During their short lives, these children had been deprived of stable emotional ties and relationships with caring adults. Freud and Dann found that the children were ignorant of the meaning of family and grew excessively upset when they were separated from one another, even for a few seconds. In addition:

> they showed no pleasure in the arrangements which had been made for them and behaved in a wild, restless, and uncontrollably noisy manner. During the first days after arrival they destroyed all the toys and damaged much of the furniture. Toward the staff they behaved either with cold indifference or with active hostility, making no exception for the young assistant Maureen who had accompanied them from Windermere and was their only link with the immediate past. At times they ignored the adults so completely that they would not look up when one of them entered the room. They would turn to an adult when in some immediate need, but treat the same person as nonexistent once more when the need was fulfilled. In anger, they would hit the adults, bite or spit. (Freud and Dann 1958:130)

Less Extreme Cases of Extreme Isolation

Other evidence of the importance of social contact comes from less extreme cases of neglect. Rene Spitz (1951) studied ninety-one infants who were raised by their parents during their first three to four months but later placed in orphanages because of unfortunate circumstances. At the time of their entry into the institution, the infants were physically and emotionally normal. At the orphanages they received adequate care with regard to bodily needs — good food, clothing, diaper changes, clean nurseries — but little personal attention. Because only one nurse was available for every eight to twelve children, the children were starved emotionally. The emotional starvation caused by the lack of social contact resulted in such rapid physical and developmental deterioration that a significant number of the children died. Others became completely passive, lying on their backs in their cots. Many were unable to stand, walk, or talk (Spitz 1951).

Such cases teach us that children need close contact with and stimulation from others to develop normally. Adequate stimulation means the existence of strong ties with a caring adult. The ties must be characterized by a bond of mutual expectation between caregiver and baby. In other words, there must be at least one person who knows the baby well enough to understand his or her needs and feelings and who will act to satisfy them. Under such conditions, the child learns that certain actions on his or her part elicit predictable responses: getting excited may cause Dad to become equally excited; crying may get Mom to soothe the child. When researchers set up experimental situations in which a parent fails to respond to his or her infant in expected ways (even for a few moments), the baby suffers considerable tension and distress (*Nova* 1986).

Meaningful social contact with and stimulation from others are important at any age. Strong social ties with caring people are linked to overall social, psychological, and physical well-being. British sociologist Peter Townsend (1962) studied the effects of minimal interaction that can characterize life for the elderly in nursing homes. The consequences for the institutionalized elderly are strikingly similar to those described by Spitz in his studies of institutionalized children:

> In the institution people live communally with a minimum of privacy, and yet their relationships with each other are slender. Many subsist in a kind of defensive shell of isolation. Their mobility is restricted, and they have little access to general society. Their social experiences are limited, and the staff leads a rather separate existence from them. They are subtly oriented toward a system in which they submit to orderly routine and lack creative occupation, and cannot exercise much self-determination. They are deprived of intimate family relationships. . . . The result for the individual seems to be a gradual process of depersonalization. He may become resigned and depressed and may display no interest in the future or things not immediately personal. He sometimes becomes apathetic, talks little, and lacks initiative. His personal and toilet habits may deteriorate. (Townsend 1962:146–147)

The work of Kingsley Davis, Rene Spitz, Anna Freud, and Peter Townsend supports the idea that a person's overall well-being depends on meaningful interaction experiences with others. On a more fundamental level, social interaction is essential to a developing sense of self. Sociologists, psychologists, and biologists agree that "it is impossible to conceive of a self arising outside of social experience" (Mead 1934:135). Yet, if the biological mechanisms involved in remembering or learning and recalling names, faces, words, and the meaning of significant symbols were not present, people could not interact with one another in meaningful ways: "You have to begin to lose your memory, if only in bits and pieces, to realize that memory is what makes our lives. Life without memory is no life at all. . . . Our memory is our coherence, our reason, our feeling, even our action. Without it we are nothing" (Bunuel 1985:22).

Individual and Collective Memory

Memory, the capacity to retain and recall past experiences, is easily overlooked in exploring socialization. Yet without memory individuals and even whole societies would be cut off from the past. On the individual level, memory is what allows people to retain their experiences; on the societal level, memory preserves the cultural past.

How memory works is still largely a mystery. The latest neurological evidence suggests that some physical trace remains in the brain after new learning takes place, stored in an anatomical entity called an engram. **Engrams**, or *memory traces* as they are sometimes called, are formed by chemicals produced in the brain. They store in physical form the recollections of experiences — a mass of information, impressions, and images unique to each person.

> It may have been a time of listening to music, a time of looking in at the door of a dance hall, a time of imagining the action of robbers from a comic strip, a time of waking from a vivid dream, a time of laughing conversation with friends, a time of listening to a little son to make sure he was safe, a time of watching illuminated signs, a time of lying in the delivery room at childbirth, a time of being frightened by a menacing man, a time of watching people enter the room with snow on their clothes. (Penfield and Perot 1963:687)

Engrams Chemically formed entities in the brain that store in physical form a person's recollections of experiences.

Scientists do not believe that engrams store actual records of past events, like films stored on videocassettes. More likely engrams store edited or consolidated versions of experiences and events, which are edited further each time they are recalled.

As we noted previously, memory has more than an individual quality; it is strongly social. First, no one can participate in society without an ability to remember and recall such things as names, faces, places, words, symbols, and norms. Second, most "newcomers" easily learn the language, norms, values, and beliefs of the surrounding culture. We take it for granted that people have this information stored in memory. Third, people born at approximately the same time and place are likely to have lived through many of the same events. These experiences, each uniquely personal and yet similar to one another, remain in memory long after the event has passed. We will use the term **collective memory** to describe the experiences shared and recalled by significant numbers of people (Coser 1992; Halbwachs 1980). Such memories are revived, preserved, shared, passed on, and recast in many forms, such as stories, holidays, rituals, and monuments.[8]

Israel is alive with memorials and reminders of the past. For example, the *Jerusalem Post* runs retrospective columns with headlines such as "No School for Jewish Children in Poland." And the roadsides in Israel are littered with remnants of Palestinian jeeps, cars, and trucks from the 1948 War of Independence (Bourne 1990).

Palestinians displaced by the 1948 and 1967 wars retain memories of their former homeland, and, like the Jews, they pass them down to those who lack personal memories. Some name their children after the cities and towns in which they had lived before the 1948 war (Al-Batrawi and Rabbani 1991). Most tell their children about the places where they used to live; they teach them to call those places home and even to respond with the name of that land if they are asked where they come from. When author David Grossman (1988) asked a group of Palestinian children in a West Bank refugee camp to tell him their birthplace, each replied with the name of a former Arab town:

> Everyone I spoke to in the camp is trained—almost from birth—to live this double life; they sit here, very much here . . . but they are also there. . . . I ask a five-year-old boy where he is from, and he immediately answers, "Jaffa," which is today part of Tel Aviv.
>
> "Have you ever seen Jaffa?"
>
> "No, but my grandfather saw it." His father, apparently, was born here, but his grandfather came from Jaffa.
>
> "And is it beautiful, Jaffa?"
>
> "Yes. It has orchards and vineyards and the sea."
>
> And farther down . . . I meet a young girl sitting on a cement wall, reading an illustrated magazine. . . . She is from Lod, not far from Ben-Gurion International Airport, forty years ago an Arab town. She is sixteen. She tells me, giggling, of the beauty of Lod. Of its houses, which were big as palaces. "And in every room a hand-painted carpet. And the land was wonderful, and the sky was always blue. . . .
>
> "And the tomatoes there were red and big, and everything came to us from the earth, and the earth gave us and gave us more."
>
> "Have you visited there, Lod?"
>
> "Of course not."
>
> "Aren't you curious to see it now?"
>
> "Only when we return." (Grossman 1988:6–7)

The Jews who went to Palestine to escape persecution and who fought to establish the state of Israel hold

Passover seder is a socialization ritual for the children of the family that promotes their identification with the Jewish historic tradition of oppression and with Israel as an ideal or nation.

©Bill Aron/Photo Researchers Inc.

Collective memory The experiences shared and recalled by significant numbers of people.

[8]A case in point is the United States Holocaust Memorial Museum in Washington, D.C. The museum was founded with the realization that surviving witnesses to the Holocaust will be dead in about ten to twenty years. The museum will help preserve Jewish collective memory of this event by housing "object survivors"— letters, diaries, identity papers, armbands, clothes—that document life in the concentration camps and ghettos (Goldman 1989).

memories that are both parallel to and different from Palestinian memories. Although members of both groups participated in many of the same historical events, their memories differ because they witnessed these events from different viewpoints. Because Israel has participated in six wars with neighboring countries, has occupied the West Bank and Gaza for more than thirty years, and is a refuge for persecuted Jews from more than eighty countries around the world, virtually everyone in Israel has memories of war and persecution. The memories may concern personal involvement in war, waiting for a loved one to return, or fleeing places where they were deemed unfit to exist.

The point is that socialization is not possible without memory. Memory is the mechanism by which group expectations become internalized in individuals and, by extension, in the whole society and by which the past remains an integral, living part of the present. The picture that any one individual holds cannot be a complete record of the past. Yet memories include the perceived reasons that things are the way they are.

Memory of past experiences allows individuals to participate in society and shapes their viewpoint. In his essay "The Problem of Generations," sociologist Karl Mannheim (1952) maintained that first impressions or early childhood experiences are fundamental to a person's view of the world. In fact, Mannheim believed that an event has more biographical significance if it is experienced early in life than if it is experienced later in life. Many of our most significant early experiences take place in groups, some of which leave a powerful and lasting impression.

The Role of Groups

In the most general sense, a **group** is two or more persons who

1. share a distinct identity (the biological children of a specific couple; members of a gymnastics team, military unit, club or organization; or persons sharing a common cultural tradition);
2. feel a sense of belonging; and
3. interact with one another in direct and/or indirect, but broadly predictable, ways.

Interaction is broadly predictable because norms govern the behavior expected of members depending on their position (mother, coach, teammate, sibling, employee) within the group. Groups vary according to a whole host of characteristics including size, degree of intimacy among members, member characteristics, purpose, duration, and the extent to which the members socialize newcomers and one another. However, sociologists identify primary groups and ingroups and outgroups as particularly powerful socialization agents.

Primary Groups

Primary groups, such as the family or a high school sports team, are characterized by face-to-face contact and strong ties among members. Primary groups are not always united by harmony and love; they can be united by hatred for another group. But in either case the ties are emotional. A primary group is composed of members who strive to achieve "some desired place in the thoughts of [the] others" and who feel allegiance to the others (Cooley 1909:24). A person may never achieve the desired place but may still be preoccupied with that goal. In this sense, primary groups are "fundamental in forming the social nature and ideals of the individual" (Cooley 1909:23). The family is an important primary group because it gives the individual his or her deepest and earliest experiences with relationships and because it gives newcomers their first exposure to the "rules of life." In addition, the family can serve to buffer its members against the effects of stressful events or negative circumstances, or it can exacerbate these effects.

Sociologists Amith Ben-David and Yoav Lavee (1992) interviewed sixty-four Israelis to learn how members of their families behaved toward one another during the SCUD missile attacks, launched by Iraq on Israel during the Persian Gulf War. During these attacks, families gathered in sealed rooms and put on gas masks. The researchers found that families varied in their response to this life-threatening situation. Some respondents reported that interaction was positive and supportive: "We laughed and we took pictures of each other with the gas masks on" or "We talked about different things, about the war, we told jokes, we heard the announcements on the radio" (p. 39).

Group Two or more people who share a distinct identity, feel a sense of belonging, and interact with one another in direct or indirect, but broadly predictable, ways.

Primary groups Social groups characterized by face-to-face contact and strong emotional ties among members.

Primary groups like the family provide meaningful interactions with others that are vital to our physical, mental, and social well-being. In the process, primary groups have an immense influence on our socialization and on how we see the world.

©Esaias Baitel/Gamma-Liaison

Other respondents reported that interaction was minimal but that a feeling of togetherness prevailed: "I was quiet, immersed in my thoughts. We were all around the radio . . . nobody talked much. We all sat there and we were trying to listen to what was happening outside" (p. 40).

Finally, some respondents reported that interaction among family members was tense: "We fought with the kids about putting on their masks, and also between us about whether the kids should put on their masks. There was much shouting and noise" (p. 39). The point is that even under extremely stressful circumstances such as war, the family can respond in ways that increase or decrease that stress.

We do not know the extent to which Israeli and Palestinian populations are composed of people who suffered the loss of important primary group figures in their lives. The historical experiences, especially those of Israelis, suggest that the proportion may be very high. For example, approximately one-third of the 8,000 Ethiopian Jews airlifted to Israel in 1984 came from one-parent families. In many cases, the other parent had died from war-related causes or starvation (*Encyclopedia Judaica Yearbook* 1987).

It seems that children whose primary group remains intact emerge in relatively good psychological condition despite widespread turmoil, violence, and destruction around them (Freud and Burlingham 1943). This finding gains indirect support from what we know about groups that turn to violence to achieve their ends. Such groups tend to draw recruits most often from populations that have suffered extreme humiliation and brutality at the hands of uniformed representatives of authority and justice (Fields 1979). For example, a high proportion of Palestinians have had their homes blown up, their villages destroyed, close family members or friends imprisoned and beaten by Israeli soldiers, or parents or siblings killed. Furthermore, research shows that both the Palestinians still living in refugee camps and the most recent Jewish émigrés to Israel are least likely to believe that the conflict can be solved by negotiation. We might speculate that living in refugee camps or the experience of being forced to migrate to Israel socialized some of the members in these groups to believe that armed struggle represents the most viable solution (Shadid and Seltzer 1988; Yishai 1985).

One very clear example of a primary group is a military unit. A unit's success in battle depends on strong ties among its members. Apparently soldiers in the primary group fight for one another, rather than for victory per se, in the heat of battle (Dyer 1985).

U.S. In Perspective

Muscular Bonding

The U.S military trains its men and women for war using techniques employed by military organizations around the world. Despite changes in technology that require less physical activity (and even fitness) than in the past, the military still relies on strategies that make the individual recruit feel inseparable from the group. One of these strategies is muscular bonding. This excerpt from *Keeping Together in Time* by William H. McNeill describes the function of muscular bonding.

> *In September 1941 I was drafted into the army of the United States and underwent basic training in Texas along with thousands of other young men. Supplies were short. We boasted a single (inoperative) anti-aircraft gun for the entire battalion, so that practical training on the weapon we were supposed to master was impossible. Consequently, whenever our officers ran out of training films and other ways of using our time, we were set to marching about on a dusty, gravelled patch of the Texas plain under the command of an illiterate noncom. A more useless exercise would be hard to imagine. Given the facts of twentieth-century warfare, troop movement in the rear was a matter of trucks and railroads. Close-order marching within range of machine guns and rifles was a form of suicide. All concerned realized these simple facts, yet still we drilled, hour after hour, moving in unison and by the numbers in response to shouted commands, sweating in the hot sun, and, every so often, counting out the cadence as we marched: Hut! Hup! Hip! Four!*
>
> *Treasured army tradition held that this sort of thing made raw recruits into soldiers. That was enough for our officers and the cadre of enlisted men who were in charge of our training. But why did young Americans not object to senseless sweating in the sun? At the time I was too busy getting used to totally unfamiliar routines and social relations to ask the question, much less reflect upon it. What I remember now, years afterward, is that I rather liked strutting around, and so, I feel sure, did most of my fellows. Marching aimlessly about on the drill field, swaggering in conformity with prescribed military postures, conscious only of keeping in step so as to make the next move correctly and in time somehow felt good. Words are inadequate to describe the emotion aroused by the prolonged movement in unison that drilling involved. A sense of pervasive well-being is what I recall; a sort of swelling out, becoming bigger than life, thanks to participation in a collective ritual.*
>
> *But such phrases are far too analytical to do justice to the experience. It was something felt, not talked about. Words, in a sense, destroy what they purport to describe because they limit and define: in this case, a state of generalized emotional exaltation whose warmth was indubitable, without, however, having any definite external meaning or attachment. The strongest human emotions—love, hate, and fear—are ordinarily triggered by encounters with other persons or particular external circumstances, and the emotion in question helps us to react successfully. But the diffused exaltation induced by drill has no apparent external stimulus. Instead, marching became an end in itself. Moving briskly and keeping in time was enough to make us feel good about ourselves, satisfied to be moving together, and vaguely pleased with the world at large.*
>
> *Obviously, something visceral was at work; something, I later concluded, far older than language and critically important in human history, because the emotion it arouses constitutes an indefinitely expansible basis for social cohesion among any and every group that keeps in time, moving big muscles together and chanting, singing, or shouting rhythmically. "Muscular bonding" is the most economical label I could find for this phenomenon, and I hope the phrase will be understood to mean the euphoric fellow feeling that prolonged and rhythmic muscular movement arouses among nearly all participants in such exercises.*

The term *bonding* has become a cliché these days, but the underlying concept is an important aspect of socialization. As you have grown up, have you been subject to any attempts at muscular bonding? What about other types of bonding? Which if any of these experiences had an important effect on you—really made you feel part of a group?

Source: *Keeping Together in Time* by William H. McNeill. Copyright ©1995 by William H. McNeill. Reprinted by permission.

Military units train their recruits always to think of the group before self. In fact, the paramount goal of military training is to make individuals feel inseparable from their unit. Some common strategies to achieve this goal include ordering recruits to wear uniforms, shave their heads, march in unison, sleep and eat together, live in isolation from the larger society, and perform tasks that require the successful participation of all unit members: if one member fails, the entire unit fails (see "Muscular Bonding"). Another key strategy is to focus the unit's attention on fighting together against a common enemy. An external enemy gives a group a singular direction and thus increases its internal cohesiveness.

Almost every Israeli can claim membership in this type of primary group because virtually every Israeli citizen, male and female, serves in the military for three years and two years, respectively. Men must serve on active duty for at least one month every year until they are sixty-four years old. (Since the *intifada,* the annual length of service has increased from thirty to sixty days.) Military training is similarly an important experience for many Palestinians, although it is less formal. Palestinian youths, especially those living in Syrian, Lebanese, Egyptian, and Jordanian refugee camps (which are outside Israeli control), join youth clubs and train to protect the camps from attack. The focus on a common enemy helps establish and maintain the boundaries of the military unit. All types of primary groups, however, have boundaries—a sense of who is in the group and who is outside the group.

Ingroups and Outgroups

Sociologists use the term **ingroup** to describe those groups with which people identify and to which they feel closely attached, particularly when that attachment is founded on hatred from or opposition toward another group. Ingroups exert their influence on our social identity in conjunction with outgroups. An **outgroup** is a group of individuals toward which members of an ingroup feel a sense of separateness, opposition, or even hatred. Outgroups also can make us conscious of where we belong. Obviously one person's ingroup is another's outgroup.

The very existence of an outgroup heightens loyalty among ingroup members and magnifies characteristics that distinguish the ingroup from the outgroup. An outgroup can unify an ingroup even when the ingroup members are extremely different from one another. For example, one could argue from a functionalist viewpoint that Israel benefits from the Palestinian presence in Israel and the Occupied Territories. The presence of non-Israelis provides a thread of unification among Israelis, who are themselves culturally, linguistically, religiously, and politically diverse. One reason for Israel's diverse population is that, since 1948, Jews from eighty different countries have settled there (see Figure 5.2). To ease communication problems caused by diversity, Israeli law requires that everyone learn Hebrew. In addition to a common language, the unifying threads are the desire for a homeland free of persecution and the ongoing conflict with an outgroup —Palestinians and Arabs in surrounding states.

Similarly, the presence of Israelis acts to unite an equally diverse Palestinian society. Palestinians come from different ethnic and religious groups, clans, and political orientations. They may be Muslim, Christian, or Druze.

Loyalty to an ingroup and opposition to an outgroup are accompanied by an us-versus-them consciousness. In Israel, the West Bank, and Gaza, this consciousness is reinforced by the fact that the work lives and personal lives of most Palestinians and Israelis are interrupted time and again by wars and military obligations. Israeli men must leave their jobs and families at least once a year to serve in the army. Palestinians

Ingroup A group with which people identify and to which they feel closely attached, particularly when that attachment is founded on hatred from or opposition toward another group.

Outgroup A group toward which members of an ingroup feel a sense of separateness, opposition, or even hatred.

Figure 5.2 Countries of Origin of Immigrants to Israel Since 1948

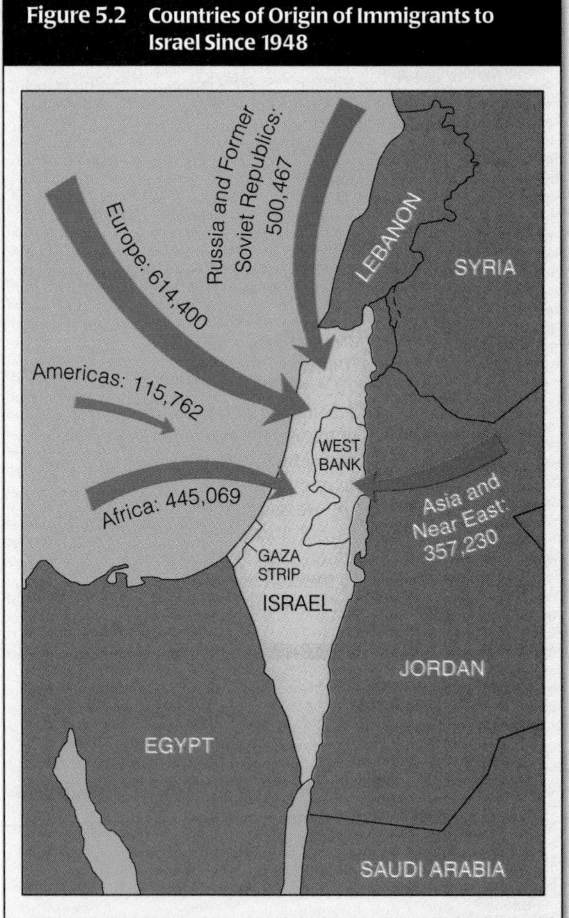

Source: Updated and adapted from *Facts About Israel* (1985).

Often members of ingroups and outgroups clash over symbols. Before the 1993 peace accord, it was illegal in Israel to display the Palestinian flag. The flag thus became the focus of many conflicts between Israelis and Palestinians.

©Stephane Compoint/Sygma

frequently sacrifice work and family when they protest the occupation with strikes, work stoppages, and business closures. Israeli officials also disrupt Palestinian work and family life when they arrest dissenters or close Palestinian businesses as punishment for protesting the occupation. Each group holds the other responsible for the state of affairs, and each seeks to control the other. To complicate matters even further, Israelis and Palestinians have little in common, other than the economic relationship they share. (In this relationship, Palestinians are concentrated in low-status jobs.) These two groups live segregated lives, for the most part, and do not share a language, religion, schools, residence, or military service.

Boundaries between the two groups are sharp; they are reinforced by residential, occupational, educational, and/or religious segregation. Because little interaction occurs between ingroup and outgroup members, they know little about one another. This lack of firsthand experience deepens and reinforces misrepresentations, mistrust, and misunderstandings between members of the two groups. Members of one group tend to view members of the other in the most stereotypical of terms. Often one of the groups has superior status, material conditions, and facilities. In this case, Israelis have the superior economic and political status. Palestinians, many of who have graduated from technical colleges or universities, can obtain only manual labor jobs and low-status service jobs.

Dr. Yorum Bilu at Hebrew University of Jerusalem designed and conducted a particularly creative research study to examine the consequences of ingroup-outgroup relations on Israel's West Bank. Dr. Bilu and two of his students asked youths ages eleven to thirteen from Palestinian refugee camps and Israeli settlements on the West Bank to keep a journal of their dreams over a specified period. Seventeen percent of Israeli children wrote that they dreamed about encounters with Arabs; 30 percent of the Palestinian children dreamed about meeting Jews:

> Among thirty-two dreams of meetings (Jews and Arabs) there is not one character identified by name. There is not a single figure defined by a personal, individual appearance. All the descriptions, without exception, are completely stereotyped; the characters defined only by their ethnic identification (Jew, Arab, Zionist, etc.) or by value-laden terms with negative connotations (the terrorists, the oppressors, etc.). . . .
>
> The majority of the interactions in the dreams indicate a hard and threatening reality, a fragile world with no defense. . . .
>
> An Arab child dreams: "The Zionist Army surrounds our house and breaks in. My big brother is taken to prison and is tortured there. The soldiers continue to search the house. They throw everything around, but do not find the person they want [the dreamer himself]. They leave the house, but return, helped by a treacherous neighbor. This time they find

Israeli and Palestinian boys took part in a 1992 summer camp called Interns for Peace. What is it that makes a group of boys feel like they belong together? What makes them feel divided?

> me, and my relatives, after we have all hidden in the closet in fright."
>
> A Jewish child dreams: ". . . suddenly someone grabs me, and I see that it is happening in my house, but my family went away, and Arab children are walking through our rooms, and their father holds me, he has a *kaffiyeh* and his face is cruel, and I am not surprised that it is happening, that these Arabs now live in my house." (Grossman 1988:30, 32–33)

Often an ingroup and an outgroup clash over symbols — objects or gestures that are clearly associated with and valued by one group. These objects can be defined by members of the other group as so threatening that they seek to eliminate them: destroying the objects becomes a way of destroying the group. Danny Rubinstein, an Israeli reporter covering the West Bank, observed that most of the clashes between Palestinian youths and Israeli soldiers are over symbols:

> Of the hundreds of clashes I have witnessed pitting Palestinian youths against Israeli military and administrative authorities in the West Bank and Gaza, most have involved symbols. Thus, for example, an ongoing battle is being waged over the Palestinian flag. Arabs hoist the flag (which is very much like the Jordanian flag), while Israeli soldiers bring it down and attempt to catch and punish the perpetrators. At times the situation takes a ridiculous turn. Some time ago, in Bethlehem, I heard an Israeli officer issue an order to close down for a week all shops on a certain street where a Palestinian flag had been hoisted on the corner utility pole the night before. I saw schoolgirls in Hebron knitting satchels modeled on the Palestinian flag and clothing stores with window displays arranged to fit its color and pattern. (Rubinstein 1988:24)

Although ingroup and outgroup clashes center around symbols, the symbol alone is not the cause. Rather, it functions as a rallying point for some issue important to both sides. For example, on September 27, 1996, Palestinians and Israelis clashed over an archeological tunnel opened alongside the Temple Mount, an artificially made plateau inside Jerusalem's Old City with a religious significance that dates back more than a thousand years for both groups. The site of the confrontation should come as no surprise as the control of Jerusalem is one of the most difficult issues that Palestinians and Israelis must face if there is to be lasting peace in the area. To complicate matters, a few weeks before the confrontation, the Israeli government lifted a freeze on settlement expansion in the West Bank.

To this point, we have examined how socialization is a product of nature and nurture. We have discussed how genetic makeup provides each individual with potentials that are developed to the extent made possible by the environment. We also have considered the importance of stimulation from caregivers in developing our genetic potential and the connection between group membership and self-awareness. Even groups to which we do not belong can have powerful influences on our sense of self. For example, an outgroup makes us clearly aware of who "they" are, which in turn re-

Symbols such as the Vietnam Veterans Memorial help to revive, preserve, and pass on important historical events.

minds us of who "we" are. Next we will examine the theories of sociologists George Herbert Mead and Charles Horton Cooley, regarding some specific way in which the self develops and in which information is transmitted to newcomers.

Symbolic Interactionism and Self-Development

Humans are not born with a sense of self; it evolves through regular interaction with others. The emergence of a sense of self depends on our physiological capacity for **reflexive thinking**—stepping outside the self and observing and evaluating it from another's viewpoint. Reflexive thinking allows individuals to learn how they come across to others and to adjust and direct behavior in ways that meet others' expectations. In essence, self-awareness emerges hand in hand with awareness of others and of their evaluations of one's behavior and appearance.

The Emergence of Self-Awareness

According to George Herbert Mead, significant symbols and gestures are the mechanisms that allow an individual to interact with others and in the process to learn about the self. A **significant symbol** is a word, gesture, or other learned sign that is used to "convey a meaning from one person to another, and that has the same meaning for the person transmitting it as for the person receiving it" (Theodorson and Theodorson 1979:430). Language is a particularly important significant symbol because "it is only through language that we enter fully into our human estate and culture, communicate freely with our fellows, acquire and share information. If we cannot do this, we will be bizarrely disabled and cut off" (Sacks 1989:8).

Symbolic gestures or signs are nonverbal cues, such as tone of voice, inflection, facial expression, posture, and other body movements or positions that convey meaning from one person to another.

As people learn significant symbols — language and symbolic gestures—they also acquire the ability to do reflexive thinking and to adjust their presentation of self to meet other people's expectations. Mead believed, however, that humans do not adhere mechanically to others' expectations. Instead a dialogue goes on continuously between two aspects of the self— the *I* and the *me*.

The *me* is Mead's term for the self as the internalized expectations of others. Before an individual acts,

Reflexive thinking Stepping outside the self and observing and evaluating it from another's viewpoint.

Significant symbol A word, gesture, or other learned sign used to convey a meaning from one person to another.

Symbolic gestures Nonverbal cues, such as tone of voice and other body movements, that convey meaning from one person to another.

the *me* takes others into account by assessing the appropriateness of the act and anticipating the responses. The *I* is the spontaneous, autonomous, creative self, capable of rejecting expectations and acting in unconventional, inappropriate, or unexpected ways. For example, a student recently expressed disappointment at receiving a failing grade on an examination. Upon seeing her score, she blurted out, "A 50! I skipped two classes to study for this stupid test!" In making this comment, she failed to anticipate its effect on the professor. Presumably the expected grade came as such a shock that her spontaneous *I* overwhelmed her calculating *me*.

Although Mead does not specify how the *I* emerges, we know that a spontaneous, creative self must exist; otherwise, human life would never change and would stagnate. Mead is more specific about how the *me* develops: through imitation, play, and games, all of which give the developing child practice with role taking.

Role Taking

Mead assumed that the self is a product of interaction experiences. He maintained that children acquire a sense of self when they become objects to themselves. That is, they are able to imagine the effect of their words and actions on other people. According to Mead, a person can see him- or herself as an object after learning to role-take. **Role taking** involves stepping outside the self and viewing its appearance and behavior imaginatively from an outsider's perspective.

Researchers have devised an ingenious method for determining when a child is developmentally capable of role taking. A researcher puts a spot of rouge on the child's nose and then places the child in front of a mirror. If the child shows no concern with the rouge, he or she presumably has not yet acquired a set of standards about how he or she ought to look; that is, the child cannot role-take or see him- or herself from another person's viewpoint. But if the child shows concern over the rouge, he or she presumably has formed some notion of self-appearance and therefore can role-take (Kagan 1989).

Role taking Stepping outside the self and viewing its appearance and behavior imaginatively from an outsider's perspective.

Play Voluntary and often spontaneous activity, with few or no formal rules, that is not subject to constraints of time or place.

Mead hypothesized that children learn to take the role of others through (1) imitations, (2) play, and (3) games. Each of these stages involves a progressively sophisticated level of role taking.

The Preparatory Stage In this stage, children have not yet developed the mental capabilities that allow them to role-take. Although they mimic or imitate people in their environment, they have almost no understanding of the behaviors that they are imitating. Children may imitate spontaneously (by mimicking a parent writing, cooking, reading, and so on), or they may repeat things that adults encourage them to say and reward them for saying. In the process of imitating, children learn to function symbolically; that is, they learn that particular actions and words arouse predictable responses from others. For example, Israeli children may be taught to respond "I will defend my homeland, Israel" to the question "What will you do when you grow up?" Similarly, Palestinian parents teach their children where they come from, even before the children learn notions of geography and understand the historical circumstances of their living arrangements. A typical exchange between a Palestinian parent and child would go something like this:

PARENT: Where are you from?

CHILD: I am from Palestine—from the city of Hebron.

PARENT: What is Israel?

CHILD: The real name for Israel is Palestine.

Both Palestinian and Israeli children, like children in nearly every culture, learn to sing patriotic songs and say prayers before they can understand the meaning of the words. Jenny Bourne, a political activist and a member of a delegation that visited the Occupied Territories, was struck by the fact that as soon as some two-year-old Palestinian children "saw the cameras come out, they were up and alert, hands outstretched as taut fingers made in unison the victory sign for our photos. No [Palestinian] child we met anywhere wanted to be photographed without that sign" (Bourne 1990:70).

The Play Stage Mead saw children's play as the mechanism by which they practice role taking. **Play** is a voluntary and often spontaneous activity, with few or no formal rules, that is not subject to constraints of time (for example, twenty-minute halves, fifteen-minute quarters) or place (for example, a gymnasium, a regulation-size field). Children, in particular, play whenever and wherever the urge strikes. If there are

Through socialization this Palestinian girl has learned to dream of living someday in a glorious Jerusalem—a city she has never seen.

©Joanna B. Pinneo/Aurora

rules, they are not imposed on participants by higher authorities (for example, rule books, officials). Participants undertake play for their amusement, entertainment, or relaxation. These characteristics make play less socially complicated than organized games (Corsaro 1985; Figler and Whitaker 1991).

In the play stage, children pretend to be **significant others**—people or characters who are important in their lives, in that they have considerable influence on a child's self-evaluation and encourage the child to behave in a particular manner. Children recognize behavior patterns characteristic of these significant persons and incorporate them into their play. For example, when a little girl plays with a doll and pretends she is the doll's mother, she talks and acts toward the doll the same way her mother does toward her. By pretending to be the mother, she gains a sense of the mother's expectations and perspective and learns to see herself as an object. Similarly, two children playing doctor and patient are learning to see the world from viewpoints other than their own and to understand how a patient acts in relation to a doctor, and vice versa.

Children's role taking can come only from what they see and hear. For the most part, Palestinian children in the West Bank and Gaza have never seen an adult male Israeli without a gun. Palestinian children's play reflects their experiences: the children pretend to be Israeli soldiers arresting and beating other Palestinian children who are pretending to be stone throwers. They use sticks and cola cans as if they were guns and tear-gas canisters (Usher 1991). One evening ABC News featured a segment on Palestinian children engaged in this type of play. When asked by the reporter which they preferred to be, soldiers or stone throwers, the children replied, "Soldiers, because they have more power and can kill."

Israeli children have had little experience with Palestinians except as manual laborers or "terrorists." Thus, it is hardly surprising that some Israeli kindergartners pretend that Israelis are Smurfs (good guys) and Palestinians portray Gargamel (a bad guy in a TV program). Israeli children, like Palestinian children, pretend to be soldiers because both men and women must serve in the Israeli military beginning at age eighteen. Israeli children pick up their fathers' guns and declare, "I'm going to kill the Arabs with this" or "Arabs are bad and must be killed." Through this type of play, children learn how they think the "enemy" views them and how they should view the enemy.

The Game Stage In Mead's theory, the play stage is followed by the game stage. **Games** are structured, organized activities that almost always involve more than one person. They are characterized by a number of constraints, including one or more of the following: established roles and rules, an outcome toward which all activity is directed, and an agreed-on starting time and place. Through games children learn to (1) follow established rules, (2) take simultaneously the role of all participants, and (3) see how their position fits in relation to all other positions.

When children first take part in games such as organized sports, their efforts seem chaotic. Instead of

Significant others People or characters who are important in an individual's life by having considerable influence on that person's self-evaluation and encouraging him or her to behave in a particular manner.

Games Structured, organized activities that usually involve more than one person and a number of constraints concerning roles, rules, time, place, and outcome.

making an organized response to a ball hit to the infield, for example, everyone tries to retrieve the ball, leaving nobody at the base to catch the throw needed to put the runner out. This chaos exists because children have not developed to the point at which they can see how their role fits with the roles of everyone else in the game. Without such knowledge a game cannot have order. Through playing games, children learn to organize their behavior around the **generalized other**—that is, around a system of expected behaviors, meanings, and viewpoints that transcend those of the people participating. "The attitude of the generalized other is the attitude of the whole community. Thus, for example, in the case of such a social group as a baseball team, the team is the generalized other insofar as it enters—as an organized process or activity—into the experience of [those participating]" (Mead 1934:119). In other words, when children play organized sports, they practice fitting their behavior into an established behavior system.

In view of this information, not surprisingly, games are the tools used in programs designed to break down barriers between Palestinian and Jewish children and adolescents. The games involve activities like

> throwing an orange into the air, calling a person's name to catch it, throwing it again with another's name, and again and again as the whoops of laughter fill the room. Then they all crowd together, take each other's hands, and turn around until they are enmeshed in a tangle of arms. Intertwined with each other, they try to unravel themselves without letting go. They talk to each other, giving advice, crouching so another can step over an arm, stooping so others can swing arms over heads, spinning around, trying to turn the snarled mess of Arab and Jewish bodies into a clean circle. (Shipler 1986:537)

Although these games seem merely fun, sociologists contend that participants are learning to see things from another perspective and to play their parts successfully in a shared activity. The participants cannot be effective unless they understand their own roles in relation to everyone else's. Although the children trying to untangle themselves may not be fully aware of it, they are learning that a Palestinian (or an Israeli) can be in positions like their own. If anyone can become untangled, participants must be able to understand everyone else's situation.

As we have learned, George Herbert Mead assumed that the self develops through interaction with others. Mead identified the interaction that occurs in play and games as important to children's self-development. When children participate in play and games, they practice at seeing the world from the viewpoint of others and gain a sense of how others expect them to behave. Sociologist Charles Horton Cooley offered a more general theory about how the self develops.

The Looking-Glass Self

Like Mead, Charles Horton Cooley assumed that the self is a product of interaction experiences. Cooley coined the term **looking-glass self** to describe the way in which a sense of self develops: people act as mirrors for one another. We see ourselves reflected in others' reactions to our appearance and behaviors. We acquire a sense of self by being sensitive to the appraisals of ourselves that we perceive others to have: "Each to each a looking glass,/Reflects the other that [does] pass" (Cooley 1961:824). As we interact, we visualize how we appear to others, we imagine a judgment of that appearance, and we develop a feeling somewhere between pride and shame: "The thing that moves us to pride or shame is not the mere mechanical reflection of ourselves but . . . the imagined effect of this reflection upon another's mind" (Cooley 1961:824).

Cooley went so far as to argue that "the solid facts of social life are the facts of the imagination." According to this logic, one person's effect on another is defined most accurately as what one person imagines the other will do and say on a particular occasion (Faris 1964). Because Cooley defined the imagining or interpreting of others' reactions as critical to self-awareness, he believed that people are affected deeply even when the image they see reflected is exaggerated or distorted. One responds to the perceived reaction rather than to the actual reaction.

On the other hand, we cannot overlook the fact that more often than not our imaginations of how other people will react and behave rests on past experiences with others. In the case of Palestinians and Israelis, for example, each group aims a number of powerful images at the other. For example, Palestinians call the Israelis "Nazis" and equate the occupation of their country and the accompanying imprisonments, beat-

Generalized other A system of expected behaviors, meanings, and viewpoints that transcend those of the people participating.

Looking-glass self A way in which a sense of self develops in which people see themselves reflected in others' reactions to their appearance and behaviors.

ings, and identification checks with the concentration camps. These labels are quite painful to Israelis, who see little similarity between concentration camps and the Palestinian situation. Israelis, on the other hand, react by defining the Palestinians as culturally primitive and incapable of managing their own affairs. They tell the Palestinians that the Israelis are responsible for turning the worthless desert land occupied previously by a backward Palestinian people into a modern, high-technology state.

Both Mead's and Cooley's theories suggest that self-awareness derives from an ability to think reflexively—to step outside oneself and view the self from another's perspective. Although Cooley and Mead describe the mechanisms (imitation, play, games) by which people learn about themselves, neither theorist addressed how a person acquires this level of cognitive sophistication. To answer this question, we must turn to the work of Swiss psychologist Jean Piaget.

Piaget's theory of cognitive development suggests that the emergence of social concern in many young people may be explained in part by the development of their ability to think abstractly and, consequently, to identify with issues and causes.

©Tony Freeman/PhotoEdit

Cognitive Development

Piaget is the author of many influential and provocative books about how children think, reason, and learn. The titles of some of his many books — *The Language and Thought of the Child* (1923), *The Child's Conception of the World* (1929), *The Moral Judgment of the Child* (1932), *The Child's Conception of Time* (1946), and *On the Development of Memory and Identity* (1967) — give some clues about the many categories of childhood thinking that Piaget investigated.

Piaget's influence reaches across many disciplines: biology, education, sociology, psychiatry, psychology, and philosophy. His ideas about how children develop increasingly sophisticated levels of reasoning stem from his study of water snails (*Limnaea stagnalis*), which spend their early life in stagnant waters. When transferred to tidal water, these lazy snails engage in motor activity that develops the size and shape of the shell to help them remain on the rocks and avoid being swept away (Satterly 1987).

Building on this observation, Piaget arrived at the concept of **active adaptation**, a biologically based tendency to adjust to and resolve environmental challenges. The theme of active adaptation runs through almost all of Piaget's writings. He believed that learning and reasoning are rooted in active adaptation. He defined logical thought, another biologically based human attribute, as an important tool for meeting and resolving environmental challenges. Logical thought emerges according to a gradually unfolding genetic timetable. This unfolding must be accompanied by direct experiences with persons and objects; otherwise, a child will not realize his or her potential ability. On the basis of cumulative experiences, a child constructs and reconstructs his or her conceptions of the world.

Piaget's model of cognitive development includes four broad stages—sensorimotor, preoperational, concrete operational, and formal operational—each characterized by a progressively more sophisticated reasoning level. A child cannot proceed from one stage to another until the reasoning challenges of earlier stages are mastered. Piaget maintained that reasoning abilities cannot be hurried; a more sophisticated level of understanding will not show itself until the brain is ready.

- *Sensorimotor stage (from birth to about age two).* In this stage, children explore the world with their senses (taste, touch, sight, hearing, and smell). The cognitive accomplishments of this stage include an understanding of the self as separate from other persons and the realization that objects and persons exist even when they are out of sight. Before this notion takes hold, very young children act as if an object does not exist when they can no longer see it.

Active adaptation A biologically based tendency to adjust to and resolve environmental challenges.

- *Preoperational stage (from about ages two to seven).* Piaget focused most of his attention on this stage. Children in this stage typically demonstrate three characteristic types of thinking. They think *anthropomorphically;* that is, they assign human feelings to inanimate objects. Thus, they believe that objects such as the sun, the moon, nails, marbles, trees, and clouds have motives, feelings, and intentions (for example, dark clouds are angry; a nail that sinks to the bottom of a glass filled with water is tired). Second, they think *nonconservatively,* a term Piaget used to signify an inability to appreciate that matter can change form but still remain the same in quantity. Third, they think *egocentrically* in that they cannot conceive how the world looks from another point of view. Thus, if a child facing a roomful of people (all of whom are looking in his or her direction) is asked to draw a picture of how a person in the back of the room sees the people, the child will draw the picture as he or she sees the people. Related to egocentric thinking is *centration,* a tendency to center attention on one detail of an event. As a result of this tendency, the child fails to process other features of a situation (see Figure 5.3).

- *Concrete operational stage (from about ages seven to twelve).* By the time children enter this stage, they have mastered these preoperational tasks but have difficulty in thinking hypothetically or abstractly without reference to a concrete event or image. For example, a child in this stage has difficulty envisioning a life without him or her in it. One twelve-year-old struggling to grasp this idea said to me, "I am the beginning and the end; the world begins with and ends with me."

- *Formal operational stage (from the onset of adolescence onward).* At this point, people are able to think abstractly. For example, they can conceptualize their existence as a part of a much larger historical continuum and a larger context.

As far as we know, this progression by stages toward increasingly sophisticated levels of reasoning is universal, but the content of people's thinking varies across cultures. As an example, all Palestinian and Israeli children learn the following rule: "If you ever see an unattended package or bag on a street or bus, don't touch it. Notify an adult immediately." Knowing the rule is a matter of safety because the package might contain a bomb. American children typically are not exposed to the same dangers and hence have no need to consider situations in which this rule can be relaxed.

New York Times reporter David Shipler, in his book *Arab and Jew,* describes his frustrations in explaining to his young children that they did not have to follow this rule when they returned to the United States. When he set some bags of newspapers on the curb to be picked up, his children reported suspicious packages outside:

> Michael [age 7] ran in another day to report a plastic cup of some sort in the street. I had seen it and asked him to throw it away. He adamantly refused to go near it, and he remained solidly unmoved by my extravagant assurances that we didn't have to worry about bombs on a quiet, tree-lined suburban street in America. (Shipler 1986:83)

From the perspective of Piaget's theory, Michael centered all of his attention on one detail (the rule) and could not respond to other aspects of the situation that made the rule irrelevant, such as geographic location.

The theories of Mead, Cooley, and Piaget all suggest that the process of social development is multifaceted and continues over time. It is important to realize that socialization is a lifelong process in which people make any number of transitions over a lifetime: from

Figure 5.3 Centration in a Child's Drawing

As they draw, young children can focus on only one detail at a time. For example, they draw the passengers and the automobile separately. When they think about the passengers or a car, they fail to consider them as a unit.

Source: From "Children's Drawing of Human Figures," by Norman H. Freeman. P. 138 in *The Oxford Companion to the Mind.* Copyright ©1987. Oxford University Press. Reprinted by permission.

single to married, from married to divorced or widowed, from childlessness to parenthood, from healthy to disabled, from one career to another, from civilian status to military status, from employed to retired, and so on. In making such transitions, people undergo resocialization.

Resocialization

Resocialization is the process of being socialized over again. In particular, it is a process of discarding values and behaviors unsuited to new circumstances and replacing them with new, more appropriate values and norms (standards of appearance and behavior). A considerable amount of resocialization happens naturally over a lifetime and involves no formal training; people simply learn as they go. For example, people marry, change jobs, become parents, change religions, and retire without formal preparation or training. However, some resocialization requires that, to occupy new positions, people must undergo formal, systematic training and demonstrate that they have internalized appropriate knowledge, suitable values, and correct codes of conduct.

Such systematic resocialization can be voluntary or imposed (Rose, Glazer, and Glazer 1979). It is voluntary when people choose to participate in a process or program designed to "remake" them. Examples of voluntary resocialization are wide-ranging — the unemployed youth who enlists in the army to acquire a technical skill, the college graduate who pursues medical education, the drug addict who seeks treatment, the alcoholic who joins Alcoholics Anonymous (AA). The twelve-step program of AA offers concrete examples of behavior that alcoholics must learn to perform if they are to free themselves from alcohol dependency.

Resocialization is imposed when people are forced to undergo a program designed to train them, rehabilitate them, or correct some supposed deficiency in their earlier socialization. Military boot camp (when a draft exists), prisons, mental institutions, and schools (when the law forces citizens to attend school for a specified length of time) are examples of environments that are designed to resocialize but that people also enter involuntarily.

In *Asylums: Essays on the Social Situation of Mental Patients and Other Inmates,* sociologist Erving Goffman writes about a setting — total institutions (with particular focus on mental institutions) — where people undergo systematic socialization. In **total institutions**, people surrender control of their lives, voluntarily or involuntarily, to an administrative staff and, as inmates, carry out daily activities (eating, sleeping, recreation) in the "immediate company of a large batch of others, all of whom are [theoretically] treated alike and required to do the same thing together" (Goffman 1961:6). Total institutions include homes for the blind, the elderly, the orphaned, and the indigent; mental hospitals; jails and penitentiaries; prisoner-of-war camps and concentration camps; army barracks; boarding schools; and monasteries and convents. Their total character is symbolized by barriers to social interaction, "such as locked doors, high walls, barbed wire, cliffs, water, forests, or moors" (p. 4).

Goffman was able to identify the general and standard mechanisms, despite their wide range, that the staffs of all total institutions employ to resocialize "inmates." When the inmates arrive, the staff strips them of their possessions and their usual appearances (and the equipment and services by which their appearances are maintained). In addition, the staff sharply limits interactions with people outside the institution to establish a "deep initial break with past roles" (p. 14).

> We very generally find staff employing what are called admission procedures, such as taking a life history, photographing, weighing, finger-printing, assigning numbers, searching, listing personal possessions for storage, undressing, bathing, disinfecting, haircutting, issuing institutional clothing, instructing as to rules, and assigning to quarters. The new arrival allows himself to be shaped and coded into an object that can be fed into the administrative machinery of the establishment. (p. 16)

Goffman maintained that the admission procedures function to prepare inmates to shed past roles and assume new ones by participating in the various enforced activities that staff members have designed, to fulfill the official aims of the total institution—whether to care for the incapable, to keep inmates out of the community, or to teach people new roles (for example, to be a soldier, priest, or nun).

In general, it is easier to resocialize people when they want to be resocialized than when they are forced to abandon old values and behaviors. Furthermore, re-

Resocialization The process of discarding values and behaviors unsuited to new circumstances and replacing them with new, more appropriate values and norms.

Total institutions Institutions in which people surrender control of their lives, voluntarily or involuntarily, to an administrative staff and carry out daily activities with others required to do the same thing.

socialization is likely to be easier if acquiring new values and behaviors requires competence rather than subservience (Rose et al. 1979). A case in point is the resocialization that takes place in medical school. Theoretically, medical students learn (among other things) to be emotionally detached in their attitudes toward patients, not prefer one patient over another (that is, patients of a particular ethnicity, gender, or age, or even level of cooperation), and provide medical care whenever it is required (Merton 1976). These attitudes are necessary for proper diagnosis and treatment, as the following accounts make clear.

An emergency room physician, Elisabeth Rosenthal, reported on an unkempt man who had been brought into the emergency room. He was uncooperative, his speech was slurred and did not make sense, he did not know the date or the name of the president of the United States, and he could not decide whether he was in a hotel or a hospital. On the basis of his appearance and behavior, he seemed to be drunk; the best treatment seemed to be to let him sleep it off. However, physicians are trained (in theory, at least) not to be influenced by stereotypes and to look beyond commonplace interpretations associated with certain physical traits.

In this case, subsequent tests revealed that the man had a massive kidney infection: "His incoherence and his lack of cooperation were caused not by intoxication but by a metabolic disturbance resulting from his infection" (Rosenthal 1989:82). For the physician, professional competence is demonstrated by learning to abandon or hold in check widely shared misconceptions about the meaning of behavior, especially the behavior of persons in different ethnic, gender, and age groups.

Michael Gorkin (1986) believes that many problems can develop between Israeli psychiatrists and Palestinian patients during therapy. (The same could be said of Palestinian psychiatrists and Israeli patients, but there are very few Palestinian psychiatrists in Israel. Consequently, most Palestinians who go to psychiatrists go to Israeli ones.) If Israeli psychiatrists do not learn to manage their stereotypes and prejudices with regard to Palestinians and to familiarize themselves with Arab culture, they may treat the client in counterproductive ways. In addition, the patient must trust the psychiatrist if therapy is to be successful.

Establishing trust is made difficult by the fact that almost all Israeli psychiatrists serve in army reserve units and must cancel therapy hours several times a year for several weeks. Many are reluctant to tell their Palestinian clients the reason for their absence; most hope that their patients will not inquire. Palestinians, of course, are aware of this commitment. Gorkin maintains that the Israeli psychiatrist must address these strains constructively with the Palestinian client in the initial sessions. He does not believe that the discussion will resolve these differences but thinks that it "sets the stage for openness in the therapeutic interaction and conveys the message that this crucial issue is not taboo" (Griffith 1977:38). The Israeli psychiatrist is likely to come to terms with prejudices and stereotypes because he or she has chosen to become a physician and because coming to terms with these beliefs demonstrates professional competence achieved through resocialization during the medical training period.

Therein lies the dilemma in finding a resolution to the Palestinian-Israeli conflict. Both sides use what they hope are resocialization measures that attempt to force the other side to change its position about land rights. Israelis deport, imprison, impose curfews, close schools, level houses, and kill. Palestinians throw stones, strike, boycott Israeli products, and kill. Both Palestinians and Israelis seem to believe that if they make life miserable enough for the other, each will gain a homeland. The problem is that if one side wins through intimidation, the other side by definition assumes a subservient position.

Summary and Implications

Humans are born with a genetic endowment. By way of their genes, parents transmit to their offspring a biological heritage common to all human beings. This heritage is such that we depend on others for a relatively long period of time. On the other hand, it presents us with a great capacity for learning—to speak (or sign) innumerable languages, perform countless movements, retain and recall past experiences, and devise and use a seemingly unlimited number of objects. The biological process of learning, inasmuch as it is tied to the central nervous system, the cerebral cortex, and other physiological equipment, is similar, if not virtually identical, for everyone.

The fact that we are born with this great learning capacity suggests that we come into the world "unfinished," lacking the information and skills needed to meet the challenges of living. To fill this gap, the unfinished person must participate in social life (Hannerz 1992). Thus, we can say that among the most significant of human biological needs is a need for social contact. Without such contact, a person cannot become a normally functioning human being. The rare instances of children raised in extreme isolation or in very restrictive and unstimulating environments, such as the cases of Anna and Isabelle, show the importance of social contact to normal development. Our strong dependence on interaction supports the view that for the individual, social contact "is a reality from which everything that matters to us flows" (Durkheim 1984:252). Thus, the need for social contact is perhaps the most important universal trait possessed by all humans.

The need for social contact indicates that we cannot speak of the individual as if he or she existed apart from others. In *Human Nature and the Social Order,* Charles Horton Cooley (1964) argues that "a separate individual is an abstraction unknown to experience" (p. 36). The few completely separate individuals we know about, such as Anna and Isabelle, did not possess the qualities that make people human when they were found.

It is obvious, then, that an individual is a product of two major sources: heredity and social interaction. The genes that parents transmit to their offspring reach back over an indefinite period of time through four biological grandparents, eight biological great-grandparents, and beyond to a common ancestor. As a result, each of us has a unique biological heritage, yet this heritage is common to all humans. In addition, through contact with a unique combination of various others (parents, grandparents, baby-sitters, peers, and so on), individuals realize their human capacities.

Socialization goes beyond the needs of creating an individual with "human qualities." It functions to link people to one another in orderly and predictable ways. Without the benefits of social interaction, newcomers fail to thrive physically and to learn the skills they need to achieve meaningful connections with others. In addition, without meaningful contact between the generations, culture (solutions to the problems of living) cannot be passed on from one generation to the next. Socialization can also be a process by which newcomers learn to think and behave in ways that reflect the interests of the teachers and to accept and fit into a system that benefits some groups more than others.

What we know about the socialization processes suggests that life does not have to unfold in a seemingly predictable fashion. Human genetic and social makeup contain considerable potential for change. First, people are not born with preconceived notions about standards of appearance and behavior. To develop standards, people must be exposed to information that leads them to expect people, behavior, and objects to be a certain way. Second, the cerebral cortex allows people to think reflexively — to step outside the self and observe and evaluate it from another viewpoint. Third, the mechanisms that teach prejudice and hatred for another group — mechanisms such as imitation, play, and games — also teach respect and understanding. The problem is that they often are used to teach children respect and understanding after the children already have learned prejudice and hatred through the same mechanisms. Finally, people can be resocialized to abandon one way of thinking and behaving for another. Preferably, the socialization process will be such that people choose to abandon old habits and, in doing so, will gain a feeling of competence and personal empowerment that comes with choice.

Finally, as noted in this chapter, each generation learns about the environment it inherits and comes to terms with it in unique ways. Table 5.1 gives some idea of the challenges facing Israel/Palestine in this regard. Although the older generations may share their personally acquired memories with later generations, these memories cannot affect the behavior of the younger generations in the same way. The continuous emergence of new generations "serves the necessary purpose of enabling us to forget," or at least to bring a fresh perspective to a situation (Mannheim 1952:294). On the other hand, we cannot assume that people of one generation cannot change the way they think about and approach their environment.

Assassinated Israeli Prime Minister Yitzhak Rabin and Palestinian President Yasir Arafat, lifelong bitter enemies, recognized the other's right to exist. They did so in an exchange of letters before the peace accord ceremony in 1993:

> **Mr. Prime Minister**
>
> The PLO recognizes the right of the State of Israel to exist in peace and security . . . [and] renounces the use of terrorism and other acts of violence. . . . (Sincerely, Yasir Arafat, Chairman, The Palestinian Liberation Organization)
>
> **Mr. Chairman . . .**
>
> The Government of Israel has decided to recognize the PLO as the representative of the Palestinian people. . . . (Yitzhak Rabin, Prime Minister of Israel)

Table 5.1 Future Generations' Population Age and Growth (estimated for 1995)

Perpetuating cultural and political identities depends on the size and socialization of future generations. Suppose you saw your interests primarily in terms of the future of Israel as a Jewish homeland. What might these numbers suggest to you? Suppose you saw your interests primarily in terms of an Arab-Palestinian homeland. What might the numbers then suggest? Suppose you saw your interests in terms of long-term peace. What might the numbers suggest then?

	Gaza	West Bank	Israel
Fertility rate (children born per woman)	7.74	5.34	1.4%
Infant mortality rate (deaths per 1,000 live births)	30.6	29.7	8.4
Population growth rate	4.55%	3.5%	1.4%
Life expectancy at birth (years)	71.09	71.42	78.14
Profile of ages			
0–14 years	52%	46%	29%
15–64 years	45%	51%	61%
65 and over	3%	3%	10%

Source: U.S. Central Intelligence Agency (1995).

Key Concepts

Use this outline to organize your review of the key chapter ideas.

- **Socialization**
  - **Nature**
  - **Nurture**
- **Collective memory**
  - **Engrams**
- **Groups**
  - **Primary**
  - **Ingroups**
  - **Outgroups**
- **Reflexive thinking**
- **Significant symbol**
- **Symbolic gestures**
- **Role taking**
  - **Play**
    - **Significant others**
  - **Games**
    - **Generalized others**
- **Looking-glass self**
- **Active adaptation**
- **Resocialization**
  - **Total institutions**

internet assignment

In this chapter, we studied the concept "collective memory" to describe the experiences shared and recalled by significant numbers of people and considered how versions of past experiences are transmitted through stories, holidays, rituals, museums, and monuments to those who were not there. Use a search engine to find a list of museums with Web sites. Browse through the list until you find a museum that attracts your interest. Why did you choose that museum? What part of past experience does the museum record? What messages are conveyed directly or indirectly about the past?

6

Social Interaction and the Social Construction of Reality

With Emphasis on Zaire (Now the Democratic Republic of the Congo)*

*On May 18, 1997, as this book was in production, rebel leader Laurent Kabila declared himself president and renamed Zaire the Democratic Republic of the Congo. To avoid confusing the Democratic Republic of the Congo with its neighbor to the west, the Republic of the Congo, we continue to use the name Zaire.

Children canoeing to school on the Ngiri River. ©Jaques Jangoux/Tony Stone Images

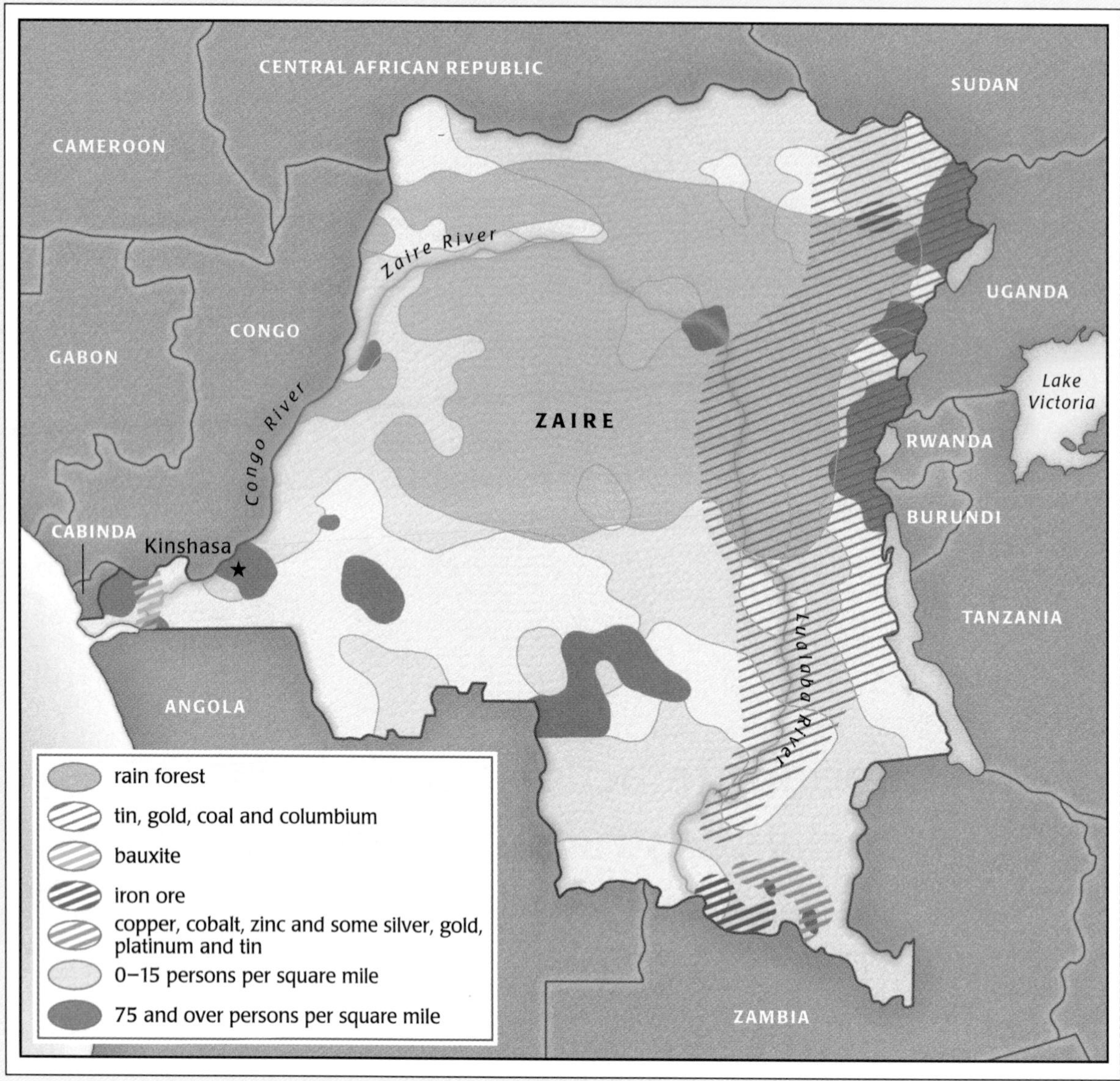

Source: Kurian (1992).

Roots of Interaction

A map of Zaire's natural resources helps explain why European countries sought to control Africa in the nineteenth and twentieth centuries. The Belgian effort to tap resources and centralize power in what is now Zaire led to 90,000 miles of road and significant river transport by the 1960s. By the mid-1990s, ruled by a dictator supported by the United States and other industrialized countries, Zaire had about 1,738 miles of passable roads. Yet massive rural-to-urban migrations are taking place within Zaire. These facts provide a context for understanding broad patterns of social interaction within Zaire and between Zaire and other countries.

Margrethe Rask, a Danish surgeon, was exposed to the virus now known as human immunodeficiency virus, or HIV, while working in a small village clinic in Zaire in 1977. The excerpt that follows highlights some of the final events in Dr. Rask's life, her last interactions with colleagues and close friends.

Grethe Rask gasped her short, sparse breaths from an oxygen bottle. . . . "I'd better go home to die," Grethe had told [her friend] Ib Bygbjerg matter-of-factly. The only thing her doctors could agree on was the woman's terminal prognosis. All else was mystery. Also newly returned from Africa, Bygbjerg pondered the compounding mysteries of Grethe's health. None of it made sense. In early 1977, it appeared that she might be getting better; at least the swelling in her lymph nodes had gone down, even as she became more fatigued. But she had continued working, finally taking a brief vacation in South Africa in early July.

Suddenly, she could not breathe. Terrified, Grethe flew to Copenhagen, sustained on the flight by bottled oxygen. For months now, the top medical specialists of Denmark had tested and studied the surgeon. None, however, could fathom why the woman should, for no apparent reason, be dying. There was also the curious array of health problems that suddenly appeared. Her mouth became covered with yeast infections. Staph infections spread in her blood. Serum tests showed that something had gone awry in her immune system; her body lacked T-cells, the [essential parts of] the body's defensive line against disease. But biopsies showed she was not suffering from a lymph cancer that might explain not only the T-cell deficiency but her body's apparent inability to stave off infection. The doctors could only gravely tell her that she was suffering from progressive lung disease of unknown cause. And, yes, in answer to her blunt questions, she would die.

Finally, tired of the poking and endless testing by the Copenhagen doctors, Grethe Rask retreated to her cottage near Thisted. A local doctor fitted out her bedroom with oxygen bottles. Grethe's longtime female companion, who was a nurse in a nearby hospital, tended her. Grethe lay in the lonely whitewashed farmhouse and remembered her years in Africa while the North Sea winds piled the first winter snows across Jutland.

In Copenhagen, Ib Bygbjerg, now at the State University Hospital, fretted continually about his friend. Certainly, there must be an answer to the mysteries of her medical charts. Maybe if they ran more tests. . . . It could be some common tropical culprit they had overlooked, he argued. She would be cured, and they would all chuckle over how easily the problem had been solved when they sipped wine and ate goose on the Feast of the Hearts. Bygbjerg pleaded with the doctors, and the doctors pleaded with Grethe Rask, and reluctantly the wan surgeon returned to the old Rigshospitalet in Copenhagen for one last chance. On December 12, 1977, just twelve days before the Feast of the Hearts, Margrethe P. Rask died. She was forty-seven years old. **(Shilts 1987:6-7)**

Why Focus on Zaire?

In this chapter, as we explore the sociological theories and concepts that sociologists use to analyze any social interaction in terms of context and content, we give particular focus to social interaction as it relates to the transmission of HIV and the treatment of AIDS. In doing so, we look closely at the central African country of Zaire (formerly the Belgian Congo) for two important reasons. First, focusing on Zaire and its relationship to other countries helps us connect the transmission of HIV to a complex set of intercontinental, international, and intrasocietal interactions. Specifically, these interactions involve unprecedented levels of international and intercontinental air travel of the privileged for pleasure and business, as well as legal and illegal migrations of the underprivileged from villages to cities and from country to country (Sontag 1989).

Second, focusing on Zaire highlights evidence that HIV existed as early as 1959—evidence in the form of an unidentified blood sample frozen in that year and stored in a Zairean blood bank. Although this hardly proves that HIV originated in Zaire, this hypothesis has received considerable support from government and health officials in Western countries.

Whether Zaire is actually the country of origin of HIV is irrelevant to our purpose. Far more important is the idea that reality is a social construction. That is, people give meaning to phenomena (events, traits, objects) that almost always emphasize some aspect of a phenomenon and ignore other aspects. For example, to say that HIV traveled from Zaire to the United States ignores the possibility that it traveled from the United States to Zaire. When we compare Western values and beliefs related to the origin of the virus and the treatment of AIDS with African values and beliefs, we realize that the Western framework is only one way of viewing the AIDS phenomenon. Moreover, by comparing the two frameworks, we can see that the meanings people give to an event have enormous consequences for the individuals involved (medical personnel, infected persons and those close to them, and noninfected persons). These meanings influence how people interact with one another and what decisions they make and actions they take to deal with HIV infection and AIDS.

We can visualize some of the interactions between Grethe Rask and her friends and colleagues, presented in the opening quote. For example, we can visualize Dr. Rask telling her friend Ib Bygbjerg, "I'd better go home to die," or asking the Copenhagen doctors whether she would die after they tell her that she is "suffering from progressive lung disease of unknown cause." Finally, when Dr. Rask decides to leave the hospital and die at home, we can imagine Ib Bygbjerg pleading, "Please come back to the hospital for more tests; maybe there is still hope."

Sociologists looking at this situation would agree that Dr. Rask's illness is the obvious and immediate reason for these **social interactions**—everyday events in which at least two people communicate and respond through language and symbolic gestures to affect one another's behavior and thinking. In the process, the parties involved define, interpret, and attach meaning to the encounter. Sociologists also assume that any social interaction reflects forces beyond the obvious and immediate. Hence, they strive to locate the interaction according to time (history) and place (culture).[1]

When sociologists study social interaction, they seek to understand and explain the forces of context and content. **Context** consists of the larger historical circumstances that bring people together. **Content** includes the cultural frameworks (norms, values, beliefs, material culture) that guide behavior, dialogue, and interpretations of events. In the case of Dr. Rask, sociologists would determine the context by asking what historical events brought Dr. Rask to Africa in the first place and what further events put her in direct contact with a deadly virus. To understand the content of her interactions, sociologists would ask how the parties involved are influenced by their cultural frameworks as

Social interactions Everyday events in which at least two people communicate and respond through language and symbolic gestures to affect one another's behavior and thinking.

Context The larger historical circumstances that bring people together.

Content The cultural frameworks (norms, values, beliefs, material culture) that guide behavior, dialogue, and interpretations of events.

[1]From a sociological viewpoint, history and culture limit the range of potential experience and point people "towards certain definite modes of behavior, feeling, and thought" (Mannheim 1952:291).

What Is the Difference Between AIDS and HIV?

Acquired immunodeficiency syndrome (AIDS) is a fatal disease that severely compromises the human body's ability to fight infections and is caused by the human immunodeficiency virus (HIV). HIV infection can be transmitted by sexual intercourse between men and between men and women; by exposure to contaminated blood or blood products; by sharing or reusing contaminated needles; and during pregnancy, childbirth, and possibly breastfeeding, from woman to child. There is no evidence that HIV infection is transmitted through casual contact, water, air, or insects.

Although persons infected with HIV may not show any clinical symptoms of AIDS for months or years, they may never become free of the virus and may infect others without realizing it. An individual is considered to have AIDS if a blood test indicates the presence of antibodies to the [HIV] virus and if he or she has one or more debilitating and potentially fatal cancers, neurological disorders, or bacterial, protozoal, or fungal infections that are characteristic of the syndrome.

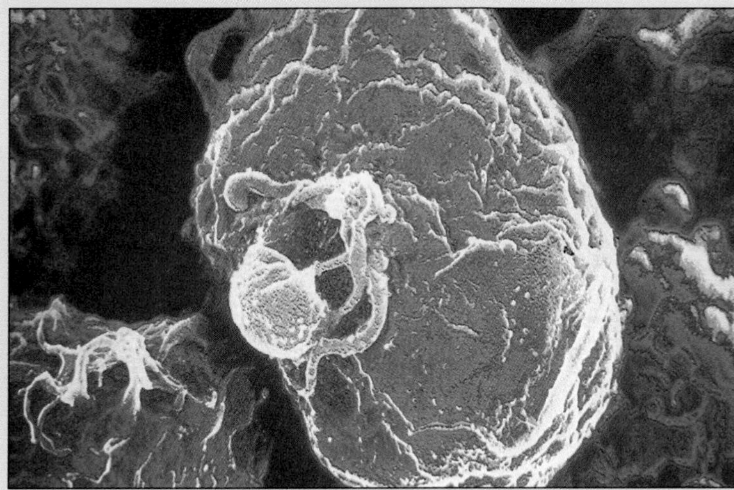

AIDS virus budding from a human lymphocyte, one of the white blood cells that form in lymphoid tissues.

Source: U.S. General Accounting Office (1987), p. 7.

they strive to define, interpret, and respond meaningfully to her condition.

Continue to think about Dr. Rask as you read this chapter. As we explore issues of context and content, we will see that an individual's seemingly unique and personal interactions are affected by history and culture. We begin by exploring the context of Dr. Rask's social interactions — the unprecedented mixing of the world's peoples and the large-scale social disruptions that have accompanied the emergence of worldwide economic and social interdependence. If we can understand these social forces, we can understand more about the transmission of viruses in general and the transmission of HIV in particular (see "What Is the Difference Between AIDS and HIV?").

The Context of Social Interaction

Emile Durkheim was one of the first sociologists to provide insights into the social forces that contributed to the rise of a "global village." In *The Division of Labor in Society* ([1933] 1964), Durkheim gives us a general framework for understanding both global interdependence and conditions that can cause large-scale social upheaval, leaving people vulnerable to phenomena like AIDS. More specifically, Durkheim's ideas provide a framework for understanding how Zaire was transformed, in less than 200 years, from a land of isolated and independent nations to a country characterized by immense social disruptions and participation in the world economy. (A **nation** is a geographical area occupied by people who share a culture and a history; a **country** is a political entity, recognized by foreign governments, with a civilian and military bureaucracy to enforce its rules.)

Nation A geographical area occupied by people who share a culture and a history.

Country A political entity, recognized by foreign governments, with a civilian and military bureaucracy to enforce its rules.

King Leopold II of Belgium, and then the Belgian government, claimed the territory now known as Zaire, exploiting the people and resources.

©Eric A. Wessman/Stock, Boston

Durkheim observes that an increase in population size and density intensifies the demand for resources. This in turn stimulates the development of more efficient methods for producing goods and services. As population size and density increase, society "advances steadily towards powerful machines, towards great concentrations of forces and capital, and consequently to the extreme division of labor" (Durkheim [1933] 1964:39). As Durkheim describes it, **division of labor** refers to work that is broken down into specialized tasks, with each task performed by a different set of persons. Not only are the tasks themselves specialized, but the parts and materials needed to manufacture products come from many geographical regions.

Because of its growing demand for resources, the West vigorously colonized much of Asia, Africa, and the Pacific in the late nineteenth and early twentieth centuries. Western governments forced local populations to cultivate and harvest crops and extract minerals and ores for export. The Belgian government claimed territory in central Africa, named it the Belgian Congo, and forced the people living there to extract rubber and mine copper. As industrialization proceeded in Europe, so did the demand for various raw materials. Over time, the world grew to depend on Zaire as a source of copper, cobalt (needed to manufacture jet engines), industrial diamonds, zinc, silver, gold, manganese (needed to make steel and aluminum dry-cell batteries), and uranium (needed to generate atomic energy and fuel the atomic bomb). The worldwide division of labor now included the indigenous people of Zaire, who mined the raw materials needed for products in distant parts of the world.

Durkheim notes that as the division of labor becomes more specialized and as the sources of materials for products become more geographically diverse, a new kind of solidarity or moral force emerges. Durkheim uses the term **solidarity** to describe the ties that bind people to one another in a society. He refers to the solidarity that characterizes preindustrial society as mechanical, and the solidarity that characterizes industrial societies as organic.

Division of labor Work that is broken down into specialized tasks, with each task performed by a different set of persons.

Solidarity The ties that bind people to one another in a society.

Mechanical solidarity Social order and cohesion based on a common conscience or uniform thinking and behavior.

Mechanical Solidarity

Mechanical solidarity is social order and cohesion based on a common conscience or uniform thinking and behavior. In this situation everyone views the world in much the same way. A person's "first duty is to resemble everybody else, [and] not to have anything personal about one's beliefs and actions" (Durkheim [1933] 1964:396). Such uniformity derives from a simple division of labor and the corresponding lack of specialization. (In other words, everyone is a jack-of-all-trades.) When everyone in a society does the same thing, they have common experiences, possess similar skills, and hold similar beliefs, attitudes, and thoughts. Therefore, a simple division of labor means that people are more alike than different. People are bound together because similarity gives rise to consensus and common conscience. In societies characterized by mechanical solidarity, the ties that bind people to one another are based primarily on kinship and religion.

As one of about 200 nations in Zaire, each of which has a distinct language and belief system, the Mbuti pygmies, a hunting-and-gathering people who live in the Ituri Forest (an equatorial rain forest) of northeastern Zaire, exhibit this type of solidarity. Their society represents the way of life that many people were forced to abandon after colonization began.

The Mbuti share a forest-oriented value system. Their common conscience derives from the fact that the forest gives them food, firewood, and materials for shelter and clothing. Anthropologist Colin Turnbull has written extensively about the Mbuti and their value system in three books, *The Forest People* (1961), *Wayward Servants* (1965), and *The Human Cycle* (1983). Excerpts from these books show the extent to which Mbuti forest-centered values permeate their life:

> For them the forest is sacred, it is the very source of their existence, of all goodness. . . . Young or old, male or female . . . the Mbuti talk, shout, whisper, and sing to the forest, addressing it as mother or father or both. (1983:30)

> It is not surprising that the Mbuti recognize their dependence upon the forest and refer to it as "Father" or "Mother" because as they say it gives them food, warmth, shelter, and clothing just like their parents. What is perhaps surprising is that the Mbuti say that the forest also, like their parents, gives them affection. . . . The forest is more than mere environment to the Mbuti. It is a living, conscious thing, both natural and supernatural, something that has to be depended upon, respected, trusted, obeyed, and loved. The love demanded of the Mbuti is no romanticism, and perhaps it might be better included under "respect." It is their world, and in return for their affection and trust it supplies them with all their needs. (1965:19)

Turnbull provides several examples of the intimacy between the Mbuti and the forest. In one instance, he came upon a youth dancing and singing by himself in the forest under the moonlight: "He was adorned with a forest flower in his hair and with forest leaves in his belt of vines and his loin cloth of forest bark. Alone with his inner world he danced and sang in evident ecstasy" (1983:32). When questioned as to why he was dancing alone, he answered, "'I am not dancing alone, I am dancing with the forest'" (1965:253).

The Mbuti, a hunting and gathering people, exemplify Durkheim's concept of mechanical solidarity. The core of their "Common conscience" is a value system centered on the forest, which provides them with all the necessities of life.

In a second instance, Turnbull asked a Mbuti pygmy whether he would like to see a part of the world outside the forest. The pygmy hesitated a long time before asking how far they would go beyond the forest. Not more than a day's drive from the last of the trees, replied Turnbull, to which the Mbuti pygmy responded with disbelief, "No trees? No trees at all?" He was highly disturbed about this and asked whether it was a good country. From the Mbuti perspective, people living without trees must be very bad to deserve that punishment. In the end, he agreed to go if they took enough food to last them until they returned to the forest. "He was going to have nothing to do with 'savages' who lived in a land without trees" (1961:248).

Finally, Turnbull notes that the pygmies are aware of the ongoing destruction of the rain forest by companies that push them farther into the forest's interior. By consensus, the pygmies do not wish to leave the forest and become part of the modern world: "The forest is our home; when we leave the forest, or when the forest dies, we shall die. We are the people of the forest" (1961:260).

Organic Solidarity

A society with a complex division of labor is characterized by **organic solidarity** — social order based on interdependence and cooperation among people performing a wide range of diverse and specialized tasks. A complex division of labor increases differences among people, and common conscience in turn decreases. Yet, Durkheim argues that the ties that bind people to one another can be very strong nonetheless. In societies characterized by a complex division of labor, these ties are no longer based on similarity and common conscience but on differences and interdependence. When the division of labor is complex and when the materials for products are geographically scattered, few individuals possess the knowledge, skills, and materials to permit self-sufficiency. Consequently, people find themselves dependent on others. Social ties are strong because people need each other to survive.

Specialization and interdependence mean that every individual contributes a small part in creating a product or delivering a service. Because of specialization, relationships among people take on a transitory, limited, impersonal, and abstract character. We relate to one another in terms of our specialized roles. We buy tires from a dealer; we interact with a sales clerk by telephone, computer, and fax; we fly from city to city in a matter of hours and are served by flight attendants; we pay a supermarket cashier for coffee; and we deal with a lab technician for only a few minutes when we give blood. We do not need to know these people personally to interact with them. Nor do we need to know that the rubber in the tires, the cobalt from which the jet engine is built, and the coffee we purchase come from Zaire. Similarly, we do not need to know whether the blood we give is kept in the United States or exported elsewhere.

When we interact in this manner, we can ignore personal differences and treat those who perform the same tasks as interchangeable. Yet, members of society "are united by ties which extend deeper and far beyond the short moments during which the exchange is made. Each of the functions that they exercise is, in a fixed way, dependent upon others. . . . [W]e are involved in a complex of obligations from which we have no right to free ourselves" (Durkheim [1933] 1964:227). In other words, because everyone is dependent on everyone else, each individual has a stake in preserving the system.

A curious feature of organic solidarity is that although people live in a state of interdependence with others, they maintain little awareness of it, possibly because of the fleeting and impersonal nature of the relationships. Because the ties with most of the people with whom we come in contact during a day are instrumental (we interact with them for a specific reason) rather than emotional, we seem to live independently of one another.

Durkheim hypothesized that societies become more vulnerable as the division of labor becomes more complex and more specialized. He was particularly concerned with the kinds of events that break down individuals' ability to connect with one another in meaningful ways through their labor, a process we take for granted until something disrupts those connections. Such events include (1) industrial and commercial crises caused by such occurrences as plant closings, massive layoffs, crop failures, technological revolutions, and war; (2) workers' strikes; (3) job specialization, insofar as workers are so isolated that few people grasp the workings and consequences of the overall enterprise; (4) forced division of labor to such an extent that occupations are based on inherited traits (race, sex) rather than on ability, in which case the "lower" groups aspire to the positions that are closed to them and seek to dispossess those who occupy such positions; and (5) inefficient management and development of workers' talents and abilities so that work for them is nonexistent, irregular, intermittent, or subject to high turnover. For example, a country might not develop enough workers (teachers, scientists, nurses) or too many workers (athletes and entertainers) for available positions, or it might fail to retrain people whose positions are vulnerable to layoff or obsolescence. These events are particularly disruptive when they arise suddenly.

Organic solidarity Social order based on interdependence and cooperation among people performing a wide range of diverse and specialized tasks.

Sociologist Emile Durkheim was particularly concerned about the kind of events that break down the ability of individuals to connect with one another in meaningful ways through their labor. Those events included war and massive layoffs. The war in Rwanda resulted in millions of people fleeing to Goma, Zaire. Massive layoffs in the United States increase the numbers of people in unemployment lines.

©B. Press/Panos Pictures (left); ©R. Crandall/The Image Works (right)

In his book *The Reckoning,* David Halberstam (1986) profiles the life of one man—Joel Goddard—after he is laid off from Ford. Goddard is married and has two children. Halberstam chronicles the changes in Goddard's life that affect his ability to connect in meaningful ways with others. For example, after Goddard loses his job, his daily contacts shift from colleagues at work to contacts with those at the unemployment office. Moreover, Goddard loses the structure to his life that comes with the routine of his job. Instead, he watches TV, fishes, or reads want ads. His ties are further disrupted when some of his friends from work move to Texas to find employment. Goddard eventually takes a job selling insurance but finds himself selling to his acquaintances. Eventually he quits, and then his wife decides to go to work. However, her success at work strains their marriage as Joel is reminded of his failures.

Zaire in Transition

From 1883 to the present, at least one of the five disruptive situations that Durkheim postulated has existed in Zaire. A brief summary of Zaire's history in the past 100 years shows the extent to which these disruptions have created a social order that severs the connections that people have to one another.

Belgian Imperialism (1883–1960)

Before 1883, inhabitants of what is now Zaire lived in villages characterized by common conscience and a simple division of labor. In 1883, however, King Leopold II of Belgium claimed the land as his private property, and millions of people were forced from their villages to work the land and mine raw materials (forced division of labor). Leopold's personal hold over the land was formally legitimized in 1885 by leaders of fourteen European countries attending the Berlin West Africa Conference. The purpose of this conference was to carve Africa into colonies. The continent was divided without regard to preexisting national boundaries, so that friendly nations were split apart and hostile nations were thrown together (see Figure 6.1).

For twenty-three years, Leopold capitalized on the world's growing demand for rubber. His reign over Zaire was the "vilest scramble for loot that ever disfigured the history of human conscience and geographical location" (Conrad 1971:118). The methods he used to extract rubber for his own personal gain involved atrocities so ghastly that in 1908 international outrage forced the Belgian government to assume administration of the Belgian Congo (see "The Essay That Mark Twain Could Not Get Published").

The Belgian government operated more humanely than Leopold, but it too forced the indigenous peoples to build roads so that minerals and crops could be transported from mines and fields across the country for export. Africans were forced to leave their villages to work the mines, cultivate and harvest the crops, and live alongside the roads and maintain them. The Belgians introduced a cash economy, imported goods from Europe that eventually became essential to native life, established a government, and built schools, hospitals, and roads. Under this system, the African people

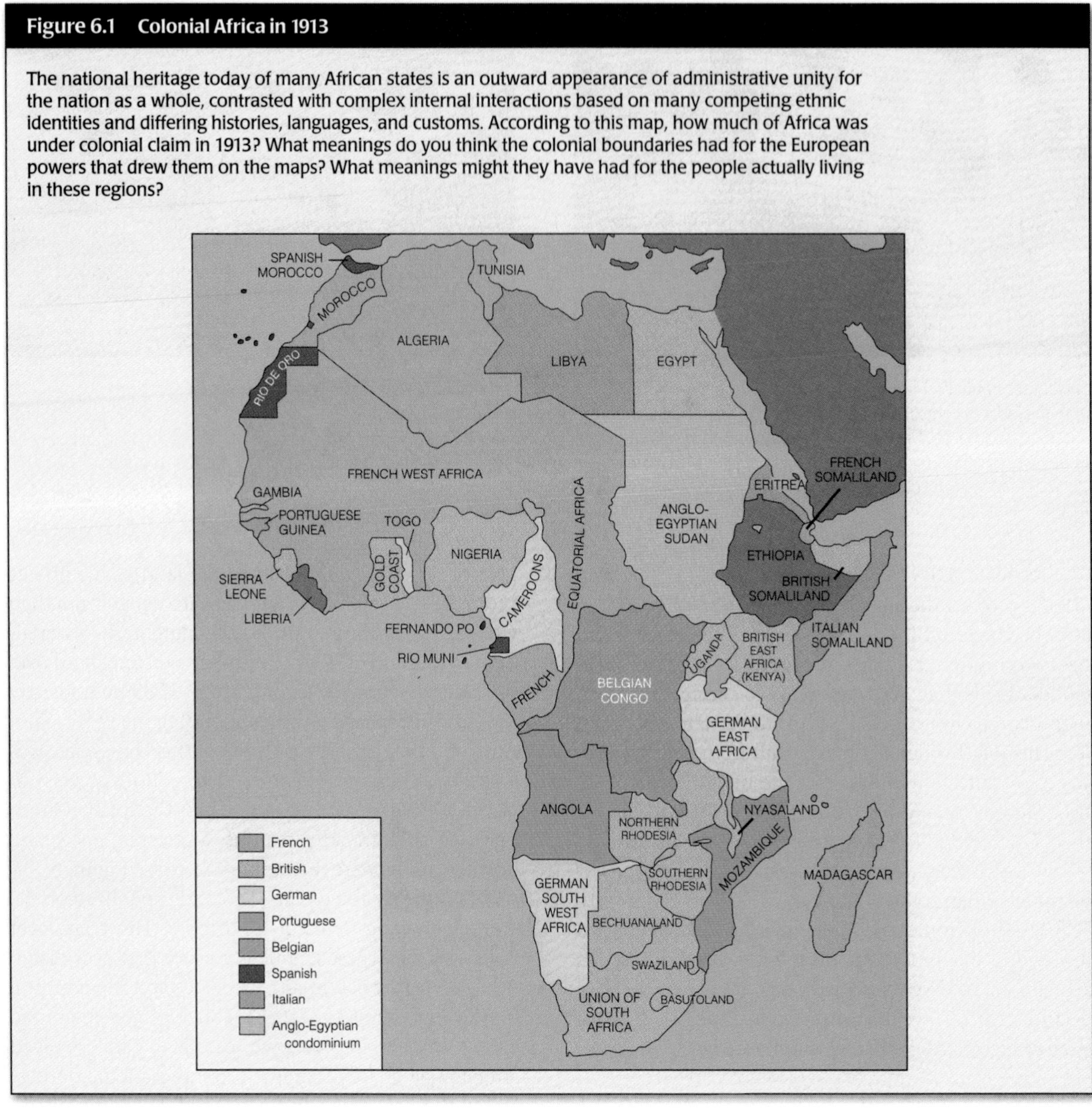

Figure 6.1 Colonial Africa in 1913

The national heritage today of many African states is an outward appearance of administrative unity for the nation as a whole, contrasted with complex internal interactions based on many competing ethnic identities and differing histories, languages, and customs. According to this map, how much of Africa was under colonial claim in 1913? What meanings do you think the colonial boundaries had for the European powers that drew them on the maps? What meanings might they have had for the people actually living in these regions?

Source: *The Times Atlas of World History* (1984).

could acquire cash in one of two ways: growing cash crops or selling their labor. They were no longer allowed to be self-sufficient, as they were before European colonization. In addition, under European colonization the African people were denied access to most professional-level and high-skilled occupations (forced division of labor; inefficient management and development of worker's talents).

The introduction of European goods and a cash economy pulled the people who inhabited the Belgian Congo into a worldwide division of labor and created a migrant labor system within the country. Since colonization, people have moved continuously from the villages to the cities, mining camps, and plantations (Watson 1970). In addition, family members have endured prolonged separations as a result of the migrant labor system.

> The migrant labor system affected Africans' lives in many fundamental ways. . . . [M]ale workers were

U.S. In Perspective

The Essay That Mark Twain Could Not Get Published

Most people associate the name Mark Twain with the novels of *Huckleberry Finn* and *Tom Sawyer.* They do not think of him as an avid critic of American and European imperialism who wrote essays and pamphlets expressing his outrage at the "Great Powers for the way they exercised their 'unwilling' missions in South Africa, China, and the Philippines" (Meltzer 1960:256). When the Congo Reform Association approached Twain in early 1905 to "lend his voice 'for the cause of the Congo natives'" (p. 257), Twain responded by writing "King Leopold's Soliloquy." In the essay, he presents a report filed by the Reverend H. E. Scrivener, a British missionary, on the plight of the people of the Belgian Congo and under King Leopold's rule. Since no magazine editor in the United States would agree to publish this essay, Congo reform groups issued it as a pamphlet and sold it for 25 cents, with proceeds going toward the relief of the Congo people. The following is excerpted from that report.

> *Soon we began talking, and without any encouragement on my part the natives began the tales I had become so accustomed to. They were living in peace and quietness when the white men came in from the lake with all sorts of requests to do this and that, and they thought it meant slavery. So they attempted to keep the white men out of their country but without avail. The rifles were too much for them. So they submitted and made up their minds to do the best they could under the altered circumstances.*
>
> *First came the command to build houses for the soldiers, and this was done without a murmur. Then they had to feed the soldiers and all the men and women—hangers on—who accompanied them. Then they were told to bring in rubber. This was quite a new thing for them to do. There was rubber in the forest several days away from their home, but that it was worth anything was news to them. A small reward was offered and a rush was made for the rubber. "What strange white men, to give us cloth and beads for the sap of a wild vine." They rejoiced in what they thought their good fortune. But soon the reward was reduced until at last they were told to bring in the rubber for nothing. To this they tried to demur; but to their great surprise several were shot by the soldiers, and the rest were told, with many curses and blows, to go at once or more would be killed. Terrified, they began to prepare their food for the fortnight's absence from the village which the collection of rubber entailed. The soldiers discovered them sitting about. "What, not gone yet?" Bang! bang! bang! and down fell one and another, dead, in the midst of wives and companions. There is a terrible wail and an attempt made to prepare the dead for burial, but this is not allowed. All must go at once to the forest. Without food? Yes, without food. And off the poor wretches had to go without even their tinder boxes to make fires. Many died in the forests of hunger and exposure, and still more from the rifles of the ferocious soldiers in charge of the post. In spite of all their efforts the amount fell off and more and more were killed. I was shown around the place, and the sites of former big chiefs' settlements were pointed out. A careful estimate made the population of, say, seven years ago, to be 2,000 people in and about the post, within a radius of, say, a quarter of a mile. All told, they would not muster 200 now, and there is so much sadness and gloom about them that they are fast decreasing.*
>
> *We stayed there all day on Monday and had many talks with the people. On the Sunday some of the boys had told me of some bones which they had seen, so on the Monday I asked to be shown these bones. Lying about on the grass, within a few yards of the house I was occupying, were numbers of human skulls, bones, in some cases complete skeletons. I counted thirty-six skulls, and saw many sets of bones from which the skulls were missing. I called one of the men and asked the meaning of it. "When the rubber palaver began," said he, "the soldiers shot so many we grew tired of burying, and very often we were not allowed to bury; and so just dragged the bodies out into the grass and left them. There are hundreds all around if you would like to see them." But I had seen more than enough, and was sickened by the stories that came from men and women alike of the awful time they had passed through. The Bulgarian atrocities might be considered as mildness itself when compared with what was done here. How the people submitted I don't know, and even now I wonder as I think of their patience. That some of them managed to run away is some cause for thankfulness. I stayed there two days and the one thing that impressed itself upon me was the collection of rubber. I saw long files of men come in, as at Bongo, with their little baskets under their arms; saw them paid their milk tin full of salt, and the two yards of calico flung to the headmen; saw their trembling timidity, and in fact a great deal that all went to prove the state of terrorism that exists and the virtual slavery in which the people are held.*

Source: From "King Leopold's Soliloquy on the Belgian Congo," in *Mark Twain and the Three R's,* by Mark Twain, pp. 47–48.

typically recruited from designated labour supply areas great distances from the centers of economic activity. This entailed prolonged family separations which had serious physical and psychological repercussions for all concerned. The populations of African towns "recruited by migration" were characterized by a heavy preponderance of men living in intolerably insecure and depressing conditions and lacking the benefits of family life or other customary supports. (Doyal 1981:114)

The Belgians did not anticipate the Africans' anger about the exploitation of their land, minerals, and people and were not prepared for the violent confrontations that took place in the late 1950s. Those in power termed the revolutions "savage" and claimed that the Africans did not appreciate the "benefits" of colonialism. When the Belgians pulled out suddenly in 1960, the Belgian Congo became an independent country without trained military officers, businesspeople, teachers, doctors, or civil servants. In fact, there were only 120 medical doctors in a country of 33 million people (Fox 1988).

Independence of Zaire (1960–present)

In the vacuum left by the Belgians, the various African ethnic groups that had been forced together to form the Belgian Congo now fought one another to obtain power (industrial and commercial crises). Civil wars raged until 1965, when Sese Mobutu took power with the help of the United States. Since that time, several power struggles have occurred between various nations within Zaire, especially in the later 1970s. To stop the rebellions, Mobutu called on mercenary forces from Morocco, Belgium, and France. To make up for the lack of skilled personnel, Mobutu invited French, Danes, Haitians, Portuguese, Greeks, Arabs, Lebanese, Pakistanis, and Indians to work as civil servants, teachers, doctors, traders, businesspeople, and researchers. It was through this invitation that Dr. Rask arrived in Zaire.

In addition to problems posed by civil war, several other major problems have made life difficult for the people of Zaire since independence. First, many cash crops were priced out of competition in a growing world economy, and a technological revolution in synthetic products reduced the demand for African raw materials (industrial and commercial crises). Second, civil wars raging in neighboring countries have caused hundreds of thousands of refugees from Sudan, Rwanda, Angola, Uganda, Congo, and Burundi to flee to Zaire. Meanwhile, Zaireans suffering from their own civil wars and economic problems sought refuge in those same countries (Brooke 1988a; U.S. Bureau for Refugee Programs 1988; U.S. Central Intelligence Agency 1995). Finally, Mobutu has diverted much of Zaire's wealth to European banks and has invested it in property outside Africa (Kramer 1993; Brooke 1988b). Some people estimate that Mobutu's personal fortune is worth as much as $5 billion (Kramer 1993).[2]

[2]Although Mobutu's latest term in office ended in 1991, presidential elections were never held. On May 16, 1997, Mobutu was finally forced out of office.

The local populations have suffered greatly and are still suffering from the ongoing massive upheaval in Zaire. These events disrupted the division of labor, and people lost an important social connection to one another. Such "change of existence, whether it be sudden or prepared, always brings forth a painful crisis" (Durkheim [1933] 1964:241). Out of economic necessity, a desire for a higher standard of living, and a need to escape war, many people have left their villages for the cities and for industrial, plantation, and mining sites. Women, children, and the elderly left behind in the villages have had little choice but to change to higher-yield and less labor-intensive crops to survive. Unfortunately, the new crops, such as cassava, are low in protein and high in carbohydrates. (Low-protein diets compromise the human immune system, making people more vulnerable to infection.) To further complicate matters, significant numbers of single women with no means of supporting themselves in the villages and rural areas have migrated to labor sites in search of employment. Because there are few employment opportunities for women at the labor sites, many are forced to survive through prostitution (inefficient management and development of workers' talent). In the meantime, when the men and women who had migrated out of the villages to find employment became sick and could no longer work, they returned home to their villages and infected an already vulnerable population with whatever diseases they carried (Hunt 1989).

The magnitude of these migrations is reflected in the population increase of Kinshasa, the capital city, which grew from 390,000 in 1950 to 5.0 million in 1993. A large portion of its population (almost 60 percent) lives in squatter slums, the largest of which is named the Cite. Here the "streets [are] stuffed with children and families living under cardboard roofs held down by rocks. Kinshasa [is] a wasteland of flooded streets and cracked sidewalks, smoldering garbage and bars catering to whores and lonely white men. There [is] an end-of-civilization atmosphere, with survivors finding shelter in the rubble" (Clarke 1988:175, 178).

As a result of this upheaval and mismanagement, Zaire fell from its position as one of the wealthiest colonies in Africa to become the poorest independent country and one of the twelve poorest countries in the world. Its gross national product per capita is an estimated $440 (U.S. Central Intelligence Agency 1995). As another indicator of how dire the situation has become, in 1960 Zaire had 90,000 miles of passable highway; by 1988 that number had dwindled to only 6,000 miles. According to the U.S. Central Intelligence

Between 1950 and 1993 Kinshasa, the capital of Zaire, grew from a city of 390,000 people to 5.0 million.

©Marc Schlossman/Panos Pictures

Agency (1995), there are 1,738 miles of paved roads in Zaire today. Obviously this impedes the transportation of medical supplies, food, and fertilizer to villages that have become dependent on these commodities (Noble 1992).

These historical events are the contextual forces that brought Dr. Margrethe Rask to a small village clinic in Zaire. Amid such chaos and poverty, Dr. Rask had to perform operations on less than a shoestring budget, with only minimal supplies. "Even a favored clinic would never have such basics as sterile rubber gloves or disposable needles. You just used needles again and again until they wore out; once gloves had worn through you risked dipping your hands in your patient's blood because that was what needed to be done" (Shilts 1987:4).

The importance of considering Zaire's history and global connections to understand the origins of AIDS is supported by the fact that disease patterns historically are affected by changes in population density and transportation patterns, both of which bring together previously isolated groups (McNeill 1976). Leading AIDS researchers believe that the transmission of HIV is indeed linked to changes in population density and transportation. Interestingly, "the medical condition which was later to be called AIDS began to be noticed in the late 1970s and early 1980s in several widely separated locations, including Belgium, France, Haiti, the United States, Zaire, and Zambia" (The Panos Institute 1989:72). Before this time, the virus may have survived in a dormant state in an isolated population with a tolerance to the virus but may have been activated when this population came into contact with another population with no tolerance. Or two harmless retroviruses, each existing in a previously isolated population, may have interacted to produce a third, lethal virus. The point is that

> a major dislocation in the social structure — love, hate, peace, war, urbanization, overpopulation, economic depression, people having so much leisure [or having no alternative source of income] they sleep with five different people a night—whatever it is that puts a stress on the ecological system, can alter the equilibrium between [people] and microbes. Such great dislocations can lead to plagues and epidemics. (Krause 1993:xii)

In view of this information about the global context of interaction, it is difficult to say who is responsible for triggering and transmitting the virus that causes AIDS. Clearly, "the foreigners introduced hitherto unknown diseases and probably aggravated some previously endemic diseases to epidemic proportions by the facilitation of transportation, forced migration of rural populations to work sites, and the creation of congested cities" (Lasker 1977:280). For example, the destruction of the Ituri rain forest in Zaire by companies from around the world brings developers into contact with previously isolated populations (such as the Mbuti) and forces many people to migrate to the city because they have lost their homes. In addition to disrupting people's lives, the commercial activities in the rain forest cause climatic changes (such as the greenhouse effect) that can alter the structure of viruses. What is important is not to determine who started the transmission (because that is impossible to ascertain) but to recognize the extent to which the world's people are interacting with one another and to become aware that the actions of one group can affect other groups.

Placing Dr. Rask's interactions in this global context helps us see how historical events bring people into interaction with one another. The opportunity for Dr. Rask to go to Zaire and practice medicine, the cir-

Zaire is considered to be one of the 12 poorest countries in the world. The poverty is exacerbated by a transportation system incapable of delivering medical supplies and goods to those in need.

©Robert Caputo/Aurora

cumstances that placed her in direct contact with patients' blood, and her subsequent illness arose from a unique sequence of historical events.

The point is that somewhere in this unprecedented mixing of people from all over the world are the conditions that facilitated both the activation and the transmission of HIV to an estimated 16.9 million adult cases of HIV worldwide (World Health Organization 1995). The importing and exporting of blood (which brings people into contact with one another in indirect ways), the development of wide-body jet aircraft, and large-scale migrations are cited as events that increased opportunities for large numbers of people from different countries and from different regions of the same country to interact with one another and to transmit the HIV infection through unprotected sexual intercourse, needle sharing, and other activities that involve blood and blood products (De Cock and McCormick 1988).

As one indicator of the amount of "indirect" interaction between blood donors and receivers, consider that the American Red Cross, the largest blood supplier in the United States, collected 5.7 million blood donations in 1994, the last year for which data are available. The Red Cross also maintains "the world's largest registry of rare blood donors and maintains a frozen supply of rare blood available for immediate shipment across the globe" (American National Red Cross 1996). In addition, the Red Cross provides the human tissue used in approximately 25 percent of transplantation surgeries.

Besides the context, the content of social interaction is important. When people interact, they identify the social status of the people they interact with. Once they determine another person's status in relation to their own, people proceed to interact on the basis of role expectations.

The Content of Social Interaction

As the division of labor has become more specialized and the sources of labor and raw materials have become more geographically diverse, the ties that connect people to one another have shifted in character from mechanical to organic, and the number of interactions people have with strangers has increased. How can people interact smoothly with people about whom they know nothing? They eliminate "strangeness" by identifying the social positions or social status of the strangers with whom they interact. Knowing a person's social status gives us some idea of the behaviors we can expect from someone in that status. It also affects how we will interact with that person.

To grasp this principle, think about when you meet someone for the first time. How do you start the interaction? You ask the stranger a question to determine his or her social status. You might ask, "What do you do?" That question sets the interaction into motion.

Social Status

In everyday language, people use the term *social status* to mean rank or prestige. To sociologists, **social status** refers to a position in a social structure. A **social structure** consists of two or more people interacting and interrelating in specific expected ways, regardless of the unique personalities involved. For example, a social structure can consist of the two statuses of doctor and patient and their relationship. Other familiar examples of statuses in a two-person social structure are husband and wife, professor and student, sister and brother, and employer and employee.

Examples of multiple-status social structures are a family, an athletic team, a school, a large corporation, and a government. Again, the common characteristic of all social structures is that it is possible to generalize about the behavior of people in each of the statuses, no matter who occupies them. Just as the behavior of a person occupying the status of a football quarterback is broadly predictable, so too is the behavior of a person occupying the status of nurse, secretary, mechanic, patient, or physician. Once we know a person's status, we have enough information to interact with the person.

Types of Status Statuses can be of two kinds—ascribed or achieved. An **ascribed status** is a position that people acquire through no fault or virtue of their own. Examples include sex, age, ethnic, and health statuses—for example male, female, African American, Native American, adolescent, senior citizen, retired, son, daughter, and disabled. An **achieved status** is a position earned (or lost) by a person's own actions or abilities. Occupational, educational, parental, and marital statuses include athlete, senator, secretary, physician, high school dropout, divorcé, and single parent.

The distinction between ascribed and achieved statuses is not always clear-cut. Often achieved statuses such as financial, occupational, and educational positions are related to ascribed statuses such as sex, ethnicity, and age. For example, physicians in the United States are disproportionately male (80 percent) and white (92 percent), whereas registered nurses are overwhelmingly female (86.2 percent) and white (90 percent) (U.S. Bureau of the Census 1992b).

Stigmas Some ascribed and achieved statuses are such that they overshadow all other statuses that a person occupies. Sociologist Erving Goffman calls such statuses **stigmas** and classifies them into three broad varieties: (1) physical deformities; (2) character blemishes due to factors such as sexual orientation, mental hospitalization, or imprisonment; and (3) stigmas of ethnicity, nationality, or religion. When a person possesses a stigma, he or she is reduced in the eyes of others from a multifaceted person to a person with one tainted status. To illustrate this point, Goffman opens *Stigma: Notes on the Management of Spoiled Identity* (1963) with a letter written by a sixteen-year-old girl born without a nose. Although she is a good student, has a good figure, and is a good dancer, no one she meets can get past the fact that she has no nose.

Every person occupies a number of statuses. For example, Dr. Rask was middle-aged, a female, a physician, a patient, and Danish. A status has meaning, however, only in relation to other statuses. For instance, the status of a physician takes on quite different meanings depending on whether the interaction is with someone who occupies the same status or a different status such as patient, spouse, or nurse. This is because a physician's role varies according to the status of the person with whom he or she interacts.

Social Roles

Sociologists use the term **role** to describe the behavior expected of a status in relationship to another status (for example, professor to student). The distinction between role and status is subtle but noteworthy: people *occupy* statuses and *enact* roles.

Social status A position in a social structure.

Social structure Two or more people interacting and interrelating in specific expected ways, regardless of the unique personalities involved.

Ascribed status A position that people acquire through no fault or virtue of their own.

Achieved status A position earned (or lost) by a person's own actions or abilities.

Stigmas Deeply discrediting statuses in that they overshadow all other statuses a person occupies.

Role The behavior expected of a status in relationship to another status.

Members of ACT-UP, a group formed to protest aggressively what its members see as a slow and insufficient response by public officials to the AIDS problem. They believe that policy makers would act more decisively if AIDS were not seen as primarily a problem for those in unpopular or socially powerless groups.

©Brooks Kraft/Sygma

Associated with every status is a **role set**, or an array of roles. For example, the status of physician entails, among other roles, the role of physician in relationship to patient, nurse, other doctors, and a patient's family members. The sociological significance of statuses and roles is that they make it possible for us to interact with other people without knowing them. Once we determine another person's status in relation to our own, we interact on the basis of role expectations attached to that status relationship.

Role expectations are socially prescribed and include both rights and obligations. The **rights** associated with a role define what a person assuming that role can demand or expect from others depending on his or her status. For example, teachers have the right to demand and expect that students will come prepared for class. The **obligations** associated with a role define the appropriate relationship and behavior that the person enacting that role must assume toward others occupying a particular status. For instance, teachers have an obligation to their students to come to class prepared.

One of the best-known descriptions of a role and its accompanying rights and obligations was given by sociologist Talcott Parsons (1975). According to Parsons, we assume a **sick role** when we are sick. Ideally, sick persons have an obligation to try to get well, seek technically competent help, and cooperate with a treatment plan. Sick persons also have certain rights: they are exempt from "normal" social obligations and are not held responsible for their illness.

Role set An array of roles.

Rights The behaviors that a person assuming a role can demand or expect from others.

Obligations The relationship and behavior that the person enacting a role must assume toward others in a particular status.

Sick role A term coined by sociologist Talcott Parsons to represent the rights and obligations accorded people when they are sick.

Social Roles and Individual Behavior Roles set general limits on how we think and act but do not imply that behavior is totally predictable. Sometimes people do not meet their role obligations, as when professors come to class unprepared or when patients do not cooperate with their physicians' treatment plans or when physicians blame their patients for getting sick (for example, thinking they have AIDS because they engaged in promiscuous and perverse sexual activity, or thinking they have lung cancer because they do not have the will to quit smoking). By definition, when people fail to meet their role obligations, other people are not accorded their role rights. When professors are unprepared, students' rights are violated; when patients do not follow treatment plans, physicians' rights are violated; when physicians blame their patients, patients' rights are violated. Moreover, the idea of role does not imply that all people occupying the same status enact the roles of that status in exactly the same way. Roles are enacted differently because of individual personalities and interpretations of how the role should be carried out. Finally, roles are enacted differently because people resolve role strain and role conflict differently.

Role strain is a predicament in which contradictory or conflicting expectations are associated with the role that a person is occupying. For example, military doctors, as physicians, have an obligation to preserve life. At the same time, they are employed to care for patients who have been placed deliberately in situations that threaten their health and lives. **Role conflict** is a predicament in which the expectations associated with two or more roles in a role set are contradictory. For example, a sick person has an obligation to want to get better and comply with treatment plans. This obligation, however, can interfere with other roles that the person holds, as in the case of a woman who finds that the side effects of a prescribed drug prevent her from being alert enough to work or care for her children.

Socially prescribed rights and obligations notwithstanding, there is always room to exercise improvisation and personal style. Yet, despite variations in how people enact roles, role is still a useful concept because, for the most part, a predictability generally exists, "sufficient to enable most of the people, most of the time to go about [the] business of social life without having to improvise judgments anew in each newly confronted situation" (Merton 1957:370). In other words, the variations usually fall within "a certain range of culturally acceptable behavior—if the performance of a role deviates very much from the expected range of behavior, the individual will be negatively sanctioned" (p. 370).

[3]In Zaire, the chief "physicians" are traditional healers. Only a small proportion are like the witch doctors of legend. (In fact, many wear business suits.) Traditional healers operate from a variety of beliefs and practices, but most approach health care from a holistic perspective.

Cultural Variations in Roles: The Patient-Physician Interaction The behaviors expected of one status in relation to another status vary across cultures. Role expectations are intertwined with norms, values, beliefs, and nonmaterial culture. In the United States, the major objective of the patient-physician interaction is to determine the exact physiological malfunction and use the available material culture and technology (tests, equipment, machines, drugs, surgery, transfusions) to treat it.

This objective is shaped by a profound cultural belief in the ability of science to solve problems. Practitioners of Western medicine are expected to use all of the tools of science at hand to establish the cause, combat the disease for as long as possible, and return the body to a healthy state. When physicians violate this norm, the violation is almost always accompanied by intense public debate.

In view of this cultural orientation, it is not surprising that in the United States physicians and their patients rely heavily on technological elements such as X-rays and CAT scans to diagnose the condition and on vaccines, drugs, and surgery to cure it. This reliance is reflected in the fact that the United States, with a population of 248.7 million (according to the 1990 census)—5 percent of the world's population—consumes an estimated 23 percent of the world's pharmaceutical supply (Peretz 1984). Given this emphasis, it is not surprising that tremendous effort is devoted to finding a technological solution to the AIDS problem.

We can contrast the U.S. physician-patient relationship with the traditional African healer-patient relationship. Although Zaire has modern health care facilities, the majority of people go to traditional healers.[3] The social interaction between the healer and patient is very different from the U.S. physician-patient social interaction.

Just as Western physicians do, traditional healers recognize the organic and physical aspects of disease. But they also attach considerable importance to other factors — supernatural causes, social relationships (hostilities, stress, family strain), and psychological

Role strain A predicament in which contradictory or conflicting expectations are associated with a person's role.

Role conflict A predicament in which the expectations associated with two or more roles in a role set are contradictory.

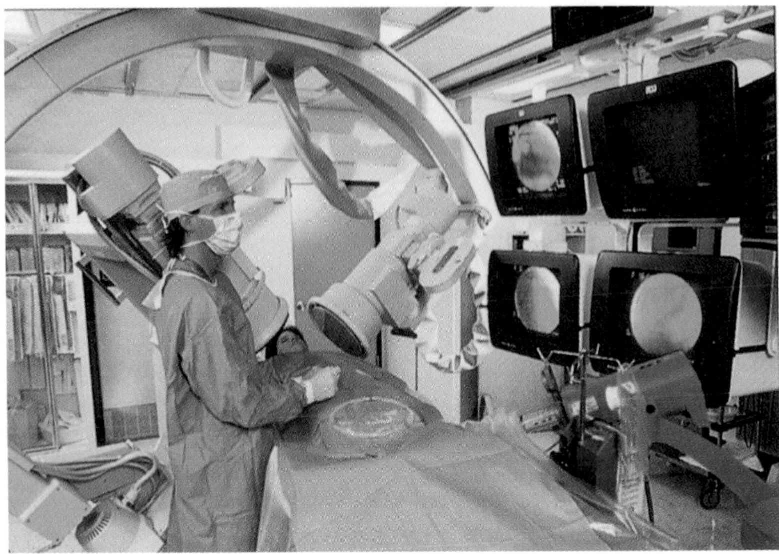

The Western approach to healing is rooted in cultural beliefs about the nature of disease and the power of science and technology to combat it.

©Bob Daemmrich/Stock, Boston

distress. This holistic perspective allows for a more personal relationship between healer and patient. Another significant difference from Western medicine is that healers concentrate on providing symptom relief instead of searching for a total cure. African healers, for example, focus on treating the debilitating symptoms of AIDS such as diarrhea, headaches, fevers, and weight loss, and they employ remedies with few side effects so as not to make people sicker (Hilts 1988). When traditional methods fail, Africans may make a trip to a hospital or clinic to consult with a doctor of Western medicine. Obviously, each system has its attractions and its weaknesses, as evidenced by the fact that Westerners suffering from incurable diseases sometimes turn to alternative treatments and medicines.[4] In fact, the U.S. National Institutes of Health named the Seattle Bastyr University Center as an alternative medicine center which will evaluate alternative treatments for HIV/AIDS.[5]

[4]Many Western doctors are showing considerable interest in traditional African medicine. There seems to be a movement toward incorporating elements of both scientific and traditional medicine into health care and learning more about the curative properties of the herbs and remedies used by healers before the plants that supply them become extinct (because of the destruction of the rain forests). Apparently, traditional medicines work faster and better for some diseases (such as hepatitis) than prescribed pharmaceuticals do (Lamb 1987).

[5]In 1992, Congress created the Office of Alternative Medicine to encourage research investigating and evaluating unorthodox medical ideas (Global Childnet 1995).

Sociologist Ruth Kornfield observed Western-trained physicians working in Zaire's urban hospitals and found that success in treating patients was linked to the foreign physician's ability to tolerate and respect other models of illness and include them in a treatment plan. For example, among some Zairean ethnic groups, when a person becomes ill, the patient's kin form a therapy management group and make decisions about administering treatments. Because many people in Zaire believe that illnesses are caused by disturbances in social relationships, the cure must involve a "reorganization of the social relations of the sick person that [is] satisfactory for those involved" (Kornfield 1986:369; also see Kaptchuk and Croucher 1986: 106–108).

The point of this discussion is not to evaluate the quality or outcome of either culture's patient-practitioner interaction. Rather, by comparing the two cultures, we can see more clearly that people think and behave in largely automatic ways because they are influenced by norms, values, beliefs, and nonmaterial culture. Those involved with Dr. Rask automatically assumed a scientific framework to define the origin and treatment of her condition. Such a conceptual framework defines illness as "a state of disease and dysfunction 'impersonally' caused by microorganisms, inborn metabolic disturbances, or physical or psychic stress" (Fox 1988:505). On this basis, we can begin to understand why Dr. Rask, other physicians, and her close friends all defined her illness as a condition contracted in Africa, requiring hospitalization, having biological origins, and related to direct contact with a patient's blood. The interactions and dialogue among Dr. Rask, other medical personnel, and friends would have been quite different if they had assumed a "nonscientific" model of illness.

The Dramaturgical Model of Social Interaction

A number of sociologists have compared roles attached to statuses with dramatic roles played by actors. Erving Goffman is a sociologist associated with the **dramaturgical model** of social interaction. In this model, social interaction is viewed as though it were theater, people as though they were actors, and roles as though they were performances presented before an audience in a particular setting. People in social situations resemble actors in that they must be convincing to others and must demonstrate who they are and what they want through verbal and nonverbal cues. In social situations, as on a stage, people manage the setting, their dress, their words, and their gestures to correspond to the impression they are trying to make or the image they are trying to project. This process is called **impression management**.

Impression Management

On the surface, the process of impression management may strike us as manipulative and deceitful. Most of the time, however, people are not even aware that they are engaged in impression management because they are simply behaving in ways they regard as natural. Women engage in impression management when they remove hair from their faces, legs, armpits, and other areas of their bodies and present themselves as hairless in these areas. From Goffman's perspective, even if people are aware that they are engaged in impression management, it can be both a constructive and a normal feature of social interaction because smooth interactions depend on everyone's behaving in socially expected and appropriate ways. If people spoke and behaved entirely as they pleased, civilization would break down. Goffman (1959) also recognized the dark side of impression management that occurs when people manipulate their audience in deliberately deceitful and hurtful ways.

Impression management often presents us with a dilemma. If we do not conceal inappropriate and unexpected thoughts and behavior, we risk offending or losing our audience. Yet, if we conceal our true reactions, we may feel that we are being deceitful, insincere, or dishonest or that we are "selling out." According to Goffman, in most social interactions the people involved weigh the costs of losing their audience against the costs of losing their integrity. If keeping our audience is important, concealment is necessary; if showing our true reactions is important, we may risk losing the audience.

In the United States, people who test positive for HIV antibodies face the dilemma of impression management. If they disclose the test results, they risk discrimination and the loss of their jobs, insurance coverage, friends, and family. If they keep this information to themselves, they may feel that they are being untrue to themselves and others who care about them. Rarely are interaction situations "either/or." Usually people compromise between the two extremes.

The dramaturgical model and the idea of impression management are useful concepts for understanding the dilemma that one partner may face if he or she suggests using a condom as a precautionary condition of sexual intercourse. This dilemma is quite different in the United States than in Zaire. In both countries, health officials recommend condom use as a way to reduce substantially the risk of HIV infection. Risk of infection extends to new sexual partners, members of high-risk groups, and even currently monogamous partners if they were sexually active, used intravenous drugs, or received a blood transfusion in the past ten years. (At the time of this writing, the interval between HIV infection and the onset of AIDS is known to be as long as ten years. In essence, an individual is going to bed not only with the other person but also with that person's past sexual partners.)

In both Zaire and the United States, the subject of condom use is a sensitive one because of the message that the condom conveys to potential sexual partners. Condom use in Zaire is associated with birth control. The taboo against birth control is strong because many

Dramaturgical model A model in which interaction is viewed as though it were theater, people as though they were actors, and roles as though they were performances presented before an audience in a particular setting.

Impression management The process by which people in social situations manage the setting and their dress, words, and gestures to correspond to the impressions they are trying to make or the image they are trying to project.

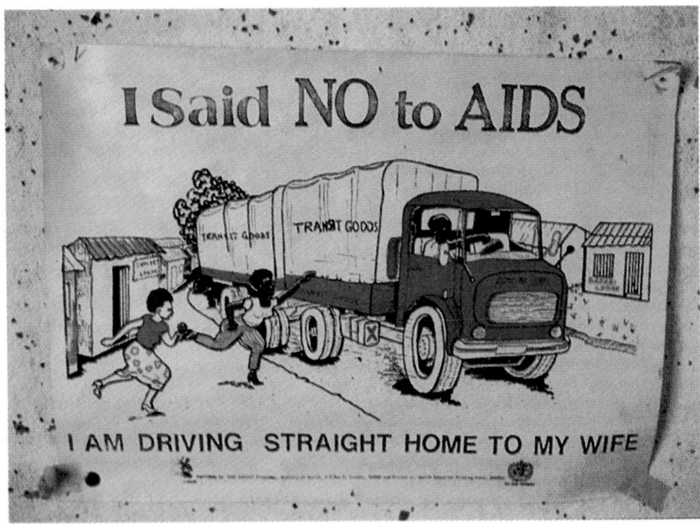

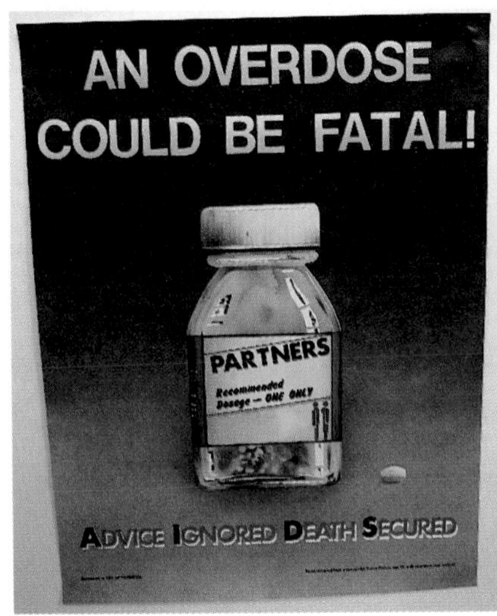

Almost every country in the world produces AIDS prevention messages tailored to the values and beliefs of people in that country. These posters are from Uganda (left) and Barbados (right).

©T. Cambre Pierce/Saba (left); ©Jeremy Hartley/Panos Pictures (right)

Africans measure their spiritual and material wealth by the number of offspring. If children survive,[6] they become economic assets to the family as well as links to ancestors (Whitaker 1988). In view of the strong pressures to have children, many Zaireans believe that condom use virtually deprives them of the approval of their families and ancestors.

Researcher Kathleen Irwin (1991) and fifteen colleagues interviewed healthy factory workers and their wives from Kinshasa, Zaire. They found that, although many respondents had heard of condoms, few actually used them. The researchers also found that, among these respondents, it is the men who decide whether to use and purchase condoms.

The widespread availability and use of female contraceptives in the United States lowers the incentives to use condoms (Urban Institute 1995). In the United States, if a person suggests using a condom during sex, he or she is implying that the partner's sexual orientation or sexual history is suspect. Advertisers try to package and market condoms in ways that counteract this message. Condoms come in all colors, are designed and manufactured in forms that stimulate greater sexual sensation, and show sexually appealing scenes on outer packages.

The point is that in both Zaire and the United States, "sexual behavior is based on long-standing cultural traditions and social values and may be very difficult to change" (U.S. General Accounting Office 1987). This fact in turn makes it difficult for any person to manipulate the meaning of a condom without offending his or her sexual partner. Therefore, many people resist using condoms, even in high-risk situations (Giese 1987).

The dramaturgical model of social interaction is useful because it helps us see how the need to convey the right impressions can work to discourage people from behaving in ways that are not in their best (or others' best) interest. Goffman uses another theater analogy — staging behavior — to identify situations in which people are most likely to engage in impression management.

Front stage The region where people take care to create and maintain expected images and behavior.

Back stage The region out of sight where individuals can do things that would be inappropriate or unexpected on the front stage.

Staging Behavior

Just as the theater has a front stage and a back stage, so does everyday life. The **front stage** is the area visible to the audience, where people take care to create and maintain expected images and behavior. The **back stage** is the area out of the audience's sight, where indi-

[6]In Zaire, the infant mortality rate is 130 per thousand; one-half of all children in Zaire die before they reach age five.

viduals can "let their hair down" and do things that would be inappropriate or unexpected on the front stage. Because back-stage behavior frequently contradicts front-stage behavior, we take great care to conceal it from the audience. Goffman uses the restaurant as an example. Restaurant employees do things in the kitchen and pantry (back stage) that they would not do in the dining areas (front stage), such as eating from customers' plates, dropping food on the floor and putting it back on a plate, and yelling at one another. Once they enter the dining area, however, such behavior stops. How often have you, as a customer, seen a server eat a scallop or a french fry from a customer's plate while in the dining area? But if you have ever worked in a restaurant, you know that this is fairly common back-stage behavior.

The division between front stage and back stage is hardly unique to restaurants but is found in nearly every social setting. In relation to the AIDS crisis, we can name a host of environments that have a front stage and a back stage, including hospitals, doctors' offices, and blood banks. Much as restaurant personnel shield diners from backstage behavior, medical personnel shield patients from backstage behavior. For example, most people know little about the blood bank industry beyond what they see when they donate, sell, or receive blood. Although the public can research such industries and learn about their inner workings, most people do not have the time to study every industry or do not know what questions to ask about the industry that provides them with goods and services.

For example, most people probably give little thought to whom the blood industry collects blood from and would never think to ask the question of where blood comes from. Thus, they would be surprised to find out that Mexicans cross the border into the United States to give blood. In 1980, a few years before HIV was discovered, *Newsweek* magazine ran a story on the thousands of poor Mexicans who earn $10 by undergoing a procedure called plasmapheresis (technicians take blood, separate the red cells from the plasma, and reinject the donor with his or her own cells minus the plasma). Because red cells are returned to donors, they may undergo the procedure up to twice a week. The article highlighted the international nature of blood collection (Clark and McGuire 1980).

The Back Stage of Blood Banks Blood bank officials found themselves in a dilemma in 1981, when officials at the Centers for Disease Control made known their suspicion that HIV was being transmitted through blood products. This revelation meant that not only the United States' blood supply but also the world's blood supply was contaminated, because the United States supplies about 30 percent of the world's blood and blood products (U.S. Bureau of the Census 1992a; *The Economist* 1981, 1983; see "Imports and Exports of Blood and Blood Products").

Until spring of 1985, U.S. blood bank officials continued to affirm publicly their faith in the safety of the country's blood supply, insisting that screening donors was unnecessary. Yet, these officials never revealed to the public the many shortcomings in production methods that could jeopardize the safety of blood. (By *shortcomings*, I do not mean negligence but rather deficiencies in medical knowledge and the level of technology.) Looking back on their policies, blood bank officials later argued that they practiced this concealment to prevent a worldwide panic. (In Goffman's terminology, they did not want to lose their audience.) Such a panic would have brought chaos to the medical system, which depends on blood products.[7] Still, the delay in implementing screening exposed many people to infection.

Here lies a potential connection to Zaire. Malaria is a common disease in Zaire, especially among children. The disease leaves its victims vulnerable to severe anemia, and blood transfusions are used to treat the anemia. As mentioned previously, the United States is a large exporter of blood products. Many countries that import this blood reexport it, and the Red Cross delivers blood products to countries in need. Therefore, some of the blood used in Zaire is likely to have originated in the United States.

Although blood bank officials announced publicly that the risk of HIV infection from blood products was one in a million, knowledge of the back-stage collection, production, and distribution of blood products leaves no doubt that the risks were in fact higher. Hemophiliacs, especially, were at risk because their plasma lacks the substance Factor VIII, which aids in clotting, or whose plasma contains an excess of anticlotting material (U.S. Department of Health and Human Services 1990). In fact, we now know that 50 percent of hemophiliacs were HIV infected from Factor VIII treatments before the first case of AIDS appeared in this group (*Frontline* 1993).

[7]Many medical treatments depend on blood products:

> Blood transfusions save the lives of premature infants and are crucial for children with Cooley's anemia and other hereditary blood disorders. A vaccine against hepatitis B is derived from blood. Injections of gamma globulin, prepared from blood, are effective in helping prevent hepatitis A, chicken pox, rabies, and other ailments. Platelet transfusions are key to some cancer treatments. (Altman 1986:A1)

U.S. in Perspective

Imports and Exports of Blood and Blood Products

To truly understand the global AIDS epidemic, we need to consider the role of exported blood products in the transmission of HIV. Figure 6.2 graphs the amount of blood products exported from the United States to selected other countries as a percentage of the receiving country's total need. The data are from 1981, the year that HIV was "discovered" and that scientists first learned that it was in the blood supply. Notice that at the time, Japan imported 98 percent of its blood products from the United States—46 million units of concentrated blood products and 3.14 million liters of blood plasma (Yasuda 1994).

Figure 6.3 maps the United States' pattern of export and import of human blood plasma around 1995.

Sources: Map: U.S. Bureau of the Census (1996a, 1996b); Bar graph: International Federation of Pharmaceutical Manufacturers Associations (1981).

Figure 6.2

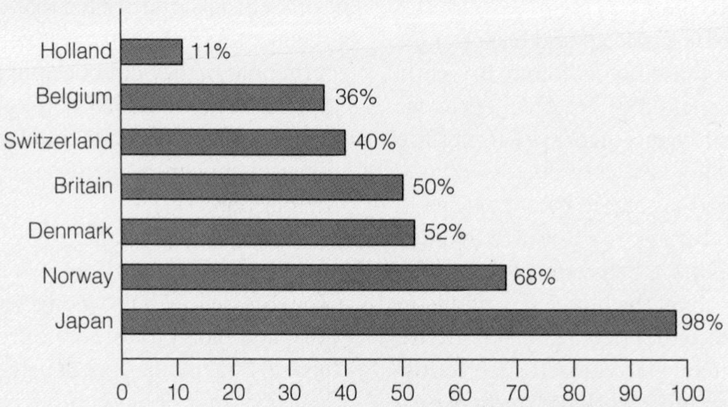

Figure 6.3

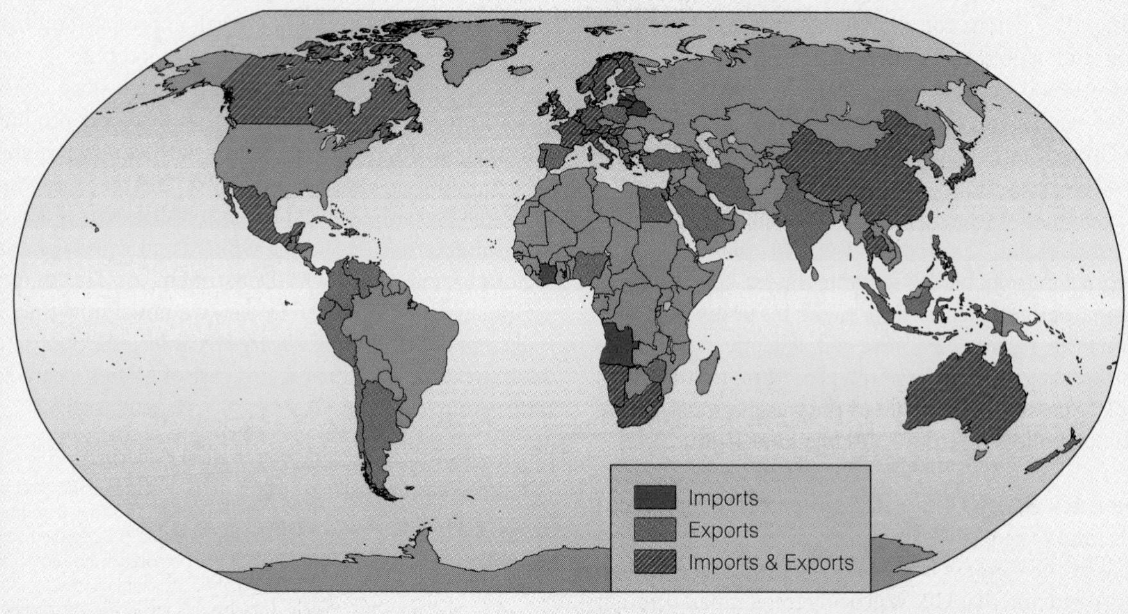

Even after blood bank officials agreed to start screening blood for HIV infection in spring 1985 (after years of debate over whether to test), they still pronounced the blood supply safe. They did not, however, announce the shortcomings of the screening tests: (1) the antibodies that the test measures may not appear in the blood for eleven months after infection with the virus, and (2) a small but unknown percentage of HIV carriers never develops detectable antibodies (Kolata 1989).[8]

In hindsight, it is easy to criticize blood bank officials' response to the situation. Yet, in fairness to the blood industry, we must acknowledge the legitimacy of their wish not to induce worldwide panic, especially very early on when no tests were available to screen blood for the infection. As the head of the New York Blood Center argued, "You shouldn't yell fire in a crowded theater, even if there is a fire, because the resulting panic can cause more deaths than the threat" (Grmek 1990:162). In addition, we must recognize that it is impossible to eliminate every element of risk. Even so, one troubling fact suggests that their decision may have been motivated by profit. Although U.S. companies tested new blood for the domestic market, they did not test blood already stored in their inventories, and they continued to supply untested blood to foreign countries for at least six months after the tests were available (*Frontline* 1993; Hiatt 1988; Johnson and Murray 1988).

To this point, we have examined a number of concepts that help us analyze the content of any social interaction. Specifically, when examining content sociologists identify the statuses of those people involved. Once statuses have been identified, they focus on the roles expected of each status in relation to the other, with special emphasis on rights and obligations. Sociologists use a dramaturgical model to think about how people enact roles. Thus, they focus on impression management (how people manage the setting, their dress, their words, and their gestures) to correspond to role expectations. Impression management is most important when people are front stage as opposed to back stage.

[8]Some of the "checks" to prevent HIV-infected people from donating blood (Zuck 1988) are as follows:

- High-risk donors practicing voluntary self-exclusion
- Confidential exclusion when donors feel pressured to give blood (After giving blood, they can tell the technician drawing blood that they wish their blood to be used for research only.)
- HIV antibody testing to detect infected units
- A national registry of ineligible donors that the Red Cross maintains and against which it checks donors (American National Red Cross 1996)

Knowing about statuses, roles, impression management, and front/back stages helps sociologists predict much of the content of social interaction. However, when we interact, we do more than identify the statuses of the people involved and act to behave in ways consistent with role expectations. We also try to assign causes to our own and others' behaviors. That is, we posit explanations for behavior, and we may act differently depending on what explanations we come up with. A theoretical approach that helps us understand how we arrive at our everyday explanations of behavior is attribution theory.

Attribution Theory

Social life is complex. As we have seen, people need a great deal of historical, cultural, and biographical information if they are to understand the causes of even the most routine behaviors. Unfortunately, it is nearly impossible for people to have this information at hand every time they want to understand the causes of behavior. For one thing "the real environment is altogether too big, too complex, and too fleeting for direct acquaintance" (Lippmann 1976:178).

Yet, despite our limitations in understanding causes of behavior, most people do attempt to determine a cause anyway, even if they rarely stop to examine critically the accuracy of their explanations. As most of us very well know, ill-defined, incorrect, and inaccurate perceptions of cause do not keep people from forming opinions and taking action. Such perceptions, however, result in actions that have real consequences. Sociologists William and Dorothy Thomas described this process very simply: "If [people] define situations as real they are real in their consequences" (Thomas and Thomas [1928] 1970:572).

Attribution theory rests on the assumption that people assign (or attribute) a cause to behavior in order to make sense of it. People usually attribute cause to either dispositional traits or situational factors. **Dispositional traits** include personal or group traits such as motivation level, mood, and inherent ability. **Situational factors** include forces outside an individual's control such as environmental conditions or bad luck. When evaluating the causes of their own behavior, peo-

Dispositional traits Personal or group traits such as motivation level, mood, and inherent ability.

Situational factors Forces outside an individual's control such as environmental conditions or bad luck.

ple tend to favor situational factors. When evaluating the causes of another's behavior, however, people tend to point to dispositional traits.[9]

A memorable example of attributing cause to dispositional rather than situational characteristics appears in Colin Turnbull's *The Lonely African.* Turnbull describes how a European farm owner reacts to an African farmhand who wears a tie without a shirt: "'I've actually got a farmhand who wears a tie—but the stupid bastard doesn't realize you don't wear a tie without a shirt!'" (1962:21). The owner attributes the behavior to a dispositional factor—the farmhand is simply stupid. If the farm owner had thought about the behavior in terms of situational factors, he would have found that the farmhand wears the tie "because it makes a bright splash of color, and is useful for tying up bundles, and refuses to wear the shirt that collects dirt and sweat and makes the Europeans smell so bad" (p. 21). Right or wrong, the attributions that people make shape the content of social interaction. In other words, the way in which people construct reality affects how they treat and deal with different individuals and groups.

Throughout history, whenever medical professionals have lacked the knowledge or technology to combat a disease, especially one of epidemic proportions, the general population has tended to hold some groups of people within or outside of the society responsible for causing the disease (Swenson 1988). In the sixteenth and seventeenth centuries, for example, the English called syphilis "the French disease," and the French called it "the German disease." In 1918, the worldwide influenza epidemic, which infected more than a billion people and killed more than 25 million, was called "the Spanish flu" even though there was no evidence to support the idea that it originated in Spain. American history holds many examples of groups blamed for bringing disease to the country, including the Irish (cholera), Italians (polio), Chinese (bubonic plague), Jews (tuberculosis), and Haitians (AIDS) (Kraut 1994).

In a similar vein, U.S. medical researchers trying to map the geographic origin and spread of AIDS have hypothesized various interaction scenarios between specific groups of people who are inferred to behave in careless, irresponsible, or immoral ways to explain how AIDS has spread transcontinentally. The hypotheses usually assume that the disease originated in Zaire and then spread to the United States via Europe, Haiti, or Cuba. Yet, no evidence supports the possibility that it did not spread transcontinentally from the United States. Two of the most prevalent hypotheses are as follows:

- The virus traveled from Zaire to Haiti to the United States. In the mid-1960s, a large number of Haitians went to Zaire to fill middle management positions in the newly independent state. In the 1970s, Mobutu sent them home. American homosexuals vacationing in Port-au-Prince brought the virus back to New York and San Francisco.
- The virus traveled from Zaire to Cuba to the United States. Cubans brought it back from Angola, which shares a long border with Zaire. In the late 1970s, the Cuban government purged the army of undesirables, including some homosexual veterans who had served in Angola. Many of these Cubans migrated to Miami.

Attributing cause to dispositional factors seems to reduce uncertainty about the source and spread of the disease. The rules are clear: if we do not interact with members of that group or behave like members of that group, then we can avoid the disease. Such rules provide us with the secure feeling that the disease cannot affect *us* because it affects *them* (Grover 1987).

In the United States, many people believe that HIV and AIDS are confined to members of a few high-risk groups, notably hemophiliacs, male homosexuals, and intravenous (IV) drug users. Dispositional explanations for the high risk among members of these groups imply that these individuals "earned" their disease as a penalty for perverse, indulgent, and illegal behaviors (Sontag 1989). As late as 1985, medical and government officials referred to AIDS as the "gay plague," even though there was overwhelming evidence that HIV infection was also transmitted through heterosexual intercourse, needle sharing, and other exchanges of blood and blood products or some other bodily fluids. In his book *And the Band Played On,* journalist Randy Shilts (1987) argues that the public health and medical research response to AIDS was delayed several years because policy makers believed that casualties were limited to unpopular and socially powerless groups such as homosexuals, IV drug users, and Haitians.

[9]Generally, people use different criteria when they attribute cause to their own behavior than when they attribute cause to another's behavior. With regard to other people's failures or shortcomings, we tend to overestimate the extent to which our own failures are due to situational factors. With regard to success, we tend to exaggerate the extent to which others' successes are caused by situational factors (a lucky break; they were in the right place at the right time). We tend to explain our own successes as resulting from dispositional factors such as personal effort and sacrifice.

Similarly, dispositional thinkers might explain the high incidence of HIV infection and AIDS among both males and females in Zaire (or other African countries) in terms of excessive sexual promiscuity or polygynous marriages[10] or in terms of rituals involving monkeys, female circumcision, and scarification.

Whereas Americans tend to regard HIV infection as originating in Africa and as being transmitted through bizarre and indulgent behaviors, Zaireans tend to view AIDS as an American disease:

> Westerners had brought AIDS to Africa with their "weird sexual propositions" — a view echoed by *La Gazette* in July 1987, which referred to Westerners coming to Africa with their "sexual perversions." "Many of the venereal diseases now found in Kenya," said an editorial in the *Kenyan Standard,* "were brought into the country by the same foreigners who are now waging a smear campaign against us." (The Panos Institute 1989:75)

In Zaire and other African countries, people often refer to the United States as the "United States of AIDS." Condoms are called "American socks," and aid in the form of condoms is known as "foreign AIDS" (Brooke 1987; Hilts 1988). People commonly believe that (1) AIDS arrived via rich American sports fans who came to Kinshasa to watch the Ali-Foreman boxing match in 1977, (2) the virus was manufactured in an American laboratory for military germ warfare and was unleashed deliberately on Africans,[11] (3) the disease came from American canned goods sent as foreign aid, and (4) the AIDS epidemic can be traced to the way in which the polio vaccine was manufactured and administered to Africans thirty years ago.

In each dispositional scenario, one can infer that those giving the disease to Africans are morally suspect, evil plotters, profit driven, or careless; these are all dispositional characteristics. Dispositional theories such as the ones just listed, whether American or African, are alike in that they define a clear culprit or scapegoat. A **scapegoat** is a person or a group that is assigned blame for conditions that cannot be controlled, threaten a community's sense of well-being, or shake the foundations of a trusted institution. Usually the scapegoat belongs to a group whose members are already vulnerable, hated, powerless, or viewed as different.

The public identification of scapegoats gives the appearance that something is being done to protect the so-called general public; at the same time, it diverts public attention from those who have the power to assign labels. In the United States, identifying AIDS as the "gay plague" diverted attention from the blood banks and the risks associated with medical treatments involving blood products. Blood bank officials maintained that the supply was safe as long as homosexuals abstained from giving blood. In Zaire, identifying AIDS as an American disease diverted attention from corrupt officials who were funneling money out of Zaire, leaving its people poor and malnourished and (by extension) vulnerable to HIV infection.

From a sociological perspective, dispositional explanations that point to a group — or characteristics supposedly inherent in members of that group — are simplistic and potentially destructive not only to the group but to the search for solutions. When the focus is a specific group and that group's behavior, the solution is framed in terms of controlling that group. In the meantime, the problem can spread to members of other groups who believe that they are not at risk because they do not share the problematic attribute of the groups identified as high-risk. This kind of misguided thinking about risk applies even to physicians and medical researchers.

Medical sociologist Michael Bloor (1991) and colleagues argue that the official statistics on the modes of transmission are influenced by researchers' beliefs about the relative riskiness of behaviors. For example, an HIV-positive male who has received a blood transfusion and who has had sexual relations with another male is placed in the transmission category "homosexual" rather than "blood recipient." This approach to classification ignores the possibility that a homosexual

Scapegoat A person or a group that is assigned blame for conditions that cannot be controlled, threaten a community's sense of well-being, or shake the foundations of a trusted institution.

[10]African officials become outraged when the Western media equate polygynous marriage arrangements with promiscuity. One villager responded as follows to this charge: "We have several wives, and we are faithful to them all, and we care for all their children until we die. You people can not even be faithful to one wife, and your children are such a nuisance to you that you send them away from home as soon as they can walk" (Turnbull 1962:28).

[11]This theory has some basis in fact. In 1969, the U.S. Defense Department discussed the theoretical possibility of the following development:

> Within the next five to ten years it would probably be possible to make a new infective micro-organism which could differ in certain important respects from any known disease-causing organisms. Most important of these is that it might be refractory to the immunological and therapeutic processes upon which we depend to maintain our relative freedom from infectious disease. (Harris and Paxman 1982:241)

male can become HIV infected through a blood transfusion or other means.

Similarly, until 1993 the official definition of AIDS did not include HIV-related gynecological disorders such as cervical cancer as one of the conditions that constituted a diagnosis of AIDS. (Also not included under the official definition of AIDS until 1993 were HIV-related pulmonary tuberculosis and recurrent pneumonia.) Under the old definition, HIV-positive persons were said to have AIDS only if they developed one of twenty-three illnesses, many of which were peculiar to the gay population with AIDS. Under the new, revised definition, HIV-positive women with cervical cancer are officially diagnosed as having AIDS. Before this change, physicians did not advise women with cervical cancer to be tested for HIV and did not give HIV-positive women with this condition the opportunity to be treated for AIDS (Barr 1990; Stolberg 1996). These two examples show that attributions about who "should" have AIDS affect the way in which the condition of AIDS is defined and influence the statistics about who has AIDS. They also illustrate that such attributions affect the content of the physician-patient interaction.

In addition to acknowledging the shortcomings related to AIDS classification and diagnosis, we also have to acknowledge that we simply do not know how many people are HIV infected worldwide or who is actually infected.

Determining Who Is HIV Infected

To obtain information on who is actually infected with HIV, every country in the world would have to administer blood tests to a random sample of its population. Unfortunately (but perhaps not surprisingly), people resist being tested.

A planned random sampling of the U.S. population sponsored by the Centers for Disease Control (CDC) was aborted after 31 percent of the people in the pilot study refused to participate, despite assurances of confidentiality (Johnson and Murray 1988). Two researchers involved with this project concluded that "it does not seem likely that studies using data on HIV risk or infection status, even with complete protection of individual identity, will be practical until the stigma of AIDS diminishes" (Hurley and Pinder 1992:625).

The United States is not the only country whose people do not want to be tested. This resistance seems to be universal. For example, at one time Zairean officials were reluctant to disclose the number of AIDS cases and infection rates to United Nations officials or to allow foreign medical researchers to test Zairean citizens because of their sensitivity to the unsubstantiated but widely held belief that Zaire was the cradle of AIDS (Noble 1989).

Random sampling, however, is the most dependable method we have of determining the number of HIV-infected persons. Until we have such information, we cannot know what factors cause a person with HIV to contract AIDS. Random blood samples would permit comparisons between the lifestyles of infected but symptom-free people and infected people who have developed AIDS or AIDS-related complex (ARC).[12] From such comparisons, we could learn which cofactors cause a healthy carrier to develop ARC or AIDS. Such cofactors might include diet, exposure to hazardous materials, or prolonged exposure to the sun or tanning booth rays—anything that might compromise the immune system in such a way as to activate a dormant infection. For example, why do some people remain HIV-infected for years without developing AIDS? Why do other people develop AIDS shortly after exposure to HIV (Altman 1986, 1995)?[13] Why is HIV absent from the bodies of some patients with AIDS-like symptoms (Liversidge 1993)? Finally, how is it that at one time HIV may have been a harmless virus? How did the virus change to become the causative pathogen of AIDS? (Some scientists speculate that HIV has been around in an inactive state for at least a century.)[14]

The point is that a person may develop AIDS as a result of other factors besides the behavior that causes a person to contract HIV infection. As long as no systematic and random sampling of populations is performed, people will continue to speculate on these factors, either overestimating or underestimating the prevalence of infection and the projected numbers of AIDS cases worldwide.

[12]AIDS-related complex (ARC) is a term applied to HIV-infected persons whose symptoms do not meet the definition of AIDS set forth by the CDC. AIDS-related complex encompasses a wide range of symptoms, including fever, rash, and bacterial or viral infections, most of which point to a diminished ability of the body's immune system.

[13]Since 1981, approximately 500,000 AIDS case have been reported and listed in the national AIDS Registry. Two-thirds have died (Stolberg 1996).

[14]Actually, the HIV-AIDS connection is not a clear-cut one. In recognition of this, more than 100 biologists from around the world have formed an organization (the Group for Scientific Reappraisal of the HIV/AIDS Hypothesis) with the purpose of rethinking this connection. This group also issues a newsletter titled *Rethinking AIDS* (Liversidge 1993).

This situation leaves people with a dilemma about how to deal with a complex health problem such as AIDS when so many unanswered questions persist. Most people do not have the time to inform themselves about all of the contextual forces underlying HIV and AIDS. Yet, this does not stop people from acting or attributing cause on the basis of limited information. Even if people do not have the time to inform themselves, they do have the option of at least being critical of their information sources; for most people, that source is television.

Public health officials believe that the media need to deliver at least one important message with regard to HIV: "It is not who you are; it is how you live and what you do" (Kramer 1988:43). For the most part, however, this has not been the message transmitted to the American public. Because television news and information shows are an important source of information for most Americans (98 percent of American households have television sets), we will examine how television producers present information in general and the AIDS phenomenon in particular.

Television: A Special Case of Reality Construction

When sociologists say that reality is constructed, they mean that people assign meaning to interaction or to some other event. When people assign meaning, they almost always emphasize some aspects of that event and ignore others. In this chapter, we have looked at some of the strategies that people use to construct reality. Consider the following points:

1. When people assign meaning, they tend to ignore the larger context.
2. When people interact with others, they assign meaning first by determining their own social status in relation to the other parties and then by drawing upon learned expectations of how people in some social statuses are to behave.
3. People create reality when they engage in impression management; that is, they manage the setting, their dress, their words, and their gestures to correspond to impressions they are trying to make.
4. People control access to the back stage so that outsiders to the back stage form opinions on the basis of the front stage.
5. People attribute cause to dispositional traits when evaluating others' behavior, and they attribute cause to situational factors when evaluating their own behavior.

People also assign meaning to events based on firsthand experiences. Television gives people access to events that they would never have the chance to experience if left to their own resources. Thus, our analysis of how people come to construct the reality they do would not be complete if we ignored the format that television, especially television news, uses to present information about what is going on in the world.

Television conquers time and space: it allows us to see what is going on in the world as soon as it happens. However, consider the following features of national and local news—television at its most serious and most informative:

- The average length of a camera shot is 3.5 seconds.
- Every three or four news items, no matter how serious, are followed by three or four commercials.
- The news of the day is often presented as a series of sensationalized images.
- Approximately fifteen news items are presented in a thirty-minute news program with approximately twelve different commercials.
- Most news coverage of events focuses on the moment; each item is presented without a context.

In *Amusing Ourselves to Death,* Neil Postman (1985) examines the format of news programs and asks how it affects the way people think about the world. Overall, he believes, this format gives viewers the impression that the world is unmanageable and that events just seem to happen. More to the point, it gives the public only the most superficial facts about events in the world. For example, most people in the United States know that AIDS exists, and they know basic facts about how the virus is transmitted and about how to reduce risk of transmission. However, a large percentage (71 percent) also admit that they do not know a lot about AIDS (U.S. Department of Health and Human Services 1992).

What characteristics of the news format produce this consequence? The brevity of camera shots is probably one of the main problems. Postman argues that with the average length of a shot being just 3.5 seconds, facts are pushed into and out of consciousness in rapid succession so that viewers do not have sufficient time to reflect on what they have seen and to evaluate it properly. Furthermore, commercials defuse the effects of any news event; they give the following message: I cannot do anything about what is happening in the

The story of Ryan White, a hemophiliac who contracted HIV, was widely publicized in the news media. How does this kind of selective coverage change people's view of a phenomenon like AIDS?

©Mary Ann Carter/Sipa Press

world, but I can do something to feel good about myself if I eat the right cereal, own the right car, and use the right hair spray.

Television is image oriented; a picture is a moment in time and, by definition, is removed from any context. This quality often causes television news to be sensationalistic. The highly publicized case of Ryan White (who died in April 1990 at age eighteen from AIDS-related respiratory failure) illustrates just how sensationalistic news reports can be. As a child, White had received HIV-infected Factor VIII while being treated for hemophilia and later was diagnosed as having AIDS. White was barred from attending Western Middle School in Kokomo, Indiana, in August 1985 after school officials learned that he had AIDS (Kerr 1990). When White returned to school in April 1986, reporters and camera crews covered the event in what the school principal termed a sensationalistic manner:

> I understand that the media has a job to do, but I think there is a fine line between informing the public and creating controversy in order for a story to keep continuing.
>
> It seems like the problems were brought out by those who jumped on the sensational. These were published and displayed on television and it created an excited atmosphere in what was really a pretty calm school situation. . . .
>
> When Ryan did come back, there were thirty kids whose parents took them out of school. That's what made the news, but there were 365 other children who stayed in the school. (Colby 1986:19)

An image-oriented format tends to ignore those historical, social, cultural, or political contexts that would make the event more understandable. Viewers are left with vivid, sensationalistic images of enraged parents, an emaciated gay AIDS patient, a skid row drug addict, a prostitute, a family home destroyed by fearful neighbors, Africans walking to an AIDS clinic, or gays protesting discrimination. Rarely, however, do television producers present AIDS as a chronic but often manageable condition or portray those with AIDS as leading responsible lives. Imagine what people's initial attitudes toward AIDS might have been if the first discussion of the disease had centered around the life of someone like Dr. Rask instead of a small group of homosexual males. Trying to understand how Dr. Rask contracted HIV would certainly have told us more about the complex social origins of the disease than did focusing on homosexual practices.

This is not to deny that television is an important tool for informing large audiences. Despite the sensationalistic coverage, most Americans know that AIDS exists and that it is related in some way to a virus, but some confusion remains about how HIV is transmitted. For example, about 29 percent of people in the United States believe that a person runs the risk of becoming infected from donating blood, and 10 percent do not know whether donating blood puts them at risk (U.S. Department of Health and Human Services 1992).

To illustrate how little context television provides, you might list the things you have learned about AIDS from this chapter that you did not learn from the media. One might argue that there is not enough time to learn about AIDS from television because so many other important events are competing for attention in the news. Still, this point does not eliminate the need for context in understanding events.

Another way to illustrate how little context television provides is to count how many events (other than sports and weather items) you remember from last night's news. If you cannot remember many events, then the information may have been presented so quickly that you could not reflect long enough to absorb it. To remedy this problem, newscasters might reduce the number of stories, increase the time given to context, and cover stories in less reactive and more reflective ways. In the case of AIDS, for example, producers could show segments explaining how HIV differs from AIDS or how the television image of AIDS differs from the experience of AIDS. With regard to this last suggestion, television producers are quite good at covering and discussing presidential campaigns as media events. They certainly could cover AIDS as a media event as well.

Summary and Implications

In this chapter, we learned that sociological concepts and theories contribute to our understanding of HIV and AIDS and, by extension, other infectious diseases. AIDS (and other infectious diseases) cannot be viewed simply as a biological event. AIDS is a social phenomenon not only because certain kinds of "high-risk" interactions are associated with the transmission of HIV. Sociology gives us a framework for conceptualizing the role of social interaction in HIV transmission and the content of interactions that occur in reaction to HIV.

The importance of considering interaction on a global scale is supported by the fact that disease patterns historically are affected by changes in population density and changes in transportation. Physician Mary E. Wilson (1996) describes the relationship between travel and the emergence of infectious diseases:

> Travel is a potent force in the emergence of disease. Migration of humans has been the pathway for disseminating infectious diseases throughout recorded history and will continue to shape the emergence, frequency, and spread of infections in geographic areas and populations. The current volume, speed, and reach of travel are unprecedented. The consequences of travel extend beyond the traveler to the population visited and the ecosystem. When they travel, humans carry their genetic makeup, immunologic sequelae of past infections, cultural preferences, customs, and behavioral patterns. Microbes, animals, and other biologic life also accompany them. Today's massive movement of humans and materials sets the stage for mixing diverse genetic pools at rates and in combinations previously unknown. Concomitant changes in the environment, climate, technology, land use, human behavior, and demographics converge to favor the emergence of infectious diseases caused by a broad range of organisms in humans, as well as in plants and animals.

As one indicator of the massive movement of human beings, consider that in 1996 an estimated 400 million people traveled on international flights (Sanchez 1994). In 1994, there were approximately 48 million foreign visitors to the United States, and 47.3 million U.S. residents traveled to a foreign country (U.S. Department of Commerce 1994).

We focused on Zaire and on Dr. Rask to show that a complex set of interactions lies behind the transmission of HIV infection. Dr. Rask represents one person with AIDS caught up in the historical events that led to worldwide economic interdependence, which in turn brought people from all over the world into contact with one another. The complexity of the interaction that brought Dr. Rask to Zaire and in contact with HIV makes it impossible to state conclusively that HIV originated in Zaire. Even if a previously isolated group in Zaire such as the Mbuti was identified as the group that harbored the virus in a dormant state, could we in good conscience define the Mbuti as the source? What about the many forces that brought them out of isolation into contact with groups having no immunity to the virus?

The historical context of interaction is only one dimension of the interaction process that sociologists examine. They also seek to understand the *content* of social interaction. Sociology offers a rich conceptual vocabulary to analyze the dialogue, actions, and reactions that take place when people interact, as well as to analyze how the parties involved in social interaction strive to define, interpret, and attach meaning to the encounter. People associated with Dr. Rask, for example, assumed a scientific framework to define the origin, significance, and treatment of her condition. As a result, Dr. Rask experienced not only a physical state of sickness but a social state in which her behavior and the behavior of the involved parties reflected the assumption

Disease patterns historically are affected by changes in population density and transportation patterns.

©Crispin Hughes/Panos Pictures

that her physiological malfunction could be understood and corrected with the available medical technologies (tests, drugs, machines, surgery, and so on).

Some of the key sociological concepts to help us analyze interaction include social status, social roles, impression management, front stage, back stage, dispositional traits, and situational factors. These concepts helped us identify a number of mechanisms that impede progress and prevent effective action in dealing with HIV and AIDS. For example:

- Expectations associated with the physician's role focus his or her attention on finding a cure for the condition without regard to side effects and to the neglect of symptom relief and patient comfort.
- Back-stage behavior plays a significant role in determining the eventual course of the disease. (How many people around the world ultimately contracted HIV because of the way blood and blood products were processed? What role did "profit" play in exposing some people to HIV who might not have otherwise been exposed?)
- Dispositional explanations that point to a group or characteristics supposedly inherent in members of that group are simplistic and potentially destructive. When the focus is the group, the solutions are framed in terms of controlling that group rather than understanding the problem and finding solutions.
- Official statistics on modes of transmission and official definitions of AIDS are influenced by researchers' beliefs about the relative riskiness of behavior and by attributions about who "should" have AIDS.
- The public identification of scapegoats diverts attention away from considering other factors that may be causing the problem. (Initially, identifying AIDS as the "gay plague" diverted attention away from the blood banks and the risks associated with medical treatments involving blood products).
- Social meanings associated with preventive measures such as condoms can cause people to resist using those measures even in clearly high-risk situations.

The global transmission of HIV infection illustrates a point about interdependence on a global scale; when something goes wrong in one part of the world, other parts of the world are affected. Our discussions of global interdependence in relation to colonization and of many countries' reliance on blood products from the United States illustrates this point; so do many other phenomena, such as global warming and illegal drug trade. Interdependence is not necessarily a negative situation. It can also lead to greater efforts to solve problems and keep the system running. As futurist John Naisbitt (1984) suggests, "If we get sufficiently interlaced economically, we will probably *not* bomb each other off the face of the planet" (p. 79). Such interdependence also means that "AIDS cannot be stopped in any country unless it is stopped in all countries" (Mahler, quoted in Sontag 1989:91); "it cannot be mastered in the West unless it is overcome everywhere" (Rozenbaum, quoted in Sontag 1989:91). These comments suggest that if AIDS policies are to be effective, they must involve worldwide effort.

Such an effort is underway. The World Health Organization (WHO) is sponsoring, directing, and coordinating a global strategy to prevent HIV and control its transmission. The Global Programme on AIDS supports national AIDS programs in 150 countries that include health education, prevention information, blood transfusion services, and cross-cultural research on human behavior and effective communication. Officials at WHO maintain that this program offers health benefits beyond the prevention of AIDS: "The global response to AIDS offers a great opportunity to accelerate the strengthening of our health care infrastructures" in general (World Health Organization 1988:15).

Key Concepts

Use this outline to organize your review of the key chapter ideas.

- **Social interaction**
  - **Context**
    - **Division of labor**
    - **Solidarity**
      - **Mechanical solidarity**
      - **Organic solidarity**
  - **Content**
    - **Social structure**
    - **Social status**
      - **Ascribed status**
      - **Achieved status**
    - **Stigma**
- **Role**
  - **Role set**
  - **Rights**
  - **Obligations**
  - **Sick role**
  - **Role strain**
  - **Role conflict**
- **Dramaturgical model**
  - **Impression management**
  - **Front stage**
  - **Back stage**
- **Attribution theory**
  - **Dispositional traits**
  - **Situational factors**

internet assignment

Skim through the table of contents in current and past issues of *Emerging Infectious Disease* for articles that focus on social factors that underlie disease emergence (including human demographics and behavior, technology, industry, travel, commerce, breakdown of public health measures, and so on). Select two or three articles to read, making note of the social dimensions of the disease(s) discussed. Skim through the list of 1996 publications from the National Center for Health Statistics, and read two or three reviews of the publications listed. Comment on the social factors connected to the physical condition or medical process covered in the review.

Emerging Infectious Diseases

http://www.cdc.gov/ncidod/EID/eid.htm
The National Center for Infectious Diseases published the peer-reviewed on-line journal *Emerging Infectious Diseases.* The goal of the journal is "to promote the recognition of new and reemerging infectious diseases and to improve the understanding of factors involved in disease emergence, prevention, and elimination."

National Center for Health Statistics 1996 Publication

http://olympus.lang.arts.ualberta.ca:8010/
The National Center for Health Statistics lists its 1996 publications covering a range of health care issues from Alzheimer's disease to vital statistics. The entire text of each publication can be accessed via downloading. Reviews of each publication, however, are available on-line.

7 Social Organizations

With Emphasis on the Multinational Corporation in India

Golden Temple, Varanasi, India. Herbert Lanks/Monkmeyer Press

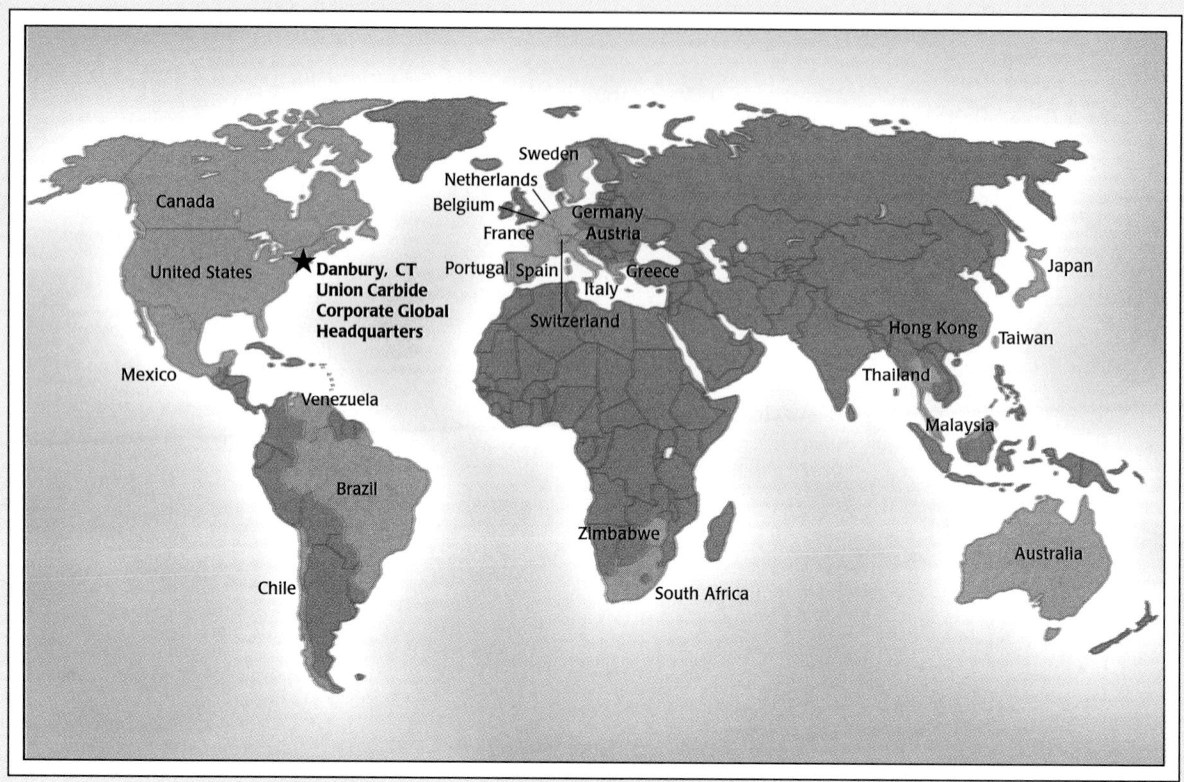

Union Carbide: A Multinational Corporation Headquartered in the United States

How many locations does Union Carbide have in the United States? How many does it have outside the United States? Can you tell from the map what parts of the world do not have Union Carbide–affiliated operations? Union Carbide is just one of many multinational corporations—by no means the largest. Does Union Carbide have a subsidiary or affiliate in India? Why do you think this is the case?

On December 3, 1984, approximately forty tons of methylisocyanate (MIC), a highly toxic, volatile, flammable chemical used in making pesticides, escaped from a Union Carbide storage tank and blanketed the densely populated city of Bhopal, India. Investigators determined that between 120 and 140 gallons of water somehow had entered the storage tank containing MIC. The combination triggered a violent chemical reaction that could not be contained. As a result, approximately 800,000 residents awoke coughing, vomiting, and with eyes burning and watering. They opened their doors and joined the "largest unplanned human exodus of the industrial age":

> Those able to board a bicycle, moped, bullock car, bus, or vehicle of any kind did. But for most of the poor, their feet were the only form of transportation available. Many dropped along the way, gasping for breath, choking on their own vomit and, finally, drowning in their own fluids. Families were separated; whole groups were wiped out at a time. Those strong enough to keep going ran three, six, up to twelve miles before they stopped. Most ran until they dropped. (Weir 1987:17)

Although exact numbers are not known, the most conservative estimates are that the chemical accident killed at least 2,500 people and injured another 250,000. Many of the injured live with the long-term and chronic side effects of their exposure, which include lung and kidney damage, visual impairment, skin diseases and eruptions, neurological disorders, and gynecological damage (Everest 1986). In 1989, Union Carbide agreed to pay $470 million to the Indian government as compensation to the victims and their families.[1] Depending on how one determines victim status,[2] this breaks down to about $3,000 per family affected. However, legal complications have delayed these payments. As of December 1994, only about 120,000 people had received payments of $110 million, with more than 300,000 cases still pending (Dalberg 1994; Hazarika 1994). The victims of Union Carbide were prevented by law from suing Union Carbide in U.S. courts **(The Economist 1994)**

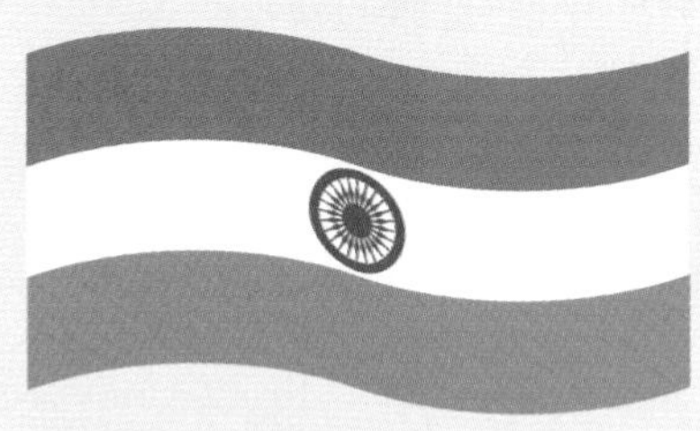

Why Focus on the Multinational Corporation in India?

In this chapter, we focus on one kind of organization—the multinational corporation. Multinational corporations, especially the 300 or so largest, are a major force in the globalization of the world economy and the prime agents behind the economic transformations taking place over the past twenty-five years (Barnet and Cavanagh 1994; McNeely 1994). In particular, we focus on the multinational corporation in India. We choose India because this country has built its economic strategy around attracting foreign investment (see "India's Economic and Growth Policy"). At the same time, the U.S. Department of Commerce has identified India as one of ten "Big Emerging Markets" (BEMs) that U.S. corporations must penetrate if they are to remain a leader in the world economy.[3] To date approximately 422 U.S. corporations have invested in India, making the United States the leading foreign-investor country (Burns 1996).

Obviously, extending business operations into foreign markets presents special issues related to organizational mission and responsibility (de la Torre 1995). To help us explore these special issues, we focus on a 1984 industrial accident involving a Union Carbide plant in Bhopal, India. The accident is considered to be one of the two worst industrial accidents in human history. Although the accident occurred in 1984, its effect on the more than 800,000 people exposed to the chemi-

India's Economic and Growth Policy

In June 1996, India's ruling party, United Front, released its economic and growth policy, which centers around and depends on foreign investment:

- Annual 7 percent growth in domestic products is targeted over the next decade to "abolish endemic poverty and unemployment."
- Rapid, labor-intensive industrialization is needed to achieve 12 percent annual growth in the industrial sector. This will require massive capital, modern technology and continued deregulation.
- India "cannot do without foreign investment" and needs at least $10 billion annually, five times current levels.
- Foreign investment in "low-priority areas" will be discouraged through government taxation and credit policies.
- If necessary, Indian companies will be protected from foreign competitors to guarantee the former a "level playing field," and competitive state-owned industries will be assisted, with the aim of making them "global giants."
- Investment in infrastructure must be increased from the current 3.5 percent to 4 percent of GDP a year to at least 6 percent. In just the next five years, India needs $200 billion to improve its power-generating capacity, oil industry, telecommunications, railroads, roads and ports.
- As the "highest priority," the fiscal deficit will be brought under 4 percent of GDP from the current 5.5 percent.
- Private ventures will be allowed into insurance, a sector formerly reserved for public-sector companies.

Source: Dahlberg (1996).

cals released are chronic and issues of compensation have yet to be resolved (Basu 1994).

The emphasis on a multinational corporation in a country such as India allows us to consider the role of multinationals in countries that are poor and have large populations.[4] These issues include (1) cultural differences between workers who come from one country and members of management who come from another country, (2) the economic conditions of the host country, and (3) the fact that the total operations of a multinational corporation are not subject to the laws of one government. This last problem—regulating the multinational corporation—is a "terribly complex and thorny issue" yet to be resolved (Keller 1986:12). In the case of Union Carbide, lawyers for that corporation argued in court that its Bhopal, India, plant is separate from its parent company and the subsidiary should be liable, not the entire corporation (Trade Environment Project 1996).

[1]On September 9, 1994, Union Carbide sold its 50.9 percent share of Union Carbide India Ltd. The sale was part of a complex deal tied to the court settlement (Lepkowski 1994).
[2]An estimated 600,000 claims have been filed against Union Carbide (Morehouse 1993).
[3]The other BEMs are the Chinese Economic Area (which includes the People's Republic of China, Hong Kong, and Taiwan), India, Indonesia, South Korea, Mexico, South Africa, Poland, and Turkey.
[4]India, with 930 million people, is the second most populous country in the world, after China. Its land area is about one-third the size of the United States. India's population increases every year by about 1.77 percent, or 17 million people (U.S. Central Intelligence Agency 1995).

Whenever there is a technology-related disaster involving the loss of human life, investigators search the scene and interview people, looking to uncover explanations. In the case of the Bhopal tragedy, some explanations that investigators found for what happened at Bhopal include technology failures, managerial and operating flaws, training deficiencies, insufficient staff, inadequate safety procedures, and widespread inattention to known risks. In the final analysis, it seemed to investigators that almost everyone involved with the Bhopal plant had in some way avoided responsibility for correcting known risks or anticipating dangers. Investigators also concluded that the disaster could have been prevented had officials made public the available knowledge about the extreme toxicity and reactivity of MIC and had they considered emergency plans to cope with an accident involving MIC (Jasanoff 1988).

These explanations and conclusions are not unique to the incident at Bhopal. They underlie other

Table 7.1 Selected Industrial Accidents of Environmental Significance, World, 1981–1996

Date	Country and Location	Origin of Accident	Products Involved	Number of Deaths	Number of Injured	Number of Evacuated
1981	Venezuela, lacoa	Explosion	Oil	145	1,000	—
1982	Venezuela, Caracas	Tank explosion	Explosives	101	1,000	—
1983	U.S., Denver	Rail accident	Nitric acid	—	43	2,000
1984	India, Bhopal	Leakage	Methyl isocyanate	2,500	50,000	200,000
1985	U.S., Institute	Fire	Aldicarbe oxide	—	140	—
1986	USSR, Chernobyl	Reactor explosion	Nuclear	8,000	300	135,000
1987	U.S., Pampa	Explosion	Toxic chemicals	3	37	—
1988	U.S., Commerce	Chemical reaction	Chlorine	—	—	20,000
1989	U.S., Prince William Sound	Spill	Oil	—	—	—
1990	Russia, Ufa	Explosion	Phenol	—	110	4,000
1991	Thailand, Bangkok	Explosion	Methyl bromide and others	3	—	6,000
1993	U.S., Milwaukee	Water supply	Protozoan cryptosporidium	40	400,000	—
1994	U.S., Torrence	Soil contamination	DDT	—	—	30 families
1996	U.S. Los Angeles	Air pollution	Particulates	5,873	—	—

Source: Organization for Economic Cooperation and Development (1987) and *New York Times Indexes* (1988, 1990, 1991, 1993, 1994, 1996).

disasters "whether they are in the form of 'mini-Bhopal's,' smaller industrial accidents that occur with disturbing frequency" in chemical plants in both developed and developing countries, or "'slow-motion Bhopals,' unseen chronic poisoning from industrial pollution that causes irreversible pain, suffering, and death" (Weir 1987:xi–xii; Trade Environment Project 1996:1). Many industrial accidents occurred in the 1980s and 1990s, posing dangers to both humans and the environment (see Table 7.1). In fact, the dominant theme of most disaster investigations is that workers at virtually every level — from maintenance workers to the chief executive officer — somehow ignore, do not receive, fail to act on, fail to enforce, or fail to pass along information that could have prevented the disaster from occurring. This finding suggests that the causes of so-called accidents go beyond individual mistakes or technological failures. Rather, it is something about the organizations themselves that halts the spread of important information and prevents responsibility for problems from falling on any one individual.

In this chapter, we examine concepts that sociologists use to analyze an **organization**, defined as a coordinating mechanism created by people to achieve stated objectives. Those objectives may be to maintain order (for example, a police department); to challenge an established order (for example, Essential Information);[5] to keep track of people (a census bureau); to grow, harvest, or process food (Pepsico); to produce goods (Sony); to make pesticides (Union Carbide); or to provide a service (a hospital) (Aldrich and Marsden 1988).

From a sociological perspective, organizations can be studied apart from the people who make them up. This is because organizations have a life that extends to some degree beyond the people who constitute them. This idea is supported by the simple fact that organizations continue on even as their members die, quit, or

[5]Essential Organizations is a nonprofit, tax-exempt organization founded in 1982 by Ralph Nader. The organization is "involved in a variety of projects to encourage citizens to become active and engaged in their communities . . . [and provides] provocative information to the public on important topics neglected by the mass media and policy makers" (Essential Information 1996).

Organization A coordinating mechanism created by people to achieve stated objectives.

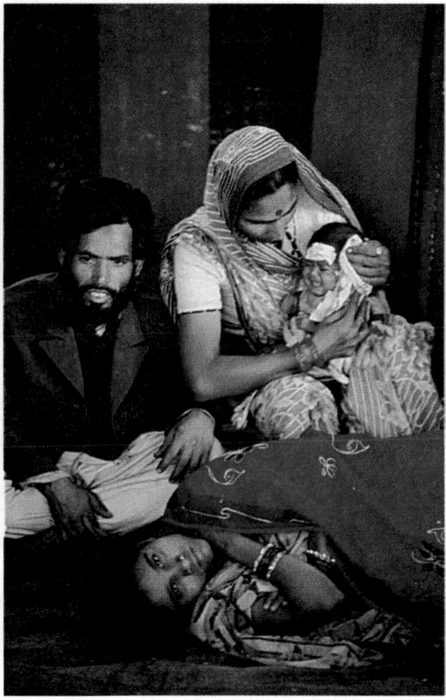

The chemical accident at Bhopal killed at least 2,500 people and injured another 250,000. Sociologists ask how the nature of organizations may contribute to tragedies like this one.

©Baldev/Sygma (left); ©Morvan/Sipa Press (right)

retire or get fired, promoted, or transferred. Organizations are a taken-for-granted aspect of life:

> Consider, however, that much of an individual's biography could be written in terms of encounters with [them]: born in a hospital, educated in a school system, licensed to drive by a state agency, loaned money by a financial institution, employed by a corporation, cared for by a hospital and/or nursing home, and at death served by as many as five organizations — a law firm, a probate court, a religious organization, a mortician, and a florist. (Aldrich and Marsden 1988:362)

Because organizations are so much a part of our lives, we rarely consider how they operate, how much power they have, and how much social responsibility they assume.

The concepts that sociologists use to study organizations can help us understand how they can be powerful coordinating mechanisms that channel individual effort into achieving goals that benefit the lives of many people. At the same time, we can use these concepts to help us see how these coordinating mechanisms can contribute to ignorance, misinformation, and failure to take responsibility for known risks. These issues of accountability are particularly relevant when they involve technologies that have the potential to cause irreparable damage to people and the environment.

Multinational corporations Enterprises that own or control production and service facilities in countries other than the one in which their headquarters are located.

The Multinational Corporation: Agent of Colonialism or Progress?

Multinational corporations (or just "multinationals") are enterprises that own or control production and service facilities in countries other than the one in which their headquarters are located. The United Nations estimates that there are at least 35,000 multinationals worldwide with 150,000 foreign affiliates (Clark 1993).[6] Multinationals are headquartered dis-

[6]According to the U.S. Internal Revenue Service (1995) in 1992, the last year for which data are available, 49,943 corporations in the United States were controlled by a "foreign person."

proportionately in the United States, Japan, and Western Europe (see "The World's Largest Industrial Corporations, 1995"). Multinationals compete against rival corporations for global market share, and they plan, produce, and sell on a multicountry and even a global scale. In addition, they recruit employees, extract resources, acquire capital, and borrow technology on a multicountry or worldwide scale (Kennedy 1993; Khan 1986; U.S. General Accounting Office 1978).

Multinationals establish operations in foreign countries for many reasons, including to obtain raw materials or make use of an inexpensive labor force (for example, a *maquila* assembly plant in Mexico, a lumber company in Brazil, a mining company in South Africa). They also establish subsidiary companies in foreign countries and employ their citizens to manufacture goods or provide services that are marketed to customers in those countries (for example, IBM Japan).

Critics maintain that multinational corporations are engines of destruction. That is, they exploit people and resources to manufacture products inexpensively. They take advantage of cheap and desperately poor labor forces, lenient environmental regulations, and sometimes nonexistent worker safety standards. According to these critics, multinational corporations represent another kind of colonialism. Advocates of multinational corporations, on the other hand, maintain that these corporations are agents of progress. They praise the multinationals' ability to transcend political hostilities, transfer technology, and promote cultural understanding.

In reality, no simple evaluation can be made that would apply to all multinationals (see Table 7.2). Obviously, at some level they "do spread goods, capital, and technology around the globe. They do contribute to a rise in overall economic activity. They do employ hundreds of thousands of workers around the world, often paying more than the prevailing wage" (Barnet and Müller 1974:151). George Keller (1996), chairman of the board and chief executive officer of Chevron Corporation, explains:

> To conduct our operations, we had to help create the necessary environment. We drilled water wells, built roads, and developed electrical power. As we made progress, the local communities grew and developed into prosperous cities. Schools and hospitals were built, and local industries emerged. (pp. 125–126)

Executive Malcolm T. Williams (1995) of Shell International Petroleum Company maintains that its Nigerian operations budget more than $20 million annually toward community development.

> This is being spent on water boreholes, school classrooms, roads, jetties, teacher sponsorship and a community healthcare scheme which is building, equipping and maintaining a network of rural clinics throughout the oil-producing areas. In addition, the company runs training skills workshops, awards more than 1,000 school and university scholarships to students from oil-producing areas each year and runs a well-established agricultural programme that dates back to the 1960s. The company believes this is an effective and worthwhile contribution but it cannot, and should not, take over the primary role and responsibility of the government in providing social infrastructure. (p. 2)

Even so, the means that multinational companies employ to achieve the maximum profit for owners and stockholders (the valued goal) are not necessarily those that alleviate a host country's problems of poverty, hunger, mass unemployment, and gross inequality (see "The Prawn Aquaculture Case" on page 200). Critics argue that, if anything, multinationals aggravate these problems because their pursuit of profit is closely related to gross social and ecological imbalances that would be obvious to anyone visiting a country such as India:

> What a curious contradiction of rags and riches. One out of every 10,000 persons lives in a palace with high walls and gardens and a Cadillac in the driveway. A few blocks away hundreds are sleeping in the streets, which they share with beggars, chewing gum hawkers, prostitutes, and shoeshine boys. Around the corner tens of thousands are jammed in huts without electricity or plumbing. . . . The stock market is booming, but babies die and children with distended bellies and spindly legs are everywhere.[7] (Barnet and Müller 1974:133–134)

How are multinationals connected with this kind of social imbalance when it seems that "without the technologies and the capital that multinationals help to introduce, developing countries would have little hope of eradicating poverty and hunger" (*Union Carbide Annual Report* 1984:107)?

Features of Modern Organizations

Sociologist Max Weber gives us one framework for understanding the two faces of organizations: organizations capable of (1) efficiently managing people, information, goods, and services on a worldwide scale and

[7]Although this description is almost twenty-five years old, it is still an accurate portrayal.

U.S. In Perspective

The World's Largest Industrial Corporations, 1995

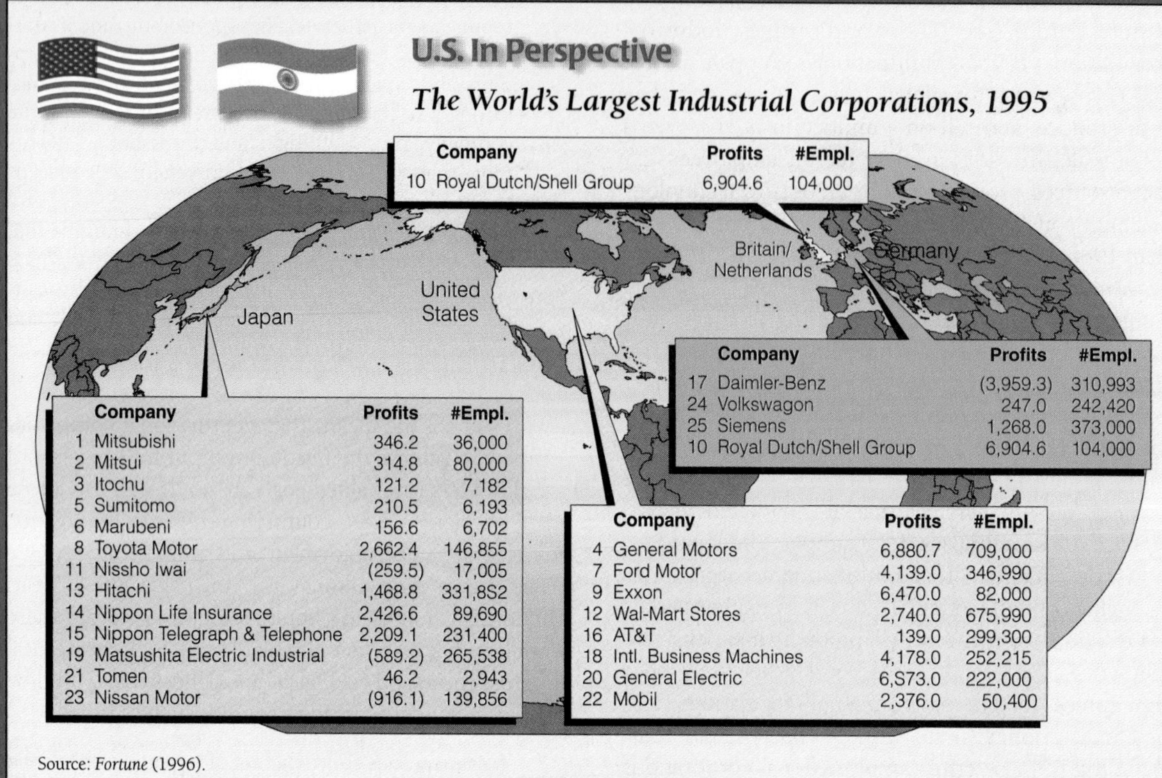

Company	Profits	#Empl.
10 Royal Dutch/Shell Group	6,904.6	104,000

Company	Profits	#Empl.
17 Daimler-Benz	(3,959.3)	310,993
24 Volkswagon	247.0	242,420
25 Siemens	1,268.0	373,000
10 Royal Dutch/Shell Group	6,904.6	104,000

Company	Profits	#Empl.
1 Mitsubishi	346.2	36,000
2 Mitsui	314.8	80,000
3 Itochu	121.2	7,182
5 Sumitomo	210.5	6,193
6 Marubeni	156.6	6,702
8 Toyota Motor	2,662.4	146,855
11 Nissho Iwai	(259.5)	17,005
13 Hitachi	1,468.8	331,8S2
14 Nippon Life Insurance	2,426.6	89,690
15 Nippon Telegraph & Telephone	2,209.1	231,400
19 Matsushita Electric Industrial	(589.2)	265,538
21 Tomen	46.2	2,943
23 Nissan Motor	(916.1)	139,856

Company	Profits	#Empl.
4 General Motors	6,880.7	709,000
7 Ford Motor	4,139.0	346,990
9 Exxon	6,470.0	82,000
12 Wal-Mart Stores	2,740.0	675,990
16 AT&T	139.0	299,300
18 Intl. Business Machines	4,178.0	252,215
20 General Electric	6,S73.0	222,000
22 Mobil	2,376.0	50,400

Source: *Fortune* (1996).

This map shows the headquarters of the twenty-five largest industrial corporations in the world in 1995. Should it be any cause for concern to the rest of the world that twenty-four of the twenty-five have their headquarters in just three countries? What factors might cause companies based in the United States, Germany, and Japan to lose touch with the needs and interests of the people in the many other countries in which they hold great economic power? What sorts of "local" concerns might they tend to overlook in those other countries?

If you were directing operations for one of these corporations, what policies might you want to establish to avoid problems, improve service, or enhance the way your company was regarded in other countries?

(2) promoting inefficient, irresponsible, and destructive actions that can affect the well-being of the entire planet. Weber's ideas about organizations are built on an understanding of rationalization and its significance in modern life.

Rationalization as a Tool in Modern Organizations

In Chapter 1, we learned that, according to Max Weber, the sociologist's main task is to analyze and explain the course and consequences of social action (actions influenced by other people, including the thoughts and feelings that lead to particular actions). We also learned that Weber classified social action into four types, according to the reasons that people pursue a goal, whatever those goals may be (for example, to increase crop yields, to make a profit, to attract a spouse). The four types are (1) traditional, (2) affectional, (3) value-rational, and (4) instrumental. Weber contended that ever since the onset of the Industrial Revolution, an individual's actions are less likely to be guided by tradition or emotion and more likely to be value-rational. He was particularly concerned about the value-rational action because, as stated in Chapter 1, the valued goal can become so all-important that people lose sight of the negative consequences that can arise from the methods used to reach that goal.

The World's Largest Industrial Corporations, 1995

Company	Headquarters In	Profits ($ Millions)	Employees
1. Mitsubishi	Japan	346.2	36,000
2. Mitsui	Japan	314.8	80,000
3. Itochu	Japan	121.2	7,182
4. General Motors	U.S.	6,880.7	709,000
5. Sumitomo	Japan	210.5	6,193
6. Marubeni	Japan	156.6	6,702
7. Ford Motor	U.S.	4,139.0	346,990
8. Toyota Motor	Japan	2,662.4	146,855
9. Exxon	U.S.	6,470.0	82,000
10. Royal Dutch/Shell Group	Britain/Netherlands	6,904.6	104,000
11. Nissho Iwai	Japan	(259.5)	17,005
12. Wal-Mart Stores	U.S.	2,740.0	675,990
13. Hitachi	Japan	1,468.8	331,852
14. Nippon Life Insurance	Japan	2,426.6	89,690
15. Nippon Telegraph & Telephone	Japan	2,209.1	231,400
16. AT&T	U.S.	139.0	299,300
17. Daimler-Benz	Germany	(3,959.3)	310,993
18. Intl. Business Machines	U.S.	4,178.0	252,215
19. Matsushita Electric Industrial	Japan	(589.2)	265,538
20. General Electric	U.S.	6,573.0	222,000
21. Tomen	Japan	46.2	2,943
22. Mobil	U.S.	2,376.0	50,400
23. Nissan Motor	Japan	(916.1)	139,856
24. Volkswagon	Germany	247.0	242,420
25. Siemens	Germany	1,268.0	373,000

Fortune (1996).

According to Weber, rationalization is a product of human technological and organizational ingenuity and proficiency, and it coincides with the specialization, the division of labor, and the mechanization that revolutionized the production process. Weber defined **rationalization** as a process whereby thought and action rooted in emotion (love, hatred, revenge, joy), superstition, respect for mysterious forces, and tradition are replaced by thought and action grounded in the logical assessment of cause and effect or the means to achieve a particular end (Freund 1968).

The thought and action guided by tradition, superstition, and emotion is different from the thought and action guided by value-rational action. We can illustrate these differences by comparing two distinctly different meanings applied to trees. One meaning is held by the Bonda, a small tribe that lives in the Orissa Mountains in India. The other is held by modern science and industry.

The Bonda reflect a culture rooted in emotion, superstition, and respect for mysterious forces. The

Rationalization A process whereby thought and action rooted in emotion, superstition, respect for mysterious forces, and tradition are replaced by thought and action grounded in the logical assessment of cause and effect or the means to achieve a particular end.

Table 7.2 Pros and Cons of Multinational Corporations (MNCs)

Frequently Heard MNC Claims of Benefits for Host Nations	Frequently Heard Criticisms of MNCs by Host Nations
Provide new products	Lead to a loss of cultural identity and traditions with the creation of new consumer tastes and demands
Introduce and develop new technical skills	Be used as channels for foreign (especially U.S.) political influence
Introduce new managerial and organizational techniques	Possess a competitive advantage over local industries
Promote higher employment	Create inflationary pressures
Yield higher productivity	Misapply host country resources
Provide greater access to international markets	Exploit host country wealth for the primary benefit of the citizens of other nations
Provide for greater accumulation of foreign exchange	Lead to loss of control by hosts over their own economies
Supplement foreign aid objectives and programs of home countries directed toward the host	Possess neither sufficient understanding nor concern for the local economy, labor conditions, and national security requirements
Serve as a point of contact for host country businesspeople and officials in the home country	Dominate key industries
Encourage the development of new ancillary or spin-off industries	Divert local savings from investment by nationals
Assume investment risks that might not have been undertaken by others	Restrict access to modern technology by centralizing research and development facilities in the home country and by employing home country nationals in key management positions
Mobilize capital for productive purposes that might have gone to other, less fruitful uses	

Source: Adapted from U.S. General Accounting Office (1978).

Bonda believe that spirits inhabit the earth, the sky, and the water; they believe that sickness, death, and poor harvests are caused by evil spirits. They are particularly respectful of spirits who live in trees and plants. Bonda priests specify which trees can and cannot be cut down. The people do not touch those trees considered to be the homes of gods and genies. Some trees are left standing if the priests believe that their removal would displease phantoms or demons and would cause them to send poor harvests, sickness, and deaths to avenge crimes against trees (Chenevière 1987). In essence, the belief that trees, plants, and animals possess souls leads people to feel reverence and respect for nature and behave accordingly toward it.

Science and technology have enabled us to break down trees and plants into various components and assign them precise functions in the larger production process. In contemporary society, for example, trees are thought of as a means to an end — a source of food, wood, rubber, quinine (a drug used to combat malaria), turpentine (an ingredient of paint thinner and solvents), cellulose (used to produce paper, textiles, and explosives), and resins (used in lacquers, varnishes, inks, adhesives, plastics, and pharmaceuticals). From a value-rational point of view, nature is something to use: "Rivers are something to dam; swamps are something to drain; oaks are something to cut; mountains are something to sell and lakes are sewers to use for corporate waste" (Young 1975:29).

Disenchantment of the world A great spiritual void accompanied by a crisis of meaning.

It is not that people in "rational" environments do not value nature on some level; rather, they place greater value on the goals of profit, employment, convenience, and global competition. Although one can argue that rationalization has released people from the bondage of superstition and tradition and has given people unprecedented control over nature. One major negative side effect of rationalization, however, is what Weber called the **disenchantment of the world** — a great spiritual void accompanied by a crisis of meaning. Disenchantment occurs when the very process of achieving a valued goal is such that it leaves people with a great spiritual void.

Weber made several important qualifications regarding value-rational thought and action. First, he used the term *rationalization* to refer to the way in which daily life is organized socially to accommodate large numbers of people, not necessarily to the way individuals actually think (Freund 1968). For example, a large chemical industry makes the products that enable millions of people to use them. The companies that make pesticides advertise them as a rational means for the quick and efficient killing of bugs in the house, in the garden, or on pets. Yet, most people who buy and use these products have no idea how the chemicals work, where they come from, or what consequences they bring except that they kill bugs. Thus, on a personal level, people deal with pesticides as if they were magic.

Second, rationalization does not assume better understanding or greater knowledge. People who live in a value-rational environment typically know little about their surroundings (nature, technology, the economy). "The consumer buys any number of products in the grocery without knowing what substances they are made of. By contrast, 'primitive' man in the bush knows infinitely more about the conditions under which he lives, the tools he uses and the food he consumes" (Freund 1968:20). Most people are not troubled by such ignorance but are content to let specialists or experts know how things work and how to make corrections when something goes wrong. People assume that if they ever need this information, they can consult an expert or go to the library and look it up.

Finally, value-rational thought is the norm; instrumental action is rare. When people are determining a goal and deciding on the means (actions) to be employed, they seldom consider and evaluate competing goals or other, more appropriate but less expedient means of reaching the stated goal. For example, people often turn to technology as the means of solving problems that they define as important. With the exception of the social critics in a society, rarely does anyone ask questions such as the four listed here to evaluate the consequences of a technology for the overall quality of life on the planet (adapted from Standke 1986, p. 66):

1. Is this technology directed toward helping achieve the highest possible human goals?
2. Does this technology use mineral and energy resources efficiently and preserve or enhance the environment?
3. Does this technology preserve or enhance "good work" for the maximum number of human beings?
4. Is this technology founded on the very best scientific and technical information in combination with the wisdom and highest values of the culture?

More often than not, people set valued goals without first considering the possible disruptive or destructive social consequences of the means or strategies used to reach them. This failure to consider such consequences is at the heart of the "destructive side" of organizations in general and of multinational corporations in particular.

Value-Rational Action: Chemical Companies in India

In the case of India, during the 1960s and 1970s the government encouraged chemical companies such as Union Carbide to locate in India (see chapter-opening map). They were supposed to be part of the Green Revolution, a plan to relieve chronic food shortages and help the country become self-sufficient in food production through agricultural technologies, including treated seeds, pesticides, and fertilizers (Derdak 1988). In addition, the chemical companies used local labor and regional raw materials and thus provided employment opportunities. The manufactured products were used to prevent malaria and other insect-borne diseases and to protect crops and harvests from insects, rodents, and diseases.

The short-term agricultural yields were indeed impressive, but the means chosen to achieve those goals have had negative long-term consequences. For exam-

This cigarette advertisement in Tokyo does not carry the kind of health warning that is legally required in the United States. When profit-making organizations are not operating under legal constraints, do they have any obligation to protect consumers from the hazards associated with their products?

©Robert Wallis

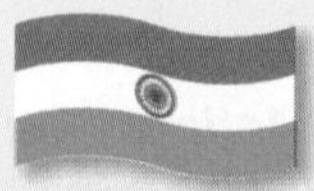

U.S. In Perspective

The Prawn Aquaculture Case

In an interview with a reporter for *Multinational Monitor,* environmental lawyer M. C. Mehta describes his work on the shrimp aquaculture case before the Supreme Court of India.

MM: Could you describe the shrimp aquaculture case which you have brought in the Supreme Court?

MEHTA: The prawn aquaculture case, filed in 1994, challenges the operation of prawn farms in the coastal areas, especially in Tamil Nadu, where they are destroying the local ecology. After four or five years, the lands where the prawn farms are located will become useless; it is very difficult to reclaim the land once it has been flooded and made into a prawn farm. First, the land becomes unfertile. Second, the land becomes saline. Third, the mangroves and other vegetation cover along the coastline are destroyed.

Community residents have undertaken a grassroots fight against the prawn companies. For the last three years, the people suffering from this activity have

This prawn farm, located in Sri Lanka, exports prawns raised there to the United States, Japan, and other countries.

©Dominic Sansoni/Panos Pictures (left); ©Anne Dowie (right)

ple, only the wealthiest Indian farmers were able to purchase the chemical technologies. In conjunction with mechanization, chemical technologies allowed the wealthier farmers to farm more efficiently. As a result, they pushed the poor, small farmer, who could not compete, off the land and out of business. In the end, millions of poor farmers migrated to the cities in search of work or deeper into woodlands and forests in search of a livelihood (Crossette 1996).

Multinationals are not charities, of course, and one could argue that they are not responsible for how the people who purchase their products use them or for unintended consequences. Nevertheless, many people question whether the multinationals and other corporations should have the right to ignore the larger long-term effects of their products on people and the environment. We can also ask whether it is acceptable for the executives and stockholders of multinational corporations to point to increases in a "poor" country's gross national product or that country's increased ability to export as evidence that multinationals are agents of progress when, by most social measures, their presence has exacerbated world poverty, world unemployment, and world inequality (Barnet and Müller 1974).

In this regard, consider the statement of Keith Richardson, the public affairs chief of B.A.T Industries,[8] the world's largest manufacturer of cigarettes. When asked whether the company did not feel some obligation to put on their cigarette packs sold in the Third World the kind of health warnings they are required to carry in Britain, the United States, and other

[8]B.A.T Industries does not use a period after the letter *T* (Moskowitz 1987:37).

agitated, held nonviolent protests, sat in satyagrahas, and many activists have been arrested on a number of occasions.

These people came to me and asked for assistance. I went to Tamil Nadu, and collected facts. On the basis of all those facts, we filed a petition in the Supreme Court of India challenging the violations committed by the prawn companies. The Supreme Court directed scientists from the National Environment Engineering Research Institute (NEERI), a prestigious institute, to go and see the situation and report to the Court.

The NEERI scientists went there and examined the situation and concluded that the prawn farms are seriously degrading the environment. Their report concludes that this activity is more harmful than beneficial.

After NEERI submitted the report, the Supreme Court issued an order saying no more agricultural land or salt pans can be converted to prawn farming.

But the mangrove forests are still being destroyed all along the coastline. The trees are being cut, and the salinity is increasing in those waters, and the people have even lost their drinking water sources.

The big multinational companies, in conjunction with Indian businessmen, have set up these prawn farms without taking any precautionary measures to control pollution. They have destroyed thousands of acres of land in this area, which is all along the nine coastal states of India.

The prawns are exported to the United States, Japan and other countries. The argument is being made that India needs foreign exchange, and that the prawn farming benefits India because it brings in foreign exchange. But I think that U.S. citizens should boycott such things, particularly from India. The prawns may be a delicacy for U.S. citizens who are eating them, but this delicacy is depriving the livelihood and bringing untold suffering to thousands of fishermen and other people who have become victims of serious pollution and environmental degradation caused by the prawn industries. The people who have lived for centuries in these areas are being displaced.

MM: What has been the reaction to the Court order?

MEHTA: The Supreme Court order is an interim one, with the final hearing on the matter set for August, 1995. Because the multinationals have a lot of money, they feel they can do anything. The rich multinationals and big businessmen housed in India have started a campaign to portray prawn farming as a very good activity. They are hiring consultants and scientists and are trying to say that prawn farming will generate employment in the country and help poor people.

Source: Interview with M. C. Mehta for *Multinational Monitor*. Copyright ©1995 Essential Information. Reprinted by permission of Multinational Monitor, PO Box 19405, Washington, DC 20036.

countries, he replied, "These are sovereign countries, and they are unenthusiastic about being told what to do by pressure groups in the U.K. and the United States. We fit in with what the government wants in each country. We let the marketplace decide. We do not try to impose" (Moskowitz 1987:40–41).

Closer to home, consider the role of automobile makers in creating American dependence on foreign oil. (Americans account for 5 percent of the world's population, yet they consume 40 percent of the gasoline in the world, two-thirds of which goes toward fuel for transportation.) Consider American consumers' preferences regarding the fuel efficiency of automobiles. When given a choice between a fuel-efficient and a gas-guzzling version of the same car model, most consumers chose the faster, less fuel-efficient model (Wald 1990). In view of the various world oil crises—most recently the Persian Gulf crisis—should corporations produce and market products that are not fuel-efficient, even if those products satisfy public wants?

The point of these examples is to show that decisions that are profitable for a corporation may not be profitable for a society because the society in question pays tremendous **externality costs**—costs that are not figured into the price of a product but that are nevertheless a price we eventually pay for using or creating a product (Lepkowski 1985). Examples of such costs include the cost of restoring contaminated and barren en-

Externality costs Costs that are not figured into the price of a product but that are nevertheless a price we eventually pay for using or creating a product.

India at a Glance

The United States has identified India as one of the ten "Big Emerging Markets" where future growth rates are posed to exceed those in the developed markets.

India is a subcontinent, nearly 2,000 miles from north to south and 1,800 miles from east to west. India's coastline is 3,800 miles long, and its area is 1.3 million square miles. Vast distances separate the most populous cities of India.

There are no national department store chains. Shopping centers are unknown, though all cities have well-known market districts. Retail sales outlets are almost always locally owned. Buying and selling is often a process of bargaining and negotiation and frequently takes place in a market. A leading manufacturer of cosmetics and personal care products sells to 200 million Indians through a network of 100,000 retail outlets across the country.

Outside the major metropolitan areas, India is an intricate network of rural villages. Poor roads make many rural districts inaccessible. Although villages may have satellite dishes and receive cable TV, moving goods is still much more difficult than broadcasting information in India.

The largest cities of India – Bombay in the west; Delhi in the north; Calcutta in the east; and Bangalore, Hyderabad, and Madras in the south – give the country a strong regional character. Just as the languages in different areas vary, so too do consumer and business preferences and marketing practices.

Population of Major Indian Cities (1991)

City	Population
Bombay	12,572,000
Calcutta	10,916,000
Delhi	8,375,000
Madras	5,361,000
Hyderabad	4,280,000
Bangalore	4,087,000

Source: India, 1991 census.

A market in Delhi, India

©Ron Levy/Liaison International

Rural women working the fields in Ladakh, India

©Robert Frerck/Woodfin Camp & Associates

PEOPLE

Population: 936,545,814 (July 1995 estimate)

Age Structure

0–14 years: 35%

15–64 years: 61%

65 years and over: 4%

Population growth rate: 1.77% (1995 estimate)

Life Expectancy at Birth

Total population: 59.04 years

Male: 58.5 years

Female: 59.61 years (1995 estimate)

Total fertility rate: 3.4 children born/women (1995 estimate)

ENVIRONMENT

Current issues: deforestation; soil erosion; overgrazing; desertifaction; air pollution from industrial effluents and vehicle emissions; water pollution from raw sewage and runoff of agricultural pesticides; tap water is not potable throughout the country; huge and rapidly growing population is overstraining natural resources.

Natural hazards: droughts; flash floods; severe thunderstorms common; earthquakes

ECONOMY

India's economy is a mixture of traditional village farming, modern agriculture, handicrafts, a wide range of modern industries, and a multitude of support services. Faster economic growth in the 1980s permitted a significant increase in real per capita private consumption. A large share of the population, perhaps as much as 40 percent, remains too poor to afford an adequate diet. Financial strains in 1990 and 1991 prompted government austerity measures that slowed industrial growth but permitted India to meet its international payment obligations without rescheduling its debt. Production, trade, and investment reforms since 1991 have provided new opportunities for Indian businessmen and an estimated 100 million to 200 million middle-class consumers. New Delhi has always paid its foreign debts on schedule and has stimulated exports, attracted foreign investment, and revived confidence in India's economic prospects. Foreign exchange reserves, precariously low three years ago, now total more than $19 billion. Positive factors for the remainder of the 1990s are India's strong entrepreneurial class and the central government's recognition of the continuing need for market-oriented approaches to economic development—for example, in upgrading the wholly inadequate communications facilities. Negative factors include the desperate poverty of hundreds of millions of Indians and the impact of the huge and expanding population on an already overloaded environment.

Major U.S. and other investments approved during the last three years include these:

Company/Country	Project	U.S. Equity ($ million)
CMS/U.S.	Thermal power generation	415
Coca Cola/U.S.	Consumer goods	70
J. Makowski/U.S.	Gas-based power	40
G.M./U.S.	Vehicles	25
Guardian/U.S.	Glass	25
Mount Everest/Singapore	Mineral water	24
Natl. Westminster Bank/U.K.	Banking	16
G.E./U.S.	White goods	15
R.R. Donelley/Singapore	Software services	15

Source: U.S. Embassy, compiled from published sources and press releases.

Five Principal U.S. Exports to India (CY–1993)

H.S. Code	Industry Sector	U.S. $ Million
8802	Aircraft	581
3100	Fertilizers	171
8411	Turbopropellers & gas turbines	167
8803	Parts of balloons, aircraft & spacecraft, etc.	117
8431	Parts for machinery	93

Five Principal U.S. Imports from India (CY–1993)

H.S. Code	Industry Sector	U.S. $ Million
7102	Diamonds	1,243
6206	Women's or girls' blouses	271
6204	Women's or girls' suits	259
7113	Articles of jewelry & parts	160
6205	Men's or boy's shirts	134

The face of a well-known Indian pop singer hovers over a family that has lived on the streets of Bombay for eighteen years. The image reminds us that despite the extreme poverty of millions of Indians, the Indian economy as a whole represents great potential for international trade.

©Steve McCurry/National Geographic Society

BEST PROJECTS FOR U.S. EXPORTS

India's market continues to offer commercial opportunities for U.S. exports of aircraft and parts; oil and gas field machinery; electronic and scientific equipment; telecommunications equipment and services; medical equipment; chemical production machinery; machine tools and metalworking equipment; computer software and services; mining equipment; computers; peripheral devices and specialized CAD (computer-aided design), CAM (computer-aided manufacturing), and CAE (computer-aided engineering) systems; port and shipbuilding equipment; specialized railway equipment; printing and graphic arts equipment; hotel and restaurant equipment; industrial process controls; environmental technologies and services; and financial services.

Source: Adapted from *Country Commercial Guides* (1997)

vironments, assisting people to cope in the face of dislocation, restoring health, and so on.

Regardless of the position we take in response to such questions, our views will be more realistic and well informed if we understand how organizations in general operate. We turn first to the systematic procedures that they follow to produce and distribute goods and services in the most efficient (especially the most cost-efficient) manner. A key concept to understanding these workings is that of the bureaucracy.

The Concept of Bureaucracy

Weber defined **bureaucracy**, in theory, as a completely rational organization—one that uses the most efficient means to achieve a valued goal, whether that goal is making money, recruiting soldiers, counting people, or collecting taxes. The following are some major characteristics of a bureaucracy that allow it to coordinate people so all of their actions center on achieving the goals of the organization:

- There is a clear-cut division of labor: each office or position is assigned a specific task toward accomplishing the organizational goals.
- Authority is hierarchical: each lower office is under the control and supervision of a higher office.
- Written rules and regulations specify the exact nature of relationships among personnel and describe the way in which tasks should be carried out.
- Positions are filled on the basis of qualifications determined by objective criteria (academic degree, seniority, merit points, or test results) and not on the basis of emotional considerations such as family ties or friendship.
- Administrative decisions, rules, regulations, procedures, and activities are recorded in a standardized format and preserved in permanent files.
- Authority belongs to the position and not to the particular person who fills the position or office. The implication is that one person can have authority over another on the job because he or she holds a higher position, but those in higher positions can have no authority over another's personal life away from the job.
- Organizational personnel treat clients as "cases" and "without hatred or passion, and hence without affection or enthusiasm" (Weber 1947:340). This approach is necessary because emotion and special circumstances can interfere with the efficient delivery of goods and services.

Bureaucracy An organization that uses the most efficient means to achieve a valued goal.

Ideal type A deliberate simplification or caricature in that the characteristics emphasized exaggerate certain aspects of a bureaucracy, which makes them the objects of comparison.

Taken together, these characteristics describe a bureaucracy as an **ideal type**—ideal not in the sense of being desirable but as a standard against which real cases can be compared. An ideal type is a deliberate simplification or caricature in that the characteristics emphasized exaggerate certain aspects of a bureaucracy, which makes them the objects of comparison (Sadri 1996). In other words, real cases can be compared against the ideal. Anyone involved with an organization realizes that the actual behavior departs from the ideal. Thus, one might ask, what is the use of listing essential characteristics if no organization exemplifies them? This list is a useful tool because it identifies important organizational features. (Note, however, that having these traits does not guarantee that things run perfectly; the rules and policies themselves can cause problems.) Rather, by comparing the actual operation with the ideal, one can determine the extent to which an organization departs from these traits or adheres to them too rigidly.

In the case of the Bhopal disaster, serious problems can be linked to either rigid adherence to or blatant departures from official rules, regulations, and procedures. The following list summarizes some of the problems:

- A refrigeration unit designed to keep MIC cool and to inhibit chemical reactions had been turned off for several months.
- Two of the plant's three main safety systems were not working.
- On the eve of the disaster, an employee who did not meet the plant's training requirements was assigned the task of cleaning out an improperly sealed pipe leading to the MIC tank. (Workers suspect that this is probably how water entered the tank.)
- The problem tank containing MIC was filled to 73 percent of capacity rather than the recommended

50 percent. The restriction was imposed so that, in the event of a reaction, there would be more time for corrective action, because a less full tank would cause pressure in the tank to rise less quickly.

- Plant operators failed to move some of the MIC in the problem tank to a spare tank, as required, because they said the spare was not empty, as it should have been.
- Training, experience, and educational qualifications of employees at the Bhopal plant were reduced sharply after the plant began to lose money.
- Instruments monitoring the chemical were unreliable.
- The Bhopal plant did not have the computer system that was present in sister plants in the United States to monitor the chemicals and alert the staff quickly to problems.
- There were no effective public warnings of the potentially disastrous effects of a chemical accident. Officials did not provide contingency plans or shelters, nor did they distribute brochures or other materials to the people in the area surrounding the plant.
- Most employees panicked as the gas escaped. They failed to behave in recommended ways that could have helped to evacuate residents in company buses (see Everest 1986; Jasanoff 1988; Shabecoff 1988a, 1988b, 1989; Weir 1987 for details on all of these problems).

This list of problems clearly shows that employees at all levels—from managers who did not hire qualified people, inspect equipment, or draft contingency plans in case of emergency to the actions of employees who routinely violated standard operating procedures — did not behave in a responsible manner. In light of this fact, sociologists ask what specific factors within organizations might be behind these problems.

Factors That Influence Behavior in Organizations

On paper, the job descriptions, relationships among personnel, and procedures for performing work-related tasks are well defined and predictable. However, the actual workings of organizations are not as predictable, because the people involved with organizations vary in the extent to which they adhere to rules and regulations. Three factors that influence how people act are the informal relationships they form with others, the way they are trained to do their jobs, and the way their performances are evaluated.

Formal Versus Informal Dimensions of Organizations

Sociologists distinguish between formal and informal aspects of organizations. The **formal dimension** consists of the official, written guidelines, rules, regulations, and policies that define the goals of the organization and its relationship to other organizations and with integral parties (for example, the government or the stockholders). This term also applies to the roles, the nature of the relationships among roles, and the way in which tasks should be carried out to realize the goals.

The **informal dimension** includes owner- or employee-generated norms that evade, bypass, do not correspond with, or are not systematically stated in official policies, rules, and regulations. In the most general sense, this term applies to behavior that does not correspond to written plans (Sekulic 1978). Examples include unwritten rules about standards of interaction, the appropriate content of conversations between employees of different ranks, and the pace at which people should work. A boss who expects employees to do work unrelated to their official job descriptions such as babysitting his or her children, running personal errands, or doing handy work around his or her house is an example of the informal dimension of organization.

Another area of informal organization is worker-generated norms that govern output or physical effort. These include informal norms against working too hard (those who do so are often called "rate busters"), working too slowly, or slacking off, as well as norms about the number and length of coffee breaks and the length of the lunch break.

Both positive and negative consequences are associated with informal norms. Informal norms about bending the rules, cutting through red tape, and han-

Formal dimension The official, written guidelines, rules, regulations, and policies that define the goals and roles of the organization and its relationship to other organizations and with integral parties.

Informal dimension Owner- or employee-generated norms that evade, bypass, do not correspond with, or are not systematically stated in official policies, rules, and regulations.

dling unusual cases or problems, for example, can increase organizational efficiency and effectiveness. Informal norms about after-work activities, friendships, and unofficial communication channels can promote loyalty and work satisfaction. On the negative side, informal norms that put the worker and public safety at risk can have destructive consequences.

At the Bhopal plant, the informal methods that workers used to monitor leaks explain, in part, how the chemical reaction went out of control at the plant on December 3, 1984. Workers could not rely on the alarm systems and gauges that monitored chemical pressure, temperature, level, flow, and composition because these monitoring devices were notoriously unreliable. Usually, the gauges did not show correct readings or did not work. When they were working, they were inadequate because the gauge range was too limited to show critical danger levels. By all accounts, the management was generally unresponsive to workers' complaints about plant equipment. Consequently, workers operated on the assumption that the instruments were inaccurate and monitored the MIC and other chemical leaks according to whether their eyes watered or burned. Their eyes burned and watered frequently, however, so they often ignored even this signal. This method of leak detection was a direct violation of official plant procedures. The Union Carbide technical manual on MIC states, "Although the tear gas effects of the vapor are extremely unpleasant, this property cannot be used as a means to alert personnel." The worker-generated norm for monitoring leaks was also used by Union Carbide workers at the West Virginia chemical production plant (Weir 1987).

The operator monitoring the gauges on the evening of the Bhopal crisis was not concerned when pressure gauges connected to the MIC tank rose substantially between 10:30 and 11:00 P.M., and he felt no need to report this change to the shift supervisor. Other operators smelled MIC and noticed that their eyes were irritated, but they maintained that this was no cause for alarm. When they reported the possibility of a leak to the plant supervisor, he decided to look into it after a tea break (Kurzman 1987).

Two other factors affect the way that people behave in organizations—specifically, whether they behave in flexible or rigid ways. One factor is how people are trained to do their jobs. The other is how the organization evaluates worker performance.

Trained incapacity The inability to respond to new and unusual circumstances or recognize when official rules and procedures are outmoded or no longer applicable.

Trained Incapacity

If an organization is to operate in a safe, creditable, predictable, and efficient manner, its members need to follow rules, guidelines, regulations, and procedures. Organizations train workers to perform their jobs a certain way and reward them for good performances. However, when workers are trained to respond mechanically or mindlessly to the dictates of the job, they risk developing what economist and social critic Thorstein Veblen (1933) called **trained incapacity,** the inability to respond to new and unusual circumstances or recognize when official rules and procedures are outmoded or no longer applicable. In other words, workers are trained to do their jobs only under normal circumstances and in a certain way; they are not trained to respond in imaginative and creative ways or to anticipate what-if scenarios so that they can perform under a variety of changing circumstances.

In her 1988 book, *In the Age of the Smart Machine,* social psychologist Shoshana Zuboff distinguishes between work environments that promote trained incapacity and those that promote empowering behavior. Zuboff's conclusions are the result of more than a decade of field research in various work environments, including pulp mills, a telecommunications company, a dental insurance claims office, a large pharmaceutical company, and the Brazilian offices of a global bank. All of these workplaces had one trait in common: the workers were learning to use computers. We focus here on the experiences of pulp mill workers.

Zuboff found that pulp mill employees who had worked in conventional mills all of their lives were overwhelmed at first by the new condition of having to run the mill while seated at computer screens. These comments illustrate their reactions:

> With computerization I am further away from my job than I have ever been before. I used to listen to the sounds the boiler makes and know just how it was running. I could look at the fire in the furnace and tell by its color how it was burning. I knew what kinds of adjustments were needed by the shades of color I saw. A lot of the men also said that there were smells that told you different things about how it was running. I feel uncomfortable being away from these sights and smells. Now I only have numbers to go by. I am scared of that boiler, and I feel that I should be closer to it in order to control it. (p. 63)

Because of advances in computer technology, many organizations now run daily production and manufacturing operations from computer terminals and control panels.

©John Coletti/Stock, Boston

> With the change to the computer it's like driving down the highway with your lights out and someone else pushing the accelerator. (p. 64)

> What strikes me as most strange, hardest to get used to, is the idea of touching a button and making a motor run. It's the remoteness. I can start it from up here, and that is hard to conceive. I can be up in the control room and touch the keyboard, and something very far away in that process will be affected. It takes a while to gain confidence that it will be OK, that what you do through the terminal actually will have the right effects. . . . It's hard to imagine that I am sitting down here in front of this terminal and running a whole piece of that plant outside. The buttons do all the work. (p. 82)

Zuboff believes that management can choose to use computers as automating tools or informating tools. *To automate* means to use the computer to increase workers' speed and consistency, as a source of surveillance (for example, by checking up on workers or keeping precise records on the number of keystrokes per minute), and to maintain divisions of knowledge and thus a hierarchical arrangement between management and workers. The pulp mill workers' comments show that this choice has resulted in trained incapacity:

> Currently, managers make all the decisions. . . . Operators don't want to hear about alternatives. They have been trained to do, not to think. There is a fear of being punished if you think. This translates into a fear of the new technology. (p. 74)

> Sometimes I am amazed when I realize that we stare at the screen even when it has gone down. You get in the habit and you just keep staring even if there is nothing there. (p. 66)

> We had another experience with the feedwater pumps, which supply water to the boiler to make steam. There was a power outage. Something in the computer canceled the alarm. The operator had a lot of trouble and did not look at the readout of the water level and never got an alarm. The tank ran empty, the pumps tripped. The pump finally tore up because there was no water feeding it. (p. 69)

On the other hand, management can choose to use computers as informating tools. *To informate* means to empower workers with knowledge of the overall production process, with the expectation that they will make critical and collaborative judgments about production tasks. The pulp mill workers who use the computer as an informating tool experience work very differently than those who use the computer as an automating tool. The following quotes illustrate this point:

> To do the job well now you need to understand this part of the mill and how it relates to the rest of the

> plant. You need a concept of what you are doing. Now you can't just look around you and know what is happening; you can't just see it. You have to check through the data on the computer to see your effects. And if you don't know what to look for in the data, you won't know what's happening. (p. 94)

> [Before automation we] never expected them to understand how the plant works, just to operate it. But now if they don't know the theory behind how the plant works, how can we expect them to understand all of the variables in the new computer system and how these variables interact? (p. 95)

> If something is happening, if something is going wrong, you don't go down and fix it. Instead, you stay up here and think about the sequence. . . . You get it done through your thinking. But dealing with information instead of things is very . . . well, very intriguing. I am very aware of the need for my mental involvement now. I am always wondering: Where am I at? It all occurs in your mind now. (p. 75)

> The computer makes your job easier . . . but it also makes things more complicated. You have to know how to read it and what it means. That is the biggest problem. What does that number actually mean? You have to know this if you want to really learn how to trust the technology. (p. 81)

Virtually all investigative reports of the Bhopal disaster point to trained incapacity as an important contributing factor in causing the runaway chemical reaction. They especially consider the training supplied after the plant began to lose money and was put up for sale. In the early days of operation, well-educated Indian supervisors who had been trained at company headquarters in the United States frequently reminded workers to wear masks, goggles, and protective clothing while handling chemicals. Over time, however, this vigilance weakened, and protective equipment fell into disuse or was not replaced when it wore out. Worker training was reduced from an intensive one-year course to a four-month course and subsequently to a thirty-day crash course. Most workers later reported that they were trained to master certain steps but not to handle the chemical in all its conditions. They had no knowledge of the production process as a whole or of the rationale behind many of the rules and procedures, as the following comments from several Bhopal operators reveal:

> I was trained for one particular area and one particular job; I don't know about other jobs. During training they just said "These are the valves you are supposed to turn, this is the system in which you work, here are the instruments and what they indicate. That's it." (Diamond 1985a:A7)

> [My three months of instrument training and two weeks of theoretical work taught me to operate only one of several MIC systems.] If there was a problem in another MIC system, I don't know how to deal with it. (Diamond 1985a:A7)

> [I] knew the pipe [leading to the MIC storage tank] was unsealed but "it was not my job" to do anything about it. (Diamond 1985b:A6)

> The management said MIC could give you a rash on your skin or irritate your eyes. They never said it could kill you. No one at this plant thought MIC could kill more than one or two people. (*New York Times* 1985a:A6)

As further evidence that workers were trained to do their jobs in a rote manner, most plant workers at all levels had little understanding of the chemicals they worked with, especially of how the chemicals might react under unusual circumstances. Union Carbide published an MIC manual stating that MIC "may cause skin and eye burns on contact. Vapors are extremely irritating and cause chest pain, coughing and choking. May cause fatal pulmonary edema. Repeated exposure may cause asthma" (*New York Times* 1985a:A6). The manual warned that the chemical was toxic, volatile, and flammable. Yet, many workers seemed genuinely surprised that MIC was so extremely dangerous. In fact, many did not receive the manual, which was written in technical English and which even English-speaking workers found difficult to read and understand.

Many newspaper, magazine, and network news reporters suggested that trained incapacity is a problem unique to underdeveloped countries such as India and that, in the case of Bhopal, the lack of what-if thinking is rooted in Hindu beliefs. Some reports suggested that inferior Indian labor and resources were behind the tragedy. Many accounts of the Bhopal crisis cite the fact that Indian law required Union Carbide to design, engineer, build, operate, and staff its Bhopal chemical plant with local labor, materials, and machines unless the company could show that local resources were not available. These accounts insinuate that Indian laws required Union Carbide to compromise safety standards. In a similar vein, many of these reports point out that foreign companies were encouraged to use manual production systems to create more jobs for the large unemployed Indian population.

Before we accept such conclusions, we need to ask how many American chemical workers are issued manuals and trained to understand the properties of the chemicals with which they work. The result of an Environmental Protection Agency (EPA) study suggests that many people in the United States who work with chemicals do not understand their hazards or know how to handle them except under routine conditions. Around the time of the Bhopal crisis, between 1980 and 1985, 6,928 chemical accidents occurred in the United States, causing 139 deaths and 1,500 injuries (Diamond 1985c). The large number of accidents reflects that in the United States, as elsewhere, standardized training programs and intensive worker training for runaway chemical reactions are rare. "The best company programs . . . include a month of classes in safety principles and two-week refresher courses yearly. The worst have no classroom or refresher training. . . . They say here's your safety gear—we wear it when things go wrong" (Diamond 1985c:D11).

Another EPA report showed that between 1961 and 1989 seventeen industrial accidents occurred in the United States in which the chemicals released exceeded the amount and toxicity of those released at Bhopal. Fortunately, these chemical releases did not have the deadly consequences of the Bhopal accident because they happened in remote regions, the wind blew the chemicals away from heavily populated areas, or the chemicals leaked in a liquid state. Liquid chemicals, although highly toxic, diffuse more slowly, so that workers have time to contain them or to allow people to be evacuated (Shabecoff 1989). The point is that if people in the United States view the Bhopal crisis as something unique to countries populated by so-called illiterate peasants, they are discouraged from analyzing their own industries and the shortcomings in the ways in which workers are trained.

So far we have looked at informal relationships between people in organizations and at the ways in which people are trained to do their jobs to understand the factors that determine how closely they adhere to organizational rules and regulations. Now we turn to a third factor: how organizations evaluate job performances.

Statistical Records of Performance

In large organizations, supervisors often compile statistics on absenteeism, profits, losses, customer satisfaction, total sales, and production quotas as a way to measure individual, departmental, and overall organizational performance. Such measures can be convenient and useful management tools because they are considered to be objective and precise and because they permit systematic comparison of individuals across time and departments. On the basis of numbers, management can reward good performances through salary increases and promotions and can take action to correct poor performances. Sociologist Peter Blau examined the problems that can occur when managers use faulty measures without taking their shortcomings into consideration.

One problem with statistical measures of performance is that a chosen measure may not be a valid indicator of what it is intended to measure, or it may measure performance by too narrow a criterion. For example, occupational safety is often measured by the number of accidents that occur on the job. On the basis of this indicator, the chemical industry has one of the lowest accident rates of all industries. This indicator, however, has been criticized as too narrow and lacking validity: chemical workers may be less likely to suffer physical injury on the job than to suffer illnesses whose symptoms go unrecognized as related to chemical exposures. Furthermore, exposure-related illnesses may take years to develop.

A second problem with statistical measures of performance is that they encourage employees to concentrate on achieving good scores and to ignore problems generated by their drive to score well. In other words, people tend to pay attention only to those areas that are being measured and to overlook those for which no measures exist. If quarterly profit is used as the main indicator of corporate performance, management may do as Bhopal management did—cut employees by 25 percent and cut costs in such critical areas as worker safety, plant maintenance, and employee training (Wexler 1989). Such single-minded efforts are most prevalent in organizations in which employee, departmental, or company performance is evaluated according to rigid measures and where strong sanctions are applied if workers do not meet target figures.

These problems associated with statistical measures of performance do not mean there should be none. On the positive side, each year since 1986 chemical companies in the United States are required to report (as a result of the Bhopal disaster) the amount and kind of chemical emissions that occurred in the past year. The report known as the *Toxic Release Inventory* shows that since 1988 the chemical industry has cut annual toxic emissions by 176 million pounds, or 47 percent (see Table 7.3).

In the case of Bhopal, no measure existed to monitor how often equipment such as refrigeration units, safety systems, and gauges failed to work. Because no measure was in place, responsibility for this dimension of operational safety obviously was never assigned to a

Table 7.3 1993 Top Increasers and Decreasers in Air/Water/Land Releases (in millions of pounds)

		Increase (in millions of pounds)		
Company	**Location**	**From**	**To**	**Chemical**
IMC-Agrico Co.	Saint James, LA	72.3	113.7	Phosphoric acid
Magnesium Corp. of America	Rowley, UT	57	67	Chlorine
Coastal Chem. Inc.	Battle Mountain, NV	.291	4.7	Ammonium nitrate
Arcadian Ohio L.P.	Lima, OH	0	3.4	Total releases
IMC-Agrico Co.	Uncle Sam, LA	5.8	11.6	Sulfuric acid
Cyprus Miami Mining Corp.	Claypool, AZ	2.8	5.6	Copper compounds
Phelps Dodge Mining Co.	Playas, NM	10.4	13.7	Copper compounds
Zinc Corp. of America	Bartlesville, OK	0	2.6	Copper, lead, and zinc compounds
Chemetals Inc.	New Johnsonville, TN	2.7	5.3	Manganese compounds
Elkem Metals Co.	Marietta, OH	3.4	4.0	Ammonia
Stone Container Corp.	Coshocton, OH	0	2.0	Formaldehyde, methanol, acetone, benzene, and acetaldehyde
American Chrome & Chemicals	Corpus Christi, TX	10	12	Chromium compounds
Inland Container Corp.	Rome, GA	.980	2.1	Methanol
Lenzing Fibers Corp.	Lowland, TN	20.4	22.3	Carbon disulfide
Doe Run Co.	Herculaneum, MO	6.1	7.8	Lead and zinc compounds
		Decrease (in millions of pounds)		
Company	**Location**	**From**	**To**	**Chemical**
Magma Copper Co.	San Manuel, AZ	22.3	0.90	Copper, zinc, lead, arsenic
Inland Steel Co.	East Chicago, IN	27.8	7.7	Manganese, chromium
Arcadian Fertilizer	Geismar, LA	35.0	18.9	Phosphoric acid, sulfuric acid
Mobile Mining and Minerals	Pasadena, TX	14.0	0.14	Phosphoric acid, sulfuric acid, ammonia
Dupont	Leland, NC	11.4	1.2	Methanol
IMC Fertilizer, Inc.	Mulberry, FL	12.7	6.8	Phosphoric acid
Mississippi Chemical Corp.	Yazoo City, MS	10.4	4.6	Ammonia
Amoco Oil Co.	Texas City, TX	4.2	0.18	Naphthalene, 1,2,4-trimethylbenzene ethylbenzene, toluene zylene (mixed isomers), chromium compounds & zinc combined
Simpson Paper Co.	Eureka, CA	4.6	0.55	Methanol
Unocal Petroleum	Kenai, AK	8.4	4.7	Ammonia, urea

U.S. Environmental Protection Agency (1996).

position in the ranks of Union Carbide or in any of the many Indian government agencies.[9] As a result, everyone who was interviewed, from the president of Union Carbide to the Bhopal police, maintained that they were not responsible for operational safety. The following quotes illustrate this point:

> But safety was the responsibility of the Indian personnel. It was "a local issue." (Warren M. Anderson, chairman, Union Carbide Corporation, quoted in Engler 1985:498)

> We expected that Union Carbide Corporation would try and palm off the blame on Union Carbide, India. But UCC cannot escape responsibility. They should have ensured that such [safety] lapses could not occur. (Kamal K. Pareek, senior project engineer,

[9]A number of Indian agencies and appointed officials monitored the Bhopal plant at some level. These include the Ministry of Petroleum and Chemicals, the Ministry of Agriculture, the Directorate General of Technical Development, the Ministry of Finance, the Department of Science and Technology, the Controller of Imports and Exports, the Chief Inspector of Explosives, the Plant Protection Advisor, the Bureau of Industrial Costs and Prices, the Ministry of Commerce, and the Ministry of Industry (Bleiberg 1987).

> Bhopal Union Carbide plant, quoted in Diamond 1985c:A1)

> Responsibility for plant maintenance, hiring and training of employees, establishing levels of training and determining proper staffing levels rests with plant management. (Union Carbide spokesperson, quoted in Diamond 1985a:A6)

> We do not design, maintain and operate plants. We only check to see that there are enough protective masks and safety guards. (Factory inspector, India Labor Department, quoted in Diamond 1985a:A6)

> It is the basic responsibility of the company to make the community aware of [the hazards]. I have not seen anything so far to show that this was done. (Arjun Singh, Madhya Pradesh government, quoted in Reinhold 1985a:A8)

As we have seen, many potential problems are associated with statistical measures of performance. To ensure that important conditions such as occupational safety are monitored, it is advisable to develop thoughtful and accurate indicators to measure them, assign responsibility to a definite position, and tie the measures to the evaluation of performance by persons occupying that position. When no such system is in place, responsibility for accident prevention never rests squarely with specific people. This diffusion of responsibility also can occur when decision makers rely on experts for advice or when the power to make decisions is concentrated in the hands of a few people at the top.

Obstacles to Good Decision Making

In his writings about bureaucracy, Weber emphasizes that power was not located in the person but in the position that a person occupied in the division of labor. The kind of power described by Weber is clear-cut and familiar: a superior gives orders to subordinates, who are required to carry out those orders. The superior's power is supported by the threat of sanctions: demotions, layoffs, firings. Sociologists Peter Blau and Richard Schoenherr (1973) recognize the importance of this form of power but identify a second, more ambiguous type — expert power — that they believe is "more dangerous than seems evident for democracy and . . . is not readily identifiable as power" (p. 19).

Expert Knowledge and Responsibility

According to Blau and Schoenherr (1973), expert power is connected to the fact that organizations are becoming increasingly professionalized. **Professionalization** is a trend in which organizations "hire" experts (such as chemists, physicists, accountants, lawyers, engineers, psychologists, or sociologists) who have formal training in a particular subject or activity that is essential to achieving organizational goals. Experts are not trained by the organization, however; they receive their training in colleges and universities. Theoretically, they are allowed to be self-directed and are not subjected to narrow job descriptions or direct supervision. Experts use the frameworks of their chosen profession to analyze situations, solve problems, or invent new technologies. From the experts' viewpoints, the information, service, or innovation they provide to the organization is technical and neutral. They do not necessarily think about or have control over the application of that information, service, or invention.

As a case in point, consider that researchers at University of California, San Francisco, invented, among other things, the hepatitis B vaccine, the basic technology for gene splicing/genetic engineering, and the human growth hormone. The genetic engineering technology, for example, allows biotech companies "to turn bacteria and yeast into chemical factories for the production of proteins such as the growth hormone and the hepatitis vaccine" (Jacobs 1996:A23). Whereas UC researchers may have been motivated to improve the human condition, the corporations marketing the technology are motivated by profit. Thus, the growth hormone invented with the goal of treating dwarfism may be sold as a solution for people dissatisfied with their height.

Blau and Schoenherr regard this arrangement between experts and organizations as problematic because it leaves nobody accountable for the actions of powerful corporations and because it complicates attempts to find individuals "whose judgments [are] the ultimate source of a given action" (pp. 20–21). This situation is complicated for two reasons. First, the recommendations and judgments of experts rest on specialized knowledge and training. The experts may understand principles of accounting, physics, biology, chemistry, or sociology, but their training for the most part is compartmentalized; they know one subject very well, but they do not know other subjects. For example, a chemist may be able to design a pesticide, but he

Professionalization A trend in which organizations hire experts with formal training in a particular subject or activity that is essential to achieving organizational goals.

Political analyst Robert Michels believed that oligarchy, or rule by the few, is the inevitable tendency of large organizations. Size alone makes it impossible to get everyone's input into organizational decisions.

©Brad Markel/Gamma-Liaison

or she has not been trained to consider the limitations of the people who use it. Similarly, a sociologist may understand the abilities and limitations of people who use a pesticide but may not understand the technology. Because the sociologist does not understand the chemistry, he or she cannot design the details of a program to educate consumers or speak knowledgeably about the social consequences of that technology.

Blau and Schoenherr emphasize that the men and women who give expert advice are decent people but that their position, training, and point of view make them unable to anticipate, plan, or control unintended consequences. In addition, decision making in large organizations is complex because no single person provides all of the input that goes into a decision. A decision is a joint product of information and judgments by a variety of experts. Often the decision maker does not understand the principles underlying an expert's recommendations and judgments. The problem is that when something goes wrong, the experts claim that they only provided the patent, information, suggestions, and recommendations, whereas management claims that it cannot predict the consequences of an invention or a service that only the experts understand.

Oligarchy Rule by the few, or the concentration of decision-making power in the hands of a few persons who hold the top positions in a hierarchy.

The Problems with Oligarchy

Oligarchy is rule by the few, or the concentration of decision-making power in the hands of a few persons who hold the top positions in a hierarchy.

> One of the most bizarre features of any advanced industrial society in our time is that the cardinal choices have to be made by a handful of men . . . who cannot have firsthand knowledge of what those choices depend upon or what their results may be. . . . [And by] "cardinal choices," I mean those which determine in the crudest sense whether we live or die. For instance, the choice in England and the United States in 1940 and 1941, to go ahead with work on the fission bomb: the choice in 1945 to use that bomb when it was made. (Snow 1961:1)

Political analyst Robert Michels believed that large formal organizations tended inevitably to become oligarchical, for the following reasons. First, democratic participation is virtually impossible in large organizations. Size alone makes it "impossible for the collectivity to undertake the direct settlement of all the controversies that may arise" (Michels 1962:66). For example, Union Carbide (headquartered in Danbury, Connecticut) employs almost 38,000 people in plants located in more than twenty-four countries (Stopford 1992). At the time of the Bhopal crisis, Union Carbide had fourteen factories, twenty-eight sales branches, and employed 9,000 workers in India alone (Diamond 1985a; Lepkowski 1992). "It is obvious that such a gigantic number of persons belonging to a unitary organization cannot do any practical work upon a system of direct discussion" (Michels 1962:65).

Second, as the world becomes more interdependent and technology becomes increasingly complex, many organizational features become incomprehensible to workers. As a result, many employees work toward achieving organizational goals that they did not define, cannot control, may not share, or may not understand. This lack of knowledge prevents workers from participating in or evaluating decisions made by executives.

A danger of oligarchy is that those who make decisions may not have the necessary background to understand the full implications of the decisions. For example, Warren Anderson, the chairman of the Union Carbide Corporation at the time of the Bhopal accident, stated, "It never entered my mind that an accident such as Bhopal could happen" (Engler 1985:495). In addition, decision makers may not consider the greater good and become preoccupied with preserving their own leadership. Guarding against these effects of oligarchy requires that the average worker and the general public be interested, attentive, and informed. As more and more people hold jobs that require them to deal with science and technology, and as our daily lives become increasingly dependent on technology, people need to understand and feel responsible for understanding what is going on around them. Otherwise, technology decisions will be made by others on their behalf, and they will be forced to accept the consequences.

The reaction of Bhopal residents to the unexpected gas leak dramatizes how an uninformed and inattentive public cannot take the precautions they need to protect themselves when a crisis occurs. Although an emergency alarm sounded at the Bhopal plant, most people living near the plant did not know what it meant; they were conditioned to hearing alarms because alarms sounded at the plant an average of twenty times a week. Many assumed that the alarm signaled a change in shift or a practice drill or that it had gone off by accident:

> We used to hear sirens go off often. . . . We thought it was routine—a change of shift or a fire in the factory. We were never alarmed. . . . We thought they were making powder, some kind of powder. . . . We were never told anything about poison, by the company or by the government. (Sabir Kahn, Bhopal resident, quoted in Reinhold 1985a:A8)

Few residents of Bhopal knew what chemicals were manufactured at the plant, how toxic they were, how to detect escaped chemicals by smell or sight, or what to do if chemicals were released accidentally from the plant:

> Several Bhopal residents said many of the people living near the plant thought it produced "Kheti Ki Dawai," a Hindi phrase meaning "medicine for the crops." Thus, [they thought] the plant's output was healthful. (Sabir Kahn, Bhopal resident, quoted in *New York Times* 1985a:A6)

One reason the chemical link at Bhopal affected so many people was that approximately 20 percent of the city's residents lives in squatter slums like this one located in Bombay, India.

©Jean-Luc Bitton/Sipa Press

The one response they knew was to run. If they had known to lie close to the ground and cover their faces with a wet cloth, most of the victims would have escaped injury or death.

In addition to this general lack of knowledge, physicians, firefighters, police, and city officials did not know the hazards that the plant presented or what precautions to take in an emergency:

> We had no inkling of what kind of emergency steps should be taken in such a situation. (Nily Chaudhuri, chairman, Central Water and Air Pollution Board, quoted in Stevens 1984:A10)

> What shocked me was to find out from the mayor of Bhopal that he didn't have the vaguest idea that this

could happen. (Stephen Solarz, U.S. congressman and chairman of the Subcommittee on Asian and Pacific Affairs, quoted in Hazarika 1984:A3)

To complicate matters even further, the physicians at the two hospitals where the victims went for treatment had no information about MIC in general or how to treat people exposed to the chemical. Physicians at one hospital—the Hemida Hospital, a 1,000-bed facility—faced 20,000 patients desperately short of medical supplies and without a firm idea of how to treat patients' symptoms (Wexler 1989). It was not that the Bhopal physicians were incompetent. Rather, they could not find out this information from Union Carbide because the company treated results from research that it commissioned on animals in 1963 and 1970 as proprietary. Moreover, almost nothing was published on MIC in medical journals as "some scientists who tried to work with it said it was too dangerous" (*Scientific American* 1995:16).

Many newspaper reporters in the United States attributed the disorder to the reaction of a poor, illiterate peasant population whom they considered incapable of understanding sophisticated technology (Everest 1986).[10] Residents of Institute, West Virginia, however, responded to a chemical leak at the local Union Carbide plant in the same disorganized way.[11] When they noticed a strong odor seeping into their houses, they did not know what to do, as evidenced by statements from various residents (Franklin 1985):

> I kept burying my head into the pillow, trying to get rid of the smell. Then I finally woke up, and I had my shoes on in no time. If it had been something really bad, it would have been too late. People would have been dead in their beds.
>
> I didn't know what it was when I saw that white cloud go up, disperse and spread out. I just locked the building, picked up my wife, and took off.
>
> The whistle was blowing and there was a terrible smell. We couldn't find anything on the radio or television, and my husband insisted that we leave. We didn't know what it was. There was a breakdown in communication.
>
> I didn't know what to do. (*New York Times* 1985b:A12)

It is needless for people, even the poorest people, to be unprepared or passive victims of the decisions of a government or a large multinational corporation. The ecological movement led by Sunderlal Bahuguna, a sixty-year-old man working to save the forests of northern India, illustrates the power of individuals. Bahuguna travels from village to village teaching the people to conserve forests, replant trees when they cut them, and protect the forests from commercial enterprises. He teaches children about the chemistry of trees and urges them to hug trees and sing a song about them: "Do not touch me with an ax, I, too, feel pain. I am your friend. I bring you fresh air. I bring you water. I always bow down before you. Why do you cut me down?" (Hutchison 1989:185). The women of some villages have hugged trees in successful efforts to prevent contractors from cutting them down and have formed human chains across access roads to prevent contractors from reaching timbering sites.

Another illustration is the work of Kishan Baburao Hazare. Hazare has helped the people of Ralegan Sidhi, a village in western India, to transform the area. At one time the region was an ecological disaster due to deforestation. Now the village is able to produce enough food to be self-sufficient and export half of what it produces (Nikore and Leahy 1993). Essentially, Hazare educated the people about inexpensive ways to fight soil erosion, such as terracing and planting trees along the hillside.

As a final example of a constructive response to seemingly impossible situations is the work of M. C. Mehta, cofounder of the India Council for Enviro-Legal Action (ICELA). Mehta has won approximately forty landmark environmental judgments including one that required more than 5,000 factories polluting the Ganges River[12] to install air and water pollution control devices and that ordered those 190 or so factories out of compliance to close until they could meet prescribed standards (Environmental Law Alliance Worldwide 1996).

To this point we have discussed a number of important concepts that help us understand how some characteristics of organizations make them coordinating mechanisms with the potential for both constructive and destructive consequences. Now we turn to Karl Marx and his concept of alienation to understand how workers are dominated so strongly by the forces of

[10]With regard to disaster coverage, Lee Wilkins (1986) argues that reporters may lack sufficient background knowledge to ask the proper questions and as a result fail to report key facts which allow readers to evaluate the event.

[11]A Union Carbide report shows 134 leaks of phosgene, MIC, or a mixture of the two chemical substances between 1980 and 1985 at the Institute plant.

[12]The Ganges River is a major river in India and one of the holiest Hindu pilgrim centers in India where every Hindu wishes their last rites to be performed.

Residents living around the Institute, West Virginia, Union Carbide plant responded in much the same way as Bhopal residents did to a chemical leak at the plant.

©A. Tannenbaum/Sygma

production that they remain uninformed or uncritical about their role in the production process.

Alienation of Rank-and-File Workers

Human control over nature increased with the development of increasingly sophisticated instruments and tools and the growth of bureaucracies to coordinate the efforts of humans and machines. Machines and bureaucratic organizations combined to extract raw materials from the earth more quickly and more efficiently and increase the speed with which necessities such as food, clothing, and shelter could be produced and distributed.

Karl Marx believed that increased control over nature is accompanied by **alienation**, a state in which human life is dominated by the forces of human inventions. Chemical substances represent one such invention; they have reduced the physical demands involved in producing goods. Fertilizers, herbicides, pesticides, and chemically treated seeds give people control over nature because they eliminate the need to fight weeds with hoes, they prevent pests from destroying crops, and they help people produce unprecedented amounts of food.

But there is a dark side to these gains. In the long run, people are dominated by the effects of this invention. Heavy reliance on chemical technologies causes the soil to erode and become less productive; it also causes insects and disease-causing agents to develop resistance to the chemicals. Chemical technologies also have altered the ways in which farmers plant crops: planting patterns have changed from many species of sustenance crops planted together to a single cash crop, planted in rows. As a result of these changes, farmers have lost knowledge of how to control insects and diseases without chemicals by interplanting a variety of flowers, herbs, and vegetables. Farmers are now economically dependent on a single crop and the chemical industry. An excerpt from the essay "A Descent into Toxic Hell in India" represents a particularly vivid example of alienation:

> I now know what Indian scientist-activist Dr. Vandana Shiva really means when she uses the phrase, "the dark side of development."
>
> The scenes that I saw were very dark indeed. We entered the industrial factory zone, drove past phar-

Alienation A state in which human life is dominated by the forces of human inventions.

Karl Marx believed that alienation occurs when the production process is divided up so that workers are treated like parts of a machine rather than as active, creative, social beings.

©Jon Feingersh/Stock, Boston

maceutical companies, chemical plants, fertilizer plants, steel plants, and came to a large, black pool—a small lake, really—that was an unlined pit for toxic waste the color of moonless midnight.

But there were more pits, and more: fourteen in all, we were told. And they went down sixty feet or more. And there it contaminated the groundwater thoroughly for miles. We started to walk along the bank of one, to take photographs, but suddenly the wind shifted and we had to stick our shirts over our noses and hurry quickly back to the road. I could only imagine the molecules of toxic substances that were entering my lungs.

The people of the nearby village don't have to imagine such things, nor did we for long, for the evidence is all over their bodies. When we drove up, a group of about twenty villagers were waiting for our arrival, which had been arranged by a local medical doctor-turned-activist. Over the course of an hour, the crowd grew and grew, people coming from their houses, bringing photographs and X-rays and children who could not walk. (Bonine 1994)

Although Marx discussed alienation in general, he wrote more specifically about alienation in the workplace.

Marx maintained that workers are alienated on four levels: (1) from the process of production, (2) from the product, (3) from the family and the community of fellow workers, and (4) from the self. Workers are alienated from the process because they produce not for themselves or for known consumers but for an abstract, impersonal market. In addition, they do not own the tools of production. Workers are alienated from the product because their roles are rote and limited—each performs a specialized task. Many workers are treated as replaceable or as interchangeable as machine parts. They are treated as economic components rather than as active, creative social beings (Young 1975). Marx believed that the conditions of work usually are such that they impair an individual's "capacity to become a multidimensional, authentic being with human qualities of compassion, reflection, judgment, and action" (Young 1975:27). As a result, no person can claim a product as the unique result of his or her labor.

Workers are alienated from their families because households and work environments are separate from one another. In other ways, households are uprooted because large-scale enterprises take over the land or force families to move to areas where work is available. When a factory is built in India, workers migrate from rural homes and gather around the factory to live in squatter settlements or slums. The population in Bhopal was 102,000 in 1961. After Union Carbide and other industries settled there in the 1960s, the population grew to 385,000 in 1971, 670,000 in 1981, and 800,000 by 1984.

> It is a familiar sight in so-called underdeveloped countries to find somewhere, in the midst of great poverty, . . . a gleaming, streamlined new factory, created by foreign enterprise. . . . Immediately outside the gates you might find a shanty town of the most miserable kind teeming with thousands of people most of whom are unemployed and do not seem to have a chance of ever finding regular employment of any kind. (Schumacher 1985:490)

At the time of the accident, approximately 20 percent of Bhopal's 800,000 residents lived in squatter settlements. The location of two of these squatter camps—directly across from the Union Carbide plant—explains why the deaths occurred disproportionately among the poorer residents. These people are paid poverty-level wages, which prohibits them from acquiring decent living quarters. In many parts of India, people are drawn to land surrounding industrial sites because they have been driven from their homes by large corporations.

The factory owners and shareholders make large profits, in part because they can find places where minimum wage, environmental protection, and workers' health and safety laws are not systematically enforced or are less stringent. In the worst case, workers handle chemicals about which "they may understand little, other than that their eyes tear, they cough harshly and they suffer recurring rashes and headaches" (Engler 1985:493). Although studies on working conditions in developing countries are imprecise and sketchy, the preliminary information suggests that a substantial number of workers are exploited without regard for health consequences. Two studies conducted in the mid-1980s found that one-third of Indians working in a DDT plant and one-fourth of all employees at battery plants were sick (Engler 1985). The World Health Organization found that of the thousands of industries in Delhi, many are "cramped, poorly lighted, ill-ventilated spaces with atmospheres full of dust, gas, vapors and fumes" and that they operate without health controls (Crossette 1989).

When the company loses money, the management at the corporate headquarters often neglects the plant, allowing equipment, employees' skills, and, sometimes, already inadequate safety standards to deteriorate. The workers are thankful for any job and hesitant about organizing to improve working conditions when such an action could mean losing their jobs and source of livelihood. In this regard, workers are alienated from the community of fellow workers because they compete for a limited number of jobs. As they compete, they fail to consider how they might unite as a force and control their working conditions. Finally, workers are alienated from themselves because "one's genius, one's skills, one's talent is used or disused at the convenience of management in the quest of private profit. If private profit requires skill, then skill is permitted. If private profit requires subdivision of labor and elimination of craftsmanship, then skill is sacrificed" (p. 28).

Summary and Implications

Organizations are powerful coordinating mechanisms that permit goods and services to be produced and delivered efficiently to millions of people. Although they manage people, information, goods, and services efficiently, organizations also can promote inefficient, irresponsible actions (see Table 7.4). This chapter introduced concepts that sociologists use to understand the two sides of organizations with special emphasis on the multinational corporation in India. We have concentrated on one case—Union Carbide and the organizational shortcomings that contributed to one of the two worst industrial accidents in history.

An analysis of the organizational issues behind the Bhopal disaster reveals that many factors within organizations can promote ignorance, misinformation, and failure to take responsibility for known risks. These factors include (1) rational decision making that emphasizes the quickest and most cost-efficient means to achieve a goal without considering the merits of other methods and goals, (2) departures from rules and regulations, (3) excessive adherence to rules and regulations, (4) reliance on experts with compartmentalized training, (5) inappropriate statistical measures of performance, (6) oligarchy, and (7) an alienated work force.

We also have learned that a complex relationship exists between an organization and the environment, especially when the organization is a multinational corporation operating in a country such as India. The case of Union Carbide shows that an organization not only draws on labor and raw materials from the environment but also affects the environment through the

Table 7.4 Organizations: What the Critics Say Versus What the Defenders Say

What the Critics Say	What the Defenders Say
1. Organizations force workers to do simplified, meaningless tasks, so robbing them of initiative, creativity, and independence. Look at the assembly line.	We can enrich jobs by deliberately building in autonomy and discretion. In Sweden, the assembly line has been modified to do just this. In any case, very few people actually work on assembly lines anymore. We are becoming a service society.
2. Organizations do not serve society or consumers well. Automobiles are unsafe, factories pollute the air, and the sheer size of organizations makes them dangerous. The multinationals dominate life in company towns and even overthrow governments.	We can control pollution, and make better products with quality-control circles and other advances pioneered by Japan and other countries. Large size produces economies of scale. Besides, only very large organizations can afford to do research on new products.
3. Organizations don't even work well. Cars break down before you get them home from the showroom, prisons don't rehabilitate, and schools turn out illiterates. We need alternative institutions.	You go too far. Cars give trouble because you want so much from them – trouble-free driving at high speeds with minimal maintenance. If we produced a serviceable car with minimum features, you wouldn't buy it. Prisons could rehabilitate if you would pay the price for vocational counseling and training. The schools do a great job – name another country in which over a third of all high-school students go on to higher education.
4. Organizations are out of control. An arrogant power elite of interlocking directors controls them, and an army of ever-increasing bureaucrats administers them. The client or citizen is powerless against organizations.	The evidence for any monolithic power elite is exaggerated or so biased as to be invalid. Client and citizen power, they are increasing. Look at the detail on contents now provided on labels at the supermarket. Look at the growth of consumer action groups.

Source: From *Organizations in Society*, by Edward Gross and Amitai Etzioni, p. 4. Copyright ©1985 by Prentice-Hall, Inc. Reprinted by permission of Prentice-Hall, Upper Saddle River, NJ.

products it manufactures, the services it provides, and the waste it disposes of.

Obviously, multinational organizations cannot be held totally responsible for negative consequences to a host country's environment as domestic industries contribute to pollution as well. In addition, an uninformed, inattentive public that consumes unthinkingly also bears responsibility for their environment. An informed and attentive public is important because it reinforces conscientious decision making on the part of management and experts. Moreover, an alert and active public is capable of responding intelligently if a crisis does occur. Scientist Kenneth Prewitt suggests that it would be helpful if the public simply understood that all technologies have unpredictable consequences and that complex problems do not have simple answers:

> A savvy population recognizes a "fix" for the unrealistic goal that it is. It is the rare problem that is solved; more often, the problem changes form. To control infectious diseases is not to solve the health problem. It is to shift attention to cancer and heart failure. And when these illnesses of late middle life are cured, the health problem will shift to senility and related illnesses of the aged. Society copes with problems; it doesn't solve them. If the public is to become actively involved in debates over the introduction of major technologies — chemical fertilizers, fluoridated water, nuclear energy, genetic engineering—then the minimum literacy necessary is appreciation of this fact. (Prewitt 1983:62–63)

The absence of an informed public is not unique to countries such as India. The United States, too, is plagued by public apathy. The problem is greater in India and other parts of the developing world, however, because such places lack watchdog agencies and because many people are so poor they will tolerate almost any conditions for a chance to make a living.

The important point is that management, the experts, the workers, and the public recognize that technology is not good or bad in itself. It carries a variety of benefits and costs depending on the uses to which it is put and on the way it is handled. Many people believe that they cannot live with technology, but they cannot live without it, either. The issue that the Bhopal crisis represents is not technology versus no technology. The real issue is the need for organizational mechanisms to coordinate the conscientious production and use of potentially hazardous materials and educated consumers and informed workers dedicated to influencing company policies beyond those aimed at the bottom line (or profit).

Economist and former U.S. Secretary of Labor Robert Reich maintains that people must evaluate corporations, particularly multinationals, in a new light. According to Reich, we cannot assume that, if corporations headquartered in a country are prosperous and profitable, the workers in that country are also prosperous. Nor can we assume that executives from foreign-owned corporations are more likely than native-born executives to make decisions that will harm the economic well-being of the host country. Why? Because "U.S.-based corporations and their shareholders can now prosper by going wherever on the globe the costs of doing business are lowest—where wages, regulations, and taxes are minimal. Indeed, managers have a responsibility to their shareholders to seek out just such business climates" (Reich 1988:79). Reich advises people to be leery of business executives who point to large trade deficits as evidence that they need wage concessions, trade protection, and other incentives if their corporations are to compete on a global scale. In the case of the United States, an unknown but significant portion of the trade deficits with other countries is the result of U.S. corporations' making products in those countries to be sold in the United States under their own brand name. For example, 40 percent of the trade that Mexico does with the United States originates from U.S. subsidiaries in Mexico (Clark 1993).

In view of this global context, Reich argues that people should not be overconcerned with the nationality of a company or trade deficits. Instead, they should focus on the quality of jobs and the quality of life that a corporation brings to a community: The mistake often made by local chambers of commerce, state governors, and federal officials when drumming up business is that they focus only on the number of jobs and overlook many of the organizational characteristics discussed in this chapter.

Key Concepts

Use this outline to organize your review of the key chapter ideas.

Organizations
Multinational corporations
Rationalization
- **Disenchantment of the world**
- **Externality costs**
- **Bureaucracy**
  - **Ideal type**

Formal dimensions of organizations
Informal dimensions of organizations
Trained incapacity
- **Informate**
- **Automate**

Oligarchy
Professionalization

internet assignment

Check out the December 1996 issue of the *Multinational Monitor* (under back issues), which features the "Corporate Hall of Shame: The Ten Worst Corporations of 1996" at http://www.essential.org/monitor/monitor.html. List the names of the ten worst corporations, and in two or three sentences, describe the actions that helped place each corporation on the list. Identify three corporations from this list whose destructive actions can be analyzed using one or more of the key concepts covered in this chapter.

Deviance, Conformity, and Social Control

With Emphasis on the People's Republic of China

The Role of Context in Defining Deviance

Deviance: The Violation of Norms

The Constructionist Approach

Structural Strain Theory

Differential Association Theory

Factors That Shape U.S. and Chinese Systems of Social Control

A view down busy Nanjing Road, Shanghai. Dan Habib/Impact Visuals

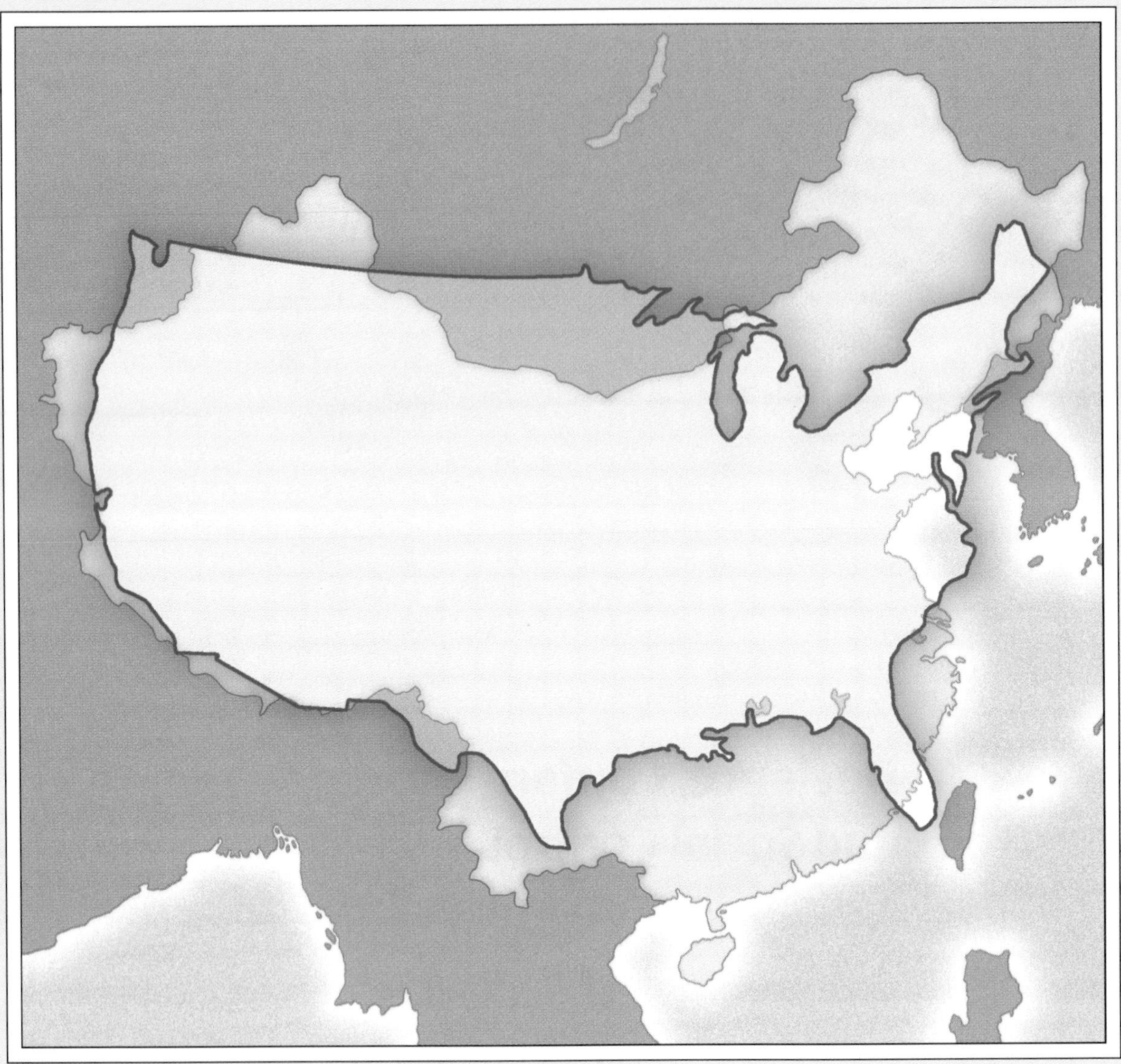

Source: U.S. Central Intelligence Agency (1995).

The Problems of Governing More Than 1.2 Billion People

China's system of social control is much more rigid than our own. One major reason for this may be the relative size of China's population in relation to its resources. More than 1.2 billion Chinese live in a space roughly the size of the United States. The United States, however, has an abundant 12 percent of the world's arable land and 4.5 percent of the world's population. China has only 7 percent of the world's arable land and 21 percent of the world's population.

Another reason may be the difficulties inherent in managing a huge and centralized society. To get an idea of the scale of Chinese society, consider these facts: As of 1994, the United States had only one city with more than 5 million people (New York), and only seven cities over 1 million. China has thirty-eight cities with more than 1 million each and four cities (not counting Hong Kong) with more than 5 million.

Or look at it this way: The central authorities of China are attempting to govern more than a billion people. In Europe and North and South America, it takes about fifty separate countries to do that.

When the Cultural Revolution began . . . my [family's house] was one of the first in the city to be ransacked.

I later found out that it was my mother's ignorance that started the ransacking. Both my grandfather and father were working in banks. They were well-known capitalists. At that time all the funds capitalists had in the banks were frozen. You couldn't withdraw anything. It was called "money made by exploitation." All the names of capitalists were listed just outside the banks. My mother didn't know about that. She went to get some money out. The bank clerks immediately called the Red Guards. They showed up in no time at my home and started to search and ransack our apartment.

[When I finally got there,] I glanced over the rooms from the corridor. Red Guards were standing everywhere, searching for things. No sign of my family. Lots of things were in shreds, smashed and torn.

My parents moved in with my mother's family. Their home had been searched and sealed too. My parents just blocked off a tiny area with a piece of cloth in the corridor. They found some wooden boards to use as a bed. When I went to see my mother, her hair had been cut and shaved by the Red Guards. (Feng Jicai 1991:58–60)

This event took place during the ten years of the Cultural Revolution (1966–1976), a period in which more than a half-million Chinese were imprisoned, tortured, humiliated, and executed. During this time, any person who held a position of authority, worked to earn a profit, showed the slightest leaning toward foreign ways, or had academic interests was subject to interrogation, arrest, and punishment. Included in this group were scientists, teachers, athletes, performers, artists, writers, private business owners, and people who had relatives living outside China, wore glasses, wore makeup, spoke a foreign language, owned a camera or a radio, or had traveled abroad (Mathews and Mathews 1983).

Contrast the events of the Cultural Revolution with Asian scholar Robert Oxman's description of China today:

All over China people are jumping into the sea. That's the new Chinese expression for going into business, making money privately in a country that used to forbid any form of capitalism. It means saving a few dollars for the simple tools to fix bicycles at a curbside shop, or running a sidewalk shoe repair stand because new soles are still cheaper than new shoes, or cutting hair at 5:30 in the morning before going to a regular job. In the country, a wife grows produce to make extra income to buy consumer goods for her family. Unleashing the natural business instincts of the Chinese people, plus the addition of huge foreign investments have fueled one of the most explosive economic take-offs in history. Around major cities, buildings, housing projects, and roads seem to emerge almost overnight. It's even been suggested that China's national bird should be the crane. Every urban horizon is filled with cranes. In Shanghai's frenzied harbor, ships from around the world compete with tiny river barges as they haul goods and move people. Once the people of China were caught in economic slow motion by a state which prized ideology above all else. Now the Chinese are rushing in fast forward in a national quest for prosperity. **(Oxman 1993a:9)**

Why Focus on the People's Republic of China?

In this chapter, we pay special attention to the People's Republic of China for one major reason. In twelve to fifteen years, China has undergone a lifetime of economic reform that is startling when contrasted with its economic policies of the Cultural Revolution of the 1960s and 1970s, which made profit-making activities criminal and which closed the country off to foreign investment and influences. During the Cultural Revolution, if a family planted small, but extra, crops to sell after the autumn harvest for money to buy food during the winter, they were criticized for taking the capitalist road. At that time the pressure against acquiring even small amounts of extra wealth drew such severe criticism that peasants accepted living in poverty rather than being equated with the so-called bourgeoisie (Bernstein 1983).

In contrast, today an estimated 220,000 foreign-invested enterprises are registered in China (U.S. Department of State 1996), up from 70,000 in 1993. Among the foreign-invested firms are the U.S.-based shoe giants Nike, Converse and Reebok; the fast-food giants PepsiCo (KFC, Pizza Hut) and McDonald's; and Procter and Gamble. In Beijing alone there are at least a hundred luxury hotels, up from nine in 1981, for foreigners and wealthy Chinese to stay in (Montalbano 1993). And in 1993, the International Monetary Fund named China the third largest economy in the world after the United States and Japan.

Because of this vast amount of economic change and freedom, which is described as one of the largest economic experiments the world has ever known (Broadfoot 1993), many investors view China as potentially the world's largest market and a "gigantic reservoir of cheap labor" (Carrel and Hornik 1994:A15)—a gold mine.

Although China has been defined as a market that must be penetrated, it has also received considerable international criticism and pressure directed at improving its human rights record. The U.S. government, for example, regularly warns the Chinese leadership about its (1) unacceptably high trade surplus with the United States (growing at a faster rate than any other trading partner's trade surplus); (2) plans to govern Hong Kong;[1] (3) infringements against copyrights, trademarks, and patents reflected in the pirating of U.S. movies, computer software, music, and books; and (4) human rights violations against political dissidents and, most recently, orphans.

China represents an interesting case for studying issues of deviance, conformity, and social control because many of the behaviors that constituted deviance during the Cultural Revolution no longer apply today. In addition, the Chinese government's ideas about what constitutes deviance and its methods of social control are often the object of international criticism.

[1]Hong Kong, a former territory of the United Kingdom, returned to China on July 1, 1997. Hong Kong is approximately six times the physical size of Washington, D.C. An estimated 50 percent of the Hong Kong population are refugees from China.

The topics of this chapter are deviance, conformity, and social control—some of the most complex issues in sociology because, as we will learn, almost any behavior or appearance can qualify as deviant under the right circumstances. **Deviance** is any behavior or physical appearance that is socially challenged and/or condemned because it departs from the norms and expectations of a group. **Conformity**, on the other hand, may be defined as behavior and appearances that follow and maintain the standards of a group. All groups employ mechanisms of **social control**—the methods used to teach, persuade, or force their members, and even nonmembers, to comply and not to deviate from norms and expectations. The controlling agent may be parents, peers, coaches, elected officials, governing bodies, or the organizational mechanisms of governments, justice systems, or international bodies.

Deviance Any behavior or physical appearance that is socially challenged and/or condemned because it departs from the norms and expectations of a group.

Conformity Behavior and appearances that follow and maintain the standards of a group.

Social control The methods used to teach, persuade, or force their members, and even nonmembers, to comply and not to deviate from norms and expectations.

As shown by the vignettes at the start of this chapter, depending on the cultural circumstances, a deviant act or appearance may be something as seemingly minor as wearing eyeglasses or withdrawing money from a bank. We can illustrate the same point within U.S. culture. When sociologist J. L. Simmons asked 180 men and women in the United States from all age, educational, occupational, and religious groups to "list those things or types of persons whom you regard as deviant," 1,154 items in all were listed.

> Even with a certain amount of grouping and collapsing, these included no less than 252 different acts and persons as "deviant." The sheer range of responses included such expected items as homosexuals, prostitutes, drug addicts, beatniks, and murderers; it also included liars, democrats, reckless drivers, atheists, self-pitiers, the retired, career women, divorcées, movie stars, perpetual bridge players, prudes, pacifists, psychiatrists, priests, liberals, conservatives, junior executives, girls who wear make up, and know-it-all professors. (Simmons 1965:223–224)

Although this study was made more than thirty years ago, Simmons's conclusions are still relevant: almost any behavior or appearance can qualify as deviant under the right circumstances. The only characteristic common to all forms of deviance is "the fact that some social audience regards them and treats them as deviant" (p. 225). In light of this fact, it is difficult to generate a precise list of deviant behaviors and appearances because something that some people consider deviant may not be considered deviant by others. Likewise, something that is considered deviant at one time and place may not be considered deviant at another. For example, wearing makeup is no longer considered a deviant behavior in China,[2] as evidenced by the fact that Avon, a U.S. cosmetics company, has recruited 18,000 salespeople to sell its products in China (WuDunn 1993). As another example, cocaine and other now-illegal drugs once were legal substances in the United States. In fact, "We unwittingly acknowledge the previous legality of cocaine every time we ask for the world's most popular cola by brand name" (Gould 1990:74). Originally Coca-Cola was marketed as a medicine that could cure various ailments. One of the ingredients used to make the drink came from the coca leaf, which is also used to produce cocaine (Henriques 1993). Finally, considering how cigarette smokers are treated today, it is interesting to note that smoking was not only accepted but glamorized in the United States just a few decades ago (see "Clearing the Air: Significant Signposts").

A society's definitions of deviance change over time. Considering how smokers are viewed today, it is interesting to note that cigarette smoking was not only accepted but glamorized in the United States just a few decades ago.

Cover drawing by Barry Blitt; ©The New Yorker Magazine, Inc.

Such wide variations in responses to what is deviant alert us to the fact that deviance exists only in relation to norms in effect at a particular time and place. The sociological contribution to understanding deviant behavior is that it goes beyond studying the individual and instead emphasizes the context under which deviant behavior occurs. Such an emphasis raises at least two general but fundamental questions about the nature of deviance. One obvious question is, How is it that almost any behavior or appearance can qualify as deviant under the "right" circumstances? It follows that any answers to this question must go beyond the deviant individual's personality or genetic makeup. Understanding context, for example, helps us see why it was once a crime to make a profit in China.

A second important question is, Who defines what is deviant? That is, who decides that a particular group,

[2]Whenever the term *China* is used in this chapter, it denotes the People's Republic of China, whose capital is Beijing. The use of *China* does not include Hong Kong (a former British colony that reverted to the People's Republic of China in 1997) or Macao (currently a Portuguese colony expected to revert to the People's Republic of China in 1999).

U.S. In Perspective

Clearing the Air: Significant Signposts

This time line shows the various public settings in which smoking is no longer acceptable. It is a story of how the meaning of a behavior shifted from something people could do in public to an activity confined to private settings.

Judging from the time line, what sorts of organizations or institutions in the United States appear to have taken the lead in reducing the acceptance of smoking? What might have motivated them to do so? What sorts of organizations or institutions appear not to have participated in a major way? Why not?

What is the recent history of tobacco regulation on your campus? Are the current rules adequate from your point of view? Are they obeyed? Why?

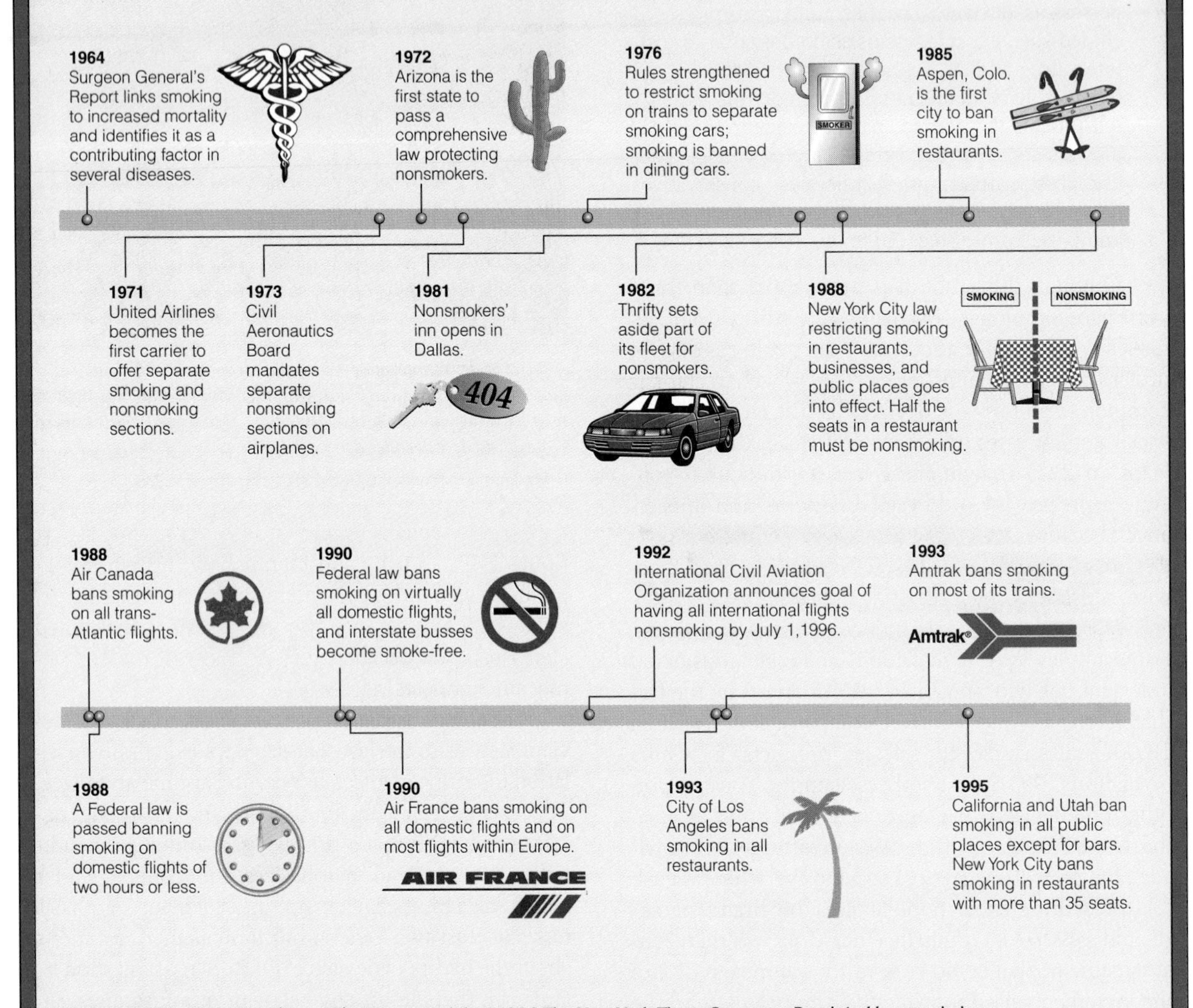

1964
Surgeon General's Report links smoking to increased mortality and identifies it as a contributing factor in several diseases.

1971
United Airlines becomes the first carrier to offer separate smoking and nonsmoking sections.

1972
Arizona is the first state to pass a comprehensive law protecting nonsmokers.

1973
Civil Aeronautics Board mandates separate nonsmoking sections on airplanes.

1976
Rules strengthened to restrict smoking on trains to separate smoking cars; smoking is banned in dining cars.

SMOKER

1981
Nonsmokers' inn opens in Dallas.

404

1982
Thrifty sets aside part of its fleet for nonsmokers.

1985
Aspen, Colo. is the first city to ban smoking in restaurants.

1988
New York City law restricting smoking in restaurants, businesses, and public places goes into effect. Half the seats in a restaurant must be nonsmoking.

SMOKING | NONSMOKING

1988
Air Canada bans smoking on all trans-Atlantic flights.

1988
A Federal law is passed banning smoking on domestic flights of two hours or less.

1990
Federal law bans smoking on virtually all domestic flights, and interstate busses become smoke-free.

1990
Air France bans smoking on all domestic flights and on most flights within Europe.

AIR FRANCE

1992
International Civil Aviation Organization announces goal of having all international flights nonsmoking by July 1, 1996.

1993
Amtrak bans smoking on most of its trains.

Amtrak®

1993
City of Los Angeles bans smoking in all restaurants.

1995
California and Utah ban smoking in all public places except for bars. New York City bans smoking in restaurants with more than 35 seats.

Source: *The New York Times* (June 30): 12xx.

The Chagan shopping center, Beijing, would be very much at home in the United States. This scene is quite remarkable when you consider that during the Cultural Revolution people were criticized for possessing foreign-made goods or appearing to have extra wealth or "bourgeoisie" status.

©Adrian Bradshaw/Saba

behavior, appearance, or person is deviant? The fact that an activity or an appearance can be deviant at one time and place and not at another suggests that it must be defined as deviant by some particular process. In the case of China, who determined that wearing eyeglasses, speaking a foreign language, and withdrawing money constituted deviant behavior and warranted such severe punishments?

In this chapter, we explore the concepts and theories that sociologists use to understand deviance, and we use them to answer these fundamental but complex questions about the social nature of deviance.

The Role of Context in Defining Deviance

It is impossible to fully understand how the Cultural Revolution disrupted Chinese lives without having experienced it (Bernstein 1982). The revolution, which began in 1966 and lasted approximately a decade, was a campaign inspired by Communist Party Chairman Mao Zedong[3] to restore revolutionary spirit to China. It was also a campaign against revisionism and against entrenched authority. Mao loosely defined *revisionism* as "an abandonment of the goals of the [Chinese communist] revolution, and acceptance of the evils of special status and special accumulation of worldly goods" (Fairbank 1987:319).[4] He loosely defined *entrenched authority* as the "Four Olds": old ideas, old culture, old customs, and old habits. At first Mao assigned the Red Guards (his name for the youths of China between the ages of nine and eighteen), and eventually the People's Liberation Army,[5] the task of finding and purging revisionists and those in entrenched positions of authority.

[3]Another common spelling is Mao T'se-tung. I chose to use *Zedong* because it corresponds to the official English spelling of Mao's name in Mainland China. The *T'se-tung* version is more common in Taiwan.

[4]The phrase "an abandonment of the goals of the revolution" refers to the 1949 revolution that resulted in the establishment of the People's Republic of China. Before this revolution, China was very poor. Between 1849 and 1949, the Chinese people were victims of every imaginable sort of exploitation by foreign countries. To survive starvation, many Chinese ate the leaves off trees and the grass off the ground, and a large portion of the population was addicted to opium. "A peasant party had come out of the hills to put an end to corruption, invasion, and humiliation; they wore straw sandals and told the truth" (Wang Ruowang 1989:40). Mao led this party; he instilled a revolutionary spirit into the masses that motivated the Chinese people to stand up to these immense problems (Strebeigh 1989). Mao was a remarkable leader and hero to his followers. The Chinese intellectual Liu Zaifu noted that "in the 1950s 'we did not believe in ourselves, but only in the all-wise, all powerful Mao'" (1989:40)

[5]Within three years, however, the Red Guards had disrupted and divided the country so deeply that Mao denounced them, claiming that they had failed to understand and implement his strategy. Thus, he banished 17 million youths to the countryside to live among the peasants and to be reeducated, and he reassigned the task to the People's Liberation Army.

"We believed deeply in Chairman Mao": The decade-long Cultural Revolution was an attempt to rekindle revolutionary spirits in China and enforce conformity to Mao's doctrine.

During the Cultural Revolution, many artifacts from China's long history (tombstones, relics, manuscripts, art objects, books, scrolls of poetry) and any other objects that suggested special status or the accumulation of worldly possessions were destroyed. Any person in a position of authority or with the slightest leaning toward foreign ways, including farmers who planted extra crops, were suspect. If someone simply remarked that a foreign-made product such as a can opener was better than its Chinese counterpart, or if someone wrapped some food or garbage in a piece of newspaper with Mao's picture, he or she was suspect (Mathews and Mathews 1983).

As Jung Chang writes in *Wild Swans: Three Daughters of China* (1991), such conditions reduced many people "to a state where they did not dare even to think, in case their thoughts came out involuntarily" (1992:6). The slightest misstep could make one a target:

> Targets might be required to stand on a platform, heads bowed respectfully to the masses, while acknowledging and repeating their ideological crimes. Typically they had to "airplane," stretching their arms out behind them like the wings of a jet. In the audience tears of sympathy might be in a friend's eyes, but from his mouth would come only curses and derisive jeering, especially if the victim after an hour or two fell over from muscular collapse. . . .
>
> To Chinese, so sensitive to peer-group esteem, to be beaten and humiliated in public before a jeering crowd including colleagues and old friends was like having one's skin taken off. (Fairbank 1987:336)

The reflections of a former Red Guard, summarized in Fox Butterfield's (1982) *China: Alive in the Bitter Sea,* some fifteen years after the event, show the intensity with which so-called revisionists and entrenched authority figures were hunted down and persecuted:

> "I was very young when the Cultural Revolution began. . . . My schoolmates and I were among the first in Peking to become Red Guards, we believed deeply in Chairman Mao. I could recite the entire book of the Chairman's quotations backward and forward, we spent hours just shouting the slogans at our teachers."
>
> Hong remembered in particular a winter day, with the temperature below freezing, when she and her faction of Red Guards put on their red armbands and made three of the teachers from their high school kneel on the ground outside without coats or gloves. "We had gone to their houses to conduct an investigation, to search them, and we found some English-language books. They were probably old textbooks, but to us it was proof they were worshipping foreign things and were slaves to the foreigners. We held a bonfire and burned everything we had found."
>
> After that, she recalled, the leader of her group, a tall, charismatic eighteen-year-old boy, the son of an

army general, whose nickname was "Old Dog," ordered them to beat the teachers. He produced some wooden boards, and the students started hitting the teachers on their bodies. "We kept on till one of the teachers start[ed] coughing blood," Hong said. . . . "We felt very proud of ourselves. It seemed very revolutionary." (Butterfield 1982:183)

The forces behind the Cultural Revolution help us understand more clearly why making money was considered a crime against the state and why such people were considered deviant.[6] The Cultural Revolution was Mao Zedong's response to the failure of an important national plan—The Great Leap Forward.

The Great Leap Forward was a plan of Mao's to mobilize the masses and transform China from a country of poverty to a land of agricultural abundance in five short years. Under this plan, Mao mobilized hundreds of thousands of people for projects ranging from killing insects to building giant dams with shovels and wheelbarrows (Butterfield 1976). The Great Leap Forward was an ill-conceived, hastily planned, widespread, sweeping reorganization of Chinese society that created economic disruption on a massive scale. The nature of this disruption is illustrated by the failure of a plan to increase fertilizer by cutting down trees, burning them, and spreading their ashes over the fields. This plan left peasants without fuel for cooking and heating. In one region peasants stopped harvesting crops and started to dig tunnels in their fields to find coal, which Communist Party officials believed was plentiful. No coal was found, however, and the crops rotted. As a result of the Great Leap Forward, 30 to 50 million Chinese died from human-made famine (Leys 1990; Liu Binyan 1993). In addition, the plan caused considerable environmental destruction. Desperate to grow food and obtain fuel, Chinese peasants destroyed more than half of the grasslands and one-third of the forests (Leys 1990).

In light of these catastrophic problems, Mao was particularly vulnerable to political attack. Mao blamed the failure of his plan on entrenched authority and on the loss of revolutionary spirit. He used the Cultural Revolution as an attempt to eliminate anyone in the Communist Party and in the masses who opposed his policies.

In view of everything that happened, why is making money acceptable in China today? Again, this question must be examined in the context of the Cultural Revolution.

[6]The circumstances leading to this revolution are not easy to summarize, however; one would have to go back at least 185 years to trace its origin.

A woman sells Mao memorabilia in contemporary China. Ironically, it is now acceptable to make money from objects celebrating the man who forbade any form of capitalism.

After Mao's death in 1976, the Cultural Revolution ended, and the new leaders were faced with a great many problems that the revolution had exacerbated and generated. The revolution took place at the expense of China's economic, technological, scientific, cultural, and agricultural development and drained the Chinese physically and mentally. The Cultural Revolution created a ten-year gap in trained workers. "'I can't honestly let any of the young doctors in my hospital operate on a patient,' said a leading surgeon. 'They went to medical school. But they studied Mao's thought, planting rice or making tractor parts. They never had to take exams and a lot of them don't know basic anatomy'" (Butterfield 1980:32). This situation applies to more than the medical profession; during the Cultural Revolution, millions of people secured positions because they were loyal to Mao, not because they were qualified (Broaded 1991).

To solve these problems, the Communist Party, under the leadership of Deng Xiaoping, needed the support of those teachers, technicians, artists, and scientists who had been struck down during the Cultural

The Importance of Ideological Commitment

In China there seem to be two groups of people: those who are ideologically sound and support the Communist Party and those who lack ideological commitment and conspire against the party. Today the ideologically sound are those who support four major political principles: (1) China is a socialist country, hence the creation of not a capitalist system but a socialist market economy; (2) power and leadership resides with the Chinese Communist Party; (3) a combination of Marxism, Leninism, and Mao Zedong thought is the guiding ideology; and (4) the state is a dictatorship by the proletariat. The principles are intentionally vague so that the party or individual leaders can claim violations against these principles when it suits their purposes (personal correspondence 1993). Any behavior that disrupts the progress or the smooth operation of the workplace, the neighborhood, or the country is conceived as an offense against the people or the country, which the Communist Party oversees. The most widespread formal sanction used in China is rehabilitation through reeducation and labor.

A single man risks his life to try to face down a line of tanks sent to suppress the 1989 demonstration in Tiananmen Square.

Obviously, Chinese who have robbed, murdered, assaulted, or raped someone have disrupted the social fabric. In addition, those who commit adultery, disobey work assignments, lose their jobs, have no honest occupation, are vagrants, disrupt a group's discipline, are expelled or drop out of school, or engage in counterrevolutionary and countersocialist activity (behaviors and appearances that challenge Communist ideology) are disruptive and in need of rehabilitation. One activity defined as counter-

Revolution. To make up for "Ten Lost Years," the Chinese leaders sent thousands of students overseas to study in the capitalist West. They allowed farmers and factory workers to keep profits from surplus crops after meeting government quotas, and they established five Special Economic Zones (SEZs — designated areas within the People's Republic of China that enjoy capitalist privileges).[7] In reversing the policies of the Cultural Revolution, Deng Xiaoping employed a rhetoric that defined the new policies as being in the best interest of the country, not individual self-interest. With regard to SEZs, Deng argued, "A few regions that have the right conditions will develop first. . . . The developed regions will then carry the developing regions until they finally reach common prosperity" (quoted in Tempest 1995:A5).

As one measure of the success of the new policies, consider that in 1994, foreign investments to China were $33.8 billion (U.S. Department of State 1996). And in 1995 China ran a $35 billion trade surplus with the United States, creating trade tensions between the two countries (Hong Kong Trade Development Council 1996). On two occasions, June 1996 and August 1996, the U.S. trade deficit with China was larger than its trade deficit with Japan. This state of economic affairs is truly remarkable when we remember that as recently as 1976 one could suffer imprisonment and hard labor simply for knowing some foreign words. On the other hand, do not interpret these economic changes to

[7]Special Economic Zones are designed to attract foreign investment and capital; foreign investors are lured there by the cheap labor force and by the potential market of more than 1 billion, 200 million Chinese consumers. The five SEZs are Hainan Island, Chantou, Shenzhen, Xiamen, and Zhuhai.

revolutionary was the series of demonstrations in Tiananmen Square by Chinese students in support of democracy,[1] eventually suppressed violently in June 1989. The students were accused of "bourgeois liberalization,"

> *a term that has never been clearly defined but, based on its usage by different sources in the media, it could perhaps be interpreted as wanton expression of individual freedom (individualism) that poses a threat to the stability and unity of the country. Such tendencies had to be curbed. In the government's view, students participated in the protests because they had led a sheltered life and were ignorant of the complexities of the reform process. Their youthful outburst did not take into consideration larger collective interests and concerns. (Kwong 1988:983–984)*

The government undertook a number of measures to persuade the students to become more knowledgeable about the complexities of life and less responsive to subversive ideas. These measures included sending them "to rural areas to teach them to endure hardship, work hard and appreciate the daily difficulties faced by China's mostly rural population" (Kristof 1989:Y1). The rationale is that proper ideological commitment can be instilled through contact with the masses and through manual labor.[2] Other measures included placing limits on the number of students entering the humanities and social sciences. In the year following the June 4 incident, almost no students were admitted to study academic areas that government officials consider to be "ideologically suspect" by nature, such as history, political science, sociology, and international studies (Goldman 1989).

Perhaps one of the most intriguing things about China is that, since 1949, the leaders in power have adhered to the following code:

> *In any circumstance and at any cost, political power must be retained in its totality. This rule is absolute, it tolerates no exception and must take precedence over any other consideration. The bankruptcy of the entire country, the ruin of its credit abroad, the destruction of national prestige, the annihilation of all efforts toward overture and modernization—none of these could ever enter into consideration once the Party's authority was at stake. (Leys 1989:17)*

This code still applies. Even today, as China undergoes massive economic growth and change, "its leaders have made it abundantly clear that they intend to preserve an unchallengeable Communist Party dictatorship, even as they pull back from trying to dominate every detail of economic and social life" (Holley 1993:H15).

[1]The movement toward democracy has been suppressed many times. In 1979 it was suppressed on the grounds that it was "unstabilizing" to China. In 1983 it was labeled a case of "spiritual pollution" (Link 1989).

[2]"A Chinese professor who at the end of the Cultural Revolution was sent to Anhui province, in central China . . . described what it was like to live there. Everything was made from mud, he recalled: the floors, the walls, even the beds and the stools the peasants sat on were constructed from pounded earth. The villagers had no wood for fuel—all the trees in the region had long ago been chopped down—so the women and children gathered grass and wheat stalks, depriving the earth of valuable natural nutrients. During the year the professor was there, he ate no meat, and the family he was quartered with had no matches, no soap, and most of the time no cooking oil, an essential part of the Chinese diet" (Butterfield 1982:16).

mean that China is moving toward establishing a capitalist economy (see "The Importance of Ideological Commitment"). While Chinese leaders have been trying to change a sluggish Soviet-style, centrally planned system to a more dynamic and flexible economy with market elements, the new economy is still under Communist control (U.S. Central Intelligence Agency 1995).

The dramatic changes in China since 1978 make it an ideal case for illustrating two major assumptions that guide sociological thinking about the nature of deviance:

1. Almost any behavior or appearance can qualify as deviant under the right circumstance.
2. Conceptions of what is deviant vary over time and place.

In Chapter 4, we examined norms, a sociological concept that underscores these two assumptions. As you recall, norms are written and unwritten rules specifying appropriate and inappropriate behaviors. The important concept of norms cannot be overlooked when discussing deviance, because it is the violation of norms that constitutes deviance.

Deviance: The Violation of Norms

Recall that in Chapter 4 we learned that some norms are considered more important than others. In that chapter, we highlighted two kinds of norms — folkways and mores — distinguished by sociologist William Graham Sumner.

Folkways and Mores

Folkways are customary ways of doing things that apply to the details of life or routine matters—how one should look, eat, greet another person, express affection toward same-sex and opposite-sex persons, and even go to the bathroom.[8] When Francisco Martins Ramos, an anthropology professor who teaches in Portugal, visited the United States, he found that folkways governing how to behave in public restrooms were different from those in his country. Ramos writes that the Portuguese newcomer

> will certainly be quite surprised with a form of cultural behavior never dreamt of: The American who urinates in public initiates conversation with the partner at his side, even if he does not know the latter! Themes of these occasional dialogues are the weather, football, politics, and so on. We can guess at the forced pleasure of the Portuguese, who heretofore has regarded urination as a necessary physical function, not as a social occasion. The public restroom! Is it an extension of the bar room? (Ramos 1993:5)

Mores are norms that people define as essential to the well-being of their group or nation. People who violate mores usually are punished severely; they are ostracized, institutionalized in prisons or mental hospitals, and sometimes executed. In comparison with folkways, however, people consider mores to be "the only way" and "the truth." Consequently, people consider mores to be final and unchangeable. The United States has a large number of mores that protect individual privacy, property, rights, and freedoms. For example, in the United States an individual has the right to marry, have children, choose a career, and change jobs and residences without appealing to a higher authority for permission. It does not matter that the country as a whole has a shortage of scientists, teachers, and nurses or that there are no physicians in some geographic areas and a glut of physicians in others. What matters is an individual's right to choose his or her occupation. It is unthinkable that Americans could not change jobs or residences whenever they wanted, for whatever reason.

In China, throughout its long history, its mores have reflected the traditional values of conformity, collectivism, and obedience to authority.[9] In China it is unthinkable that an individual could marry, have a baby, or obtain housing without first obtaining approval from the Communist Party–controlled work unit or neighborhood committee (Oxman 1993b). Until recently, a person could not select an occupation; purchase a train ticket, bicycle, or television; secure a hotel room; obtain employment; or buy food without written approval from a unit or a committee.

One indicator of the dominant mores which govern U.S. policy toward China can be found in the official position that imposing sanctions on China for human rights violations and property rights infringements would make it more difficult for the United States to influence commercial, social, and political reform in China. In addition, if the U.S. imposed sanctions on China, such a policy would give economic advantages to Japan and Europe (Stevenson 1996; see "The President Renews MFN Status for China"). China, on the other hand, values its economic relationship with the United States but places the preservation of social and political stability above this relationship even if that means confronting the United States (Tyler 1994b) as it has when it tested nuclear weapons, sold arms to the "enemies" of the United States, and staged military exercises close to Taiwanese territory.

Usually people abide by established folkways and mores because they accept them as "good and proper, appropriate and worthy" (Sumner 1907:60). For the great majority of people, "the rule to do as all do suf-

Folkways Customary ways of doing things that apply to the details of life or routine matters.

Mores Norms that people define as essential to the well-being of their group or nation.

[8]Fox Butterfield (1982) describes several aspects of life — sleeping arrangements, living arrangements, and departures—in which Chinese folkways differ from those of Americans. In regard to sleeping arrangements, the Chinese are surprised to find that most American children have their own rooms, because Chinese are accustomed to sleeping with their entire family. According to another Chinese folkway, children live at home with their parents until they marry. Butterfield describes a response of "Oh, I'm so sorry" made by a Chinese reporter when he learned that an American woman to whom he was talking lived alone. He assumed that she lived alone because her parents had died or because she had been forced to take a job that separated her from them geographically. As to departures, when guests leave, the Chinese host walks them out to their vehicle and then stands and waves until the visitors are out of sight.

[9]In fact, the Communist Party, which was formed in the 1920s and eventually came to power under Mao's leadership in 1948, sought to maintain these traditional values but to shift them away from the family to the party and party-led institutions. At the same time, the party wanted to get rid of many old mores that were based on a disdain for physical labor, allegiance to the family, religious beliefs, and reliance on personal networks and to substitute mores that reflected new values including a love for physical labor, loyalty to the party and its leader, and atheism. The Communists aimed to create a new style of person ("socialist man"), and they built upon some old mores and values and introduced a new set of values, mores, and sanctions to discourage deviance and enforce their own rule (personal correspondence 1993).

U.S. In Perspective

The President Renews MFN Status for China

The White House
Office of the Press Secretary
May 31, 1996

Statement by the Press Secretary

The President announced today that he has renewed Most-Favored-Nation (MFN) trade status for China this year. The President has taken this step because he believes that continued engagement with China is the best way to help it become a constructive force for stability and prosperity in Asia and to advance important American interests. . . .

Far from giving China a special deal, renewal of MFN confers on it a trading status equal to that enjoyed by most other nations. Simply put, it gives China normal trade status. It is not being granted as a favor to China.

The President is renewing MFN for China because he believes it advances critical U.S. interests at a time when China is at a critical turning point. Maintaining our overall relationship with China enables the U.S. to engage China in the months and years ahead, to enhance areas of cooperation and to pursue American interests in areas where we differ. That engagement can help determine whether China becomes a destabilizing threat or a constructive force in Asia and in the world.

Substantial U.S. interests are at stake and renewal of MFN best advances those interests. Revoking MFN would raise average tariffs on Chinese imports from 5 percent to 45 percent. It would effectively sever our economic relationship with China, undermining our capacity to influence China in a broad range of areas, including human rights, nonproliferation, trade, Taiwan relations and others. It is a clumsy and counterproductive instrument that would set us down the wrong path. It would reverse three decades of bipartisan China policy and would seriously weaken our influence not only in China, but throughout Asia.

Revoking MFN would also undermine America's economic interests. U.S. exports to China support 170,000 American jobs and have been growing at a rate of 20 percent a year. Chinese retaliation would imperil or eliminate these jobs, exclude American companies and workers from future business in one of the world's most dynamic markets and give an open field to our competitors.

Furthermore, revoking MFN would only set back efforts to promote human rights and democracy in China. Whether by telephone, fax, E-mail or daily contact, Chinese citizens are receiving a greater understanding of American ideals of personal, political and economic freedom. This influx of new ideas and information has helped fuel China's transformation over the last twenty years. Revoking MFN would cut those links and set back a dialogue that is feeding China's development for the next century.

At the same time, the U.S. will continue to stand with those who are fighting for freedom and human rights in China, as we did last month in cosponsoring a resolution in the U. N. Human Rights Commission condemning China's human rights record. We continue to have serious concerns about China's human rights practices and we will continue to press China on these matters. But revoking MFN is not the right way to make fundamental human rights progress.

Revocation of MFN status would not advance the interests of Taiwan or Hong Kong. The economies of both of these countries are enormously dependent on continued trade between the U.S. and China. That trade has contributed to the prosperity—and in turn—the establishment of a strong democracy on Taiwan. For Hong Kong, a strong and vibrant economy is an important way of supporting its autonomy as it moves toward the transition to Chinese sovereignty. That is why even those in Hong Kong who are most critical of China support renewal of MFN.

Engagement with China does not mean acquiescence in Chinese policies or practices we oppose. The President has demonstrated that he is prepared to use sanctions and other means at his disposal to promote America's goals regarding China, whether it is protecting U.S. intellectual property, deterring dangerous proliferation or promoting human rights. These are the right tools to use in advancing U.S. interests. Revocation of MFN is not.

In what ways does the press secretary's statement touch on issues of social control across international boundaries? Is the United States encouraging "deviance" in China?

fices." Recall from Chapter 5 that socialization is the process by which most people come to learn and accept as natural the ways of their culture. Because socialization begins as soon as a person enters the world, there is little opportunity to avoid exposure to the culture's folkways and mores.

If we compare the ways in which life is structured for four-year-olds in Chinese preschools and American preschools, we can see that different but important cultural lessons are incorporated into their daily activities. Even though it is impossible to generalize about preschools in countries as large and as diverse as the United States and China, there are some broad, outstanding differences. For the most part, Chinese preschoolers are taught to suppress individual impulses, play cooperatively with other children, and attune themselves to group enterprises. In contrast, American preschoolers are taught to cultivate individual interests and compete with other children for success and recognition by the teacher.

Quite different mores govern behavior in American and Chinese preschools. Whereas Chinese preschool teachers emphasize structured activities and social-mindedness, teachers in the United States are more likely to teach pupils self-direction, freedom of choice, and individuality.

©Forrest Anderson/Gamma-Liaison (left); ©Paul Conklin/Monkmeyer Press (right)

Socialization as a Means of Social Control: Preschool in China and in the United States

Professors Joseph Tobin and Dana Davidson and researcher David Wu filmed daily life in Chinese and U.S. preschools to learn how teachers in each system socialize children to participate effectively in their respective societies.[10] (It is worth noting that the great majority of preschool children whom they filmed in both countries seemed happy and productive.) The researchers found that in comparison with U.S. preschools, Chinese preschools are highly structured and socially minded: Chinese teachers discipline their four-year-old students "by stopping them from misbehaving before they even know they are about to misbehave" (Tobin, Wu, and Davidson 1989:94), and they promote loyalty to the group. The bathroom scene described here is an example of the extent to which Chinese children are taught to follow instructions and attune themselves to group enterprises:

> It is now 10:00, time for children to go to the bathroom. Following Ms. Wang, the twenty-six children walk in single file across the courtyard to a small cement building toward the back of the school grounds. Inside there is only a long ditch running along three walls. Under Ms. Wang's direction and, in a few cases, with her assistance, all twenty-six children pull down their pants and squat over the ditch, boys on one side of the room, girls on the other. After five minutes Ms. Wang distributes toilet paper, and the children wipe themselves. Leaving the toilet, again in single file, the children line up in front of a pump, where two daily monitors are kept busy filling and refilling a bucket with water that the children use to wash their hands. (pp. 78–79)

When the researchers showed U.S. parents and teachers the film portraying daily life in Chinese preschools, Americans were particularly disturbed by the bathroom scene and asked why children were forced to go to the bathroom in this manner.[11] To this question, one Chinese educator replied:

> Why not? Why have small children go to the bathroom separately? It is much easier to have everyone go at the same time. Of course, if a child cannot wait,

[10]Actually, the authors studied preschools in three countries: China, the United States, and Japan.

[11]The Chinese system seems harsh from an American point of view, but it ensures that children wipe themselves properly and that their hands are washed. This method has some practical benefits because many diseases are spread through contact with bodily substances, including saliva, mucus, and feces.

> he is allowed to go to the bathroom when he needs to. But, as a matter of routine, it's good for children to learn to regulate their bodies and attune their rhythms to those of their classmates. (p. 105)

In the United States, preschoolers also undergo discipline, but they are more likely to be disciplined after they do something wrong or get out of hand. In comparison with their Chinese counterparts, U.S. teachers encourage self-direction, freedom of choice, independence, and individuality—qualities that often depend on a supply of material items. For example, American children typically use as much paper as they want; they start a drawing, decide they do not like it, and crumple up the paper. When they play house, store, or firefighter, they use costumes, plastic dishes, children's versions of household appliances, plastic food items, and so on. American teachers also encourage children to choose from a number of activities. A typical exchange between a preschool teacher and his or her students follows. It is difficult to imagine this exchange taking place in a Chinese classroom:

> Who would like to paint? Michelle. Mayumi. Nicole. Okay, you three get your smocks from your cubbies and you can paint. [The teacher holds up a wooden block.] Who wants to do this? Mike? Okay, that's one. Stu, that makes two. Billy is three. . . . Here's a puzzle piece. You want to start on the puzzles? Okay? [The teacher holds up a toy frying pan.] Who wants to start in the house? Lisa, Rose, Derek. Go ahead to the housekeeping corner. Kerry, what do you want to do? The Legos? You're going to work on the radio, Carl? That's fine. Who is going to come over to the book corner to read this book? It's called *Stone Soup*. Okay, come on with me. (p. 130)

Reaction to Socialization of Another Culture

Both the Chinese and the Americans who watched the films were disturbed by the other country's system of handling preschool. Comments by Chinese viewers showed their clear preference for their own way of structuring early education. They maintained that their form of discipline expresses care and concern, and they regard U.S. preschools as chaotic, undisciplined, and promoting self-centeredness. As one Chinese viewer remarked, "There are so many toys in the classroom that children must get spoiled. When they have so much, children don't appreciate what they have" (p. 88). On the other side, the U.S. viewers criticized the Chinese preschools for being rigid, totalitarian, too group-oriented, and overrestrictive, "making children drab, colorless, and robot-like" (p. 138). The people in each country are uncomfortable with the other's system because the lessons that are taught in the country clash with the prevailing mores or ideas about which behaviors are essential to their own country's well-being.

The point is that early socialization experiences prepare children to fit within the existing system. Each society tries to prepare its people to fit into and accept their respective environments. In the case of preschool socialization, most Chinese preschoolers begin kindergarten having learned a great deal about the need for group cooperation. Most American preschoolers graduate knowing that they will continue to be evaluated and rewarded on their individual performance. Even so, primary socialization experiences such as those that take place during preschool are uneven at best. Not all preschools are alike, and some children do not attend preschools. Even among those who attend, some children do not internalize (take as their own and accept as binding) the values, norms, and expectations that they are taught. Therefore, all societies establish other mechanisms of social control to ensure conformity.

Ideally, conformity is voluntary. That is, people are internally motivated to maintain group standards and feel guilty if they deviate from them. As we have seen, during the Cultural Revolution it was considered deviant to wear glasses, use makeup, speak a foreign language, or break or destroy items that displayed Mao Zedong's picture. Many Chinese conformed to these rules on their own and punished themselves if they violated them even if they did so by accident. The memories of one Chinese man illustrate this point:

> As a boy, I did not know what a god looked like, but I knew that Mao was the god of our lives. When I was six, I accidentally broke a large porcelain Mao badge. Fear gripped me. In my life until that moment, the breaking of the badge seemed the worst thing I had ever done. Desperate to hide my crime, I took the pieces and threw them down a public toilet. For months I felt guilty. (Author X 1992:22)

In this case, the author's own guilt is a sign of voluntary conformity. Often, however, if conformity cannot be achieved voluntarily, people employ various means to teach, persuade, or force others to conform.

Mechanisms of Social Control

Ideally, socialization brings about conformity, and, ideally, conformity is voluntary. But when conformity cannot be achieved voluntarily, other mechanisms of social control are used to convey and enforce norms and ex-

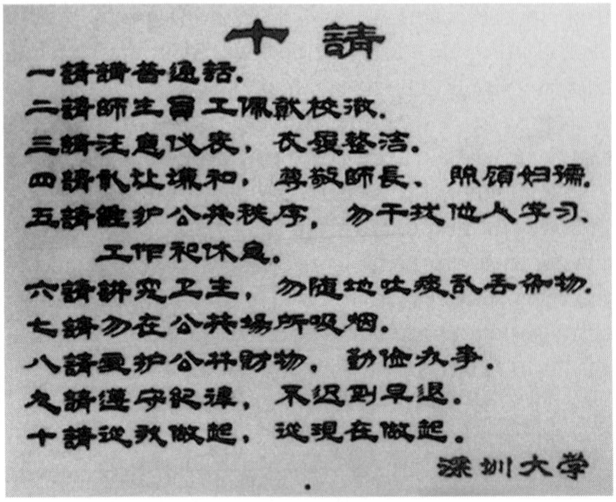

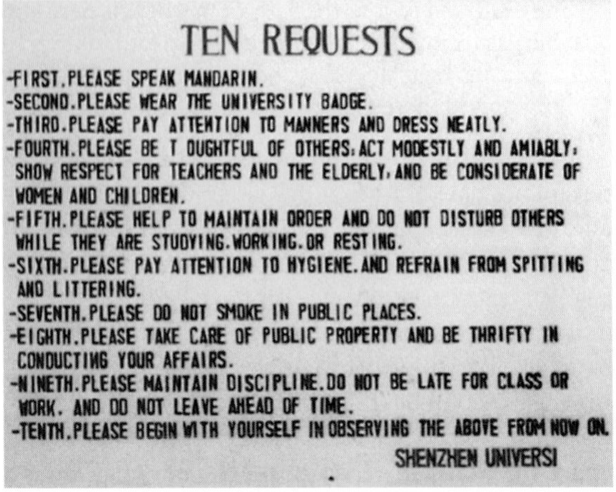

Formal sanctions include written rules governing appropriate behavior, as illustrated by these signs posted in both Chinese and English at China's Shenzhen University.

pectations. Such mechanisms are known as **sanctions** — reactions of approval and disapproval to behavior and appearances. Sanctions can be positive or negative, formal or informal. A **positive sanction** is an expression of approval and a reward for compliance; such a sanction may take the form of applause, a smile, or a pat on the back. A **negative sanction** is an expression of disapproval for noncompliance; the punishment may be withdrawal of affection, ridicule, ostracism, banishment, physical harm, imprisonment, solitary confinement, or even death.

Informal sanctions are spontaneous and unofficial expressions of approval or disapproval; they are not backed by the force of law. The following incident involving the son of a well-known psychologist, Sandra Bem, shows how informal sanctions usually are applied. Bem and her husband raised their children not to be limited by those "attributes and behaviors the culture may have stereotypically defined as inappropriate" for one sex or the other (Monkerud 1990:83). Their efforts to go against norms and expectations are challenged through informal sanctions at every step of the way:

> At age four, Bem's son, Jeremy, wore barrettes to nursery school. One day a boy repeatedly told him that "only girls wear barrettes." Jeremy tried to explain that wearing barrettes didn't make one a boy or girl: only genitalia did. Finally, in frustration, he pulled down his pants to show the boy that having a penis made him a boy. The boy responded, "Everybody has a penis; only girls wear barrettes." (p. 83)

This incident also shows how early socialization to norms works. Even at the age of four, children use sanctions to enforce the norms that they have learned from adults, teachers, and other sources such as television.

Formal sanctions are definite and systematic laws, rules, regulations, and policies that specify (usually in writing) the conditions under which people should be rewarded or punished and that define the procedures for allocating rewards and imposing punishments. Examples of formal positive sanctions include medals, cash bonuses, and diplomas. Formal negative sanctions include fines, prison sentences, the

Sanctions Reactions of approval and disapproval to behavior and appearances used to convey and enforce norms and expectations.

Positive sanction An expression of approval and a reward for compliance.

Negative sanction An expression of disapproval for noncompliance.

Informal sanctions Spontaneous, unofficial expressions of approval or disapproval that are not backed by the force of law.

Formal sanctions Definite and systematic laws, rules, regulations, and policies that specify (usually in writing) the conditions under which people should be rewarded or punished and that define the procedures for allocating rewards and imposing punishments.

death penalty, or high-pressure waterhoses, rubber bullets, or tear gas to disperse demonstrators (Xu 1995). Sociologists use the word **crime** to refer to deviance that breaks the laws of society and is punished by formal sanctions. People in every society have different views of what constitutes crime, what causes people to commit crimes, and how to handle offenders (see "The Extent of Social Control in China").

Defining Deviance

According to sociologist Randall Collins (1982), Emile Durkheim presented one of the most sophisticated sociological theories of deviance. Durkheim ([1901] 1982) argued that, although deviance does not take the same form everywhere, it is present in all societies. He defined deviance as those acts that offend collective norms and expectations. The fact that always and everywhere there are people who offend collective sentiments led him to conclude that deviance is normal as long as it is not excessive and that "it is completely impossible for any society entirely free of it to exist" (p. 99). According to Durkheim, deviance will be present even in a "community of saints in an exemplary and perfect monastery" (p. 100). Even in seemingly perfect societies, acts that most persons would view as minor may offend, create a sense of scandal, or be treated as crimes.

Durkheim drew an analogy to the "perfect and upright" person. Just as such a person judges his or her smallest failings with a severity that others reserve for the most serious offenses, so do societies that are supposed to contain the most exemplary people. Even among such exemplary individuals, some act or appearance will offend simply because "it is impossible for everyone to be alike if only because each of us cannot stand in the same spot. It is also inevitable that among these deviations some assume a criminal character" (p. 100). What makes an act or appearance criminal is not so much the character or the consequences of that act or appearance but the fact that the group has defined it as something dangerous or threatening to its well-being. Wearing eyeglasses, for example, is clearly not a behavior that is harmful to others. However, as we have learned, this behavior was defined as a clear indicator of other threatening behaviors, such as the crimes of revisionism and entrenched authority, that were not so easily observable.

Durkheim maintained that deviance is functional for society for at least two reasons. First, the ritual of punishment (exposing the wrongdoing, determining a punishment, and carrying it out) is an emotional experience that serves to bind together the members of a group and establish a sense of community. Durkheim argued that a group that went too long without noticing crime or doing something about it would lose its identity as a group. In evaluating Durkheim's argument, consider your reaction when you learn that your government has criticized the Chinese system. On the other hand, consider how people in China might react when its officials criticize something about the U.S. system or another system. The merits of the respective criticisms aside, the act of exposing "wrongdoing" functions to generate patriotic feelings in many people as the indirect message is "Be glad you live here and not there!"

Second, deviance is functional because it is useful in making necessary changes and preparing people for change. It is the first step toward what will be. Nothing would change if someone did not step forward and introduce a new perspective or new ways of doing things. Almost every invention or behavior is rejected by some group when it first comes into existence.

Durkheim's theory offers an intriguing explanation for why almost anything can be defined as deviant. Yet, Durkheim did not address an important question: Who decides that a particular activity or appearance is deviant? One answer can be found from labeling theory.

Labeling Theory

In *Outsiders: Studies in the Sociology of Deviance,* sociologist Howard Becker[12] states the central thesis of labeling theory: "All social groups make rules and attempt, at some times and under some circumstances, to enforce them. When a rule is enforced, the person who is supposed to have broken it may be seen as a special kind of person, one who cannot be trusted to live by the rules agreed on by the group. He is regarded as an outsider" (1963:1).

As Becker's statement suggests, labeling theorists operate under the assumption that rules are (1) socially constructed and (2) not enforced uniformly or consis-

[12]A number of sociologists — Frank Tannenbaum (1938), Edwin Lemert (1951), John Kitsuse (1962), Kai Erikson (1966), and Howard Becker (1963) — are linked to the development of what is conventionally called labeling theory. The scholar most frequently associated with labeling theory, however, is Howard Becker.

Crime Deviance that breaks the laws of society and is punished by formal sanctions.

The Extent of Social Control in China

In China, "social control is everywhere and involves everyone" (Clark and Clark 1985:109), including relatives, friends, colleagues, employers, and everyone a person comes in contact with (Xu 1995). Thus, "each person has a social duty to participate in group activities and to 'help' others in the collective living arrangement. The mandate to 'help' means assuming responsibility for others and correcting their faults" (Rojek 1985:119).

This billboard in Beijing is part of the Chinese effort to control population growth through a policy of one child per family.

©Forrest Anderson/Gamma-Liaison

Theoretically, every Chinese belongs to a work unit and a neighborhood committee headed by Communist Party members. Until recently, almost every important area of life was supervised by the unit or the committee. The work unit issued job assignments; determined salary; distributed ration coupons and other goods (light bulbs, contraceptives, bicycles, television sets); granted permission to travel, change jobs, and change residences. The neighborhood committee is part of the "chain" connecting its members to local leaders and the police, and up the various levels of leadership between it and to the central authority structure in Beijing (Farley 1995). It still scrutinizes requests to marry and have children, and it determines whether a worker or his or her children may take college entrance examinations. The work unit maintains a confidential file[1] on every person, which contains information on education, work, class background as far back as three generations, the party's evaluation of the person (as an activist or as a counterrevolutionary), and any political charges made against him or her by informants. Members of the work unit are encouraged to report wrongdoings committed by other members. Whereas Americans are hesitant to report their suspicions of misconduct such as child abuse, spousal abuse, or drug use because they believe in a person's right to live without interference, the Chinese typically are afraid not to voice such suspicions because, by remaining silent, they are viewed as accomplices to such acts.

The neighborhood committee monitors life outside the workplace. It enforces birth control policies, monitors contacts between its members

---

tently. Support for the first assumption comes from the fact that definitions of what is deviant vary across time and place. If this is the case, then people must decide what is deviance. The second assumption is supported by the fact that some people break rules and escape detection whereas others are treated as offenders even though they have broken no rules. Labeling theorists maintain that whether an act is deviant depends on whether people notice it and, if they do notice, on whether they label it as a violation of a rule and subsequently apply sanctions. Such contingencies suggest that violating a rule does not make a person deviant. That is, from a sociological point of view, a rule breaker is not deviant (in the strict sense of the word) unless someone notices the violation and decides to take corrective action (see Figure 8.1).

Labeling theorists suggest that for every rule a social group creates, there are four categories of people: conformists, pure deviants, secret deviants, and the falsely accused. The category that one belongs to depends on whether a rule has been violated and on whether sanctions are applied.

Conformists are people who have not violated the rules of a group and are treated accordingly. **Pure deviants**, on the other hand, are people who have broken the rules and are caught, punished, and labeled as outsiders. As a result, the rule breaker takes on the **master status of deviant**, an identification that "proves to be more important than most others. One will be identified as a deviant first, before other identifications are

Conformists People who have not violated the rules of a group and are treated accordingly.

Pure deviants People who have broken the rules and are caught, punished, and labeled as outsiders.

Master status of deviant An identification marking a rule breaker first as a deviant and then as having any other identification.

and outsiders, settles domestic problems, scrutinizes each household's activities, investigates disputes among neighbors, deals with petty crime, and educates members about new policy and law. The committee has the right to search living quarters without the occupants' consent. As in the work unit, members are encouraged to inform on other members' wrongdoings. "When society is so tightly organized, where can one run to [when one unintentionally or intentionally violates the rules]?" (Wu Han 1981:39). Recently multinational corporations have employed neighborhood committees to pass out product samples to its members and to survey them about product preferences (Farley 1995).

Among the aspects of Chinese life that are controlled by work units and neighborhood committees, perhaps the most widely publicized outside of China is procreation. Because of its huge population, China has tried to impose a limit of one child per couple in hopes of slowing population growth so that it will stabilize at 1.2 billion by the year 2000 (Tien 1990). Each province and each city is assigned a yearly quota with regard to how many children can be born, and the neighborhood committee or work unit determines which couples will be included in that quota. In essence, the work unit and the neighborhood committee decide which married couples can have a baby, determine when they can start trying, oversee contraceptive use, and even record women's menstrual cycles. Female officials check to see that IUDs are in place. If a woman misses several periods and has not been authorized to try to conceive, she is persuaded to have an abortion (Ignatius 1988). As one Chinese official told an American journalist, "We don't force her. . . . We talk to her again and again until she agrees" (Hareven 1987:73). Couples who have only one child and sign an agreement to have no more children are rewarded with positive sanctions: salary bonuses and other financial incentives, educational opportunities, housing priority, and extra living space. If couples request permission to have a second child, they are asked to wait at least four years; if they have two children, they are persuaded not to have any more. Various kinds of economic sanctions or penalties are imposed on couples who have more than one child.[2]

[1]In China, access to what people living in the United States would consider in the public domain such as demographic statistics is considered classified information. Chinese students and professors interested in such information must make a formal request. Such requests are entered into their permanent file (Rorty 1982).

[2]The policy of one child per couple has been more successful in urban areas than in rural areas. Chinese peasants have resisted government efforts because they believe that more children bring more happiness, that sons are more effective laborers in the fields than daughters, and that sons offer security to parents in old age. (When a male child marries, his wife moves in with him and his parents, and the couple supports them in their old age.)

made" (Becker 1963:33). We must remember that, although pure deviants undeniably violate rules, rule enforcers "select" the people they apprehend and punish.

Consider how highway patrol officers choose, from among all of the cars speeding along a highway, which drivers to pull over. A study of vehicles' and drivers' characteristics on a stretch of the New Jersey Turnpike showed that fewer than 5 percent of the vehicles observed were late-model cars with out-of-state license plates, driven by black males. Yet, 80 percent of the arrests made on that stretch of highway fit this profile (Belkin 1990). The drivers who are pulled over for speeding assume the status of deviant; those not stopped assume the status of secret deviant. As a second example, consider how drug enforcement agents make a decision to stop someone in an airport because they suspect him or her of smuggling drugs. A substantial number of people who are stopped by drug enforcement agents are stopped because they fit a profile of a drug courier and a drug swallower. The profile includes the following traits: having dark skin and/or a foreign appearance; being obviously in a hurry; having an exotic hairstyle or wearing brightly colored clothing; purchasing a one-way ticket; paying cash for a ticket; changing flights at the last minute; flying to or from Detroit, Miami, or another large city; and flying to or from the Caribbean, Nigeria, Jamaica, or other known drug-supplying regions (Belkin 1990). Although not everyone stopped is found to be a drug smuggler, the profile directs law enforcement officials' attention away from those who do not fit the profile, which increases the chances that those drug smugglers not fitting the profile remain as secret deviants.

Secret deviants are people who have broken the rules but whose violation goes unnoticed, or, if it is no-

Secret deviants People who have broken the rules but whose violation goes unnoticed, or, if it is noticed, no one reacts to enforce the law.

Are people who use marijuana for medical purposes criminals because their behavior leads to harmful consequences or because some powerful segments of society have defined the behavior as illegal?

©Mark Richards/PhotoEdit

ticed, no one reacts to enforce the law. Becker maintains that "no one really knows how much of this phenomenon exists," but [he is] convinced that the "amount is very sizable, much more so than we are apt to think" (1963:20). For example, a 1994 U.S. Bureau of Justice survey of crime victims documented that 42.4 million crimes were committed against U.S. residents twelve years of age and older. Of these 42.4 million crimes, only about 38 percent of the victims reported the crime to police (U.S. Bureau of Justice 1996a, 1996b; see Table 8.1).

The **falsely accused** are people who have not broken the rules but who are treated as if they have done so. The ranks of the falsely accused include those who are the victims of eyewitness errors, perjury, and cover-up. They include innocent suspects who make false confessions under the pressure of interrogation. In their book, *In Spite of Innocence,* sociologist Michael L. Radelet and philosopher Adam Bedau reviewed more than 400 cases of innocent people convicted of capital crimes and found that 56 had made false confessions. It appears that some innocent suspects admit guilt, even

Falsely accused People who have not broken the rules but who are treated as if they have done so.

Figure 8.1 Labeling Theory

Major Assumptions
1. Rules are socially constructed.
2. Rules are not enforced uniformly or consistently.
3. Deviants are those whose behavior people have noticed, *and* labeled as such, and applied sanctions.

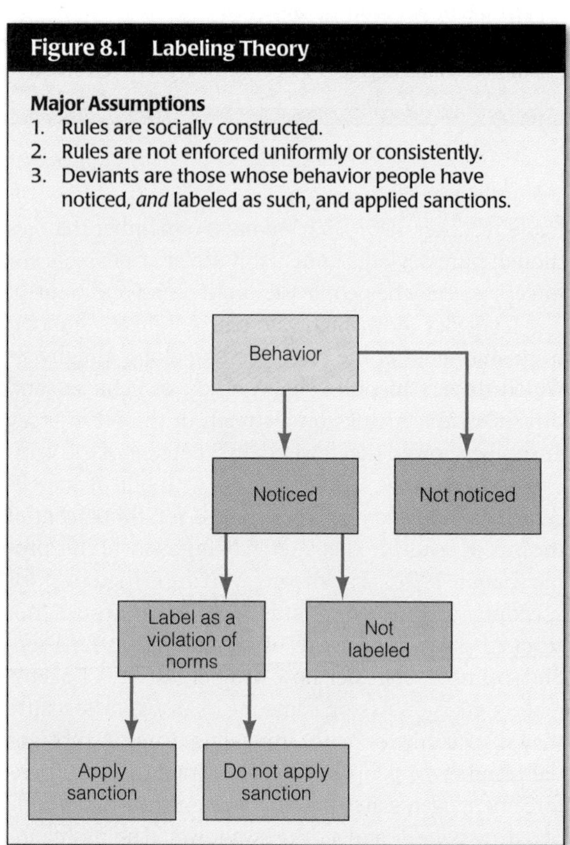

Table 8.1 Percentage Distribution of Victimizations and Whether Reported to Police

Sector and Type of Crime	Number of Victimizations	Percentage of Victimizations Reported to the Police
All crimes	39,582,880	36.5
All personal crimes	10,268,280	43.0
Crimes of violence	9,866,200	43.2
Crimes of theft	22,883,060	27.3
Property crimes	10,035,060	34.2

Source: U.S. Bureau of Justice Statistics (1996a, 1996b).

to heinous crimes, to escape the stress of interrogation (Jerome 1995). As with the phenomenon of secret deviants, no one knows how often people are falsely accused, but it probably happens more often than we think. In any case, the status of accused often lingers even if the person is cleared of all charges. Such cases lead us to ask a larger question: Under which circumstances are people most likely to be falsely accused?

The Circumstances of the Falsely Accused

Sociologist Kai Erikson (1966) identifies a particular situation in which people are likely to be falsely accused of a crime: when the well-being of a country or a group is threatened. The threat can take the form of an economic crisis (an economic depression or recession), a moral crisis (family breakdown, for example), a health crisis (say, AIDS), or a national security crisis (such as war). At such times, people need to define a seemingly clear source of the threat. Whenever a catastrophe occurs, it is common to find someone to blame for it. Identifying the threat gives an illusion of control. The person blamed is likely to be someone who is at best indirectly responsible, someone in the wrong place at the wrong time, and/or someone who is viewed as different.

This defining activity can take the form of a **witch-hunt**, a campaign against subversive elements with the purpose of investigating and correcting behavior that undermines a group or a country. In actuality, a witch-hunt may not accomplish this goal because the real cause of a problem is often complex and may lie beyond the behavior of a targeted person or group. Often the people who are defined as the problem are not in fact the cause of the threat but are used intentionally to represent the cause of a complicated situation. For example, as happened during the Cultural Revolution, sometimes the seemingly most insignificant acts were classified as crimes against a group or country to divert the public's attention from the shortcomings of those in power, unite the public behind a cause, or take people's attention away from a disruptive event such as massive job layoffs.[13]

The internment of more than 110,000 people of Japanese descent (80 percent of whom were American citizens) living on the West Coast of the United States during World War II is another example of a situation in which a group was targeted in conjunction with a crisis. Japanese Americans were forced from their homes and taken to desert prisons surrounded by barbed wire and guarded with machine guns (Kometani 1987). There was no evidence of anti-American activity on the part of Japanese Americans. Yet, the wartime hysteria, combined with long-standing prejudices, led to the shipping of men, women, and children to concentration camps.

The existence of the falsely accused underscores the fact that the study of deviance must go beyond looking at people identified or labeled as rule breakers. The roles of rule makers and rule enforcers must be examined as well.

Rule Makers and Rule Enforcers

Sociologist Howard Becker (1973) recommends that researchers pay particular attention to the rule makers and rule enforcers and to how they achieve power and

[13]Another example is the timing of the 1993 debate over whether gays should be allowed to serve in the U.S. military. Sociologists might ask, Is it a coincidence that gays became the focus of media attention just when the Clinton administration and Congress were planning such large cuts in the defense budget?

Witch-hunt A campaign against subversive elements with the purpose of investigating and correcting behavior that undermines a group or a country.

Conflict theorists emphasize that not all rules and rule breakers are treated equally. A sharp financial dealer like Michael Milken (shown here doing community service as part of his punishment) may inflict far more harm on people than street thieves (right). Yet the white-collar felon is often treated far less harshly by the criminal justice system.

©Bart Bartholomew/Black Star (left); ©Rick Friedman/Black Star (right)

then use it to define how others "will be regarded, understood, and treated" (p. 204). This topic, of course, interests not only labeling theorists but also conflict theorists. According to conflict theorists, those with the most wealth, power, and authority have the power to create laws and crime-stopping and -monitoring institutions. Consequently, we should not be surprised to learn that law enforcement efforts focus disproportionately on crimes committed by the poor and other powerless groups rather than on those committed by the wealthy and other politically powerful groups. This uneven focus gives the widespread impression that the poor, the uneducated, and minority group members are more prone to criminal behavior than are people in the middle and upper classes, the educated, and majority group members. Crime exists in all social classes, but the type of crime, the extent to which the laws are enforced, access to legal aid, and the power to shape laws to one's advantage vary across class lines (Chambliss 1974). In the United States, for example, police efforts are directed at controlling crimes against individual life and property (crimes such as robbery, assault, homicide, and rape) rather than against white-collar and corporate crime.

White-collar crime Crime committed on the job by well-respected people with high social status.

Corporate crime Crime committed by a corporation as it competes with other companies for market share and profits.

White-collar crime consists of "crimes committed by persons of respectability and high social status in the course of their occupations" (Sutherland and Cressey 1978:44). **Corporate crime** is crime committed by a corporation as it competes with other companies for market share and profits. Usually white-collar and corporate crimes, such as the manufacturing and marketing of unsafe products, unlawful disposal of hazardous waste, tax evasion, and money laundering, are handled not by the police but by regulatory agencies that have little power and minimal staff. Escaping punishment is easier for white-collar crimes because offenders are part of the system: they occupy positions in the organization that permit them to carry out illegal activities discreetly. In addition, white-collar and corporate crime "is directed against impersonal — and often vaguely defined — entities such as the tax system, the physical environment, competitive conditions in the market economy, etc." (National Council for Crime Prevention in Sweden 1985:13). These crimes are without victims in the usual sense because they are "seldom directed against a particular person who goes to the police and reports an offense" (p. 13).

Deviance is a consequence not of a particular behavior or appearance but of the application of rules and sanctions to a so-called offender by others (see "The Presumption of Guilt or Innocence"). For these reasons, sociologists are concerned less with rule violators

U.S. In Perspective

The Presumption of Innocence

"China is oriented toward crime control not due process" (Xu 1995:84). In China a defendant is presumed to be guilty, not innocent until proved guilty as in the U.S. system. Before he or she appears in court, the defendant will have been scrutinized already by a work unit or neighborhood committee. If the work unit or committee members cannot correct the defendant's outlook or behavior, they will refer the case to the control committee, which can either assign a punishment or refer the case to the courts. If a case is referred to the courts, it is presumed that the offender is guilty and that some punishment is warranted. The court appearance is likely to be just a sentencing hearing, as evidenced by the fact that approximately 99 percent of persons whose cases are heard by the Supreme People's Court are convicted. Theoretically, a Chinese offender can appeal a decision, but the court system allows only one appeal (Chiu 1988).

A man accused of counterrevolutionary activity is brought to trial before being executed. By the time that his case reached the trial stage, he had already been presumed guilty.

©Agence Vu

In theory, a defendant in the United States is assumed innocent until proved guilty; the burden of proving guilt rests with the prosecution. Furthermore, sentences and punishments can be reduced through plea bargaining, shock probation (sudden and unexpected release from prison), or plea of insanity or self-defense, or simply because prisons are overcrowded. In 1990, 300,000 criminal cases were processed in New York City alone. Only 10 percent, or 30,000 cases, were tried to completion. The remaining 270,000 cases ended in dismissals, reduction in charges, or plea bargaining (Shipp, Baquet, and Gottlieb 1991). Finally, defendants who can afford the fees are able to purchase the best legal defense and thus have the greatest chance of being found innocent or receiving lighter sentences.

The role of lawyers in each society dramatically reflects the basic differences between Chinese and American views on how defendants should be handled. In China, lawyers are paid by the state and must remit any fees to the state. Although the Chinese lawyer is trained to protect the legitimate rights of clients, those rights are secondary to the rights and interests of the state and the people. In addition, Chinese lawyers cannot use the impression management strategies that American lawyers employ to convince a jury of a client's innocence when in fact the client is guilty (Lubman 1983). If Chinese lawyers followed such techniques, they would be accused of conspiring with an enemy of the state.

than with those persons who make the rules and those who support and enforce the rules. Any effort to explain deviance is complicated by the fact that the powerful play an important role in defining what is deviant and establishing the sanctions that should be applied to correct such behavior. This leaves us with the question of how the powerful are able to get the public to accept such definitions and apply the recommended sanctions. The work of social psychologist Stanley Milgram gives us one answer to this question.

Obedience to Authority

When Stanley Milgram (1974) conducted his research for *Obedience to Authority,* he was interested in learning about how people in positions of authority manage to get people to accept their definitions of deviance and to conform to orders about how to treat people classified as deviant. His study gives us insights about how events like the Holocaust, the Cultural Revolution, and the systematic rape of women by soldiers could have taken place. The fact that such atrocities required the cooperation of a large number of people raises important questions about people's capacity to obey authority.

> The person who, with inner conviction, loathes stealing, killing, and assault may find himself performing these acts with relative ease when commanded by authority. Behavior that is unthinkable in an individual who is acting on his own may be executed without hesitation when carried out under orders. (p. xi)

Milgram designed an experiment to see how far people would go before they would refuse to conform to an authority's orders. The findings of his experiment have considerable relevance for understanding the conditions under which rules handed down by authorities are enforced by the masses.

The participants for this experiment were volunteers who answered an ad placed by Milgram in a local paper. When each participant arrived at the study site, he or she was greeted by a man in a laboratory jacket who explained to the participant and another apparent volunteer the purpose of the study: to test whether the use of punishment improves the ability to learn. (Unknown to the subject, the other apparent volunteer was actually a confederate—someone who works in cooperation with the experimenter.) The participant and the confederate drew lots to determine who would be the teacher and who would be the learner. The draw was fixed, however, so that the confederate was always the learner and the real volunteer was always the teacher.

The learner was strapped to a chair, and electrodes were placed on his or her wrists. The teacher was placed where he or she could not see the learner, in front of an instrument panel containing a line of shock-generating switches. The switches ranged from 15 to 450 volts and were labeled accordingly from "slight shock" to "danger, severe shock." The experimenter explained that when the learner made a first mistake, the teacher was to administer a fifteen-volt shock and increase the voltage with each subsequent mistake. In each case, as the strength of the shock increased, the learner expressed increased discomfort. One learner even said that his heart was bothering him. After that statement, there was complete silence. When the volunteers expressed concern about the learner's safety, the experimenter firmly told them to continue administering the shock. Although many of the volunteers protested against administering such severe shocks, a substantial number obeyed and continued to administer shocks, "no matter how vehement the pleading of the person being shocked, no matter how painful the shocks seemed to be, and no matter how much the victim pleaded to be let out" (1987:567).

The results of Milgram's experiments are especially significant when one considers that no penalty was imposed on the participants if they refused to administer a shock. Obedience in this situation was founded simply on the firm command of a person with a status that gave minimal authority over the subject. If this level of obedience is possible under the circumstances of Milgram's experiments, one can imagine the extent to which it is possible in a situation in which disobedience brings severe penalties or negative consequences.

In view of this need to understand more about the role that rule makers play in molding deviance, we turn to the constructionist approach, which concentrates on a particular type of rule maker—the claims maker.

Constructionist approach A sociological approach that focuses on the process by which specific groups, activities, conditions, or artifacts become defined as problems.

The Constructionist Approach

The **constructionist approach** focuses on the process by which specific groups (for example, illegal immigrants or homosexuals), activities (for example, child abuse or drug swallowing), conditions (teenage pregnancy, infertility, pollution), or artifacts (song lyrics, guns, art, eyeglasses) become defined as problems. In particular, constructionists examine the claims-making activities that underlie this process.[14] Claims-making activities include "demanding services, filling out forms, lodging complaints, filing lawsuits, calling press conferences, writing letters of protest, passing resolutions, publishing exposés, placing ads in newspapers, supporting or opposing some governmental practice or policy, setting up picket lines or boycotts" (Spector and Kitsuse 1977:79). Claims makers are people who artic-

[14]Sociologist Joel Best (1989) assembled thirteen articles, written by various sociologists, in a single volume called *Images of Issues: Typifying Contemporary Social Problems.* Articles express a constructionist viewpoint. The titles include "Horror Stories and the Construction of Child Abuse," "Dark Figures and Child Victims: Statistical Claims about Missing Children," "AIDS and the Press: The Creation and Transformation of a Social Problem," and "The Surprising Resurgence of the Smoking Problem." In spite of the diversity of topics, all thirteen authors take a constructionist approach, a method with strong roots in symbolic interaction but that also appeals to conflict theorists.

ulate and promote claims and who tend to gain in some way if the targeted audience accepts their claims as true.

Claims Makers

Claims makers include but are not limited to victims of discrimination, government officials, professionals (medical doctors, scientists, professors), and pressure groups (any group that exerts pressure on public opinion and government decision makers in order to advance or protect its interests). The success of a claims-making campaign depends on a number of factors, including access to the media, available resources, and the claims maker's status and skills at fund raising, promotion, and organization (Best 1989). During the 1989 Tiananmen Square incident in Beijing,[15] for example, government officials were engaged in claims making activities as they controlled the information that the troops received about the demonstration. Essentially, the soldiers were denied access to newspapers and television for about a week. When soldiers and students became friendly with one another, the government sent in replacements, some of whom came from the North and were not fluent in standard Chinese (Calhoun 1989). Thus, the replacements could not communicate with the demonstrators. Obviously, the government officials, because of their positions, were better able than the students to control the soldiers' understanding of the situation.

According to sociologist Joel Best, when constructionists study the process by which a group or behavior is defined as a problem to the society, they focus on who makes the claims, whose claims are heard, and how audiences respond. Constructionists are guided by one or more of the following questions: What kinds of claims are made about the problem? Who makes the claims? Which claims are heard? Why is the claim made when it is made? What are the responses to the claim? Is there evidence that the claims maker has misrepresented or inaccurately characterized the situation? In answering this last question, constructionists examine how claims makers characterize a condition. Specifically, they pay attention to any labels that claims makers attach to a condition, the examples they use to illustrate the nature of the problem, and their orientation toward the problem (describing it as a medical, moral, genetic, educational, or character problem).

Labels, examples, and orientation are important because they tend to evoke a particular cause and a particular solution to a problem (Best 1989). For example, to call AIDS a moral problem is to locate its cause in the goodness or badness of human action and to suggest that the solution depends upon changing evil ways. To call it a medical problem is to locate its cause in the biological workings of the body or mind and to suggest that the solution rests with a drug, a vaccine, or surgery. Similarly, a claims maker who uses the example of a promiscuous homosexual male to illustrate the nature of the AIDS problem sends a much different message about AIDS than does a claims maker who uses hemophiliacs or HIV-infected children as examples (see "Claims About the Dangers of Legal and Illegal Substances" and "Grounds for Divorce and Annulment").

To this point, we have examined several sociological concepts — socialization, norms (folkways and mores), and mechanisms of social control — and how they relate to deviance. In addition, we have also examined Durkheim's theory of deviance for insights about how any behavior or appearance can come to be defined as deviant. We have looked at labeling theory and the constructionist approach for insights about the role that rule makers play in shaping deviance. Now we turn to the theory of structural strain, which gives us insights about how society creates deviance by virtue of its valued goals and the opportunities it offers people to achieve those goals.

Structural Strain Theory

Robert K. Merton's theory of structural strain takes into account three factors: (1) culturally valued goals defined as legitimate for all members of society, (2) norms that specify the legitimate means of achieving these goals, and (3) the actual number of legitimate opportunities available to people to achieve the culturally valued goals. According to Merton, **structural strain**, or anomie, is any situation in which (1) the valued goals

Structural strain A condition that occurs when the valued goals have no clear boundaries, when it is not clear whether the legitimate means that society provides will lead to the goals, and when the legitimate opportunities for meeting the goals are closed to a significant portion of the population.

[15]The Tiananmen Square incident is the most well-known event in a series of demonstrations that took place in China between April 15 and June 4, 1989. In all, eighty-four cities and up to 3 million students joined by workers, teachers, even police and soldiers were involved in demonstrations against those in power. The image that remains with us is that of a young Chinese man facing down a tank. Although no official figures are available, hundreds are estimated to have been killed and thousands injured.

U.S. In Perspective

Claims About the Dangers of Legal and Illegal Substances

We use the constructionist approach to examine the accuracy of claims that alcohol and cigarettes are legal because they are less dangerous substances than cocaine and other now-illegal drugs. This claim is one reason that drug enforcement officials generate profiles of people likely to be drug carriers and drug swallowers but not people likely to be tobacco or alcohol vendors who sell these products to minors.

In his article "Taxonomy as Politics: The Harm of False Classification," biologist Stephen Jay Gould (1990) describes a war of words between two claims makers—former Surgeon General C. Everett Koop and Illinois Representative Terry Bruce—that began after Koop announced that nicotine was no less addicting than heroin or cocaine:

> *Representative Terry Bruce (D-IL) challenged [Koop's] assertion by arguing that smokers are not "breaking into liquor stores late at night to get money to buy a pack of cigarettes." Koop properly replied that the only difference resides in social definition as legal or illegal: "You take cigarettes off the streets and people will be breaking into liquor stores. I think one of the things that many people confuse is the behavior of cocaine and heroin addicts when they are deprived of the drug. That's the difference between a licit and an illicit drug. Tobacco is perfectly legal. You can get it whenever you want to satisfy the craving." (p. 75)*

Koop's hypothesis that cigarette smokers deprived of their drug would behave like cocaine and heroin addicts is supported by events in Italy, where the government has a monopoly over the manufacture and distribution of cigarettes. In 1992, approximately 13 million smokers were prevented from purchasing cigarettes because of a strike by the workers who distributed them. The forced abstinence resulted in the following behaviors: (1) panic buying and hoarding of the cigarettes still on the shelves when the strike began, (2) a run on nicotine gum and patches, (3) a sevenfold increase in traffic into Switzerland, (4) an underground market in cigarettes, and (5) robbery of cigarette vendors at gunpoint (Cowell 1992).

Gould argues that illegal drugs such as cocaine should be legalized and controlled, like alcohol and tobacco, and their use strongly discour-

have unclear limits (that is, people are unsure whether they have achieved them), (2) people are unsure whether the legitimate means that society provides will lead to the valued goals, and (3) legitimate opportunities for meeting the goals are closed to a significant portion of the population. The rate of deviance is likely to be high under any one of these situations. Merton uses the United States, a country where all three conditions exist, to show the relationship between structural strain and deviance.

Structural Strain in the United States

In the United States, most people place a high value on the culturally valued goal of economic affluence and social mobility. In addition, Americans tend to believe that all people, regardless of the circumstances in which they are born, can achieve monetary success. Such a viewpoint suggests that success or failure results from personal qualities and that persons who fail have only themselves to blame. Merton does not believe that the same proportion of people in all social classes accept the cultural goal of monetary success, but he does think that a significant number of people across all classes do so. He argues that "Americans are bombarded on every side" with the message that this goal is achievable, "even in the face of repeated frustration" (1957:137).

According to Merton, considerable structural strain exists in the United States because this culturally valued goal has no clear limits. He believes that there is no point at which people can say they have achieved monetary success. "At each income level . . . Americans want just about 25 percent more (but of course this 'just a bit more' continues to operate once it is obtained)" (p. 136).

Merton also maintains that structural strain exists in the United States because the legitimate means of achieving affluence and mobility are not entirely clear; the individual's task is to choose a path that leads to success. That path might involve education, hard work, or natural talent. The problem is that school, hard work, and talent do not guarantee success. With regard to education, for example, many Americans believe that the diploma (especially the college diploma) in itself entitles them to a high-paying job. In reality, however, the diploma is only one component of many needed to achieve success.

Finally, structural strain exists in the United States because too few legitimate opportunities are available

aged. He would even accept, reluctantly, a policy in which alcohol and tobacco had the same illegal status as drugs such as cocaine. Gould (1990) cannot find any rationale, however, for "an absurd dichotomy (legal versus illegal) that encourages us to view one class of substances with ultimate horror as preeminent scourges of life . . . while the two most dangerous and life-destroying substances by far, alcohol and tobacco, form a second class advertised in neon on every street corner of urban America" (p. 74).

Gould's claim is supported by data from the National Institute on Drug Abuse, an agency of the U.S. government, and the Robert Wood Johnson Foundation. The National Institute on Drug Abuse estimated for every one cocaine-related death in 1987, there were 300 tobacco-related and 100 alcohol-related deaths (Reinarman and Levine 1989). A more recent study sponsored by the Robert Wood Johnson Foundation (1992) found that substance abuse, which includes smoking, drinking, and drugs, causes 500,000 premature deaths each year in the United States. Cigarettes account for 419,000. The study also found that alcohol-related conditions account for 25 to 40 percent of all general hospital admissions. A third study published in the *Journal of the American Medical Association* estimated that each year 400,000 premature deaths are due to tobacco, 100,000 to alcohol, and 20,000 to the illegal drugs.

Gould (1990) does not deny that all such drugs mentioned are dangerous, but he finds that the illegal status for one set of substances and the legal status for another "permit a majority of Americans to live well enough with one, while forcing a minority to murder and die for the other" (p. 75). He argues that illegal status causes more social and medical problems than it solves and that illegal status itself causes crime. For example, the responses to the prohibition of alcohol in the United States between 1920 and 1933 are remarkably similar to those associated with cocaine's illegal status. The responses include murder, payoffs, bribes, and international intrigue. There is one argument, seldom articulated, in favor of keeping drugs like cocaine illegal: prohibition might keep the cocaine problem from reaching the tragic dimensions of the alcohol and tobacco problems.

to achieve desired goals. Although all Americans are supposed to seek financial success, they cannot all expect to achieve it legitimately; the opportunities for achieving success are closed to many people, especially those in the lower classes. For example, many young African American men living in poverty believe that one seemingly sure way to achieve success is through sports. The opportunities narrow rapidly, however, as an athlete advances: fewer than 2 percent of the athletes who play college basketball, for instance, even have a chance at the professional ranks.

Merton believed that people respond in identifiable ways to structural strain and that their response involves some combination of acceptance and rejection of the valued goals and means (see Figure 8.2). He identified the following five responses — only one of which is not a deviant response.

- **Conformity** is the acceptance of the cultural goals and the pursuit of these goals through legitimate means.
- **Innovation** involves the acceptance of the cultural goals but the rejection of legitimate means to obtain these goals. For the innovator, success is equated with winning the game rather than playing by the rules of the game. After all, money may be used to purchase the same goods and services whether it was acquired legally or illegally. Merton argues that when the life circumstances of the middle and upper classes are compared with those of the lower classes, the lower classes clearly are under the greatest pressure to innovate, though no evidence suggests that they do so (for example, white-collar and corporate crime are forms of innovation).
- **Ritualism** involves the rejection of cultural goals but a rigid adherence to the legitimate means of those goals. This response is the opposite of innovation; the game is played according to the rules despite defeat. Merton maintains that this re-

Conformity The acceptance of the cultural goals and the pursuit of these goals through legitimate means.

Innovation The acceptance of the cultural goals but the rejection of legitimate means to obtain these goals.

Ritualism The rejection of cultural goals but a rigid adherence to the legitimate means of those goals.

U.S. In Perspective

Grounds for Divorce or Annulment

This box provides another way to look at how U.S. society varies in its attitude toward alcohol, other drugs, and another behavior often called "deviant." The map shows which states define alcoholism, drug addiction, and adultery as definite grounds for divorce or annulment. Making such behaviors grounds for divorce is one way of controlling the behaviors without making them strictly illegal. What patterns do you see in the data?

Practically speaking, the illegality of drugs like marijuana and cocaine is currently maintained based on federal law. Suppose someone proposed that grounds for divorce should also be standardized at the national level rather than being controlled by the states. What arguments would you make for or against such a proposal?

Suppose someone proposed that a spouse's addiction to prescription drugs should be grounds for divorce. Could a case be made for such a law?

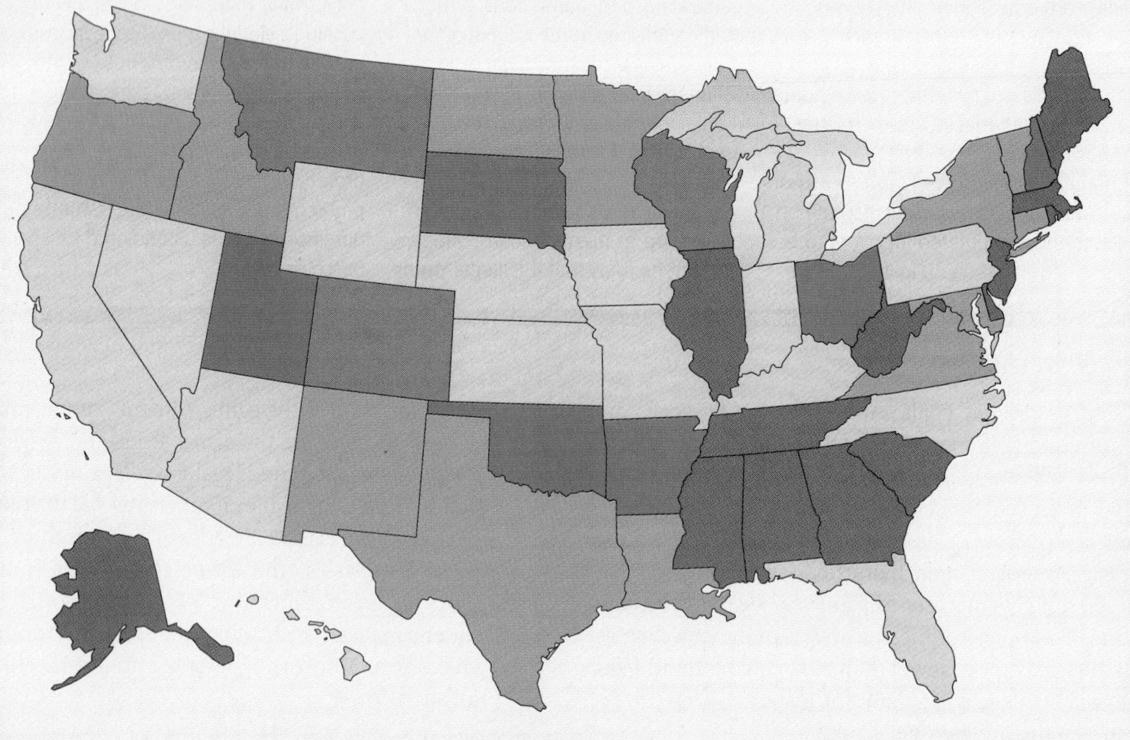

	Grounds for Divorce		
	Adultery	Alcoholism	Drug Addiction
	Yes	Yes	Yes
	Yes	Yes	No
	No	Yes	Yes
	Yes	No	No
	No	No	Yes
	No	No	No

Source: *The World Almanac and Book of Facts 1996* (p. 729)

Figure 8.2 Merton's Typology of Responses to Structural Strain

Merton believed that people respond to structural strain and that their response involves some combination of acceptance and rejection of valued goals and means.

Mode of Adaptation	Goals	Means
Conformity	+	+
Innovation	+	–
Ritualism	–	+
Retreatism	–	–
Rebellion	+/–	+/–

+ Acceptance/achievement of valued goals or means

– Rejection/failure to achieve valued goals or means

Source: Adapted from Merton (1957:140), "A Typology of Modes of Individual Adaptations."

sponse can be a reaction to the status anxiety that accompanies the ceaseless competitive struggle to stay on top or to get ahead. Ritualism finds expression in the clichés "Don't aim high and you won't be disappointed" and "I'm not sticking my neck out." Ritualism can also be the response of people who have few employment opportunities open to them. If one wonders how it is possible for a ritualist to be defined as deviant, consider the case of a college graduate who can find only a job bagging groceries at $5.15 an hour. Most people react as if this person is a failure even though he or she may be working full-time.

- **Retreatism** involves the rejection of both cultural goals and the means of achieving these goals. The people who respond in this way have not succeeded by either legitimate or illegitimate means and thus have resigned from society. According to Merton, retreatists are the true aliens or the socially disinherited — the outcasts, vagrants, vagabonds, tramps, drunks, and addicts. They are "in the society but not of it."
- **Rebellion** involves the full or partial rejection of both goals and means and the introduction of a new set of goals and means. When this response is confined to a small segment of society, it provides the potential for subgroups as diverse as street gangs and the Old Order Amish. When rebellion is the response of a large number of people who wish to reshape the entire structure of society, the potential for a revolution is great.

Structural Strain in China

In China, one source of structural strain can be traced to the actual number of legitimate opportunities (one) open to married couples to achieve the culturally valued goal of population control. As pointed out in "The Extent of Social Control in China," since 1979 China has tried to impose a limit of one child per couple (means) in hopes of slowing population growth (Sturm and Zhang 1994). The one-child policy poses a problem because in China there is a cultural preference for boys, especially among the people living in the countryside. One reason sons are valued over daughters is that sons and their families are expected to care for parents in old age. Thus, couples who manage to produce a son can be confident that someone will care for them later in life. The legitimate means open to a couple in China include these: obtain permission to have a baby, accept the sex of the child (that is, not aborting female fetuses or killing a daughter), report the birth and sex of the child to appropriate agencies, and practice birth control to avoid conceiving other children.

Merton's typology of responses to structural strain can be used to describe the reactions of couples (Figure 8.3). Those most likely to be conformists are those whose first child is a healthy son. Conformists would also include those who have no preference as to the sex of their child and/or those who are firmly committed to upholding the laws related to birth control because they see them as critical to China's quality of life (Bolido 1993:6). The majority of people in China can be classified as conformists.

Innovators accept the culturally valued goal of population control but reject the package of legitimate means to obtain this goal. Upon learning she is expecting a child, the woman may undergo ultrasound to learn the sex of the fetus. If it is a girl, the couple may decide to abort the baby. Or upon the birth of a girl baby, the parents may kill or arrange to have a midwife kill the infant. Such practices are blamed for the so-

Retreatism The rejection of both cultural goals and the means of achieving these goals.

Rebellion The full or partial rejection of both goals and means and the introduction of a new set of goals and means.

Figure 8.3 Merton's Typology Applied to China's One-Child Policy

Culturally valued goal: population control
Culturally valued means: one-child limit
Sources of structural strain: preference for boys

Response	Goal: Population Control	Means: One-Child Policy	Examples
Conformists	+	+	Couples with no preference "Ideologically sound" couples
Innovators	+	–	Couples who abort female or unhealthy babies Couples who abandon babies
Ritualists	–	+	Couples who follow the rules but reject policies
Retreatists	–	–	Couples who reject the idea of population control and hide the "extra" babies
Rebels	+/–	+/–	Couples who replace official goals and means with new goals and means

+ Acceptance/achievement of valued goals or means

– Rejection/failure to achieve valued goals or means

called missing girls problem; the 1990 census in China showed that for every 114 boys born there are 100 girls (Oxman 1993c). A second category of innovators are those who abandon an unhealthy and/or female child or abandon a second child born in violation of family planning regulations. These children fill the country's sixty-seven state-run orphanages (Tyler 1996b).[16]

Ritualists reject the cultural goal of population control, but they adhere to the rules. They do not agree with the government policies, but they are afraid they will be punished if they do not follow the rules. For example, one survey found that 78 percent of rural women in China considered two children ideal. Although most complied with the government's policies, they practiced reversible contraception because in the event of a policy change they would have a second child (Remez 1991).

Retreatists, on the other hand, reject the one-child goal and reject the legitimate means open to them. This category of deviants may include women and men who reject population control and are afraid to conceive out of fear that their firstborn may not be the desired sex. Retreatists might also include those couples who keep having children until they have a boy and hide the birth of the baby girls from party officials. In fact, Sterling Scruggs of the United Nations Population Fund argues that there is good evidence that many of the so-called missing girls include those who have not been reported (Oxman 1993c). In a sense, these parents are in society but, because of their secret, "not of it," as Merton would say.

Rebels reject the cultural goal of population control and reject the legitimate means; instead, they introduce new goals and new means. One could argue that this response applies to couples who are part of China's fifty-five ethnic minority populations that make up 9 percent (91.2 million) of the total population. The majority of these ethnic groups are exempt from the one-child policy. The government permits them to have two children, and some groups are permitted to have three and in special cases four children. For these groups, the

[16]The annual death rates in some orphanages are higher than 20 percent, and in some institutions the death rate is between 57.2 and 72.5 percent (Schell 1996). Human rights groups argue that the high mortality rate reflects a deliberate policy to minimize the number of abandoned children. Chinese officials argue that the high orphan mortality rate reflects the poor health of the abandoned children and the shortage of electricity needed to keep buildings warm (Tyler 1996a, 1996c).

cultural goal is to increase the size of the minority populations, and the means of achieving that goal is to exempt them from the one-child policy. This option has prompted many couples to claim ethnic minority status (Tien et al. 1992).

Differential Association Theory

Sociologists Edwin H. Sutherland and Donald R. Cressey advanced a theory of socialization called **differential association** to explain how deviant behavior, especially delinquent behavior, is learned. It refers to the idea that "when persons become criminal, they do so because of contacts with criminal patterns and also because of isolation from anticriminal patterns" (1978:78). The contacts take place within **deviant subcultures**—groups that are part of the larger society but whose members adhere to norms and values that favor violation of the larger society's laws. That is, people learn techniques of committing crime from close association and interaction with people who engage in and approve of criminal behaviors.

Sutherland and Cressey maintain that impersonal forms of communication such as television, movies, and newspapers play a relatively small role in the genesis of criminal behavior. If we accept the premise that criminal behavior is learned, then criminals constitute a special type of conformist. That is, they conform to the norms of the group with which they associate. Furthermore, the theory of differential association does not explain how a person makes contact with deviant subcultures in the first place (unless, of course, the person is born into such a subculture). Once contact is made, however, the individual learns the subculture's rules for behavior in the same way that all behavior is learned.

Sociologist Terry Williams studied a group of teenagers, some as young as fourteen, who sold cocaine in the Washington Heights section of New York City. These youths were recruited by major drug suppliers because, as minors, they could not be sent to prison. Williams (1989) argues that the teenagers were susceptible to recruitment because they saw little chance of finding well-paying jobs and because they perceived drug dealing as a way to earn money that would enable them to pursue a new life.

To become a successful drug dealer, a youth must learn a number of skills. Williams (1989) describes some of these skills:

> Each week, Max is fronted [consigned] three to five kilos (in 1985, this would have a street value of $180,000 to $350,000) to distribute to the kids. The quantity he receives varies according to the quantity he has previously sold, how much he has on hand, and how much he is committed to deliver both to the crew and directly to other customers.
>
> The quality, variety and amount of cocaine each crew member receives, then, is determined by Max and by his suppliers. The kids might ask for more but whether or not they get more depends on how much they sold from their last consignment. Personal factors are also involved. (p. 34)

Williams's findings suggest that once teenagers get involved in drug networks, they learn the skills to do their jobs in the same way that everyone learns to do a job. Success in a "deviant" job is measured in much the same way that success is measured in mainstream jobs: pleasing the boss, meeting goals, and getting along with associates.

In China, notions of differential association are the basis of the philosophy underlying rehabilitation: a deviant individual, whether a thief or a revisionist, becomes deviant because of "bad" education or associations with "bad" influences (see "Is There a Rationale for Control?"). The way to correct these influences is to reeducate deviant individuals politically, inspire them to support the Communist Party, and teach them a love of labor. Particular attention is paid to the 5 million or so university or advanced technical institute students, the selected and elite few (3 percent of middle or high school students). As one Beijing University professor commented in an interview with Asian scholar and *MacNeil/Lehrer Newshour* special correspondent Robert Oxman (1994b), "It is very challenging job here on the campus how to learn those advanced technology and also very interesting ideas, value systems from the outside [and at] the same time to keep our own traditions" (p. 7). In other words, the Chinese government must prepare new generations of young people for coping with global interdependence and for running the country's economy and other affairs. In doing this, students associate with people and ideas that challenge Chinese ways of doing things. To counteract such exposures,

Differential association A theory of socialization that explains how delinquent behavior is learned. It refers to the idea that "when persons become criminal, they do so because of contacts with criminal patterns and also because of isolation from anticriminal patterns" (Sutherland and Cressey 1978:78).

Deviant subcultures Groups that are part of the larger society but whose members adhere to norms and values that favor violation of the larger society's laws.

U.S. In Perspective

Is There a Rationale for Control?

The Chinese believe in the malleability of human beings and the social basis for deviant behavior (Zhou 1995:84). As a result, they support early intervention and have a definite rehabilitation program for offenders, consisting of thought reform and reeducation through labor:

> *The Chinese believe that it is important to intervene early, before deviance has become extreme or caused too much damage. To refrain from correcting minor expressions of deviance on the grounds that such correction interferes with the individual's civil rights strikes the Chinese as analogous to withholding early treatment of a disease on the grounds that the disease has a right to develop until it proves to be lethal. This is particularly true since society (through bad education) is at least partially to blame for the deviance; it thus has a corresponding duty to correct it. (Bracey 1985:142)*

An offender's work unit, neighborhood committee, school, or parents can submit requests to local "control committees," composed of party members who evaluate the application and who can sentence a person to reeducation through labor for up to four years without judicial review.

Each city and each province maintains labor camps and prisons. Offenders are placed in one of six types of correctional facilities: (1) prisons, (2) labor reform battalions located in rural areas, (3) reeducation-through-labor battalions, (4) forced job placement battalions composed of ex-prisoners, (5) detention centers for the not-yet-rehabilitated, and (6) juvenile facilities (Mosher 1991). The purpose of incarceration is to rehabilitate or "rehumanize" offenders. Rehabilitation programs are designed to teach offenders a productive skill and to make their outlook socially minded, patriotic, and law-abiding. Offenders labor in mines, farms, or factories or on large infrastructure projects such as road building and irrigation. A significant number of prisoners produce goods for export. They also undergo "thought reform," which involves self-criticism and intensive study of Communist principles. In sum, "Enormous effort is invested in the attempt to drive a convicted criminal into a hell of personal despair and loneliness, from which the only escape is by genuinely turning over a new 'moral leaf' and wholeheartedly accepting the political line of the Party" (Bonavia 1989:164).

China's treatment of political dissidents has drawn considerable criticism. In fact, the U.S. House of Representatives passed a resolution asking the International Olympic

Communist ideology is officially espoused at the universities. Also, Chinese students are required to take a course in Marxism, and campus security remains tight (Oxman 1994b).

We close the chapter with a discussion of the factors that influence the Chinese and U.S. systems of social control. They include population size, the age of the two countries, and the degree of internal turmoil.

Factors That Shape U.S. and Chinese Systems of Social Control

The size of the Chinese population is one major reason for the rigid system of social control. More than 1.2 billion Chinese—roughly one person of every five people alive in the world—live in a space roughly the size of the United States. After subtracting deserts and uninhabitable mountain ranges, the habitable land area is about half that of the United States. Although almost 21.3 percent of the world's population lives in China, China has only 7 percent of the world's agriculture land (Han Xu 1989). Even though China manages to feed most of its population well, approximately 65 million people live in a state of absolute poverty; that is, they lack the resources to meet basic needs of food and shelter. Another 290 million live in a state of substantial deprivation. World Bank poverty specialist Alan Piazza (1996) maintains that "most Americans could not begin to comprehend what 65 million Chinese have to endure and what that level of deprivation is all about" (p. 47). On the average, there are 18,000 births and 7,000 deaths per day in China. If these patterns persist, the total population could increase by another 100 million by the year 2000. To complicate the situation, the population has expanded rapidly in the past forty-five years, from approximately 500 million in 1949 to 1.2 billion in 1995 (U.S. Central Intelligence Agency 1995).

Such rapid growth strains the country's ability to house, clothe, educate, employ, and feed its people and can overshadow industrial and agricultural advance-

Committee to deny Beijing the privilege of hosting the 2000 Olympic Games (Oxman 1993d). Many critics in the United States believe that China should be denied Most Favored Nation status (a trading status that has been given to all but about twelve countries in the world) until it improves its human rights record.

The human rights watch group Asia Watch maintains that 1,300 people arrested during Tiananmen Square remain in prison. Chinese officials say that the figure is seventy. Because China gives no monitoring group such as the International Committee of the Red Cross access to its prisons, accounts of human rights abuses such as torture, solitary confinement, and other physical and mental abuses are based on anecdotal evidence.[1] Chinese officials deny that such abuses take place and further state that such practices are against the law.

Critics also claim that China uses many of its 10 million prisoners in labor camps to manufacture exports. However, the U.S.–China Business Council claims that prisoner-made exports cannot possibly exceed 1 percent of all exports. China has responded to the charge by opening five labor camps believed to be sites of such manufacturing activities (Oxman 1994a).

In the United States, on the other hand, no clear rationale, such as moral reform, underlies the formal sanctions. In other words, there is little agreement about the purpose of corrections: Should it be rehabilitation? Restitution? Revenge? Moreover, no coherent philosophy guides the operation of the 1,185 state and 61 federal prisons (Lilly 1991). American politicians and criminal justice personnel debate continually not only the purpose of prisons but also the effectiveness of the various methods, which may include an uneven array of educational, vocational, and therapeutic programs. Currently, the overwhelming emphasis is placed on ensuring that criminals serve time by establishing mandatory punishment schedules free of judicial discretion (Xu 1995).

[1]It is worth noting that in China, execution is one method of education. Its purpose is reflected in the Chinese saying "Kill a chicken to scare the monkeys." In an anticrime campaign, the Chinese government executed 10,000 people between 1983 and 1986. "The victims of such instruction might be murderers; but they might also be pimps, arsonists, prostitutes, gamblers, procurers, rapists, white collar criminals, thieves, muggers or members of Chinese secret societies. In other words, disrupters of life. A suitable candidate for capital punishment is anyone who goes against the grain; and unlawful assembly, or airing allegedly seditious material, as occurred this past May and June, is going against the grain" (Theroux 1989:7).

ments. When one considers that approximately 1 billion people live under approximately fifty independent governments in Europe, North America, and South America (Fairbank 1989), the magnitude of China's population problem appears even more overwhelming. China's large population explains in part why jobs and housing are assigned and why until recently nearly everything was rationed. Rice and other grains, cooking oil, soap, light bulbs, bicycles, and television sets all have been rationed. Rice, for example, was purchased with ration coupons. Even if a bowl of rice was purchased in a restaurant, it counted as part of a person's monthly allotted ration. In light of the rapid economic growth and the accompanying abundance of goods and food, the rationing system has become obsolete (Oxman 1993b).

The Chinese government must create work for many people, especially the large number of people awaiting employment. (Theoretically, in China there are no unemployed people — only those awaiting employment.) Currently, 583.6 million people (160 million in urban areas and 420 million in rural areas) constitute the workforce in China. However, approximately 120 million people in the rural work force (which is growing by 20 million a year) are surplus workers (Chinese Embassy 1996). As many as 100 million surplus rural workers known as the "floating population" have migrated to the cities in search of work but find only part-time low-paying jobs. Consequently, there is a considerable amount of "make-work." For example, "every park employs dozens of ticket takers, some to collect the ten cents for a ticket, some to collect the ticket, and some to collect the stub" (Mathews and Mathews 1983:34).

Living space is another problem. About 10 percent of housing in China is owned privately. Usually housing is allocated by the government or work units. Space is tight. Family members sleep in the same room, and most households share bathroom and kitchen facilities with other households (Butterfield 1982).

A second reason for the difference between Chinese and U.S. social control systems is rooted in history. Most Americans are immigrants or descendants of immigrants from many nations and cultures who have settled in the United States over the past 200 years. Geographic separation from their native lands enabled them to break with tradition and either retain or discard elements of their native cultures (Fairbank 1989). In addition, the United States is a land with abundant resources and an impressive ratio of people to resources. The material abundance, combined with geographic separation from native cultures, permitted those who immigrated to the United States to create a society in which, in theory, citizens live wherever they can afford to and where they can manage their own lives.

The Chinese, on the other hand, not only are members of the most populous country in the world but also belong to the longest-continuing civilization, which dates back some 3,700 years. This long and complex history has at least four strong and persistent traits:

1. The Confucian[17] system of ethics—which emphasizes order, justice, harmony, personal virtue and obligation, devotion to the family (including the spirits of one's ancestors), and respect for tradition, age, and authority—assigns everyone an unalterable role and place in society.
2. The system of family responsibility makes each member responsible for the conduct of other family members.
3. An imperial tradition gives rulers supreme authority over the lives of the people—historically, those in power have been above the law — and their power has not been constrained by institutionalized checks and balances.
4. No regime in China has ever relinquished its power without first resorting to bloodshed (Fairbank 1987, 1989).

A third issue that explains the rigid system of social control in China also relates to history. Since the beginning of the twentieth century, the Chinese people have suffered through several wars of foreign aggression and four revolutionary civil wars.[18] The most recent civil war, the Cultural Revolution (1966–1976), ended less than twenty years ago; the last civil war in the United States took place more than 125 years ago. As a people, Americans are unfamiliar with the division, chaos, and destruction that war and revolution can bring. Often, strict control is the only way to restore order and ensure the stability needed to rebuild a society whose citizens have opposed each other violently.[19]

Summary and Implications

Deviance is a complex concept because (1) definitions of deviance change over time and place; (2) not everyone who commits a deviant act is caught, and not everyone who is punished actually committed the crime of which he or she is accused; and (3) rule making and rule enforcing affect how some behaviors or appearances come to be defined as deviant and others do not.

For these reasons, sociologists maintain that deviance is not the result of a particular behavior or appearance but a consequence of time, place, and the application of rules and sanctions by others to an "offender." The case of China's dramatic shift in attitude toward profit-making activities illustrates that examining a specific behavior or the personality of someone defined as deviant cannot give us a true understanding about the nature of deviance. Examining the social and historical context surrounding a behavior or appearance adds significant insights into the causes of deviant behavior.

A knowledge of existence of secret deviants and falsely accused also adds tremendous and important implications for the study of deviance. Their existence helps us build the case that the causes of deviant be-

[17]Confucius lived between 551 and 479 B.C., a time of constant war, destruction, and chaos. It is not surprising, therefore, that his teachings emphasize rules that support order and harmony.

[18]The four revolutionary civil wars include the Republican Revolution of 1911, the Nationalist Revolution of 1925–1928, the Kuomintang-Communist Revolution of 1945–1949, and the Great Proletarian Cultural Revolution of 1966–1976.

[19]Mao Zedong, the charismatic Chinese leader who mobilized the masses in two of these civil wars, stated, "A revolution is not a dinner party, or writing an essay, or painting a picture, or doing embroidery; it cannot be so refined, so leisurely and gentle, so temperate, kind, courteous, restrained and magnanimous. A revolution is an insurrection, an act of violence by which one class overthrows another" (Mao Zedong 1965:28).

Political purges directed by Communists are not unique to the Cultural Revolution. Whenever China has faced serious economic problems or whenever the authority of Chinese officials has been questioned, campaigns have been launched to hunt down and purge those who opposed them.

havior go beyond studying the personal characteristics of the officially accused (pure deviants and the falsely accused) and must include the behavior of rule makers and rule enforcers. Researchers who compare prison populations with a population that has never been imprisoned to discover traits that distinguish the two learn little about what causes people to break rules. Such studies do not identify the characteristics that make criminals but only those that cause a person to end up in prison. This is the case because, as the U.S. Bureau of Justice Statistics (1996a, 1996b) victimization studies show, the so-called never-imprisoned population includes a large number of people who have committed the same crimes as those who are imprisoned but who have escaped detection or conviction in part because victims fail to report crimes. To complicate matters even further, the prison population can contain innocent people. The contribution of sociology to the study of deviance is that it reminds us to consider factors other than the individual defined as deviant. By considering the larger context, we will be better able to understand the true nature of deviance and design policies to change not just the individual but the larger society.

Key Concepts

Use this outline to organize your review of the key chapter ideas.

- **Deviance**
  - **Pure deviants**
  - **Master status of deviant**
  - **Secret deviants**
  - **Falsely accused**
    - **Witch-hunts**
- **Conformity**
- **Social control**
  - **Sanctions**
    - **Positive sanctions**
    - **Negative sanctions**
    - **Formal sanctions**
    - **Informal sanctions**
- **White-collar crime**
- **Corporate crime**
- **Constructionist approach**
  - **Claims makers**
- **Structural strain**
  - **Conformity**
  - **Innovation**
  - **Ritualism**
  - **Retreatism**
  - **Rebellion**
- **Differential association**

internet assignment

In this chapter, we learned that it is just as important to study the rule makers as the rule breakers. Skim the *Special Report to Congress: Cocaine and Federal Sentencing Policy.* It consists of eight chapters and three appendices, which can be accessed through the URL http://www.acsp.uic.edu/lib/ussc/chapter 1.htm. This report seeks to determine whether federal sentencing guidelines and rules related to cocaine offenses are fair and effective. Mandatory minimum sentences currently are based on the specific quantity of drug being distributed. Under current policies, crack cocaine is treated differently from powder cocaine such that possession of 500 grams of powder triggers a five-year minimum mandatory sentence, whereas possession of only 5 grams of crack triggers the same sentence.

Based on the information presented in this report, which groups are likely to be convicted for powder versus crack cocaine? Is crack cocaine different enough from powder cocaine to warrant the harsher penalty? What does this report say about the nature of deviance?

9

Social Stratification

With Emphasis on South Africa

Capetown, South Africa. ©Milner/Sygma

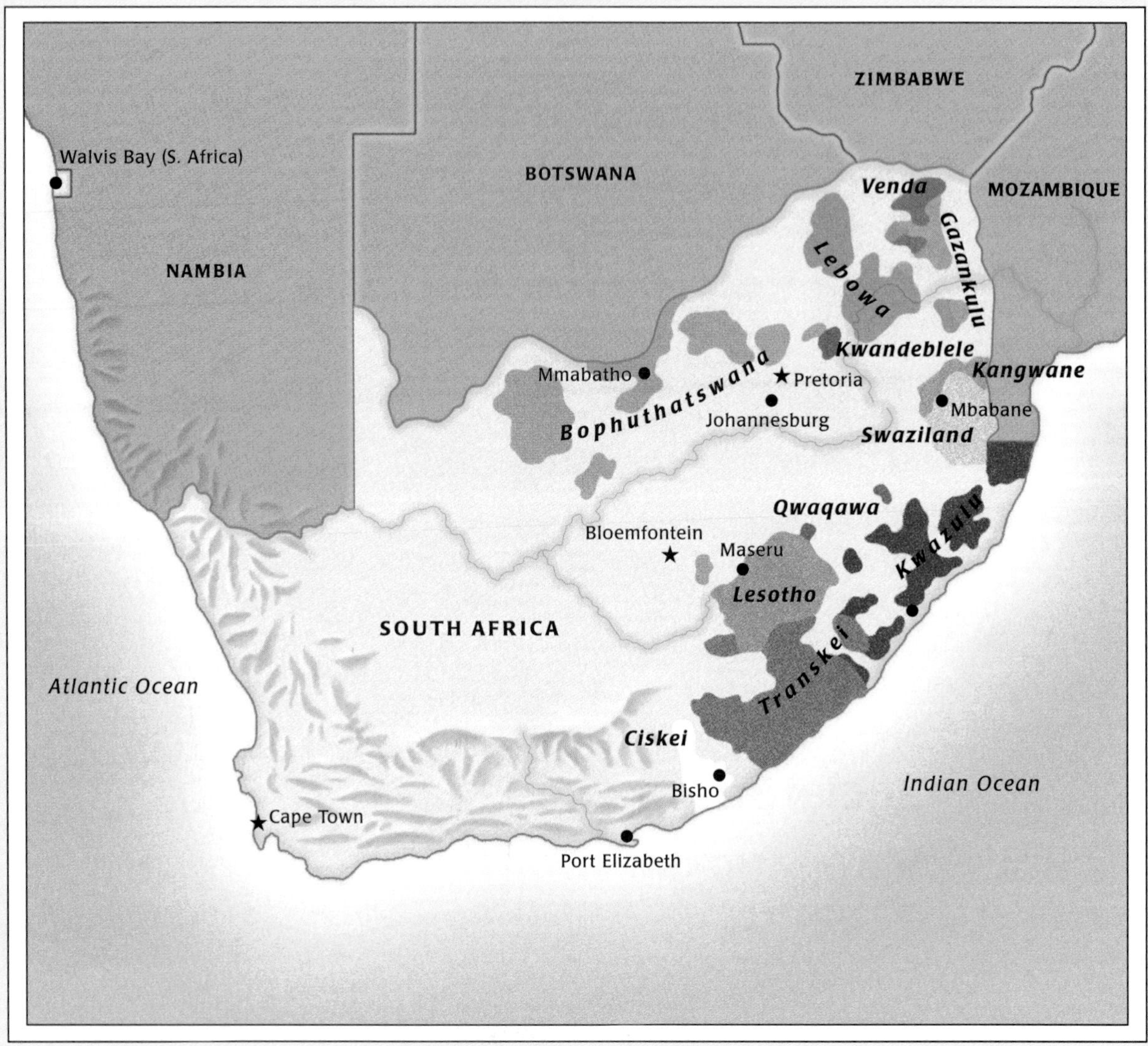

Source: Central Intelligence Agency Maps

The Geography of Apartheid

South African apartheid was an attempt to create a society in which a person's race determined his or her rights and privileges in life. It divided the country into land for whites and land for blacks. The Land Acts of 1913 and 1936 mandated that ten "homelands" be set aside, one for each of the ten major African ethnic groups. As shown, one homeland could be composed of several dozen pieces of land. The irregular shapes reflect the all-white government's decision to give the best land to the whites. By the 1960s, South Africa's policy of racial stratification had made it a moral pariah. By the 1990s, it was clear that the policies of apartheid were a dead end politically, economically, and socially. Studying apartheid shows how sociologists look at problems of social stratification. It also offers points of comparison with issues of caste and class in the United States.

***P**reamble*

Constitution of South Africa, as adopted 8 May 1996

We, the people of South Africa,

Recognize the injustices of our past;

Honour those who suffered for justice and freedom in our land;

Respect those who have worked to build and develop our country; and

Believe that South Africa belongs to all who live in it, united in our diversity.

We therefore, through our freely elected representatives, adopt this Constitution as the supreme law of the Republic so as to—

Heal the divisions of the past and establish a society based on democratic values, social justice and fundamental human rights;

Lay the foundations for a democratic and open society in which government is based on the will of the people and every citizen is equally protected by law;

Improve the quality of life of all citizens and free the potential of each person;

Build a united and democratic South Africa able to take its rightful place as a sovereign state in the family of nations.

May God protect our people.

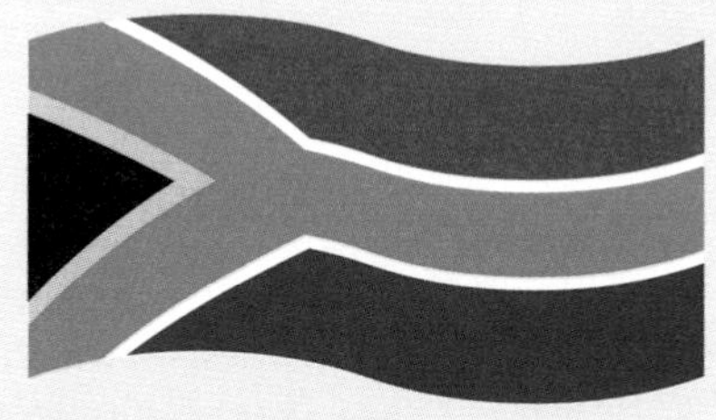

Why Focus on South Africa?

In this chapter, we focus on South Africa, a country that is attempting to change its system of social stratification from one in which 14 to 17 percent of the population (the whites) controlled the fate of the remaining 83 to 86 percent to a multiracial democracy. When Nelson Mandela became president in 1994, he inherited a country characterized by enormous economic disparities, created over 300 years, between whites and nonwhites. Mandela promised that in the five years of his administration, his government would work to address this disparity by building more than one million low-income houses, creating jobs, investing heavily in health and education, bringing electricity to 20 million people, purchasing and redistributing farmland and residential property, and taking control of the mining industry so that wealth beneath the land will go to the people of South Africa (Keller 1993). Revenue for these projects would come not from redistributing the existing wealth but from new wealth generated from foreign and domestic investment.

In the realm of foreign investment, the United States has put together a $600-million trade, investment, and development package to help South Africa achieve its goals (U.S. Central Intelligence Agency 1995). In addition, since the Bush and Clinton administrations lifted economic sanctions on South Africa, the United States has reestablished itself as South Africa's top trading partner and the country with the greatest number of corporations (208 as of June 1995) with subsidiaries in South Africa (Menaker 1995).

For many Americans, the interest in seeing the new South Africa succeed goes beyond economics. One senior Clinton administration official described the interest, even excitement, as connected

> to the broad sense of ownership that we all feel over what has gone on in South Africa. I've said in the past, every American student who protested his or her university's involvement in South Africa; every stockholder who lobbied to get their corporation out of South Africa; any American corporation that stayed in South Africa because they believed that that was the best way to fight apartheid—to create jobs, to

overcome racial discrimination; the pension fund managers who decided not to invest in South Africa; the members of Congress who allowed themselves to be arrested outside of the South African Embassy; the African Americans who, in their churches on Sunday, for years and years finished the service saying, "Free Mandela"—all of these people feel an ownership over this issue, and it is one we are trying hard as we can to foster because ownership means involvement and commitment. (Office of the Press Secretary 1994).

Social stratification is the systematic process by which people are divided into categories that are ranked on a scale of social worth. The categories in which people are placed affect their **life chances**, a critical set of potential social advantages including "everything from the chance to stay alive during the first year after birth to the chance to view fine art, the chance to remain healthy and grow tall, and if sick to get well again quickly, the chance to avoid becoming a juvenile delinquent — and very crucially, the chance to complete an intermediary or higher educational grade" (Gerth and Mills 1954:313).

This chapter examines several important and interrelated features of social stratification, including the kinds of social categories into which people are divided, the symbols that designate a person's category (for example, skin color, type of car owned, occupation), types of stratification systems, reasons for stratification systems, and the effects of stratification on life chances.

Social stratification The systematic process by which people are divided into categories that are ranked on a scale of social worth.

Life chances A set of potential social advantages including "everything from the chance to stay alive during the first year after birth to the chance to view fine art, the chance to remain healthy and grow tall, and if sick to get well again quickly, the chance to avoid becoming a juvenile delinquent—and very crucially, the chance to complete an intermediary or higher educational grade" (Gerth and Mills 1954:313).

Ascribed characteristics Attributes that people have at birth, develop over time, or possess through no effort or fault of their own.

Achieved characteristics Attributes that are acquired through some combination of choice, effort, and ability.

Status value A situation in which persons who possess one feature of a characteristic are regarded and treated as more valuable or more worthy than persons who possess other categories.

Social Categories

Every society in the world divides its people into categories — whether formally by the government and other institutions or informally in the course of social interaction. Almost any criterion can be used (and at one time or another has been used) to categorize people: hair color and texture, eye color, physical attractiveness, weight, height, occupation, sexual preferences, age, grades in school, test scores, and many others. Two major kinds of criteria are used to categorize people: ascribed and achieved characteristics.

Ascribed characteristics are attributes that people (1) have at birth (such as skin color, sex, or hair color), (2) develop over time (such as baldness, gray hair, wrinkles, retirement, or reproductive capacity), or (3) possess through no effort or fault of their own (national origin or religious affiliation that was "inherited" from parents).

Achieved characteristics, on the other hand, are acquired through some combination of choice, effort, and ability. In other words, people must act in some way to acquire these attributes. Some examples of achieved characteristics include occupation, marital status, level of education, and income. Ascribed and achieved characteristics seem clearly distinguishable, but such is not always the case. Debate continues, for example, over whether sexual preference is achieved or ascribed. Some people argue that homosexuality is genetically based behavior over which people have no control; others insist that it is learned and ultimately a matter of choice.

Sociologists are interested in those ascribed and achieved characteristics that take on social significance and **status value**: when that occurs, persons who possess one feature of a characteristic (white skin versus brown skin, blond hair versus dark hair, professional athlete versus high school teacher) are regarded and treated as more valuable or more worthy than persons who possess other categories (Ridgeway 1991).[1] Sociologists are particularly interested in studying different social values attached to ascribed characteristics because these are attributes over which people have no control.

A classic demonstration of how ascribed characteristics can affect people's access to valued resources involves a third-grade class in Riceville, Iowa. In 1970, teacher Jane Elliot conducted an experiment in which she divided her students into two groups according to a physical attribute — eye color — and rewarded them accordingly. She did this to show her class how easy it is for people to (1) assign social worth, (2) explain behavior in terms of an ascribed characteristic such as eye color, and (3) build a reward system around this seemingly insignificant physical attribute. The following excerpt from the transcript of the program "A Class Divided" (*Frontline* 1985) shows how Elliot established the ground rules for the classroom experiment:

ELLIOT: It might be interesting to judge people today by the color of their eyes ... would you like to try this?

CHILDREN: Yeah!

ELLIOT: Sounds like fun, doesn't it? Since I'm the teacher and I have blue eyes, I think maybe the blue-eyed people should be on top the first day. . . . I mean the blue-eyed people are the better people in this room. . . . Oh yes they are, the blue-eyed people are smarter than brown-eyed people.

BRIAN: My dad isn't that . . . stupid.

ELLIOT: Is your dad brown-eyed?

BRIAN: Yeah.

ELLIOT: One day you came to school and you told us that he kicked you.

BRIAN: He did.

ELLIOT: Do you think a blue-eyed father would kick his son? My dad's blue-eyed, he's never kicked me. Ray's dad is blue-eyed, he's never kicked him. Rex's dad is blue-eyed, he's never kicked him. This is a fact. Blue-eyed people are better than brown-eyed people. Are you brown-eyed or blue-eyed?

BRIAN: Blue.

ELLIOT: Why are you shaking your head?

[1]South Africa's constitution names several statuses that cannot take on status value. They are race, gender, sex, pregnancy, marital status, ethnic or social origin, color, sexual orientation, age, disability, religion, conscience, culture, language, and birth.

Teacher Jane Elliot's now-classic experiment showed how easy it is for people to assign social worth, explain behavior, and build a reward system around a seemingly insignificant physical attribute — eye color.

©Anne Dowie

BRIAN: I don't know.

ELLIOT: Are you sure that you're right? Why? What makes you sure that you're right?

BRIAN: I don't know.

ELLIOT: The blue-eyed people get five extra minutes of recess, while the brown-eyed people have to stay in. . . . The brown-eyed people do not get to use the drinking fountain. You'll have to use the paper cups. You brown-eyed people are not to play with the blue-eyed people on the playground, because you are not as good as blue-eyed people. The brown-eyed people in this room today are going to wear collars. So that we can tell from a distance what color your eyes are. [Now], on page 127 — one hundred twenty-seven. Is everyone ready? Everyone but Laurie. Ready, Laurie?

CHILD: She's a brown-eye.

ELLIOT: She's a brown-eye. You'll begin to notice today that we spend a great deal of time waiting for brown-eyed people. (*Frontline* 1985:3–5)

Once Elliot set the rules, the blue-eyed children eagerly accepted and enforced them. During recess, the children took to calling each other by their eye colors, and some brown-eyed children got into fights with blue-eyed children who called them "brown-eye." The teacher observed that these "marvelous, cooperative, wonderful, thoughtful children" turned into "nasty, vicious, discriminating little third-graders in a space of fifteen minutes" (p. 7).

This experiment illustrates on a small scale how categories and their status value are reflected in the distribution of valued resources. Because the category to which one is assigned determines status and is related to life chances, it is essential that we examine the shortcomings of classification schemes.

One major shortcoming of any classification scheme is that not all people in a society fit neatly in the categories designated as important. For example, the third-grade teacher whose experiment we described divided her students into just two categories: the blue-eyed and the brown-eyed. Such a classification scheme, however, does not accommodate people with green eyes, hazel eyes, gray eyes, or mixed-color eyes (one blue and one brown). This shortcoming in the classification scheme leaves us unsure about what to do with someone who does not fit into any of the designated categories. Typically, when people do not fit a category, others find ways to make them fit. In the third-grade class, brown-eyed people were required to wear collars to make it absolutely clear who belonged to that category.

Such a strategy is not unique to that third-grade classroom experiment. Strategies to make people fit into categories are a part of all classification schemes—even those for gender, age, and race. For example, there is no perfect dividing line to separate people into the categories "male" and "female." A small but significant number of babies are born hermaphrodites; that is, they have both male and female reproductive organs. In the United States, parents must choose what sex to put on a hermaphrodite child's birth certificate. Given the importance of sexual categories, physicians tell parents that it is in the child's best interest to undergo a sex-clarifying operation. Because most people accept without questioning their society's category system (see Chapter 11 on gender), it is difficult to see clearly those categories as social constructions that everyone is more or less made to fit into and/or that the majority of people work to fit into. To demonstrate the many shortcomings associated with classification schemes, we will critique one major form of categorizing people—race.

Classifying People into Racial Categories

Most people in the United States equate race with groups of people who possess certain distinctive and conspicuous physical traits. Moreover, racial categories represent "natural, physical divisions among humans that are hereditary, reflected in morphology, and roughly but correctly captured by terms like Black, White, and Asian (or Negroid, Caucasoid, and Mongoloid)" (Haney López 1994:6). This three-category classification scheme, however, has many shortcomings, which immediately becomes evident when we imagine using it to classify the more than 5.6 billion people in the world. If we attempt this task, we would soon learn that three categories are not enough. This shortage of categories has a historical and social basis:

> The idea that there exist three races and that these races are "Caucasoid," "Negroid," and "Mongoloid," is rooted in the European imaginations of the Middle Ages, which encompassed only Europe, Africa, and the Near East. . . . The peoples of the American continents, the Indian subcontinent, East Asia, Southeast Asia and Oceania—living outside the imagination of Europe . . . — are excluded from the three major races for social and political reasons, not for scientific ones. Nevertheless, the history of science has been the history of failed efforts to justify these social beliefs. (Haney López 1994:13–14)

Adding more categories, however, would not ease the task of classifying the world's billions of people because racial classification rests on the fallacy that clear-cut racial categories exist. Why is this a fallacy? First, many people do not fit clearly into a racial category because no sharp dividing line distinguishes, for example, black skin from white skin or curly hair from wavy hair. This lack of a clear line, however, has not discouraged people from trying to devise ways to make the line seem clear-cut. A hundred years ago in the United States, for example, churches "had a pinewood slab on the outside door and a fine toothed comb hanging on a string" (Angelou 1987:2). People would go into the church if they were no darker than the pinewood and if they could run the comb through their hair without it snagging. At one time in South Africa, the state board that oversaw racial classification used a pencil test to classify individuals as white or black. If a pencil placed in the person's hair fell out, the person was classified as white (Finnegan 1986).

A South African woman classified as Colored explains how racial classification worked as late as 1989:

> Under South African law, I am officially considered Colored. But so is my light-skinned sister with brown hair and my brother who has kinky hair and skin even blacker than mine. The state determines what color you are. At sixteen, you have to fill out some forms, attach a photograph and send them to a state authority where your race will be decided. Differences in color are noted by official subdivisions. For example, I am a Cape Colored whereas my sister is

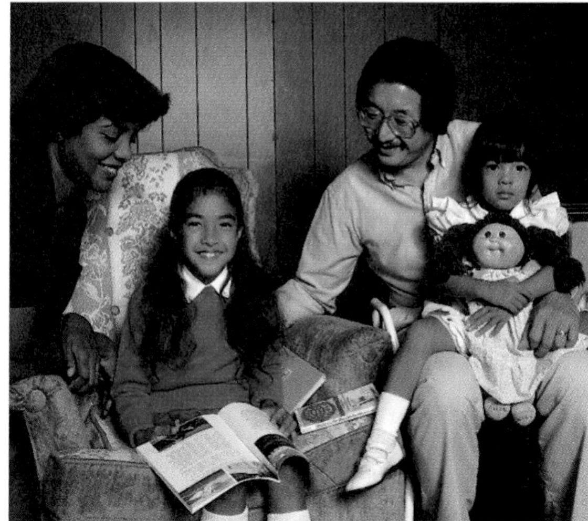

One reason that racial classification schemes are highly problematic is that many people around the world have mixed ancestry.

©Bernard Gotfryd/Woodfin Camp & Associates (left); ©Rieder/Monkmeyer Press (right)

> called Indian Colored. Of course, many people categorized as Colored are of mixed Black and white ancestry. In fact, if you can prove having had a white grandparent or parent and are yourself very light skinned, you can even make an application to be reclassified from Colored to white. (Chapkis 1986:69)

Despite the official status of classification in South Africa, the passage just cited shows that people often disagreed with and sometimes even formally appealed their racial assignments. Even if the system of formal classification completely disappears, the legacy of classification will remain for some time to come.

A second problem with trying to classify people according to any racial scheme is that millions of people in the world have a mixed ancestry and possess the physical traits of more than one race. In addition, some people belong to one race in one society but would belong to another if they lived elsewhere. For example, racial categories in South Africa are different from those used in the United States. In South Africa the four officially recognized racial categories are white,[2] African, Colored, and Asian.

Coloreds, Asians, and Africans are regarded by white South Africans as "black." Coloreds are those of mixed descent, the offspring of sexual and marital unions between white settlers and the indigenous peoples of South Africa or indentured servants from India and Malaya. The Asians, primarily Hindus of Indian descent, trace their South African residency to the late nineteenth century, when their ancestors arrived as indentured servants to work the sugar cane fields of Natal Province. Africans are divided further into ten different ethnic categories of which the largest are the Zulu, the Xhosa, and the Sotho.

For all practical purposes, in South Africa "Colored" is a miscellaneous category for those who do not fit easily into the other three groups (Sparks 1990). In the United States, however, there is no parallel category. According to the rule of hypodescent, descendants of a black-white union or a nonwhite-white union are not assigned to special categories but are usually classified arbitrarily as members of the subordinate group. At one time in the United States, people were considered black if they were known to be $^1/_{32}$ black. Consequently, in the United States, "blackness is a taint . . . [and] light-skinned people have more status" (Poussaint 1987:7).

A third shortcoming in systems of racial classification is that racial categories and guidelines for placing people in racial categories are often vague, contradictory, unevenly applied, and thus subject to change. To illustrate, consider that for the 1990 census, coders were instructed to classify as "white" those who said they were "white-black" but to classify as "black" those who said they were "black-white." Likewise, the U.S.

[2]Whites, who constitute about 14 percent of the population, consist of two distinct groups: the Afrikaners (descendants of the early Dutch, German, and Huguenot farmers who settled in South Africa in the late 1600s and who speak Afrikaans) and the descendants of British settlers, who speak English. Afrikaners make up about 60 percent of the white population, whereas English speakers make up about 40 percent of the white population.

National Center for Health Statistics has changed its guidelines for recording race and ethnicity on birth and death certificates. Before 1989, for example, a child was designated as white if both parents were white; if only one parent was white, the child was classified according to the race of the nonwhite parent; if the parents were different nonwhite races, the child was assigned the father's race; and if one parent's race was unknown, the infant was assigned the race of the parent whose race was known. After 1989, the rules for classifying newborns changed: now an infant's race is the same as the mother's (Lock 1993)—as if identifying the mother's race would present no challenges.

Sociologists are particularly interested in any classification scheme that incorporates the belief that certain important abilities (such as athletic talent or intelligence), social traits (such as criminal tendencies or aggressiveness), and cultural practices (dress, language) are passed on genetically or the belief that some categories of people are inferior to others by virtue of genetic traits and therefore should receive less wealth, income, and other socially valued items. Sociologists are interested in classification schemes because they have enormous consequences for society and the relationships among people.

For sociologists, one important dimension of stratification systems is the extent to which people "are treated as members of a category, irrespective of their individual merits" (Wirth [1945] 1985:310). In this vein, they examine how "open" or "closed" a stratification system is.

"Open" and "Closed" Stratification Systems

Despite wide variations among forms of stratification systems, each falls somewhere on a continuum between two extremes: a **caste system** (or "closed" system, in which people are ranked on the basis of traits over which they have no control) and a **class system** (or "open" system, in which people are ranked on the basis of merit, talent, ability, or past performance). No form of stratification exists in any pure form. Three characteristics most clearly distinguish a caste from a class system of stratification: (1) the rigidity of the system (how difficult it is for people to change their category), (2) the relative importance of ascribed and achieved characteristics in determining people's life chances, and (3) the extent to which there are restrictions on social interaction between people in different categories. Caste systems are considered closed because of how rigid they are—rigid in the sense that ascribed characteristics determine life chances and social interaction among people in different categories is restricted. In comparison, class systems are open in the sense that achieved characteristics determine life chances and no barriers exist to social interaction among people in different categories.[3]

Caste system A scheme of social stratification in which people are ranked on the basis of traits over which they have no control.

Class system A scheme of social stratification in which people are ranked on the basis of merit, talent, ability, or past performance.

Caste Systems

When people hear the term *caste*, what usually comes to mind is India and its caste system, especially as it existed before World War II. India's caste system, now outlawed in its constitution, was a strict division of people into four basic categories with 1,000 subdivisions. The strict nature of the caste system is shown by some of the rules that specified the amount of physical distances that were to be maintained between people of different castes: "a Nayar must keep 7 feet (2.13 m) from a Nambudiri Brahmin, an Iravan must keep 32 feet (9.75 m), a Cheruman 64 feet (19.5 m), and a Nyadi from 74 to 124 feet (22.6 to 37.8 m)" (Eiseley 1990:896).

Most sociologists use the term *caste* not to refer to one specific system but to designate any scheme of social stratification in which people are ranked on the basis of physical or cultural traits over which they have no control and that they usually cannot change. Whenever people are ranked on the basis of such traits, they are part of a caste system of stratification.

The rank of any given caste is reflected in the public esteem accorded to those who belong to it, the power wielded by members of that caste, and the opportunities of members of that caste to acquire wealth and other valued items. People in lower castes are seen as innately inferior in intelligence, personality, morality, capability, ambition, and many other traits. Conversely, people in higher castes consider themselves to

[3]Class and caste systems are ideal types. We learned in Chapter 7 that an ideal type is ideal not in the sense of having desirable characteristics but as a standard against which "real" cases can be compared. Actual stratification systems depart in some way from the ideal types.

Table 9.1 Life Chances and Racial Classification in South Africa

	White	Asian	Colored	African
Percentage of the total population	15.0	3.0	9.0	73.0
Infant mortality rate	12/1,000	18/1,000	52/1,000	110/1,000
Percentage of babies born underweight				
Urban (%)	16.0	35.0	49.0	28.0
Rural (%)	—	—	—	43.0
Number of pupils per teacher	18	25	29	42
Government expenditures per pupil	$1,700	$1,100	$600	$220
Literacy rate	99/100	69/100	62/100	50/100
Life expectancy in years	70	65	—	59
	White	**Asian**	**Colored**	**African**
Access to				
Piped water in home (%)	99.7	—	78.9	17.5
Electricity (%)	99.8	100.0	86.2	36.5
Drive car to work (%)	82.0	67.7	25.7	8.1
Occupation				
Male				
Professional/managerial (%)	41.7	32.0	9.7	7.2
Laborer/semiskilled (%)	0.8	17.3	41.4	31.2
Female				
Professional/managerial (%)	43.8	23.4	13.3	15.3
Laborer/semiskilled (%)	0.8	22.2	52.0	31.1

Sources: Wilson and Ramphele (1989), *The World Almanac and Book of Facts 1991* (1990), South Africa Labour Research Unit (1994)

be superior in such traits. Moreover, caste distinctions are treated as if they are absolute; that is, their significance is never doubted or questioned (especially by those who occupy higher castes), and they are viewed as unalterable and clear-cut (as if everyone is supposed to fit neatly into a category). Finally, heavy restrictions are placed on the social interaction among people in higher and lower castes. For example, marriage between people of different castes is forbidden. South Africa's system of apartheid is a classic example of caste system of social stratification.

Apartheid: A Caste System of Stratification

Apartheid was a system of laws in which everyone in South Africa was put into a racial category and issued an identity card denoting his or her race. The system was one in which skin color overwhelmingly determined a person's life chances (see Table 9.1). Under apartheid, a person is not just born a baby; a person is born a black, colored, Indian, or white baby—and that label profoundly affects every aspect of a person's life (Mabuza 1990), determining where, and with whom he or she can "live, work, eat, travel, play, learn, sleep, and be buried" (Roberts 1994:54).

Apartheid (the Afrikaans word for "apartness") was practiced in South Africa for hundreds of years. But it did not become official policy until 1948, when the conservative white Nationalist Party, led by D. F. Malan, won control of the government. Once in power, the Nationalists passed hundreds of laws and acts mandating racial separation in almost every area of life and severely restricting blacks' rights and opportunities. The ultimate result has been legalized political, social, and economic domination by whites (14 percent of the population) over nonwhites (86 percent of the population).

It is virtually impossible to give an overview of the effects of apartheid, a system that has inflicted enduring misery, economic damage, and daily indignities on South Africa's nonwhite population (Wilson and

Apartheid A system of laws in which everyone in South Africa was put into a racial category and issued an identity card denoting his or her race.

Ramphele 1989). Although apartheid has been dismantled on paper, the challenge lies in dismantling its effects on relationships between whites and nonwhites and on life chances for nonwhites. To fully grasp how apartheid laws determined life chances, we turn to an overview of these policies. As you read this overview, do not try to memorize names, dates, and specific purpose of the acts. Instead, think about how every aspect of life in South Africa was regulated according to skin color. And consider the inevitable challenges that accompany the dismantling of apartheid.

Apartheid Policies Apartheid policies centered on one aim: maintaining separate black (that is, mixed-race, Asian, and African) and white areas. The Transvaal Province Law of 1922 (paragraph 267) stated that a nonwhite "should only be allowed to enter urban areas, which are essentially the white man's creation, when he is willing to enter and to minister to the needs of the white man and should depart therefrom when he ceases so to minister" (Wilson and Ramphele 1989:192). Other legislation — the Separate Amenities Act, the Group Areas Act, the Land Acts, and the Population Registration Act—represent additional ways in which a white-ruled government worked to achieve this aim.

The Separate Amenities Act of 1953, abolished on October 16, 1990,[4] authorized the creation of separate and unequal (in quantity and quality) public facilities — parks, trains, swimming pools, libraries, hotels, restaurants, hospitals, waiting rooms, pay telephones, beaches, cemeteries, and so on — for whites and nonwhites. In areas officially designated for "whites only," local governments constructed only the most basic and necessary public facilities (such as toilets and bus stations) for blacks "ministering to the needs of whites" and prohibited them from using all other public facilities (for example, parks and libraries).

The Land Acts of 1913 and 1936, repealed on June 5, 1991, reserved 85 percent of the land for five million whites (14 percent of the population) and the remaining 15 percent of the land for the rest of the population. These acts mandated that ten homelands in the eastern half of the republic be set aside, one for each of the ten major African ethnic groups (as defined by white South Africans). The homelands consisted of land that was arid, underdeveloped, and lacking in natural resources. Needless to say, the Africans had no influence in determining the location or type of land that would constitute their homelands. For example, a homeland may be composed of as many as two dozen pieces of land, separated by large, white-owned plantations, farms, and industries (Liebenow 1986; see the chapter-opening map).

Regardless of where they were born or where they had lived most of their lives, each African was declared a citizen, not of South Africa but of one of the ten homelands. Four of these homelands were declared independent (Bophuthatswana, Transkei, Ciskei, and Venda), and the government hoped that the other six eventually would accept independent status. However, no other country in the world recognized the independence of any of the homelands because they were established only to keep black South Africans from living in white areas while meeting the low-wage employment needs of white South Africans.

The logic underlying the Land Acts was that if all Africans had citizenship in a homeland, they had no

During the same period as the 1953 Separation Amenities Act, the United States had its own version—the "Jim Crow" laws—in a number of southern states that legalized racial segregation of all public facilities.

[4]Despite the 1990 legislation, many in power have created ways to avoid the legal obligation to integrate public facilities. Residents of white areas, for example, can treat nonwhites as "nonresidents" and charge them fees as high as $200 a year to use public facilities such as the library (Wren 1991).

claim to economic and political rights when they were outside their homeland. Thus, the South African government could deny them a vote and treat them as guest workers and noncitizens:

> The homelands have a more sinister side. On the one hand, they are a labor pool for industry; on the other, they are a rubbish heap for the people industry has no use for—the old, the sick, the very young. By law, if you cannot work you must go to your homeland, and there you are dumped. Sometimes there are "resettlement camps" of tents or huts for groups who are moved [by the government] to a countryside that they do not know. Sometimes people are just left on the land, and told to survive off relatives. Schools are bad or nonexistent, medical facilities the same. Some people survive. Many die. The homelands policy is a policy of polite extermination. Gas chambers are not needed when people can simply starve to death. (Seidman 1978:111)[5]

The Group Areas Act of 1950, repealed on June 5, 1991, designated separate living areas for whites and nonwhites. Under this act, Africans were permitted to work in white areas, but they had to be housed in adjacent townships or in single-sex (usually male) closed compounds called hostels. Buses brought blacks from the townships or compounds into white areas to work and then took them home after work. Most blacks spent three or more hours a day traveling to and from work:

> Well, let's take a common scenario of a middle-aged black man living in a township 30 miles from his job. He lives there because that's where he's been put. About 4 A.M. he gets up so he can make the two-hour trek into the city each day. He rides a bus or train that makes our worst rush-hour scene look peaceful by comparison.
>
> When the workday is over, the man must leave the city by a certain time because of his curfew and take his two-hour trek back home again.
>
> Or let's say the man happens to be a miner. He'll live in a crowded dormitory and make about $5000 a year, while white miners live in rent-free homes and make around $20,000 a year for often doing the same work. It's against the law for the black miner to bring his wife and children to live with or near him or for them to visit him. Except for his two short visits home a year, he stays at the mine and works at a very hazardous job.
>
> Now what about his wife? Well, she can't live with her husband if he works at the mines, and she can't live with her family if she's a domestic. If she works for a white family, they often will insist that she live in housing quarters behind their home. If so, her kids have to be left back in the homeland with grandma or auntie. Occasionally, she too will be permitted to travel back home to visit her family. (Lambert 1988:29–30)[6]

Under the Group Areas Act, every square foot of land was officially classified as reserved for whites, Coloreds, Asians, or Africans. Figure 9.1 shows a map of the "apartheid city model" that was drawn up as part of governmental guidelines about how areas would be divided and arranged for planned racial separation. Anyone found living in the wrong area could be resettled or forced out. In addition, the government had the right to change the classification of an area. It is estimated that, between 1960 and 1983, the South African government "resettled" more than 3.5 million nonwhites to places where they did not choose to go (Wilson and Ramphele 1989).

South Africa's pass laws restricted the free movement of blacks. According to these laws, which were abolished in 1986, blacks were required to carry passbooks that contained information about their racial, residential, and employment status. Without their passbooks, blacks were not permitted to be present in areas designated as white (85 percent of the land). Between 1916 (when statistics on pass law violations first were recorded) and 1986, seventeen million Africans were prosecuted for pass law violations. One white employer explained:

> We protested the pass laws, too. They were dreadful. If our garden man left our garden to water some plants on the other side of the hedge and if he didn't have his pass in his working shirt (not wanting to get it dirty and spoiled because without it he was a lost soul), he could have been carted off by any policeman who happened to walk by. We wouldn't know where he was. We wouldn't know what had happened. We'd ring up the police stations, and if we happened to strike somebody a bit simpatico, we might learn his whereabouts. This used to happen all the time in the suburbs. It was driving us mad. Every evening you used to see lines of these people handcuffed to one another being dragged to the nearest

[5]Many of the people sent to the homelands were born in the cities and had no strong family ties in the homelands. There were no public facilities to care for them and they were not wanted (*Weekend World* 1977).

[6]A South African government official explained the purpose of this labor system: "The African labor force must not be burdened with superfluous appendages such as wives, children, and dependents who could not provide service" (African National Congress 1996).

Figure 9.1 The Apartheid City

This is how South African urban planners laid out their "ideal of geographic racial and economic stratification under apartheid." (The abbreviation CBD stands for "commercial and business district.") Apart from the distinction between whites and Africans, what class distinctions does the map define within races? What do the areas marked "SSSS" suggest to you about how well such a map would actually comprehend the human realities?

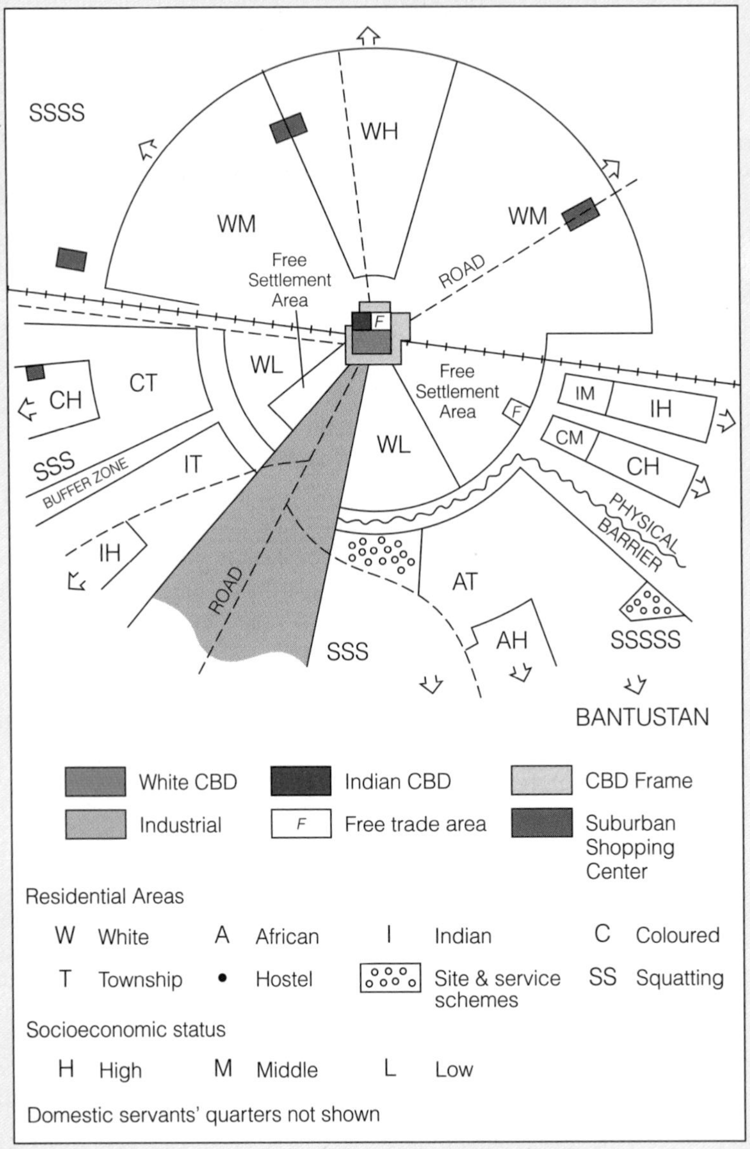

Source: From "The Apartheid City" by David Simon. P. 313 in *Atlas of World Development*. Copyright © 1994 Developing Areas Research Group. Reprinted by permission of the Developing Areas Research Group of the Royal Geographical Society (with the Institute of British Geographers) and John Wiley and Sons Ltd.

police station. We succeeded in getting that done away with. Of course, the passes still exist, but there is usually less harassment.

When I began thinking, I was distressed to have to write a piece of paper. This was before they issued passbooks. I had an African man servant whom we absolutely adored. He had worked for us for over twenty-five years. We had built very beautiful servants' quarters for him and our other help. We all loved him. And this man was really noble. He was old

The Population Registration Act required that everyone in South Africa be classified by race and that they carry an identification card denoting their race.

©Reuters/Corbis-Bettmann

> enough to be my father, and yet he had to come to me if he wanted to go out. I would have to scribble, "Please pass Amos until six o'clock—or eight o'clock—tonight." (Crapanzano 1985:131–132)

The Population Registration Act of 1950, repealed on June 18, 1991, required that everyone in South Africa be classified by race and issued an identification card denoting his or her race. Classification boards composed of white social workers ruled on questionable race identity cases and heard appeals from those who wished to change their racial classification. It is not known how many people applied for such a change:

> Deliberations take months, sometimes years, with the board paying careful attention to skin tint, facial features, and hair texture. In one typical twelve month period, 150 coloreds were reclassified as white; ten whites became colored; six Indians became Malay; two Malay became Indians; two coloreds became Chinese; ten Indians became coloreds; one Indian became white; one white became Malay; four blacks became Indians; three whites became Chinese.
>
> The Chinese are officially classified as a white subgroup. The Japanese, most of whom are visiting businessmen, are given the status of "honorary whites." Absurd? Yes, if all this were part of an antiutopian novel. But in real life it is a chilling part of the most complex system of human control in the world. (Lamb 1987:320–321)

In light of this history, it is not surprising that South Africa's new constitution seeks to protect rights denied to nonwhites under apartheid (see "The U.S. Bill of Rights and South Africa's Bill of Rights").

Apartheid is a classic example of a caste system of stratification: people are categorized according to physical traits such as skin color, ascribed characteristics determine life chances, and there are many barriers to social interaction among people who belong to different racial categories. Now we turn to a system of stratification that is at the opposite end on the continuum—a class system.

Class Systems

In class systems of stratification, "people rise and fall on the strength of their abilities" (Yeutter 1992:A13). Class systems of stratification contain economic and occupational inequality, but that inequality is not systematic. By "not systematic," we mean that there is no connection between a person's sex, race, age, or ascribed characteristics and the inequality that exists. A class system differs from a caste system in that inequality that exists in a society can be explained by differences with regard to merit, talent, ability, and past performance and not to attributes over which people have no control, such as skin color or sex. In class systems, people assume that they can achieve a desired level of education, in-

U.S. In Perspective

The U.S. Bill of Rights and South Africa's Bill of Rights

The following is an excerpt from the 1996 Constitution of South Africa, Bill of Rights, as adopted May 8, 1996. The document was adopted six years after the then South African President F. W. de Klerk traveled to the United States to convince its government and people that South Africa was beginning a sincere and irreversible effort to dismantle apartheid. During that visit, he emphasized that the principles of the U.S. political system (such as the Bill of Rights) deserved to be emulated.

As you read the South African Bill of Rights, consider the ways in which it is similar to and goes beyond the U.S. Bill of Rights. The South African Bill of Rights also alerts us to the various areas of life in which life chances are affected by a country's system of social stratification. What are some of those areas? Which Bill of Rights would you rather live under? Why?

Everyone is equal before the law and has the right to equal protection and benefit of the law.

Everyone has inherent dignity and the right to have their dignity respected and protected.

Everyone has the right to life.

Everyone has the right to freedom and security of the person, which includes the right –
a. not to be deprived of freedom arbitrarily or without just cause;
b. not to be detained without trial;
c. to be free from all forms of violence from both public and private sources;
d. not to be tortured in any way; and
e. not to be treated or punished in a cruel, inhumane or degrading way.

Everyone has the right to bodily and psychological integrity, which includes the right –
a. to make decisions concerning reproduction;
b. to security in and control over their body; and
c. not to be subjected to medical or scientific experiments without their informed consent.

No one may be subjected to slavery, servitude or forced labor.

Everyone has the right to privacy, which includes the right not to have –
a. their person or home searched;
b. their property searched;
c. their possessions seized; or
d. the privacy of their communications infringed.

Everyone has the right to freedom of conscience, religion, thought, belief or opinion.

Everyone has the right to freedom of expression, which includes –
a. freedom of the press and other media;
b. freedom to receive and impart information and ideas;
c. freedom of artistic creativity; and
d. academic freedom and freedom of scientific research.

Everyone has the right, peacefully and unarmed, to assemble, to demonstrate, to picket and to present petitions.

Everyone has the right to freedom of association.

Every citizen is free to make political choices.

No citizen may be deprived of citizenship.

Everyone has the right to freedom of movement.

Every citizen has the right to choose their trade, occupation or profession freely.

Everyone has the right to fair labor practices.

Everyone has the right –
a. to an environment that is not harmful to their health or well-being; and
b. to have the environment protected, for the benefit of present and future generations, through reasonable legislative and other measures that –
i. prevent pollution and ecological degradation;

Social mobility Movement from one class to another.

Vertical mobility A change in class status that corresponds to a gain or loss in rank or prestige.

Downward mobility A change in social class that corresponds to a loss of rank.

Upward mobility A change in social class that corresponds to a gain of rank.

Intragenerational mobility Movement (upward or downward) during an individual's lifetime.

come, and standard of living through hard work. Furthermore, people can change their class position upward during their own lifetimes, and their children's class position can be different (and ideally higher) from their own.

Movement from one class to another is termed **social mobility**. There are many kinds of social mobility in class systems. **Vertical mobility** exists if the change in class status corresponds to a gain or loss in rank or prestige, as when a medical student becomes a physician or when a wage earner loses a job and goes on unemployment. A loss of rank is **downward mobility**, whereas a gain in rank is **upward mobility**. **Intragenerational**

ii. promote conservation; and
iii. secure ecologically sustainable development and use of natural resources while promoting justifiable economic and social development.

No one may be deprived of property except in terms of law of general application, and no law may permit arbitrary deprivation of property.

Everyone has the right to have access to adequate housing.

Everyone has the right to have access to—
a. health care services, including reproductive health care;
b. sufficient food and water; and
c. social security, including, if they are unable to support themselves and their dependents, appropriate social assistance.

Every child has the right—
a. to a name and a nationality from birth;
b. to family care, parental care or appropriate alternative care when removed from the family environment;
c. to basic nutrition, shelter, basic health care services and social services;
d. to be protected from maltreatment, neglect, abuse or degradation;
e. to be protected from exploitative labor practices;
f. not to be required or permitted to perform work or provide services that—
i. are inappropriate for a person of that child's age; or
ii. place at risk the child's well-being, education, physical or mental health, or spiritual, moral or social development.

Everyone has the right—
a. to a basic education, including adult basic education; and
b. to further education, which the state must take reasonable measures to make progressively available and accessible.

Everyone has the right to use the language and to participate in the cultural life of their choice, but no one exercising these rights may do so in a manner inconsistent with any provision of the Bill of Rights.

Persons belonging to a cultural, religious or linguistic community may not be denied the right, with other members of their community, to—
a. enjoy their culture, practice their religion and use their language; and
b. form, join and maintain cultural, religious and linguistic associations and other organs of society.

Everyone has the right of access to—
a. any information held by the states; and
b. any information that is held by another person and that is required for the exercise or protection of any rights.

Everyone has the right to administrative action that is lawful, reasonable and procedurally fair.

Everyone has the right to have any dispute that can be resolved by the application of law decided in a fair public hearing in a court or, where appropriate, another independent and impartial forum.

Everyone who is arrested for allegedly committing an offense has the right—
a. to remain silent;
b. to be informed promptly—
i. of the right to remain silent; and
ii. of the consequences of not remaining silent;
c. not to be compelled to make any confession or admission that could be used in evidence against that person;
d. to be brought before a court as soon as reasonably possible, but not later than 48 hours after the arrest, but if that period expires outside ordinary court hours, to be brought before a court on the first court day after the end of that period;
e. at the first court appearance after being arrested, to be charged or to be informed of the reason for the detention to continue, or to be released; and
f. to be released from detention if the interests of justice permit, subject to reasonable conditions.

Note: The South African Bill of Rights also includes a detailed list of rights that apply to all phases of justice after arrest. The complete document can be accessed via the following URL address: gopher://gopher.anc.org.za/00/anc/misc/sacon96.txt.

mobility is movement (upward or downward) during an individual's lifetime. **Intergenerational mobility** is a change in rank over two or more generations.

In contrast to caste systems, distinctions between classes are not always clear. In principle, people can move from one class to another. It is often difficult to identify a person's class through observation alone. That is, a person may drive a BMW and wear designer clothes — symbols of solid middle-class status — but may work at a low-paying job, live in a low-rent apartment, and have exceeded his or her line of credit. Or a person may change class position through marriage, graduation, inheritance, or job promotions but may not assume the lifestyle of that class immediately (see "Caste Versus Class").

Does the United States Have a Class System?

Before we answer this question, let us recall that in a true class system, ascribed characteristics do not determine social class. Although considerable economic inequality may be present, equality of opportunity exists.

Intergenerational mobility A change in rank over two or more generations.

Simply dismantling the apartheid laws will not reverse the enormous disparities between black and white South Africans that have developed over 300 years.

©A. Tannenbaum/Sygma

On paper, the United States has a class system. The 1776 Declaration of Independence, for example, states:

> We hold these Truths to be self-evident, that all Men are created equal, that they are endowed by their Creator with certain unalienable Rights, that among these are Life, Liberty, and the Pursuit of Happiness—that to secure these Rights, Governments are instituted among Men, deriving their just powers from the Consent of the Governed, that whenever any Form of Government becomes destructive of these Ends, it is the Right of the People to alter or abolish it, and to institute new Government, laying its foundation on such Principles.

Although this document supports the "American creed," we must remember that this document was signed in 1776, but the right to vote was denied on the basis of race (until 1870) and sex (until 1919). The ideas with regard to human rights verbalized in the Declaration of Independence, however, has made this a remarkable, inspiring document for all people.

However much Americans want to believe that they live in a true class system, evidence indicates otherwise. The chances for economic and occupational success are more often than not connected to forces over which people have little control, such as social background, ascribed characteristics, and massive restructuring of the economy. To put it bluntly, economic inequality follows a clear pattern in the United States. Although the pattern of inequality in the United States is not anywhere near as systematic as that which existed under apartheid and that exists today in South Africa, it is striking nonetheless.

Income Inequality in the United States The U.S. Bureau of the Census (1992) classifies households ac-

U.S. in Perspective

Caste Versus Class

India-born novelist Bharati Mukherjee, who now lives in the United States, was asked, "What does America mean to you as an idea?" Her answer clarifies the differences between class and caste systems of stratification:

> *What America offers me is romanticism and hope. I came out of a traditional society where you are what you are, according to the family that you were born into, the caste, the class, the gender. Suddenly, I found myself in a country where—theoretically, anyway—merit counts, where I could choose to discard that part of my history that I want, and invent a whole new history for myself. . . .*
>
> *[America represents] that capacity to dream and then try to pull it off, if you can. I think that the traditional societies in which people like me were born really do not allow the individual to dream. To dream big. (Mukherjee, quoted in Tucher 1990:3–4)*

Is Mukherjee's vision of the United States accurate? Is the United States a class system?

This painting by John Trumbull shows the first U.S. Congress in 1776. What do you not see in this picture? About how many years would pass before you could expect to see those missing groups in a picture of U.S. legislators?

A detail from John Trumbull, *The Declaration of Independence, July 4, 1776*. Yale University Art Gallery. Trumbull Collection.

cording to income categories. Table 9.2 shows how income is distributed across the 97.1 million households in the United States. The table shows that 14.3 percent of all households are in the lowest income category, whereas 12.5 percent are in the highest income category. The presence of inequality as shown in Table 9.2 by itself is not enough evidence to support a claim that the United States is not a class system.

When we examine income profiles for households classified as white, black, and Hispanic, however, we see that almost 30 percent (28.9 percent) of black households and 20 percent of Hispanic households fall in the lowest income categories (see Table 9.3 and "Income Inequality in Chicago"). Such differences suggest that the United States is not a class system in the

Table 9.2 Money Income by Household, Percentage of 97.1 Million Households in Each Income Category, 1993

Income Category	Percentage
Under 10,000	14.3
$10,000–14,999	9.2
15,000–24,999	16.9
25,000–34,999	14.7
35,000–49,999	16.3
50,000–74,999	16.1
75,000+	12.5
Median income	$31,241

Source: U.S. Bureau of the Census (1995).

Table 9.3 Money Income by Household, Percentage Distribution of Households Classified as White, Black, and Hispanic in Each Income Category, 1993

	Households Classified As		
Income Category	White	Black	Hispanic
Under $10,000	12.2	28.9	20.1
$10,000–14,999	8.9	11.8	12.4
$15,000–24,999	16.6	19.2	21.5
$25,000–34,999	14.9	13.8	16.5
$35,000–49,999	17.0	12.0	13.4
$50,000–74,999	17.0	9.3	10.8
$75,000+	13.4	5.1	5.4
Median income	32,960	19,533	22,886

Source: U.S. Bureau of the Census (1995).

U.S. in Perspective

Income Inequality in Chicago

This map shows ZIP code areas of downtown Chicago. Within each ZIP code, two figures are given: in blue, the percentage of whites living in that ZIP code; in red, the percentage of "low-income" households (income below $15,000 per year). The table lists the same values shown on the map as well as the values for several adjacent ZIP codes. It lists the data in order of percentages of whites in a given ZIP code. Is there a pattern to the data? What is it? Do any ZIP codes seem to depart remarkably from the pattern?

Sources: U.S. Postal Service (1996) and U.S. Bureau of the Census (1996).

Percentages of Whites and Low-Income Households in ZIP Code Areas of Downtown Chicago

Zip Code	Whites (%)	Households with Income Less Than $15,000 (%)
60624	00.9	50.9
60612	14.6	57.0
60623	23.7	39.7
60651	26.5	32.9
60616	32.3	42.4
60608	39.8	44.1
60647	39.9	38.3
60605	40.0	61.6
60622	43.4	46.8
60607	46.3	53.5
60604	60.6	63.5
60606	72.6	88.2
60603	73.7	25.4
60602	73.7	26.0
60614	79.4	23.6
60601	81.8	25.4
60611	86.3	20.4

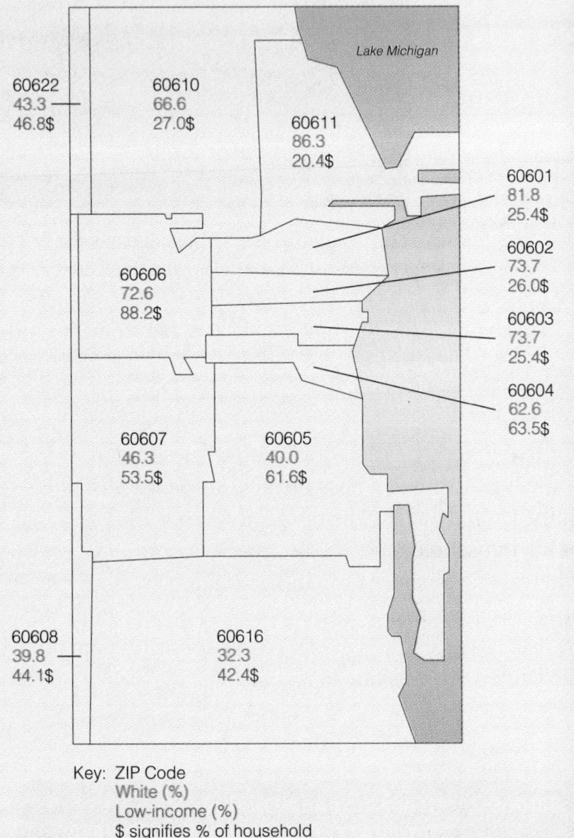

true sense of the word. If it were a true class system, the percentages of households in each income category would be the same across all three groups.

The systematic character of inequality in the United States is even more obvious when we compare median incomes of various types of families classified as white, black, and Hispanic households. Table 9.4 shows that of these household types, female-headed households in general have the lowest median income ($17,443). When we look at female-headed families classified as black and Hispanic, we see the median incomes are $11,909 and $12,047, respectively. In view of these findings, we should not be surprised to learn that some groups are concentrated more heavily than others in the lowest-paying and least prestigious occupational categories.

Table 9.4 Median Income of Families Classified as White, Black, and Hispanic, 1993

Type of Family	White	Black	Hispanic	All
Married couple	43,675	35,218	28,454	43,005
Wife in paid labor force	51,630	44,805	35,973	51,204
Wife not in paid labor force	30,878	22,207	20,721	30,218
Male householder, wife absent	28,269	19,476	21,717	26,467
Female householder, husband absent	20,000	11,909	12,047	17,443
Two earners	48,332	37,124	32,172	47,424

Source: U.S. Bureau of the Census (1995).

Occupational Inequality in the United States

Occupations are considered to be segregated according to ascribed characteristics when some occupations (librarian, secretary, physician, lawyer, chief executive officer) are filled primarily by people of a particular race, ethnicity, sex, or age. The occupational categories shown in Tables 9.5 and 9.6 are examples of those with the highest concentrations by race and by sex.

Occupational segregation is a problem when certain groups are concentrated in the low-paying, low-ranking jobs. When we compare the annual income of nonfamily, single-person households classified as black, white, and Hispanic, the impact of occupational segregation becomes evident (see Figure 9.2). The fact that income and occupation are connected to race and ethnicity (an ascribed characteristic) means that we must conclude that the United States is at best a mixture of class and caste systems of stratification.

Mixed Systems: Class and Caste

Systems of stratification are usually a combination of class and caste. In the United States, virtually every occupation contains members of different ethnic, racial, age, and sex groups. At the same time, however, some groups such as women and blacks are severely over- or underrepresented in some occupations. Moreover, in the United States, a person's ascribed characteristics can overshadow his or her achievements in such a way that "no amount of class mobility will exempt a person from the crucial implications of . . . birth" (Berreman 1972:399). For example, even though a person holds a high-ranking occupation, others may question or overlook that person's talent, merit, and accomplishments if they have dark skin or female reproductive organs. The interrelations between caste and class

> can be brought readily to mind by thinking of the relative advantages and disadvantages which accrue in Western class systems to persons who occupy such occupational statuses as judge, garbage man, stenographer, airline pilot, factory worker, priest, farmer, agricultural laborer, physician, nurse, big businessman, beggar, etc. The distinction between class and birth-ascribed stratification can be made clear if one imagines that he encounters two Americans, for example, in each of the above mentioned occupations, one of whom is white and one of whom is black. This quite literally changes the complexion of the matter. . . . Obviously something significant has been added to the picture of stratification in these examples which is entirely missing in the first instance—something over which the individual generally has no control, which is determined at birth, which cannot be changed, which is shared by all those of like birth, which is crucial to social identity, and which vitally affects one's opportunities, rewards, and social roles. The new element is race (color), caste, ethnicity (religion, language, national origin), or sex. (Berreman 1972:385–386)

The box "How Ideas of Race Affect Other Perceptions" examines how the element of race affects the ways in which people view and evaluate accomplishments, appearance, and identity.

Two examples demonstrate how class and caste stratification can operate together in the United States. Consider how positions are filled and how players relate to one another on a coed softball team. Usually, females are assigned to the least central positions (those requiring the least amount of involvement in completing a play). Also, when a player hits the ball toward a female player, at least one male comes over to "help out." Rarely does a female charge over to aid a male teammate. Finally, when a female player comes to the plate to bat, the four outfielders move in to the infield; when a male player bats, they stay in the outfield. (A female might be a good singles hitter, but it is difficult to hit a single when there are nine infielders.)

Table 9.5 Occupations in Which People Who Declare Themselves Black and Hispanic Are Disproportionately Under- and Overrepresented in 1994*

Underrepresented

Occupation	Percentage of All Employed	
	Black	Hispanic Origin
Managerial and professional	7.1	4.5
Scientists	3.8	1.6
Physicians	4.2	5.2
Dentists	3.7	4.5
Pharmacists	2.6	4.1
College professors	5.0	2.9
Lawyers and judges	3.3	3.0
Authors	2.8	2.9
Technical writers	4.0	0.2
Artists†	2.7	3.2
Dental hygienists†	1.1	3.3
Airplane pilots and navigators	1.5	0.4
Sales occupations	7.1	6.8
Waiters and waitresses	5.5	7.6

Overrepresented

Occupation	Percentage of All Employed	
	Black	Hispanic Origin
Maids and housemen	27.9	20.2
Nurses' aides, orderlies, and attendants	29.3	8.9
Cleaners and servants (private households)	20.0	31.0
Postal clerks (except mail carriers)	24.2	5.9
Pressing machine operators	18.7	21.6
Short-order cooks	17.9	16.8
Textile sewing machine operators	19.4	24.1
Farm workers†	8.6	26.8

*According to the 1990 census, blacks constitute 12.1 percent, persons of Hispanic origin constitute 9.0 percent, and whites constitute 80.3 percent of the U. S. population.

† 1991 figures

Source: Adapted from U.S. Bureau of the Census (1995).

Table 9.6 Examples of Occupations in Which Women Are Disproportionately Under- and Overrepresented in 1994

Underrepresented

Occupation	Percentage of All Employed
Engineers	8.3
Dentists	13.3
Clergy	11.1
Firefighting and fire prevention	2.1
Mechanics and repairers	4.5
Construction trades	2.2
Transportation and material-moving occupations	9.4
Forestry and logging operations	7.0
Airplane pilots and navigators	2.6

Overrepresented

Occupation	Percentage of All Employed
Registered nurses	93.8
Speech therapists	94.6
Prekindergarten and kindergarten teachers	98.1
Dental hygienists†	99.8
Licensed practical nurses	95.1
Secretaries, stenographers, typists	98.0
Receptionists	96.4
Financial records processing (bookkeepers)	91.4
Eligibility clerks, social welfare	80.6
Data entry keyers	83.8
Teachers' aides	90.3
Child care (private household)	97.3
Cleaners and servants (private)	95.8
Dental assistants	96.6
Textile sewing	86.0

* According to the 1990 census, females constitute 51 percent of the total population in the United States.

† 1991 figures

Source: Adapted from the U.S. Bureau of the Census (1995).

We could make the case that women play the least central positions and are helped by men in adjacent positions because they lack the skills and experience to field, throw, and hit the ball adequately. When talent and skill are the only criteria for assigning positions to men and women, a class system of stratification is at work. When there is a predictable relationship between an ascribed characteristic such as sex and an achievement such as shortstop, this is a clue for sociologists to investigate the source of this pattern. Often the reasons for the pattern can be traced to the fact that an undetermined, yet significant, amount of the male-female dif-

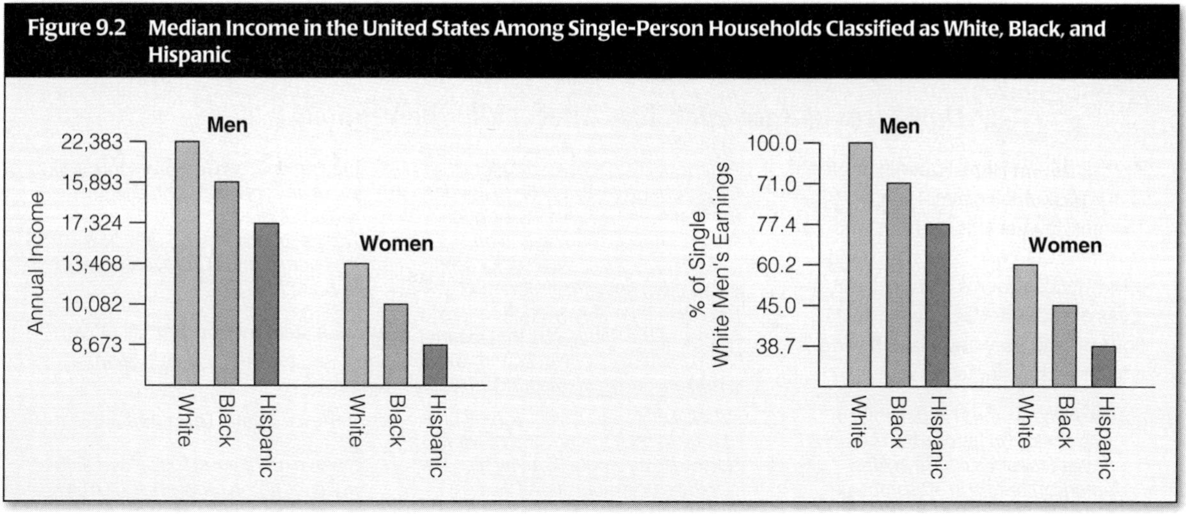

Figure 9.2 Median Income in the United States Among Single-Person Households Classified as White, Black, and Hispanic

Source: U.S. Bureau of the Census (1995).

ference in talent and ability is imposed externally; that is, they are by-products of the ways in which males and females are socialized.

Starting in infancy, parents elicit more active and more physical behavior from sons than daughters. They also channel their children toward sex-appropriate sports. Girls are guided into predominantly non-contact sports, often individual sports, that require grace, flowing movements, flexibility, and aesthetics, such as gymnastics, tennis, and swimming. By contrast, boys are encouraged to participate in team sports that involve contact, lifting, throwing, catching, and running. Because boys and girls participate in different kinds of athletic activities, they develop different skills and styles. Finally, the number of organized teams for each sex (from T-ball to professional) shows that more human and monetary resources are devoted to male than to female athletic development.

In the case of coed softball, a class system of stratification is operating on one level because those who have the most ability and experience are assigned to the most central positions. Yet, on another level a caste system is also at work, because social practices contribute to differences between males' and females' talent and ability.

Caste and class systems of stratification operate together in professional sports as well. Sociologists have long noted that black athletes are concentrated in positions that require strength, speed, and agility, whereas whites are concentrated in central leadership, "thinking," and playmaking positions (Loy and Elvogue 1971; Medoff 1977). In professional baseball, for instance, whites tend to play in the infield positions (including pitcher and catcher), whereas blacks tend to play in the outfield. Because no on-field position in professional baseball excludes black athletes completely, some element of class stratification must be at work. At the same time, however, coaches, who are predominantly white, obviously assign blacks to non-central positions and whites to leadership positions. This hypothesis is supported by the fact that most elementary and high school athletes play on predominantly white or black teams, which means that black athletes have experience in playing all positions. Sports sociologists believe that black athletes are removed systematically from positions of leadership and are assigned to the less central positions as they advance from the elementary to high school to college to the professional levels. This practice continues after their on-field professional career is ended: relative to the number of white athletes in the three major professional sports, a smaller percentage of black athletes become head coaches or general managers (see Table 9.7).

From a sociological point of view, any person who explains these differences as due to biological differences between men and women and across race and ethnic groups is, in the succinct words of social psychologist E. A. Ross ([1908] 1929), "too lazy" to trace these differences to the social environment or historical conditions.

Clearly, social stratification is an important feature of a society, one with significant consequences for the life chances of the advantaged and disadvantaged alike. In the remainder of this chapter, we will examine various theories that seek to explain why stratification occurs and the forms it takes.

U.S. in Perspective

How Ideas of Race and Class Affect Other Perceptions

Historian John Hope Franklin, professional basketball player Isaiah Thomas, lawyer Otis Graham, and journalist Clarence Page have made insightful comments on how their ideas of race affects people's definitions of who they are, what they do, and what they should look like.

> *It's often assumed I'm a scholar of Afro-American history, but the fact is I haven't taught a course in Afro-American history in 30-some-odd years. They say I'm the author of 12 books on black history, when several of those books focus mainly on whites. I'm called a leading black historian, never minding the fact that I've served as president of the American Historical Association, the Organization of American Historians, the Southern Historical Association, Phi Beta Kappa, and on and on.*
>
> *The tragedy . . . is that black scholars so often have their specialties forced on them. My specialty is the history of the South, and that means I teach the history of blacks and whites. (Franklin, quoted in Megurl 1990:13)*

> *When [Larry] Bird makes a great play, it's due to his thinking, and his work habits. It's all planned out by him. It's not the case for blacks. All we do is run and jump. We never practice or give a thought to how we play. It's like I came dribbling out of my mother's womb.*
>
> *Magic [Johnson] and Michael Jordan and me, for example, we're playing only on God-given talent, like we're animals, lions and tigers, who run around wild in a jungle, while Larry's success is due to intelligence and hard work. (Thomas, quoted in Berkow 1987:D27)*

About a steal that basketball player Larry Bird made in game 5 of the playoffs in the 1987 Eastern Conference finals, Thomas said:

> *It was a great basketball play. I didn't put enough zip on the ball, and [Bill] Laimbeer didn't step up to get it. Meanwhile, this white guy on the other team who is supposed to be very slow, with little coordination, who can't jump, all of a sudden appears out of nowhere, jumps in,*

Are you inclined to guess that one or the other of these players has more "God-given talent"? Do you expect that the other has more of some other quality? In what ways may your beliefs about race and athletic ability affect how you interpret what you see?

©Dennis MacDonald/PhotoEdit

Theories of Stratification

One theory—functionalism—seeks to explain why resources are distributed unequally in society. A second set of theories deals with identifying the various strata within society.

A Functionalist View of Stratification

Sociologists Kingsley Davis and Wilbert Moore (1945) took a functionalist perspective and asked how stratification — the unequal distribution of social rewards — contributes to maintain order and stability in society. They claimed that social inequality is the device by which societies ensure that the most functionally important occupations are filled by the best-qualified people.

Davis and Moore concede that it is difficult to document the functional importance of an occupation, but they suggest two somewhat vague indicators: (1) the degree to which the occupation is functionally unique (that is, few occupations can perform the same function adequately) and (2) the degree to which other occupations depend on the one in question. In view of these indicators, garbage collectors, although functionally important to sanitation, need not be rewarded highly because little training and talent are required to do that job. Even though we depend on garbage collec-

grabs the ball, leaps up in the air as he's falling out of bounds, looks over the court in the space of two or three seconds, picks out a player cutting for the basket, and hits him with a dead-bullet, picture-perfect pass to win the game. You tell me this white guy—Bird—did that with no God-given talent? (Thomas, quoted in Berkow 1987:D27)

Look at Eldrick "Tiger" Woods, the golf prodigy. His mother, from Thailand, is half Thai, a quarter Chinese, and a quarter white. His father is half black, a quarter Chinese and a quarter American Indian. Yet, since age six, when he first was profiled in Ebony *magazine, he has been called a "black" golf prodigy. In the spring of 1995 when he became, at age nineteen, the fourth "black" golfer to play the Masters Tournament in Augusta, Georgia, while still enrolled in Stanford, he refused to settle for that label. It was an injustice to all of his other heritages, he said, to call him simply "black." He was not ashamed to be black. But it was not all that he was proud of, either.*

What does he put on forms? "Asian," he told Sports Illustrated. *(Page 1996:285)*

No one objected when white actors, models, and consumers wore the cosmetic lenses, but when black talk-show host Oprah Winfrey wore them on TV, there was an immediate avalanche of attacks from both whites and blacks who could not understand why a black person would wear green contacts. White people seemed to be threatened by the notion that black people could actually avail themselves of cosmetic advances and appropriate beauty characteristics that white people had theretofore defined as exclusively their own. Black audiences, too, looked at rich, powerful, and famous Oprah and feared that she was somehow about to "buy" herself out of the black race and leave us bereft of one more black heroine and role model. In the end, when the host held her ground on her black identity, black and white viewers wised up and realized that the ever dedicated and down-to-earth Winfrey wasn't going anywhere she didn't belong. Colored contacts were not going to change her. (Graham 1995:230)

tors to maintain sanitary environments, many people are able to do the work.

Davis and Moore argue that society must offer extra incentives to induce the most talented individuals to undergo the long and arduous training needed to fill the most functionally important occupations. They specify that the incentives must be great enough to prevent the best-qualified and most capable people from finding less functionally important occupations as attractive as the most important occupations.

Davis and Moore concede that the efficiency of a stratification system in attracting the best-qualified people is weakened when (1) capable individuals are overlooked or not granted access to training, (2) elite groups control the avenues of training (as through admissions quotas), and (3) parents' influence and wealth (rather than the ability of their offspring) determine the status that their children attain. Yet Davis and Moore believe that the system adjusts to such inefficiencies. For example, when there are shortages of personnel to fill functionally important occupations, the society must increase people's opportunities to enter those occupations, allowing those who were previously denied entry. If this is not done, the society as a whole will suffer and will be unable to compete with other societies.

A functionalist might argue that such an "adjustment" is reflected in white South Africans' recent moves to repeal the system of job reservation, which restricted each race to certain types of jobs and prohibited the advance of a nonwhite over a white in the same occupation. To compete effectively (especially after the internationally imposed antiapartheid sanctions ended in 1993), the South African economy needs an increasingly skilled work force, which cannot be maintained by the white population alone. In fact, an article in the February 1994 issue of *World Trade* named the nearly 50 percent unemployment rate, coupled with a shortage of middle managers and professionals in South Africa, as liabilities for foreign companies thinking of investing there (Jones 1994). Even though the long-standing job and mobility restrictions placed on nonwhite workers have been repealed, it will take many years to train those who were denied the educational opportunities to obtain such positions. Thus, the functionalist argument that society will adjust and that all will work out in the end introduces a critical moral question: Should the life chances of nonwhites be tied to the needs of the dominant (white) group in society? It also introduces another more general question: Is social inequality the way to ensure that the most important occupations are filled by the most qualified people?

Critique of the Functionalist Perspective

The publication of "Some Principles of Stratification" prompted a number of responses that took issue with

Table 9.7 Minority Employment in Manager and Coach Positions: Major League Sports

Sport	Number of Minorities in Manager and Coach Positions	Total Manager and Coach Positions	Percentage of Minorities
Major League Baseball (28 teams)			
18% of the players classified as black; 18% as Latino			
Manager	4	28	14.3
Coaches	40	157	25.5
National Football League (30 teams)			
68% of players classified as black			
Manager	10	30	33.3
Head coaches	2	30	6.7
Assistant coaches	85	385	22.1
National Basketball Association (27 teams)			
79% of players classified as black			
General manager	7	27	25.9
Head coaches	5	27	18.5
Assistant coaches	25	59	42.3

Source: Plaschke (1995).

the fundamental assumption underlying the Davis and Moore theory — that social inequality is a necessary and universal device that societies use to ensure that the most important occupations are filled by the best qualified people. Two especially insightful critiques are Melvin M. Tumin's (1953) "Some Principles of Stratification: A Critical Analysis" and Richard L. Simpson's (1956) "A Modification of the Functional Theory of Social Stratification."

Neither Tumin nor Simpson believes that a position commands great social rewards simply because it is functionally important or because the available personnel are scarce. Some positions command large salaries and bring other valued rewards even though their contribution to the society is questionable. Consider the salaries of athletes, for example. For the 1995 season, the average salary of a major league professional baseball player was $1,099,875. In 1993, 40 percent of the 650 athletes who played for the twenty-six major league teams were paid $1 million or more (Bodley 1996; Chass 1992, 1993). Elementary and secondary teachers, on the other hand, were paid an average of $34,098 per year (*The World Almanac and Book of Facts 1994* 1993). This difference in pay raises the question of whether professional athletes and entertainers are more essential to society than teachers — or whether other forces are at least equally important in defining occupational rewards and status.

Critics of functionalism also question why a worker should receive a lower salary for the same job just because the person is of a certain race, age, sex, or national origin. After all, the workers are performing the same job, so functional importance is not the issue. This question relates to issues connected with pay equity. For example, why do females working as registered nurses make on average of $0.98 for every dollar a male registered nurse makes? On the other hand, why do female mechanics and repairers earn an average of $1.06 for every dollar their male counterparts earn?

Critics of the functionalist perspective on social stratification also point to the "comparable worth" debate. Advocates of comparable pay for comparable work ask whether women who work in predominantly female occupations (registered nurse, secretary, day-care worker) should receive salaries comparable to those earned by men who work in predominantly male occupations that are judged to be of roughly comparable worth (housepainter, carpenter, automotive mechanic). For example, assuming comparable worth, why should a female working at a child day-care center get paid an average weekly salary of $205.98 while a male working in an auto supply store makes $464 (U.S. Department of Labor 1995; see Figure 9.3 and Table 9.8)?

In addition, Tumin and Simpson argued that it is very difficult to determine the functional importance of an occupation, especially in societies characterized by a complex division of labor. The specialization and interdependence that accompany a complex division of labor imply that every individual contributes to the whole operation. In light of this interdependence, one

Figure 9.3 Two Approaches to Fairness

Pay Equity
When men and women work in the same firm in the same occupation, they must not be paid differently.

Comparable Worth
When occupational categories are agreed to be *equivalently valuable within a firm*, then compensation must be equivalent across those categories *at that firm.*

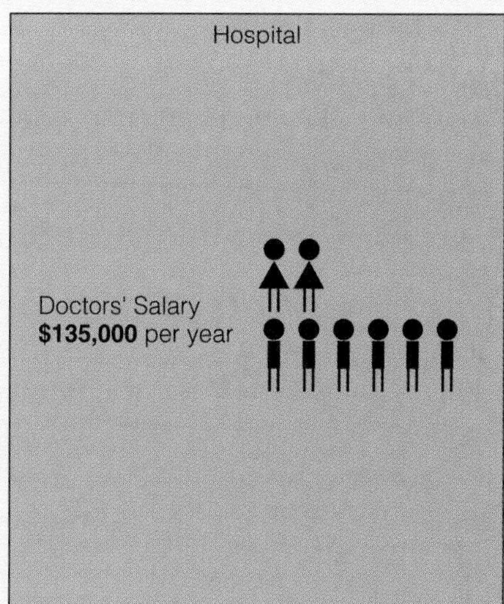

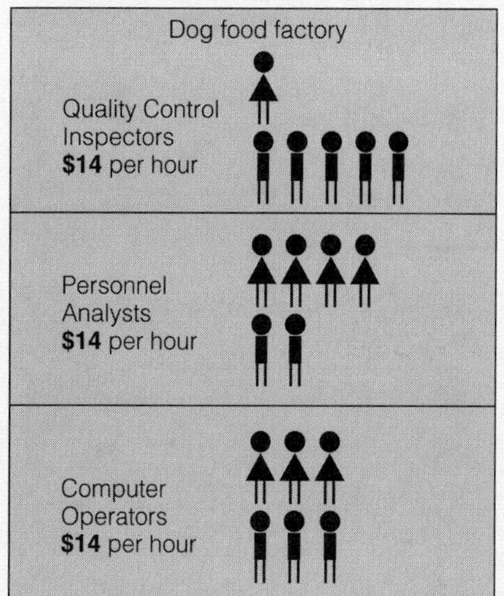

What happens (that is, what interpretation can we attach to the situation?) when pay differentials show up between equivalent firms whose chief difference is the gender composition of the workers?

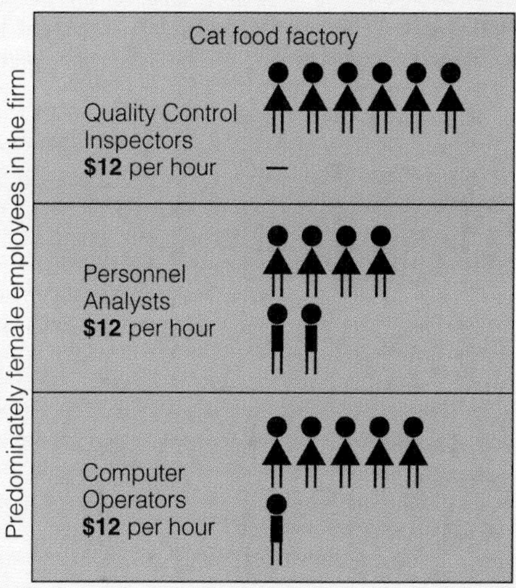

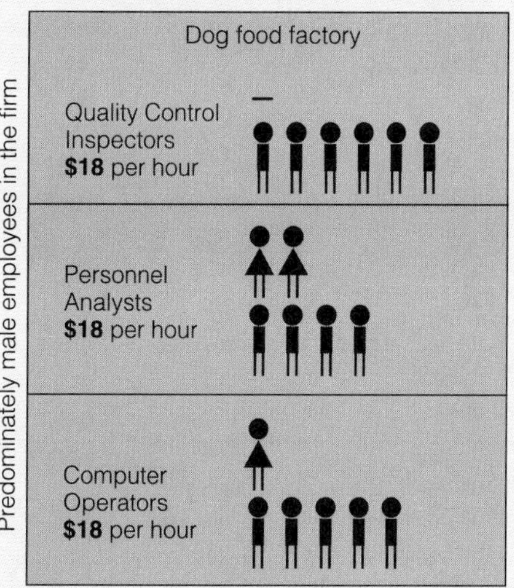

Alternatively. . .
What can we interpret from data that show that certain occupations or industries contain disproportionate shares of either male or female employees, and there is a pay differential? Does the pay differential go with the occupation or industry or with the gender of the worker?

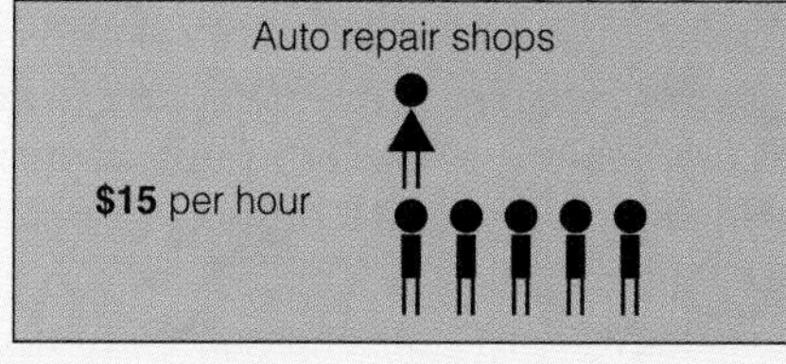

Source: U.S. Bureau of the Census (1995b).

Table 9.8 Average Weekly Earnings for Employees Working in Predominantly Male and Predominantly Female Occupations

Field of Work	Male (%)	Average Weekly Earnings
Coal mining	94.3	864.42
Heavy construction equipment rental	85.6	554.06
Automotive repair shops	84.4	415.01
General building contractors	84.2	532.32
Trucking and warehousing	84.0	482.22
Sanitary services	83.6	535.52
Auto and home supply stores	83.0	318.69
New and used car dealers	82.4	464.20
Surveying services	79.1	450.67
Automotive dealers	79.0	398.94
Electric, gas, and sanitary services	77.6	734.58
Disinfecting and pest control services	77.0	362.12
Automotive dealers and service stations	75.0	359.97
Local and suburban transportation	73.2	418.47
Agricultural services	61.4	288.19
Field of Work	**Female (%)**	**Average Weekly Earnings**
Child day care services	93.1	205.98
Home health care services	92.4	314.88
Beauty shops	90.6	229.04
Offices and clinics of dentists	88.4	344.60
Intermediate care facilities	88.0	261.14
Nursing and personal care facilities	86.2	285.90
Health services	81.9	407.55
Offices and clinics of medical doctors	80.1	406.79
Hospitals	79.9	491.32
Social services	77.9	263.03
Veterinary services	76.9	247.10
Apparel and accessory stores	75.0	184.51
Variety stores	73.0	187.97
Physical fitness facilities	70.5	152.15
Miscellaneous personal services	69.6	145.37

Source: U.S. Department of Labor (1995).

could argue that every individual makes an essential contribution. "Thus to judge that the engineers in a factory are functionally more important to the factory than the unskilled workmen involves a notion regarding the dispensability of the unskilled workmen, or their replaceability, relative to that of the engineers" (Tumin 1953:388). Even if engineers, supervisors, and CEOs have the more functionally important positions, how much inequality in salary is necessary to ensure that people choose these positions over unskilled ones? In the United States, for example, the average salary of the CEOs of 424 of the largest U.S. corporations in 1993 was $3.3 million—170 times the average salary of a factory worker (Crystal 1995). Are such high salaries really necessary to make sure that someone chooses the job of CEO over the job of factory worker? Probably not. But such high salaries have been justified as necessary to recruit the most able people to run a corporation in the context of a global economy. It is unclear whether such salaries accurately reflect the CEO's contribution to society relative to that of the factory worker. Even though unskilled workers might be replaced more easily than engineers or CEOs, an industrialized society depends on motivated and qualified people in all positions. Moreover, should low-skill workers be denied a livable wage simply because they can be easily replaced? This question becomes especially critical when we consider that 4.5 million families with at least one person working live below the poverty line (Rockefeller Foundation 1996c).

Finally, both Tumin and Simpson argue that the functional theory of stratification implies that a system of stratification evolves as it does to meet the needs of the society. In evaluating such a claim, one must look at whose needs are being met by the system. In the case of apartheid, the needs of whites unquestionably were met at the expense of the needs of blacks. We can see this claim supported by the way in which electricity is still distributed to the people of South Africa (Passell 1994). South Africa produces 60 percent of all of the electricity used throughout the African continent. But 66 percent of all South Africans (mostly nonwhites) do not have electricity in their homes. Those without electricity face enormous difficulties in obtaining the fuel to cook food and to warm and light their houses. As a result, people without electricity[7] must resort to gathering wood:

> In the high grassland areas of KwaZulu [one of the ten African homelands], for example, the average distance walked in collecting one headload [of fuel] was a little over 5 miles (8.3 km) and the average time taken in collecting the one load was 4.5 hours.
>
> Not only is the collecting of firewood exhausting, time-consuming, and dangerous but it has serious ecological consequences. Each household uses between three and four tons of wood a year. . . . Over the relatively brief span of the past 50 years, 200 of the 250 forests in KwaZulu have disappeared. (Wilson and Ramphele 1989:44)

This example presents a clear challenge to the idea that stratification functions to meet the needs of a society. In light of the environmental consequences resulting from inequality in distribution of electricity, it is clear that only one segment of South Africa's society benefited—the whites.

Analyses of Social Class

Although sociologists use the term *class* to refer to one form of stratification, they also use **class** to denote a category that designates a person's overall status in society. Sociologists consider social class to be an important factor in determining life chances. Sociologists, however, are preoccupied with two questions: (1) How many social classes are there, and (2) what constitutes a social class? For some answers to these questions, we turn to the works of Karl Marx and Max Weber.

Karl Marx and Social Class

Karl Marx viewed every historical period as characterized by a system of production that gave rise to specific types of confrontation between the exploiting and exploited classes in society. Consequently, Marx was interested in relationships between various social classes that make up a society. He gave several answers to the question, How many social classes are there? In *The Communist Manifesto* ([1848] 1996), he named two: the bourgeoisie and the proletariat. In *Capital: A Critique of Political Economy* (1909), he named three social classes: wage laborers, capitalists, and landlords. In *The Class Struggles in France 1848–1850* ([1895] 1976), he named at least six: the finance aristocracy, the bourgeoisie, the petty bourgeois, the proletariat, landlords, and peasants. According to French sociologists Raymond Boudon and François Bourricaud (1989), a careful reading of Marx's writings suggests that he believed "that the number of classes to be defined depends on the reason why we want to define them" (p. 341). The fact that Marx paid so much attention to class and the class divisions in society underscores his belief that the most important engine of change is class struggle. A brief overview of these works clarifies this point.

In *The Communist Manifesto,* written with Friedrich Engels in 1848, Marx described how class conflict between two distinct classes propels society from one historical epoch to another. Over time free men and slaves, nobles and commoners, barons and serfs, and guildmasters and journeymen have confronted each other. Marx observed that the rise of factories and mechanization as a means of production created two modern classes: the bourgeoisie (the owners of the means of production) and the proletariat (those who must sell their labor to the bourgeoisie). In light of this historical theme, it is appropriate that in *The Communist Manifesto* Marx focused on the two social classes he believed would usher society out of capitalism and into another era.

In *Capital: A Critique of Political Economy,* Marx (1909) named three classes: wage laborers, capitalists,

[7]Since Nelson Mandela came to power in April 1994, 525,000 homes (almost all in impoverished black townships) have been wired for electricity. The plan is to wire 300,000 homes each year until 2000. Even at that pace, more than 17 million people will still be without electricity in South Africa at the start of the new millennium (Drogin 1996).

Class A category that designates a person's overall status in society.

and landlords. Each class is composed of people whose revenues or income "flow from the same common sources" (p. 1032). For wage laborers, the source is wages; for capitalists, profit; for landowners, ground rent. Marx acknowledged that the boundaries separating landowners from capitalists are not clear-cut. In this three-category classification scheme, for example, Henry Ford, the founder of Ford Motor Company, is both a landlord (because he owned a rubber plantation in Brazil) and a capitalist (because he owned the factories and the machines and purchased the labor). Marx also acknowledged that each of the three classes can be subdivided further. The category of landowners, for example, can be divided into owners of vineyards, farms, forests, mines, fisheries, and so on. In this case Marx was interested in distinguishing people according to their sources of income, therefore a three-category social class scheme made sense.

The Class Struggles in France 1848–1850 ([1895] 1976) is a historical study of an event in progress—the 1848 revolutions against several European governments (Germany, Austria, France, Italy, and Belgium), with special emphasis on France. In this book, Marx sought to describe and explain "a concrete situation in its complexity" (Boudon and Bourricaud 1989:341). He described the 1848 revolution as a struggle for the necessities of life and as "a fight for the preservation or annihilation of the bourgeois order" (Marx [1895] 1976:56). The latter consisted of a finance aristocracy, which lived in obvious luxury among masses of starving, low-paid, unemployed people.

Marx outlined the major factors that triggered the 1848 revolution and explained why he believed it failed. The widespread discontent was fueled by two world economic events. One was the potato blight and the bad harvests of 1845 and 1846, which raised the already high level of frustration among the people. The resulting rise in the cost of living caused bloody conflict in France as well as on the rest of the continent. The other event was a general commercial and industrial crisis which resulted in an economic depression and a collapse of international credit. The revolutions were centered in the cities, where the Industrial Revolution had created a proletariat from persons who had migrated there in search of work. Generally, the workers were paid very low wages, lived in squalor, and lacked the most basic necessities.

Although the faces of those who ruled the French government changed as a result of the 1848 revolution, the exploitive structure remained. In the end, the workers were put down by "unheard of brutality" (Marx [1895] 1976:57). Marx believed that the uprising failed because, even though the workers displayed unprecedented bravery and talent, they were "without chiefs, without a common plan, without means and for the most part, lacking weapons" (p. 56). Also, the revolution failed because the "other" classes did not support the proletariat when they moved against the finance aristocracy.

It is difficult to apply Marx's ideas about social class in a total way because, as he made clear in *The Class Struggles in France 1848–1850,* the reality of class is very complex. He left us, however, with some useful ideas with which to approach social class. First is the idea that conflict between two distinct classes propels us from one historical epoch to another. South Africa clearly contained two distinct classes: those designated as white and those designated as something else. In the United States, on the other hand, one area that has received considerable attention is the class division between skilled and unskilled workers. Unskilled workers in the United States are vulnerable because in the capitalistic economy corporations have transferred (and still are transferring) low-skilled jobs out of the country. Class conflict is an important and impending agent of social change in both South Africa and the United States.

A second important idea left to us by Marx is the concept of viewing social class in terms of the sources of income. Approaching social class in this way carries our understanding of social class beyond the simple notion of occupation (or relationship to the means of production). Often, however, such information on income is very hard to acquire. This information is not available for the South African population; at best, it is incomplete for the United States. In the United States the Federal Reserve sponsors the Survey of Consumer Finances, conducted every six years or so. Unfortunately, the various sources of income are available for only five income groups. Households with annual incomes of $50,000 or more are grouped into one category. Therefore, we cannot determine sources of income for the wealthiest Americans.

Finally, Marx's ideas remind us that the conditions that lead to a successful revolt by an exploited class against the exploiting class are multifaceted and complex. He recognized that exploitive conditions can trigger uprisings but observed that other factors such as a well-thought-out plan, effective leadership, the support of other classes, and access to weapons determine the success or failure of the revolt.

To illustrate this point, consider that two of the best-known demonstrations against apartheid occurred in 1960 and 1976. In 1960, the residents of

The Survey of Consumer Finances does not really let us determine sources of income for the wealthiest Americans. It treats all households with incomes of over $50,000 as one category. Might this fact reflect the class interests of the wealthiest Americans—not making facts about class in the United States better known and understood?

©Steve Starr/Stock, Boston

Sharpville, an African township, marched to the police station without their passes to protest the pass laws (U.S. Department of State 1996). The police fired into the crowd, killing 69 persons and wounding 180. Although the police response initiated riots throughout South Africa, in the end the white government won and declared all antiapartheid organizations illegal. In 1976, 20,000 Soweto schoolchildren marched in protest against the use of Afrikaans as the language of classroom instruction. The police opened fire; hundreds of children were killed and thousands were wounded. This brutal response set off riots throughout the country. The point is that nonwhite South Africans have always protested apartheid, but only since the mid-1980s has the other class—whites—taken steps to end the system.

What factors other than protests by the exploited class have contributed to the movement to end apartheid? First, the United Nations stripped South Africa of its member status in 1974. Then in 1976 the United States and Europe applied new and more forceful economic sanctions while resistance from within increased. Also, the African continent and South Africa lost their strategic importance with the collapse of communism and the end of the Cold War between the United States and the Soviet Union. As a result, the United States no longer needed to support the white South African government to counter communist influences within South Africa and surrounding countries.

Max Weber and Social Class

Although Karl Marx did not consistently specify an exact number of social classes in society, he clearly stated that a person's social class was based on his or her relationship to the means of production. Max Weber, like Marx, did not specify how many social classes exist. For Weber, though, the basis for a social class was not the means of production; rather, it was the marketplace. According to Weber, class situation is ultimately market situation. It is based on the chances of acquiring goods and services, obtaining a well-paying job in the marketplace, and finding inner satisfaction.

According to Weber ([1947] 1985), people's class standing depends on their marketable abilities (work experience and qualifications), their access to consumer goods and services, their control over the means of production, and their ability to invest in property and other sources of income. Persons completely unskilled, lacking property, and dependent on seasonal or sporadic employment constitute the very bottom of the class system. They form the **"negatively privileged" property class**. Those at the very top — the **"positively privileged social class"**—monopolize the purchase of the highest-priced consumer goods, have access to the most socially advantageous kinds of education, control the highest executive positions, own the means of production, and live on income from property and other investments. Between the top and the bottom of the ladder is a continuum of rungs.

Weber states that class ranking is complicated by status groups and parties, of which there are many different kinds. He defines **status group** as a plurality of persons held together by virtue of a common lifestyle, formal education, family background, or occupation and "by the level of social esteem and honor accorded to them by others" (Coser 1977:229). This definition suggests that wealth, income, and position are not the only factors that determine an individual's status group. "The class position of an officer, a civil servant or a student may vary greatly according to their wealth and yet not lead to a different status since upbringing and education create a common style of life" (Weber 1982:73). In South Africa, English-speaking whites form a status group distinct from Afrikaans-speaking white Afrikaners. The two groups practice what sociologist Diana Russell calls a voluntary apartheid: "speaking different languages, attending different schools, living in different areas, and voting for different political parties" (Russell 1989:4).

Political parties are organizations "oriented toward the planned acquisition of social power [and] toward influencing social action no matter what its content may be" (Weber 1982:68). Parties are organized to represent class, status, and other interests; they exist at all levels (within an organization, a city, a country). The means employed to obtain power can include violence, vote canvassing, bribery, donations, the force of speech, suggestion, and fraud. In South Africa, the best-known antiapartheid organization is the African National Congress (ANC), founded in 1912. Nelson Mandela, the most famous ANC member, was convicted of treason and sentenced to life in prison in 1964 for acts of sabotage (that is, damaging property so as to disrupt the functioning of society) directed against apartheid. At his trial, Mandela (1990) explained why sabotage was necessary:

> The initial plan was based on careful analysis of the political and economic situation of our country. We believed that South Africa depended to a large extent on foreign capital and foreign trade. We felt that planned destruction of power plants and interference with rail and telephone communications, would tend to scare away capital from the country, make it more difficult for goods from the industrial areas to reach the seaports on schedule, and would in the long run be a heavy drain on the economic life of the country, thus compelling the [white] voters of the country to reconsider their position. (pp. 26–27)

A second example of a political party oriented toward the acquisition of power is the "Third Force," South African security forces who worked to sabotage the transition from apartheid to democracy. The acts of sabotage included infecting Johannesburg prostitutes with AIDS and supplying the bitter rivals of the ANC Party, the Inkatha Freedom Party, with assault weapons to help escalate the fighting among black factions. The Third Force also employed death squads, terrorist bombs, and torture to control antiapartheid sympathizers and destabilize the country (Drogin 1995).

An example of a political party in the United States is political action committees (PACs), unions, corporations, and other interest groups that contribute to the political candidates and parties who are likely to support their position on key issues. Table 9.9 shows top contributors to the Republican and Democratic parties between June 1, 1995, and June 30, 1996.

Weber's conception of social class enriches that of Marx. Weber views class as a continuum of rungs on a

"Negatively privileged" property class Persons completely unskilled, lacking property, and dependent on seasonal or sporadic employment constitute the very bottom of the class system.

"Positively privileged social class" Persons at the very top of the class system who monopolize the purchase of the highest-priced consumer goods, have access to the most socially advantageous kinds of education, control the highest executive positions, own the means of production, and live on income from property and other investments.

Status group A plurality of persons held together by virtue of a common lifestyle, formal education, family background, or occupation and "by the level of social esteem and honor accorded to them by others" (Coser 1977:229).

Political parties Organizations "oriented toward the planned acquisition of social power [and] toward influencing social action no matter what its content may be" (Weber 1982:68).

African National Congress leader, antiapartheid activist, and now president of South Africa, Nelson Mandela was sent to prison as a young man in 1964 (left) and was finally released twenty-seven years later (right).

©Selon Mention/Sipa Press (left); ©Jacques Witt/Sipa Press (right)

social ladder, with the top and the bottom rungs being the positively privileged class and the negatively privileged property class. Weber argues that a "uniform class situation prevails only when completely unskilled and propertyless persons are dependent on irregular employment" (1982:69). We cannot speak of a uniform situation with regard to the other classes because class standing is complicated by such elements as occupation, education, income, the status groups to which people belong, differences in property, consumption patterns, and so on.

Weber's idea of top and bottom rungs, with everyone else somewhere between, inspires us to compare the situation of the wealthiest person with that of the very poor. The box "Measuring Income Inequality Within Countries" discusses various population groups in what the United Nations calls high-income economies. Among other things, it shows the wealthiest in society (the wealthiest 10 and 20 percent) have annual household incomes that are anywhere from four to nine times that of the poorest 20 percent of the population.

Weber's ideas about social class also draw our attention to the negatively privileged classes. The proportion of negatively privileged persons tells us something important about the extremes of inequality in a society. Although some people of all racial groups in South Africa have high incomes, income distribution is clearly related to race. Table 9.10 shows the average annual per capita income by race and the percentage of households in each racial group in various income categories. Fifty percent of white households have incomes of more than 8,000 rand (about $25,000 in U.S. dollars), as compared with less than 5 percent of Asian, Colored, and African households. Inequality is even more extreme in the homelands, where 80 percent of households are in a state of dire poverty (Wilson and Ramphele 1989). The plight of the negatively privileged in South Africa can be traced directly to apartheid.

The existence of a negatively privileged property class in the United States can also be traced to structural factors—in particular, to changes in the occupational structure. This may come as a surprise to some Americans who attribute mobility, whether upward or downward, to individual effort and do not consider changes in the occupational structure as the cause. Many Americans may not recognize that some groups are affected by changes in the occupational structure more strongly than others. In *The Truly Disadvantaged* (1987) and other related articles and books (Wilson 1983, 1991, 1994), sociologist William Julius Wilson describes how structural changes in the U.S. economy helped create what he termed, in his 1990 presidential address to the American Sociological Association, the "ghetto poor." A number of economic transformations

Table 9.9 Top Fifty PACs' Contributions to Candidates, January 1, 1995, to June 30, 1996

Rank	Political Action Committee		
1.	Democratic Republican Independent Voter Education Committee	L	$1,584,710
2.	Association of Trial Lawyers of America Political Action Committee	T	$1.552,975
3.	International Brotherhood of Electrical Workers Committee on Political Education	L	$1,321,600
4.	American Federation of State County & Municipal Employees	L	$1,311,222
5.	UAW-V-CAP (UAW Voluntary Community Action Program)	L	$1,293,775
6.	National Education Association Political Action Committee	L	$1,213,230
7.	Dealers Election Action Committee of the National Automobile Dealers Association	T	$1,204,475
8.	Laborers' Political League	L	$1,163,800
9.	Build Political Action Committee of the National Association of Home Builders	T	$1,134,349
10.	United Parcel Service of America Inc Political Action Committee (UPSPAC)	C	$1,125,531
11.	American Telephone & Telegraph Company Political Action Committee (AT&T PAC)	C	$1,019,183
12.	American Medical Association Political Action Committee	T	$1,018,505
13.	Active Ballot Club. a Dept of United Food & Commercial Workers Int'l Union	L	$1,007,901
14.	NRA Political Victory Fund	T	$963,368
15.	American Institute of Certified Public Accountants Effective Legislation	T	$958,675
16.	Realtors Political Action Committee	T	$952,008
17.	Machinists Non-Partisan Political League	L	$929,275
18.	Carpenters Legislative Improvement Comm, United Brotherhood of Carpenters	L	$847,256
19.	Transportation Political Education League	L	$840,000
20.	American Federation of Teachers Committee On Political Education	L	$839,514
21.	American Maritime Officers, AFL-CIO Voluntary Political Action Fund	L	$827,960
22.	National Beer Wholesalers' Association Political Action Committee (NBWA PAC)	T	$812,242
23.	American Bankers Association BANKPAC	T	$782,350
24.	CWA-COPE Political Contributions Committee	L	$699,430
25.	American Dental Political Action Committee	T	$689,721
26.	Lockheed Martin Employees Political Action Committee	C	$681,750
27.	United Steelworkers of America Political Action Fund	L	$674,650
28.	National Association of Life Underwriters Political Action Committee	T	$638,500
29.	Sheet Metal Workers International Association Political Action League (PAL)	L	$621,750
30.	National Committee for an Effective Congress	M	$604,270
31.	United Association Political Education Committee	L	$595,000
32.	Committee On Letter Carriers Political Education	L	$570,009
33.	Employees of Northrop Grumman Corporation Political Action Committee (ENGPAC)	C	$568,700
34.	Air Line Pilots Association Political Action Committee	L	$565,000
35.	Ernst & Young Political Action Committee	M	$550,365
36.	Philip Morris Companies Inc Political Action Committee (AKA PHIL-PAC)	C	$549,730
37.	Federal Express Corporation Political Action Committee "FEPAC"	C	$543,000
38.	Seafarers Political Activity Donation (SPAD)	L	$538,950
39.	Associated General Contractors Political Action Committee	T	$538,550
40.	Union Pacific Fund for Effective Government	C	$517,290
41.	National Association of Retired Federal Employees Political Action Committee	T	$516,100
42.	Team Ameritech Political Action Committee	C	$506,390
43.	National Restaurant Association Political Action Committee	T	$489,783
44.	Ironworkers Political Action League	L	$474,965
45.	RJR Political Action Committee RJR Nabisco Inc. (RJR PAC)	C	$448,400
46.	American Hospital Association Political Action Committee (AHAPAC)	T	$436,976
47.	American Crystal Sugar Political Action Committee .	V	$434,525
48.	Tenneco Inc. Employees Good Government Fund	C	$431,475
49.	General Electric Company Political Action Committee	C	$423,050
50.	Independent Insurance Agents of America Inc Political Action Committee	T	$422,057

C = Corporate
L = Labor
M = Non-Connected
T = Trade/Membership/Health
V = Cooperative
W = Corporation without stock

Source: Federal Election Commission (1996).

U.S. in Perspective

Measuring Income Inequality Within Countries

Each year the United Nations publishes the most recent statistics available for every country in the world. One rough measure of the inequality within a country is the distribution of total household income in a given year. This table focuses on high-income countries. Based on the total household income within each country, it shows the percentage of that income that was earned by the poorest 20 percent of households, the richest 20 percent, and the richest 10 percent.

What countries show the widest income gaps between their poorest and wealthiest citizens? What countries show the narrowest gaps? What governmental policies might lead to such differences? What other factors might be at work? Is great income inequality desirable? Why?

Percentage Share of Household Income in High-Income Economies

	Lowest 20 Percent of Population	Highest 20 Percent of Population	Highest 10 Percent of Population
Saudi Arabia	—	—	—
Spain	6.9	40.0	24.5
Ireland	—	—	—
Israel	6.0	39.6	23.5
Singapore	5.1	48.9	33.5
Hong Kong	5.4	47.0	31.3
New Zealand	5.1	44.7	28.7
Australia	4.4	42.2	25.8
United Kingdom	5.8	39.5	23.3
Italy	6.8	41.0	25.3
Kuwait	—	—	—
Belgium	7.9	36.0	21.5
Netherlands	6.9	38.3	23.0
Austria	—	—	—
United Arab Emirates	—	—	—
France	6.3	40.8	25.5
Canada	5.7	40.2	24.1
Denmark	5.4	38.6	22.3
Germany, Fed. Rep.	6.8	38.7	23.4
Finland	6.3	37.6	21.7
Sweden	8.0	36.9	20.8
United States	4.7	41.9	25.0
Norway	6.2	36.7	21.2
Japan	8.7	37.5	22.4
Switzerland	5.2	44.6	29.8

Source: Adapted from *World Development Report* (1990), Table 30.

Table 9.10 Annual Household Income, by Race (in Rands*)

	Average	Less than 4,999	5,000–14,999	15,000–49,999	50,000
White	25,344	9.3	5.8	31.9	61.0
Indian	11,112	3.6	13.5	49.7	33.7
Colored	5,196	8.9	30.1	53.7	6.3
African	2,520	30.2	45.0	22.7	2.1
Urban	3,192	—	—	—	—
Rural	1,920	—	—	—	—

*A Rand is equivalent to $3.8 U. S.

Source: South Africa Labour Development Research Unit (1994).

have taken place, including the restructuring of the American economy from a manufacturing-based economy to a service- and information-based economy; a labor surplus that began in the 1970s, marked by the entry of women and the large baby boom segment into the labor market; a massive exodus of jobs from the cities to the suburbs; and the transfer of low-skilled manufacturing jobs out of the United States to offshore locations (see Chapter 2). These changes are major forces behind the emergence of the ghetto poor or urban underclass, a "heterogeneous grouping of families and individuals in the inner city that are outside the mainstream of the American occupational system and that consequently represent the very bottom of the economic hierarchy" (Wilson 1983:80).

Wilson (in collaboration with sociologist Loic J. D. Wacquant; Wacquant and Wilson 1989) looks at Chicago as a case in point. (Actually, the point applies to every large city in the United States—Los Angeles, New York, Detroit, and so on.) In 1954, Chicago was at the height of its industrial power. Between 1954 and 1982, the number of manufacturing establishments within the city limits dropped from more than 10,000 to 5,000, and the number of jobs dropped from 616,000 to 277,000. This reduction, in conjunction with the outmigration of stably employed working-class and middle-class black families, fueled by new access to housing opportunities outside the inner city, had a profound impact on the daily life of people left behind. The exodus of the stably employed resulted in the closing of hundreds of local businesses, service establishments, and stores. According to Wacquant (1989), the single most significant consequence of these historical and economic events was the "disruption of the networks of occupational contacts that are so crucial in moving individuals into and up job chains . . . [because] ghetto residents lack parents, friends, and acquaintances who are stably employed and can therefore function as diverse ties to firms . . . by telling them about a possible opening and assisting them in applying [for] and retaining a job" (Wacquant 1989:515–516).

The ghetto poor are the most visible and most publicized underclass in the United States which leads many Americans to associate poverty with minority groups. However, approximately 50 percent of the population classified as living in poverty is white.[8] In addition, demographers William P. O'Hare and Brenda Curry-White (1992) estimate that approximately 736,000 rural residents can be classified as an underclass. More than two-thirds (70 percent) of the less visible rural poor are white (Rockefeller Foundation 1996b). Like their urban counterparts, the rural underclass is concentrated in geographic areas with high poverty rates. They too have been affected by economic restructuring, which includes the decline of farming, mining, and timber industries and the transfer of routine manufacturing out of the United States.

The rural and urban underclass represent two distinct segments of the population that live below the poverty line, set at about $15,029 for a family of four in 1994. On the basis of this definition, almost 32 million Americans (or 13 percent of the population) live below the poverty line. Included in this 32 million are 12 million people whose income is less than half the amount officially defined as the poverty level. However, the definition of poverty encompasses diverse groups of people, some of whom might not be considered really poor (such as graduate students and retired people who have assets but who live on a low fixed income). Still, it is

[8]In numerical terms, people classified as white represent the largest category of people living in poverty. The percentage of "whites" living in poverty in 1994, however, was 11.7 percent compared with 30.6 percent of "blacks" and 30.7 percent of "Hispanics" (O'Hare 1996).

Although individual effort is one important variable in upward and downward mobility, this crowded unemployment office illustrates the reality that structural changes in the economy can significantly affect people's ability to find and keep well-paying jobs.

©Alon Reininger/Woodfin Camp & Associates

important to point out that two out of every three poverty-level households[9] are headed by women. For many of these women "their only 'behavioral deviancy' is that their husbands or boyfriends left them" (Jencks 1990:42); in the case of many older women, their husbands have died. For the most part, two reasons account for women's poverty: (1) the economic burden of children and (2) women's disadvantaged position in the labor market. These issues will be explored further in Chapter 11.

Summary and Implications

In this chapter, we have examined the workings of social stratification (the systematic division of people into categories). More importantly, we have learned that classification schemes have a profound impact on people's life chances. Some forms of social stratification affect life chances more strongly than others. Caste systems such as apartheid have a decided impact because in such systems life chances are determined by characteristics over which people have no control. Class systems, although not models of equality, permit life chances to be enhanced on the basis of individual effort. Class and caste systems are end points on a continuum.

South African society still approaches a caste system because in that country people's life chances and access to scarce and valued resources are clearly connected to race. As much as we would like to believe that the United States represents a pure class system, the evidence tells us that it is more castelike than we care to acknowledge. At the same time, the United States is classlike in the sense that every occupation contains people of different ethnic, racial, age, and sex groups. On the other hand, there is no question that some ethnic, racial, age, and sex groups are concentrated in the low-status occupations.

The most intriguing and most problematic feature of stratification systems in general involves the criteria used to rank people, especially when ascribed characteristics are the important criteria. How can ranking

[9] A Rockefeller Foundation (1996a) survey found that the average person in the United States estimated the cutoff for poverty-level income for a family of four to be approximately $3,000 higher than the federal measure.

systems exist in which people who belong to one category of an ascribed characteristic (such as white skin or blue eyes) are treated as more valuable or worthy than people who belong to other categories? Jane Elliot, the third-grade teacher who separated her students by eye color and rewarded them accordingly, gives one answer: "This is not something I can do alone" (*Frontline* 1985:20). By this statement, Elliot meant that the experiment could not have worked without the cooperation of the people on top. Her observation suggests that people cooperate in maintaining systems of stratification. But why do people cooperate? One answer comes from a blue-eyed member of Jane Elliot's class, now an adult, as he recalled the experience. "Yeah, I felt like I was—like a king, like I ruled them brown-eyes, like I was better than them, happy" (p. 13). The feeling of being better translated into an unexpected result. Elliot explains:

> The second year I did this exercise I gave little spelling tests, math tests, reading tests two weeks before the exercise, each day of the exercise and two weeks later and, almost without exception, the students' scores go up on the day they're on the top, down the days they're on the bottom and then maintain a higher level for the rest of the year, after they've been through the exercise. We sent some of those tests to Stanford University to the Psychology Department and they did, sort of an informal review of them, and they said that what's happening here is kids' academic ability is being changed in a 24-hour period. And it isn't possible but it's happening. Something very strange is happening to these children because suddenly they're finding out how really great they are and they are responding to what they know now they are able to do. And it's happened consistently with third graders. (p. 17)

One clear answer to the question of why the people on top cooperate to maintain a system of stratification is that they benefit (whether they know it or not) from the system of stratification and the way in which rewards are distributed. In the third-grade class, the blue-eyed children benefited from the system that distributed rewards on the basis of eye color. In South Africa, whites clearly have benefited from a system that rewards people on the basis of race.

In comparison with South Africa, it is more difficult to see the inequality that exists in the United States. Perhaps this is so because we believe that the United States is a model of equal opportunity because (1) we can find examples of people from all racial, ethnic, sex, and age groups who do achieve rewards; (2) we have no obvious laws governing ascribed characteristics and life chances; and (3) we believe that anyone can transcend his or her environment through hard work.

The case of South Africa reminds us how difficult it is for people to give up their privileges and to put into practice a new ranking and reward system. South African President F. W. de Klerk traveled to the United States in 1990 to meet with President Bush and other government leaders and to convince the American people that the political changes in South Africa over the past year represented a sincere, irreversible effort to dismantle apartheid. During his visit, he emphasized repeatedly that the principles underlying the American political system (such as the Bill of Rights and the Constitution) deserved to be emulated (Wren 1990). Earlier in this chapter, we noted that these documents, remarkable as they are, also are frustrating in that they allow for many injustices to be interpreted as being in agreement with these documents in principle. Hundreds of times throughout U.S. history, members of some group or other have been defined as less than human or as not completely human; as a result, they have been denied equitable access to justice, freedom, health care, housing, jobs, and education. Moreover, there are cases substantiating that one group's rights have been upheld at the expense of another, less powerful group. The point is that adopting a set of principles does not guarantee that they will be put into practice. Gunnar Myrdal calls this situation "an American dilemma" in a book by that name (1944). The dilemma is that a substantial gulf exists between the so-called American creed and actual conduct. In addition, even if these principles were implemented immediately, the people who were penalized in the past would continue to be penalized.

The sociological perspective is valuable in that it enables us to see how social stratification systems are connected to life chances. When we know what is going on we have an obligation to work to change things. However, people usually are not so clear-sighted. The case of South Africa shows that it took pressure from outside (in the form of economic sanctions and cultural isolation) and from within (in the form of mass demonstrations, strikes, and bloodshed) to push South African whites to take the first steps toward dismantling apartheid and toward creating a multiracial democracy. The success of South Africa's attempts to dismantle apartheid will hinge on whether it can dismantle the legacy of this policy — the profound social and economic inequalities that now keep South Africans apart.

Key Concepts

Use this outline to organize your review of the key chapter ideas.

Life chances
Social stratification
- **Caste**
  - **Apartheid**
  - **Ascribed characteristics**
- **Class**
  - **Achieved characteristics**

Status value
Social mobility
- **Vertical mobility**
  - **Downward mobility**
  - **Upward mobility**
- **Horizontal mobility**
- **Intergenerational mobility**
- **Intragenerational mobility**

Social class
- **"Negatively privileged" property class**
- **"Positively privileged" property class**
- **Status group**
- **Political parties**
- **Urban underclass**

internet assignment

In Chapter 9 we learned that however much Americans want to believe that upward or downward social mobility is tied to individual effort (or lack of effort), structural factors are often at work. A U.S. Bureau of Labor Statistics report found that between January 1991 and December 1993, 4.5 million workers were displaced from jobs they had held for three or more years. And if we consider workers who held their jobs less than three years, some nine million workers were displaced over the same period – many because of changes in the occupational structure. The report can be found at ftp://stats.bls.gov/pub/news.release/disp.txt. Read the report and answer the following questions. (1) How does the Bureau of Labor Statistics define displaced workers? (2) What are the top reasons displaced workers lose their jobs? (3) What category of worker is at highest risk of job displacement? (4) What geographic region of the United States has the highest number of displaced workers?

Race and Ethnicity

With Emphasis on Germany

Brandenburg Gate, Berlin. ©Dave Bartruff/Artistry International

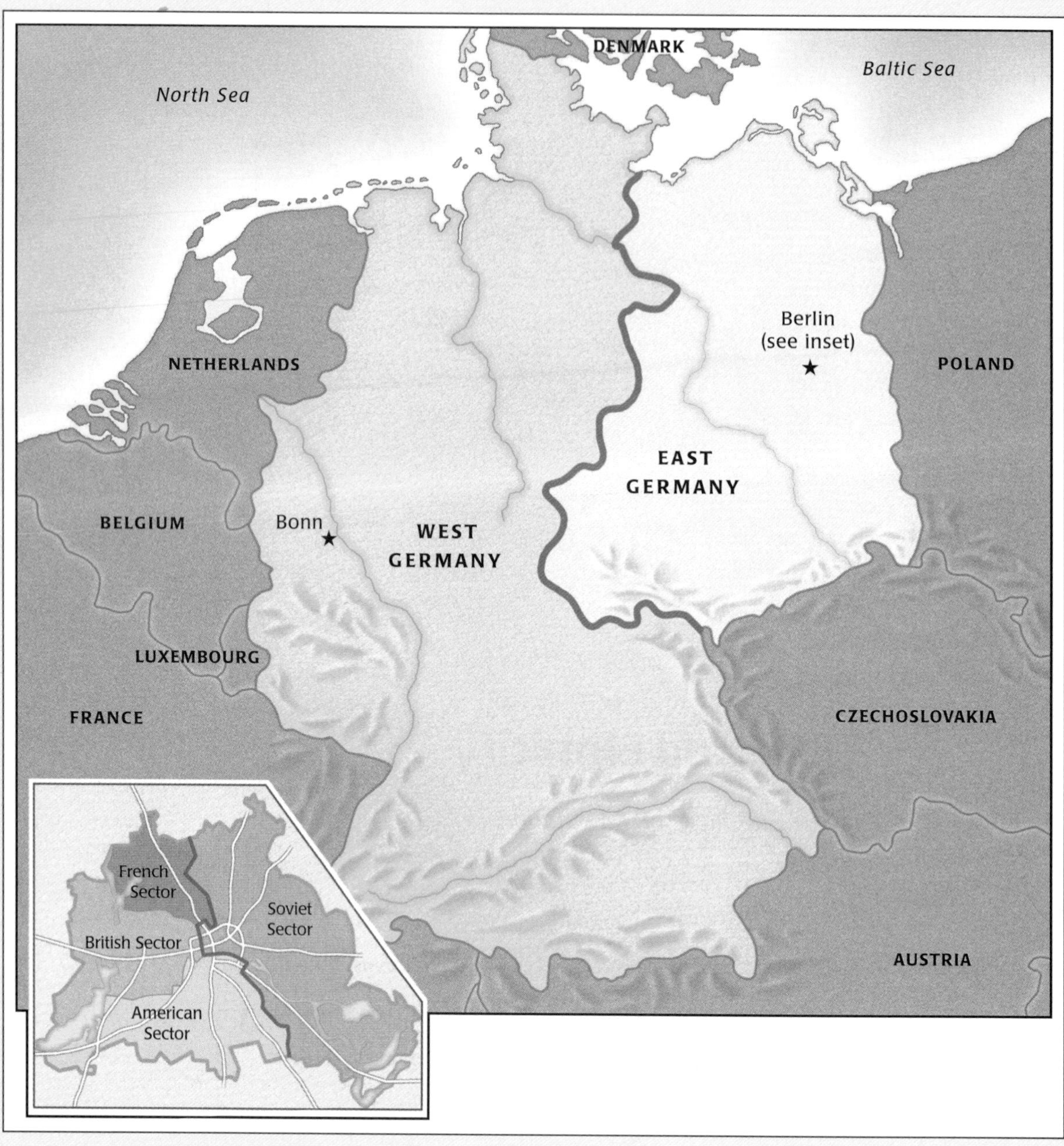

Source: Adapted from *The Times Atlas of World History* (1984).

Race, Ethnicity, and the Berlin Wall

At the end of World War II, Germany was divided into East Germany under the Soviet Union and West Germany under the Allied powers. Berlin was also divided. Between 1945 and 1961, 3 million East Germans and 9 million from former German territories migrated to West Germany, providing labor for West Germany's rapid postwar reconstruction. In 1961, when 3,000 migrants per day were joining the East-West flow, the Soviets sealed off East Germany and East Berlin from West Germany and West Berlin with over 970 miles of wall and fortified fences.

Suddenly West Germany faced a labor shortage. In response, it began to recruit "guest workers" from many countries. Although it wanted their labor, it tried to rule out their settling in Germany or gaining rights of German citizenship. Today about 8 percent of the people living in Germany are "foreigners," including nearly 2 million Turks, though many of them have lived in Germany all their lives.

Reunited Germany is a context in which to consider some problems of race and ethnicity that are relevant to the United States as well. What happens when a government attempts to define race or ethnicity in a clear-cut fashion?

Ender Bsaran is twenty-five now. He looks no different from most young Germans on his street. He has the latest German haircut—long in back, and short and spiky on top. . . . [W]ith . . . his baggy jeans and the pack of Marlboros sticking out of his shirt pocket, it's hard to think of him as "foreign." He likes to say that his family is "more of a European family," although his father goes to the mosque every day and sits with the men after prayers and listens to the gossip. . . . He wants you to know that his mother is nothing at all like the squat, ruddy Turkish women you see on the streets of Hemshof, wrapped in the bulky coats and gabardine head shawls that they brought from Turkey. He likes Turkey, but he doesn't think that either he or his mother belongs in Turkey. (Kramer 1993:56)

I'm German, and I'm dark. But then not all that dark either. I've often looked in the mirror and asked myself what distinguishes me, what makes me so different in the eyes of others. Inside I am German because of my German environment, school, my home—just German. And yet it was always made clear to me that is exactly what I am not. But why? It's all based on externalities. (Wiedenroth 1992:165)

[Piri] Thomas's parents moved from Puerto Rico to Spanish Harlem, where Piri and his three siblings were born. Once in the United States, however, the family faced the peculiar American necessity of defining itself as White or Black. . . . The Thomas family—hailing from Puerto Rico of mixed Indian, African, and European antecedants—considered themselves White and pursued the American dream, eventually moving out to the suburbs in search of higher salaries and better schools for the children. Yet in their bid for Whiteness, the family gambled and lost, because even while the three other children and Piri's mother were fair, Piri and his father were dark skinned. Babylon, Long Island proved less forgiving of Piri's dark skin than Spanish Harlem. In the new school, the pale children scoffed at Piri's claim to be Puerto Rican rather than Black, taunting Piri for "passing for Puerto Rican because he can't make it for white," and proclaiming "[t]here's no difference . . . [h]e's still black." Piri's morphology shattered not only the family's White dream, but eventually the family itself. (Haney López 1994:39–40)

My friend Alena's father, a white American, met her mother in Thailand when he was stationed there while in the military. Alena has many of her mother's physical characteristics: dark skin, dark hair, big eyes, and full lips. Many people try to classify her as Hawaiian or Filipino. When people ask, "Where are you from?" or "What are you?" she usually answers, "American." It is obvious that she has been raised in America. She dresses and talks like an American. When she was a child, her paternal grandmother took care of her a lot, and all of her friends are white. She knows very little Thai and is Catholic (her mother is Buddhist). (Northern Kentucky University student 1996)

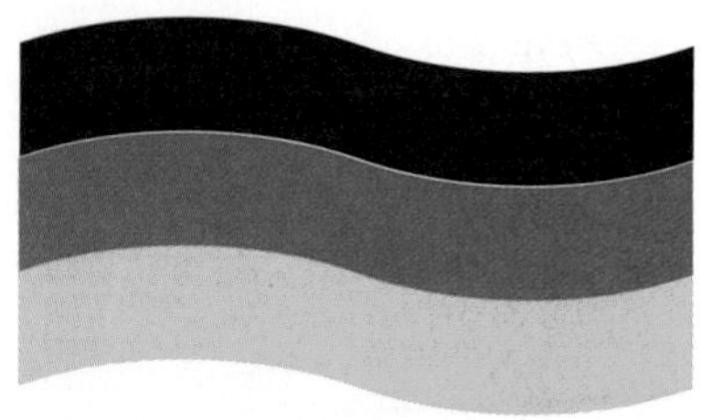

Why Focus on Germany?

In this chapter, we focus on Germany because it shares something in common with the United States: each country's government classifies the people within its borders according to race and/or ethnicity. On the surface, determining someone's race and ethnicity may seem like a relatively simple task because most people believe that race is something that is easily observable and assume that everyone knows his or her race. In the United States, the Census Bureau classifies everyone within its borders as belonging to *one* of four official categories: (1) White, (2) Black, (3) American Indian, Eskimo, Aleut, or (4) Asian or Pacific Islander. It further classifies people as belonging to one of two official ethnic categories: (1) Hispanic or (2) non-Hispanic. It releases data on U.S. population that suggest this classification scheme works. As we will learn, this system can make life particularly difficult for those who fit certain categories and those who do not fit or look like they fit into the official categories.

The German government also is concerned with the race and/or ethnicity of people within its borders. Under German law, people who can prove German ancestry are entitled to citizenship regardless of their country of birth. It does not matter if they cannot speak German or if they know nothing about German culture; the criterion for German citizenship is biological. This law makes life especially difficult for those who have lived in Germany all of their lives but who are not biologically German or who do not "look" German. The "foreigners" who live in Germany—even those who have lived and worked in the same town for thirty or more years and those who were born there and know only the German language—are considered guests (Thränhardt 1989).

The flaws of the system of racial and ethnic classification used in Germany and the United States become especially evident when we identify the people who do not fit easily into one of the official categories. Such cases tell us that race and ethnicity are not easily definable characteristics immediately evident on the basis of physical or other easily observable characteristics. Rather, such cases tell us that ideas of race and ethnicity are defined and maintained by people through a complex array of formal and informal mechanisms. We will use the cases of Germany and the United States to explore the hypothesis that race and ethnicity are not biological facts but are products of racial and ethnic classification systems. We also explore the consequences of classification on the lives of people.

Classifying People by Race

Every ten years, the U.S. government attempts to count the number of people living under its jurisdiction and classify them according to race. On the surface, a reading of the Census Bureau data suggests that virtually everyone in the United States belongs to one of four broad racial categories: (1) white, (2) black, (3) American Indian, Eskimo, Aleut, or (4) Asian and Pacific Islander (see Table 10.1). I recently asked students in a race and gender class whether they knew of someone who might not fit neatly into only one category. Of the seventy students in this class, nineteen responded in the affirmative. Here are two of those responses:

> My friend Debra's parents are both of mixed ancestry: her mother is Native American, Portuguese, and black; her father is French and black. Debra has the darkest skin of anyone in her family, but that does not save her from being teased by her friends for trying to be a white girl. Debra has light skin and fine hair, which led many people to assume that she is really not "black."

> Three years ago my good friends Cathy and Sam found out they were going to have their first baby. They spent the next nine months preparing to make everything perfect for the new arrival. Cathy eventually gave birth to a blue-eyed, blond-haired baby boy whom they named Michael. The hospital nurse who prepared the birth certificate told the parents that Michael would be classified according to the race of his father, black. Cathy was outraged by this idea, afraid that her son would be an object of discrimination. How will people react when Michael checks

Table 10.1 U.S. Population by Race

According to the U.S. Census Bureau, everyone in the United States can be classified into four major categories plus one vague category labeled "other race." It is significant that the fifth category is not labeled "multiracial," or "mixed race," or even "race unknown."

Racial Category	Population
White	199,827,064
Black	29,930,524
American Indian, Eskimo, or Aleut	2,015,143
Asian or Pacific Islander	7,226,986
Other race	9,710,156

Source: U.S. Bureau of the Census (1996).

"black" as his race on the various forms and applications that will come his way in the future? Will he be viewed as an impostor if he claims his black heritage?

These two examples raise an important question: How is it that racial categories are treated as mutually exclusive when we can identify many cases in which people have complex racial histories? Maybe race is not a biological fact, an inherited trait like eye or hair color. Perhaps, the fact everyone in the United States seems to fit into a single racial category is the result of the system of racial classification, not biological reality.

When most people meet someone, however, they do not think to learn the facts of that person's ancestry; instead, they search for the visible clues that have come to be associated with a race and proceed to classify that person accordingly. In "Racelessness," Cecile Ann Lawrence (1996) alerts us to the insanity of a racial classification system that labels people like retired general Colin Powell, professional golfer Tiger Woods, law professor Lani Guinier,[1] and the early sociologist W. E. B. Du Bois, all of whom have mixed ancestry, as "black" (see Chapter 1).

Such a case tells us that **race** is not an easily observed characteristic immediately evident on the basis of physical clues but is a category defined and maintained by people through a complex array of formal and informal mechanisms. One of those mechanisms is the rules the Census Bureau uses to ensure that people fit into one of the four official racial categories (see Figure 10.1).

Classifying People by Ethnicity

In addition to a question about race, the Census Bureau asks U.S. residents several ethnicity-related questions, including questions about Hispanic origin, ancestry, place of birth, and language (including a self-rating of ability to speak English). Although the U.S. Census Bureau asks a variety of questions related to ethnicity, the only ethnic categories it recognizes officially are (1) Hispanic/Spanish and (2) non-Hispanic origin. Contrary to popular belief, "Hispanic" is not a race. People classified as Hispanic can be of any race. How is this the case? The history of Latin America (and the Americas, for that matter) is intertwined with that of Asia, Europe, the Middle East, and Africa. As a result of this interconnected history, the countries of Latin America are populated not by a homogeneous group known as Hispanics but by native- and foreign-born persons, immigrant and nonimmigrant residents, and persons from every conceivable ancestry, not just Spanish (see "The Complexity of North American and 'Hispanic' Origins").

As with race, the Census Bureau codes and manipulates data on ethnicity such that no problems appear to be associated with categorizing the population of the United States into Hispanic/non-Hispanic categories. The official definition declares Hispanic to be "a person of Mexican, Puerto Rican, Cuban, Central or South American or other Spanish culture or origin, regardless of race." According to this definition, baseball legend, Ted Williams (whose mother is of Mexican ancestry) and Walt Disney (who was born in Spain) should be classified as Hispanic, as should Vanna White, Raquel Welch, and Linda Carter (Toro 1995). The Hispanic population would include someone born in Costa Rica, of Chinese ancestry,[2] who later immigrated to the United States. It would also include as Hispanic the following case:

> A Filipina with a Spanish surname is filling out the Census form. She was born to parents who moved to the United States shortly after World War II. She is aware that "Spanish surname" used to be the name of today's "Hispanic" classification, and that the Philip-

[1]In 1993, President Bill Clinton nominated Lani Guinier, a woman of Jewish and black ancestry, to head the Civil Rights division of the Justice Department. Later he withdrew the nomination because conservatives criticized her ideas about how black voters could have more influence in elections.

[2]In the article "Distribution of the Overseas Chinese in the Contemporary World," Poston and Yu (1990) estimate that approximately 220,000 Chinese live in the Americas, excluding Canada and the United States.

Race A term that refers to a group of people who possess certain distinctive physical characteristics.

What "race" is Gregory Howard Williams? The left photo shows Williams's grandmother with his father around 1920. The photo above shows Williams, far left, with his immediate family in 1992. How meaningful or valid is it for a government to routinely categorize people—or for people to categorize themselves—by racial constructs that ignore the extent and biological realities of racial blending? If distinctions of race are drawn solely on the basis of outward appearances, what real purpose do they serve?

Courtesy of Dr. Gregory H. Williams, author of *Life on the Color Line: The True Story of a White Boy Who Discovered He Was Black.*

pines were a Spanish colony for centuries before becoming a U.S. dependency after the Spanish-American War in 1898. She is a Catholic, but neither she nor any member of her family speaks Spanish. She knows some Chicanos at work, but her close friends are all Filipino. (Toro 1995:1262)

The variety of people included in the category Hispanic reminds us that classifying people according to ethnicity is not an easy task. When sociologist Raymond Breton and his colleagues studied ethnicity in Toronto, they asked respondents 167 questions to determine their ethnicity (Breton, Isajiw, Kalback, and Reitz 1990), including questions about language spoken at home, the importance of ethnic background, the ethnic composition of the neighborhood in which they lived as children and as adults, whether they subscribed to ethnic magazines, and their parents' ethnic background. The point is that no single indicator can be used as a mark of ethnicity.

Ethnic blending "Inter-ethnic unions (interbreeding) and shifts in ethnic affiliation [such that] most ethnic communities are either amalgams of different peoples or have absorbed significant numbers of other groups through conquest, the expansion of national boundaries, or acculturation" (Hirschman 1993:549–550).

To further complicate ethnic classification, Charles Hirschman (1993) points out that there is no such thing as mutually exclusive or clear-cut categories. There has always been **ethnic blending**—"inter-ethnic unions (interbreeding) and shifts in ethnic affiliation [such that] most ethnic communities are either amalgams of different peoples or have absorbed significant numbers of other groups through conquest, the expansion of national boundaries, or acculturation" (pp. 549–550). Thus, practically everyone in any society belongs to multiple ethnic categories.

Figure 10.1 U.S. Census Bureau Flowchart of Procedures for Recording Race

Consider this diagram taken from the U.S Census Bureau's (1994) interviewing manual. It is a flowchart of decisions that census enumerators and interviewers must make when they encounter problem cases—respondents who say they are more than one race or name a race not listed as a response. Notice that the flowchart directs interviewers to classify the so-called problem respondent as belonging to one race.

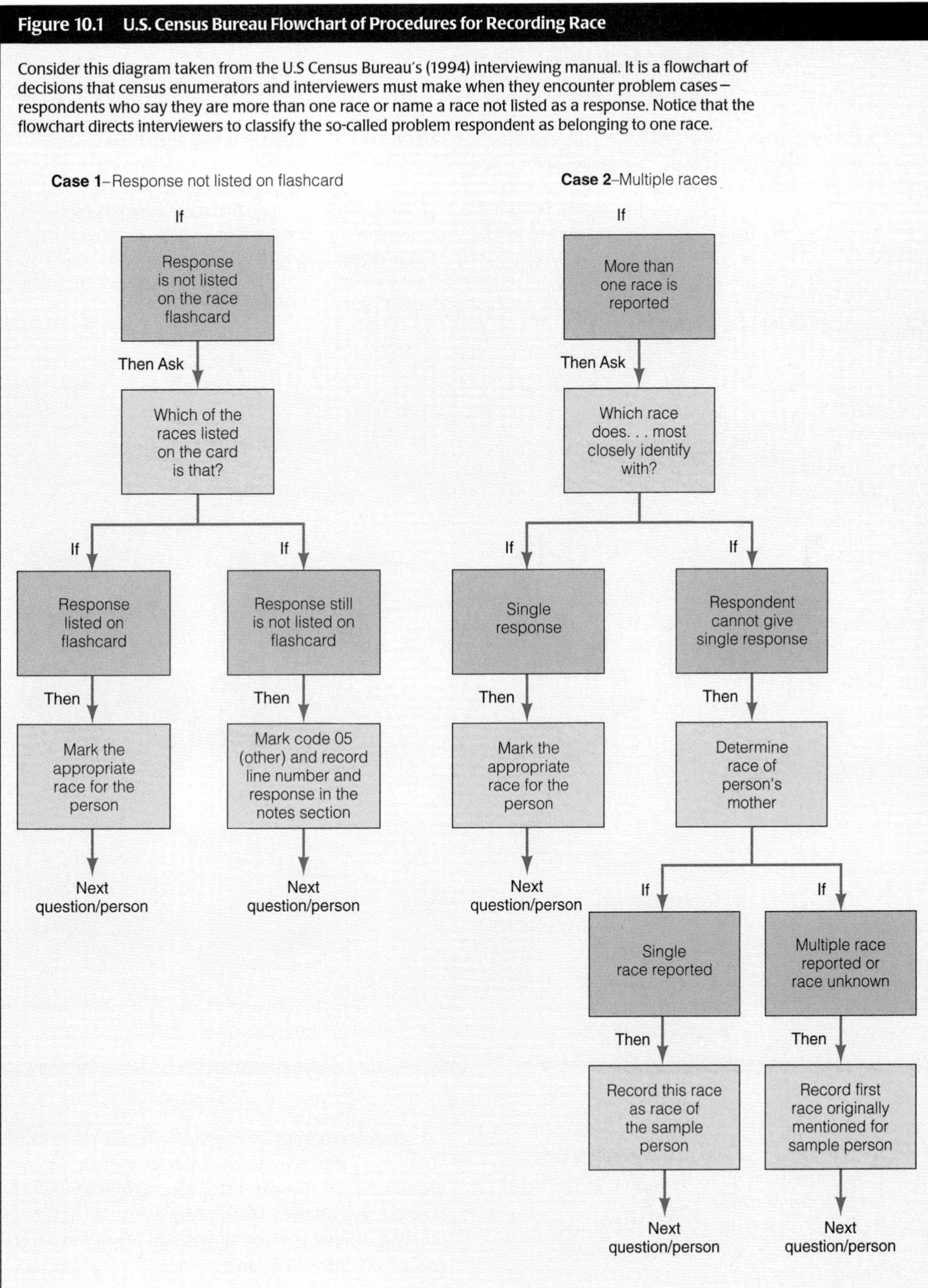

U.S. In Perspective

"Hispanic" Origin

The U.S. Bureau of the Census (1996) sums up the racial origins of U.S. Hispanics in what looks like neat and simple fashion:

Hispanic Origin	
White	11,402,291
Black	645,928
American Indian, Eskimo, or Aleut	148,336
Asian or Pacific Islander	232,684
Other race	94,708

The map and notes shows estimates of the numbers of people migrating to the Americas from other regions of the world. They document the multiracial origins of the immigrant American peoples. With regard to Latin America, the map and tables suggest a wider range of ethnic and linguistic origins than the word *Hispanic* implies, in addition to many indigenous peoples.

Based on these data, what are some possible ethnic and linguistic origins of a person who has immigrated to the United States from South or Central America?

Modern emigration to the Americas (in millions)

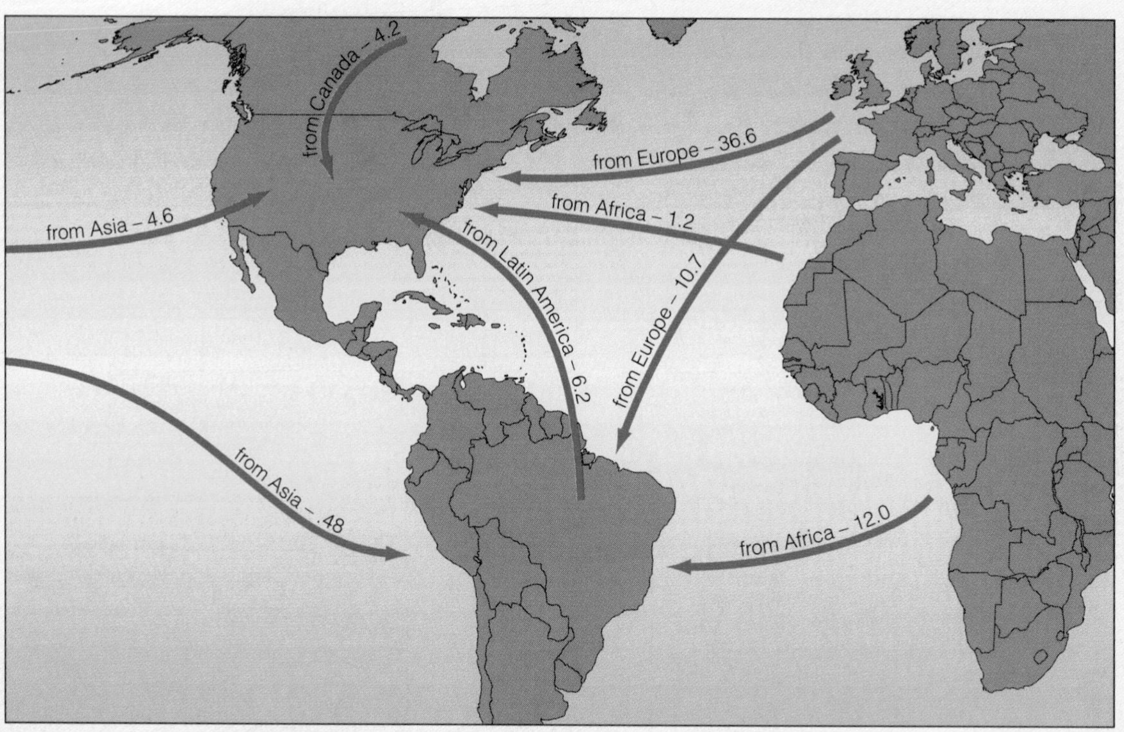

Sources: Segal (1993), Chaliand and Rageau (1995); *Encyclopedia of Latin American History and Culture* (1995); Stalker (1994).

Ethnicity A term that is used to classify people according to any number of attributes, including national origin, ancestry, distinctive and visible cultural traits (religious practice, dietary habits, style of dress, body ornaments, or languages) and/or socially important physical characteristics.

Sociologist Paul D. Starr (1978) notes that in the absence of distinctive skin color and other physical characteristics, people determine **ethnicity** on the basis of any number of other imprecise attributes — language, dress, jewelry, tattoos and other body scars, modes of expression, and residence. In determining someone's race and ethnicity, people rarely look beyond the most visible characteristics. Most people fail to consider the details of another person's life.

If race and ethnicity are such vague categories, perhaps the most appropriate definition of a **racial** or **ethnic group** is that its members believe (or outsiders believe) that they share a common national origin, cultural traits, or distinctive physical features. It does not matter whether this belief is based on reality. The point is that membership in an ethnic or racial group is a matter of social definition, an interplay of self-definition and others' definitions.

Even though racial and ethnic classification schemes are problematic, many people argue that they know a white, black, Asian, or Arabic person when they see one. If they meet someone who does not fit the image, they say, "But you don't look like someone of African/Asian/Arabic descent." That is because their vision of human variety is limited by the images of people portrayed on airline posters, magazine ads, and television sitcoms (Houston 1991). Sometimes people will go to extremes to create an ethnic and racial group that fits an ideal image. The United States, however, is not the only country in the world that divides, even forces, its population into racial and ethnic categories that are not clear-cut. In Chapter 9, we considered South Africa's system of racial classification. Here we consider Germany's classification system.

Racial and Ethnic Classification in Germany

Germany is the most economically powerful country in the European Union. Germany, like the United States, is a country dependent on foreign labor. Yet, despite this dependence, Germany is officially a nonimmigration country (Joppke 1996). Its constitution "does not seek to increase its national population through naturalization of foreigners" (Holzner 1982:67). The foreigners who live in Germany, even those who have lived and worked in the same town for thirty years and those who were born there and know only the German language, are considered "guests" or "temporary labor migrants" (Faist and Häubermann 1996).[3]

Under German law, people who can prove German ancestry are entitled to citizenship regardless of their country of origin. It does not matter if they cannot speak German or if they know nothing about German culture; the criterion for German citizenship is biological. Theoretically, then, 25 percent of the U.S. population (the descendants of the more than 7 million Germans who have immigrated over the past three centuries) could qualify for German citizenship (Department of State 1996). The political changes in the former Soviet Union and Eastern Europe since 1988[4] have put this citizenship law to the test. Between 1990 and 1991, over 600,000 ethnic Germans from countries such as Poland, Romania, and the former Soviet Union settled in Germany (U.S. Bureau for Refugee Programs 1992).

This citizenship law makes life especially difficult for those who do not look German or who have lived in Germany all their lives but who are not biologically German. Afro-Germans fit into this first category. On the one hand, Afro-Germans[5] qualify for citizenship because they are of German ancestry. On the other hand, they are perceived to be foreign because they have dark skin and other physical features not associated with German ancestry. Comments from some Afro-German women illustrate their marginal status:

> Everywhere the same thing — in the job market, in the search for a place to live. Always I have to identify myself, prove, two times, three times, that I'm German, prove my right to exist. "Oh, we thought you

Racial or **ethnic group** A group whose members believe (or outsiders believe) that they share a common national origin, cultural traits, or distinctive physical features.

[3]According to Faist and Häubermann (1996), most of the guest workers of the 1960s have acquired an "unlimited residence permit." Some "temporary labor migrants" can apply for citizenship after ten to fifteen years of residence, provided they have "proven their willingness to assimilate to life, language, and culture of the German demos" (Kurthen 1995:930).

[4]In 1988, when Mikhail Gorbachev was named president of the Soviet Union, he instituted policies of glasnost ("openness"), perestroika ("economic restructuring"), and noninterference in Eastern Europe. Gorbachev also announced plans to withdraw a half-million troops and 10,000 tanks from Eastern Europe over two years. These policies led to the overthrow of hard-line Communist regimes in Poland, East Germany, Czechoslovakia, Bulgaria, and Romania. (In addition, they led to the breakup of the Soviet Union into separate and independent countries.) On November 9, 1989, the Berlin Wall was dismantled. The subsequent reunification of East Germany (with a population of 17 million) with West Germany (with a population of 63 million) followed less than a year later, on October 3, 1990.

[5]Afro-Germans is a term that is "not immediately comprehended by most people who hear it. Virtually no Americans and indeed few Germans have ever heard the term at all" (Adams 1992:234). This term applies to more than the "occupation babies" (babies born to white German mothers and black servicemen from the United States and from the African countries who fought in World War II). It also applies to the products of unions between men and women whose paths crossed as a result of the European colonization of the African continent.

One problem with trying to classify people according to a racial scheme is that millions of people in the world have mixed ancestry and possess the physical traits of more than one race.
©Janice Rubin/Black Star

> were a foreigner." Foreigners are different; they are singled out; they are — as I said — "also-people." (Wiedenroth 1992:167)

> Given the black-white matrix in people's minds — you are placed on the nonwhite side and you are classified as an "also-person." After all, Blacks are "also" people. (Wiedenroth 1992:166)

> My color is black, therefore I'm perceived as a foreigner — African or American. I'm always being asked how come I speak German so well, where I come from, etc. This quizzing gets on my nerves. Most of the time I answer provocatively that I'm German. In spite of my unequivocal answer they continue: How? Why? (Adomako 1992:199)

In addition to Afro-Germans, who are "German" but do not look German, a second group of people is affected by Germany's citizenship laws: those 2.6 million guest workers recruited to work in what was then West Germany between 1961 and 1973. When the Soviet Communists erected the Berlin Wall in 1961, West German employers lost a major labor pool.[6] To make up for the shortage, the German government established labor recruitment offices in Turkey, Yugoslavia, Italy, Greece, Portugal, and Spain. From these offices, officials screened male and female job applicants. Those who possessed the needed occupational skills, had no police records, and enjoyed good health were admitted to West Germany as guest workers to live in employer-provided work site housing. They were admitted on the assumption that they were temporary workers and would eventually return to their home countries (Castles 1986; Herbert 1995). Many stayed, however.

Today, united Germany's population is 81.3 million. This figure includes 5.3 million foreigners, most of whom live in large cities (Strasser 1993). The most numerous, poorest, and most visibly different foreigners are the Turks (including 400,000 Kurds), who make up one-third of the foreign-born population (Jones and Pope 1993; Martin and Miller 1990). The Turks, along with North Africans, sub-Saharan Africans, Pakistanis, and Persians (virtually all of whom are Muslims), are considered "rejected foreigners" and are targets of prejudice and discrimination (Safran 1986; see Table 10.2 and Figure 10.2). The irony is that many people from these groups have lived in Germany all of their lives and have more in common with the West Germans than the West Germans have in common with the East Germans (see "East Meets West"). To help us understand these dynamics, we turn to the concept of minority groups.

Minority Groups

Minority groups are subgroups within a society that can be distinguished from members of the dominant groups by visible and identifying characteristics, including physical and cultural attributes. Members of such subgroups are regarded and treated as inherently different from those in dominant groups. For these reasons they are systematically excluded, consciously or unconsciously, from full participation in society and denied equal access to positions of power, prestige, and

Minority groups Subgroups within a society that can be distinguished from members of the dominant groups by visible and identifying characteristics, including physical and cultural attributes.

[6]Between 1945, when World War II ended, and 1961, 3 million refugees from East Germany and 9 million from former German territories migrated to West Germany. These migrants provided the labor needed to offset the labor loss due to war casualties and to fuel postwar economic reconstruction. In 1961, when the East-West flow of migrants reached an average of 3,000 per day, the Soviet Communists erected the Berlin Wall—103 miles of heavily guarded, 10-foot-high, steel-fortified concrete—to seal off East Berlin from West Berlin and to stop the flow of people from East to West (McFadden 1989). Also, the Soviets helped construct fortified fences and walls along the additional 860-mile border that separated East Germany from West Germany.

In 1961, Communist authorities decided to stop the flow of people from the East to the West by erecting the Berlin Wall (left). Nearly thirty years later, demonstrators were eager to have a hand in taking down the infamous wall while East German border guards looked on placidly (right).

UPI/Corbis-Bettmann (left); Reuters/Corbis-Bettmann (right)

wealth. Thus, members of minority groups tend to be concentrated in inferior political and economic positions and isolated socially and spatially from members of the dominant groups.

On the basis of these characteristics, many groups can be classified as minorities, including some racial, ethnic, and religious groups, women, the very old, the very young, and the physically different (for example, visually impaired people or overweight people). Although we focus on ethnic and racial minorities in this chapter, the concepts that follow can be applied to any minority.

Table 10.2 Attitudes Toward Foreigners of Various Ethnic Groups Who Live in Germany

Category	Nationality of Foreigner	Attitude Toward Foreigner
Noble foreigners	British, French, Americans, Swedes	Positive
Foreigners	Spaniards, Yugoslavs, Greeks	Neutral
Strange foreigners	Portuguese, Italians, Vietnamese	Neutral, with tendency toward negative
Rejected foreigners	North Africans, Black Africans, Pakistani, Persians, Turks	Rejected by substantial parts of the population

Source: Adapted from Thränhardt (1989:13).

Figure 10.2 Foreign Migrations into Germany, 1961–1995 (in thousands)

This map shows major flows of immigration into Germany. The numbers (in thousands) are net gains of people from various countries between 1961 and 1995. Do any of the numbers surprise you? Are you surprised not to see substantial migrations from certain other countries not shown in these flows? What might explain their absence?

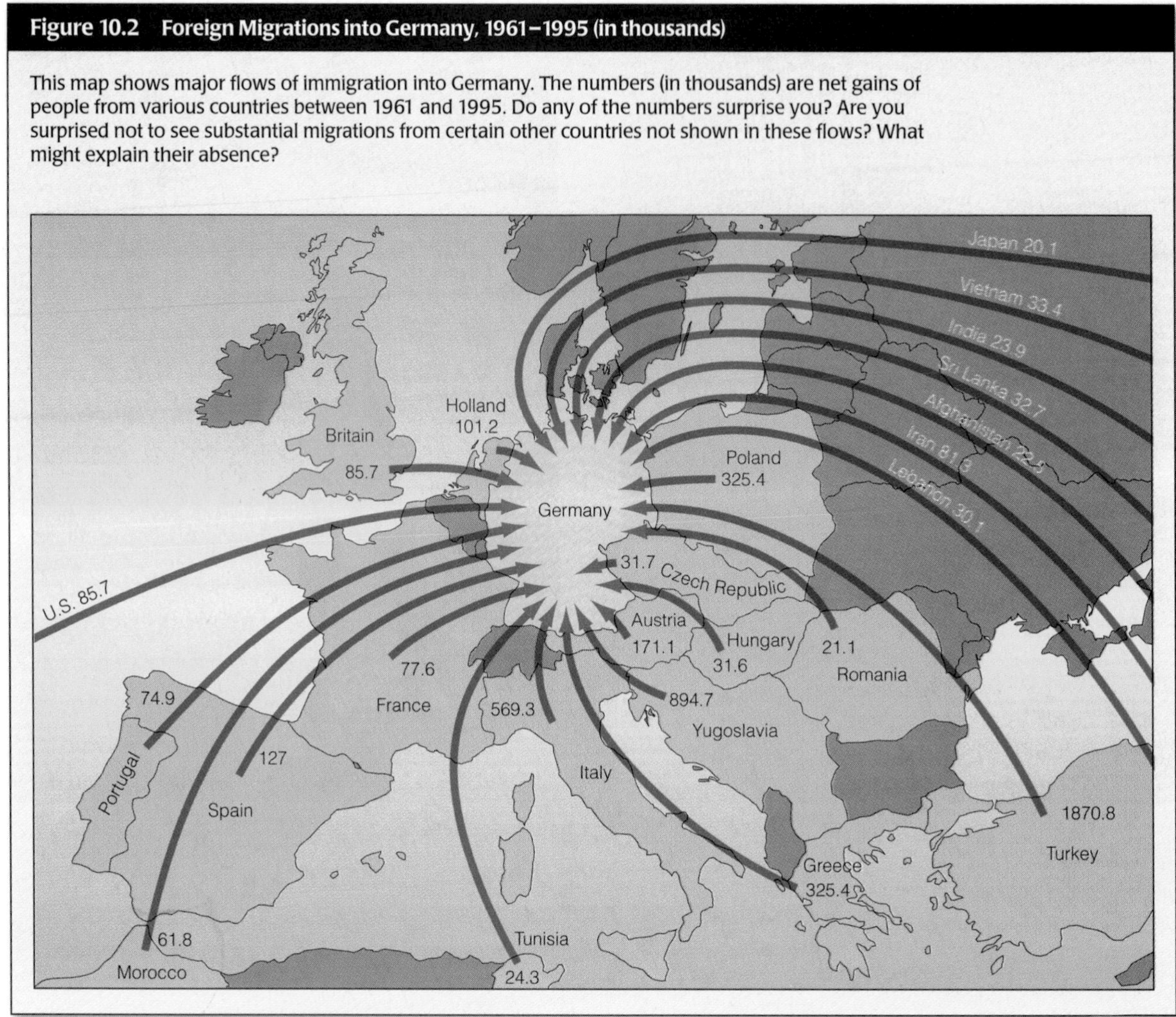

Source: U.S. Central Intelligence Agency (1995); Glassner (1996).

Sociologist Louis Wirth (1945) made a classic statement on minority groups, identifying a number of essential traits characteristic of all minority groups. First, membership is involuntary: as long as people are free to join or leave a group, no matter how unpopular the group, they do not by virtue of that membership constitute a minority. This first trait is quite controversial because the meaning of "free to join or leave" is unclear. For example, if a very light-skinned person of African and German descent can pass as German, is he or she "free" to leave the African connections in his or her life? Second, minority status is not necessarily based on numbers; that is, a minority may be the numerical majority in a society. The key to minority status, then, is not size but access to and control over valued resources. South Africa is an obvious example: roughly 85 percent of the population (Asian Indians, blacks, and people of mixed race) are controlled by the 14 percent of the population who are white. A third characteristic, and the most important, is nonparticipation by the minority group in the life of the larger society. That is, minorities do not enjoy the freedom or the privilege to move within the society in the same way as members of the dominant group do. Sociologist Peggy McIntosh (1992) identifies a number of privileges that most members of the dominant group take for granted, including the following:

- I can, if I wish, arrange to be in the company of people of my race [or ethnic group] most of the time.

Although many Turkish people have been living in Germany for forty years, they have not been assimilated into German society.
©Regis Bossu/Sygma

- If I should need to move, I can be pretty sure of renting or purchasing housing in an area which I can afford and in which I would want to live.
- I can go shopping alone most of the time, fairly well assured that I will not be followed or harassed by store detectives.
- I can be late to a meeting without having the lateness reflect on my race [or ethnicity].
- Whether I use checks, credit cards, or cash, I can count on my skin color not to work against the appearance that I am financially reliable. (pp. 73–75)

The final and most troublesome characteristic of a minority group is that people who belong to such a group are "treated as members of a category, irrespective of their individual merits" (Wirth 1945:349) and often irrespective of context. In other words, people outside the minority group focus on the visible characteristics that identify someone as belonging to a minority. This visible characteristic becomes the focus of interaction as in the scene described here in which a woman can only focus on the physical features of the people she encounters to the neglect of what is going on around her:

> "It obsesses everybody," declaimed my impassioned friend, "even those who think they are not obsessed. My wife was driving down the street in a black neighborhood. The people at the corners were all gesticulating at her. She was very frightened, turned up the windows, and drove determinedly. She discovered, after several blocks, she was going the wrong way on a one-way street and they were trying to help her. Her assumption was they were blacks and were out to get her. Mind you, she's a very enlightened person. You'd never associate her with racism, yet her first reaction was that they were dangerous." (Terkel 1992:3)

The characteristics that Wirth identifies as associated with minority group status indicate that minorities stand apart from the dominant culture. Some people argue that minorities stand apart because they do not wish to assimilate into mainstream culture. To assess this claim, we turn to the work of sociologist Milton M. Gordon, who has written extensively on assimilation.

Perspectives on Assimilation

Assimilation is a process by which ethnic and racial distinctions between groups disappear. There are two main types of assimilation. One form is absorption assimilation. The other form is called melting pot assimilation.

Absorption Assimilation

In this form of assimilation, members of a minority ethnic or racial group adapt to the ways of the dominant group, which sets the standards to which they must adjust (Gordon 1978). According to Gordon, absorption assimilation has at least seven levels. That is, an ethnic or racial group is completely "absorbed" into the dominant group when it goes through all of the following levels:

1. The group abandons its culture (language, dress, food, religion, and so on) for that of the dominant group (an action known as acculturation).
2. The group enters into the dominant group's social networks and institutions (structural assimilation).

Assimilation A process by which ethnic and racial distinctions between groups disappear.

East Meets West

The following excerpt from the *New York Times Magazine* points out the inevitable strains between East and West Germans that arise from a forty-five-year separation. It also raises the question of whether Turks who have lived in West Germany all their lives have more in common with West Germans than do East Germans.

> *It has been forty years. I think it is time to stand back from this social experiment that we call the divided Germany and assess the results.*
>
> *True, an experiment was the last thing the allies had in mind when they agreed to that border running through the middle of Europe. Call it an involuntary experiment, then, born of the pressures of victory, in which the allies functioned as principal researchers and the Germans as extraordinarily cooperative white mice.*
>
> *Perhaps it would clarify matters if I mention a related scientific area, one in which I have some expertise as an amateur; I am speaking of research on twins.*
>
> *Let us assume we are dealing with two hell-raising twins who share a criminal past. Through allied efforts, they are finally forcibly separated and sent to two extremely different boarding schools. One twin grows up in the bracing climate of Western values; first with difficulty and then with growing enthusiasm, he learns to appreciate as basic values democracy, capitalism and individual freedom; and he develops great respect for the Western principal researcher.*
>
> *The other twin has quite a different fate. He is often beaten and brutalized, and finally learns just as assiduously the basic values of Eastern culture: "solidarity," "social commitment," "passion for socialism," and of course "eternal friendship" for the Eastern principal researcher.*
>
> *Let us further assume that a wall is constructed between the twins and an odd system of visitation rights is established. The Western twin can move in any direction he likes, including east; he can visit his brother on the other side, chat with him and compare experiences, bring him presents, then return to the Western half to have dinner in a French restaurant.*
>
> *The Eastern twin, on the other hand, has some freedom of movement north and south, and to the east has access to an almost unlimited recreational area (which, however, has only recently come to be considered a place worth visiting). But access to the West is blocked. There is the famous forbidden door, a good 1,400 kilometers wide. It still opens legally only in exceptional cases, and after a wait of at least two years; anyone who does not wish to wait that long must take his life in his hands and jump, or dig.*
>
> *Let us further assume that the twin in the West, thanks to assistance such as the Marshall Plan and the Western market economy, gradually gets rich. His twin brother, meanwhile, not only has to pay war debts to the far poorer principal researcher in the East; he also has to adopt the researcher's inefficient economic system.*
>
> *At least one result we can safely predict—the twin in the East will fall victim to a psychological law. Every wall in the world, whether German or Chinese, begs to be overcome. Also, because he finds himself in the awkward situation of having to wait for his Western sibling to come to him, a certain reproachful attitude forms. "That guy over there really could come to see me more often," thinks the Eastern twin. "At the very least, he could write or phone regularly. And he could be a bit more generous, because it's turned out that he has the better deal. Not that he has done anything to* deserve *it, by the way; it was pure luck that he happened to be*

3. The group intermarries and procreates with those in the dominant group (marital assimilation).
4. The group identifies with the dominant group (identification assimilation).
5. The group encounters no widespread prejudice from members of the dominant group (attitude receptional assimilation).
6. The group encounters no widespread discrimination from members of the dominant group (behavior receptional assimilation).
7. The group has no value conflicts with members of the dominant group (civic assimilation).

Gordon advances a number of hypotheses about how the various levels of assimilation relate to one another. First, he maintains that acculturation is likely to take place before the other six levels of assimilation. Gordon also states, however, that even if acculturation is total (as in the cases of Ender Bsaran and some Afro-Germans), it does not always lead to the other levels of assimilation.

Gordon proposes that a clear connection exists between the structural and marital levels of assimilation. That is, if the dominant group permits people from ethnic and racial minority groups to join its social cliques, clubs, and institutions on a large enough scale, a substantial number of interracial or interethnic marriages are bound to occur:

> If children of different ethnic backgrounds belong to the same play-group, later the same adolescent cliques, and at college the same fraternities and sororities; if the parents belong to the same country club and invite each other to their homes for dinner; it is completely unrealistic not to expect these children, now grown, to love and to marry each other, blithely oblivious to previous ethnic extraction. (pp. 177–178)

living on the right side of the Elbe at the right time.

"But now his success has gone to his head. Instead of sharing, he acts as though he had all the talent and claims he works harder. He can't fool me; I know him, we started out under the same roof, and he's just as industrious and just as lazy as I am. He's gotten pretty arrogant, even self-righteous. Actually, he's still living off the misery of the world's poor, whom he exploits, but he doesn't want to hear that.

"Well, if he doesn't have a conscience, at least he could show a little family feeling, a little interest in his relatives in the East, at least listen to us—is that asking too much? He's gained an awful lot of weight, by the way; even if he is rich, that doesn't look like happiness to me."

Meanwhile, the Western twin is concocting a monologue of his own. He feels pressured by what he calls his relative's "eternal posture of expectation."

"I can see that the poor guy behind his wall doesn't have an easy time of it," thinks the Western twin. "But these demands, these unspoken reproaches really cast a pall over our relationship. God knows I'm happy to give him things, but it's no fun bringing a present when the other person always expects you to. Those people over there seem to think that cars and color televisions grow on trees. But we're not born *with a Mercedes; you have to earn it; you have debts, interest payments—concepts my brother knows only by hearsay. . . .*

"I'd like to explain this, but he won't listen; he just talks and talks. Of course, it's not his fault that he always has to stand in line to buy oranges; but at least he should admit that he bet on the wrong horse, that the socialist economic system is a disaster—no one's criticizing him personally. The problem is—and it's so German *of him—he takes any kind of criticism personally. Instead of agreeing with me, he tries to spoil my success. He claims to be an idealist—I'm glad to let him call himself that, because the poor fellow has a lot to compensate for. . . . But he accuses me of being a conspicuous consumer and a 'conformist'—a compliment I'm glad to return by pointing out that his so-called 'socialist' or even 'revolutionary' virtues are all a pose; I know him, after all.*

"Sometimes I feel downright relieved when the visit is over; there's an unpleasant tension between us that we really should talk about some day. Next time."

And what happens when the Eastern twin comes west? Every immigrant is expected to declare allegiance to the Western way of life, of course, but I can hardly think of anyone to whom crossing the border has not brought culture shock. Far from softening this effect, the common language exacerbates it, because it stimulates a commonality that daily experience does not bear out.

The Western twin finds it hard to understand the complaints of his Eastern brother, now safely "back home." He's never satisfied, he always finds something to criticize. What's all this whining about how "cold" it is in the West, about the lack of "real friendship" and "coziness"? Suddenly, the Western twin begins to suspect that his brother has taken on a whole raft of official socialist virtues, even when he claims to be a sworn anti-Communist; after three beers, this difficult relative begins to dream of being back behind the wall, where a word was a word and the promise "I'll give you a call" really meant something.

Source: "If the Wall Came Tumbling Down," by Peter Schneider, pp. 27, 61–62, in the *New York Times Magazine* (June 25). Copyright © 1989 by The New York Times Company. Reprinted by permission.

Of the seven levels of assimilation, Gordon believes the structural level is the most important because if it occurs, the other levels of assimilation inevitably will follow. Yet, even though the structural level is the most important, it is very difficult to achieve in practice. Members of ethnic or racial minorities, such as Turks, Afro-Germans, and African Americans, are denied easy and comfortable access to the dominant group's networks and institutions. In fact, all of the important and meaningful primary relationships that are close to the core of personality and selfhood are confined largely to people of the same racial or ethnic group:

> From the cradle in the sectarian hospital to the child's play group, the social clique in high school, the fraternity and religious center in college, the dating group within which he [or she] searches for a spouse, the marriage partner, the neighborhood of his [or her] residence, the church affiliation and the church clubs, the men's and the women's social and service organizations, the adult clique of "marrieds," the vacation resort, and then, as the age cycle nears completion, the rest home for the elderly and, finally, the sectarian cemetery — in all these activities and relationships which are close to the core of personality and selfhood—the member of the ethnic group may if he wishes follow a path which never takes him across the boundaries of his [or her] ethnic structural network. (p. 204)

This scenario especially characterizes the primary group relations of **involuntary minorities**, ethnic and racial groups that did not choose to be a part of a country. These groups were forced to become part of a

Involuntary minorities Ethnic and racial groups that were forced to become part of a country by slavery, conquest, or colonization.

country by slavery, conquest, or colonization. Native Americans, African Americans, Mexican Americans, and native Hawaiians are examples of involuntary minorities. Unlike **voluntary minorities**, whose members come to a country expecting to improve their way of life, members of involuntary minorities have no such expectations. Their forced incorporation involves a loss of freedom and status (Ogbu 1990).

Melting Pot Assimilation

Assimilation need not be a one-sided process in which a minority racial and ethnic group disappears, or is absorbed, into the dominant group. Ethnic and racial distinctions also can disappear in another way known as **melting pot assimilation** (Gordon 1978). In this process, the groups involved accept many new behaviors and values from one another. This exchange produces a new cultural system, which is a blend. Melting pot assimilation is total when significant numbers of people from each ethnic and racial group take on cultural patterns of the other, enter each other's social network, intermarry and procreate, and identify with the blended culture.

The melting pot concept can be applied to the various African ethnic groups imported to the United States as slaves. They were "not one but many peoples" (Cornell 1990:376), who spoke many languages and came from many cultures. Slave traders capitalized on this diversity: "Advertisements of new slave cargoes frequently referred to ethnic origins, while slaveowners often purchased slaves on the basis of national identities and the characteristics they supposedly indicated" (Cornell 1990:376; see also Rawley 1981). Although slave owners and traders acknowledged ethnic differences among Africans, they treated Africans from various ethnic groups as belonging to one category of people — slaves. Because slave traders sold and slave owners purchased individual human beings, not ethnic groups, this treatment had the effect of breaking down ethnic concentrations. In addition, slave owners tended to mix together slaves of different ethnic origins in order to decrease the likelihood of the slaves' plotting a rebellion. To communicate with each other, the slaves invented pidgin and Creole languages. In addition to inventing a new language, the slaves created a common and distinctive culture based on kinship, religion, food, songs, stories, and other features. The harsh conditions of slavery, in combination with the mixing together of people from many ethnic groups, encouraged slaves to borrow aspects of each other's cultures and to create a new, blended culture.

In Germany (which calls itself a nonimmigration country) and the United States (which calls itself a melting pot culture), assimilation has been mainly a one-sided process in that newcomers are expected to adapt to the dominant culture. Sociologist Ruth Mandel believes that the status of the labor migrant in Germany makes assimilation difficult:

> The very term by which these people are commonly referred, "Gastarbeiter," guest workers, underlies the ambiguity of their status . . . guests are by definition temporary, and are expected to return home. Guests are bound to rules and regulations of hosts. Whatever the intentions, guests rarely feel "at home" in foreign environs. The second half of the compound word, "arbeiter," worker, refers to the economic and use value of the migrant determined solely in relation to his or her labor. (1989:28)[7]

When most people think of assimilation, they do consider it a process of mutual exchange in which members of the dominant group form and identify with a blended culture (Opitz 1992a). This is not to say that the dominant culture has not been shaped and influenced by racial and ethnic minorities; rather, the dominant group more often than not fails to acknowledge others' contributions to the society in which they live. Stratification theory is one major approach to understanding forces that work against assimilation (absorption or melting pot) between dominant and minority groups.

Stratification Theory and Assimilation

Stratification theory is guided by the assumption racial and ethnic groups compete with one another for scarce and valued resources. As a whole, the dominant group retains the advantage because its members are in a po-

Voluntary minorities Racial and ethnic groups that come to a country expecting to improve their way of life.

Melting pot assimilation A process of cultural blending in which the groups involved accept many new behaviors and values from one another.

[7]German political leaders have made some attempts to call international labor migrants something other than "guest workers." One such term is *ausländische Mitbürger,* or "foreign co-citizens." Yet, in the words of sociologist Czarina Wilpert (1991), "In any other language this would be a contradiction in terms, and even in German it highlights the ambiguity of belonging to the collective of foreign settlers in Germany" (p. 53).

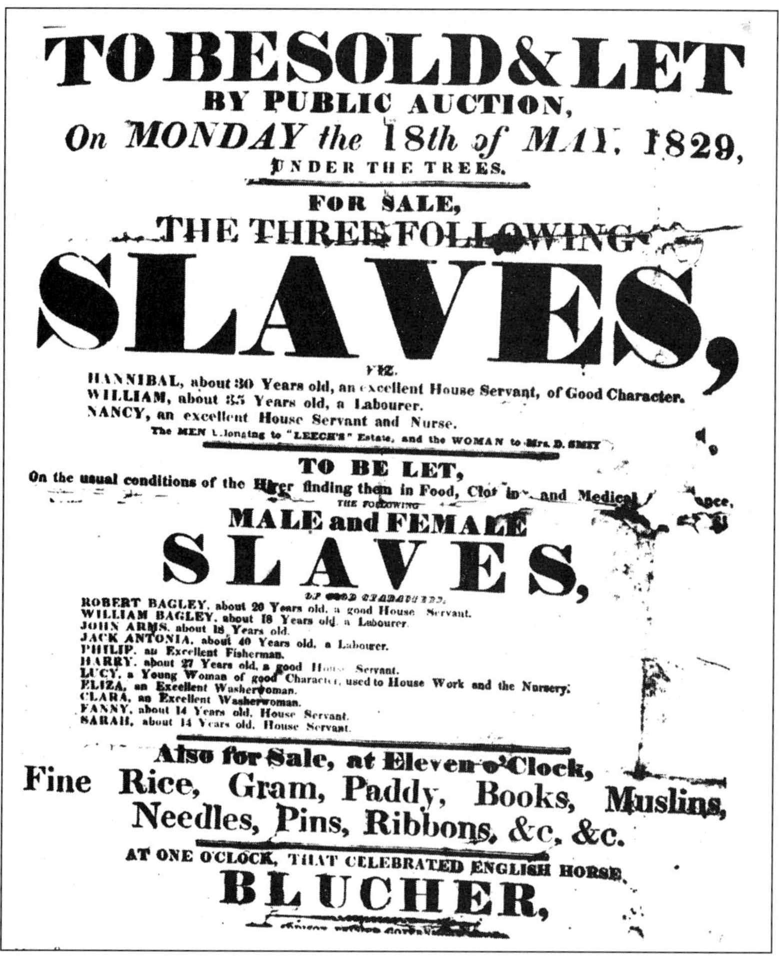

Throughout history, members of ethnic or racial groups have become involuntary minorities through conquest, annexation, or slavery.

Corbis/Bettmann

sition to preserve the system that gives them their advantages. Stratification theorists focus on the mechanisms employed by people in the dominant group to preserve inequality (Alba 1992). These include racist ideologies, prejudice and stereotyping, discrimination, and institutional discrimination.

Racist Ideologies

An **ideology** is a set of beliefs that are not challenged or subjected to scrutiny by the people who hold them. Thus, ideologies are taken to be accurate accounts and explanations of why things are as they are. On closer analysis, however, ideologies are at best half-truths, based on misleading arguments, incomplete analysis, unsupported assertions, and implausible premises. They "cast a veil over clear thinking and allow inequalities to persist" (Carver 1987:89–90). One such ideology is racism. People who follow this ideology believe that something in the biological makeup of an ethnic or racial group explains and justifies its subordinate or superior status.

Racist ideologies are structured around three notions: (1) people can be divided into categories on the basis of physical characteristics; (2) a close correspon-

Ideology is a set of beliefs that are not challenged or subjected to scrutiny by the people who hold them.

U.S. In Perspective

Is There Such a Thing as Black Racism?

Black racism. For some blacks it is a laughable oxymoron. ("How can the victims of racism be racist?") For some whites it is an excuse for doing little to reduce the inequalities that still plague the country. ("See, blacks are just as bad as they say we are.")

The recent diatribe against Catholics, Jews and homosexuals (among others) by Khalid Abdual Muhammad, a member of the hierarchy of the Nation of Islam, has focused attention on the disturbing question of how much racism permeates black America. Indeed, the public fascination with the Nation of Islam—a fringe group with negligible political and economic power—is partly explained by white fear that its views are shared by other blacks. Looking at the cleaning women, the bank tellers, partners in their law firm, some whites silently ask, "Do you agree with Louis Farrakhan?"

Until recently, the question of how blacks view whites has been largely unexplored territory for scholars and sociologists. Why have sociologists not probed the black psyche to see how much racial animus resides there? Perhaps because they are afraid of what they might find; perhaps because they feel that since whites are the dominant group, their views are more important; perhaps because they feel that blacks have no opinions that whites are bound to respect.

But recent work by pollsters, sociologists and political scientists has uncovered evidence of specific negative attitudes that blacks may harbor toward whites in general and Jews in particular. Seeking responses to a series of assertions (Jews tend to stick together more than other Americans; Jews wield too much power on Wall Street; Jews are more willing than others to use shady practices to get what they want), a 1992 survey by the Anti-Defamation League of B'nai B'rith found that blacks are more than twice as likely as non-Jewish whites to hold stereotypical views of Jews.

Preliminary finds in a study to be published soon by the National Conference of Christians and Jews found that blacks feel that whites think themselves superior, and that they do not want to share power and wealth with nonwhites. The poll found that black antipathy toward whites is far greater than that of other minorities, such as Latinos and Asian-Americans.

"Of course, there's black racism," said Roger Wilkins, a professor of history at George Mason University, who is black. "I think that any time that you say that a whole group of people is this or that, fill in the blanks, and the blanks are all negative adjectives, and then you go on to prove your conclusions by telling stories that support your use of those adjectives, that's racism."

But, as another black writer, Ellis Cose, points out in his 1993 book, The Rage of a Privileged Class *(HarperCollins), the harboring of negative views by blacks does not tell the whole story. Surveys of blacks also find that many give American whites, both Jew and gentile, high marks for intelligence and for creating a democratic society. Viewed in this light, the results of the surveys may be less an affirmation of racism than an expression of envy and a plea to share power.*

Racism among blacks, whatever its extent, has been like magma bubbling under black nationalism. Like other nationalist movements—Zionists, the Quebecois in Canada, Serbs seeking a "Greater Serbia"—those embraced by some blacks have used ethnic solidarity as a political organizing tool.

dence exists between physical traits and characteristics such as language, dress, personality, intelligence, and athletic capabilities; and (3) physical attributes such as skin color are so significant that they explain and determine social, economic, and political inequalities. Any racial or ethnic group may use racist ideologies to explain their own or another group's behavior. (See "Is There Such a Thing as Black Racism?") One example of a racist ideology is the hypothesis offered by former Los Angeles police chief Daryl Gates to explain why so many blacks have died from restraining chokeholds: their "veins and arteries do not open up as fast as they do on normal people" (Dunne 1991:26). The fact behind these statistics is that many police officers, black and white, tend to handle black suspects more harshly than they do their white counterparts.

German Nazis rationalized the extermination and persecution of Jews, Gypsies, Poles, Russians, blacks, and others "deemed unfit to breed" (including disabled, homosexual, or mentally ill persons) with the racist ideology that the Aryan race was superior to other races. According to this ideology, the Jews were responsible for Germany's defeat in World War I and did not even belong to a race of people; they were "nonhuman; blood suckers, lice, parasites, fleas . . . organisms to be squashed or exterminated by chemical means" (Fein 1978:283).

The premise of racial superiority is also at the heart of other rationalizations used by one group to dominate another. Sociologist Larry T. Reynolds (1992) observes that race as a concept for classifying humans is a product of the 1700s, a time of widespread European exploration, conquest, and colonization that did not begin to subside until the end of World War II. Racist ideology also supported Japanese annexation and domination of Korea, Taiwan, Karafuto (the southern half of the former Soviet island Sakhalin), and the Pacific Islands before World War II. Both the Japanese

For some mainstream black politicians, black anger toward whites presents both opportunity and temptation. Just as white politicians often take advantage, some discreetly and others more cynically, of racist attitudes among whites, so black politicians are aware of hostility among their constituents. And when they tap into it, they confront a question of conscience: When do they cross the line into demagoguery?

A call for unity is by implication a call for exclusivity. The desire to maintain that solidarity makes it difficult for many blacks to denounce black racism. But the censure of Mr. Muhammad by people like the Rev. Jesse Jackson, Benjamin Chavis, executive director of the National Association for the Advancement of Colored People, and Representative Kweisi Mfume of Maryland, chairman of the Congressional Black Caucus, shows that difficult does not mean impossible. Indeed, black officials have shown more willingness to move beyond solidarity than have members of the United States Senate who met with silence the remarks of Senator Ernest F. Hollings when he jokingly implied in December, while commenting on trade talks in Switzerland, that African heads of state are cannibals. A Hollings aide later said the Senator meant no offense.

There are those who argue that even if blacks espouse racist views, it makes little difference since they lack the power to turn those views into action. Such a rationale was articulated by Mr. Farrakhan at his recent press conference: "Really, racism has to be coupled with a sense of real power."

Mr. Farrakhan may have a point. For all the racial bombast by the Nation of Islam and other strident black nationalists, America has had no episodes of large-scale coordinated attacks by blacks against whites. But individual acts like the beating of the trucker Reginald O. Denny during the Los Angeles riots demonstrate that black bigotry can maim people and poison that atmosphere.

There is no denying that black bigotry is fed by social and economic conditions. Even discounting, as some whites do, the country's history of slavery, segregation and discrimination, there are trends that insure black resentment of whites will not diminish soon, and that black demagogues will find allies among the disaffected.

Black poverty rates remain stubbornly high. Progress toward closing the income gap between the races stalled in 1975, according to census data, and since then the gap has widened. (For instance, male black college graduates who entered the job market in 1971 earned on average 2 percent more than their white counterparts, but by 1989 the same blacks were earning 25 percent less on average than white men who graduated in 1971.)

Blacks remain the most segregated group, according to the census. Whether it is the result of whites making blacks feel unwelcome, or of declining faith in integration among some blacks, black Americans remain huddled in their corners, afraid that any attempt to reach out will tar them with labels: bleeding heart, Uncle Tom. "We don't live next door to each other," said Lani Guinier, the law school professor whose nomination as Assistant Attorney General for civil rights was withdrawn in a controversy over her voting-rights theories. "No one talks to each other. Racial stereotypes fester in isolation."

Source: "Behind a Dark Mirror: Traditional Victims Give Vent to Racism," by Steven A. Holmes, p. 4E in the *New York Times* (February 15).

and Europeans used racial schemes to classify the people they encountered; the idea of racial differences became the "cornerstone of self-righteous ideology," justifying their right by virtue of racial superiority to exploit, dominate, and even annihilate conquered peoples and their cultures (Lieberman 1968).

Prejudice and Stereotyping

A **prejudice** is a rigid and usually unfavorable judgment about an outgroup (see Chapter 5) that does not change in the face of contradictory evidence and that applies to anyone who shares the distinguishing characteristics of that group. Prejudices are based on **stereotypes**—exaggerated and inaccurate generalizations about people who are members of an outgroup—and they are applied to everyone who possesses that distinguishing characteristic on sight. Many Germans stereotype Turks[8] as backward people who come from villages where "their houses are huts, the streets are unpaved, . . . and where there are goats, sheep, and chickens, and fields of corn, wheat, and melons" (Teraoka 1989:111).

[8]For example, if we know that someone comes from Turkey, it does not follow that he or she is a Turk in an ethnic, linguistic, or religious sense. Turkey's population includes 3 to 10 million Kurds, Arabs (who migrated from Egypt and North Africa in the 1930s), Armenians, Greeks, and Jews. Even the so-called Turkish population contains at least three distinct regional groups.

Prejudice A rigid and usually unfavorable judgment about an outgroup that does not change in the face of contradictory evidence and that applies to anyone who shares the distinguishing characteristics of that group.

Stereotypes Exaggerated and inaccurate generalizations about people who are members of an outgroup.

Among the victims of the Nazi ideology of a "pure race" were Gypsies like those being held at this concentration camp in Belzec, Poland. Historians estimate that between 20 and 50 percent of European Gypsies died in the "Gypsy Holocaust."

Archives of Mechanical Documentation, Warsaw, Poland. Courtesy of the United States Holocaust Memorial Museum.

Stereotypes "give the illusion that one [group] knows the other" and "confirms the picture one has of oneself" (Crapanzano 1985:271–272). Germans who claim that Turks are backward are saying that they, by definition, are not. In addition, Germans who hold such stereotypes about Turks fail to see Turks in Germany as a highly heterogeneous group with regard to social origin, cultural norms, and length of residence in Germany (Teraoka 1989).

Stereotypes are supported and reinforced in a number of ways. One way is through **selective perception**, in which prejudiced persons notice only those behaviors or events that support their stereotypes about an outgroup. In other words, people experience "these beliefs, not as prejudices, not as prejudgments, but as irresistible products of their own observations. The facts of the case permit them no other conclusion" (Merton 1957:424).

Columnist Clarence Page (1990) reflects in an editorial on the fact that stereotypes about black teenagers and violence can cause people to fear contact with any black teenager:

> As much as everyone tells us how cute and handsome our little boy is now (and they're right!), I also know that someday he will be a teen-ager and many adults have an outright fear of black male teen-agers.
>
> It's getting to be an old story. See a group of small black children coming toward you on the street and you want to scoop the little darlings up and hug them. See the same kids approaching a few years later as teen-agers in sneakers and fashionably sculpted haircuts and, depending on who you are, you very well might want to cross the street, thinking to yourself about how you can't be too careful.
>
> I dread the experiences of a family friend who is trying to help his seventh-grade son cope with his transformation from the cuddly cuteness of childhood to the imposing presence of a full-sized African American teen-ager. Formerly greeted with delight, he now is sometimes greeted with apprehension.
>
> Positive thinker that I am, I cannot help but think negatively about the damage done by the negative signals of fear we send young black males, simply because some have sinned. (p. A10)

Stereotypes also persist in another way: when prejudiced people encounter a minority person who contradicts the stereotype, they see that person as an ex-

Selective perception The process in which prejudiced persons notice only those behaviors or events that support their stereotypes about an outgroup.

ception. The fact that they have encountered a minority group member who is "different" only serves to reinforce their stereotypes. In addition, prejudiced people use facts to support their stereotypes. A prejudiced person can point to the small number of black or other minority quarterbacks, pitchers, and baseball managers as evidence that they do not possess the leadership qualities to be successful at such positions. It does not occur to the prejudiced person that prejudice and discrimination by owners and general managers may explain the small numbers instead. In a similar vein, a prejudiced person may point to the fact that "black" athletes dominate the sports such as basketball and track as evidence of natural leaping ability and quickness. At the same time, this person does not use the same kind of logic to explain why "white" athletes dominate sports like gymnastics, golf, and hockey.

Finally, prejudiced individuals keep stereotypes alive when they evaluate the same behavior differently at different times, depending on the person who exhibits that behavior (Merton 1957). For example, incompetent behavior on the part of racial and ethnic minority members often is attributed to innate flaws in their biological makeup, whereas incompetence exhibited by someone from the dominant group is almost always treated as an individual issue. Similarly, prejudiced people treat certain ideas, when expressed by members of a minority group, as more threatening than when those ideas are expressed by members of a dominant group. Spike Lee drew considerable criticism for advocating violence as a means for solving problems when he ended his film *Do the Right Thing* with two quotes—one by Martin Luther King, rejecting violence as evil and ultimately ineffectual, and the other by Malcolm X: "I don't even call it violence; when it's self-defense, I call it intelligence." From Lee's point of view, such criticism meant only that "we're not allowed to do what everyone else can. The idea of self-defense is supposed to be what America is based on. But when black people talk about self-defense, they're militant. When whites talk about it, they're freedom fighters" (McDowell 1989:92).

Discrimination

In contrast to prejudice, **discrimination** is not an attitude but a behavior. It is intentional or unintentional unequal treatment of individuals or groups on the basis of attributes unrelated to merit, ability, or past performance. Discrimination is behavior aimed at denying members of minority groups equal opportunities to achieve valued social goals (education, health care, long life) and/or blocking their access to valued goods and services. The U.S. Commission on Civil Rights (1981) describes some examples of common discriminatory actions:

- Personnel officers whose stereotyped beliefs about women and minorities justify hiring them for low level and low paying jobs exclusively, regardless of their potential experience or qualifications for higher-level jobs
- Teachers who interpret linguistic and cultural differences as indications of low potential or lack of academic interest on the part of minority students
- Guidance counselors and teachers whose low expectations lead them to steer female and ethnic minority students away from "hard" subjects, such as mathematics and science, toward subjects that do not prepare them for higher-paying jobs
- Parole boards that assume minority offenders to be more dangerous or more unreliable than white offenders and consequently more frequently deny parole to members of minorities than to whites convicted of equally serious crimes (p. 10)

Robert K. Merton explores the relationship between prejudice (the attitude) and discrimination (the behavior). He distinguishes between two types of individuals: the nonprejudiced (those who believe in equal opportunity) and the prejudiced (those who do not). Merton asserts that people's beliefs about equal opportunity are not necessarily related to their conduct, as he makes clear in his four-part typology (see Figure 10.3).

Nonprejudiced nondiscriminators (all-weather liberals) accept the creed of equal opportunity, and their conduct conforms to that creed. They represent a "reservoir of culturally legitimized goodwill" (Merton 1976:193). In Germany, for example, some religious, industrial, and political groups have formed a coalition and have launched Action Courage, a national campaign encouraging Germans to challenge antiforeign and antiethnic remarks, to intervene and protect foreigners from discrimination, and to wear a button bear-

Discrimination Intentional or unintentional unequal treatment of individuals or groups on the basis of attributes unrelated to merit, ability, or past performance.

Nonprejudiced nondiscriminators (all-weather liberals) Persons who accept the creed of equal opportunity, and their conduct conforms to that creed.

Figure 10.3 A Typology of Ethnic Prejudice and Discrimination

	Attitude Dimension: Prejudice and Nonprejudice	Behavior Dimension: Discrimination and Nondiscrimination
Type I: Unprejudiced nondiscriminator	+	+
Type II: Unprejudiced discriminator	+	–
Type III: Prejudiced nondiscriminator	–	+
Type IV: Prejudiced discriminator	–	–

* + Attitude/behavior supports equal opportunity
– Attitude/behavior rejects equal opportunity

Source: Adapted from Merton (1976:192).

ing the word courage to symbolize their commitment to these goals (Marshall 1992). People who act to realize these goals would qualify as nonprejudiced nondiscriminators in Merton's typology.

Unprejudiced discriminators (fair-weather liberals) believe in equal opportunity but engage in discriminatory behaviors because it is to their advantage to do so or because they fail to consider the discriminatory consequences of some of their actions. In one example, unprejudiced persons decide to move out of their neighborhood after a black family moves in because they are afraid that property values might start to decrease. In another instance, a white personnel officer tells friends and neighbors (who also are likely to be white) about a job opening. This word-of-mouth method of recruiting reduces the chances that a minority candidate will learn of the opening.

Examples of unprejudiced discriminators specific to Germany include those who fail to challenge young fans for yelling "Jew to Auschwitz" at referees who make "bad" calls or those who join in singing songs whose rhyming words note that the Christmas season is the time "to burn a Turk" (Marshall 1992:A13).

Prejudiced nondiscriminators (timid bigots) do not accept the creed of equal opportunity but refrain from discriminatory actions primarily because they fear the sanctions they may encounter if they are caught. Timid bigots do not often express their true opinions about racial and ethnic groups; instead, they use code words. In *Race: How Blacks and Whites Think and Feel About the American Obsession,* Studs Terkel (1992) cites several examples of speaking in code, including this one in Chicago: during the black mayor's first campaign, his white opponent's slogan was "Before It's Too Late."

Prejudiced discriminators (active bigots) reject the notion of equal opportunity and profess a right, even a duty, to discriminate. They derive significant social and psychological gains from the conviction that anyone from the ingroup (including the village idiot) excels anyone from the hated outgroup (Merton 1976:198). Active bigots are most likely to believe that they "have the moral right" to destroy the people whom they see as threatening their values and way of life. Of the four categories in Merton's typology, prejudiced discriminators are the most likely to commit hate

Unprejudiced discriminators (fair-weather liberals) Persons who believe in equal opportunity but engage in discriminatory behaviors because it is to their advantage to do so or because they fail to consider the discriminatory consequences of some of their actions.

Prejudiced nondiscriminators (timid bigots) Persons who do not accept the creed of equal opportunity but refrain from discriminatory actions primarily because they fear the sanctions they may encounter if they are caught.

Prejudiced discriminators (active bigots) People who reject the notion of equal opportunity and profess a right, even a duty, to discriminate.

Ku Klux Klan members in the United States (left) and neo-Nazis in Germany (right) exemplify active bigotry based on a racist ideology.

©Bob Daemmrich/Sygma (left); ©Adenis/Sipa Press (right)

crimes, actions aimed at humiliating minority-group people and destroying their property or lives (see Table 10.3).

The government of Saxony (one of fifteen states in Germany) published a profile of 1,244 persons suspected of committing hate crimes between 1991 and 1992. Ninety-six percent were males; two-thirds were age eighteen or younger; 50 percent had been drinking before the incident; 90 percent of the hate crimes occurred near the suspects' homes (within twelve miles); more than 98 percent of the suspects had less than a tenth-grade education. Surprisingly, only 20 percent of the suspects were unemployed or school dropouts. Eighty percent were first-time offenders. Approximately 30 percent justified their violent action with firm right-wing ideological views (Marshall 1993). A second systematic investigation of 1,398 police files related to antiforeigner offenses committed between January 1991 and April 1992 produced similar findings (Krell, Nicklas, and Ostermann 1996).

Institutionalized Discrimination

Sociologists distinguish between individual and institutionalized discrimination. **Individual discrimination** is any overt action on the part of an individual that depreciates someone from the outgroup, denies outgroup opportunities to participate, or does violence to lives and property. **Institutionalized discrimination**, on the other hand, is the established and customary way of doing things in society — the unchallenged rules, policies, and day-to-day practices that impede or limit minority members' achievements and keep them in a subordinate and disadvantaged position. It is the "systematic discrimination through the regular operations of societal institutions" (Davis 1978:30). Institutional discrimination can be overt or subtle. It is overt when laws and practices are designed with the clear intention of keeping minorities in subordinate positions, as in the 1831 Act Prohibiting the Teaching of Slaves to Read. Institutionalized discrimination can also be subtle, as when the discriminatory consequences of a practice are neither planned nor intended. One might even argue that the way in which the dominant group implements affirmative action programs is a subtle form of institutionalized discrimination. As essayist and English professor Shelby Steele (1990) observes, affirmative action programs, as they typically have been carried out, have some troubling side effects because the quality that earns minorities preferential treatment is an implied inferiority.

Individual discrimination Any overt action on the part of an individual that depreciates someone from the outgroup, denies outgroup opportunities to participate, or does violence to lives and property.

Institutionalized discrimination The established and customary way of doing things in society—the unchallenged rules, policies, and day-to-day practices that impede or limit minority members' achievements and keep them in a subordinate and disadvantaged position.

Table 10.3 Hate Crimes in the United States

In 1995, 7,947 hate crime incidents involving 10,469 victims and 8,433 known offenders (which represents known offenders for 62 percent of hate crimes) were reported to the FBI.

Intimidation Motivation (%)		**Nature of Hate Crime (%)**	
Racial bias	61	Intimidation	41
Religious bias	16	Damage/destruction/vandalism of property	23
Sexual orientation	13	Simple assault	18
Ethnicity/national origin	10	Aggravated assault/Murder	13
Victim (classified as)		**Offender (classified as)**	
White	1,226	White	4,991
Black	2,988	Black	2,253
American Indian/Alaskan Native	41	American Indian/Alaskan Native	45
Asian/Pacific Islander	355	Asian/Pacific Islander	211
Hispanic	516	Multiracial group	318
Multiracial group	221	Unknown	615

Source: Federal Bureau of Investigation (1996).

> However this inferiority is explained—and it is easily enough explained by the myriad deprivations that grew out of our oppression—it is still inferiority.
>
> The effect of preferential treatment—the lowering of normal standards to increase black representation—puts blacks at war with an expanded realm of debilitating doubt. (1990:48–49)

The key words in Steele's observations are "have been carried out." As a case in point, some of my students have reported cases in which they know of employers who have deliberately hired incompetent or unqualified minorities in the name of affirmative action as a way of "proving" they are less qualified than white applicants. There is no way of knowing the extent to which this kind of "affirmative action" hiring takes place, but it would certainly make for an interesting research project.

One wonders whether this side effect of implied inferiority would exist if blacks and other minorities were recruited for jobs and college scholarships in the same way they are recruited for college sports. Few people seem to be disturbed when black athletes are recruited to play at predominantly white colleges, possibly because school officials, coaches, recruiters, and students believe that the black athlete's presence on campus benefits the school. If the same energy and positive attitude were applied to recruiting black college students or job applicants, some of the troublesome features of the affirmative action programs might be alleviated. In many ways, the on-the-field integration of major league baseball is a model (although not a perfect one) for affirmative action programs. Syndicated columnist William Raspberry (1991) pointed this out on the *MacNeil/Lehrer Newshour:*

> The Major Leagues [until 1948 had] been an exclusive preserve of white men, skilled white men, but white men. And it was also very clear then that there were a lot of black baseball players as good as many of those whites who were in the big leagues. But the rules said you couldn't be a major league ballplayer if you were black. Branch Rickey made the breakthrough in hiring Jackie Robinson to say we're going to bust this up. . . . But once the decision was made to bust it up, a couple of things happened. One had to change . . . the ways baseball looked for and recruited ballplayers. You couldn't just look to the minor leagues anymore because blacks weren't there either. They had to look in some places they weren't accustomed to looking and had to use some techniques they weren't accustomed to using to find prospects. That's affirmative action in one sense and I buy it absolutely. But there was another piece of it. Nobody supposed that pretty good black baseball players who because they had been denied opportunity to play in the big leagues before then should have been given special breaks and brought into the big leagues with lesser skills than their white counterparts. The assumption was, and we bought it, black and white, that if we were given the opportunity to use our skills, and to develop our skills, you didn't have to cut us any breaks.

Raspberry's point is that people lose sight of the real purpose of affirmative action policies: to "bust up"

U.S. In Perspective

U.S. Department of Justice Press Release

For Immediate Release
Thursday, October 31, 1996

MONTGOMERY REALTY COMPANY TO PAY $30,000 FOR ALLEGEDLY STEERING BLACKS AND WHITES TO DIFFERENT AREAS BECAUSE OF THEIR RACE

Case Stems from Justice Department Nationwide Testing Program

Washington, D.C.—A Montgomery, Alabama real estate company that was sued by the Justice Department for allegedly refusing to refer African-Americans to properties in predominantly white parts of town will pay $30,000 in damages, the Justice Department announced.

The case, brought last January, stems from the Justice Department's highly successful nationwide fair housing testing program. Under the program, trained pairs of African-Americans and whites posing as prospective tenants inquire about the availability of rental units. By comparing the experiences of the testers, investigators discover whether minorities were treated less favorably than whites.

The agreement, filed today in U.S. District Court in Montgomery, resolves Justice Department allegations that rental agents at Hamilton Realty Company based in Montgomery, steered African-Americans toward properties located in predominantly minority areas of the city and whites toward predominantly white areas. "Steering" occurs when a rental agent discloses different properties to applicants with the intent to concentrate persons in different areas according to race.

"Housing discrimination is usually subtle but it inflicts deep wounds on its victims and serves to segregate our society," said Assistant Attorney General for Civil Rights, Deval L. Patrick. "All Americans should have the ability to live in the neighborhood of their choice regardless of the color of their skin."

Under the agreement, the company, which owns and manages approximately 200 rental properties in the Montgomery area, will create a $30,000 fund to compensate any identified victims of the alleged discriminatory practice. Any money not paid to identified victims will be paid to the government as a civil penalty.

The suit alleged that Hamilton Realty rental agents did not tell African-Americans about properties located in predominantly white areas of Montgomery, while whites were told. It also claimed that agents would steer African-Americans toward properties located in predominantly minority areas of Montgomery and whites toward predominantly white areas.

"This lawsuit and the Consent Order entered today signal our continuing commitment to the principle that equal rights in housing for all people is a basic right of citizenship," said Redding Pitt, U.S. Attorney in Montgomery.

The Justice Department's testing program has produced 33 federal cases in nine states—Ohio, Michigan, California, South Dakota, Indiana, Missouri, Florida and Virginia—resulting in more than $3 million dollars in settlements. At any given time, the Justice Department is conducting testing in about a dozen cities.

U.S. Department of Justice (1996).

a flawed hiring system in which not everyone who was qualified had a chance to apply or be considered. Instead, the attention focuses on whether the person who is hired is qualified. No one seems to remember that the hiring system that affirmative action replaced was a program that benefited the friends and acquaintances of those in power.

Two examples of institutionalized discrimination in Germany are laws that limit the proportion of foreign children in German classrooms to 20 to 30 percent and laws that limit the percentage of foreigners living in an area to 9 percent (Wilpert 1991). Another example of institutionalized discrimination, in both the United States and Germany, is the selective law enforcement practice in which one racial or ethnic group is more likely than another to be arrested or to receive stiffer penalties for breaking the law.

Of the two levels of discrimination, institutionalized discrimination is the more difficult to identify, condemn, hold in check, and punish, because it can exist in society even if most members are not prejudiced. Institutionalized discrimination cannot be traced to the motives and actions of particular individuals; discriminatory actions result from simply following established practices that seem on the surface to be impersonal and fair or part of the standard operating procedures (see "U.S. Department of Justice Press Release").

To this point, we have looked at barriers that exist in society to prevent racial and ethnic minorities from adapting and/or assimilating. These barriers include racist ideology, dependence on foreign labor, prejudice, and discrimination (individual and institutional). Although we have viewed these barriers in a general way, we have not examined how they operate in everyday interaction. In *Stigma: Notes on the Management of Spoiled Identity*, sociologist Erving Goffman (1963) gives us such a framework.

Social Identity and Stigma

When people encounter a stranger, they make many assumptions about what that person ought to be. Judging by an array of clues such as physical appearance, mannerisms, posture, behavior, and accent, people anticipate the stranger's social identity—the category (for example, male or female, black or white, professional or blue-collar, under forty or over forty, American born or foreign born) to which he or she belongs and the qualities that they believe, rightly or wrongly, to be "ordinary and natural" (Goffman 1963:2) for a member of that category.

Goffman was particularly interested in social encounters in which one of the parties possesses a stigma. We learned in Chapter 6 that a **stigma** is an attribute that is defined as deeply discrediting in that person's society. A stigma is considered discrediting because it overshadows all other attributes that a person might possess.

Thus, the important quality underlying all stigmas is that those who possess them are not seen by others as multidimensional, complex persons but as one-dimensional beings. When many Germans encounter a Turkish woman wearing a head scarf, for example, they do not see the scarf in all its complexities—as "signifying the relation between the wearer and any number of things, such as her male relatives, her personal religio-political views, her financial means, or her region of origin" (Mandel 1989:30). Instead, they see only a foreigner, pointing to the scarf as proof that Turks cannot integrate into German society and using it as evidence to prove why Turks should be sent home. In other words, this single attribute overshadows any other attribute the woman might possess.

The same dynamic applies to how most Germans view people of Afro-German ancestry. Many Germans see only one attribute — dark skin — and fail to acknowledge other attributes Afro-Germans possess such as the ability to speak German. Two Afro-German women explain how their dark skin dominates the course of interaction:

> I have dark skin, too, but I am a German. No one believes that, without some further explanation. I used to say that I was from the Ivory Coast, in order to avoid further questions. I don't know that country, but to me it sounded so nice and far away. And after this answer, I didn't get any more questions either. Germans are that ignorant. I could tell people any story I wanted to, the main thing was that it sounded foreign and exotic. But no one ever believes that I am German. When I respond to the remark, "Oh, you speak German so well" by saying "So do you," people's mouths drop open. (Emde 1992:109–110)

> It often happens with me that people have their own expectations and ignore what I say. When I tell them that I grew up here [in Germany] and have spent my entire life here, the question still might come afterward: "Yes, and when are you going back?" Crazy. Now and then I have the feeling of not belonging anywhere; on the other hand, I've grown up here, speak this language, actually feel secure here, and can express myself as I want. I share a background with these people here even if they don't accept me. "Yes, I'm German," I say, perhaps out of spite, to shake up their black-and-white thinking. (Opitz 1992b:150)

According to Goffman, it is important to remember that the discrediting trait itself is not the stigma. Rather, the stigma consists of the set of beliefs about the trait. Furthermore, from a sociological point of view, the person who possesses the trait is not the problem; the problem is how others react to the trait. Thus, Goffman maintained that sociologists should not focus on the attribute that is defined as a stigma. Instead, the focus should be on interaction — specifically, interaction between the stigmatized and normals. Goffman did not use the term *normal* in the literal sense of "well adjusted" or "healthy." Instead, he used it to refer to those people who are in the majority or those who possess no discrediting attributes. Goffman's choice of this word is unfortunate because some readers forget how Goffman intended it to be used.

Stigma An attribute defined as deeply discrediting because it overshadows all other attributes that a person might possess.

Mixed contacts "The moments when stigmatized and normals are in the same 'social situation,' that is, in one another's immediate physical presence, whether in a conversation-like encounter or in the mere co-presence of an unfocused gathering" (Goffman 1963:12).

Mixed Contact Between the Stigmatized and the Dominant Population

In keeping with this focus, Goffman (1963) wrote about **mixed contacts**, "the moments when stigmatized and normals are in the same 'social situation,' that is, in one another's immediate physical presence, whether in a conversation-like encounter or in the mere co-presence of an unfocused gathering" (p. 12). According to Goffman, when normals and the stigmatized interact, the stigma comes to dominate the course of interaction. First, the very anticipation of contact can cause normals and stigmatized individuals to try to avoid one another. One reason is that interaction threatens their sense of racial and ethnic "purity" or loyalty.

Sometimes the stigmatized and the normals avoid each other to escape the other's scrutiny. Persons of the same race may prefer to interact with each other so as to avoid the discomfort, rejections, and suspicions they encounter from people of another racial or ethnic group. ("I have a safe space. I don't have to defend myself or hide anything, and I'm not judged on my physical appearance" [Atkins 1991:B8].)

Another reason that stigmatized persons and normals make conscious efforts to avoid one another is that they believe widespread social disapproval will undermine any relationship. With regard to interracial relationships, some people believe that racist attitudes will destroy even the "most perfect and loving interracial relationships [because] racism waits like a cancer, ready to wake and consume the relationship at any, even the most innocuous, time" (Walton 1989:77).

An article in *U.S. News and World Report* magazine profiled the friendship of two six-year-old girls, one black (Jennifer) and one white (Phoebe), to demonstrate how societal pressures can interfere with the development of an interracial relationship. Whenever Jennifer accompanies Phoebe and her parents on outings, she worries that strangers will think she is adopted (Buckley 1991). Consequently, Jennifer cannot relax and enjoy Phoebe's company because she feels pressured to explain why she is with Phoebe. The lesson of this example is that for this six-year-old, interracial relationships are so rare that she cannot imagine people thinking she is Phoebe's friend.

Goffman observed a second pattern that characterizes mixed contacts: upon meeting each other, each party is unsure how the other views him or her or will act toward him or her. Thus, the two parties are self-conscious about what they say and about their behavior:

> How was I to know which whites were good and which were bad? What litmus test could I devise? I distanced myself from everyone white, watching, listening, for hints of latent prejudice. But there were no formulas to follow. (McClain 1986:36)

Sociologist Erving Goffman wrote about a pattern that characterizes "mixed contacts." Each party is unsure how the other views him or her, making both parties self-conscious about what they say and do in each other's presence.

> When you're black, you cannot help but second-guess everything. Does this person like me because I'm black or because I'm me? Does this person hate me because I'm black or because I'm a jerk? You're always second-guessing, unless it's another black person. (King 1992:401)

For the stigmatized, the source of the uncertainty is not that everyone they meet will view them in a negative light and treat them accordingly. Rather, the chance that they might encounter prejudice and discrimination is great enough to give them reason to be cautious about all encounters. According to the Kolts Report (a report written in response to the Rodney King beating, issued by Special Counsel James C. Kolts, a retired superior court judge, and his staff), there were 62 "problem" deputies out of 8,000 in the Los Angeles County Sheriff's Department. The investigators concluded that "nearly all deputies treat nearly all individuals, most of the time, with at least minimally acceptable levels of courtesy and dignity" (*Los Angeles Times* 1992:A18). Although only a small proportion of deputies (less than 1 percent) were identified as "problems," the cases of mistreatment were "outrageous enough and frequent enough to poison the well in some communities" (p. A18).

Similarly, the majority of German citizens do not discriminate against Turks and other dark-skinned people in conscious or overt ways. Yet, there are about 6,000 skinheads in Germany who identify with Nazi ideology. The government believes that they are responsible for as many as five extremely brutal attacks per day against foreigners (Kramer 1993), a number large enough to make life dangerous for minorities in Germany.

A third pattern characteristic of mixed contacts is that normals often define accomplishments by the stigmatized, even minor accomplishments, "as signs of remarkable and noteworthy capacities" (Goffman 1963:14) or as evidence that they have met someone from the minority group who is an exception to the rule. In *Two Nations: Black and White, Separate, Hostile, Unequal,* Andrew Hacker (1992) writes that whites attending professional meetings or panel discussions tended to applaud longer and more strenuously at the introduction of black participants and at the end of their remarks. Although Hacker acknowledges that in some instances the applause is well deserved, he also maintains that many whites do this to let the black speakers know that they are "in the company of friendly whites" (p. 56).

In addition to defining the accomplishments of the stigmatized as something unusual, normals also tend to interpret the stigmatized person's failings, major and even minor (such as being late for a meeting or cashing a bad check), as related to the stigma.

A fourth pattern characteristic of mixed contacts is that the stigmatized are likely to experience invasion of privacy, especially when people stare:

> I am tired of walking into restaurants, especially with a group of African-Americans, and having the patrons and proprietors act as if they were being visited by Martians.
>
> I'm tired of being out with my eight-year-old nephew and his best friend and never being completely at ease because I know they will act like little boys. Their curiosity will be perceived as criminal behavior. (Smokes 1992:14A)

If the stigmatized show their displeasure at such treatment, normals often treat such complaints as exaggerated, unreasonable, or much ado about nothing. Their argument is that everyone suffers discrimination in some way and that the stigmatized do not have the monopoly on oppression. They announce that they are tired of the complaining and that perhaps the stigmatized are not doing enough to help themselves (Smokes 1992). Normals are apt to respond:

> I'm tired of them. (Smokes 1992:14A)

> Why can't they put past discrimination behind them? I think they've been whining too long, and I'm sick of it. (O'Connor 1992:12Y)

> My grandparents, when they came here from Italy, didn't ask anyone to speak Italian for them. They became American, and they never asked for anything. (Barrins 1992:12Y)

Such responses suggest that normals do not understand the large social context and historical forces that shape interaction. The discussion thus far may lead you to believe that members of racial and ethnic minorities are passive victims who are at the mercy of the dominant group. This is not the case, however; minorities respond in a variety of ways to being treated as members of a category.

Responses to Stigmatization

In *Stigma: Notes on the Management of Spoiled Identity,* Goffman describes five ways in which the stigmatized respond to people who fail to accord them respect or who treat them as members of a category. One way is to attempt to correct that which is defined as the failing, as when people change the visible cultural characteristics that they believe represent barriers to status and belonging. A person may undergo plastic surgery or do other things to alter the shape of nose, eyes, or lips, or may enroll in a school to change an accent. In her research for "Medicalization of Racial Features: Asian American Women and Cosmetic Surgery," Eugenia Kaw (1991) found that among Asian American women who choose to undergo cosmetic surgery, the most often requested kind of surgery is "double-eyelid" to make their eyes look bigger. Klaw found that these women were motivated to correct the perception they believed many people associated "'small, slanty' eyes and 'flat' nose" with a perception that they are passive, dull, and nonsocial. The direct attempt to correct that which is defined as a failure is not unique to Asian American women. An Afro-German woman describes her efforts to change her physical appearance:

> When I was about thirteen I started to straighten my "horse hair" so that it would be like white people's hair that I admired so much. I was convinced that with straight hair I would be less conspicuous. I would squeeze my lips together so that they appeared less "puffy." Everything, to make myself beautiful and less conspicuous. (Emde 1992:103)

Turks who change their religion from Islam to Christianity, who Germanize their names, or who give up wearing head scarves to fit into German society represent direct attempts to correct the stigma. Often these kinds of responses are not easy because such persons

Protesting stigmatization: This Turkish woman is wearing a yellow star identifying herself as a Turk. The star recalls the Nazi period, when German Jews were forced to wear the Star of David.

may be considered traitors to their racial or ethnic group (Safran 1986). People may experience conflicting reactions to the visible changes they make, too:

> My wife is twenty-five years old, she is beginning to dress German now, she is becoming more and more beautiful in her German clothes. . . . [M]y friends grumble because my wife dresses so modern, they want her, if she's going to wear German clothes, to wear at least a headscarf and long skirts; why does everyone have to see from far away that my wife is a Turk[?] (Teroaka 1989:110)

The stigmatized also may respond in a second way. Instead of taking direct action and changing the "visible" attributes that normals define as failings, they may attempt an indirect response. That is, they may devote a great deal of time and effort to trying to overcome the stereotypes or appear as if they are in full control of everything around them. They may try to be perfect—to always be in a good mood, to outperform everyone else, or to master an activity ordinarily thought to be beyond the reach of or closed to people with such traits.

> [You have to be on guard all the time;] there is no way to get away from it. Because if you do something like close the door to your room, people will start saying, "Is she being angry? Is she being militant?" You can't even afford to be moody. A white girl can look spacey and people will say, "Oh, she's being creative." But if you walk around campus with anything but a big smile on your face, they'll wonder, "Why is she being hostile?" (Anson 1987:92)

> Don't ever do anything bad, because people are always looking for you to do something bad. You not only have to be good, you have to be perfect. If you do something bad, it's not a mark against yourself but a mark against the entire race. (p. 129)

In another type of indirect response, the stigmatized take issue with the way that normals define a particular situation, or they take action to change the way normals respond to them. In 1955, Rosa Parks, a black seamstress from Montgomery, Alabama, refused to give her seat on the bus to a white person. Her actions, which challenged a law requiring her to do so, triggered a boycott of Montgomery buses and was revolutionary in sparking the civil rights movement.

Another example is the declared goal of the African American Marketing and Media Association. As the organization's president announced, one goal of the organization is to combat stereotypes in advertising (*New York Times* 1990). Although such actions to change the way normals respond are in one way very direct, they fall into Goffman's category of indirect responses because the stigmatized person does not try to change his or her traits but rather attempts to change the way that normals respond to those traits.

Sometimes the stigmatized respond in a third way: they use their subordinate status for secondary gains, including personal profit or "an excuse for ill success that has come [their] way for other reasons" (Goffman 1963:10). If a black person, for example, levels a charge of racism and threatens to file a lawsuit in a situation in which he or she is justly sanctioned for poor work, academic, or other performance, that person is using his or her status for secondary gains. A fourth response is to view discrimination as a blessing in disguise, especially for its ability to build character or for what it teaches a person about life and humanity.

The hopeful symbolism of candles graces a demonstration by young Germans against xenophobia, or fear of foreigners.
©Rainer Unkel/Saba

Finally, the stigmatized can condemn all of the normals and view them negatively:

> You build up these perceptions of whites, that whites are mean and vile, never trust a white person. (Anson 1987:127)

> This coldness here in Germany makes me sick; I am still homesick today, sometimes even more strongly than five years ago; homesickness is an illness, and the illness can only be cured in Turkey; but in Anatolia there is no work for me, no gain, no possibility to move up in life, to live like a human being, with a house and a steady income: I must stay in Germany for now, I must live with this illness; in this cold; the coldness here in Germany, that is its people. (Teroaka 1989:107)

Summary and Implications

We began the chapter with cases of people who did not fit clearly into one racial or ethnic category yet were forced into a category: (1) Ender Bsaran, born and raised in Germany, who is still a "Turk"; (2) an Afro-German who is "German" yet does not "look German" and is thus treated as an outsider; (3) Piri Thomas, born in Puerto Rico of mixed Indian, African, and European ancestry, who finds that in the United States he is "black"; and (4) Alena, the daughter of a white U.S. serviceman and a Thai mother who was born and raised in the United States and is treated as "foreign." These cases tell us that pure race or ethnicity categories cannot exist if only because people from different "races" can reproduce offspring. The offspring of parents classified as belonging to different racial categories challenge the validity of those categories. For some reason the obvious fact that the offspring belong to more than one category has been lost to the strange idea that one parent's (usually the nonwhite or, in the case of Germany, the non-German one) genetic contribution offsets the other parent's contribution. However, when the "white" parent's physical features dominate the child is able to "pass" as white or German. Whatever the case "we renounce the reality of our real families, and we replace it with the unreal reality of a social construct" (Scales-Trent 1995:63).

The goal of the chapter is to show the absurdity of racial and ethnic classification, not to advocate new systems of racial classification with clearer dividing lines. This goal should also not be confused with a desire to show that because there is no such thing as race or ethnicity, "race" and "ethnicity" no longer matter. The important point is to see that people construct classification schemes that carry real and problematic consequences. Classification schemes are problematic because they are used to assign social significance and status value to categories over which people have no control. When that occurs, people who possess one variety of a characteristic are regarded as more valuable and worthy of reward than people who possess other varieties. We used the concepts minority group, prejudice, discrimination, and stigma to analyze the consequences of racial/ethnic classification.

The larger implications of the information in this chapter suggests that the diversity curriculum aimed at teaching about the separate histories[9] and distinct cultures of so-called racial and ethnic groups misses the mark in a deeply profound way. Rather, we need a curriculum that shows the interdependence among peoples and considers how we came to see each other in terms of such categories. This curriculum would challenge us to see connections in our own family histories and come to terms with a society that puts everyone in a position in which they (or their ancestors) have had to reject some part of their heritage.

[9]The statement does not mean that history has not adequately covered the experiences of those outside the mainstream or dominant group.

Key Concepts

Use this outline to organize your review of the key chapter ideas.

- **Race/ethnic group**
- **Minority groups**
  - **Involuntary minorities**
  - **Voluntary minorities**
- **Assimilation**
  - **Absorption assimilation**
  - **Melting pot assimilation**
- **Discrimination**
  - **Individual discrimination**
  - **Institutionalized discrimination**
- **Ideology**
  - **Racist ideology**
- **Prejudice**
  - **Stereotypes**
  - **Selective perception**
- **Discrimination–prejudice typology**
  - **Nonprejudiced discrimination**
  - **Prejudiced nondiscrimination**
  - **Prejudiced discrimination**
    - **Hate crimes**
  - **Nonprejudiced nondiscrimination**
- **Stigma**
  - **Social identity**
  - **Mixed contacts**

internet assignment

Browse the Web site "Interracial Voice" (http://www.webcom.com/~intvoice), which posts letters to the editor and articles that focus on the U.S. racial classification system clarifying the need for a new "racial" category for interracial individuals. List two or three examples of writing that dramatize the flaws of the U.S. system. Then go to the U.S. government site, which posts a critique of the current system of racial classification and the proposed suggestions for changing the way the Census Bureau asks race and ethnicity questions (http://www1.whitehouse.gov/WH/EOP/OMB/html/fedreg/race-ethnicity.html). In light of the information presented in this chapter, do the proposed suggestions correct the flaws? Explain.

Gender

With Emphasis on the Former Yugoslavia

Sarajevo before the Bosnian war. ©Fridmar Damm/Leo de Wys, Inc.

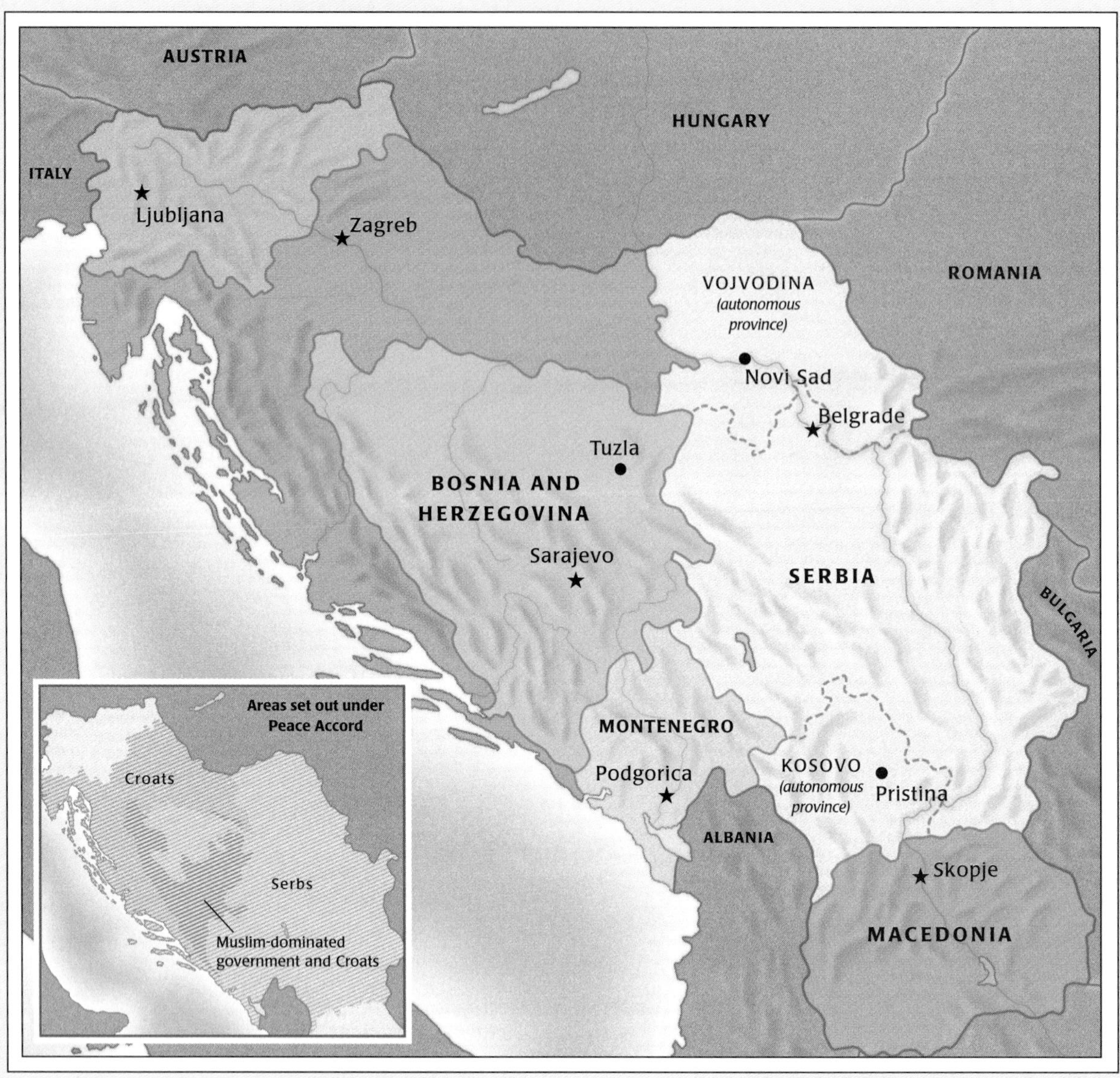

At War with Itself

Perhaps it is at times of war that societies traditionally seem to have fallen back on the clearest-cut distinctions between male and female gender stereotypes. In the old romantic view of things, the men "marched off to war" and the women "kept the home fires burning." (In actuality, of course, armies often trample through those home fires and the people who try to keep them going.)

So it has been in the former Yugoslavia, wracked by interethnic warfare since 1990 that has centered mainly in Bosnia and Herzegovina. The peoples of the former Yugoslavia have had little chance to air the kind of public debate over gender-based behaviors and opportunities that has risen elsewhere in the world. Still, recent events and conditions in that former country illuminate many gender issues and help show how gender differences in modern societies are more accurately seen as social constructs than as biological givens.

In The Balkan Express: Fragments from the Other Side of War, *Croatian writer Slavenka Drakulić (1993a) reflects on an article that appeared in a local newspaper entitled "Will You Come to My Funeral?"*

> It was about the younger generation and how they feel about the war. I remember an answer by Pero M., a student from Zagreb [Croatia].
>
> "Perhaps I don't understand half of what is going on, but I know that all this is happening because of the fifty or so fools who, instead of having their sick heads seen to, are getting big money and flying around in helicopters. I'm seventeen and I want a real life, I want to go to the cinema, to the beach . . . to travel freely, to work. I want to telephone my friend in Serbia and ask how he is, but I can't because all the telephone lines are cut off. I might be young and pathetic-sounding, but I don't want to get drunk like my older brother who is totally hysterical or to swallow tranquilizers like my sister. It doesn't lead anywhere. I would like to create something, but now I can't."
>
> And then he said to the reporter interviewing him something that struck me:
>
> "Lucky you, you are a woman, you'll only have to help the wounded. I will have to fight. Will you come to my funeral?"
>
> This is what he said in the early autumn of 1991. I could almost picture him, the street-wise kid from a Zagreb suburb, articulate, smart, probably with an earring and a T-shirt with some funny nonsense on it, hanging out in a bar with a single Coke the entire evening, talking about this or that rock group. The boy bright enough to understand that he might die and that there is nothing that he could do about it. But we—me, you, that woman reporter—we are women and women don't get drafted. They get killed, but they are not expected to fight. After all someone has to bury the dead, to mourn and to carry on life and it puts us in a different position in the war. **(1993a:133–134)**

Why Focus on the Former Yugoslavia?

In this chapter, we consider the topic gender, which sociologists define as social distinctions based on culturally conceived and learned ideas about appropriate behavior and appearance for males and females. In exploring the concept of gender, we focus on the former Yugoslavia, which has split into five separate countries. Four of the six republics of the former Yugoslavia—Croatia, Slovenia, Bosnia-Herzegovina, and Macedonia—began to declare independence about two years after the fall of the Berlin Wall in 1989. The two remaining republics—Serbia and Montenegro—call themselves the new Federal Republic of Yugoslavia. Although the former Yugoslavia is now five countries, it is impossible to speak about one of these countries without considering the other four. For all practical purposes, the fate of one is intertwined with that of the others.

We pay special attention to the former Yugoslavia because it offers an interesting contrast with the United States. In the United States, the question of gender differences, specifically the position of women in society relative to that of men, is subject to constant analysis and debate. Specifically, the debate evolves around the connection between gender and life chances (opportunities to experience and achieve a wide range of potential outcomes including the chance to live a long life, to stay out of poverty, to recover from illness, to go to school, and so on). In view of the ongoing, severe economic upheaval that has taken place in the former Yugoslavia over the past decade and in light of the wars that have been fought since 1991, gender is not considered a pressing issue there. In fact, gender was never an issue in Yugoslavia in the same way as in the United States. Officially, at least, the question of inequality between women and men was resolved under Communism. This does not mean, however, that there is no connection between gender and life chances in the countries of the former Yugoslavia.

In this chapter, we explore the basic concepts sociologists use to analyze the connection between gender and life chances. In outlining this connection, sociologists distinguish between sex (a biologically based classification scheme) and gender (a socially constructed phenomenon). They also focus on the extent to which society is gender polarized — that is, organized around the male-female distinction. In addition, sociologists seek to explain gender stratification and the mechanisms by which people learn and perpetuate their society's expectations about appropriate behavior and appearances for males and females.

Distinguishing Sex and Gender

Although many people use the words *sex* and *gender* interchangeably, the two terms do not have the exact same meaning. Sex is a biological concept, whereas gender is a social construct. In the following section, we will pursue this distinction further because it will help illustrate how the social differences between males and females develop.

Sex as a Biological Concept

A person's **sex** is determined on the basis of **primary sex characteristics**, the anatomical traits essential to reproduction. Most cultures divide the population into two categories — male and female — largely on the basis of what most people consider to be clear anatomical distinctions. Like race (see Chapter 10), however, even biological sex is not a clear-cut category, if only because a significant (but unknown) number of babies are born **intersexed**. This is a broad term used by the medical profession to classify people with some mixture of male and female biological characteristics.[1]

If some babies are born intersexed, why, then, is there no intersexed category? There is no category because parents of intersexed children collaborate with physicians to assign their offspring to one of the two recognized sexes. Intersexed infants are treated with surgery and/or hormonal therapy. The rationale underlying medical intervention is the belief that the condition "is a tragic event which immediately conjures up visions of a hopeless psychological misfit doomed to live always as a sexual freak in loneliness and frustration" (Dewhurst and Gordon 1993:A15).

Adding a third category to include the intersexed would not do justice to the complexities of biological sex. French endocrinologists Paul Guinet and Jacques Descourt estimate that on the basis of variations in the appearance of external genitalia alone, the category "true hermaphrodite" may contain as many as ninety-eight subcategories (Fausto-Sterling 1993). This variation tells us that "No classification scheme could [do] more than suggest the variety of sexual anatomy encountered in clinical practice" (p. A15).

The picture becomes even more complicated when we consider that a person's primary sex characteristics may not match the sex chromosomes. Theoretically one's sex is determined by two chromosomes: X (female) and Y (male). Each parent contributes a sex chromosome: the mother contributes an X chromosome and the father an X or a Y depending on which one is carried by the sperm that fertilizes the egg. If this chromosome is a Y, then the baby will be a male. Al-

Sex A biological concept determined on the basis of primary sex characteristics.

Primary sex characteristics The anatomical traits essential to reproduction.

Intersexed A broad term used by the medical profession to classify people with some mixture of male and female biological characteristics.

[1]The intersexed group includes three very broad categories: true hermaphrodites, persons who possess one ovary and one testis; male pseudohermaphrodites, persons who possess testes and no ovaries, but some elements of female genitalia; and female pseudohermaphrodites, persons who have ovaries and no testes but some elements of male genitalia.

though we cannot possibly know how many people's sex chromosomes do not match their anatomy, the results of mandatory "sex tests" of female athletes over the past twenty-five years have shown us that such cases exist and that a few women are disqualified from each Olympic competition and other major international competitions because they "fail" the tests (Grady 1992).[2]

Perhaps the most highly publicized after-the-fact case is that of Spanish hurdler Maria José Martinez Patino, who although "clearly a female anatomically, is, at a genetic level, just as clearly a man" (Lemonick 1992:65). Upon giving her the test results, track officials advised her to warm up for the race but to fake an injury so as not to draw the media's attention to her situation (Grady 1992). Patino lost her right to compete in amateur and Olympic events but subsequently spent three years challenging the decision. The International Amateur Athletic Federation (IAAF) restored her status after deciding that her X and Y chromosomes gave her no advantage over female competitors with two X chromosomes (Kolata 1992; Lemonick 1992).[3]

In addition to primary sex characteristics and chromosomal sex, **secondary sex characteristics** are used to distinguish one sex from another. These are physical traits not essential to reproduction (breast development, quality of voice, distribution of facial and body hair, and skeletal form) that result from the action of so-called male (androgen) and female (estrogen) hormones. I use the term "so-called" because, although testes produce androgen and ovaries estrogen, the adrenal cortex produces androgen and estrogen in both sexes (Garb 1991). Like primary sex characteristics, none of these physical traits has any clear dividing lines to separate males from females. For example, biological females have the potential for the same hair distribution as biological males — follicles for a mustache, a beard, and body hair. Moreover, females produce not only estrogen but also androgen, a steroid hormone that triggers hair growth.

Given this information, we must ask, Why is it that women seem to have different patterns of facial and body hair growth than men? Before we answer this question, we must consider the interrelationship between sex in the physical and reproductive sense and the concept of gender.

Gender as a Social Construct

Whereas sex is a biological distinction, **gender** is a social distinction based on culturally conceived and learned ideas about appropriate appearance, behavior, and mental or emotional characteristics for males and females (Tierney 1991). The terms **masculinity** and **femininity** signify the physical, behavioral, and mental or emotional traits believed to be characteristic of males and females, respectively (Morawski 1991).

To grasp the distinction between sex and gender, we must note that no fixed line separates maleness from femaleness. The painter Paul Gauguin pointed out this ambiguity in his observations about Maori men and women, which he recorded in a journal that he kept while painting in Tahiti in 1891. These observations are influenced by the norms regarding femininity around the turn of the century:

> Among peoples that go naked, as among animals, the difference between the sexes is less accentuated than in our climates. Thanks to our cinctures and corsets we have succeeded in making an artificial being out of woman. . . . We carefully keep her in a state of nervous weakness and muscular inferiority, and in guarding her from fatigue, we take away from her possibilities of development. Thus modeled on a bizarre ideal of slenderness . . . our women have nothing in common with us [men], and this, perhaps, may not be without grave moral and social disadvantages.

Secondary sex characteristics Physical traits not essential to reproduction (breast development, quality of voice, distribution of facial and body hair, and skeletal form) that result from the action of so-called male (androgen) and female (estrogen) hormones.

Gender A social distinction based on culturally conceived and learned ideas about appropriate appearance, behavior, and mental or emotional characteristics for males and females.

Masculinity The physical, behavioral, and mental or emotional traits believed to be characteristic of males.

Femininity The physical, behavioral, and mental or emotional traits believed to be characteristic of females.

[2]There has been only one case in Olympic history in which a man competed as a woman. In 1936, Nazi officials forced a German male athlete to enter the women's high jump competition. Three women jumped higher than he did (Grady 1992).

[3]Medical researchers hypothesize that a mutation in the Y chromosome in the case of anatomical females (who are genetically male) or a mutation in the X chromosome in the case of anatomical males (who are genetically female), respectively, suppresses or fails to suppress the production of excess testosterone.

At the turn of the twentieth century the artist Paul Gauguin observed that among Maori men and women the differences between the sexes was less accentuated such that there was something virile in the women and something feminine in the men.

Paul Gauguin, *Tahitian Women Inside a Hut.* The Hermitage Museum, St. Petersburg, Russia. Photo © Scala/Art Resource, NY.

> On Tahiti, the breezes from forest and sea strengthen the lungs, they broaden the shoulders and hips. Neither men nor women are sheltered from the rays of the sun nor the pebbles of the sea-shore. Together they engage in the same tasks with the same activity. . . . There is something virile in the women and something feminine in the men. ([1919] 1985:19–20)

Often we attribute differences between males and females to biology, when in fact they are more likely to be socially created. In the United States, for example, norms specify the amount and distribution of facial and body hair appropriate for females: it is acceptable for women to have eyelashes, well-shaped eyebrows, and a well-defined triangle of pubic hair but not to have hair above their lips, under their arms, on their inner thighs (outside the bikini line), or on their chin, shoulders, back, chest, breasts, abdomen, legs, or toes. Most men, and even women, do not realize that women work to achieve these cultural standards and that their compliance makes males and females appear more physically distinct on this trait than they are in reality. We lose sight of the fact that significant but perfectly normal biological events — puberty, pregnancy, menopause, stress — contribute to the balance between two hormones, androgen and estrogen. Changes in the proportions of these hormones trigger hair growth that departs from societal norms about the appropriate amount and texture of hair for females. When women grow hair as a result of these events, they tend to think something is wrong with them instead of seeing it as natural. A "female balance" between androgen and estrogen is one in which a woman's hair is consistent with these norms.[4]

The extreme measures taken by some women to eliminate facial and body hair are reflected in reports of physicians who intervened after women were harmed by commercial treatments. For example, X-ray treatments, popular between 1899 and 1940, caused ad-

[4]Although "excess" hair sometimes warns a doctor to check for the presence of an underlying endocrine disorder such as an ovarian tumor, 99 percent of the time such hair is not associated with any pathology.

Jennifer Miller, a bearded lady for the Coney Island Sideshow, decided when she was 20 years old that shaving was about shame and hiding and that the only reason to shave would be to conform to an ideal standard of femininity.

© Deanne Fitzmaurice/San Francisco Chronicle

verse side effects including wrinkling, scarring, discoloration, and cancerous growths.[5] A depilatory containing thallium acetate, a highly toxic substance that was popular in the 1960s, caused paralysis (Ferrante 1988).

Just as women strive to meet norms for facial and body hair, they work to achieve the ideal standards of feminine beauty as portrayed in such places as magazines and television. In *How We Survived Communism and Even Laughed*, Drakulić (1992) powerfully describes the profound effect of magazines such as *Vogue* on the thinking of people who read them, although these magazines were rarely available in most Eastern European countries before the end of the Cold War. Just holding the magazine was "almost like holding a pebble from Mars or a piece of a meteor that accidentally fell into your yard" (p. 27). One Budapest woman, an editor of a scientific journal, told Drakulić in an interview that *Vogue* magazine "makes me feel so miserable I could almost cry. Just look at this paper—glossy, shiny, like silk. You can't find anything like this around here. Once you've seen it, it immediately sets not only new standards, but a visible boundary. Sometimes I think that the real Iron Curtain[6] is made of silky, shiny images of pretty women dressed in wonderful clothes, of pictures from women's magazines" (p. 27).

Drakulić observes that even though women in the West are bombarded on all sides with these images, they still notice them. She maintains that the average Western woman "still feels a slight mixture of envy, frustration, jealousy, and desire while watching this world of images" (p. 28) and proceeds to buy the products that promise that image. For the average Eastern European woman, who studied "every detail with the interest of those who had no other source of information about the outside world" (p. 28), the images created hatred for the reality that surrounded them. Although Eastern European women are affected by the ideal image of the women "imported" from the West, gender is not considered a pressing social issue there.

The Question of Gender in the Former Yugoslavia

One reason that gender is not an issue in the former Yugoslavia in the same sense as in the United States may be that there is virtually no feminist voice there. In the broadest sense of the word, a **feminist** is a man or woman who actively opposes gender scripts (learned patterns of behavior expected of males and females) and believes that men's and women's self-image, aspirations, and life chances should not be constrained by those scripts (Bem 1993). For example, a man should be free to choose to stay home and take care of the chil-

Feminist A man or woman who actively opposes gender scripts (learned patterns of behavior expected of males and females) and believes that men's and women's self-image, aspirations, and life chances should not be constrained by those scripts.

[5]The severity of damage caused by X-ray treatments is illustrated by the case of Mrs. A. E. C., one of many reported in the *Journal of the American Medical Association:*

> Mrs. A. E. C., age thirty-two, with a history of treatment five years ago by the Tricho [X-ray] system in Boston. She had fifteen treatments on her chin for [superfluous hair]. She was treated every two weeks, and three areas on the underside of her chin were treated each time. Twice she was treated by the bookkeeper when the nurse was not in. Three years ago red blotches began to appear and they have continued. She shows definite signs of telangiectasis [a chronic dilation of blood vessels causing dark red blotches] and atrophy over an area of about three inches in diameter just below the sides of the chin. (American Medical Association 1929:286)

[6]The "Iron Curtain" is a term coined by British Prime Minister Winston Churchill to describe the division between Communist Eastern Europe and non-Communist Western Europe after World War II.

Croatian writer Slavenka Drakulić helped found Yugoslavia's first feminist group in 1979.

©Filip Horvat/Saba

dren rather than pursuing a full-time career; a female athlete should be able to develop her physique beyond what is considered feminine. Unfortunately for many people, the term *feminist* evokes extremely negative images and stereotypes of mannish-looking women who hate men and who find vocations such as mother and wife oppressive and unrewarding. In the former Yugoslavia, even more than in the United States, very few men or women will declare themselves feminists, even if they live according to the basic feminist principle (Drakulić 1993b).

In the former Yugoslavia (and throughout Eastern Europe, for that matter), only a handful of feminists talk and write about gender issues. The unofficial feminist organizations that exist are concentrated in three cities—Belgrade, Zagreb, and Ljubljana—and they are "small and without money or institutional support" (Drakulić 1993b:127; see also Ramet 1991). Colleges and universities in the former Yugoslavia offer no women's studies or gender studies programs. Not until 1979 did Slavenka Drakulić (the best-known feminist of her country) and about thirty other persons, including some male journalists and intellectuals, found Yugoslavia's first feminist group (Drakulić 1993b; Kinzer 1993), and feminism still has not gained strength as a movement. In fact, Drakulić's book *The Deadly Sins of Feminism,* the first feminist book produced in Yugoslavia, was not published until 1984 (Drakulić 1990).

By contrast, feminist literature in the United States has a relatively long history. The feminist scholarship that has been published by men and women, especially since 1960, is vast and wide-ranging (Komarovsky 1991). Drakulić is quite popular in the United States; she has appeared on talk shows, and her two books of essays, *How We Survived Communism and Even Laughed* and *The Balkan Express,* sold quite well. They are not available in Croatia, where the most vocal critics consider her a "cheap and defective" writer who took up feminism as a way to "rape Croatia" (Kinzer 1993:4Y). In some newspaper editorials, Drakulić and other feminists have been labeled "witches" (Kirka 1993).

Another reason that organized feminism fails to thrive in Yugoslavia, and in Eastern Europe in general, is that the Communist system as implemented has failed to meet people's basic needs and has made life hard for everyone. Thus, it is difficult to view men as advantaged or as able to achieve their aspirations: "It's hard to see them as an opposite force, men as a gender . . . perhaps because everyone's identity is denied" (Drakulić 1992:109). In comparison with the former Yugoslavia, the United States has an impressive ratio of people to resources, abundant job opportunities, and a wide choice of services and products. Because opportunities for mobility are relatively plentiful in the United States, mobility and the factors that enhance or prohibit mobility are an important question here.

Finally, there is almost no feminist voice in the former Yugoslavia because the official position is that "the woman question" ceased to exist after World War II. The end of that war also marked the end of the Yugoslavian National Liberation War against Fascist Germany. During this war, the Yugoslav Communist Party formed and coordinated the activities of the Anti-Fascist Women's Front of Yugoslavia, which supported male-dominated resistance groups. This organization "brought together some 2,000,000 women during the war; approximately 100,000 took part in regular partisan military units, and of that number 25,000 were killed and 40,000 badly wounded" (Milic 1993:111). Many other women, through their veterans' and political activities, attained leading positions in the party and local administrative bodies, or became national heroes.

Many women who participated in the national struggle for freedom found it a liberating experience. It gave them opportunities to experience something beyond the closed world of the traditional patriarchal family, which limited their life choices to marriage or domestic service. After the war they could not return to this world.

Table 11.1 Characteristics of Selected Nations and Nationalities* within the former Yugoslavia

Characteristic	Albanians	Croats	Hungarians	Macedonians
Sex ratio (no. of women per 100 men)	109	93	90	102
Number of births per 1,000	36.4	16.2	12.4	18.5
Highly educated (% total population)	1.2	3.3	1.7	3.7
Highly educated men (%)	1.8	4.4	2.2	4.9
Highly educated women (%)	0.5	2.3	1.2	2.5
Women's share of highly educated population (%)	21.4	36.6	38.7	33.7
Labor force participation (%)				
Women	15.3	40.7	35.0	48.5
Men	66.8	66.6	37.9	70.2
Directors (%)				
Women	0.3	0.5	0.4	0.4
Men	10.9	2.5	1.6	2.8
Women's share of directorial positions	4.2	12.6	13.0	9.5
Characteristic	**Montenegrins**	**Muslims**	**Serbs**	**Slovenes**
Sex ratio	102	101	98	92
Number of births per 1,000	17.6	23.1	15.4	15.7
Highly educated (% total population)	5.0	1.6	3.1	3.4
Highly educated men (%)	8.5	2.3	4.2	4.7
Highly educated women (%)	2.1	0.9	2.0	2.3
Women's share of highly educated population (%)	23.4	29.0	33.5	35.3
Labor force participation (%)				
Women	35.8	25.9	47.7	58.8
Men	65.7	69.3	73.2	69.6
Directors (%)				
Women	0.7	0.4	0.5	1.0
Men	4.5	1.6	2.2	3.5
Women's share of directorial positions	7.9	9.1	11.6	20.7

*Data based on 1981 census, the last year for which census data are available.

Source: Adapted from Darville and Reeves (1992:281).

When a new Communist government was formed after the war, its leaders made women's rights a part of official doctrine, but with a twist. Government leaders took the position that, because women's liberation was achieved through the pursuit of a higher goal — national liberation — their equality would come with the pursuit of other more important goals, such as class struggle and economic stability. This assumption underlay all efforts to address "the woman question." The Communists passed legislation that gave women the right to vote, equal pay for equal work, access to the salaried labor force, publicly funded child care, the right to abortion, and paid maternity leave, among other things. The Communist Party set a 30 percent quota to encourage women to participate in the political arena.

In spite of this kind of legislation, however, "women remained subordinated and segregated in all walks of life" (Milic 1993:111). Drakulić argues, for example, that, despite the quota for political participation, women never achieved an independent, secure voice in the political sphere (Drakulić 1993b; see Table 11.1). As evidence of this claim, she points to the fact that after the 1990 elections, the proportion of women in government declined from 30 to somewhere between 5 and 10 percent (depending on the republic).

Critics argue that the overall economic and social gains made under Communism have eroded steadily since Tito's death in 1980. The gains allowed Yugoslavia to be the most liberal and economically well-off Communist country. They gave people the feeling that the Communist plan was working. The gains were

The Breakup of Yugoslavia

Here is some background information on the breakup of Yugoslavia. As you read, consider why the various political factions would be concerned about the "kinds" of babies born.

A woman in Sarajevo visits the graves of loved ones during a cease-fire.

©Chris Rainier

One can argue that the breakup of Yugoslavia in the early 1990s actually began with the death of Marshal Josip Tito, Yugoslavia's "President-for-Life," in 1980. Tito established a Communist government in 1945 after uniting various ethnic factions (Serbs, Croats, Muslims, and others) who were at war with each other to defeat the German army occupying the country at that time.[1]

While Tito was in power, he used a careful, if often brutal, balance of reward and repression to realize his ideal of ethnic tolerance in multicultural society (C. Williams 1993). People who claimed to be Serb, Croat, or other ethnic category first and Yugoslav second risked arrest for being a nationalist (Ignatieff 1993). In addition, Tito distributed political power according to ethnicity so that each ethnic group dominated a specific republic or autonomous region.

Under this arrangement, the Serbs, Croats, Slovenes, Macedonians, Montenegrins, and Muslims were considered nations. It is worth noting that Tito gave Muslims nation status in the late 1970s to put to rest the Serb-Croat conflict over whether the Muslims were Serbian or Croatian. The answer would have allowed either Serbs or Croats to achieve majority status in Bosnia (Curtis 1992).

Tito's death created a leadership vacuum that proved particularly problematic in 1989, the year in which the Berlin Wall was dismantled. By that time, Yugoslavia's economy had deteriorated to the point of collapse.[2] With the end of the Cold War and the overthrow of Communist governments in Eastern Europe, Yugoslavia was no longer able to play the Soviet and American blocs against each other to gain economic aid (Borden 1992). This economic and political atmosphere intensified the forces of fragmentation within Yugoslavia.

In 1990 Slovenia, Bosnia, Macedonia, and Croatia elected non-Communist leaders who favored market reforms. Serbia and Montenegro, on the other hand, elected Communist leaders by decisive margins. Particularly interesting was Serbia's election of Slobodan Milosevic, a fervent nationalist who claimed the Yugoslav government had shortchanged Serbian interests ever since Tito came to power (Borden 1992). He used the government-controlled media to advocate a policy of uniting all of the Serbs of Yugoslavia into one state. Before and after his 1990 election, Milosevic focused his attention on Kosovo, an autonomous region within Serbia in which 92 percent of the population were Albanian and 8 to 10 percent were Serbian. Milosevic sent the national army into the region to prevent Kosovo from declaring its independence from Serbia and subsequently abolished its autonomous status.

Soon after this event, Slovenia and Croatia, the two wealthiest republics, declared their independence. Because the Serbian members dominated the votes in the collective federal presidency, Milosevic was able to use the Yugoslavian People's Army—minus the Croats and Slovenes, who had been dismissed—to forcibly preserve Yugoslavia's territorial integrity. The ensuing war with Slovenia lasted about ten days. The war with Croatia lasted approximately six months, until a U.N.–monitored cease-fire was declared. By that time the Serbian-controlled armed forces had managed to take control of 25 percent of Croatia's territory.

In Bosnia, the multiethnic parliament, not wishing to be part of the Serbian-dominated "new" Yugoslavia, declared its sovereignty. A referendum was placed before the Bosnian people and passed with 99 percent of the vote in favor. A significant number of Bosnian Serbs and Croats, however, boycotted the election. The Serbs threatened to secede to become part of Serbia if the Bosnian government declared independence, while the Croats (17 percent of the population) warned that they would secede to become part of Croatia if Bosnia did *not* declare independence.

After the referendum passed, the European Community and the United

States recognized Bosnia as a new state. Serbs opposed to independence took control of roads leading into Sarajevo, Bosnia's capital and a 600-year-old city with an ethnically and culturally mixed population of 600,000, making it impossible for anyone to enter or leave. The so-called Yugoslav national army entered Bosnia to take control of the cities and towns that border Serbia. Until late 1995, Serbian armed forces surrounded Sarajevo, slowly strangling the city with the goal of making Sarajevo part of Greater Serbia. Despite the relentless attack by Serbian forces, approximately 90,000 ethnic Serbs remained in Sarajevo with Muslims, "Yugoslavs,"[3] Croatians, and other ethnic groups who refused to be part of the nationalist policies (Glenny 1992). The Serb leadership made a concerted effort to kill as many educated Muslims as possible so that in the event that an independent Bosnia and Herzegovina remained when the war ended, it would be bereft of people who could make it work (Rieff 1992).

At first Croatians and Bosnians (including some Bosnian Serbs) fought as allies against the Serbian-dominated armies. The Bosnia-Croatia alliance collapsed, however, as nationalistic Croats from both Croatia and Bosnia sought to claim their share of Bosnian territory. Nationalistic Bosnian Serbs and Croats declared Serbian and Croatian states within Bosnia. Each boasted military units supported with arms, money, and recruits from Serbia and Croatia, respectively. Between them, the Serbs and Croats seized or surrounded so much of Bosnia's territory that the Bosnian Muslims were rendered practically stateless. Bosnia, a new country with no army, weapons, or military traditions, was no match for the Croatian and Serbian forces. The Bosnian army—which is predominantly Muslim but contains significant numbers of Croats, Serbs, and other ethnic groups opposed to the formation of ethnic states—was particularly affected by a U.N.–mandated arms embargo to the region. The U.N. did impose widespread internationally backed sanctions against the new Yugoslavia (Serbia and Montenegro) to stop the flow of arms to Bosnian Serbs. On August 14, 1994, Yugoslavia stopped its support of the Bosnian Serbs, and the U.N. eased sanctions. On November 21, 1995, Croat, Bosnian, and Serb leaders met in Dayton, Ohio, and initialed a peace agreement that was later signed in Paris on December 14.

Ethnic Composition of the Three Largest Republics of the Former Yugoslavia

Republic	Number	Percentage
Bosnia-Herzegovina	4,365,639	—
Muslims	1,900,000	44
Serbs	1,450,000	33
Croats	750,000	17
Yugoslavs	250,000	6
Others	15,639	0.4
Croatia	4,703,941	
Croats	3,500,000	74
Serbs	700,000	15
Yugoslavs	400,000	9
Others	103,941	2
Serbia	9,721,177	
Serbia Proper	5,753,825	
Serbs	5,500,000	96
Muslims	125,000	2.2
Gypsies	50,000	0.9
Croats	40,000	0.7
Others	38,825	0.6
Kosovo	1,954,747	
Albanians	1,630,000	83
Serbs	250,000	12.8
Muslims	40,000	2.0
Gypsies	30,000	1.5
Others	4,747	0.2
Vojvodina	2,012,605	
Serbs	1,400,000	70
Hungarians	450,000	22
Croats	100,000	5
Romanians	50,000	2.5
Others	12,605	0.6

From *Yugoslav Survey* 32 (March 1990–91), 5. The numbers for republics and autonomous provinces are actual; the numbers for the populations groups are estimates.

[1]In 1974 Tito instituted a government structure to succeed him after his death. Executive power was to be held by a committee with one representative from each of the six republics and the two autonomous regions within Serbia. The chairmanship was to rotate among the members. This created a power vacuum, however, because no one person was in charge, and no one was able to enforce Tito's ideal of ethnic tolerance in a multicultural society.

[2]At that time, at least 1 million people were unemployed, and one-third of the workforce lived "below poverty." Inflation was at four figures, and the standard of living had declined to the level of thirty years earlier.

[3]"Yugoslavs" are people who claim no national (ethnic) identification. That is, instead of identifying themselves as Croat, Muslim, Serb, and so on, they identify themselves with the country.

possible because post–World War II Yugoslavia was unique among Eastern European countries. It resisted alignment with the Soviet Union and maintained friendly, if distant, relations with the United States. Yugoslavia was unique in that the Communist government permitted its people to travel and work outside the country.[7] As a result, Yugoslavia benefited from money that its people earned as guest workers (in countries such as West Germany and Austria) and sent home. Yugoslavia's economy also depended heavily on tourism, a service industry sensitive to seasonal fluctuations and political and economic events (Curtis 1992).[8]

Yugoslavia's prosperity — built on remittances from Yugoslav guest workers working abroad, tourism, and loans from the International Monetary Fund — began its decline in 1979 when the world recession forced many guest workers to return home. The accompanying declines in living standards, together with inflation and unemployment, left some groups more vulnerable than others, making clear who benefited most under the Communist system. The gains made by women under Tito were especially vulnerable, as evidenced by the disappearance of women from political life. In addition, newly elected leaders in all of the former republics are conservative. In the atmosphere of rapid change and severe economic upheaval connected with the fall of Communism, severe economic hardship, and the wars, gender issues still are considered irrelevant to the larger national concerns that must be solved first.

In light of this history, very little literature exists on gender issues in Yugoslavia, or in Eastern Europe, for that matter. The first anthology of essays devoted to the status of women in Eastern Europe was just published in the United States in 1993 (Katzarova 1993).

To this point, we have made a distinction between sex and gender, and we have explored the extent to which gender is an important social issue in the United States and not in the former Yugoslavia. Although sociologists acknowledge that there are no clear biological markers to distinguish males from females, they would not argue that biological differences do not exist. Sociologists, however, are interested in the extent to which differences are socially induced. To put it another way, they are interested in the actions men and women take to accentuate differences between them. As we see in the next section, these actions lead to gender polarization.

Gender polarization "The organizing of social life around the male-female distinction," so that people's sex is connected to "virtually every other aspect of human experience, including modes of dress, social roles, and even ways of expressing emotion and experiencing sexual desire" (Bem 1993:192).

Gender Polarization

In *The Lenses of Gender,* Sandra Lipsitz Bem (1993) defines **gender polarization** as "the organizing of social life around the male-female distinction," so that people's sex is connected to "virtually every other aspect of human experience, including modes of dress, social roles, and even ways of expressing emotion and experiencing sexual desire" (p. 192). To understand how just about every aspect of life is organized around this distinction, we consider research by Alice Baumgartner-Papageorgiou.

In a paper published by the Institute for Equality in Education, Baumgartner-Papageorgiou (1982) summarizes the results of a study of elementary and high school students, in which she asked the students how their lives would be different if they were members of the opposite sex. Their responses reflect culturally conceived and learned ideas about sex-appropriate behaviors and appearances. The boys generally believed that as girls their lives would change in negative ways. Among other things, they would become less active and more restricted in what they could do. In addition,

[7]Slavenka Drakulić (1993a) describes how the people of Yugoslavia were caught between Eastern and Western Europe:

> People in the West always tend to forget one key thing about Yugoslavia, that we had something that made us different from the citizens of the Eastern bloc: we had a passport, the possibility to travel. And we had enough surplus money with no opportunity to invest in the economy (which was why everyone who could invested in building weekend houses in the mid-sixties) and no outlet but to exchange it on the black market for hard currency and then go shopping. Yes, shopping to the nearest cities in Austria or Italy. We bought everything — clothes, shoes, cosmetics, sweets, coffee, even fruit and toilet paper. I remember times when my mother who lives in a city only a short drive from Trieste would go there every week to get in stores what she couldn't get here. Millions and millions of people crossed the border every year just to savor the West and to buy something, perhaps as a mere gesture. But this freedom, a feeling that you are free to go if you want to, was very important to us. It seems to me now to have been a kind of a contract with the regime: we realize you are here forever, we don't like you at all but we'll compromise if you let us be, if you don't press too hard. (p. 135)

[8]In 1988, 9 million tourists visited Yugoslavia. Two republics — Slovenia and Croatia — benefited the most from guest worker programs and tourism. The government transferred about 25 percent of the two republics' earnings to other less prosperous republics.

they would become more conscious about tending to their appearance, finding a husband, and being alone and unprotected in the face of a violent attack:

- I would start to look for a husband as soon as I got into high school.
- I would play girl games and not have many things to do during the day.
- I'd use a lot of make-up and look good and beautiful. . . . I'd have to shave my whole body.
- I'd have to know how to handle drunk guys and rapists.
- I couldn't have a pocket knife.
- I would not be able to help my dad fix the car and truck and his two motorcycles. (pp. 2–9)

The girls, on the other hand, believed that if they were boys they would be less emotional, their lives would be more active and less restrictive, they would be closer to their fathers, and they would be treated as more than sex objects:

- I would have to stay calm and cool whenever something happened.
- [I could sleep later in the mornings] since it would not take [me] very long to get ready for school.
- My father would be closer because I'd be the son he always wanted.
- I would not have to worry about being raped.
- People would take my decisions and beliefs more seriously. (pp. 5–13)

These beliefs about how the character of one's life depends on one's sex seem to hold even among the college students enrolled in my introductory sociology classes. In the fall 1993 semester, I asked students to take a few minutes to write about how their lives would change as members of the other sex. The men in the class believed they would be more emotional and more conscious of their physical appearance and that their career options would narrow considerably. Here are some of their responses:

- I would be much more sensitive to others' needs and what I'm expected to do.
- I wouldn't always have to appear like I am in control of every situation. I would be comforted instead of always being the comforter.
- People would put me down for the way I look.
- I would be more emotional.
- I would worry more about losing weight instead of trying to gain weight.
- I probably wouldn't really feel any different, but people would see me as a female and respond accordingly. If I stayed in the construction program, I would have to fight the belief that men are the only real construction workers.
- My career options would narrow. Now I have many career paths to choose from, but as a woman I would have fewer.
- I would have to be conscious of the way I sit.

Notice that the first two responses suggest that the "feminine" traits would in some ways be a plus (being "more sensitive to others' needs" and being "comforted instead of always being the comforter"). It is important to realize that men as well as women can feel constrained by their gender roles.

The women in the class believed that as men they would have to worry about asking women out and about whether their major was appropriate. They also believed, however, that they would make more money, be less emotional, and be taken more seriously. Here are some of their responses:

- I would worry about whether a woman would say "yes" if I asked her out.
- I would earn more money than my female counterpart in my chosen profession.
- People would take me more seriously and not attribute my emotions to PMS.
- My dad would expect me to be an athlete.
- I'd have to remain cool when under stress and not show my emotions.
- I think that I would change my major from "undecided" to a major in construction technology.

These comments by high school and college students show the extent to which life is organized around male-female distinctions. They also show that students' decisions about how early to get up in the morning, what subjects to study, whether to show emotion, how to sit, and whether to encourage a child's athletic development are gender-schematic decisions. Decisions and viewpoints about any aspect of life are **gender-schematic** if they are influenced by a society's polarized definitions of masculinity and femininity rather than

Gender-schematic A term describing decisions that are influenced by a society's polarized definitions of masculinity and femininity rather than on the basis of other criteria such as self-fulfillment, interest, ability, or personal comfort.

Table 11.2 Selected Gender-Related Statistics for the United States and Yugoslavia

	United States	Yugoslavia
Ratio of females to males 60 years and older	138/100	141/100
Life expectancy (female to male)	+7.1 years	+5.9 years
Average age at first marriage		
Males	25.2	26.2
Females	25.3	22.2
Ratio of females to males in selected occupational groups		
Administrative/managerial	61/100	15/100
Clerical services	183/100	138/100
Production/transportation	23/100	23/100
Agricultural	19/100	88/100
Adults who smoke (%)		
Males	30	57
Females	24	10

Adapted from United Nations (1991:22, 26, 67, 104).

on the basis of other criteria such as self-fulfillment, interest, ability, or personal comfort. For example, college students make gender-schematic decisions about possible majors if they ask, even subconsciously, what is the "sex" of the major and, if it matches their own sex, consider it a viable option or, if it does not match, reject it outright (Bem 1993; see "Gender Schemes in Educational Choices").

Even sexual desire between men and women is organized around male-female characteristics unrelated to reproduction. Bem (1993) argues that

> neither women nor men in American society tend to like heterosexual relationships in which the woman is bigger, taller, stronger, older, smarter, higher in status, more experienced, more educated, more talented, more confident, or more highly paid than the man, they do tend to like heterosexual relationships in which the man is bigger, taller, stronger, and so forth, than the woman. (p. 163).

The negative consequences of channeling sexual desire according to age differences is evident when we consider that the average woman outlives her spouse by about nine years in the United States and by about seven years in the former Yugoslavia. In the United States, the average life expectancy for women is seven years longer than that of men; in the former Yugoslavia, the average life expectancy for women before the war was six years longer than that of men. (The difference in life expectancy can be explained in part by the fact that men tend to do the most hazardous jobs in society.) In both countries, men tend to marry younger women (see Table 11.2 and Figure 11.1). This practice, in combination with differences in life expectancy, means that women can expect to live a significant portion of their lives as widows.

Social emotions Internal bodily sensations experienced in relationships with other people.

Not only is sexual desire between men and women influenced strongly by gender-polarized ideas, but emotions toward persons of the same sex are also influenced. In Chapter 4, we learned that **social emotions** are internal bodily sensations that we experience in relationships with other people and that feeling rules are norms specifying appropriate ways to express those sensations. When I asked students in my class to comment on social emotions or "internal bodily sensations" that they had felt and expressed toward someone of the same sex, most indicated that other people made them feel uncomfortable and defensive about such feelings (see "Expressing Affection Toward Same-Sex Friends").

A society's feeling rules are so powerful that they even affect how people solve problems. For example, when human evolutionists discovered petrified footprints believed to be 3.5 million years old, they inferred that the prints belonged to a man and a woman, not to two women, two men, or an adult and a child. Ian Tattersall (1993), curator of the American Museum of Natural History, explains the logic underlying this conclusion:

> We know that [the people who left the footprints] were walking side by side because even though the

U.S. In Perspective

Gender Schemes in Educational Choices

This chart shows the number of U.S. degrees in each category and the percentage of degrees in each category awarded to males and females. Note that 55 percent of degrees in all fields were awarded to women. If gender had no bearing on choice of major, 55 percent of degrees in each category would be awarded to women. Which fields are most disproportionately female? Disproportionately male? Why do you think 10 percent more women than men received a degree in 1993–1994?

Earned Degrees Conferred, 1993–1994

		Bachelor's Degree	
	Men (%)	**Women (%)**	**Total**
Agriculture and natural resources	65	35	18,070
Architecture and related programs	64	36	8,975
Area, ethnic, and cultural studies	35	65	5,573
Biological/life sciences	49	51	51,383
Business management	52	49	246,654
Communications	41	59	51,164
Communications technologies	55	45	663
Computer and information sciences	71	29	24,200
Education	22	78	107,600
Engineering	84	16	62,220
Engineering-related technologies	91	9	16,005
English language and literature	34	66	53,924
Foreign languages and literatures	30	70	14,378
Health professions	18	82	74,421
Home economics	12	88	15,522
Law and legal studies	52	48	2,171
Liberal/general studies	65	35	33,397
Library science	8	92	62
Mathematics	54	46	14,396
Multi-interdisciplinary studies	36	64	25,167
Parks, recreation, & leisure studies	51	49	11,470
Philosophy and religion	65	35	7,546
Physical sciences	65	35	18,400
Precision production trades	73	27	420
Protective services	62	38	23,009
Psychology	26	74	69,259
Public administration and services	22	88	17,815
ROTC and military technologies	84	16	19
Social sciences and history	54	46	133,680
Theological studies	76	24	5,434
Transportation and material moving	89	11	3,923
Visual and performing arts	40	60	49,053
Other and unclassified by field	49	51	3,302
All fields	45	55	1,169,275

Source: U.S. Department of Education (1995).

Figure 11.1 Projected Numbers of Males and Females by Age Cohort for the Year 2000, Croatia and the United States

Notice that in the United States and Croatia, there are more males between the ages of zero to four. For the category fifty-five to fifty-nine, however, there are more females than males, and the differences between number of males and females continues to increase for each age category. Look at the graph for the United States. How does it compare?

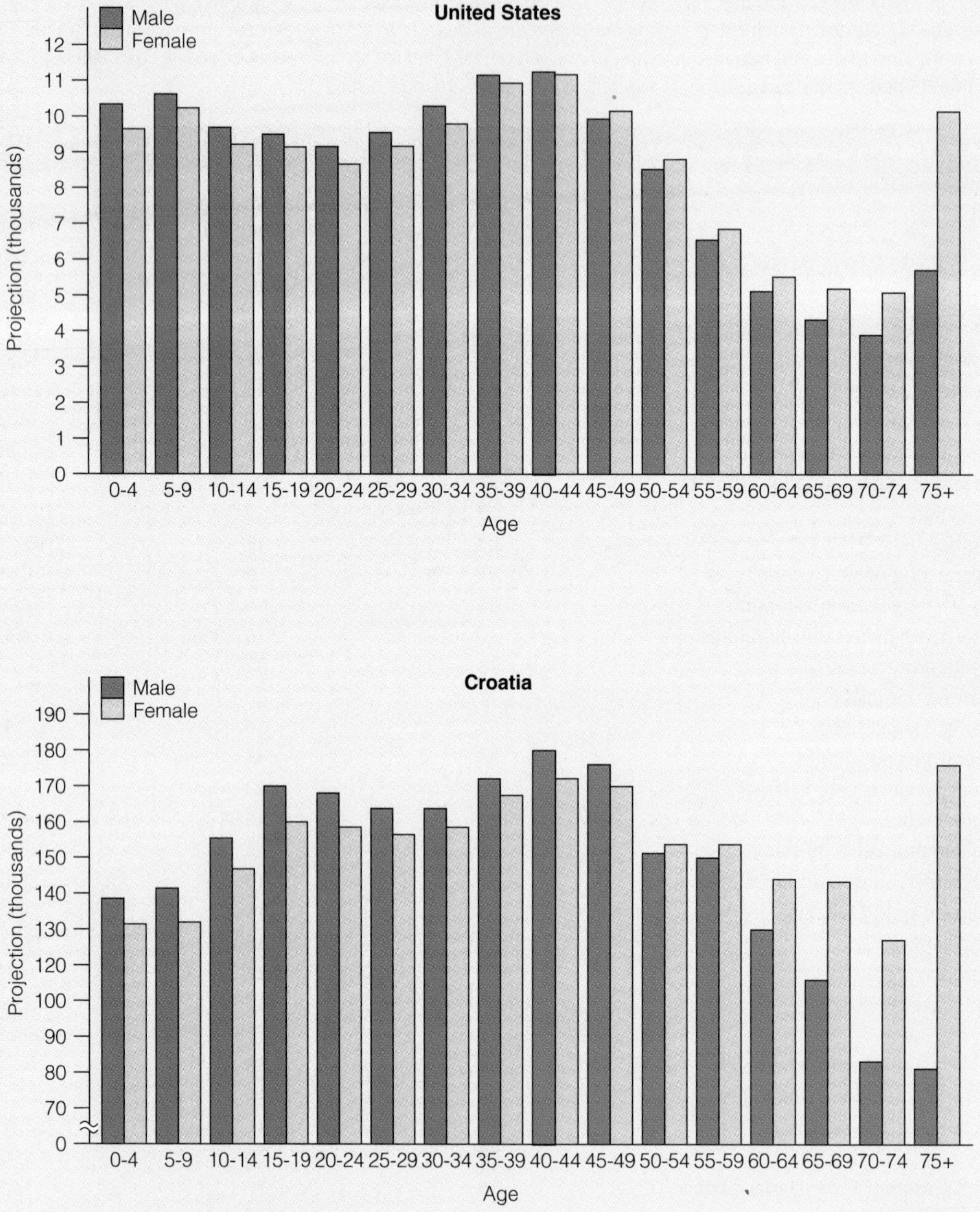

United States

Male
Female

Projection (thousands)

0 1 2 3 4 5 6 7 8 9 10 11 12

0-4 5-9 10-14 15-19 20-24 25-29 30-34 35-39 40-44 45-49 50-54 55-59 60-64 65-69 70-74 75+

Age

Croatia

Male
Female

Projection (thousands)

0 70 80 90 100 120 130 140 150 160 170 180 190

0-4 5-9 10-14 15-19 20-24 25-29 30-34 35-39 40-44 45-49 50-54 55-59 60-64 65-69 70-74 75+

Age

Source: U.S. Bureau of the Census (1991).

Expressing Affection Toward Same-Sex Friends

The comments of Introduction to Sociology students show that their relationships with same-sex friends, specifically expressing affection, are influenced by norms specifying appropriate ways to express such sensations.

I have noticed that some people struggle with how to show affection. I coached a sixth-grade boys volleyball team. On this team was a little boy who happened to be a very tactile kid. When he talked to me he would always grab my hand or arm and shake or swing it according to how intense he felt about what he was telling me. When people on the team did something good I think he wanted to give them a big hug, but instead he ended up slapping them around.

My best friend and I are really close. Sometimes when we haven't seen each other for a long time we run to each other and hug and kiss. Our boyfriends look at us like we're crazy. I explain to my boyfriend that she is like my sister and that is the way I love her, not the same way I love him. It's sad that two friends can't be close without others looking at them as if they were weird.

I don't know of any guys that I am good friends with that I would consider touching. For the most part, I think my friends would think I was gay if I did. It is all right for two guys to be friends but I can't picture two guys holding hands and not having some sexual feelings for each other. Because I am an athlete, people always ask me why do guys pat each other on the butt and my response is—"I don't know."

Feeling rules in the United States tend to restrict physical contact between members of the same sex, particularly men, to specific situations such as sports. Even though it is common to see male athletes hugging, many people would regard this gay couple's affectionate behavior as inappropriate.

This discussion about "feeling rules" has opened up ideas I've held most of my life. I have few friends, but feel very close to the ones I do have. I never hesitate to show affection, albeit sometimes in a subdued manner. Sometimes I can feel tension if I hug someone or pat them on their back, but that's usually for the first time. I feel affection is the easiest way to transfer positive energy. I enjoy giving and receiving affection.

Unfortunately, in our society people think that if two guys touch and they are not playing sports, then they're gay. I noticed some people's reactions in the class when you asked us about the way we express positive feelings toward someone of the same sex and (mostly guys) acted as if you had said something gross and immoral. In other cultures, men do hold hands, even in public; it's accepted and almost expected to hug and touch each other. A lot of people my age are so homophobic that they are afraid to express their true selves around same-sex friends. It's sick.

I don't know who I love more, my mom, dad, sister or my husband. I know my sister Marcia is my favorite person to be around, because she's most like me. In ideas, mannerisms, thinking, we just click. I have such strong feelings for her, I just want to show her how much I love her every time I see her. I hug and kiss her on the cheek when we part. But since my brother David asked if we were lesbians, she shies away from my affection.

There is an unwritten rule my friends and I follow: men shouldn't touch each other. We say, "If you're going to touch me—make it hurt."

Source: Introduction to Sociology class, fall 1993, Northern Kentucky University.

individuals are of different size, because their footprint sizes are different, their stride lengths are matched. They must have been walking together. And if they were walking together, the footprints are so close that they must have been in some kind of bodily contact with each other. What the nature of that contact was we don't know. We have chosen to put the arm of the male around the shoulder of the female. It's a bit anthropomorphic, but we couldn't think of a less, a more non-committal, if you want, kind of a gesture. (p. 13)

The point is that Tattersall could not imagine someone putting his or her arm on the shoulder of a same-sex person. In the United States physical contact between same-sex persons is reserved for specific situations. Men can give full-body hugs during a sports contest but cannot hold hands or put their hands on each other's shoulders while walking down the street. As of 1993, soldiers seen holding hands with someone of the same sex are subject to investigation (Lewin 1993). These norms against touching someone of the same sex are so powerful that some museum curators assume they existed 3.5 million years ago and construct exhibits that reflect such norms.

The information on gender and gender polarization suggests that one's sex has a profound effect on life chances—determining how long one can expect to live, the major subject one chooses to study in college, and whether one dates a shorter or taller person. A person's sex then is an important variable in determining his or her position in a society's system of social stratification.

Gender Stratification

Recall that in Chapter 9 we learned that social stratification is the system societies use to place people in categories. When sociologists study stratification, they examine how the category people are placed in affects their perceived social worth and their life chances. They are particularly interested in how persons who possess one category of a characteristic (male reproductive organs versus female reproductive organs) are regarded and treated as more valuable than persons who possess the other category. Sociologist Randall Collins (1971) offers a theory of sexual stratification to analyze this phenomenon.

Economic Arrangements

Randall Collins (1971) offers a theory of sexual stratification based on three assumptions: (1) people use their economic, political, physical, and other resources to dominate others; (2) any change in the way that resources are distributed in a society changes the structure of domination; and (3) ideology is used to justify one group's domination over another. In the case of males and females, males in general are physically stronger than females. Collins argues that, because of differences in strength between men and women, the potential for coercion by males exists in every encounter of males with females.[9] He maintains that the ideology of sexual property—which he defines as the "relatively permanent claim to exclusive sexual rights over a particular person" (p. 7)—is at the heart of sexual stratification and that for the most part women historically have been viewed and treated as men's sexual property.

Collins believes that the extent to which women are viewed as sexual property and subordinate to men historically has depended, and still depends, on two important and interdependent factors: (1) women's access to agents of violence control, such as the police, and (2) women's position relative to men in the labor market. On the basis of these factors, Collins identifies four historical economic arrangements: low-technology tribal societies, fortified households, private households, and advanced market economies.

The four economic arrangements that Collins identifies are ideal types; the reality is usually a mixture of two or more types. (Note that his theory does not account for the type of communistic self-management that characterizes Yugoslavia or Communist economies in general.) Each arrangement is characterized by distinct relationships between men and women. We begin with the first type, characteristic of low-technology tribal societies.

Low-technology tribal societies include hunting-and-gathering societies with technologies that do not permit the creation of surplus wealth, or wealth beyond what is needed to meet the basic needs (food and shelter). In such societies, sex-based division of labor is minimal because all members must contribute if the group is to survive. Some evidence, however, shows that in hunting-and-gathering societies, women perform more menial tasks and work longer hours than men. Men hunt large animals, for example, while women gather most of the food and hunt smaller animals. Because there is almost no surplus wealth, marriage between men and women from different families does little to increase a family's wealth or political power. Consequently, daughters are not treated as "property" in the sense that they are not used as bargaining chips to achieve such aims.

Fortified households include preindustrial arrangements in which there is no police force, militia,

Low-technology tribal societies Hunting-and-gathering societies with technologies that do not permit the creation of surplus wealth.

Fortified households Preindustrial arrangements in which there is no police force, militia, national guard, or other peacekeeping organization; the household is an armed unit, with the head of the household its military commander.

[9]Please note the word *potential*, which means "possible." Strength differences give men the potential to control women with physical force. Obviously most men do not take advantage of this potential.

national guard, or other peacekeeping organization. Therefore, the household is an armed unit, and the head of the household is its military commander. Fortified households "may vary considerably in size, wealth, and power, from the court of a king or great lord . . . down to households of minor artisans and peasants" (Collins 1971:11). All fortified households, however, have a common characteristic: the presence of a nonhouseholder class consisting of propertyless laborers and servants. In the fortified household, "the honored male is he who is dominant over others, who protects and controls his own property, and who can conquer others' property" (p. 12). Men treat women as sexual property in every sense: daughters are bargaining chips for establishing economic and political alliances with other households; male heads of household have sexual access to female servants; and women (especially in poorer households) bear many children, who eventually become an important source of labor. In this system, women's power depends on their relationship to the dominant men.

Private households emerge with a market economy, a centralized, bureaucratic state, and the establishment of agencies of social control that alleviate the need for citizens to take the law into their own hands. Thus, private households exist when the workplace is separate from the home, men are still heads of households but assume the role of breadwinner (as opposed to military commander), and women remain responsible for housekeeping and child rearing. Men, as heads of households, control the property; it is a relatively new practice in the United States, for example, to put a house or credit in the names of both husband and wife. Moreover, men monopolize the most desirable and important economic and political positions.

Collins states that a decline in the number of fortified households, the separation of work from home, smaller family size, and the existence of a police force to which women can appeal in cases of domestic violence give rise to the notion of romantic love as an important ingredient in a marriage. In the marriage market, men offer women economic security because they dominate the important, high-paying positions. Women offer men companionship and emotional support and strive to be attractive—that is, to achieve the ideals of femininity, which may include possessing an eighteen-inch waist or removing most facial and body hair. At the same time, they try to act as sexually inaccessible as possible because sexual access is something they offer men in exchange for economic security.

Advanced market economies offer widespread employment opportunities for women. Although women are far from being men's economic equals, some women now can enter relationships with men with more than an attractive appearance; now they can offer an income and other personal achievements. Having more to offer, women can demand that men be physically attractive and meet the standards of masculinity. This situation may explain why more commercial attention has been given in the past decade to males' appearance — body building, hair styles, male skin and cosmetic products.

Collins's classic theory of sexual stratification suggests that men and women cannot be truly equal in a relationship until women are men's economic equals (see Table 11.3). To reach this goal—still unrealized—fathers must share equally in household and child rearing responsibilities.

Household and Child-Rearing Responsibilities

In the United States, research shows consistently that even among couples in which the spouses earn the same income and share egalitarian ideals, the husbands spend considerably less time than the wives in preparing meals, taking care of children, shopping, and performing other household tasks (Almquist 1992). The same is true of the division of household labor in prewar Yugoslavia and other Eastern European countries (United Nations 1991). On the other hand, in 1990, the last year for which data are available, Yugoslavia allowed men and women 105 to 210 paid days for parental leave (Drakulić 1990). In the United States, it was not until August 1993, when the Family Leave Act was passed, that men and women who had worked for an employer for at least one year could take up to sixty days of unpaid leave without the fear of losing their job and other benefits.[10] In addition to having equality

[10]The Family Leave Act is not as straightforward as it might appear. For example, it excludes persons who are employed at work sites with fewer than fifty employees. And employers may deny leave to salaried employees who are in the top 10 percent of pay categories.

Private households The economic arrangements that exist when the workplace is separate from the home, men are still heads of households but assume the role of breadwinner (as opposed to military commander), and women remain responsible for housekeeping and child rearing.

Advanced market economies Economic arrangements that offer widespread employment opportunities for women as well as for men.

Table 11.3 Women's Nonagricultural Wage as a Percentage of Men in Fifty-five Selected Countries*

This table presents one way to measure the relative economic equality of men and women in various countries. Based on your general knowledge of the countries, are you surprised to see certain countries among the most equal? Are you surprised to see certain countries among the least equal? Why are they surprising? Is this a measure of economic opportunity for women? If so, by this standard, how well does the United States earn its position as "the land of opportunity"?

Country	Women's Nonagricultural Wage as Percentage of Men's	Country	Women's Nonagricultural Wage as Percentage of Men's
Tanzania	92.0	Germany	75.8
Vietnam	91.5	United States	75.0
Australia	90.8	Mexico	75.0
Sri Lanka	89.8	Belgium	74.5
Iceland	89.6	Uruguay	74.5
Sweden	89.0	Swaziland	73.0
Norway	86.0	Central African Rep.	72.6
Bahrain	86.0	Singapore	71.1
Kenya	84.7	Spain	70.0
Colombia	84.7	United Kingdom	69.7
Turkey	84.5	Hong Kong	69.5
Jordan	83.5	Ireland	69.0
Costa Rica	83.0	Thailand	68.2
Denmark	82.6	Switzerland	67.6
Hungary	82.0	Luxembourg	65.2
Mauritius	81.3	Argentina	64.5
France	81.0	Ecuador	63.7
New Zealand	80.6	Canada	63.0
Italy	80.0	Bolivia	62.3
Egypt	79.5	Philippines	60.8
Zambia	78.0	Cyprus	60.8
Greece	78.0	Chile	60.5
Poland	78.0	Syrian Arab Rep.	60.0
Austria	78.0	China	59.4
Finland	77.0	Korea, Rep. of	53.5
Netherlands	76.7	Bangladesh	42.0
Portugal	76.0	Average	74.9
Brazil	76.0		

*Data are for latest available year.

Source: United Nations (1994); government data from national consultants and Psacharopoulos and Tzannatos (1992).

in the household and child-rearing responsibilities, women must also have access to agents of violence control.

Access to Agents of Violence Control

Collins (1971) argues that women must have access to agents of violence control if they are to be men's equals. In other words, when women find themselves in a disadvantaged situation, they must be able to neutralize men's physical strength with support from the outside. Women's access to violence control is particularly an issue in wartime. To illustrate, we consider the case of Bosnia. In all countries, wars typically have been fought primarily, or at least disproportionately, by men. In most of these conflicts, raping of women has been a more than incidental occurrence — whether as a byproduct of war's savagery or as a deliberate instrument of warfare. These observations lead to questions about the relationship between war rape and gender roles.

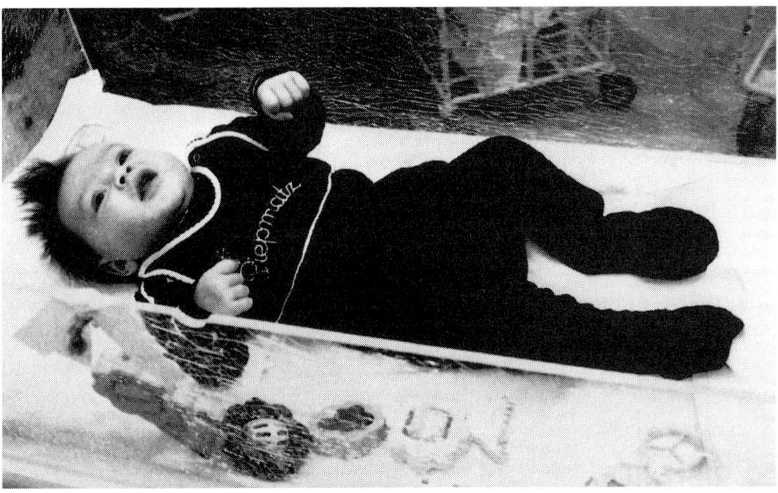

Among the casualties of war are the orphans produced by rape of "enemy" women. This baby was conceived after her seventeen-year-old Bosnian Muslim mother was repeatedly raped in a Serb-run detention center. The mother refused to see her after her birth.

AP/Wide World Photos

Although soldiers on all sides of the war in Bosnia raped women, the Serbian soldiers were under orders to rape non-Serbian women and even Serbian women who opposed the establishment of an ethnically pure Greater Serbia. This systematic mass rape is part of the Serbian policy of ethnic cleansing, a euphemism for forcing people of ethnic heritage different from the dominant group out of a territory.

According to Jeri Laber (1993), who investigated war crimes against women in Bosnia and Croatia, the rapes seemed to follow several different patterns. One pattern occurred when Serbian military units invaded a village or town populated primarily by Muslims or Croatians. Rape was used as a weapon to terrorize non-Serbs so they would flee their homes or sign documents indicating that they were leaving voluntarily. Other rapes occurred in Serbian-operated detention centers or camps, where hundreds or thousands of women were held. Still other rapes took place in Serbian-controlled houses, schools, or hotels.

The Serbian soldiers' rape of women resembled other war-related rapes. First, the rape was often a public event: it occurred in front of witnesses. The witnesses may have been other villagers, neighbors, or the woman's family. It was public so that the enemy, especially the woman's male family members, were humiliated (Swiss and Giller 1993). The rape was used to symbolize ultimate control and was carried out with extreme cruelty. "Rape by a conquering soldier destroys all remaining illusions of power and property for men of the defeated side" (Brownmiller 1975:38). Second, more than one soldier and as many as twenty participated; all women were vulnerable, regardless of age. Third, the soldiers who raped often killed their victims because they were disgusting to them. One Serbian soldier commented, "I only remember that I was twentieth, that her hair was a mess, that she was disgusting and full of sperm, and that I killed her in the end" (Mladjenovic 1993:14). In the Bosnian war, some women reported knowing the men who raped them.

The Serbian soldiers also had been ordered to forcibly impregnate non-Serbian women to produce *chetnicks* (babies). Some women were held in camps until they were far enough along in their pregnancies to make abortion dangerous (Mladjenovic 1993). One woman interviewed by United Nations human rights workers described the horror of the camps:

> It is a nightmare that cannot be talked about, or described, or understood. Sometimes I think I will go crazy and that the nightmare will never end. Every night in my dreams I see the face of Stojan, the camp guard. He was the most ruthless among them. He even raped ten-year-old girls, as a delicacy. Most of those girls didn't survive. They murdered many girls, slaughtered them like cattle. (Mirsada 1993:32)

It is difficult to determine how many women have been raped in this war. Official statistics never have been collected on the number of women killed or injured as a result of wartime rape; these wars in the former Yugoslavia are no exception.

Women of all ethnic groups in the former Yugoslavia who were raped during the war lack access to agents of violence. Not until 1993, in the wake of the massive and systematic rapes in Bosnia, did the United Nations pass a resolution identifying rape as a war crime and call for an international tribunal to prosecute those who ordered, committed, or did nothing to stop rape (Swiss and Giller 1993). The problem is that it is difficult to have rape recognized as a war crime unless there is clear proof that high-ranking military officials ordered the rapes. Otherwise rape is considered an individual act (Brew 1994). In a 1993 article in the *Journal of the American Medical Association,* Shana Swiss and Joan Giller specified medical technologies that might be used to establish rapists' identities, given that many of the raped women did not know their attackers. These include collecting and storing sperm collected from the genital tract, taking placental tissue (from abortions or delivery), and drawing blood samples from mother and child. DNA from these specimens could be matched later against DNA obtained from the alleged perpetrator's blood and hair follicles. The point is that this medical technology can be used for violence control if a potential rapist realizes that a rape can be traced to him.

We have seen that a number of sources and expressions of inequality based on gender exist. Even where there is a physical basis for these inequalities (as in the case of physical strength), the inequalities themselves are social rather than biological realities. Even in the case of war rape, soldiers are not acting out a biological imperative. Their behavior reflects, among other things, the way their social reality has been redefined so that women are considered suitable targets of aggression and rape is considered acceptable under the circumstances.

Mechanisms of Perpetuating Gender Expectations

As mentioned at the onset of this chapter, people vary in the extent to which they conform to their society's gender expectations. This fact, however, does not prevent us from using gender expectations to evaluate our own and other behavior. For many people, failure to conform (whether that failure is deliberate or reluctant) is a source of intense confusion, pain, and pleasure. This leads sociologists to identify the mechanisms by which we learn and perpetuate a society's gender expectations. To address this issue, we examine three important factors: socialization, situational constraints, and ideologies.

Socialization

In Chapter 5, we learned that socialization is a learning process that begins immediately after birth and continues throughout life. By this process, "newcomers" develop their human capacities, acquire a unique personality and identity, and internalize, or take as their own and accept as binding, the norms, values, beliefs, and language they need to participate in the larger community. Socialization theorists argue that an undetermined but significant portion of male-female differences are products of the ways in which males and females are treated.

Child development specialist Beverly Fagot and her colleagues observed how toddlers in a play group interacted and communicated with each other and how teachers responded to the children's attempts to communicate with them at age twelve months and at age twenty-four months (Fagot et al. 1985). Fagot found no real sex differences in the interaction styles of twelve-month-old boys and girls: all of the children communicated by gestures, gentle touches, whining, crying, and screaming. The teachers, however, interacted with the toddlers in gender-polarized ways. They were more likely to respond to girls when the girls communicated in gentle, "feminine" ways and to boys when the boys communicated in assertive, "masculine" ways. That is, the teachers tended to ignore assertive acts by girls and to respond to assertive acts by boys. Thus, by the time these toddlers reached two years of age, the differences in their communication styles were quite dramatic.

Fagot's findings may help explain the differing norms governing body language for males and females. According to women's studies professor Janet Lee Mills, norms governing male body language suggest power, dominance, and high status, whereas norms governing female body language suggest submissiveness, subordination, vulnerability, and low status. Mills argues that these norms are learned and that people give them little thought until someone breaks them, at which point everyone focuses on the rule breaker. Such norms can prevent women from conveying a sense of security and control when they are in positions that demand these qualities, such as lawyer, politician, or physician. Mills suggests that women face a dilemma: "To be successful in terms of femininity, a woman needs to be passive, ac-

These Bosnian boys, dressed like fighters and brandishing toy machine guns, are learning through socialization that fighting in war is part of being a male.
AP/Wide World Photos

commodating, affiliative, subordinate, submissive, and vulnerable. To be successful in terms of the managerial or professional role, she needs to be active, dominant, aggressive, confident, competent, and tough" (Mills 1985:9).

Children's toys figure prominently in the socialization process along with the ways in which adults treat children. Barbie dolls, for example, have been on the market for more than thirty years and currently are available in sixty-seven countries. Executives at Mattel, the company that created Barbie, are studying Eastern Europe as a potential new market. The company considers Barbie to be an **aspirational doll**—that is, the doll is a role model for the child. Barbie accounts for approximately half of all toy sales by Mattel (Boroughs 1990; Cordes 1992; Morgenson 1991; Pion 1993). Ninety-five percent of girls between ages three and eleven in the United States have Barbie dolls, which come in several different skin colors. Market analysts attribute Mattel's success to the fact that "they generally have correctly assessed what it means to a little girl to be grown-up" (Morgenson 1991:66).

In *How We Survived Communism and Even Laughed,* Slavenka Drakulić (1992) describes how the standard of beauty in Yugoslavia, at least since World War II, has come from Hollywood. She tells how the faces and bodies of Hollywood women such as Rita Hayworth, Ava Gardner, and Brigitte Bardot, who appeared in a fashion magazine that their parents forbade them to read, were the models that she and her friends used for paper dolls, along with paper outfits and accessories. Drakulić recalls that she and her friends "painted [the dolls'] little lips and nails bright red, and dressed them in tight sexy dresses" (p. 61). Reflecting on this childhood activity, Drakulić remarks:

> Sometimes I think that at that early age I learned everything about my sex from these paper dolls. . . . Later on, it took me—and our whole generation of women—years and years of hard work to unglue ourselves from those paper idols; to break through into another dimension, away from the dolls of our childhood, to which we were constantly reduced. (pp. 61–62)[11]

[11]There is some debate over whether Barbie is a "bimbo" or a "feminist." Some critics say that in real life Barbie's figure would measure 36-18-33 and that she is quite materialistic. Accessories range from "toe nail polish to a pink RV camper to haul around her never-quite-big-enough wardrobe" (Cordes 1992:46). Others argue that Barbie's résumé is quite impressive and that she allows little girls to dream of what they can be. Over the years Barbie has been marketed as a fashion model, ballerina, stewardess, teacher, fashion editor, medical doctor, Olympic athlete, TV news reporter, corporate executive, and animal rights volunteer (*Harper's* 1990).

Aspirational doll A doll considered to be a role model for a child.

Janet Lee Mills is famous for this pose, which she captions, "Could you say no to this woman?" In it she violates many traditional female behaviors with relaxed posture: arms, legs positioned away from body; direct, confrontive eye contact; no affiliative smile.

All photos courtesy of Janet Lee Mills.

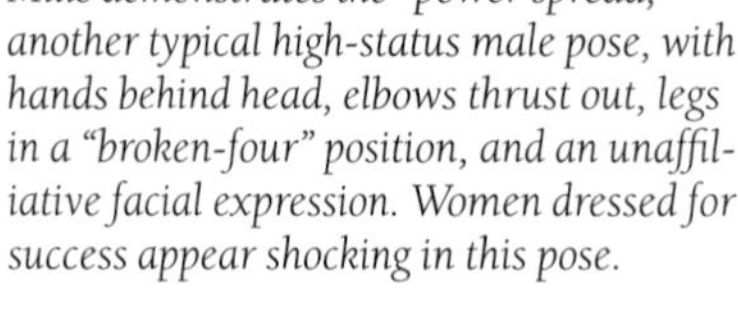

Mills demonstrates the "power spread," another typical high-status male pose, with hands behind head, elbows thrust out, legs in a "broken-four" position, and an unaffiliative facial expression. Women dressed for success appear shocking in this pose.

Mills recruited Richard Friedman, Assistant to the President at University of Cincinnati, to model for these photographs. Friedman does not appear shocking in the same pose, because many male executives conduct business from a similar position.

Ah, but doesn't Mills look feminine in this typically feminine pose with canted head, affiliative smile, ankles crossed, and hands folded demurely?

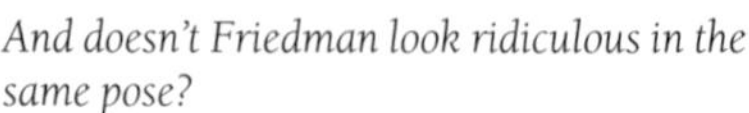

And doesn't Friedman look ridiculous in the same pose?

Power is often wielded in postural alignment, gesture, and use of objects. Mills holds power, along with papers on which attention is focused; whereas Friedman signals deference with lowered gaze and constricted body.

Now Friedman holds power, along with papers, with lower limbs spread away from his body; whereas Mills signals deference with an attentive pose, submissively bowed head, and a hand-to-mouth gesture of uncertainty.

In this typical office scene, Friedman holds power with an authoritative stance, one hand in pocket and the other at midchest, straight posture, and head high; whereas Mills is submissive with canted head, smile, arms and hands close to her body. Note Friedman's wide, stable stance and Mills's unstable stance. Many women tend to slip into a similar posture when talking to a shorter male authority figure.

Now the tables are turned. Friedman defers to authority in a feminine, subordinate posture with scrunched-up spine, constricted placement of arms and legs, canted head, and smiling attentiveness.

Structural or Situational Constraints

Situational theorists agree with socialization theorists that the social and economic differences between men and women are not explained by something in their biological makeup. In their view, the causes of these differences are structural or situational constraints. **Structural constraints** are the established and customary rules, policies, and day-to-day practices that affect a person's life chances. An example is occupations segregated by sex, so that women tend, more often than men, to be concentrated in low-paying, low-ranking, dead-end jobs. (Recall the discussion in Chapter 9 of occupations in which women are disproportionately underrepresented and overrepresented and of the inequality in men's and women's earnings.) Even when women are in professional and management positions, they are concentrated in specialties and fields that handle children and young adults, that involve supervising other women, and/or that are otherwise considered feminine (a professor of social work versus a professor of mathematics or computer sciences).

These structural differences reinforce expectations about gender. The different social and physical demands and skills required to perform the jobs held by men and women function toward "channeling their motivations and their abilities into either a stereotypically male or a stereotypically female direction" (Bem 1993:135). This point does not preclude the fact that men and women may limit their job search to positions that are considered "sex-appropriate." On the other hand, considerable evidence supports the hypothesis that once women are hired, management steers male and female employees into different "gender-appropriate" assignments and offers them different training opportunities and chances to move into better-paying positions.

A case in point is Lucky Stores, Inc., which operates 188 stores in northern California. The company lost a class action sex discrimination suit because (among other things) it failed to post job openings but filled them instead at the management's discretion. Lawyers for Lucky Stores argued that women were concentrated into part-time, low-paying, dead-end jobs in the delicatessen and bakery rather than jobs in stocking and receiving because women preferred such work. Once the store began to post job openings, however, the percentage of women in entry-level management positions jumped from 12 to 58 percent (Gross 1993).

Structural constraints The established and customary rules, policies, and day-to-day practices that affect a person's life chances.

The case of Lucky Stores shows that women are no less highly motivated than men to seek advancement in the workplace. Many other cases also serve to support this point. The implication is clear: if we remove structural constraints to advancement, women will seek to improve their position.

Sociologist Renee R. Anspach's research illustrates vividly how one's position in a social structure can channel behavior in stereotypically male or female directions. Anspach spent sixteen months conducting field research (observing and holding interviews) in two neonatal intensive care units (NICUs). Among other things, she found that nurses (almost all of whom were female) and physicians (usually male) used different criteria to answer the question, "How can you tell if an infant is doing well or poorly?" Physicians tended to draw on so-called objective (technical or measurable) information and immediate perceptual cues (skin color, activity level) obtained during routine examination:

> Well, we have our numbers. If the electrolyte balance is OK and if the baby is able to move one respirator setting a day, then you can say he's probably doing well. If the baby looks gray and isn't gaining weight and isn't moving, then you can say he probably isn't doing well.
>
> The most important thing is the gestalt. In the NICU, you have central venous pressure, left atrial saturations, temperature stability, TC (transcutaneous) oxymeters, perfusions (oxygenation of the tissues)—all of this adds in. You get an idea, when the baby looks bad, of the baby's perfusion. The amount of activity is also important—a baby who is limp is doing worse than one who's active. (Anspach 1987:219–220)

Although immediate perceptual and measurable signs were also important to the nurses, Anspach found that the nurses also considered interactional clues such as the baby's level of alertness, ability to make eye contact, and responsiveness to touch:

> I think if they're doing well they just respond to being human or being a baby. . . . Basically emotionally if you pick them up, the baby should cuddle to you rather than being stiff and withdrawing. Do they quiet when held or do they continue to cry when you hold them? Do they lay in bed or cry continuously or do they quiet after they've been picked up and held

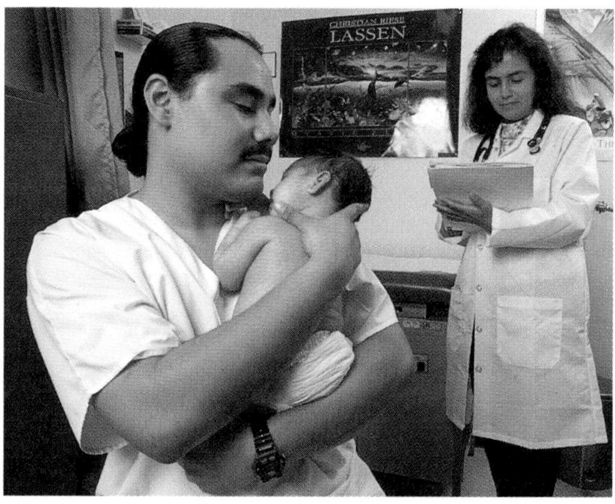

One's position in the social structure can channel behavior in stereotypically male or female directions. The job description of nurse, for example, requires the person in that position to interact more with patients than physicians do. Consequently the nurse is more likely to consider interactional clues in evaluating a patient's medical condition.

©Anne Dowie

> and fed. . . . Do they have a normal sleep pattern? Do they just lay awake all the time really interacting with nothing or do they interact with toys you put out, the mobile or things like that, do they interact with the voice when you speak? (p. 222)

Anspach concluded that the differences between nurses' and physicians' responses to the question "How can you tell if an infant is doing well or poorly?" could be traced to their daily work experiences. In the division of hospital labor, nurses interact more with patients than do physicians. Also, doctors and nurses have access to different types of knowledge about infants' condition, which correspond to our stereotypes of how females and males manage and view the world. Because physicians have only limited amounts of daily interaction and contact with infants, they tend to rely on perceptual and technological (measurable) cues. Nurses, on the other hand, are in close contact with infants throughout the day; consequently, they are more likely to consider interactional cues as well as perceptual and technological ones.

Anspach (1987) suggests that one's position in the division of labor "serves as a sort of interpretive lens through which its members perceive their patients and predict their futures" (p. 217). Her findings suggest that when physicians make life-and-death decisions about whether to withdraw or continue medical care, they should collaborate with NICU nurses so that they can consider interactional as well as technological and immediate perceptual cues. Anspach's findings also suggest that if nurses' experiences and opinions counted more in medical diagnosis, we might see a corresponding increase in the prestige and salary associated with this largely female position.

Sexist Ideology

In Chapters 2 and 10, we learned that ideologies are ideas that support the interests of the dominant group but that do not stand up to scientific investigation. They are taken to be accurate accounts and explanations of why things are as they are. On closer analysis, however, we find that ideologies are at best half-truths, based on misleading arguments, incomplete analysis, unsupported assertions, and implausible premises.

Sexist ideologies are structured around three notions:

1. People can be classified into two categories, male and female.
2. There is a close correspondence between a person's primary sex characteristics and characteristics such as emotional activity, body language, personality, intelligence, the expression of sexual desire, and athletic capability.
3. Primary sex characteristics are so significant that they explain and determine behavior and social, economic, and political inequalities that exist between the sexes.

Sexist ideologies are so powerful that "almost everyone has difficulty believing that behavior they have always associated with 'human nature' is not human nature at all but learned behavior of a particularly complex variety" (Hall 1959:67). One example of a sexist ideology is the belief that men are prisoners of their hormones, making them powerless in the face of female nudity or sexually suggestive dress or behavior. Another example is the belief that men are not capable of forming relationships with other men that are as meaningful as those that women form. Since the 1980s, dozens of books were written by men in response to these stereotypes (Shweder 1994).

We might also add a fourth characteristic about sexist ideology: people who behave in ways that depart from ideals of masculinity or femininity are considered deviant, in need of fixing, and subject to negative sanctions ranging from ridicule to physical violence. This ideology is reflected in military policy toward homosexuals. According to a U.S. Department of Defense (1990) directive:

> Homosexuality is incompatible with military service. The presence of such members adversely affects the ability of the Armed Forces to maintain discipline, good order, and morale; to foster mutual trust and confidence among the members; to ensure the integrity of the system of rank and command; to facilitate assignment and worldwide deployment of members who frequently must live and work under close conditions affording minimal privacy; to recruit and retain members of the military services; to maintain the public acceptability of military services; and, in certain circumstances, to prevent breaches of security. (p. 25)

No scientific evidence, however, supports this directive. In fact, it seems that whenever Pentagon researchers (with no links to the gay and lesbian communities and with no ax to grind) found evidence that ran contrary to this directive, high-ranking military officials refused to release the information or found the information unacceptable and directed researchers to rewrite the reports. For example, when researchers found that sexual orientation is unrelated to military performance and that men and women known to be gay or lesbian displayed military suitability that is as good as or better than that of men and women believed to be heterosexual,[12] U.S. Deputy Undersecretary of Defense Craig Alderman, Jr., wrote the researchers that the "basic work is fundamentally misdirected" (Alderman 1990:108). He explained that the researchers were to determine whether there was a connection between being a homosexual and being a security risk, not to determine whether homosexuals were suitable for military service. Although the researchers found no data to support a connection between sexual orientation and security risks, Alderman maintained that the findings were not relevant, useful, or timely. (See "Ideology to Support Military Policy.") The research that Alderman dismissed would have gone unnoticed if Congressman Gerry Studds and House Arms Subcommittee Chairwoman Patricia Schroeder had not insisted it be released.

This example shows the role that ideologies play in setting policy. In this case, the ideologies are that homosexuality is incompatible with military service and that being homosexual represents a security risk to the United States. The case of the military also alerts us to the fact that other variables, such as a person's sexual orientation, interact with biological sex to affect the experience of being male or female differently. To illustrate this interaction, we turn to the work of sociologists Floya Anthias and Nira Yuval-Davis, who have written about the interconnection among gender, race and ethnicity, and country (the state).

Gender, Ethnicity, Race, and the State

Ethgender refers to people who share (or are believed by themselves or others to share) the same sex, race, and ethnicity. This concept acknowledges the combined (but not additive) effects of gender, race, and ethnicity on life chances. Ethgender merges two ascribed statuses into a single social category. In other words, a person is not a Croat and a woman but a Croatian woman; a person is not a Chinese and a man but a Chinese man (Geschwender 1992). To complicate matters, the country or state that people of a particular ethgender inhabit (and their legal relationship to the state—as citizen, refugee, or temporary worker) has a significant effect on their life chances. We use the term *state* here to mean a governing body organized to manage and control specified activities of people living in a given territory.

Everyone has some "legal" relationship to the state, whether as a citizen by birth or naturalization, a refugee, a temporary worker, an immigrant, a permanent resident, or an illegal alien. Sociologists Floya Anthias and Nira Yuval-Davis (1989) give special attention to women, their ethnicity, and the state. They argue, "Women's link to the state is complex," and women "are a special focus of state concerns as a social category with a specific role (particularly human reproduction)" (p. 6). In the broadest sense, reproduction includes biological reproduction, especially in relation to the birth of children who become the state's citizens and future labor force. Anthias and Yuval-Davis maintain that the state's policies and discourse reflect its concerns about the kinds of babies (that is, their race and/or ethnicity) to which women give birth and about the ways in which the babies are socialized. They identify five areas of women's lives over which the state may choose to exercise control. One should not conclude,

Ethgender People who share (or are believed by themselves or others to share) the same sex, race, and ethnicity.

[12]"Believed to be homosexual" is an important consideration because the military currently follows a "don't ask, don't tell policy" and because some recruits remain in the "closet."

Ideology to Support U.S. Military Policy

People who oppose the presence of gays and lesbians in the military stereotype homosexuals as sexual predators who are waiting to pounce on a heterosexual person while he or she is showering, undressing, or sleeping. Opponents seem to believe that any same-sex person is attractive to a gay or lesbian person. But as one gay ex-midshipmen notes, "Heterosexual men have an annoying habit of overestimating their own attractiveness" (Schmalz 1993:B1). The excerpt from Pentagon research included here shows that there is no evidence to support such a stereotype.

Those who resist changing the traditional policies support their position with statements of the negative effects on discipline, morale, and other abstract values of military life. Buried deep in the supporting conceptual structure is the fearful imagery of homosexuals polluting the social environment with unrestrained and wanton expressions of deviant sexuality. It is as if persons with nonconforming sexual orientations were always indiscriminately and aggressively seeking sexual outlets. All the studies conducted on the psychological adjustment of homosexuals that we have seen lead to contrary inferences. The amount of time devoted to erotic fantasy or to overt sexual activity varies greatly from person to person and is unrelated to gender preference. In one carefully conducted study, homosexuals actually demonstrated a lower level of sexual interest than heterosexuals.

Homosexuals are like heterosexuals in being selective in their choice of partners, in observing rules of privacy, in considering appropriateness of time and place, in connecting sexuality with the tender sentiments, and so on. To be sure, some homosexuals are like some heterosexuals in not observing privacy and propriety rules. In fact, the manifold criteria that govern sexual interest are identical for homosexuals and heterosexuals, save for only one criterion: the gender of the sexual partner.

Age, gender, kinship, class membership, marital status, size and shape, social role, posture, manners, speech, clothing, interest/indifference signaling, and other physical and behavioral criteria are all differentiating cues. They serve as filters to screen out undesirable or unsuitable potential sex partners. With such an array of cues, many (in some cases, all) potential objects of interest are rejected. For most people, only a small number of potential partners meet the manifold criteria. Whether in an Army platoon or in a brokerage office, people are generally selective in their choice of intimate partners and in their expression of sexual behavior. Heterosexuals and homosexuals alike employ all these variables in selecting partners, the only difference being that the latter include same-gender as a defining criterion, the former include opposite-gender.

Source: Sarbin and Karols (1990:37).

however, that women accept the policies and programs without resistance that the state directs at them. In fact, women often work to modify these policies.

1. Women as Biological Reproducers of Babies of a Particular Ethnicity or Race As factors that can underlie a state's population control policies, Anthias and Yuval-Davis (1989) name "fear of being 'swamped' by different racial and ethnic groups" or fear of a "demographic holocaust"—that is, a particular racial or ethnic group will die out or become too small to hold its own against other ethnic groups (p. 7). Such policies can range from physically limiting numbers of a particular racial or ethnic group deemed undesirable to actively encouraging the "right kind" of women to produce more children. Policies that limit numbers include immigration control (limiting or excluding members of certain ethnic groups from entering a country and subsequently producing children), physical expulsion (which includes ethnic cleansing), extermination, forced sterilization, and massive birth control campaigns. Policies that encourage women to reproduce the "right kind" of babies include ideological mobilization (appeals to a woman's duty to her country), tax incentives, maternal leave, and other benefits.

Even before the wars in Yugoslavia and its eventual breakup (see "The Breakup of Yugoslavia"), various leaders pointed to the "fear of being swamped" by other ethnic groups as a reason for their women to bear children. Most notably, Serbian leaders such as Slobodan Milosevic pointed to the high birthrate among Albanians and Muslims as a serious threat to Serbian autonomy and quality of life. Similarly, some Croatian leaders of parties such as the Croatian Democratic Union have asked that the right to an abortion be made illegal and have asked Croatian women to bear at least three, but ideally five, children (Drakulić 1990; Enloe 1993).

2. Women as Reproducers of the Boundaries of Ethnic or National Groups In addition to implementing policies intended to encourage or discourage women in "having children who will become members of the various ethnic groups within the state" (Anthias and Yuval-Davis 1989:9), states also implement policies that define the "proper ways" to reproduce offspring. Examples include laws prohibiting sexual relationships with men or women of another race or ethnicity, laws specifying legal marriage if the child is to be recognized as legitimate, and laws connecting the child's ethnic and legal status to the ethnicity of the mother and/or father.

Although these laws apply to both men and women, the woman generally pays a heavier social price when the law is broken. For example, in *Wake Up Little Susie: Single Pregnancy and Race Before Roe v. Wade,* Rickie Solinger (1992) documents the options open to unmarried girls and women who faced pregnancy between 1945 and 1965, which included:

> futilely appealing to a hospital abortion committee, [which at that time were not concerned about the question of when human life begins but about punishing single mothers]; being diagnosed as neurotic, even psychotic by a mental health professional; [being] expelled from school (by law until 1972); [becoming] unemployed; [enrolling] in a Salvation Army or some other maternities home; [and being] poor, alone, ashamed, threatened by the law. (p. 4)

Solinger argues that the policies and programs encouraged white women to place their babies up for adoption but encouraged black women to keep their babies and to prevent them from having more.

Today, by contrast, many school districts across the United States offer on-site day care, private tutoring, and special classes to pregnant girls. Still, considerable political debate revolves around the questions of whether single mothers should receive welfare, especially after they have a second child, and whether it might be to society's benefit to bring the stigma of the past back (L. Williams 1993). For the most part, these debates rarely focus on stigmatizing the fathers of these children.

3. Women as Transmitters of Social Values and Culture The state can institute policies that either encourage women to be the main socializers of their offspring or leave socialization in the hands of the state. Examples include tying welfare payments to nonemployment (or under welfare reform to employment) so that mothers are forced to stay home with the children, instituting liberal or restrictive maternity leave policies, and subsidizing day-care centers, providing opportunities to enroll children in preschools. Sometimes state leaders become concerned that children of particular ethnic or racial groups are not learning the cultural values and/or language they need to succeed in the dominant culture. This concern motivates them to fund programs that expose particular kinds of children to the necessary personal, social, and learning skills.

4. Women as Signifiers of Ethnic and Racial Differences Political leaders often use various images of women to symbolize the most urgent issues they believe the state faces. In wartime, the state is represented as "a loved woman in danger or as a mother who lost her sons in battle" (Anthias and Yuval-Davis 1989:9–10). Men are called to battle to fight and protect the women and children. Often the leaders present the image of a woman who meets the culture's ideal of femininity and who belongs to the dominant ethnic group. Sometimes political leaders use veiled language to evoke images of women of a certain ethnic or racial group as the source of a country's problems (for example, Albanian women who produce many children; African American welfare mothers with no economic incentives to practice birth control). Often with a check of the facts, such images are unfounded or there is no evidence to support such generalizations. In "Fertility Among Women on Welfare: Incidence and Determinants," sociologist Mark R. Rank (1989) maintains that "it is impossible to calculate with any precision the fertility rate of women on public assistance" (p. 296) because the data available have serious flaws. "There is no way of judging whether the fertility rate of women on welfare is high or low" (p. 296) relative to the fertility rate of other women.

5. Women as Participants in National, Economic, and Military Struggles States implement policies governing the roles that women and men can assume in crises, notably in war. Historically, women have played supportive and nurturing roles, even in situations in which they have been exposed to great risks. In most countries, women are not drafted; they volunteer to serve. If they are drafted, the state defines acceptable military roles. If women do fight, they often do so as special units or in an unofficial capacity. In February 1994, Bosnian-Serb leaders announced that "The entire able-bodied population will be mobilized, either into military or labor units, and special women's units will be formed" (Kifner 1994:A4).

Regardless of their official roles in the war, women are affected by war. It is estimated that since World War II, civilians have suffered 80 percent of the deaths and casualties in war (Schaller and Nightingale 1992).

Women are killed, taken prisoner, tortured, and raped. Even so, they often are not trained formally and systematically to fight. As a result, women occupy a different position than men during war.

The fact that women's combat roles are limited does not mean that women are incapable of combat, however. A small number of Serbian female battalions fought in Bosnia, and we hear occasional accounts of women "warriors," such as the profile of a Sarajevan female sniper that aired on the Canadian Broadcasting Corporation (CBC) in 1993:

> They say nobody loves their city the way the people of Sarajevo do, and when I saw it all, the way they are destroying the city, destroying people in it, I knew it had to be like this. The first person I shot at was a soldier; he was also a sniper. I shot at the very last second because he was aiming at me, so if I had waited another second, we would not be sitting here talking now. There is no time to think about it. . . .
>
> If I counted all of my victims I doubt I could preserve my sanity. It's not easy to pull the trigger but when I do I save at least ten other lives. The others shoot children at play, men and women standing in the bread line, or civilians walking down the street. I don't think I kill; I try to save as many lives as possible, the lives of civilians, the lives of innocent people, lives in general. (Canadian Broadcasting Corporation 1994)

Through its military institutions, the state even establishes policies that govern male soldiers' sexual access to women outside military bases, both in general and in times of war. The Serbs, for example, captured some women and sent them to places resembling concentration camps in which many were raped, but they also kept other women in brothel-like houses and hotels. We also know that during World War II, Japanese military authorities forcibly recruited[13] between 60,000 and 200,000 women, mostly Korean but also Chinese, Taiwanese, Filipino, and Indonesian to work as sex slaves in army brothels in the war zone (Doherty 1993; Hoon 1992). They referred to them as "comfort women."

In *Let the Good Times Roll: Prostitution and the U.S. Military in Asia,* Saundra Pollock Sturdevant and Brenda Stoltzfus (1992) examine "the sale of women's sexual labor outside U.S. military bases" (p. vii). They present evidence that the U.S. military helps regulate prostitution; that retired military officers own some of the clubs, massage parlors, brothels, discotheques, and hotels; and that the military provides the women with medical care so as to prevent the spread of sexually transmitted diseases between the soldiers and the women. In 1993, thousands of Filipino women who live near the Subic Bay Naval Base filed a class action suit against the United States, arguing that the United States has moral and legal responsibilities to support the estimated 8,600 children fathered by U.S. servicemen stationed at Subic Bay.[14]

In evaluating the role of the military in women's lives, we must consider that many poor women who live near the bases see a relationship with a U.S. recruit as their only way out of poverty and a desperate situation. This means that the Subic Bay situation is more complicated than the media present it. To clarify this point, I will use a personal example told to me by a student in one of my classes.

A young male student, perhaps twenty-three or twenty-four years old, stopped me after a class in which we had discussed the Subic Bay Base closing. He told me that when he was in the Navy and the ship he was on docked at Subic Bay, it seemed as if the entire town turned out to welcome the ship. Local women were everywhere. Whereas many recruits visited prostitutes for one-night stands, others fell in love with local women. Usually the commanding officers did everything they could to discourage a permanent relationship. Although some might criticize the officers' actions, the officers were perhaps anticipating the difficulties ahead for the recruit if he took the woman as his wife and continued life "at sea," which would leave her alone. Unfortunately, the complexity of the relationship between the military and the local populations is overlooked in media accounts focusing on the base closings. The same can be said of media reports on the sensitive subject of date rape. The media tends to present the issue in very simplistic terms.

Gender and Date Rape

In 1987, Mary P. Koss, Christine A. Gidycz, and Nadine Wisniewski published their findings from a nationwide survey about college students' sexual experiences. The

[13]*Recruit* is the word the Japanese used to describe how they managed to bring Korean women to war zones. Documents show that the Japanese acquired the women through civilian agents, village raids, and applying pressure on Korean school administrations to provide girls (Hoon 1992).

[14]Under current U.S. immigration law, Filipino children of servicemen are not permitted to migrate to the United States with sponsorship from any American, as can the children of servicemen stationed in South Korea, Thailand, Cambodia, and Laos (Lambert 1993).

researchers found that 57 percent of female college students said they had experienced some form of sexual victimization (from fondling to forced anal or oral intercourse) during the past academic year. Fifteen percent of female college students said they had been victims of rape, and 12 percent said they had been victims of attempted rape. The researchers also found that most of the women knew the perpetrators. Because slightly more than one-fourth of the female student population reported that they had experienced a rape or an attempted rape, the researchers concluded that rape takes place more frequently than we are led to believe by official statistics such as Uniform Crime Reports or the National Crime Survey.

Koss and her colleagues used the term **hidden rape** to refer to rape that goes unreported. Media accounts of this research coined the catchier if less precise terms *date rape, acquaintance rape,* and *campus rape.* The last of these terms is especially imprecise because Koss did not ask whether the rapists were also college students or attended the same institution as the victim. A number of articles appearing in popular magazines such as *Time* and *Vogue* suggest that an epidemic of "date rape" is occurring, especially on college campuses.

Sociologists G. David Johnson, Gloria J. Palileo, and Norma B. Gray (1992) observed that between 1987, when Koss's study was published, and 1991, the media changed their focus from discovering and labeling a social problem on campus to debating whether so-called date rape is not actually something women claim after a "bad sexual experience" or when they do not hear again from their partner. In January 1991, these researchers replicated the Koss study. To do so, they surveyed a sample of 1,177 male and female college students attending a southern university. They hoped (among other things) to clarify a number of unanswered questions clouding this debate, including the following:

1. Is there an epidemic of date rape on U.S. college campuses?
2. Were the female respondents defining "forcible" sexual experiences as rape or were the researchers defining such experiences as rape? In Koss's study, for example, female students were asked, "Have you ever had sexual intercourse when you didn't want to because a man threatened you or used some degree of physical force (twisting your arm, holding you down, etc.) to make you?" Critics point out that women who say yes to this question are not asked whether they considered that sexual experience to be rape, even though the experience, as worded, constitutes the legal definition of rape.
3. Is so-called date rape something we might attribute to miscommunication rather than sexual aggression? Critics of the Koss findings often focused on this question. "This criticism emphasizes the ambiguity of sexual speech and nonverbal communication: When does 'no' really mean no, and when does it mean maybe, or even yes?" (Johnson et al. 1992:38).

Hidden rape Rape that goes unreported.

Johnson and his colleagues found that the prevalence of rape at the campus they studied was strikingly similar to that recorded in Koss's national study in 1987. This fact suggests that date rape is not an epidemic, in the strict sense of the word, on the campus they studied.

With regard to the second question, approximately half (51 percent) of the 149 women in this study who said they had experienced forced sexual relations said the experience was not rape; 12 percent said it was rape; 37 percent said that "some people would think it was close to rape" or that "many would call it rape."

The researchers also found considerable miscommunication between men and women. Figure 11.2 shows the percentage of female respondents who said no to sex when they meant yes. Slightly more than one-third of the women reported that they never say no when they mean yes, and 66 percent reported that they have said no when they meant yes. This finding shows a significant number of female university students miscommunicate their sexual intentions.

> We expected this outcome, but were surprised by the magnitude of the miscommunication. That one in six females at a college campus today always says "no" when she means "yes" indicates the presence of a significant problem. Clearly the ambiguity of sexual communication is a significant barrier for the achievement of nonexploitive sexual relationships between men and women. (Johnson et al. 1992:41)

It is important to note that the researchers did not ask the women if they ever say yes when they would like to say no.

In interpreting the data, we must be careful about the conclusions we draw. For one thing, Johnson and his colleagues did not ask women who said they had experienced force whether they had said no when they

Figure 11.2 Frequency That Female Respondents Say No to Sex When They Mean Yes

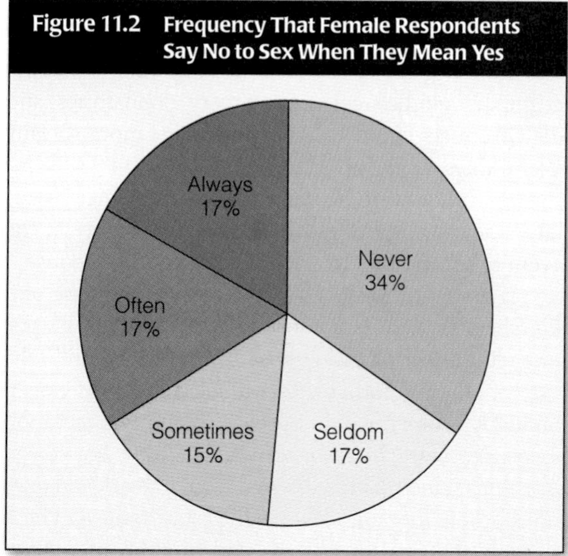

Source: Adapted from Johnson et al. (1992:44).

meant yes in those instances. Also, we cannot conclude that the women were defining as rape those instances when they reported that they had said no but meant yes. Finally, on the basis of questions asked in the Koss and Johnson studies, we cannot conclude that women are crying "rape" the morning after having "bad sex."

These qualifications aside, we are still left with the question of why 66 percent of the female respondents at this university at one time or another miscommunicated their intentions to their dates. For one answer to this question, we turn to a classic ethnographic study of a U.S. high school and its students, conducted by Jules Henry (1963). Although the study is more than thirty years old, Henry's observations are still relevant today.

Henry argues that sending mixed signals is one strategy women can use to communicate their sexual interests while maintaining their "reputation" and not appearing "easy." The woman gives double messages (saying no while indicating in other subtle ways that she means yes or saying yes when she would like to say no out of fear she will lose her partner), because the feeling rules are such that she cannot communicate her desires directly without being considered "loose" or a "slut." Sixty-six percent of women miscommunicating their sexual intentions, even if it is at only one university, is cause for alarm. This finding suggests that women are not confident enough, for whatever the reason, to let their partners know their true feelings.

Summary and Implications

In this chapter, we distinguished between sex (a biological distinction) and gender (a social construction). We have considered the problems associated with gender—culturally conceived and learned ideas about appropriate behavior and appearance for males and females. If we simply think about the men and women we encounter everyday, we quickly realize that people of the same sex vary in the extent to which they meet their society's gender expectations. Some people conform to gender expectations; others do not. This variability, however, does not stop most people from using their society's gender expectations to evaluate their own and others' behavior and appearances in "virtually every other aspect of human experience, including modes of dress, social roles, and even ways of expressing emotion and experiencing sexual desire" (Bem 1993:192).

Sociologists find gender a useful concept, not because all people of the same sex look and behave in uniform ways but because a society's gender expectations are central to people's lives whether they conform rigidly or resist. For many people, failure to conform to gender expectations, even if they fail deliberately or conform only reluctantly, is a source of intense confusion, pain, and/or pleasure (Segal 1990). For example, Pero M. (see the opening vignette) is confused and pained because, as a male, he is expected (required) to serve his country. From his point of view, women are lucky because their sex puts them in a different, safer position than his. On the other hand, we might speculate that Ila Borders, the first female to pitch in a men's NAIA collegiate baseball game, experienced considerable pleasure from accomplishing a feat that goes against gender expectations. In 1997, Borders was invited to try out with the St. Paul Saints, a minor league team. Yet, at the same time, Borders finds it annoying that she must continually remind people that she is "not a bruiser" and likes being a girl, using lipstick, and wearing feminine clothes (Stevenson 1994:B10).

Unfortunately, many people equate discussions of gender and gender polarization with the position that women should abandon traditional vocations such as housewife and mother in favor of careers. Furthermore, many critics point to "gender awareness" and the corresponding push toward sexual equality as the cause of family breakdown.

From a sociological perspective, gender awareness and the goal of sexual equality should strengthen the family. How can this be? The information in this chapter speaks to this question. Simply consider the matter of conveying sexual intentions. If women did not feel constrained about communicating sexual concerns and interests (or lack of interest) and if men were more sensitive to these gender constraints on women, perhaps there would be fewer unwanted pregnancies and abortions. Whether a person is pro-choice or pro-life, he or she would agree that 1.5 million abortions[15] per year is an unacceptable, even alarming, number. The number of abortions suggests that honest communication between women and men about the consequences of sex does not take place for a significant number of couples.

If men and women did not feel constrained to select partners according to age (that is, the typical woman is several years younger than the typical man), the sex ratio would be more balanced, especially in old age. There would be fewer widows, or, at the very least, women would spend fewer years alone.

If men and women did not feel constrained to meet artificial standards of beauty, the personal energy and financial resources channeled toward obtaining cosmetics, clothes, cosmetic surgery, depilatories, and diet products could be spent in other, more socially useful ways.

If women were paid a salary equal to men's, perhaps they would be less vulnerable to poverty in the event of separation, divorce, or the death of a spouse.

If men's and women's self-images, aspirations, and life chances were less constrained by gender scripts, men could feel free to choose to stay home and take care of children rather than pursue a full-time career. Similarly, women would not face the no-win situation associated with choosing between a family and career. As things stand today, if married women work and have no children, they are selfish; if they stay home and raise a family, they are considered "underemployed" at best; if they work and raise a family, people wonder how they can possibly do the job right; if they are divorced with children, they are responsible for the breakdown of the family unit; if they do not marry, they are considered spinsters.

[15]The number is especially high when we consider that there are about 4.0 million births each year. For every forty live births, fifteen abortions are performed.

Key Concepts

Use this outline to organize your review of the key chapter ideas.

- **Sex**
  - **Primary sex characteristics**
  - **Secondary sex characteristics**
- **Gender**
  - **Gender polarization**
  - **Social emotion**
  - **Gender-schematic decisions**
  - **Femininity**
  - **Masculinity**
- **Intersexed**
- **Feminist**
- **Sexual stratification**
  - **Low-technology tribal societies**
  - **Fortified households**
  - **Nonhouseholder class**
  - **Private households**
  - **Advanced market economies**
- **Perpetuating sexual stratification**
  - **Socialization**
  - **Situational constraints**
  - **Sexist ideologies**
  - **Institutionalized discrimination**
- **Hidden rape**
- **Ethgender**
- **State**

internet assignment

Although our discussion in this chapter has focused on the inequality of women relative to men, this does not mean that men suffer no disadvantages in the labor market. On the one hand, men as a group, especially white males, have an economic advantage relative to women in the labor market. On the other hand, men are concentrated in the most hazardous occupations. The Bureau of Labor Statistics publishes data on fatal and nonfatal occupation-related illnesses and injuries. (*Note:* Depending on the browser you are using, some tables may be difficult to read.)

ftp://stats.bls.gov/pub/news.release/osh2.txt

Document: Characteristics of Injuries and Illnesses Resulting in Absences from Work, 1994

ftp://stats.bls.gov/pub/news.release/osh.txt

Document: Workplace Injuries and Illnesses in 1994

ftp://stats.bls./gov/pub/news.release/cfoi.txt

Document: National Census of Fatal Occupational Injuries

What type of injuries and illnesses are most likely to be underreported in the Bureau of Labor Statistics Survey? Which eight industries accounted for more than 30 percent of the 6.3 million nonfatal illnesses and injuries? How many industries are male dominated? How many are female dominated? What proportion of nonfatal illnesses and injuries are specific to men? Review the tables in "Workplace Injuries and Illnesses in 1994." Are there some injuries and illnesses for which females are overrepresented? How do you account for this?

Now review Table 4 in "National Census of Fatal Occupational Injuries." Of the 6,588 fatalities in 1994, how many were men? Why do you think men have such high rates of injury, illness, and fatality?

LEE
U.S. TOP
LEVIS
STAROUP
315
LOJAS
CASA
FIOS
Lapidação de
Diamantes Antuerpia
LENTES DE CONTATO
PERFUMAR
245

Population and Family Life

With Emphasis on Brazil

Street scene in São Paolo, Brazil. ©Paulo Frieman/Sygma

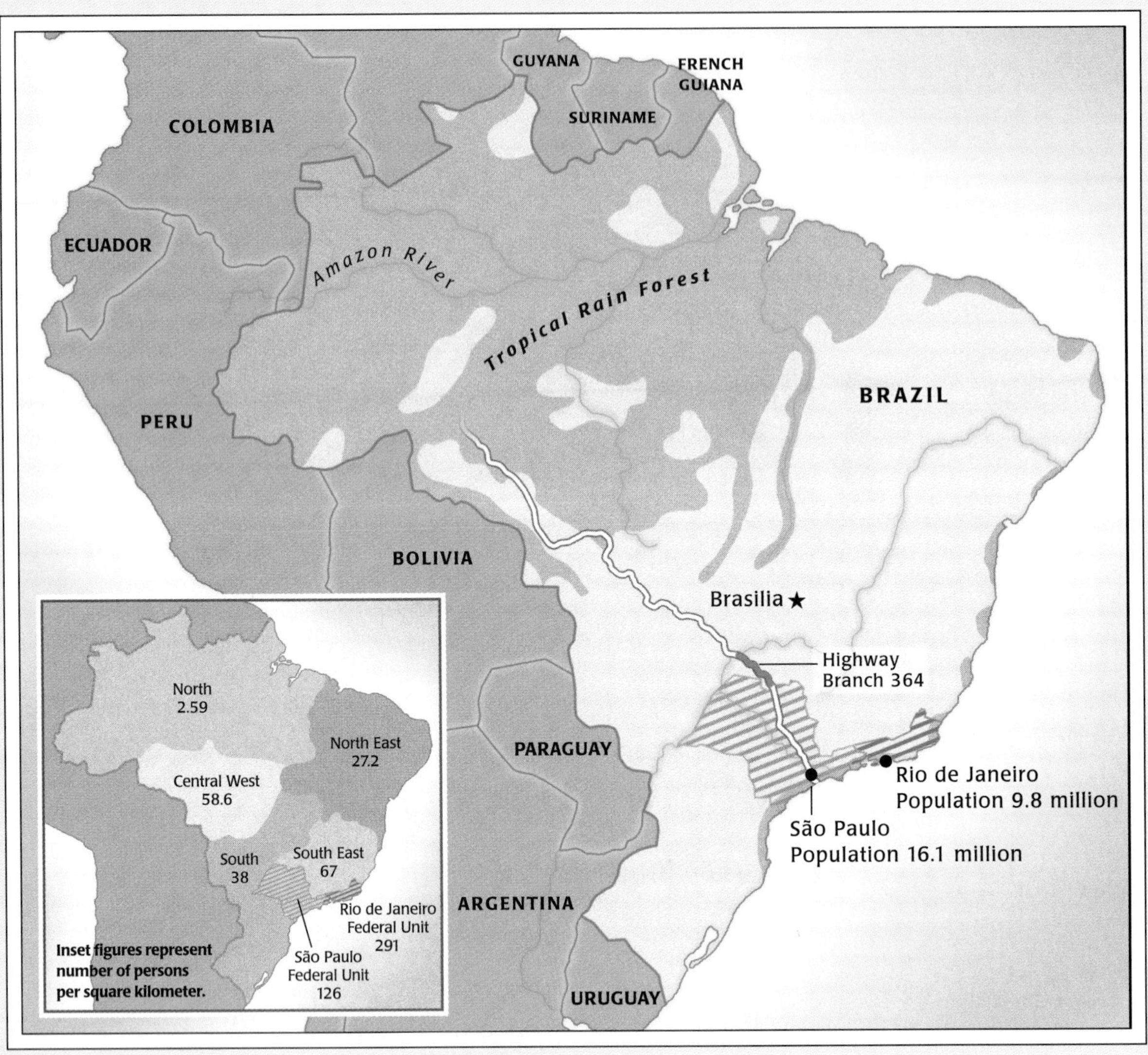

Road to Disaster?

Highway BR364 was built by the Brazilian government in 1980 to link its densely populated southeast region to its northern Amazon territories. The result was a sudden movement of poor people and economic "developers" into the rain forests and a sudden, appalling wave of forest destruction. Did the rapidly growing population of Brazil make this event inevitable? Is the virtual disappearance of the rain forest also unavoidable?

This chapter explores whether rapid population growth threatens the viability of life for families and for whole societies. Is humanity simply overwhelming the carrying capacity of Mother Earth or are other factors at work? For example, with more than 50 percent of the cultivable land in Brazil owned by less than 1 percent of the country's landowners—and much of that land lying idle—was it "inevitable" for the rain forests to be destroyed?

The chapter will also lead to us to consider carefully what the concept "family" means.

There were nine children before me and twelve after me. . . . Out of this total of twenty-two, seventeen lived, but four died in infancy, leaving thirteen still to hold the family fort.

Mine was a difficult birth, I am told. Both mother and son almost died. . . . After my birth Mother was sent to recuperate for some weeks and I was kept in the hospital while she was away. I remained there for some time, without name, for I wasn't baptized until my mother was well enough to bring me to church. **(Brown 1992:85)**

One of the common experiences of people in their early forties, which they seem to need to talk about, is having parents who are in their sixties and seventies. . . . Ten years ago, when we were in our early thirties, we talked about our children—about pregnancies and births, bottles versus breast-feeding, how to get "them" to sleep through the night, and then about toilet training and schools. We were primarily parents, and our own parents were secondary subjects. They were just grandparents and in-laws who were or were not helpful or demanding, visiting, vacationing, or whatnot. Only in the last four or five years, I realize, have I been telling and hearing stories about parents, usually with friends, but sometimes with people I have just met, if they are my age. **(Sayre 1983:124)**

There were a couple of years' worth of thinking that went into the decision to have another child. We already had a school-age son, and for a long time I thought we'd just have one child, because working and raising a family and trying to have a normal kind of life, too, was kind of hard work. But after Erik was in school, and I saw him developing into a really neat little person, I kept trying to decide in my own mind whether to do it again. . . . I felt if we were going to add to the family, this would be the time. **(Sorel 1984: 60–61)**

Rachel's day is long. She rises "when the sky is beginning to lighten," cooks breakfast, gets the children off to school and cleans the house. Then she sets off for the holding, two miles away up and downhill, where she tethers the animals to graze and gets down to planting, digging or weeding her corn, casava and cowpeas, eating a snack in the field. On her way home she gathers whatever firewood she finds. Then she fetches water, half an hour's walk away, with the return journey uphill. As the sun begins to set, she cooks the evening meal of [thick] maize porridge in a pot balanced on three stones, until it is as stiff as bread dough: "You are stirring solidly for an hour," she complains, "and it gets harder and harder until at the end the sweat is pouring off you." Getting the maize ground is another chore: twice a week she must trek two miles to the nearest neighbor who possesses a handmill. **(Harrison 1987:438–439)**

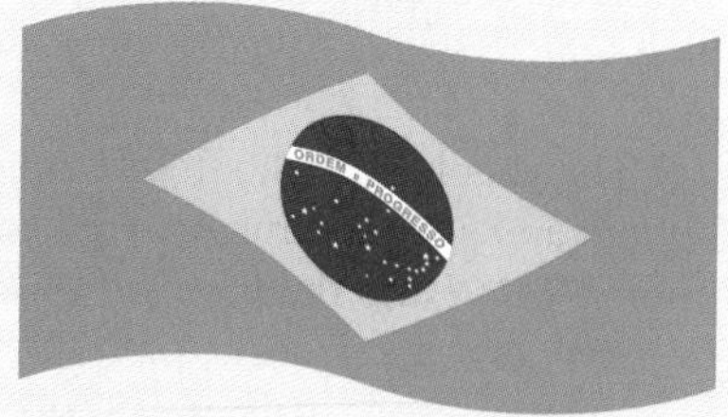

Why Focus on Brazil?

In this chapter, we focus on Brazil, which has the sixth largest population in the world and is the most populous country in South America with 155 million people (U.S. Department of State 1996). Approximately 32 percent of Brazil's population is less than fifteen years old; 5.2 percent is over sixty-five years old. In contrast, 22 percent of the U.S. population is less than fifteen years old; 12.6 percent is over sixty-five. These data tell us that Brazil has a relatively young population, whereas the United States has a relatively old population. To put it another way, in Brazil one in twenty people is over sixty-five compared with one in eight people in the United States. Obviously this difference in age structure affects family composition, the nature of family relationships, and the family-related issues considered most pressing in each country.

In addition to age structure, we will identify and compare other demographic characteristics that affect family composition, relationships, and issues in the United States and Brazil. Those demographic characteristics include these:

1. Births, including the number of children born to a woman and the spacing between those births. The number can range from none to a child born in every year of a woman's reproductive cycle.
2. Deaths, including how and when (infancy through old age) a family member dies
3. Employment, including how much and what kind of work each family member must do to sustain the family's standard of living

Population and family life are paired together in this chapter because the factors that affect a country's population size, distribution, and age-sex composition also affect the "typical" family's composition, relationships, and experiences. Obviously, in comparison with the United States, fewer people in Brazil, a country in which one in twenty people is over sixty-five, experience the need to talk about having parents in their sixties and seventies. Similarly, more men and women in Brazil will have to face and come to terms with a child's death, as 58 of every 1,000 live infants born dies before the age of one each year. In the United States, the comparable figure is 9 of every 1,000 live infants born. Finally, in a society such as Brazil where 73 percent of the rural population lives in absolute poverty, we would expect that many people would work a "long day" simply trying to secure basic resources their families need to survive.

The sociologists who study population size, distribution, and age-sex structure are known as **demographers**. We devote the first half of the chapter to an overview of the key demographic concepts and theories. In the second half, we will consider key demographic events: family composition, relationships, and experiences. One event that is central to demographic analysis is the Industrial Revolution.

Demographers The sociologists who study population size, distribution, and age-sex structure.

Doubling time The estimated number of years required for a country's population to double in size.

Infant mortality The number of deaths in the first year of life for every 1,000 live births.

The Industrial Revolution

As we learned elsewhere in this textbook (see Chapters 1, 6, and 7), the Industrial Revolution was not something unique to Western Europe and the United States. It was an event that forced people from even the most seemingly isolated and remote regions of the planet into a worldwide division of labor that continues through today (see "Company Towns: The Brazilian Experience"). Its effect is not uniform but varies according to country and region of the world.

We can classify the countries of the world into two broad categories with regard to industrialization: the mechanized rich and the labor-intensive poor. Comparable but misleading dichotomies include developed and developing, industrialized and industrializing, and First World and Third World. They are misleading names because they suggest that a country is either industrialized or is not. The dichotomy implies that a failure to industrialize is what makes a country poor, and it camouflages the fact that as Europe and North America plunged into industrialization, they took possession of Asia, Africa, and South America and then established nonindustrial economies there that were oriented to their industrial needs. The point is that labor-intensive poor countries were part of the Industrial Revolution from the beginning.

The World Bank, the United Nations, and other international organizations use a number of indicators to distinguish between mechanized rich and labor-intensive poor countries, including these:

- **Doubling time**, the estimated number of years required for a country's population to double in size
- **Infant mortality**, the number of deaths in the first year of life for every 1,000 live births

- **Total fertility**, the average number of children women bear over their lifetime
- **Per capita income**, the average income that each person in a country would receive if the country's gross national product were divided evenly
- The percentage of the population engaged in agriculture
- **Annual per capita consumption of energy**, the average amount of energy each person consumes over a year. When per capita energy consumption is low, it suggests that, for the vast majority of people, work is labor-intensive rather than machine-intensive. In labor-intensive work, considerable physical exertion is required to produce food and goods.

When the term **labor-intensive poor** (or an equivalent term) is used to characterize a country, it means that country differs markedly on these and other indicators from countries considered to be industrialized (Stockwell and Laidlaw 1981; see Table 12.1, p. 370). According to these measures, approximately 107 countries are low- or lower-income economies (labor-intensive poor), and 40 are high-income economies (mechanized rich) (Lutz 1994; see Figure 12.2, p. 371).

The official statistics used to identify labor-intensive poor and mechanized rich countries are counts of how many people have experienced an event. For example, infant mortality is an annual count of how many infants die for every 1,000 born. A high infant mortality rate tells us that for many families, the death of an infant is a common experience.

Industrialization and Brazil

Brazil has a powerful economy (the eighth largest in the world) that is likely to surpass the economies of Canada, Italy, and Great Britain before the year 2000. Moreover, after the United States, it is the world's largest exporter of agricultural products. Its major exports include soybeans, coffee, transport equipment, footwear, orange juice, iron ore, and steel products. The benefits that we associate with industrialization, however, have bypassed the majority of Brazilians, who live at or below subsistence level. Brazil is a country in which poverty is widespread and chronic (not the result of some temporary misfortune). Sixty percent of its population can be classified as poor (Calsing 1985).

Large changes are taking place in the world's labor-intensive poor countries. Behind the crowded downtown street scene from São Paulo, Brazil (see p. 362) is a national economy that is right now about to outrank the economies of Canada, Italy, and Great Britain.

The World Bank classifies Brazil as a country with a middle-income to upper-middle economy. Yet, as many as 60 percent of Brazil's people live in extreme poverty. The unequal distribution of wealth is reflected by the fact that the percentage share of household income for the most affluent 20 percent of the population is 63 percent, whereas the percentage share for the least affluent 40 percent of the population is only 8 percent (The World Bank 1990). In Brazil the effects of industrialization vary across and within its five major regions:

- The North (Amazon Basin covering half the country with vast reaches of largely uninhabited tropical forest)
- The Northeast (semiarid scrubland prone to periodic drought and massive flooding, heavily settled and poor)
- The South (rich farmland and pasture lands and large modern cities with a large and relatively prosperous population)
- The Southeast (huge, densely populated urban centers, including the city and state of São Paulo)
- The Central West (home of one of the earth's major ecological frontiers and some of the largest cattle ranches in South America, sparsely populated). The major city in this region (the Central West of Brazil) is Cuiaba, which is the site of a satellite tracking station maintained by NASA (Beresky 1991).

Total fertility The average number of children women bear over their lifetime.

Per capita income The average income that each person in a country would receive if the country's gross national product were divided evenly.

Annual per capita consumption of energy The average amount of energy each person consumes over a year.

Labor-intensive poor A term used to characterize a country that differs markedly from industrialized countries on indicators such as doubling time, infant mortality, total fertility, per capita income, and annual per capita consumption of energy.

Company Towns: The Brazilian Experience

The need to provide accommodation at new sites of economic activity has long been integral to the development process, from pithead villages, industrial and philanthropic settlements (Pullman, Port Sunlight) to socialist planning in eastern Europe (Nova Huta, Dunaujvaros). Single enterprise communities continue to be established in the Third World, most commonly where resource endowments requiring substantial capital investment occur in environments lacking established settlement. Such "company towns" are particularly associated with agribusiness, lumbering, mining, and industries dependent on bulky raw materials or cheap energy. Their precise form varies from temporary camps exploiting ephemeral resources such as timber, or at project construction sites; tracts of company-built workers' housing grafted onto existing settlements; to complete "new towns" set into existing urban networks or at the resource frontier.

Figure 12.1a The Location of Company Towns in Brazil

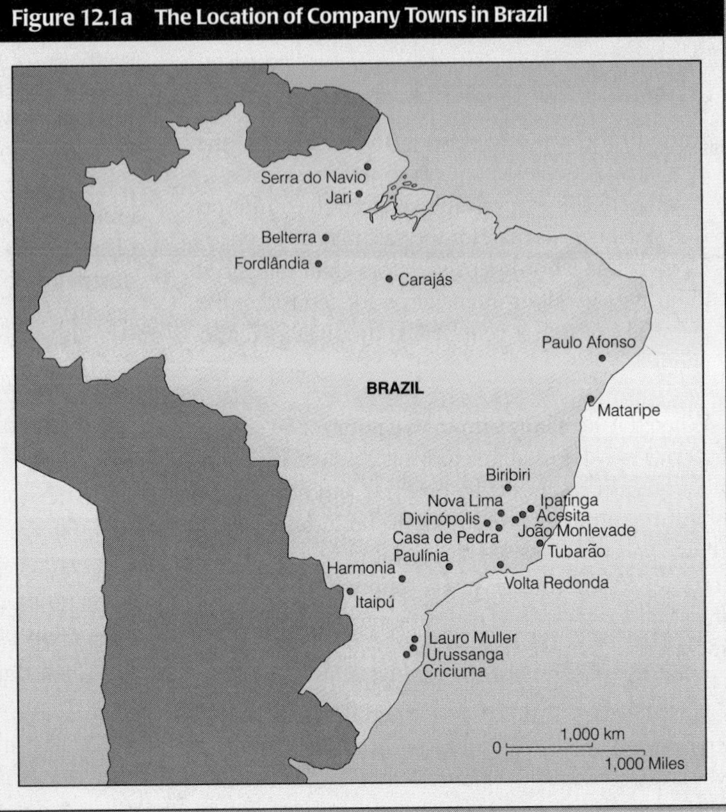

In Brazil the earlier company towns were associated with nineteenth-century mines and water-powered textile mills (Figure 12.1a). In [the region of Brazil called] Minas Gerais, the little cotton town of Biribiri was laid out at a remote waterfall in the Serra do Espinhaço, consisting of the mill, church, store, schoolhouse and a neat square of 40 houses. The British-owned St. John del Rey gold-mining company, established in 1834, built a town for its workers at Nova Lima where, until the 1950s, there was social segregation between British managers and foremen, and Brazilian laborers. Railway building fostered the construction of company housing adjacent to railyards and workshops, as at Divinópolis.

The (Brazilian) steel industry has created several company towns where plant size, coupled with raw material requirements, compelled companies to build steel mills and substantial settlements at green field sites. The iron ores of Minas Gerais generated steel towns in the Doce Valley, at Monlevade (1935), Acesita (1944) and Ipatinga (1956). Brazil's class company town is Volta Redonda, established at a green field site in the Paríba valley in 1941, to service the country's first coke-fired steelworks. The (company) Cia. Siderúrgica Nacional, initially built 2,000 houses, adding a further 2,000 by 1952. It also provided churches, hotels, cinemas, sports facilities, hospitals, primary and secondary schools and a technical school to train employees. The town is socially zoned, with workers' housing near the plant, and more substantial properties for technicians and managers on the hillsides. Population grew from 200 in 1941 to over

Approximately 25 percent of Brazil's people live in the Southeast, which is dominated by the metropolises of São Paulo[1] and Rio de Janeiro — two of the world's twenty largest cities and centers of multinational commercial, industrial, and agricultural activity (see the chapter-opening map). As many as 30 percent of the people in these two cities live in urban squatter settlements called *favelas*,[2] in makeshift dwellings constructed from cardboard, metal, or wood, with inadequate sewers, running water, and electricity.

[1]According to Latin American Studies scholar Warren Dean (1991), "São Paulo has to be explained. The rest of Brazil and indeed the rest of Latin America has not experienced that city's and state's breakthrough of sustained economic development based on diversified and technologically complex industrial production" (p. 649).

[2]Many of the people who live in these squatter settlements have been pushed off the surrounding plantation land because of mechaniza-

30,000 in 1954 and 180,000 in 1980. In addition, the company built small settlements at its iron and coal mines (Casade Pedra, Lauro Muller, etc.).

Other company towns have been associated with manganese (Serra do Navio), paper (Harmonia), oil refining (Matairpe and Paulínia), and hydro-electric projects (Paulo Afonso and Itaipú). In Amazonia there have been abortive plantation settlements at Fordiândia, Belterra and Jari.

The domination of these settlements by a single enterprise influences their economy, population and townscape. The company provides employment, controls rents, zones lands, and supplies essential utilities, social services, recreational facilities, and sometimes retail provision. Populations have grown rapidly, and tend to be immigrant, youthful and relatively affluent; employment opportunity is male-dominated. Town plans are simple, uniform in layout and date, with standardized housing which varies only in size and site with employee status.

Brazil's most recent company town is at the iron mine of Carajás (Figure 12.1b). Its sponsor, the Cia. Vale do Rio Doce, had built earlier towns at its mines in Minas Gerais and port of Tubarão (Espírito Santo). Carajás was established in 1980, for a planned workforce of 2,000 to be housed in single men's flats and family houses, with the company providing essential services. The townsite is screened from the mine, and includes leisure areas and protected forest reserves. Though carefully planned, Carajás demonstrates some of the difficulties engendered by company towns, being located in an ecologically sensitive area, offering very little employment for women, and attracting spontaneous migrants to unplanned adjacent settlements.

Figure 12.1b Layout of the Company Mining Town of Carajás, Pará

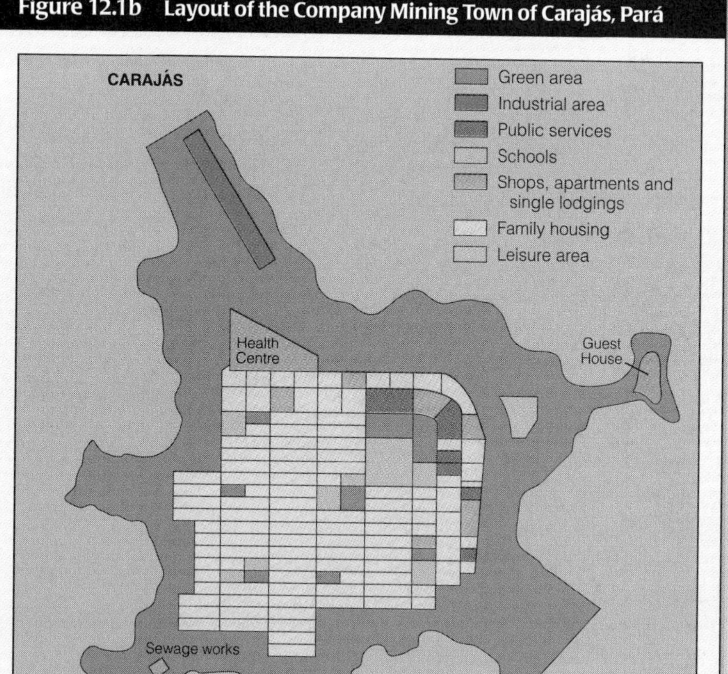

When a private company founds an isolated company town, what implications does that have for the people who will need to come to work there? What sorts of advantages and disadvantages is a company town likely to present to them? What sorts of social ties may be broken? What new ones reinforced? What sorts of personal opportunities might be diminished?

Source: From "Company Towns: The Brazilian Experience" by John Dickenson in *Atlas of World Development* edited by Tim Unwin. New York: John Wiley & Sons Ltd. Copyright © Developing Areas Research Group. Reprinted by permission of the Developing Areas Research Group of the Royal Geographical Society (with the Institute of British Geographers) and John Wiley & Sons, Ltd.

tion and have come to the cities in search of a better life. Latin American Studies scholar Thomas G. Sanders (1988) maintains that outsiders stereotype *favelas* as centers of crime, prostitution, and extreme family disorganization but that most studies show the opposite. The vast majority of *favelas* dwellers are "honest, employed, hardworking, and have high aspirations for their children. They live in *favelas* because of low cost, location near their workplaces, and often lack of viable alternative housing solutions" (p. 5).

About one-third of Brazil's people, most of whom are the descendants of slaves who worked the Portuguese sugarcane plantations between the sixteenth and nineteenth centuries, live on the drought-stricken and exhausted land in the northeastern part of the country. Even after slavery was abolished in 1888, the plantation agricultural system, which was oriented toward export, continued. Today the most productive

Table 12.1 Demographic Differences Between Labor-Intensive Poor and Mechanized Rich Countries

Labor-intensive poor countries are very different from mechanized rich countries with regard to indicators used by international organizations to classify countries. Data are presented for the ten most populous countries in the world. Compare the U.S. figures with those for any of the labor-intensive poor countries. What contrasts do you see? How might each contrast affect the people living in that country?

	Population Doubling Time (years)	Infant Mortality (per 1,000 births)	Total Fertility (children born per woman)	Per Capita GNP Income (in U.S. $)	Annual per Capita Consumption of Energy (in kilograms of oil equivalent)
Labor-Intensive Poor					
Nigeria	23	72.6	6.3	1,250	109
Bangladesh	29	104.6	4.4	1,040	70
Pakistan	23	99.5	6.4	1,930	389
Brazil	37	58.0	2.4	5,580	1,589
India	34	76.3	3.4	1,360	324
Indonesia	40	65.0	2.7	3,090	207
China (mainland)	53	52.1	1.8	2,500	593
Mechanized Rich					
United States	89	7.9	2.1	25,850	11,236
Germany	(–)	6.3	1.5	16,580	5,683
Japan	217	4.3	1.6	20,200	6,263

Source: U.S. Central Intelligence Agency (1995).

land in the Northeast is still used to grow sugarcane, soybeans, and other export crops, leaving the peasants with no land on which to grow subsistence crops.

Over the years landless peasants from the Southeast and Northeast have migrated to overcrowded cities in search of work or have sought land in the sparsely populated interior of the country. The interior houses the Amazon forest, the largest tropical jungle and rain forest in the world. The rate of migration to the Amazon region accelerated when the government began constructing a network of highways and roads in the 1960s to connect the Amazon region with the rest of the country. The network of roads opened the land to foreign and Brazilian investors, those living on the fringes of the cities, and the landless and unemployed from the Northeast. To convert forest to pasture and farmland, settlers cut the trees and other vegetation, allowed them to dry, and then set them afire during the dry season, which runs from June to October. (Brazilian and U.S. scientists, monitoring this practice via satellite, counted 170,000 fires in 1987.) The ash from the burning trees and other foliage acts as a fertilizer for a few years, but after that time the land no longer can support crops. The land then is abandoned to cattle ranchers, whose herds graze it for a few years until it is totally exhausted (Simons 1988).

Although isolated geographically from the rest of the country, the Amazon region was not uninhabited; it was occupied by indigenous people, rubber tappers, nut gatherers, and others whose forest-centered livelihood was disrupted by the highway construction and subsequent human migration. The forest dwellers, especially the indigenous people, suffered (and still suffer) cultural and physical extinction at the hands of mining companies that extract mineral wealth buried beneath the rain forest, lumber companies in search of rare jungle trees, cattle ranchers grazing their herds on deforested land, government development projects, and land-hungry peasants. Although the government has set aside land for these peoples, the forest dwellers are under pressure to abandon their language and culture and enter the dominant society. Unfortunately, if they enter society, they do so as landless peasants, low-paid laborers, or beggars (Caufield 1985).

The background information on Brazil sets the context for thinking about industrialization's wide-ranging and uneven consequences. To help us grasp these consequences, we turn to a model that outlines historical changes in birth- and death rates in Western Europe and the United States, especially as these changes are affected by the Industrial Revolution. This model is the theory of the demographic transition.

Figure 12.2 The Highest per Capita Incomes Are Found in the North

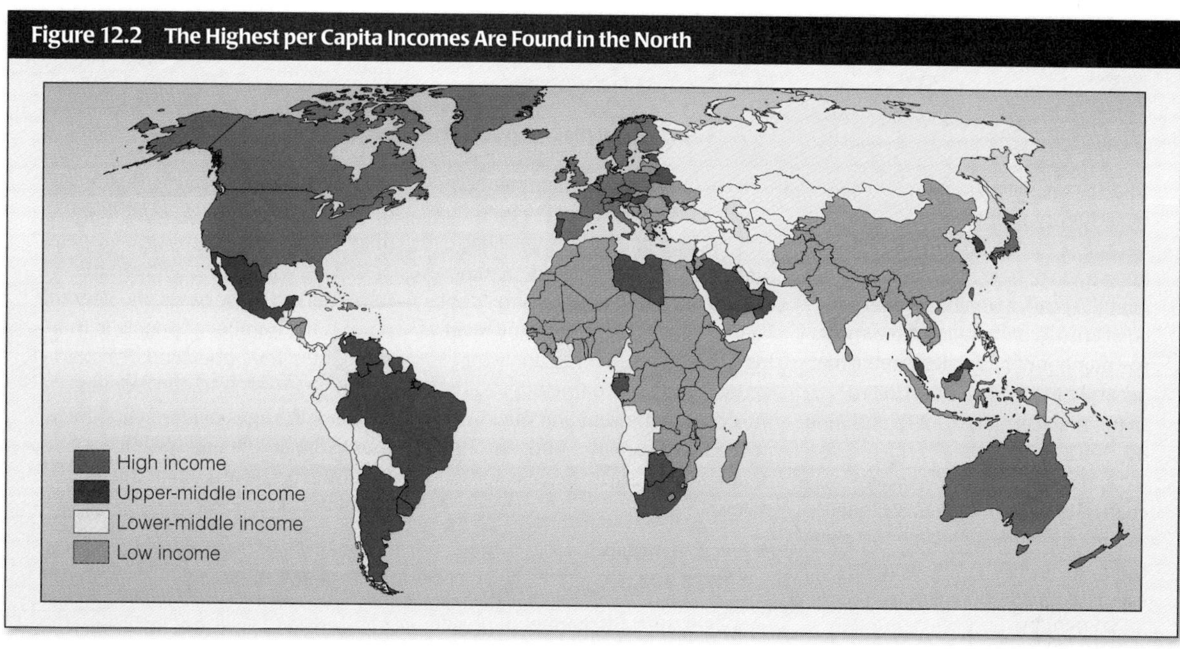

Source: Adapted by John R. Weeks in 1996 from data in the World Bank (1994), Table 1.

After we describe the model, we discuss how well it applies to non-Western countries such as Brazil.

The Theory of Demographic Transition

In the 1920s and early 1930s, demographers observed birth- and death rates in various countries and noticed a pattern: both birth- and death rates were high in Africa, Asia, and South America; death rates were declining while birthrates remained high in Eastern and Southern Europe; and birthrates were declining and death rates were low in Western Europe and North America. Demographers observed that Western Europe and North America had the following sequence with regard to birth- and death rates (see "How Demographers Measure Change").

1. Birth- and death rates were high until about the middle of the eighteenth century, at which time death rates began to decline.
2. As the death rates decreased, the population grew rapidly because there were more births than deaths. The birthrates began to decline around 1800.
3. By 1920, both birth- and death rates had dropped below 20 per 1,000 (see Figure 12.3).

On the basis of these observations, demographers put forth the theory of the demographic transition: a country's birth- and death rates are linked to its level of industrial or economic development.

This model documents the general situation and should not be construed as a detailed description of the experiences of any single country. Even so, we can say that all countries have followed or are following the essential pattern of the demographic transition, although they differ with regard to the timing of the declines and the rate at which their populations increase after death rates begin to decline. The theory of the demographic transition includes more than this pattern; it is also an explanation of the events that caused birth- and death rates to drop in the mechanized rich countries (with the exception of Japan). The factors that underlie changes in birth- and death rates in the labor-intensive poor countries, however, are fundamentally different from those that caused such changes in the so-called developed or industrialized countries.

Stage 1: High Birth- and Death Rates

For most of human history — the first 2 to 5 million years — populations grew very slowly, if at all. World population remained below 1 billion until A.D. 1800, at which point it began to grow explosively. By 1930, the world's population had increased to 2 billion; during

U.S. In Perspective

How Demographers Measure Change

To discuss change, demographers must specify a time period (a year, a decade, a century) over which they keep track of how many times an event (a birth, a death, a move) occurs. The accompanying table shows the number of births and deaths that occurred in Brazil and the United States between July 1, 1991, and June 30, 1992.

The simplest way to express change is in absolute terms—that is, to state the number of times an event occurred. (In 1991, there were an estimated 3,955,000 births in Brazil and 3,553,000 births in the United States.) Expressing change in absolute terms is not very useful, however, for making comparisons between countries that have different population sizes. Consequently, for comparative purposes demographers calculate *rates* of births, deaths, and migrations, usually per 1,000 people in the population. Rates are calculated by dividing the number of times an event occurs by the size of the population at the onset of the year and then multiplying that figure by 1,000. The 1991 death rate for Brazil is calculated as follows:

$$\frac{1{,}107{,}000}{158{,}202{,}000} \times 1{,}000 = 7$$

When the total number of people in the population is used as the denominator, the rates are called *crude rates.* Sometimes demographers wish to know how many births, deaths, or moves occur within a specific segment of the population (among males or among females ages fifteen to forty-four). In these cases, the denominator is the number of people in that segment of the population. For example, there are 35,356,000 women between the ages of fifteen and forty-four in Brazil. The age-specific birthrate for these women thus is calculated as follows:

$$\frac{3{,}955{,}000}{35{,}356{,}000} \times 1{,}000 = 112$$

Source: Data adapted from U.S. Central Intelligence Agency (1992:47, 358).

	Mid-1992 Population	Births	Deaths	Births per 1,000 Population	Deaths per 1,000 Population
Brazil	158,202,000	3,955,000	1,107,000	25	7
United States	254,521,000	3,553,000	2,291,000	14	9

the next fifty years, it increased by another 2.5 billion. During the 1980s, the world's population increased by about 750 million to its present size of 5.4 billion. Demographers speculate that growth until 1800 was slow because **mortality crises**—frequent and violent fluctuations in the death rate caused by war, famine, and epidemics—were a regular feature of life.

Stage 1 is often referred to as the stage of high potential growth: if something happened to cause the death rate to decline—for example, improvements in agriculture, sanitation, or medical care—population would increase dramatically. In this stage, life is short and brutal; the death rate is almost always above 50 per 1,000. When mortality crises occur, the death rate seems to have no limit. Sometimes half of the population is affected, as when the Black Plague struck Europe, the Middle East, and Asia in the middle of the fourteenth century and recurred for approximately 300 years. It is estimated that within twenty years of its onset, the plague killed up to three-fourths of the people in the affected populations.[3]

Mortality crises Frequent and violent fluctuations in the death rate caused by war, famine, and epidemics.

[3]The medieval Italian writer Giovanni Boccaccio ([1353] 1984) recorded his impressions of the event:

> It began in both men and women with certain swellings either in the groin or under the arm-pits [and] began to spread indiscriminately over every part of the body; and after this, the symptoms of the illness changed to black . . . almost all died after the third day (p. 728).

Figure 12.3 The Demographic Transition

The theory of the demographic transition can be represented by a graph of the historical changes in birth- and death rates among the mechanized rich countries.

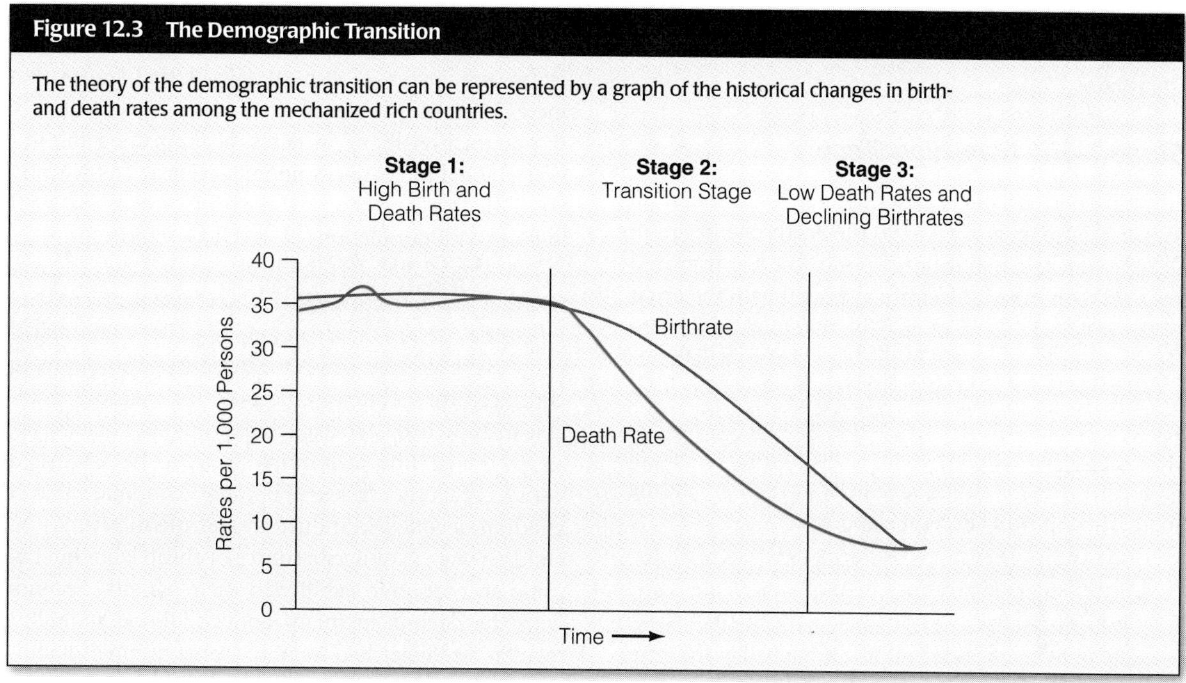

Another mortality crisis, which has not received as much attention as the Black Plague, affected the indigenous populations of North America when the Europeans arrived in the fifteenth century. A large proportion of the native population died because they had no resistance to diseases such as smallpox, measles, tuberculosis, and influenza, which the colonists brought with them. Others simply were killed by colonists because they refused to work as slaves on plantations. Historians debate what proportion of the native population died as a result of this contact; estimates range between 50 and 90 percent.

The point is that in stage 1 average life expectancy at birth remained short—perhaps between twenty and thirty-five years — with the most vulnerable groups being women of reproductive age, infants, and children under age five. (The high infant and child mortality rate pulled down the average life expectancy at birth; many people managed to live well beyond age thirty.) It is believed that women gave birth to large numbers of children and that the crude birthrate was about 50 per 1000, the highest rate possible for human beings. Families remained small, however, because one infant in three died before reaching age one, and another died before reaching adulthood. If the birthrate had not remained high, the society would have become extinct. Demographer Abdel R. Omran (1971) estimates that in societies in which life expectancy at birth is thirty years, each woman must have an average of seven live births to ensure that two children survive into adulthood. She must bear six sons to ensure that at least one survives until the father reaches age sixty-five (if the father lives that long).

In Western Europe before 1650, high mortality rates were associated closely with food shortages and famines. Even when people did not die directly from starvation, they died from diseases that preyed on their weakened physical state. Thus, Thomas Malthus ([1798] 1965), a British economist and an ordained Anglican minister, concluded that "the power of population is so superior to the power in the earth to produce subsistence for man, that premature death must in some shape or other visit the human race" (p. 140). According to Malthus, **positive checks** served to keep population size in line with the food supply. He defined positive checks as events that increase mortality, including epidemics of infectious and parasitic disease, war, and famine. Malthus believed that the only moral way to prevent populations from growing beyond a size that could not be supported by the food supply was delayed marriage and celibacy. He regarded any other

Positive checks Events that increase mortality, including epidemics of infectious and parasitic disease, war, and famine.

method—such as infanticide, homosexuality, or sterility caused by sexually transmitted diseases — as immoral.

Stage 2: The Transition Stage

Around 1650, mortality crises became less frequent in Western Europe, and by 1750 the death rate began to decline slowly in that region. The decline was triggered by a complex array of factors associated with the onset of the Industrial Revolution. The two most important factors were (1) increases in the food supply, which improved the nutritional status of the population and increased its ability to resist diseases, and (2) public health and sanitation measures, including the use of cotton to make clothing and new ways of preparing food. The following excerpt elaborates:

> The development of winter fodder for cattle was important; fodder allowed the farmer to keep his cattle alive during the winter, thereby reducing the necessity of living on salted meats during half of the year. . . . [C]anning was discovered in the early nineteenth century. This method of food preservation laid the basis for new and improved diets throughout the industrialized world. Finally, the manufacture of cheap cotton cloth became a reality after mid-century. Before then, much of the clothes were seldom if ever washed, especially among the poor. A journeyman's or tradesman's wife might wear leather stays and a quilted petticoat until they virtually rotted away. The new cheap cotton garments could easily be washed, which increased cleanliness and fostered better health. (Stub 1982:33)

Contrary to popular belief, advances in medical technology had little influence on death rates until the turn of the twentieth century, well after improvements in nutrition and sanitation had caused dramatic decreases in deaths due to infectious diseases.

Over a hundred-year period, the death rate fell from 50 per 1,000 to below 20 per 1,000, and life expectancy at birth increased to about fifty years of age. As death rates declined, fertility remained high. It may even have increased temporarily, because improvements in sanitation and nutrition enabled women to carry more babies to term. With the decrease in the death rate, the **demographic gap**—the difference between birthrates and death rates—widened and population size increased substantially. **Urbanization**, an increase in the number of cities and the proportion of the population living in cities, accompanied the unprecedented increases in population size. (As recently as 1850, only 2 percent of the world's population lived in cities with populations of 100,000 or more.)

Around 1880, fertility began to decline. The factors that caused birthrates to drop are unclear and subject to debate among demographers. But one thing is certain: the decline was not caused by innovations in contraceptive technology, because the methods available in 1880 had been available throughout history. Instead, the decline in fertility seems to be associated with several other factors. First, the economic value of children declined in industrial and urban settings, as children were no longer a source of cheap labor but became an economic liability to their parents. Second, with the decline in infant and childhood mortality, women no longer had to bear a large number of children to ensure that a few survived. Third, a change in the status of women gave them greater control over their reproductive lives and made childbearing less central to women's lives. Scholars disagree, however, about the specific conditions under which women are able to control their reproductive lives.

Demographic gap The difference between birthrates and death rates.

Urbanization An increase in the number of cities and the proportion of the population living in cities.

Stage 3: Low Death Rates and Declining Birthrates

Around 1930, both birth- and death rates fell below 20 per 1,000, and the rate of population growth slowed considerably. Life expectancy at birth surpassed seventy years, an unprecedented age. The remarkable successes in reducing infant, childhood, and maternal mortality rates were such that accidents, homicides, and suicide have become the leading causes of death among young people. Because the risk of dying from infectious diseases is reduced, people who would have died of infectious diseases in an earlier era survive into middle age and beyond, when they face the elevated risk of dying from degenerative and environmental diseases (heart disease, cancer, strokes, and so on). For the first time in history, persons fifty years of age and over account for more than 70 percent of the annual deaths. Before this stage, infants, children, and young women accounted for the largest share of deaths (Olshansky and Ault 1986).

As death rates decline, disease prevention becomes an important issue. The goal is to live not only a long life but a quality life (Olshansky and Ault 1986; Omran 1971). As a result, people become conscious of the link between their health and their lifestyle (sleep, nutrition, exercise, and drinking and smoking habits). In addition to low birth- and death rates, stage 3 is distinguished by an unprecedented emphasis on consumption (made possible by advances in manufacturing and food production technologies).

Theoretically, all of the mechanized rich countries are in stage 3 of the demographic transition. At one time some sociologists and demographers maintained that the so-called developing countries would follow this model of development as they industrialized. However, the nature of industrialization in labor-intensive poor countries is so fundamentally different from the version that occurred in the mechanized rich countries that these countries are unlikely to follow the same path.

The "Demographic Transition" in Labor-Intensive Poor Countries

One reason that the nature of industrialization in labor-intensive poor countries was so fundamentally different from that of mechanized rich countries is that most labor-intensive poor countries were once colonies of mechanized rich countries. The mechanized rich countries established economies oriented to their own industrial needs, not the needs of the countries they colonized. In the case of Brazil, the Portuguese forced the indigenous peoples to grow crops and mine metals and materials for export to the mother country, but they did not allow the people to develop and establish their own native industries. For more than a century following its independence from Portugal in 1822, Brazil possessed a one-crop, export-oriented economy that was dominated first by sugarcane, then by rubber, and then by coffee. As is usually the case with one-product export economies, Brazil's economy experienced cycles of boom and bust. The rubber boom is illustrative. Between 1900 and 1925, the Brazilian city of Manaus in the Amazon was the rubber capital of the world, supplying 90 percent of the world's demand. The profits were reaped by a handful of Amazon rubber barons, who controlled huge plantations on which Indians and peasants from northeastern Brazil worked under conditions resembling slavery. The rubber boom came to an end when Henry Wickham, a British businessman, smuggled 70,000 rubber plant seedlings out of Brazil to plant in Singapore. Eventually, Asian rubber priced Brazilian rubber out of the world market (Revkin 1990). Even today, multinational corporations employ workers in developing countries to do low-skill, low-paying, labor-intensive work whose products are exported to people in the developed countries.

The fact of colonization helps explain why the model of the demographic transition does not apply to labor-intensive poor countries. When compared with mechanized rich countries, labor-intensive poor countries differ on several characteristics: they have a faster decline in death rates, relatively high birthrates despite declines in the death rate, a more rapid increase in population size, and greater (unprecedented) levels of rural-to-urban and rural-to-rural migration.

Death Rates

The decline in death rates in the labor-intensive poor countries occurred much faster than that in the mechanized rich countries; only 20 to 25 years (rather than 100 years) elapsed before the death rate fell from 50 per 1,000 to less than 10 per 1,000. Demographers attribute the relatively rapid decline to cultural diffusion (see Chapter 4). That is, the labor-intensive poor countries imported some Western technology—pesticides, fertilizers, immunizations, antibiotics, sanitation practices, and higher-yield crops—which caused an almost immediate decline in the death rates. Brazil, for example, imported DDT to eradicate the mosquitoes that carried diseases such as yellow fever and malaria. In addition, the Brazilian government has developed antidotes for snake, spider, and scorpion bites and has distributed them to clinics around the country. Among other things, the government is also working to stamp out Chagas' disease and schistosomiasis, serious disorders caused by parasites in the blood (Nolty 1990).

The swift decline in death rates has caused the populations in developing countries to grow very rapidly. Some demographers believe that developing countries may be caught in a **demographic trap**—the point at which population growth overwhelms the environment's carrying capacity:

> Once populations expand to the point where their demands begin to exceed the sustainable yield of local forests, grasslands, croplands, or aquifers, they begin directly or indirectly to consume the resource

Demographic trap The point at which population growth overwhelms the environment's carrying capacity.

As the example of Brazil demonstrates, the issue of population growth is not simply a question of sheer numbers but also involves the carrying capacity of the land.

©Alain Keeler/Sygma

> base itself. Forests and grasslands disappear, soils erode, land productivity declines, water tables fall, or wells go dry. This in turn reduces food production and incomes, triggering a downward spiral. (Brown 1987:28)

Nepal (between China and India) and Costa Rica (in Central America) represent two cases of the demographic trap. Because of population pressures, the peasants of Nepal have been forced to cultivate steep and forested hillsides, and the women have no choice but to collect feed for livestock and wood for cooking and heating from these sites. As the forests recede, the daily journey for fodder and fuel grows longer and cultivation becomes more difficult. As a result, family incomes have dropped and diets have deteriorated. In fact, the malnutrition rates in the Nepal villages are correlated strongly with deforestation rates (Durning 1990).

In Costa Rica, a country once covered by tropical forest, the demographic trap is fueled by population growth and land policies of the past two decades, which favor a small number of cattle ranch owners (perhaps 2,000) and disregard the needs of peasants who have become landless under these policies. About half the nation's cultivable land is used to raise cattle, an industry that requires little labor and hence provides few employment opportunities for landless peasants. "The rising tide of landlessness has spilled over into expanding cities, onto the fragile slopes, and into the forests, where families left with little choice accelerate the treadmill of deforestation" (Durning 1990:146).

Keep in mind that the point at which population growth overwhelms the environment's carrying capacity is usually not caused by population pressures alone. In 1980, the Brazilian government built highway BR364 to link São Paulo with the Amazon state of Rondônia. Significant numbers of landless peasants were offered 100 acres of free land in the Amazon forest provided they clear the land and construct a house. Unfortunately, much of the land turned out to be sandy and unfit to sustain crops; millions of acres of forest land were destroyed without easing the land problem. Technically, one could argue that the population growth in the Amazon overwhelmed its capacity to feed people. On the other hand, critics of the plan argue that government officials sent peasants blindly into the jungle, unprepared for how difficult it would be to make a living. In addition, the 100-acre plots were divided without reference to terrain (many plots consisted of rocky hills and many had no source of water). Most importantly, the chronic landlessness is caused by the system of land distribution, not by a lack of land: 50 percent of the cultivable land in Brazil is owned by less than 1 percent of all landowners, and much of that land remains idle. Thus, in the final analysis, it is land distribution policies, not population pressures per se, that send peasants into the hillsides, forests, and grasslands (Sanders 1986).

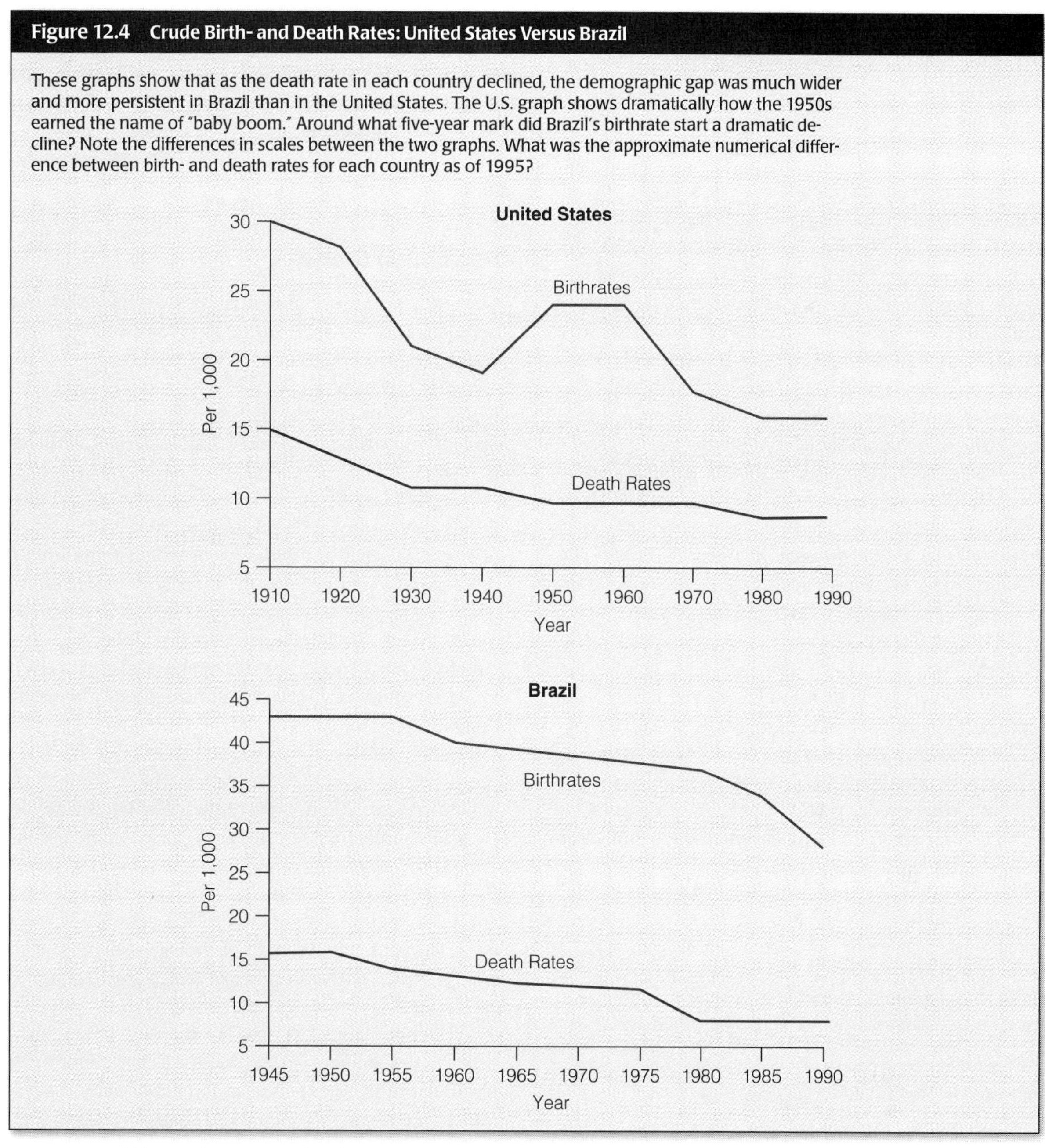

Figure 12.4 Crude Birth- and Death Rates: United States Versus Brazil

These graphs show that as the death rate in each country declined, the demographic gap was much wider and more persistent in Brazil than in the United States. The U.S. graph shows dramatically how the 1950s earned the name of "baby boom." Around what five-year mark did Brazil's birthrate start a dramatic decline? Note the differences in scales between the two graphs. What was the approximate numerical difference between birth- and death rates for each country as of 1995?

Source: *Statistical Abstracts of Latin America* (1989).

Birthrates

Birthrates, although still high, are beginning to show signs of decline in most developing countries. The demographic gap remains wide, however (see Figure 12.4). It is not clear exactly which factors have caused total fertility to decline in developing countries. Sociologist Bernard Berelson (1978) identifies some important "thresholds" associated with industrialization and declines in fertility. These include the following conditions:

1. Less than 50 percent of the labor force is employed in agriculture.
2. At least 50 percent of persons between the ages of five and nineteen are enrolled in school.

3. Life expectancy is at least sixty.
4. Infant mortality is less than 65 per 1,000 live births.
5. Eighty percent of the females between the ages of fifteen and nineteen are unmarried.

Most of these conditions have been met in Brazil. Although total fertility is still high, it has declined over the past two decades from 5.75 children to 2.39 children per woman (Brooke 1989; U.S. Central Intelligence Agency 1996). The factors identified by Berelson may indeed be responsible for this decline, along with other factors such as changes in government family policy and the near-universal access to television. It is true that during the 1960s, the Brazilian government established pronatalist policies designed to increase the size of the population, with particular emphasis on increasing the population in its Amazon states. Those who had children received tax breaks and maternity bonuses (U.S. Department of the Army 1983). Finally, however, the government abandoned these policies because they only served to increase the number of people living in already densely populated regions of the country.

Wider access to television has played some role in declining fertility in Brazil. Since 1960, the percentage of households with television sets has risen from 5 to 72 percent, and many Brazilians now are able to watch programs that encourage new norms. For example, the popular Brazilian soap operas feature small, consumer-oriented families. When large families are part of the drama, they usually are depicted as poor and miserable (Brooke 1989).

Population pyramid A series of horizontal bar graphs, each representing a different five-year age cohort.

Cohort A group of people who share a common characteristic or life event.

Expansive pyramids Population pyramids characteristic of labor-intensive poor countries that are broadest at the base with each successive bar smaller than the one below it, showing that the population is increasing in size and composed disproportionately of young people.

Constrictive pyramids Population pyramids characteristic of some European societies, most notably Switzerland and western Germany, that are narrower at the base than in the middle, showing that the population is composed disproportionately of middle-aged and older people.

The speed at which death rates declined in relation to birthrates in developing countries is only one of several important differences between the mechanized rich countries and the labor-intensive poor countries. Two other important differences are (1) the rate at which population is increasing and (2) migration, the movement of people from one area to another.

Population Growth

The rate at which populations increase is tied to a number of factors, including the sex-age composition. Generally, population size increases faster in countries with a disproportionate number of young adults (men and women of reproductive age) than in countries with a disproportionate number of middle-aged and older people.

A population's age and sex composition is commonly represented by a **population pyramid**, a series of horizontal bar graphs, each of which represents a different five-year age cohort. (A **cohort** is a group of people who share a common characteristic or life event; in this case it includes everybody born in a specific five-year period.) Two bar graphs are constructed for each cohort, one for males and another for females; the bars are placed end to end, separated by a line representing zero. Usually, the left-hand side of the pyramid depicts the number or percentage of males that make up each age cohort and the right-hand side depicts the number or percentage of females. The graphs are stacked according to age; the age 0–4 cohort forms the base of the pyramid and the 80+ age cohort is at the apex (see Figure 12.5). The population pyramid allows us to view the relative sizes of the age cohorts and to compare the relative number of males and females in each cohort.

The population pyramid is a snapshot of the number of males and females in the various age cohorts at a particular time. Generally, a country's population pyramid approximates one of three shapes — expansive, constrictive, or stationary. **Expansive pyramids**, characteristic of labor-intensive poor countries, are triangular; they are broadest at the base, and each successive bar is smaller than the one below it (see Figure 12.5). The relative sizes of the age cohorts in expansive pyramids show that the population is increasing in size and composed disproportionately of young people.

Constrictive pyramids, which characterize some European societies, most notably Switzerland and western Germany, are narrower at the base than in the middle. This shape shows that the population is com-

Figure 12.5 Population Pyramids

The two graphs show (a) the expansive pyramid characteristic of Brazil and (b) the stationary pyramid characteristic of the United States.

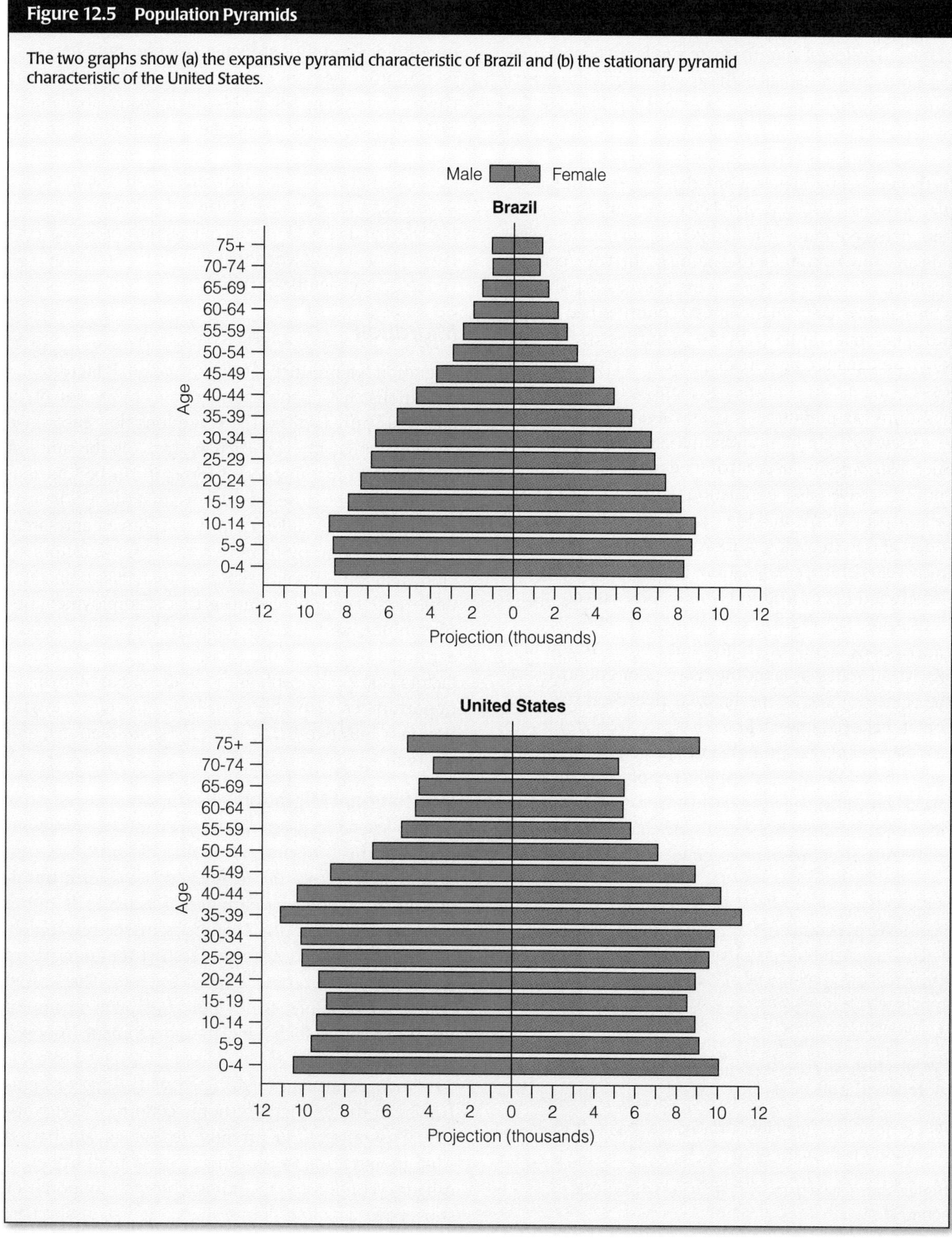

posed disproportionately of middle-aged and older people. **Stationary pyramids**, which are characteristic of most developed countries, are similar to constrictive pyramids except that all of the age cohorts in the population are roughly the same size and fertility is at replacement level (see Figure 12.5).

Demographers study age-sex composition to place birth- and death rates in a broader context. For example, between 1990 and mid-1991, the world's population increased by an estimated 90 million people. A significant proportion of this annual increase can be attributed to two countries—the People's Republic of China and India. Even though 90 countries had higher birthrates than India and 134 countries had higher birthrates than China (U.S. Bureau of the Census 1991), these two countries accounted for more than one-third (33 million) of the increase; China's population grew by 16 million, and India's grew by 17 million. Such a large share can be attributed to the fact that both countries have large populations to begin with (1.15 billion people live in China and 870 million live in India) and to the fact that more than 50 percent of the Indian and Chinese populations are of childbearing age—between fifteen and forty-nine (U.S. Bureau of the Census 1991).

Age-sex composition helps us understand in part why death rates in labor-intensive poor countries are the same as those in mechanized rich countries (and sometimes even lower). Brazil has an official death rate of 7 per 1,000; the United States has an official death rate of 9 per 1,000. Thirteen percent of the U.S. population, however, is over sixty-five, compared with five percent of Brazil's population. If the chances of survival were truly equal in the two countries, the death rate should be substantially higher in the mechanized rich countries because a greater percentage of the population is older and thus at higher risk of death.

Another reason death rates are higher in the United States than in Brazil is that in many labor-intensive poor countries, especially in rural areas, an unknown number of deaths are never registered. Consequently, there is a gap between statistics and reality. According to researchers Marilyn K. Nations and Mara Lucia Amaral (1991), "for all official purposes death occurs in Brazil only when the event is registered by surviving family members" (p. 207).

Stationary pyramids Population pyramids characteristic of most developed countries that are similar to constrictive pyramids except that all of the age cohorts in the population are roughly the same size and fertility is at replacement level.

Push factors The conditions that encourage people to move out of an area.

Pull factors The conditions that encourage people to move into a particular area.

Emigration The departure of individuals from one country.

Immigration The entrance of individuals into a new country.

Migration

Migration is a product of two factors: **push factors**, the conditions that encourage people to move out of an area, and **pull factors**, the conditions that encourage people to move into a particular area. "On the simplest level [and in the absence of force] it can be said that people move because they believe that life will be better for them in a different area" (Stockwell and Groat 1984:291). Some of the most common push factors include religious or political persecution, discrimination, depletion of natural resources, lack of employment opportunities, and natural disasters (droughts, floods, earthquakes, and so on). Some of the most common factors that pull people into an area are employment opportunities, favorable climate, and tolerance. Migration falls into two broad categories: international and internal.

International Migration International migration is the movement of people between countries. Demographers use the term **emigration** to denote the departure of individuals from one country and the term **immigration** to denote the entrance of individuals into a new country. Unless their countries are severely underpopulated, most governments restrict the numbers of foreign people who are allowed to enter. The United States, for example, had a policy of open migration until 1822, at which time the government imposed restrictions.

Today, consulates associated with American embassies in the immigrants' home countries screen potential immigrants. Currently, foreigners are barred from immigrating to the United States by thirty-eight different conditions, including criminal, immoral, or subversive activities; physical or mental disabilities; and economic factors suggesting that the individual would be a drain on the welfare system or would compete unfairly with American workers for jobs.

Three major flows of intercontinental (and, by definition, international) migration occurred between 1600 and the early part of the twentieth century: (1) the massive exodus of European peoples to North America, South America, Asia, and Africa to establish colonies and commercial ventures, in some cases eventually displacing native peoples and establishing independent countries (this occurred in the United States, Brazil, Argentina, Canada, New Zealand, Australia, and South Africa); (2) the smaller flow of Asian migrants to East Africa, the United States (including Hawaii, which did not become a state until 1959), and Brazil, where they provided cheap labor for major transportation and agricultural projects; and (3) the forced migration of some 11 million Africans by Spanish, Portuguese, French, Dutch, and British slave traders to the United States, South America, the Caribbean, and the West Indies. In all the Americas, Brazil imported the greatest number of African slaves (Skidmore 1993).

These three migration flows are responsible for the ethnic composition of Brazil: "There are few other [countries] on earth where such a wide spread of skin tones, from whitest white to yellow to tan to deepest black, are grouped under one nationality" (Nolty 1990:6). The mixing process began when the Portuguese sailors who colonized Brazil intermingled (often forcibly) with native women and African slave women. After slavery was abolished in 1888, the Brazilian government encouraged Europeans and Asians to immigrate into the country to replace slave labor. Between 1884 and 1914, a large number of Japanese and Italians immigrated to Brazil as indentured servants; they worked on large plantations until they had earned enough to buy their way out of servitude.

Internal Migration In contrast to international migration, internal migration is movement within the boundaries of a single country — from one state, region, or city to another. Demographers use the term **in-migration** to denote the movement of people into a designated area and the term **out-migration** to denote the movement of people out of a designated area.

One major type of internal migration is the rural-to-urban movement (urbanization) that accompanies industrialization.[4] However, urbanization in labor-intensive poor countries is substantially different from that which occurred in mechanized rich societies:

> The world has never seen such extremely rapid urban growth. It presents the cities, especially in the developing countries, with problems new to human experience, as well as old problems — urban infrastructure, food, housing, employment, health, education — in new and accentuated forms. Many developing countries will have to plan for cities of sizes never conceived of in currently developed countries. High population growth in developing countries, whatever other factors enter the process, is inseparable from this phenomenon. (Rusinow 1986:9)

The unprecedented growth of urban areas is caused by a number of factors. First, the cities in the mechanized rich countries grew in closer proportion to the number of jobs created during their industrialization process a century and a half ago. Many Europeans who were pushed off the land were able to emigrate to sparsely populated places like North America, South America, South Africa, New Zealand, and Australia. If those who fled to other countries in the eighteenth and nineteenth centuries had been forced to make their livings in European cities, the conditions would have been much worse than they actually were:

> Ireland provides the most extreme example. The potato famine of 1846–1849 deprived millions of peasants of their staple crop. Ireland's population was reduced by 30 percent in the period 1845–1851 as a joint result of starvation and emigration. The immigrants fled to industrial cities of Britain, but Britain did not absorb all the hungry Irish. North America and Australia also received Irish immigrants. Harsh as life was for these impoverished immigrants, the new continents nonetheless offered them a subsistence that Britain was unable to provide.
>
> . . . [T]here are no longer any new worlds to siphon off population growth from the less industrialized countries. (Light 1983:130–131)

In Brazil, the problem of urbanization is compounded by the fact that many migrants who come to

[4]Ninety-four cities in the world have populations exceeding 2 million. Of these cities, thirty-nine are in mechanized rich nations and fifty-five are in labor-intensive poor countries. By the year 2000, thirty-four additional cities will be added to this list; all but four of these new cities will be in developing countries. The proportion of the population that lives in the largest cities (2 million or more) varies considerably by country. Almost 25 percent of Mexico's population lives in Mexico City; about 20 percent of Brazil's population lives in São Paulo and Rio de Janeiro. In contrast, only 9 percent of the U.S. population lives in its largest city, New York City.

In-migration The movement of people into a designated area.

Out-migration The movement of people out of a designated area.

the cities are from some of the most economically precarious segments of Brazil. In fact, most rural-to-urban migrants are not pulled into the cities by employment opportunities but are forced to move there because they have no alternatives. In the northeast region of Brazil, peasants are pushed out because of droughts and floods, because land is deforested and exhausted, and because the land is concentrated in the hands of a few people. When the migrants come to the cities, they face not only unemployment but also a shortage of housing and a lack of services (electricity, running water, waste disposal). One of the most distinguishing characteristics of cities in labor-intensive poor countries is the prevalence of slums and squatter settlements, poorer and larger than even the worst slums in the mechanized rich countries. A survey of living conditions in São Paulo, the largest city in Brazil and the third largest in the world, sponsored by the International Institute for Labour Studies found that between 31.0 and 42.3 percent of families had housing needs. That is, their homes were constructed with makeshift materials or they lacked enough space for ordinary family activities such as sleeping, cooking, bathing and laundry. An estimated 35.7 percent of families were classified as "bad," "very poor," or "poor" with regard to labor market status. An estimated 74 percent of the population in São Paulo were found to be lacking in at least one of four areas: (1) housing, (2) education, (3) employment, and/or (4) income (SEADE Foundation 1994).

Another type of massive internal migration — rural-to-rural—is taking place in many labor-intensive poor countries. We know very little about the extent of this migration (except that it may involve as many as 370 million people worldwide) or its impact on the communities that the migrants enter and leave. We do know, however, that in this kind of migration peasants move to increasingly more marginal or fragile lands. Generally, the move does not improve their social and economic status. In other words, peasants do not move to find better jobs than those they have; they move to find any kind of work or to search for land (Feder 1971).

In Brazil, an estimated 10.5 million workers without land "migrate all over the country . . . invading any empty patch which may seem unclaimed" (Cowell 1990:137). Some of the most desperate migrate into regions of the Amazon states. These peasants do not move into the wilderness, clear the ground of trees and rocks, and become prosperous farmers, however. More often than not, they barely grow enough food to survive. This situation is not unique to Brazil. Sociologist Alfredo Molano (1993) estimates that in Colombia, settlers — "fleeing violence, political persecution, or economic deprivation — go to the forests looking for another way of finding subsistence" (p. 43). The desperate activities of migrants in search of wood for cooking and land to grow food and set up households results in a level of deforestation that surpasses the number of acres taken by the lumber industry (Semana 1993).

Industrialization, whatever its form, has been a major influence on the basic structure of family life. In this section, we have identified various trends such as the relatively rapid decline in death rates coupled with relatively slow decline in birthrates that obviously affect family size and composition. Likewise, an increase in life expectancy is associated with changes in childbearing experiences (a decrease in infant and maternal mortality, for example) and the increased probability that parents will live to see their children into adulthood. In light of the fact that changes affect the structure of the family we should not be surprised to learn that definitions and ideas of what the typical family is like reflect these changes.

Industrialization and Family Life

The sheer diversity of experiences makes it impossible to make blanket statements about how industrialization affects family life. Consider the differences among three Brazilian families that sociologist Hélène Tremblay (1988) stayed with as part of her 60-country, 118-family research circuit to learn about how families around the world live. Excerpts from her diaries about how each family begins the day give some insights into the impossibility of generalizing about family life in a specific country. For example, the Yanomami family of Amazonia lives in a *shabono,* a communal hut that houses fifteen fire sites, one for each family unit. The family consists of a mother (one of three wives, two of whom live elsewhere), a father, and three children (ages eighteen months, four years, and seven years).

> Mother plucks a few bananas hanging over her head and throws them on the coals along with some palm leaves stuffed with nutmeats. Breakfast will soon be ready.
>
> After feeding her children [she] joins the other women on their way to fish. [Father] is off hunting with the other men, taking advantage of the final days of the dry season. (p. 3)

Family Portraits

Can a single image represent all that we mean by "family"? What ideas of "family" are conveyed by these pictures? The examples shown here are just a few of the many types of arrangement. Many sociologists are greatly interested in the concept of family and problems of families, in all their complexity. At an undergraduate level, this takes the form of a course often called Marriage and Family.

Among the 118 photographs on board the Voyager 1 *and* Voyager 2 *spacecraft was this one entitled "Family Portrait."*

©Nina Leen/Life Picture Service

An African mother and her children (absent the father, who lives and works in the city).

©Betty Press/Woodfin Camp & Associates

A Mexican family celebrating a birthday.

©Chip and Rosa Maria de la Cueva Peterson

A father with his adopted children.

©J. Patrick Forden/Sygma

Lesbian parents with their child.

©Chris Maynard/Gamma Liaison

The Mariano de Souza Caldas family of Guriri lives in a mud hut on a government-issued land plot. The smell is appalling because they live just fifty meters from the city garbage dump. The family consists of a pregnant mother, a father, and ten children who range in age from eleven months to seventeen years old. The father

> gently pushes his sleeping son and gets out of bed. He moves the baby's crib, which is blocking the doorway . . . and heads for the well at the bottom of the yard. . . . He showers and rinses his mouth . . . will milk the cows . . . and will spend all morning combing the dumps in search of items that he can recycle for money. (p. 23)

The Aratanha family of Rio de Janeiro lives in a fenced-in, highly guarded high rise with many amenities (soccer field, tennis courts, a gym). The family consists of a mother and father (both of whom are physicians) and two children (ages eighteen months and three years). They have a maid who helps care for the children. The father "gets out of bed quickly without waking his wife . . . [and] leaves very early to avoid the traffic jams. . . . [The mother is] grateful for her new car. It saves her half an hour every morning" (p. 31).

Another reason that it is difficult to generalize about how industrialization affects family life is that industrialization and the changes that are thought to accompany it may run up against some political, cultural, or historical events that alter the course of events away from a predicted or hypothesized outcome. For example, birthrates are expected to decline with industrialization and urbanization. Yet, sociologists and demographers are still trying to understand the forces underlying decisions by American couples to have relatively large families after World War II (between 1946 and 1963—a period of dramatic industrialization and the years that produced the cohort known popularly as the baby boomers).

In addition, government policies may have a strong influence on birthrates. For example, after the Romanian government legalized abortion in 1957, there were approximately 4,000 abortions for every 1,000 live births (about 80 percent of pregnancies ended in abortion), and the birthrate was about 15 per 1,000. In 1967, President Nicolai Ceausescu, in an effort to increase the population and the number of future workers, drastically restricted the sale of contraceptives and placed tight restrictions on abortions. Under Ceausescu's policies, Romanian women were required to undergo regular pregnancy tests to ensure that all pregnancies were carried to term, and they were given access to contraceptives and abortions only if they were over forty-five years of age or already had four living children. Physicians who administered illegal abortions faced twenty-five years in prison and even the death penalty. Under these policies, the birthrate climbed to 27.4 per 1,000 (Burke 1989; van de Kaa 1987). By 1983, however, the birthrate had tumbled to 14 per 1,000, and Ceausescu announced more restrictions and closer monitoring.

In the sections that follow, we will consider how the changes in life expectancy, which includes changes in birth and death rates, lead to changes in family function and composition and in the status of children. We will also consider how these factors lead to changes in the division of labor between men and women and to changes in the reasons people marry. Finally, in the summary and implication section we will consider how these demographic factors affect the definition of family.

The Consequences of Long Life

Since the turn of the century, the average life expectancy at birth has increased by twenty-eight years in the mechanized rich countries and twenty years (or more) in the labor-intensive poor countries. In *The Social Consequences of Long Life,* sociologist Holger R. Stub (1982) describes at least four ways in which increases in life expectancy have altered the composition of the family since 1900. First, the chances that children will lose one or both parents before they reach sixteen years of age has decreased sharply. There was a 24 percent chance of such an occurrence in 1900; today the probability is less than 1 percent. At the same time, parents can expect that their children will survive infancy and early childhood. In 1900, 250 of every 1,000 children born in the United States died before age one; 33 percent did not live to age eighteen. Today only 10 of every 1,000 children born die before they reach age one; fewer than 5 percent die before reaching age eighteen.

Second, the potential length of marriages has increased. Given the mortality patterns in 1900, newly married couples could expect their marriage to last an average of twenty-three years before one partner died (if we assume they did not divorce). Today, assuming that they do not divorce, newly married couples can expect to be married for fifty-three years before one partner dies. This structural change may be one of several factors underlying the high divorce rates today. At the turn of the century, death nearly always intervened before a typical marriage had run its natural course.

Changing Patterns of Marriage and Divorce in the United States

This table contains a lot of information about changes since 1970 in the proportions of their lives that U.S. men and women spend as married or single persons. Try to summarize the main implications of the table in one or two sentences. What is happening to us with regard to the number of years of our lives we are spending alone?

As of 1988, roughly what percentage of their grown-up lives did the average U.S. male and female spend unmarried? Do you expect that percentage to continue to increase between now and, say, the year 2010? Why?

If the percentage of the average life spent unmarried did continue to increase, what effects might that have on population growth? On overall consumption? On consumer demand for particular kinds of goods or services? On how people choose to socialize or spend their time?

Is the change in the average life spent unmarried likely to affect our ideas about "family," especially the figures for women alone?

Changing Patterns of Marriage and Divorce in the United States: 1970, 1988

Category	Males 1970	Males 1988	Females 1970	Females 1988
Life expectancy at birth	67.2	70.7	74.8	78.5
Number of those ever marrying of those surviving to age 15 (%)	96.0	83.5	96.5	87.9
Average age at first marriage	23.4	27.5	21.8	25.1
Marriages ending in divorce (%)	37.3	42.7	35.7	43.2
Marriages ending in widowhood (%)	18.7	18.3	45.3	39.3
Marriages ending in own death (%)	44.0	39.0	19.0	17.6
Widowed persons remarrying (%)	27.0	17.3	10.3	6.3
Divorced persons remarrying (%)	86.0	78.3	80.2	72.3
Average duration of a marriage	26.5	24.5	26.8	24.8
Amount of life spent never married (%)	35.0	46.2	30.6	39.1
Amount of life spent currently married (%)	58.4	44.8	50.5	41.2
Amount of life spent widowed (%)	3.1	2.8	12.6	10.0
Amount of life spent divorced (%)	3.6	6.2	6.3	9.6

Source: Schoen and Weinick (1993), Table 1.

Now, many marriages run out of steam with decades of life remaining for each spouse. When people could expect to live only a few more months or years in an unsatisfying relationship, they would usually resign themselves to it. But the thought of twenty, thirty, or even fifty more years in an unsatisfying relationship can cause decisive action at any age (Dychtwald and Flower 1989:213). According to Stub, divorce dissolves today's marriages at the same rate that death did 100 years ago. (See "Changing Patterns of Marriage and Divorce.")

Third, people now have more time to choose and get to know a partner, settle on an occupation, attend school, and decide whether they want children. Moreover, an initial decision made in any one of these areas is not final. The amount of additional living time enables individuals to change partners, careers, or educational and family plans, a luxury not shared by their turn-of-the-century counterparts. Stub (1982) argues that the midlife crisis is related to long life because many people "perceive that there yet may be time to make changes and accordingly plan second careers or other changes in life-style" (p. 12).

Finally, the number of people surviving to old age has increased. (In countries where fertility is low or declining, the proportion of old people in the population

— not merely the number — is also increasing.) In 1970, about 25 percent of people in their late fifties had at least one surviving parent; in 1980, 40 percent had a surviving parent. Even more astonishing, in 1990, 20 percent of people in their early sixties and 3 percent of people in their early seventies had at least one surviving parent (Lewin 1990). Although it has always been the case that a small number of people have lived to age eighty or ninety, "There is no historical precedent for the aging of our population. We are in the midst of a new phenomenon" (Soldo and Agree 1988:5).

Much has been written about the growing numbers of people over age sixty-five in the world, especially about issues of caring for the disabled and the frail elderly. Yet, the emphasis on this segment of the older population should not obscure the fact that the elderly are a rapidly changing and heterogeneous group. They differ according to gender, age (a thirty-year difference separates persons age sixty-five from those in their nineties), social class, and health status. Most older people today are in relatively good health; in the United States, only 5 percent of persons age sixty-five and older live in nursing homes (Eckholm 1990).

In most countries, including the United States, most disabled and frail elderly persons are cared for by female relatives (Stone, Cafferata, and Sangl 1987; Targ 1989). In the United States, approximately 72 percent of the caregivers are women. Of this 72 percent, 22.7 percent are spouses, 28.9 percent are daughters, and 19.9 percent are daughters-in-law, sisters, grandmothers, or some other female relative or nonrelative (Stone et al. 1987). The major issue faced by even the closest of families is how to meet the needs of the disabled and frail elderly so as not to "constrain investments in children, impair the health and nutrition of younger generations, impede mobility," or impose too great a psychological, physical, and/or time-demand stress on those who care for them. This problem may be even more intense in developing countries where the presence of older dependents "may alter the level of subsistence of other members in the family" (United Nations 1983:570). In subsistence cultures, even in those that display considerable respect toward the elderly, "the decrepit elderly may be seen as too much of a burden to be supported during periods of deprivation and may be killed, abandoned, or forsaken" (Glascock 1982:55). Although many programs exist and are being developed to support caregivers and address the issues associated with caring for the disabled and frail elderly, most of these programs are still in the pilot stages. "It is going to take as many as fifteen or twenty years to sort out what works" (Lewin 1990:A11).

The Status of Children

The technological advances associated with the Industrial Revolution decreased the amount of physical exertion and time needed to produce food and other commodities. As human muscle and time became less important to the production process, children lost their economic value. In nonmechanized, extractive economies, children are an important source of cheap and unskilled labor for the family. This fact may explain in part why the states in Amazonian Brazil have some of the highest total fertility rates found anywhere in the world (Butts and Bogue 1989). The total fertility rate in rural Acre is 8.03, a level that demographers have defined as approximating the human capacity to reproduce. (See "The Economic Role of Children in a Labor-Intensive Poor Country.")

Middle- and upper-class children in mechanized rich economies are likely to live in settings that strip them of whatever potential economic contribution they might make to the family economy (Johansson 1987). In these economies, the family's energies shift away from production of food and other necessities and toward the consumption of goods and services.

The U.S. Department of Agriculture estimates that the average family earning an annual income of at least $56,700 will spend about $346,980 to house, feed, clothe, transport, and supply medical care (not covered by insurance) to a child until he or she is twenty-two years old (*American Demographics* 1996), with annual expenses ranging between $5,100 and $10,510 (Wark 1995). These figures represent a basic sum; the costs grow even higher when extras (summer camps, private schools, sports, music lessons) are included. Even if children go to work when they reach their teens, usually they use that income to purchase items for themselves and do not contribute to household expenses. Demographer S. Ryan Johansson (1987) argues that in the mechanized rich economies, couples who choose to have children bring them into the world to provide intangible, "emotional" services — love, companionship, an outlet for nurturing feelings, enhancement of dimensions of adult identity.

Urbanization and Family Life

Urbanization encompasses two phenomena: (1) the migration of people from rural areas to cities so that an increasing proportion of the population comes to live

The Economic Role of Children in a Labor-Intensive Poor Country

Children often are an important source of cheap labor, especially for nonmechanized industries. The childhood experiences of Chico Mendes, the son of an Amazon rubber tapper, illustrate the economic role of children in the kind of labor-intensive work that takes place in rubber plantations.

Chico was an important Brazilian grassroots environmentalist and union activist dedicated to preserving the Amazon and to improving rubber tappers' economic standard of living as compared with those of cattle ranchers and rubber barons. He was murdered in 1988 by Amazon cattle ranchers. Indians and rubber tappers are now working together against those who misuse or otherwise destroy the resources of the forests.

> *Childhood for Chico Mendes was mostly heavy work. . . . If all his siblings had lived to adulthood, Chico would have had seventeen brothers and sisters. As it was, conditions were so difficult that by the time he was grown, he was the oldest of six children—four brothers and two sisters.*
>
> *When he was five, Chico began to collect firewood and haul water. A principal daytime occupation of young children on the* seringal *has always been lugging cooking pots full of water from the nearest river. . . . Another chore was pounding freshly harvested rice to remove the hulls. A double-ended wooden club was plunged into a hollowed section of tree trunk filled with rice grains, like an oversize mortar and pestle. Often two children would pound the rice simultaneously, synchronizing their strokes so that one club was rising as the other descended.*
>
> *By the time he was nine, Chico was following his father into the forest to learn how to tap. . . . Well before dawn, Chico and his family would rise. . . . [T]hey would grab the tools of their trade—a shotgun, . . . a machete, and a pouch to collect any useful fruits or herbs found along the trail. They left before dawn, because that was when the latex was said to run most freely. . . .*
>
> *As they reached each tree, the elder Mendes grasped the short wooden handle of his rubber knife in two hands, with a grip somewhat like that of a golfer about to putt. . . . It was important to get the depth [of the cut] just right. . . . A cut that is too deep strikes the cambium [generative tissue of the tree] and imperils the tree; a cut that is too shallow misses the latex-producing layer and is thus a waste. . . .*
>
> *Chico learned how to position a tin cup, or sometimes an empty Brazilian nut pod, just beneath the low point of the fresh cut on a crutch made out of a small brand. . . . The white latex immediately began to dribble down the slash and into the cup. . . .*
>
> *Chico quickly adopted the distinctive, fast stride of the rubber tapper. . . . The pace has evolved from the nature of the tapper's day. The 150 and 200 rubber trees along an* estrada *are exasperatingly spread out. A simple, minimal bit of work is required at each tree, but there is often a 100-yard gap between trees. Thus, a tapper's morning circuit can be an 8 to 11-mile hike. And that is just the morning. Many tappers retrace their steps in the afternoon to collect the latex that has accumulated from the morning cuts.*
>
> *By the time Chico and his father came full circle on an* estrada *and returned to the house, it would be close to midday, time for a lunch. . . . Then, they would retrace their steps on the same trail, to gather the latex that had flowed from the trees during the morning. Only rarely would they return home before five o'clock. By then they would be carrying several gallons of raw latex in a metal jar or sometimes in a homemade, rubber-coated sack.*
>
> *The very best rubber is produced when this pure latex is immediately cured over a smoky fire.*
>
> *The smoking of the latex was done in an open shed that was always filled with fumes from the fire. The smokier the fire, the better. Usually, palm nuts were added to the flames to make the smoke extra thick. Chico would ladle the latex onto a wooden rod or paddle suspended over a conical oven in which the nuts and wood were burned. As the layers built up on the rod, the rubber took on the shape of an oversize rugby football. The smoking process would continue into the evening. After a day's labor of fifteen hours or more, only six or eight pounds of rubber were produced. The tappers often developed chronic lung diseases from exposure to the dense, noxious smoke.*
>
> *Toward December, with the return of the rainy season, the tappers stopped harvesting latex and began collecting Brazil nuts.*
>
> *During the rainy season, the Mendeses often crouched on their haunches around a pile of Brazil nut pods, hacking off the top of each one with a sharp machete blow, then tossing the loose nuts onto a growing pile. For tappers in regions with Brazil nut trees, the nuts can provide up to half of a family's income. One tree can produce 250 to 500 pounds of nuts in a good season and some tappers' trails pass enough trees to allow them to collect more than three tons of nuts each season. While that might sound like a potential windfall, even as late as 1989, tappers received only three or four cents a pound for the harvest—which later sold for more than $1 a pound at the export docks.*
>
> *When the Mendeses were not harvesting latex or nuts, they tended small fields of corn, beans, and manioc.*

Source: From *The Burning Season* by Andrew Revkin, pp. 69–76.

The effects of industrialization on family life are too complex to be captured by sweeping generalizations. Brazil, for example, includes regions where children remain sources of cheap labor in a nonmechanized economy. Yet it also includes cities where middle-class children, like their counterparts in the United States, play the role of consumers in a mechanized rich country.

©Stephanie Maze/Woodfin Camp & Associates (left); ©Paulo Fridman/Sygma (right)

in cities (see "The Growth of Major Urban Centers") and (2) a change in the ties that bind people to one another. (See the discussion of mechanical and organic solidarity in Chapter 6.) In mechanized rich countries, people who live in urban settings usually are not part of self-sufficient economic units composed of family members. They have opportunities to make contact with people outside the family network, and they come to depend on people and institutions other than the family (the workplace, hospitals and clinics, counseling services, schools, day-care centers, the media) to meet various needs. People's relationships with most of the individuals they meet in the course of a day are transitory, limited, and impersonal. This does not mean that the family ceases to be important to people's lives, however. Instead, it is important in a different way: it becomes less an economic unit and more a source of personal support. Most sociological research on the American family shows that the great majority of people have some family members living near them, that they interact with them regularly, and that they define these interactions as meaningful and important (Goldenberg 1987).

Urbanization in itself, however, does not necessarily reduce the economic function of the family. Janice Perlman (1967), in her studies of the urban poor in Brazil, found that the majority of migrants to Rio de Janeiro (67 percent) came to the city in the company of relatives or went to the homes of relatives upon arrival and that many make every effort to live as close to one another as possible. Thus, Perlman found that some neighborhoods were dominated by large kin networks. The research also shows that family networks are extremely useful to people living in precarious financial conditions. Perlman found that 63 percent of the people she studied had gotten their first job through friends or kin.

Anthropologist Claudia Fonseca (1986) also learned the significance of family relationships in Brazil. She studied 68 mothers from a shantytown of 750 squatters living on an empty lot in a middle-class neighborhood in southern Brazil. She found that 50 percent of the mothers had sent a child away to live under the care of another person while they recovered from illness or financial setbacks. Most often, this caretaker was a relative.

Judith Goode (1987) has written extensively about Colombian street children, and Thomas Sanders (1987a, 1987b) has written about Brazilian street children. Both groups of children are frequently misperceived by government officials and middle-class people as abandoned by their parents, as delinquents, and as a reflection of widespread family pathology among squatters. Both writers conclude, however, that street children (usually between the ages of ten and fifteen) are from poverty-stricken families "who are trying to maximize their household incomes by utilizing the money-earning capacities of their children in such ac-

U.S. In Perspective

The Growth of Major Urban Centers

Most people in the United States are probably still accustomed to thinking of Europe and the United States as leading cultural forces in the world. Yet we also associate culture and innovation with great cities.

Consider this table and map showing the world's fifteen largest cities, ranked by size. In what regions and countries of the world are the largest cities concentrated today? Is this strong reason to expect that one or more of these parts of the world have become, or are about soon to become, great centers of world culture? Why?

Suppose you were asked to develop a system for classifying or ranking these cities with respect to characteristics other than sheer size. What characteristics would you want to study? Why those particular characteristics?

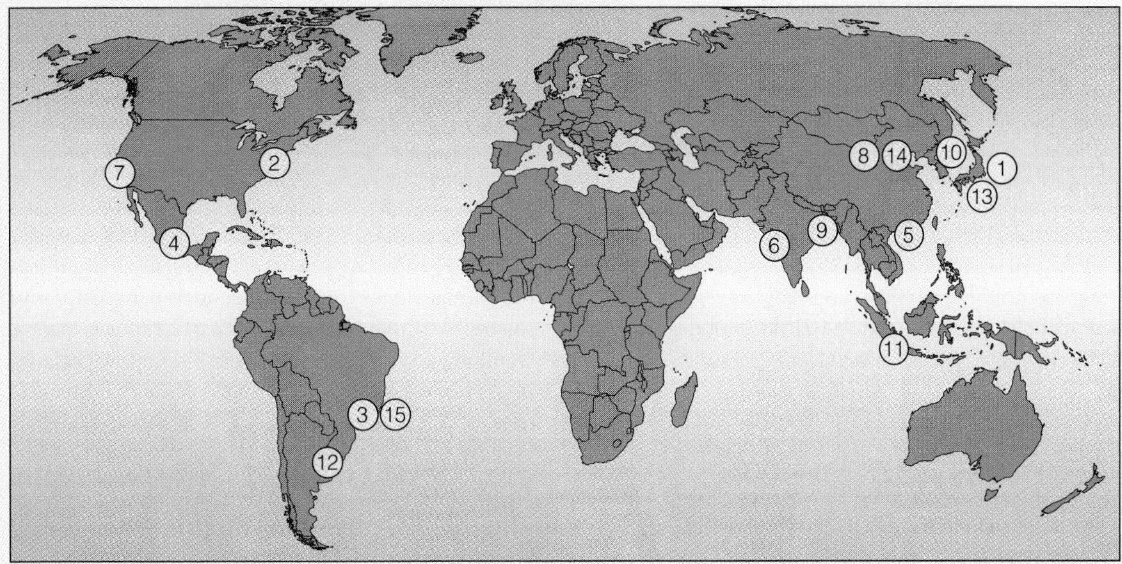

Rank	City, Country	Population 1994 (thousands)	Annual Growth Rate 1990–1995 (%)	Percentage Increase Between 1975 and 1995	Population of City as Percentage of Total Population*
1	Tokyo, Japan	26,518	1.4	35.7	21.2
2	New York City, U.S.	16,271	0.3	2.8	6.2
3	São Paulo, Brazil	16,110	2.0	66.0	10.1
4	Mexico City, Mexico	15,525	0.7	39.2	16.9
5	Shanghai, China	14,709	2.3	31.8	1.2
6	Bombay (Mumbia), India	14,496	4.2	120.1	1.6
7	Los Angeles, U.S.	12,232	1.6	39.0	4.7
8	Beijing, China	12,030	2.6	44.7	1.0
9	Calcutta, India	11,485	1.7	48.0	1.3
10	Seoul, South Korea	11,451	1.9	71.2	25.7
11	Jakarta, Indonesia	11,017	4.4	138.9	5.7
12	Buenos Aires, Argentina	10,914	0.7	20.3	31.9
13	Osaka, Japan	10,585	0.2	7.7	8.5
14	Tianjin, China	10,376	2.9	73.5	0.9
15	Rio de Janeiro, Brazil	9,817	0.8	25.6	6.2

*Denotes percentage of the total population of the country in which the corresponding city is located.

While media headlines about street children in Brazil focus on the violent gangs, most street children try to find work to help increase their families' household incomes.

©Sean Sprague/Panos Pictures

tivities as shining shoes, carrying packages, watching cars, and begging" (Sanders 1987b:1). Goode (1987) writes:

> Attempts to classify types of income for street children often distinguish between those activities which are legal, such as performing services as shoeshine boys, street vendors, and car watchers; those which are more marginal, such as begging and singing on buses; and those which are illegal and anti-social, such as purse-snatching, picking pockets, and other forms of theft. (p. 6)

Division of Labor Between Men and Women

In 1984, sociologist Kingsley Davis published "Wives and Work: The Sex Role Revolution and Its Consequences." In this article, Davis identifies "as clear and definite a social change as one can find" (p. 401): between 1890 (the first year for which reliable data exists) and 1980, the proportion of married women in the labor force rose from less than 5 percent to more than 50 percent. Davis found that this pattern holds true for virtually every industrialized country except Japan.[5] He attributes this change to the Industrial Revolution and its effect on the division of labor between men and women (see Figure 12.6). It is important to keep in mind that Davis's theory is intended to explain how industrialization in the mechanized rich countries affected the division of labor between males and females. It does not systematically address how industrialization is affecting the division of labor between males and females in labor-intensive poor countries.

[5]Japan followed an accelerated model: it underwent a rapid decline in mortality followed by an equally rapid decline in fertility. The rapid decline in fertility was due to high rates of abortion.

The Division of Labor and the Breadwinner System

Before industrialization — that is, for most of human history — the workplace was the home and the surrounding land. The division of labor was based on sex. In nonindustrial societies (which includes two major types, the hunting-and-gathering and the agrarian), men provided raw materials through hunting or agriculture, and women processed these materials. Women also worked in agriculture and provided some raw materials by gathering food and hunting small animals, and they took care of the young.

The Industrial Revolution separated the workplace from the home and altered the division of labor between men and women. It destroyed the household economy by removing economic production from the home and taking it out of the women's hands:

> The man's work, instead of being directly integrated with that of wife and children in the home or on the surrounding land, was integrated with that of nonkin in factories, shops, and firms. The man's eco-

Figure 12.6 Percentage of Married Women in the U.S. Labor Force, 1890–1992

Sources: Adapted from U.S. Bureau of the Census as reported by Davis (1984:398–399) and U.S Bureau of the Census (1993:400).

> nomic role became in one sense more important to the family, for he was the link between the family and the wider market economy, but at the same time his personal participation in the household diminished. His wife, relegated to the home as her sphere, still performed the parental and domestic duties that women had always performed. She bore and reared children, cooked meals, washed clothes, and cared for her husband's personal needs, but to an unprecedented degree her economic role became restricted. She could not produce what the family consumed, because production had been removed from the home. (Davis 1984:403)

Davis called this new economic arrangement the "breadwinner system." From a historical point of view, this system is not typical. Rather, it is peculiar to the middle and upper classes and is associated with a particular phase of industrialization — from the point at which agriculture loses its dominance to the point where only 25 percent of the population still works in agriculture. In the United States, "the heyday of the breadwinner system was from about 1860 to 1920" (p. 404). This system has been in decline for some time in the United States and other industrialized countries, but it tends to recur in countries undergoing a particular phase of development.

Davis asks, "Why did the separation of home and workplace lead to this system [in most mechanized rich countries]?" The major reason, he believes, is that women had too many children to engage in work outside the home. This answer is supported by the fact that family size, in the sense of the number of living members, reached its peak from the mid-1800s to the early 1900s. This occurred because infant and childhood mortality declined while the old norms favoring large families persisted.

The Decline of the Breadwinner System The breadwinner system did not last long because it placed too much strain on husbands and wives and because a number of demographic changes associated with industrialization worked to undermine it. The strains stemmed from several sources. Never before had the roles of husband and wife been so distinct. Never before had women played less than a direct, important role in producing what the family consumed. Never before had men been separated from the family for most of their waking hours. Never before had men had to bear the sole responsibility of supporting the entire family. Davis regards these events as structural weaknesses in the breadwinner system. In view of these weaknesses, the system needed strong normative controls to survive: "The husband's obligation to support his family, even after his death, had to be enforced by law and public opinion; illegitimate sexual relations and reproduction had to be condemned, divorce had to be punished, and marriage had to be encouraged by making the lot of the 'spinster' a pitiful one" (p. 406).

Davis maintains that the normative controls collapsed because of the strains inherent in the breadwinner system and because of demographic and social changes that accompanied industrialization. These changes included decreases in total fertility, increases in life expectancy, increases in the divorce rate, and increases in opportunities for employment perceived as suitable for women.

Declines in Total Fertility The decline in total fertility began before married women entered the labor force. This fact led Davis to conclude that the decline itself changed women's lives in such a way that they had the time to work outside the home, especially after the children entered school. During the 1880s, total fertility among white women was approximately 5.0; in the 1930s, it averaged 2.4; during the 1970s, it averaged 1.8. Not only did the number of children decrease, but reproduction ended earlier in women's lives and the births were spaced closer together than in earlier years. (The median age of mothers at the last birth was forty in 1850; by 1940, it had fallen to 27.3.) Davis attributes the changes in childbearing patterns to the forces of industrialization, which changed children from an economic asset to an economic liability, and to the "desire to retain or advance one's own and one's children's status in a rapidly evolving industrial society" (p. 408).

Increased Life Expectancy In view of the relatively short life expectancy in 1860 and the age at which women had their last child (age forty), the average woman was dead by the time her last child left home. By 1980, given the changes in family size, spacing of children, and age of last pregnancy, the average woman could expect to live thirty-three years after her last child left home. As a result, child care came to occupy a smaller proportion of a woman's life. In addition, although life expectancy has increased for both men and women, on the average women can expect to live longer than men. In 1900, women outlived men by about 1.6 years on the average; in 1980, they outlived men by approximately seven years. Yet, because brides tend to be three years younger on the average than their husbands, married women can expect to live past their husbands' death for an average of ten years. In addition, the distorted sex ratio caused by males' earlier death decreases the probability of remarriage. Although few women think directly about their husbands' impending death as a reason for working, the difference in mortality remains a background consideration.

Increased Divorce Davis traces the rise in the divorce rate to the breadwinner system, specifically to the shift of economic production outside the home:

> With this shift, husband and wife, parents and children, were no longer bound together in a close face-to-face division of labor in a common enterprise. They were bound, rather, by a weaker and less direct mutuality — the husband's ability to draw income from the wider economy and the wife's willingness to make a home and raise children. The husband's work not only took him out of the home but also frequently put him into contact with other people, including young unmarried women [who have always worked] who were strangers to his family. Extramarital relationships inevitably flourished. Freed from rural and small-town social controls, many husbands either sought divorce or, by their behavior, caused their wives to do so. (pp. 410–411)

Davis notes that an increase in the divorce rate preceded married women's entry into the labor market by several decades. He argues that once the divorce rate reaches a certain threshold (above a 20 percent chance of divorce), more married women seriously consider seeking employment to protect themselves in case of divorce. When both husband and wife are in the labor force, the chances of divorce increase even more. Now both partners interact with people who are strangers to the family. Moreover, they live in three different worlds, only one of which they share.

Increased Employment Opportunities for Women

Davis believes married women are motivated to seek work by changes in childbearing experiences, increases in life expectancy, the rising divorce rate, and the inherent weakness of the breadwinner system. This motivation was realized as opportunities to work increased for women. With improvements in machine technology, productivity increased, the physical labor required to produce goods and services decreased, and wages rose in industrialized societies. (Chapter 2 explains how industrialization proceeds in labor-intensive poor countries.) As industrialization matures, there is a corresponding increase in the kinds of jobs perceived as suitable for women (traditionally, nursing, clerical and secretarial work, and teaching).

The two-income system is not free of problems, however. First, the system at present "lacks normative guidelines. It is not clear what husbands and wives should expect of each other. It is not clear what ex-

As reflected in the Ms. Foundation for Women's "Take Our Daughters to Work" campaign, the growing importance of employment and careers to women in industrialized societies represents a profound social change as well as an economic one. Here, a nine-year-old girl is spending a day at work with her mother, a state supreme court justice.

©Marty Lederhandler/AP/Wide World Photos

wives and ex-husbands should expect, or children, cohabitants, friends, and neighbors. Each couple has to work out its own arrangement, which means in practice a great deal of experimentation and failure" (Davis 1984:413). Second, although the two-income system gives wives a direct role in economic production, it also requires that they work away from the home, a situation that makes child care difficult. Third, even in the two-income system, the woman is still primarily responsible for domestic matters. Davis maintains that women bear this responsibility because men and women are unequal in the labor force. (On the average, women earn sixty-six cents for every dollar earned by males.) As long as women make an unequal contribution to the household income, they will do more work around the house.

The problems of the new system are stressed further by the fact that a large percentage of married women (almost 40 percent) do not work outside the home. Davis believes that this lack of participation reflects the psychosocial costs of employment to married women. The dilemma women face can be summarized as follows: if married women work and have no children, they are selfish; if they stay home and raise a family, they are considered underemployed at best; if they work and raise a family, people wonder how they can possibly do either job right.

Summary and Implications

We have waited until the end of the chapter to offer a definition of the family because the challenge of defining the family is complicated by historic events that affect its composition, size, and function (see Table 12.2). Simply consider the ways in which changes in the division of labor brought on by declines in total fertility and increased life expectancy (especially increases in female life expectancy relative to men) and the subsequent effect on family size and composition. The challenge of defining the family is further complicated by the fact that there is an amazing variety of family arrangements worldwide — a variety reflected in the

Table 12.2 Effects of Industrialization on Key Episodes in Family Life

Key Episode	Before Industrialization	Mature Phase of Industrialization
Childbearing experiences		
Birthrates (annual)	30–50 births/1,000 people	Fewer than 20 births/1,000 people
Total fertility	9	2
Spacing	Entire reproductive life (15–49)	Short segment of reproductive life
Maternal mortality (annual)	Over 600 deaths/100,000 births	30 deaths/100,000 births
Estimated lifetime	1 in 21	1 in 10,000
Chance of dying from pregnancy-related causes	—	
Chances of Survival		
Death rate (annual)	50+ deaths/1,000 people	Fewer than 10 deaths/1,000 people
Life expectancy	20–30	70+
Major causes of death	Epidemic sources, deficiency diseases, malnutrition, parasite diseases	Degenerative and human-made
Nature of Labor		
Livelihood	Labor-intensive agriculture	Information and services
Source of power	Human muscles and other sources of power	Machines and created power
Economy	Subsistence-oriented	Consumption-oriented
Residence	Rural	Urban

numerous norms that specify how two or more people can become a family. These include norms that govern the number of spouses a person can have, the way a person should select a spouse, the ideal number of spacing of children, the circumstances under which offspring are considered legitimate (or illegitimate), the ways in which people trace their descent, and the nature of the parent-child relationship over the child's and parent's lives. In light of this variability, we should not be surprised to learn that it is difficult to construct a definition of family and that "no general theory or universal model of the family can be formulated" (Behnam 1990:549).

Most official definitions of family emphasize blood ties, adoption, or marriage as criteria for membership, and the function of procreation and socialization of offspring. The U.S. Bureau of the Census (1993) uses the term *family* to mean "a group of two or more persons related by birth, marriage, or adoption and residing together in a household" (p. 5). This definition has been in effect since 1950. Between 1930 and 1950, however, the definition of family revolved around a head of the household and reflected living arrangements in a more rural America:

> a private family comprises a family head and all other persons in the house who are related to the head by blood, marriage, or adoption, and who live together and share common housekeeping arrangements. The term "private household" is used to include the related family members (who constitute the private family) and the lodgers, servants, or hired hands, if any who regularly live in the home. (U.S. Bureau of the Census 1947:2)

Until the mid-1970s, the official definition of family in Brazil included the following criteria:

> (1) [that] family is synonymous with legal marriage, (2) marriage lasts until a spouse dies; (3) the husband is the breadwinner and the sole earner; (4) the wife is a full-time home-maker and her work has no [formal] economic value; (5) the husband is the legal head of the family. (Goldani 1990:525)

The Brazilian government's official definition of what constitutes a family has changed several times in the past twenty-five years such that this definition of family had been abandoned constitutionally. For example, the 1988 constitution recognized a stable union between a man and a woman as a family and children living with one parent as a family.

Changes in the official definition of family reflect dramatic changes in the size and composition of families that have occurred in both Brazil and the United States in the past twenty to thirty years (see Tables 12.3 and 12.4). Both the contemporary Brazilian and U.S.

Table 12.3 Distribution (Percent) of Family and Nonfamily Households in the United States*

Total Households	1960 (53,021,000)	1992 (95,669,000)
I. Family household		
A. Nuclear		
Married couple with children	52.18	26.46
Married couple with no children	22.00	21.64
Father-child(ren)	2.33	3.16
Mother-child(ren)	8.48	12.22
B. Extended†		
Grandparent(s), grandchildren, both parents	N/A	0.52
Grandparent(s), grandchildren, mother only	N/A	1.82
Grandparent(s), grandchildren, father only	N/A	0.15
Grandparent(s), grandchildren, no parents	N/A	0.91
II. Nonfamily household		
One person living alone	14.88	25.06
Institutions‡	3.61	3.48
III. Other living arrangements§	6.71	3.46

*Percentages do not total 100 because of inconsistencies and overlaps in the categories defined by the U.S. Bureau of the Census; N/A means "not available."

†Only those extended family categories used by the U.S. Bureau of the Census.

‡Institutions include those living in correctional institutions, nursing homes, and juvenile centers.

§Other living arrangements include college dormitories, military quarters, emergency shelters, and other unrelated persons living together.

Sources: U.S. Bureau of the Census (1961:40; 1993:64–65).

Table 12.4 Distribution (Percent) of Family and Nonfamily Arrangements in Brazilian Households

Family and Nonfamily Household	1960*	1984†
I. Family household	92.8	93.2
A. Nuclear	68.9	70.3
Married couple with children	54.1	46.5
Married couple with no children	8.5	12.8
Father or mother child–family	6.3	11.0
B. Extended	22.6	14.1
Married couple with children, unmarried and relatives	13.1	6.6
Married couple, no children and other relatives	2.9	1.6
Other extended family (no married couple, only father or mother child–family and relatives)	6.6	5.9
C. Complex	1.5	8.7
Married couple with children unmarried and non-relatives	1.5	8.7
II. Nonfamily household	7.2	6.8
Men or women living alone	5.3	5.5
Other nonfamily	1.9	1.3
Total percentage	100	100
Absolute values in thousands	13,532	31,075

*For the 1960 census the family-household classification did not include maids and lodgers. At other times only maids are not included.

†1984 is the last year for which data are available. The rural population of the Central West region was not considered in 1976, nor was the North region considered in 1984.

Source: Goldani (1990:527). Original sources: Brazilian Population Censuses, tapes of 1 percent in 1970 and 75 percent in 1980; and Household Survey, 1976 and 1984 tapes.

definitions reflect changes in the division of labor between men and women.

Still, these definitions fail to capture certain demographic realities that are beyond any one individual's control. For example, according to the official definitions of family, two elderly widows who live together and offer each other emotional support and financial commitment do not qualify as a family, nor would a couple who "adopt" an elderly man and commit to care for him. Likewise, older people who choose to assume caregiver roles to people in need (foster grandchildren, the sick and dying) do not qualify as family. On the surface, this may not seem like a big deal until we consider that official definitions of family can be used to deny family-related benefits to people who do not fit the definition of family. Such benefits include insurance coverage, housing, time off from work, inheritances (especially in the absence of a will), and opportunities to cash in individual retirement accounts without penalty to help a "family member" in need.

In this vein, a number of recent court decisions in the United States have ruled that criteria other than socially recognized bonds of blood, marriage, or adoption must be considered in defining whether a person belongs to a family and thus is entitled to family related benefits. These factors are part of any healthy family environment and include exclusivity, longevity, emotional support and financial commitment (Gutis 1989a, 1989b). The point is that, at least with regard to official definitions of family, it is not especially useful to think about the family in terms of specific membership (that is, son, daughter, mother, father) or a specific function such as procreation or socialization. Instead, we would do better to take a contextual view of the family and consider how key demographic events are transforming its character in fundamental ways (the transformation may have positive or negative consequences). The official definition, however, drives policies that can alleviate or exacerbate those consequences.

Key Concepts

Use this outline to organize your review of the key chapter ideas.

- **Demography**
- **Mechanized rich**
- **Labor-intensive poor**
  - **Annual per capita consumption of energy**
  - **Doubling time**
  - **Infant mortality**
  - **Total fertility**
  - **Per capita income**
- **Theory of demographic transition**
  - **Demographic gap**
- **Demographic trap**
- **Population pyramid**
  - **Cohort**
  - **Expansive pyramid**
  - **Stationary pyramid**
  - **Constructive pyramid**
- **Migration**
  - **International migration**
    - **Emigration**
    - **Immigration**
  - **Internal migration**
    - **In-migration**
    - **Out-migration**
  - **Push factors**
  - **Pull factors**
- **Mortality crisis**
- **Urbanization**

internet assignment

Go to Pennsylvania State University's Population Research Institute Web site (http://www.pop.psu.edu/Demography/demography.html). Browse the links connected with this Population Research Institute and/or the links to other Web sites. Select a document that adds to information in this chapter. Write a brief description of the information, and explain how it further clarifies a demographic concept covered in this chapter.

Education

With Emphasis on Public Education in the United States

College graduation ceremony. ©William Strode/Woodfin Camp & Associates

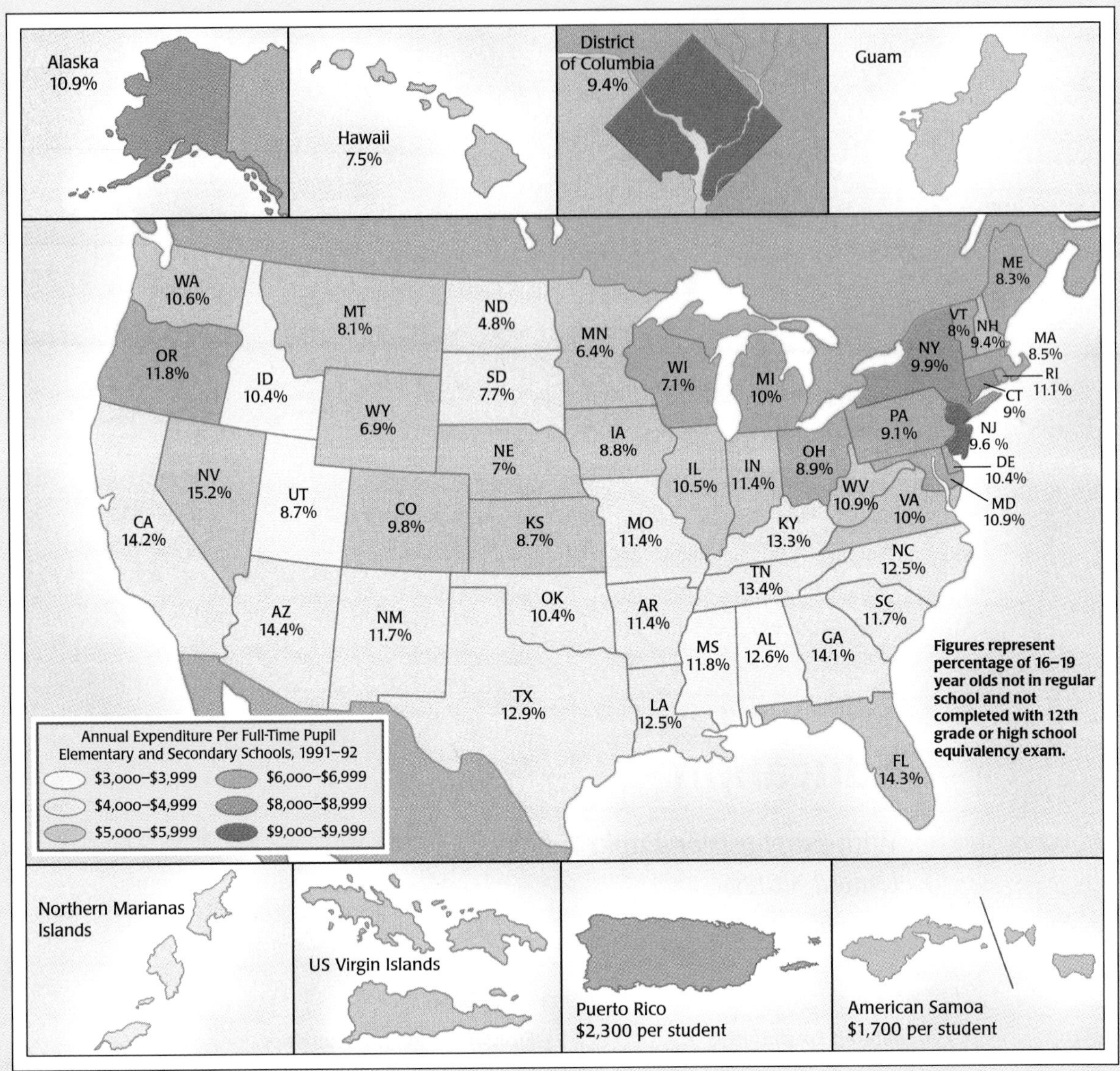

Sources: U.S. Bureau of the Census (1990) and U.S. Department of Education (1995).

Problem or Solution?

Some Americans blame many of their country's problems on its system of public education. Many also protest that we cannot improve the system by "throwing money" at it.

The color coding on this map shows how the states and U.S. territories vary in their level of educational funding per student. The numbers on the map tell what percentage of people ages sixteen to nineteen was not in regular school and had not completed twelfth grade or a high school equivalency exam in 1990. Based on these data, what relationship do you see between level of funding and the success of U.S. education?

This chapter explores the sociology of education. It asks, How well does the U.S. system of education serve its students? How well does it serve its broader social purposes? What sociological problems can education be expected to help solve? What problems can it not solve?

The Congress declares that the National Education Goals are the following:

By the year 2000

- *All children in America will start school ready to learn.*
- *The high school graduation rate will increase to at least 90 percent.*
- *All students will leave grades 4, 8, and 12 having demonstrated competency over challenging subject matter including English, mathematics, science, foreign languages, civics and government, economics, arts, history, and geography.*
- *Every school in America will ensure that all students learn to use their minds well, so they may be prepared for responsible citizenship, further learning, and productive employment in our nation's modern economy.*
- *The nation's teaching force will have access to programs for the continued improvement of their professional skills and the opportunity to acquire the knowledge and skills needed to instruct and prepare all American students for the next century.*
- *United States' students will be first in the world in mathematics and science achievement.*
- *Every adult American will be literate and will possess the knowledge and skills necessary to compete in a global economy and exercise the rights and responsibilities of citizenship.*
- *Every school in the United States will be free of drugs, violence, and the unauthorized presence of firearms and alcohol and will offer a disciplined environment conducive to learning.*
- *Every school will promote partnerships that will increase parental involvement and participation in the social, emotional, and academic growth of children.*

These national goals, to be achieved by the year 2000, were first announced by President George Bush in September 1989 at the Governors Conference. In April 1994, President Bill Clinton signed the Goals 2000 Act, which sets aside $700 million in federal funds for states and school districts that implement programs aimed at meeting guidelines related to these goals.

Notice the preoccupation with preparing students to compete in a global economy and with creating persons prepared for responsible citizenship. These preoccupations suggest that the system of education functions to teach children (and adults) the skills they need to adapt to their environment and to make decisions in their everyday lives that benefit humankind. In the United States, the ongoing belief that the school system does not achieve these functions drives efforts to reform or restructure the system. Ironically, while schools have been treated as the primary source of many of the country's problems, they are also treated as the solution to those problems.

Why Focus on the United States?

In this chapter, we give special emphasis to public education in the United States, the system in which 89 percent of students are enrolled. Throughout U.S. history, it seems as if public education has always been in a state of crisis and under reform. Between 1880 and 1920, for example, the public was concerned about whether the schools were producing qualified workers for the growing number of factories and businesses and whether they were instilling a sense of national identity (that is, patriotism) in immigrant and ethnically diverse student population. When the United States was involved in major wars—World War I, World War II, Korea, and Vietnam—the public was concerned about whether the schools were turning out recruits physically and mentally capable of defending the American way of life.

When the Soviet *Sputnik* satellites were launched in the mid-1950s, Americans were forced to consider the possibility that the public schools were not educating their students as well as the Soviets were in mathematics and the sciences. In the 1960s, civil rights events forced Americans to question whether the public schools were offering children from less advantaged ethnic groups and social classes the knowledge and skills to compete economically with children in more advantaged groups. Beginning in the late 1970s and continuing throughout the 1980s and the 1990s, many critics, at home and abroad, maintain that the U.S. system of education is not adequate for meeting the challenges associated with being part of a global economy. Many employers claim that they are unable to find enough workers with a level of reading, writing, mathematical, and critical thinking skills needed to function adequately in the workplace. The comments of IBM chairman Louis V. Gerstner in a March 27, 1996, interview illustrate this view:

> [T]here are an awful lot of wonderful public schools in this country that produce outstanding students, but there aren't enough of them. And the vast majority of students that come out of school today simply do not have the basic skills in subjects such as math, reading, being able to compute, communicate. . . . We're not looking for schools to teach vocational skills. I can teach engineers and other people who come to IBM to work with machines, how to do marketing, how to do finance, but we're really not equipped to teach 'em to read and write. (*Online NewsHour* 1996b)

What Is Education?

In the broadest sense, **education** includes those experiences that train, discipline, and develop the mental and physical potentials of the maturing person. An experience that educates may be as commonplace as reading a sweater label and noticing that it was made in Taiwan or as intentional as performing a scientific experiment to learn how genetic makeup can be altered deliberately through the use of viruses. In view of this definition and the wide range of experiences it encompasses, we can say that education begins when people are born and ends when they die.

Sociologists make a distinction, however, between formal and informal education. **Informal education** occurs in a spontaneous, unplanned way. Experiences that educate informally occur naturally: they are not designed by someone to stimulate specific thoughts or interpretations or to impart specific skills. Informal education takes place when a child puts her hand inside a puppet, then works to perfect the timing between the words she speaks for the puppet and the movement of the puppet's mouth.

Formal education is a purposeful, planned effort intended to impart specific skills and modes of thought. Formal education, then, is a systematic process (for example, military boot camp, on-the-job training, programs to stop smoking, classes to overcome fear of flying) in which someone designs the educating experiences. We tend to think of formal education as consisting of enriching, liberating, or positive experiences, but it can include impoverishing and narrowing occurrences (such as indoctrination or brainwashing) as well. In any case, formal education is considered a success when the people instructed internalize (or take as their own) the skills and modes of thought that those who design the experiences seek to impart. This chapter is concerned with a specific kind of formal education—schooling.

Schooling is a program of formal and systematic instruction that takes place primarily in classrooms but also includes extracurricular activities and out-of-classroom assignments. In its ideal sense, "[e]ducation must make the child cover in a few years the enormous distance traveled by mankind in many centuries" (Durkheim 1961:862). More realistically, schooling is the means by which those who design and implement

Education Those experiences that train, discipline, and develop the mental and physical potentials of the maturing person.

Informal education Education that occurs in a spontaneous, unplanned way.

Formal education A systematic, purposeful, planned effort intended to impart specific skills and modes of thought.

Schooling A program of formal and systematic instruction that takes place primarily in classrooms but also includes extracurricular activities and out-of-classroom assignments.

Schooling can serve different purposes. Training in the use of computers, for example, may be designed primarily to equip students with the skills they need to adapt to their environment. But it can also be used to broaden students' intellectual horizons and encourage creativity and independent thinking. Have you been asked to use computers in any of your previous education? Apart from the "computer skills," were the assignments also designed to broaden your intellectual perspective or social opportunities? How?

©James Wilson/Woodfin Camp & Associates

programs of instruction seek to pass on the values, knowledge, and skills they define as important for success in the world. What constitutes an ideal education — the goals that should be achieved, the material that should be covered, the best techniques of instruction — is elusive and debatable. Conceptions vary according to time and place; they differ according to whether schools are viewed primarily as mechanisms by which the needs of a society are met or the means by which students learn to think independently, thus becoming free from the constraints on thought imposed by family, peers, culture, and nation.

Social Functions of Education

Sociologist Emile Durkheim believed that education functions to serve the needs of society. In particular, schools function to teach children the skills they need to adapt to their environment. To ensure this end, the state (or other collectivity) reminds teachers "constantly of the ideas, the sentiments that must be impressed" on children if they are to adjust to the society in which they must live. Educators must achieve a sufficient "community of ideas and sentiments without which there is no society." Otherwise, "the whole nation would be divided and would break down into an incoherent multitude of little fragments in conflict with one another" (Durkheim 1968:79, 81). Such logic underscores efforts to use the schools as mechanisms for meeting the needs of society, whether they be to strip away identities of ethnicity and social origin to implant a common national identity, transmit values, train a labor force, take care of children while their parents work, or to teach young people to drive.

According to another quite different conception, education is a liberating experience that releases students from the blinders imposed by the accident of birth into a particular family, culture, religion, society, and time in history. Schools therefore should be designed to broaden students' horizons so they will become aware of the conditioning influences around them and will learn to think independently of any authority. When schools are designed to achieve these goals, they can function as agents of change and progress.

These aims are not necessarily contradictory if what benefits the group also liberates the individual. For example, democracies and the free market system require an informed public that is capable of independent thought. Most sociological research, however, suggests that schools are more likely to be designed to meet the perceived needs of society rather than to liberate minds. In spite of this intention, however, a significant percentage of the population in every country seems to be **functionally illiterate**—that is, they do not possess the level of reading, writing, and calculating skills needed to adapt to the society in which they live. In fact, to many critics of the U.S. educational system, illiteracy in the United States has reached crisis proportions. To evaluate the "literacy crisis," we need to define illiteracy, to distinguish among kinds of literacy, and to ask how the United States compares to other countries when it comes to preparing its students to be educated workers and citizens.

Illiteracy in the United States

In the most general and basic sense, **illiteracy** is the inability to understand and use a symbol system, whether it is based on sounds, letters, numbers, pictographs, or some other type of character. Although the term *illiteracy* is used traditionally in reference to the inability to understand letters and their use in reading and writing, there are as many kinds of illiteracy as there are symbol systems—computer illiteracy, mathematical illiteracy (or innumeracy), scientific illiteracy, cultural illiteracy, and so on.

If we confine our attention merely to languages, of which there are thought to be between 3,500 and 9,000 (including dialects), we can see that the potential number of literacies is overwhelming (Ouane 1990) and that people cannot possibly be literate in every symbol system. If a person speaks, writes, and reads in only one language, by definition he or she is illiterate in perhaps as many as 8,999 languages. Yet, such a profound level of illiteracy rarely presents a problem because usually people need to know and understand only the language of the environment in which they live.

This point suggests that illiteracy is a product of one's environment—that is, people are considered illiterate when they cannot understand or use the symbol system of the surrounding environment in which they wish to function. Examples include not being able to use a computer, access information, read a map to find a destination, make change for a customer, read traffic signs, follow the instructions to assemble an appliance, fill out a job application, or comprehend the language of those around them.

In the United States (and in all countries, for that matter), some degree of illiteracy has always existed, but conceptions of what people needed to know to be considered literate have varied over time. At one time, people were considered literate if they could sign their names and read the Bible. At other times, a person who had completed the fourth grade was considered literate. The National Literacy Act of 1991 defines literacy as "an individual's ability to read, write, and speak English and compute and solve problems at levels of proficiency necessary to function on the job and in society, to achieve one's goals, and to develop one's knowledge and potential" (U.S. Department of Education 1993a:3).

In 1988, the U.S. Congress requested that the Department of Education define literacy in the context of the new economic order and attempt to estimate how many Americans are illiterate. The project involved a representative sample of 26,091 adults (13,600 were interviewed, and 1,000 were surveyed in each of eleven states.) In addition, 1,100 federal and state prison inmates were interviewed. The project was completed in 1993. The researchers found that 21 to 23 percent of those contacted "demonstrated skills in the lowest level of prose, document, and quantitative proficiencies." Another 25 to 28 percent performed at level 2.[1] All of these respondents fell below the level 3 criteria: they could not (1) write a brief letter explaining an error made on a credit card bill, (2) identify information from a bar graph, or (3) use a calculator to find the difference between the regular price and the sale price of an item in an advertisement.

Functionally illiterate A term describing people who do not possess the level of reading, writing, and calculating skills needed to adapt to the society in which they live.

Illiteracy The inability to understand and use a symbol system, whether it is based on sounds, letters, numbers, pictographs, or some other type of character.

[1]According to the authors of this study:

> The approximately 90 million adults who performed in Levels 1 and 2 did not necessarily perceive themselves as being "at risk." Across the literacy scales, 66 to 75 percent of the adults in the lowest level and 93 to 97 percent in the second lowest level described themselves as being able to read or write English "well" or "very well." Moreover, only 14 to 25 percent of the adults in Level 1 and 4 to 12 percent in Level 2 said they get a lot of help from family members or friends with everyday prose, document, and quantitative literacy tasks. It is therefore possible that their skills, while limited, allow them to meet some or most of their personal and occupational literacy needs. (U.S. Department of Education 1993a:XV)

Figure 13.1 Estimated Percentage of the Population at Each Proficiency Level on Three Adult Literacy Scales, by Selected Countries, 1994

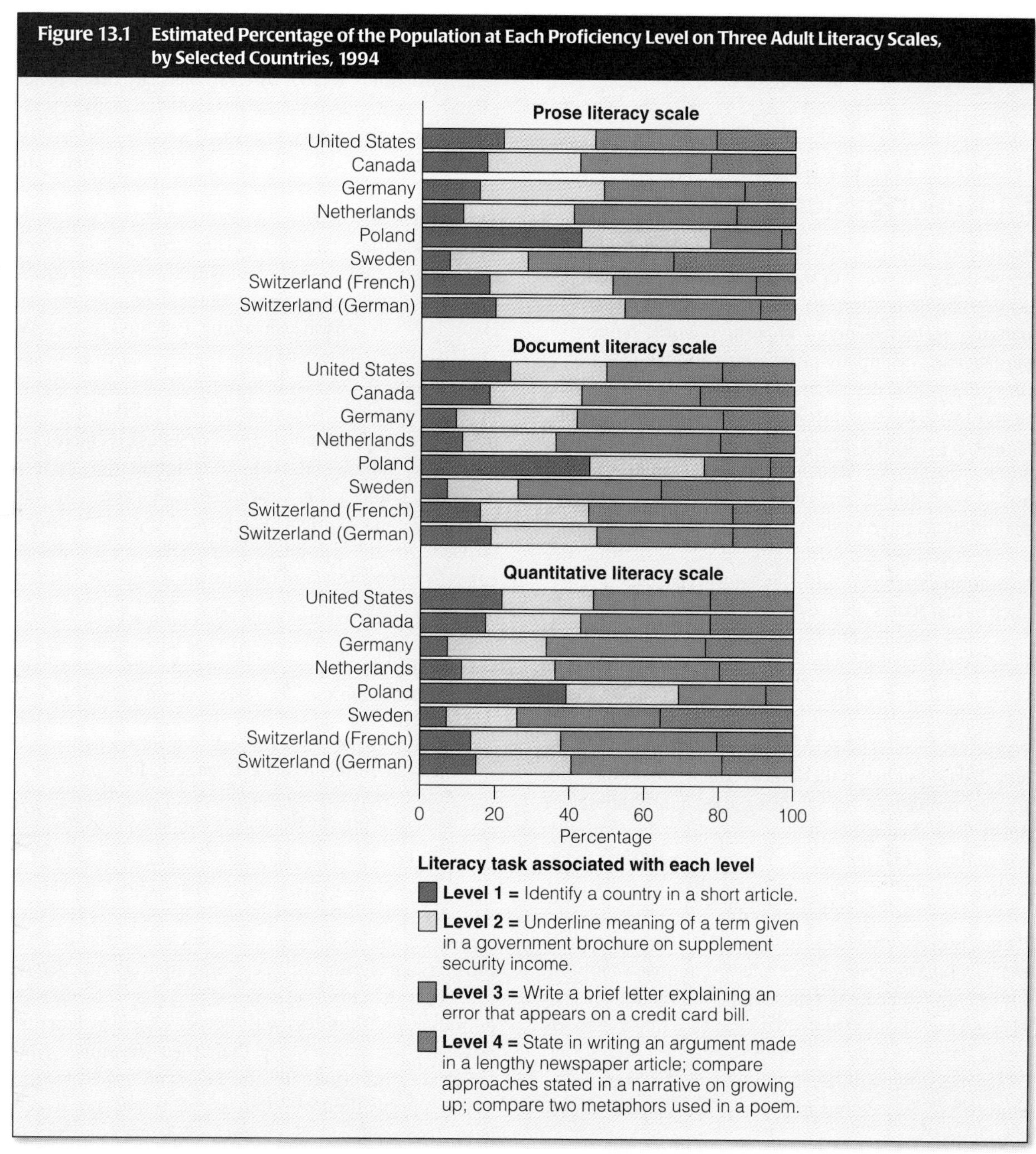

Sources: U.S. Department of Education (1996); Organization for Economic Co-operation and Development and Statistics Canada (1995).

The fact that almost half of the adult population could function at the two lowest levels and that the United States has one of the highest concentration of adults in the lowest literacy levels compared with selected European countries (see Figure 13.1) leads social critics, most notably U.S. government officials and business leaders, to point to the schools as one major source of the problem.

The high estimates of illiteracy even among those who completed high school led people to ask how so many students could attend school for at least ten to twelve years without acquiring enough reading, writing, and problem-solving skills to deal effectively with the work-related problems encountered in the new kinds of entry-level jobs. This situation has prompted many critics to examine the ways in which schools in countries considered as major economic competitors to the United States, most notably European and Pacific Rim countries, educate their students.

Insights from Foreign Education Systems

Sociologists John W. Meyer and David P. Baker (1996) argue that one legacy of the Reagan-Bush presidencies is internationalization of U.S. educational policy. Apparently an increasingly competitive world economic climate and accompanying concerns about world position prompted critics to compare the U.S. system of education with those of foreign countries on a host of attributes, including amount of homework, parental support, teachers' salaries, classroom environment, and the amount of time watching television. The assumption underlying such comparisons is that there is a connection between the way a society structures the education experience and education-related outcomes such as test scores, literacy levels, and work skills.

Meyer and Baker (1996) point out that the International Education Association (IEA) has collected cross-national data on educational achievement since the 1950s. Yet, it has only been in the past ten to fifteen years that the media has focused on the findings. Meyer and Baker maintain that, more often than not, the media emphasizes that the test scores of U.S. students are "not especially high" or are "very poor" in comparison with students in other countries. Usually the media suggests that U.S. scores are lower than they actually are because the media reports rank ("eleventh of fourteen countries," "last," or "next to last"). It fails to report that average test scores for many countries ranking lower and higher are not measurably different from the U.S. average. For example, whereas U.S. ninth graders averaged 67 percent on the IEA science test, that score is not measurably different from Spain (67.5) or Scotland (67.9), two countries that technically rank above the United States.

In addition, the media and education critics focus considerable attention on how U.S. students compare with students in selected countries—those considered to be major economic competitors (Germany, Japan) or countries expected to be major economic competitors (China, South Korea). And as Meyer and Baker point out, a disproportionate focus is placed on tests in which U.S. students do less well than on tests in which they do better than students in other countries. The fact that students in twenty countries score measurably lower than U.S. nine-year-olds in reading attracts much less attention than the fact that U.S. nine-year-olds rank ninth out of ten countries in mathematics. Consequently, it should come as no surprise that the cross-national research focuses on the United States in comparison with those countries in which its students do less well (see Table 13.1).

In this regard, one of the most systematic and well-designed studies comparing U.S. academic activities and mathematics achievement with those of two countries considered as economic competitors was done by Harold W. Stevenson, Shin-ying Lee, and James W. Stigler (1986). They compared mathematics achievement as well as the classroom and home environments of kindergartners, first-graders, and fifth-graders in three cities: Minneapolis (United States), Sendai (Japan), and Taipei (Taiwan). They found that, across all three grades, the Taiwanese, but especially the Japanese, consistently outperformed their U.S. counterparts. At the fifth-grade level, however, the differences in test scores were the most striking:

> The highest average score of an American fifth-grade classroom was below that of the Japanese fifth-grade classroom with the lowest average score. In addition, only one Chinese [Taiwanese] classroom showed an average score lower than the American classroom with the highest average score. Equally remarkable is the fact that the lowest average score for a fifth-grade American classroom was only slightly higher than the average score for the best first-grade Chinese classroom. (p. 694)

After thousands of hours of classroom observation and after interviews with both mothers and teachers, Stevenson and his colleagues concluded that Americans devote significantly less time to academic activities either in school or at home and that U.S. parents help their children less with homework. U.S. parents, however, are more likely than their Taiwanese and Japanese counterparts to rate the quality of education at the schools their children attend as good or excellent: 91 percent of U.S. parents, 42 percent of Chinese parents, and 39 percent of Japanese parents rate the quality as good or excellent and are satisfied with the qualities of education (see Table 13.2). Stevenson (1992) maintains that one reason U.S. parents rate quality high is that their school system gives them no clear guidelines about what academic skills children in various grades should possess. Thus, they have no baseline by which to judge their child's performance.

A more recent study led by the U.S. Department of Education compared the math and science scores of

Table 13.1 International Distribution of Academic Achievement Relative to the United States, 1991–1992

This table is better than many other public reports on the quality of American education. It does not exaggerate the differences between U.S. and foreign students. In some tables we only see one column of data "rank," which does not tell us whether the differences between the United States and the countries ranking higher or lower are really significant. In addition, the media that filter the data before we see it often fail to give headline attention to areas in which U.S. students do very well. Rather than giving us the entire picture, they give us selected parts of it that exaggerate the failure of U.S. students.

What measurable differences do you see in this table between the U.S. and other countries? What specific strengths or weaknesses does the table suggest in U.S. education?

	Number of Countries Performing:				
Subject and Age	Measurably Higher than the U.S.	Not Measurably Different from the U.S.	Measurably Lower than the U.S.	Number of Countries in the Study	U.S. Rank
Mathematics (1991)					
9-year-olds	7	2	0	10	9
13-year-olds	12	1	1	15	13
Science (1991)					
9-year-olds	0	7	2	10	3
13-year-olds	5	7	1	14	11
Reading (1991–92)					
9-year-olds	1	1	20	23	2
14-year-olds	3	15	6	23	9

Source: U.S. Department of Education (1996).

more than 500,000 students in forty-one countries. The findings showed that students in Asian countries and city-states (Singapore, South Korea, Japan, and Hong Kong) did the best, followed by students in Eastern Bloc countries. U.S. students scored slightly below average in math and slightly above average in science. It seems that these differences in scores can be traced to the content of the U.S. curriculum (see "The Content of Math Curriculum").

In light of the findings from international comparisons, Meyer and Baker (1996) maintain that the challenge lies with deciding what features of U.S. education should be reformed. Although many reforms have been initiated in response to the perceived standing of U.S. students relative to foreign students

> much of the need for reform is put forward as if the current American situation is the result of sloth and ineptitude (or, more precisely, suboptimization of effort by students, teachers, administrators, and parents). This view may understate the extent to which the American educational system—with all of what are now considered its great defects—has its own substantial roots in the country's educational and political history. (p. 126)

In the next sections of this chapter, we examine the historical background of the U.S. system of education and then consider some of its general and distinguishing characteristics.

The Development of Mass Education in the United States

The United States was the first country in the world to embrace the concept of mass education. In doing so, it broke with the European view that education should be limited to an elite few (for example, the top 5 percent or those who could afford it). In 1852, Massachusetts legislators passed a law making elementary school mandatory for all children, and within sixty years all of the states had passed compulsory attendance laws.

A number of factors other than the legal mandate encouraged parents to comply with attendance laws, however. First, as the pace of industrialization increased, jobs moved away from the home and out of neighborhoods into factories and office buildings. As mechanization increased, apprenticeship opportuni-

Table 13.2 Comparison of Chinese (Taiwanese), Japanese, and American Kindergartners, First-Graders, and Fifth-Graders

Homework	**Americans**	**Taiwanese**	**Japanese**
Minutes per day doing homework*			
First-graders	14 min.	77 min.	37 min.
Fifth-graders	46 min.	114 min.	57 min.
Minutes doing homework (Saturday)			
First-graders	7 min.	83 min.	37 min.
Fifth-graders	11 min.	73 min.	29 min.
Time parent spent helping with homework			
Fifth-graders	14 min.	27 min.	19 min.
Percentage who possess a desk at home			
Fifth-graders	63%	95%	98%
Percentage of parents who purchase workbooks for children to get extra practice			
Fifth-graders (math)	28%	56%	58%
Fifth-graders (science)	1%	51%	29%
Classroom			
Percentage of time devoted to academic activities			
First-graders	69.8%	85.1%	79.2%
Fifth-graders	64.5%	91.5%	87.4%
Hours per week devoted to academic interests			
First-graders	19.6 hrs.	40.4 hrs.	32.6 hrs.
Fifth-graders	19.6 hrs.	40.4 hrs.	32.6 hrs.
Total hours per week spent in school			
First-graders	30.4 hrs.	44.1 hrs.	37.3 hrs.
Fifth-graders	30.4 hrs.	44.1 hrs.	37.3 hrs.
Percentage of time a child known to be in school was not in classroom			
Fifth-graders	18.4%	<0.2%	<0.2%
Proportion of time teacher spends imparting information (all grades)	21% or 6 hrs.	58% or 26 hrs.	33% or 12 hrs.
Mother's Perceptions			
Evaluation of child's achievement in math on a scale of 1 to 9†	5.9	5.2	5.8
Evaluation of child's intellectual ability†	6.3	6.1	5.5
Percentage who believe school is doing a good or excellent job educating children	91%	42%	39%
Percentage very satisfied with child's academic performance	40%	<6%	<6%
Evaluation of what is more important to success—ability or effort?	Ability	Effort	Effort
Percentage who gave child assistance in math	8%	2%	7%

*Mothers' estimates.

†With 1 being much below average and 9 being much above average.

Source: Data from "Mathematics Achievement of Chinese, Japanese, and American Children," by Harold W. Stevenson, Shin-ying Lee, and James W. Stigler in *Science*. Copyright © 1986 by the AAAS. Reprinted by permission of Harold W. Stevenson and AAAS.

ties gradually disappeared. As family farms and businesses disappeared, parents could no longer train their children because familiar skills were becoming obsolete. Thus, with the home and work environments independent of each other, parents were no longer available to oversee their children. Second, a tremendous influx of immigrants to the United States between 1880 and 1920 created a large labor pool—a surplus—from which factory owners could draw workers, thereby eliminating the need for child laborers. This combination of events created an environment in which there was no place for children to go except to school.

At least two prominent features of early American education have endured to the present: (1) textbooks modeled after catechisms and (2) single-language instruction.

Textbooks

The most vocal early educational reformers such as Benjamin Rush, Thomas Jefferson, and Noah Webster believed that schools were an important mechanism by which a diverse population could acquire a common culture. They believed that the new "perfectly homogeneous" American was one who studied at home (not abroad) and who used American textbooks. To use Old World textbooks "would be to stamp the wrinkles of decrepit age upon the bloom of youth" (Webster, quoted in Tyack 1966:32).

The first textbooks in the United States were modeled after catechisms, short books covering religious principles written in question-and-answer format. Each question had one answer only, and in repeating the answer, the student strictly adhered to the question's wording (Potter and Sheard 1918). This format discouraged readers from behaving as active learners "who could frame questions, interpret materials, and reflect on the significance of what was presented. . . . No premium was placed on generating and inventing ideas or arguing about the truth or value of what others had written (Resnick 1990:18). The reader's job was to memorize the "right" answers for the questions. With this as the model, not surprisingly, textbooks tend to be written in such a way that the primary reason to read them is to find the "right" answers to the accompanying questions.

The influence of the catechisms on learning today is evident whenever students are assigned to read a chapter and answer the list of questions at the end. Many students discover that they do not need to read the material to answer the questions; they can simply skim the text until they find key words that correspond to those in the question, and then they copy the surrounding sentences.

In the past few years, educators have questioned the value of textbooks as a learning tool. Although we know that some schools have abandoned textbooks, especially at the elementary level, we do not know the scope of this trend. It seems that educators in the field of elementary reading are leading the way as they increasingly abandon traditional textbooks modeled after catechisms in favor of children's literature books and daily writing projects (Richardson 1994). Historical novels and edited volumes that include essays, various perspectives, and so-called "real" or authentic literature are among the alternatives to textbooks.

Single-Language Instruction

The "peopling of America is one of the great dramas in all of human history" (Sowell 1981:3). It involved the conquest of the native peoples, the annexation of Mexican territory along with many of its inhabitants (who lived in what is now New Mexico, Utah, Nevada, Arizona, California, and parts of Colorado and Texas), and an influx of millions of people from practically every country in the world. School reformers, primarily people of Protestant and British background, saw public education as the vehicle for "Americanizing" a culturally and linguistically diverse population, instilling a sense of national unity and purpose, and training a competent workforce. As Benjamin Rush argued, "Let our pupil . . . be taught to love his family but let him be taught, at the same time, that he must forsake and even forget them, when the welfare of his country requires it" (Rush, quoted in Tyack 1966:34). To meet these nation-building objectives, people had to abandon their native languages and learn to speak a common language. Consequently, students were taught in English, the language of the established elite.

Although the United States is hardly unique in pressuring its people to abandon their native tongues and learn to speak a common language, it is probably the only country in the world that places so little emphasis on learning at least one other language. Education critic Daniel Resnick believes that the absence of serious foreign language instruction contributes to the parochial nature of American schooling. The almost exclusive attention to a single language has deprived students of the opportunity to appreciate the connection between language and culture and to see that language is a tool that enables them to think about the world.

U.S. In Perspective

The Content of U.S. Math Lessons

This graph shows that 87 percent of eighth-grade U.S. math lessons observed were rated as low in quality. No lessons were rated as high in quality. In November 1996, *Newsweek* education reporter Pat Wingert explained in an interview with *MacNeil/Lehrer Newshour* (*Online Newshour* 1996a) correspondent Elizabeth Farnsworth why the ratings were so low relative to Japan and Germany, two countries regarded as major economic competitors with the United States:

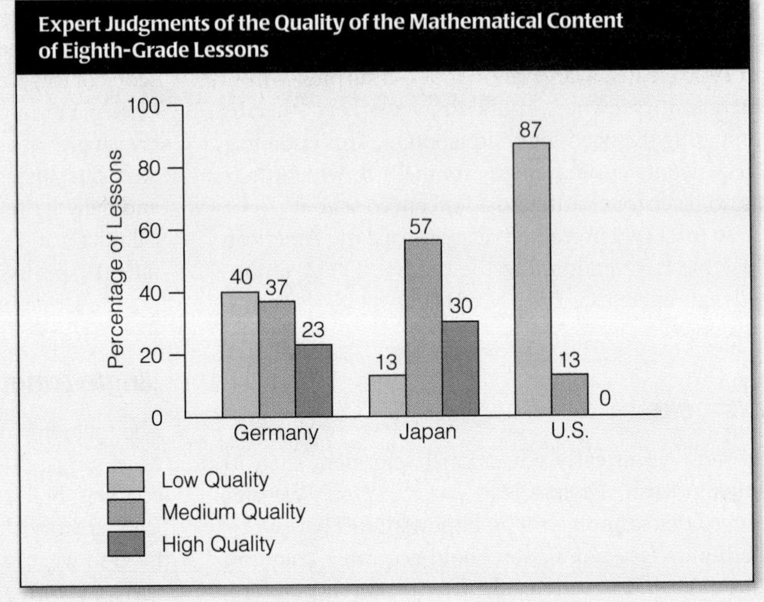

Source: Third International Mathematics and Science Society Study; unpublished tabulations, Videotape Classroom Study, University of California–Los Angeles, 1996.

FARNSWORTH: What did this study find American schools are doing wrong?

WINGERT: Basically, we seem to be doing a lot of things wrong. This study says that American teachers spend more time teaching math and science. They give out more homework, which means they're correcting more homework. They're covering more topics in a year than your typical teachers around the world. But they're teaching in a very dry way. They present a problem to the class and say, "I'm going to try to do this kind of fraction problem." Then they have kids do fraction problems at their desks. Then they give them fraction problems for homework.

That's not the most effective way to do it. What [the researchers] saw in the videotapes in Japan, for example, was that the Japanese teachers were teaching the way that American reform documents have been saying for years that we should be doing it. They give kids problems in a more realistic way. You know: "You have this piece of property. You have to divide it in half. How would you do it?"

And then they have the kids sit there and really struggle. It's not an easy answer. It's something that they may have to work out for 20 minutes, not sure they have the right answer. Then they have a class discussion where kids will get up and say, "I think this is the way to do it." That kind of discussion gets kids emotionally involved in the answer. Brain research says the more emotional you are about something that you're learning, the better you're going to remember it. So those kids are getting

Resnick states that the focus on a single language "has cut students off from the pluralism of world culture and denied them a sense of powerfulness in approaching societies very different from their own" (1990:25). It also denies those students from non-English-speaking heritages the means of reflecting on and fully appreciating their ancestors' lives.

In the United States, English became the language people needed to learn to speak if they were to enter the mainstream of society and have a chance at upward mobility. In this respect, English was positioned against the language and even regional accents of parents and grandparents. English was the language "through which one's ethnic self was converted into one's acquired American personality . . . [and] as an instrument through which one could hide and mask ethnic (and regional) origins" (Botstein 1990:63). Noah Webster, an early influential education reformer, wrote in

excited. They're fighting over their answer. Then the teacher finally gets up and says, "Well, I think this is the best way. Now watch while I do this." And all the kids that have been sitting there struggling suddenly say, "He's right! That is the right way." And they remember it as a result.

It was very frustrating to the researchers to realize, when they talk to the teachers in American classrooms, that the teachers are very aware of what math reform documents had said and thought that they were doing it. Yet, when they videotaped, the researchers [saw] they're not doing it. [The teachers] don't realize they're not doing it.

So what do we do now to get them to understand that we still have big changes to make? How do we get American teachers now to say, after spending all this time doing all this homework, we're still not doing better? How do we keep them on track to fight this?

FARNSWORTH: You have an example to show us, don't you?

WINGERT: Right. [Wingert shows an example.] This example was handed out by the Department of Education. This shows how a typical math problem is taught in the United States. And as you can see, basically once you learn the method or the fast way of doing it, it doesn't really draw you in. Once you know the method, you just do it.

[Wingert shows a second example.] [In a better] question, you have to read what the question is about, and then you have to sit there and look at the graph and struggle. Several of us were looking at this before the show and had a discussion among us. Was this the right answer? Is this the right answer? Is this what they're trying to say? You have to sit there and think about it, so you get drawn in more. You have to think more, apply more.

FARNSWORTH: Is part of the problem there's too much being covered and not enough in this different way?

WINGERT: Right. Most math teachers follow their textbooks, and most textbooks are written in a way so that teachers in California and teachers in Texas and teachers in New York will all buy them. The result is we're packing way too much into each textbook. We're covering way too many topics. What the Japanese would do, for example, is start teaching kids fractions and [continue teaching] fractions every day for multiple days. They would build on that knowledge. In a typical American classroom kids might get a little fractions this year in the fall and maybe a little more in the spring. By spring, of course, they've forgotten some of the stuff that they learned in the fall. Then next year they might learn a little more, but the teacher has to take them back to teach them again. We keep going over the same ground again and again. We're wasting time.

FARNSWORTH: Does this imply mostly changes in teacher training?

WINGERT: Yes.

FARNSWORTH: There may be too much content, and there need to be changes in the way teachers approach it?

WINGERT: There needs to be reorganization of the content. It needs to be done in a more efficient way that builds on knowledge. Teachers need to be retrained. Teachers need to see videotapes or see real live teachers teaching the way that they should be teaching. It's very hard to change, just like it's hard to change the way that you're parenting. So teachers need to see real-life examples. They need to see a teacher in front of them, doing it the right way.

Does this interview give any clues about other changes, outside the classroom – for example, at administrative levels or among publishers – that might lead to improvements in U.S. schools?

Source: From "Education Report Card" (November 21), *Online Newshour*.

the preface of his spelling book that the United States must promote a "uniformity and purity of language" and "demolish" those odious distinctions of provincial dialects which are subject to reciprocal ridicule in different states" (quoted in Tyack 1966:33). The national memory of learning English, of breaking with the past, remains with subsequent generations in their ambivalent attitudes not just toward language acquisition but toward the meaning of learning (Botstein 1990).

Fundamental Characteristics of Contemporary American Education

A number of characteristics distinguish American education from other systems of education. They include the availability of college, the lack of a uniform curriculum, funding that varies by state and community, the belief that schools can be the vehicle for solving a variety of social problems, and ambiguity of purpose and value. This section discusses these characteristics.

The Availability of College

One of the most distinctive features about the United States is that in theory anyone can attend a college if he or she has graduated from high school or has received a GED. As a result, the United States has the world's highest postsecondary enrollment ratio. Sixty-two percent of 1995 high school graduates enrolled in college in fall 1994 (U.S. Department of Education 1995).[2] This figure varies depending on racial classification. Whereas 63 percent of high school graduates classified as white enrolled in college, only 54 percent of high school graduates classified as Hispanic and 51 percent classified as black enrolled.

To understand another indicator of relatively easy access to college, consider that 20 percent of U.S. four-year colleges and universities accept students regardless of what courses they took, what grades they earned in high school, or what scores they received on ACT or SAT tests. Seventy-four percent of colleges and universities offer remedial courses in reading, writing, and mathematics for those students who lack skills necessary to do college-level work. Thirty percent of all freshman students who entered college in 1989 took one or more remedial courses (U.S. Bureau of the Census 1993).

In most other countries, education beyond high school is available to a smaller percentage of the population (see "How Many Get Their Bachelor's Degree?"). In England and Wales, sixteen-year-olds who make low scores on the General Certificate of Secondary Education Test cannot take the advanced courses required for college. Similarly, scores on the French baccalaureate examination determine who can attend college (Chira 1991). From this perspective, the educational opportunities in the United States are admirable. A college education is open to everyone (who can find money to pay the tuition) regardless of previous educational failures; it is not reserved for a privileged few (see Table 13.3). This policy reflects the American belief in equal opportunity to compete at whatever point in life one wishes to enter the competition: "We may not all hit home runs, the saying goes, but everyone should have a chance at bat" (Gardner 1984:28).

[2]This does not mean that the 62 percent who enroll out of high school go on to complete college.

At the same time, the unrestricted right to a college education seems to be connected with a decline in the value of a high school education and the effort put into achieving that level of education. As high school enrollments increased over time to include the middle class and eventually the poor and minority groups, the perceived value of the high school diploma declined and the perceived need for a college degree increased. As college enrollments increased and as a college degree came to be defined as important for success, the high school was no longer the last stop before young people entered the mainstream labor force. Consequently, high schools were subject to less pressure to make sure that their graduates had acquired the skills necessary for literacy in the workplace. As a result, the high school curriculum became less important and less rigorous (Cohen and Neufeld 1981).

Defenders of U.S. public schools argue that a less rigorous curriculum is a necessary consequence of trying to educate all citizens: if everyone is to pass through, the standards must be reduced. Even if we accept this logic, the consequences of such a policy are clear: the high school diploma loses its value, and the goal of achieving equality through compulsory and free education is undermined. Mass education accomplished through using social promotion (passing students from one grade to another on the basis of age rather than academic competency), awarding high school diplomas to people with fifth-grade reading abilities, and issuing certificates of achievement to those who fail minimum competency tests is not the same as educating everyone. True equality in education is achieved only if everyone has an opportunity to earn a degree that is valued.

Differences in Curriculum

No uniform curriculum exists in the United States. Each of the fifty states sets broad curriculum requirements for kindergarten through high school; each school interprets and implements these requirements. Consequently, even with state guidelines, the textbooks, assignments, instructional methods, staff qualifications, and material covered vary across the schools within each state. In addition to curriculum differences across states, students in the same school are usually grouped or "tracked" on the basis of tests or past performance. For example, students enrolled in standard diploma (versus college preparatory, honors, or advanced) tracks often take fewer mathematics courses and different kinds of mathematics (general math in-

U.S. In Perspective

How Many Get Their Bachelor's Degree?

This graph shows the percentages of males and females earning bachelor's or equivalent degrees in selected countries as of 1991. Spend some time considering what you can learn from it about specific countries that interest you and about comparisons between the United States and other countries. Here are some questions you might ask:

1. Which countries educate lower percentages of college graduates than does the United States? Which educate higher percentages? What do the more striking differences suggest?
2. Is it necessarily a good thing for a country to have the highest percentage of college graduates? Are there ways in which a high percentage might reflect or conceal important weaknesses in a country's overall educational system?
3. What differences do you see in the proportion of men versus women completing college in various countries? What might account for these differences in several countries? Would you consider an equal proportion of male and female graduates ideal? How so? How perhaps not?
4. The text says that "the United States has the world's highest postsecondary *enrollment* ratio." Is that claim inconsistent with the data in this graph? Why?

Bachelor's Degree Recipients as a Percentage of Population in the Appropriate Age Group for Selected Countries, by Sex, 1991

Country
Australia
Austria
Belgium
Canada
Denmark
Finland
France
Germany
Ireland
Italy
Japan
Netherlands
New Zealand
Norway
Spain
Sweden
Switzerland
Turkey
United States
0 10 20 30 40 50
Percentage
Male
Female

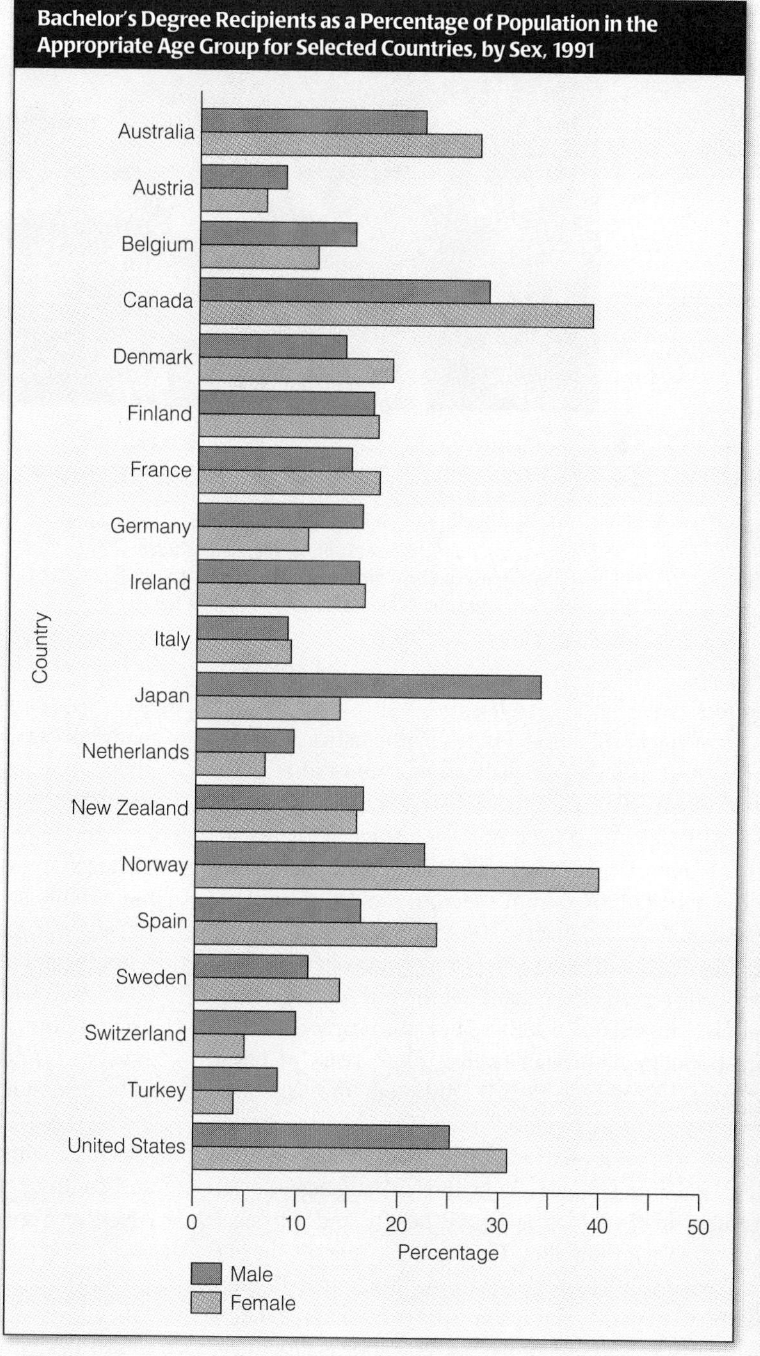

Sources: U.S. Department of Education (1996); Organization for Economic Co-operation and Development, unpublished data.

As the United States has made it easier for people to enter college, it has made secondary schools less rigorous. Is this a wise trade-off? Should high school diplomas be based on stricter and more specific standards of accomplishment with regard to verbal and mathematical literacy? For example, should all students be required to pass algebra? How might such requirements have affected you and other students you knew in high school?

stead of algebra) to meet a state's mathematics requirement. And they take English composition rather than creative writing to meet the state English requirement.

Although most countries also track students at some point in their careers, all students are exposed to a core curriculum, even if everyone cannot assimilate the material. The Japanese, for example, believe that a standardized curriculum is the only way to ensure that everyone has an equal chance at the rewards that education brings (Lynn 1988). "They put nearly their entire population through twelve tough years of basic training" (Rohlen 1986:38). Although the Japanese have three tracks at the secondary level — academic, specialized vocational, and comprehensive — all Japanese students take classes in foreign language, social studies, mathematics, science, health and physical education, and fine arts. The Japanese schools do not have gifted or self-paced learning programs (Rohlen 1986). They do not teach some students general mathematics and others algebra; everyone is taught algebra. In elementary school, all Japanese children learn to read music and play a wind instrument and a keyboard instrument (Rohlen 1986). Likewise, in Germany all students take a foreign language, German, physics, chemistry, biology, mathematics, and physical education.

Another source of differences in curriculum in the United States is that some minority group members are more likely to take or be assigned to less demanding classes. For example, in 1994, 61.6 percent of high school graduates classified as white took algebra II compared with 43.7 percent of those classified as black. Only 39.2 percent of students classified as American Indian took algebra II. However, 67 percent of high school graduates classified as Asian took algebra II (U.S. Department of Education 1996).

E. D. Hirsch, Jr. (1993), author of a best-selling and controversial[3] book, *Cultural Literacy: What Every American Needs to Know,* maintains that the nonuniform

[3]Hirsch's book is controversial because it attempts to identify and define concepts that every American should know. According to Hirsch, "this common knowledge or collective memory allows people to communicate, to work together, and to live together. It forms the basis for communities, and if shared by enough people, it is a distinguishing characteristic of a national culture" (Hirsch, Kett, and Trefil 1993:ix).

Table 13.3 Range of Tuition at Four-year Institutions,* 1995–1996

About how much per year did most private and public college students pay for their 1995–1996 tuition? How much more would you add for other expenses such as books and transportation? Consider your own college tuition. Where do you fit on this chart? How do you compare with your peers?

	Number of Colleges	Proportion of Total Enrollment (%)
Private Institutions		
$20,000 or more	36	3.2
19,000–19,999	29	2.6
18,000–18,999	20	1.8
17,000–17,999	29	2.6
16,000–16,999	32	2.8
15,000–15,999	32	2.8
14,000–14,999	61	5.4
13,000–13,999	72	6.4
12,000–12,999	87	7.7
11,000–11,999	92	8.1
10,000–10,999	115	10.2
9,000– 9,999	97	8.6
8,000– 8,999	81	7.1
7,000– 7,999	79	7.0
6,000– 6,999	67	5.9
5,000– 5,999	62	5.5
4,000– 4,999	71	6.3
3,000– 3,999	36	3.2
2,000– 2,999	22	1.9
1,000– 1,999	6	0.5
Less than $1,000	7	0.6
Total	1,133	100.0
Public Institutions		
$5,500 or more	10	1.9
5,000–5,499	11	2.1
4,500–4,999	17	3.2
4,000–4,499	29	5.4
3,500–3,999	60	11.3
3,000–3,499	66	12.4
2,500–2,999	71	13.3
2,000–2,499	99	18.6
1,500–1,999	151	28.3
1,000–1,499	10	1.9
Less than $1,000	9	1.7
Total	533	100.0

*Includes only those institutions that provided final or estimated 1994–95 tuition and fees by September 1, 1995.

Source: The College Board (1996).

and diverse nature of the U.S. curriculum has created one of the most unjust and inegalitarian school systems in the developed world.

> It happens that the most egalitarian elementary-school systems are also the best. . . . The countries that achieve these results tend to teach a standardized curriculum in early grades. Hungarian, Japanese, and Swedish children have a systematic grounding in shared knowledge. Until third graders learn what third graders are supposed to know, they do not pass on to fourth grade. No pupil is allowed to escape the knowledge net. (Hirsch 1989:32)

Surprisingly, little evidence indicates that tracking students into remedial or basic courses contributes to intellectual growth, corrects academic deficiencies, prepares students for success in higher tracks, or increases interest in learning. Instead, the special curricula exaggerate and widen differences among students and perpetuate beliefs that intellectual ability varies according to social class and ethnic group (Oakes 1986a, 1986b).

Differences in Funding

A 1993 report on education in twenty-four countries sponsored by the Organization for Economic Co-operation and Development (OECD) based in Paris concluded that no country has greater educational disparities between rich and poor than the United States (Sanchez 1993). In the United States, schools differ with regard not only to curriculum requirements but also to funding. Elementary and secondary schools receive approximately 7 percent of their funding from the federal government, 45.6 percent from the state government, and the rest from the local sources, primarily property taxes (U.S. Department of Education 1993b). Heavy reliance on state revenue is problematic because the less wealthy states generate less tax revenue than do the wealthier states. Of the twenty-four countries studied, the United States is third in spending per student at the secondary level and first at the primary level (Celis 1993a). On the other hand, the size of school funding disparities between the poorest and wealthiest states is considerably greater in the United States than that in other industrialized countries.

For example, as in the United States, the twelve Canadian provinces spend different amounts of money per student; expenditures range from a high of $6,500 per year per student to a low of $3,000. Thus, a difference of about $3,500 separates the richest province from the poorest province. In the United States, state-level expenditures range from a high of $9,415 per student to a low of $3,761; differences of $5,654 separate the richest and poorest states (U.S. Department of Education 1995).

Likewise, an even heavier reliance on local revenue is also problematic because it causes funding disparities among schools within states. In this regard, the courts in at least twenty-eight states have evaluated (or are in the process of evaluating) claims that methods of financing have helped create unequal school systems within the state or have ruled that the methods of school financing are unconstitutional (Celis 1992, 1993b). A dramatic case in point is Kentucky. In response to a 1986 lawsuit filed by sixty-six, mostly rural, school districts that challenged the state's system of funding education, the Kentucky Supreme Court declared on June 8, 1989, that it found the entire state system of public education deficient and unconstitutional. The court declared that every aspect of the public school system should be reconsidered and a new system created no later than April 15, 1990. "The court concluded that a school system in which a significant number of children receive an inadequate education or ultimately fail is inherently inequitable and unconstitutional" (Foster 1991:34). To remedy this inequity, the court made the academic success of all students a constitutional obligation and required the state legislature to devise a system to ensure that every student was "learning at the highest level of which he or she is capable" (Foster 1991:36). In 1990, the legislature responded by passing the Kentucky Education and Reform Act (KERA), which set into motion the restructuring of education's rules, roles, and relationships. The sociological significance of KERA is that the state recognized that inequality could only be corrected if there was a complete rethinking and overhaul of the way education is delivered to students.

Education-Based Programs to Solve Social Problems

The United States uses education-based programs to address a variety of social problems including parents' absence from the home, racial inequality, drug and alcohol addictions, malnutrition, teenage pregnancy, sexually transmitted diseases, and illiteracy. Although all countries have education-based programs that address social problems, the United States is unique in that education is viewed as the primary solution to many of its problems. In the United States

> [t]he process became familiar: discover a social problem, give it a name, and teach a course designed to remedy it. Alcoholism? Teach about temperance in

every school. Venereal disease? Develop courses in social hygiene. Youth unemployment? Improve vocational training and guidance. Carnage on the highways? Give driver education classes to youth. Too many rejects in the World War I draft? Set up programs in health and physical education. (Tyack and Hansot 1981:13)

The importance of these programs notwithstanding, the schools alone cannot solve such complex problems. Other programs must be implemented in concert with education-based programs. For example, an estimated 350,000 infants each year are born with drug- and alcohol-related problems. A disproportionate number of these babies are born to poor women who have had virtually no access to prenatal care (Blakeslee 1989; Dorfman 1989). When these children reach school age, they will present special problems to the school system. In *The Broken Cord,* Michael Dorris summarizes the concerns of teachers who work with children affected by alcohol before birth, including "difficulty staying on task, distracting other children, poor use of language, inability to structure their work time, and a constant need for monitoring and attention" (1989:241).[4]

The root causes of this totally preventable problem are connected, in part, to a health-care system that uses heroic measures to save low–birth-weight babies, many of whom are born to drug- and alcohol-addicted mothers, but that does not provide adequate prenatal care.

Ambiguity of Purpose and Value

As we noted earlier, in comparison with the people of other countries, Americans tend to be ambivalent about the purpose and value of an education. In general, the U.S. public supports without question mass education and the right to a college education despite past academic history. Yet, for many Americans elementary school, high school, and college are merely something to be endured; students count the days until they are "out." Ernest Boyer (1986) interviewed hundreds of students from public schools around the United States and found no one who could articulate why he or she was in school:

> The most frequent response was, "I have to be here." They know it's the law. Or, "If I finish this, I have a better chance at a job." The "this" remains a blank. Or, "I need this in order to go to college." Or, "This is where I meet my friends." Not once in all our conversations did students mention what they were learning or why they should learn it. (p. 43)

Most Americans tend to equate education with increased job opportunities even though, since the late 1960s and early 1970s, the country has produced college graduates faster than the economy could absorb them, at least into the kinds of jobs the college-educated expect (Guzzardi 1976). In 1990 (the last year for which figures are available), approximately 122.6 million workers were in the labor force. Of these workers, 99.3 million held jobs that did not require a college degree. Slightly less than 20 percent of American workers were employed in jobs that required a college degree (Shelley 1992). Because approximately 29 million workers have had four or more years of college and only 23.2 million jobs require a college degree, about one college graduate in five is underemployed. This gap between the number of college-educated workers and the number of jobs requiring a college education is reflected in a survey of college degree recipients who graduated in 1990: 44 percent reported that they did not believe a degree was required for the job they had obtained in 1991. This figure is up from 37 percent in 1985 (U.S. Department of Education 1993b; see Figure 13.2).

These findings do not mean that level of education is unrelated to occupation or income. Rather, they indicate that a large proportion of college graduates are underemployed, if only because there are not enough high-skill jobs available to absorb the increasing number of graduates. In view of this trend, "job" seems a narrow criterion by which to evaluate an education. Yet, high school and college students commonly evaluate their courses, especially general requirements, as useless because "I will never use it in the real world"—in particular, on the job.

Open college enrollments, diverse and special curricula, unequal funding, the problem-solving burden, and a national ambivalence toward education explain in part the dropout rate, the high number of functional illiterates, and the United States's poor academic showing relative to its European and Pacific Rim counterparts. In the next section, we look more closely at the role of the classroom environment.

[4]Dorris maintains that

> the most striking characteristic of children with fetal alcohol syndrome is their lack of imagination: not so much the ability to make up a fictional story, but the basic act of foreseeing a possible consequence in one's own life. Most of us do it all the time: if I do x today, then y will occur tomorrow. If I work at a job, I will earn money. If I earn money, I can purchase more options for myself. If I go to sleep early, I will be rested when I wake up. If I eat my lunch at nine in the morning, I will be hungry at two. (Dorris 1989:245)

Figure 13.2 Percentage of 1989–90 Bachelor's Degree Recipients Indicating That a Four-Year College Degree Was Not Required for Their Job, 1991

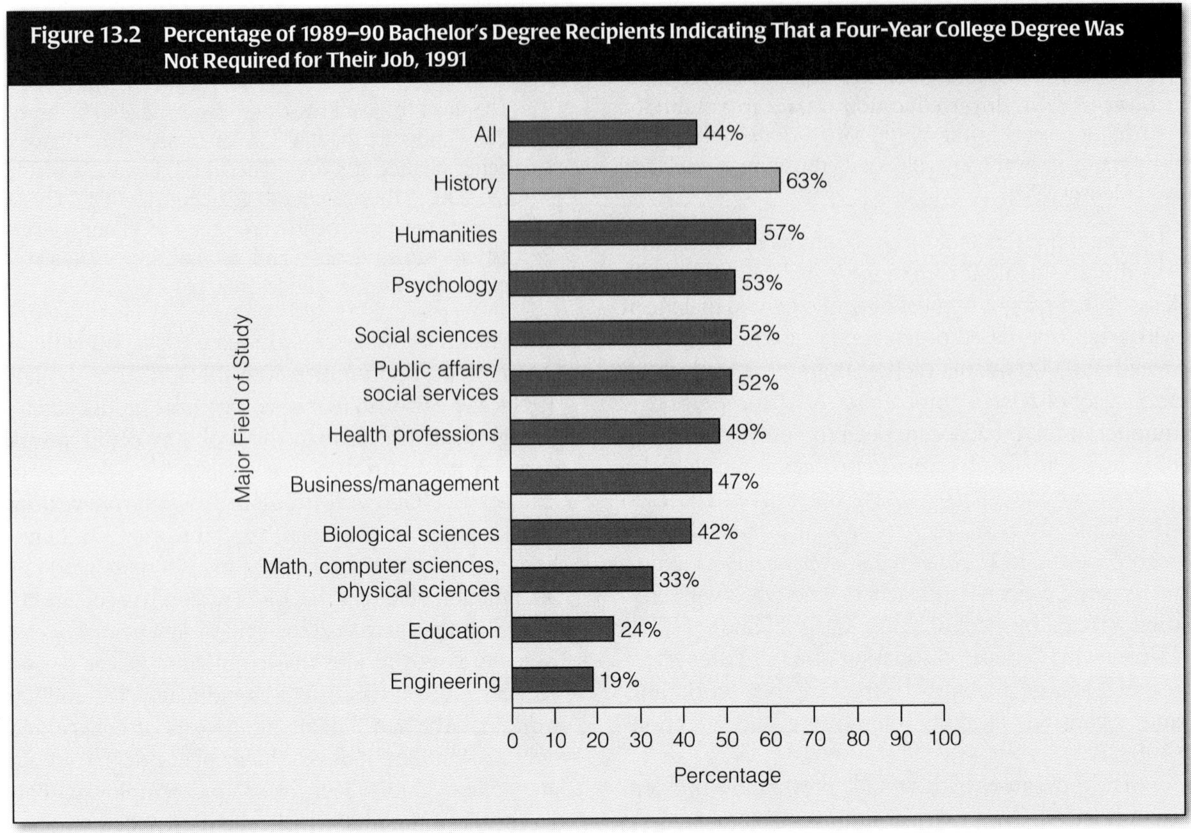

Source: U.S. Department of Education (1993b:18).

A Close-Up View: The Classroom Environment

The classroom is where schooling takes place. In this section, we examine what goes on in the classroom: the curriculum to which many students are exposed, the practice of tracking, how students are tested, and the problems that teachers face. The section focuses on practices that lead to boredom and failure and that undermine the time and energy devoted to academic pursuits. Certainly some schools in the United States have stimulating classroom environments. The problem is that there are not enough of them.

Formal curriculum The various subjects such as mathematics, science, English, reading, physical education, and so on.

Hidden curriculum All the things that students learn along with the subject matter.

The Curriculum

Teachers everywhere in the United States teach two curricula simultaneously—a formal curriculum and a hidden curriculum. The various academic subjects—mathematics, science, English, reading, physical education, and so on—make up the **formal curriculum**. Students do not learn in a vacuum, however. As teachers instruct students and as students complete their assignments, other activities are going on around them. Social anthropologist Jules Henry (1965) maintains that these other activities are important and that they are the hidden curriculum. The **hidden curriculum** is all the things that students learn along with the subject matter. The teaching method, the types of assignments and tests, the tone of the teacher's voice, the attitudes of classmates, the number of students absent, the frequency of the teacher's absences, and the number of interruptions during a lesson are examples of things going on as students learn the formal curriculum. These so-called extraneous events function to convey messages to students not only about the value of the subject but about the values of society, the place of learning in their lives, and their role in society.

Hidden Curriculum: The Case of Spelling Baseball

Henry (1965) uses typical classroom scenes (acquired from thousands of hours of participant observation) such as a session of "spelling baseball" to demonstrate the seemingly ordinary process by which a hidden curriculum is transmitted and to show how students are exposed simultaneously to the two curricula. Although Henry observed this scene in 1963, his observations hold even today.

> The children form a line along the back of the room. They are to play "spelling baseball," and they have lined up to be chosen for the two teams. There is much noise, but the teacher quiets it. She has selected a boy and a girl and sent them to the front of the room as team captains to choose their teams. As the boy and girl pick the children to form their teams, each child chosen takes a seat in orderly succession around the room. Apparently they know the game well. Now Tom, who has not yet been chosen, tries to call attention to himself in order to be chosen. Dick shifts his position to be more in the direct line of vision of the choosers, so that he may not be overlooked. He seems quite anxious. Jane, Tom, Dick, and one girl whose name the observer does not know, are the last to be chosen. The teacher even has to remind the choosers that Dick and Jane have not been chosen.
>
> The teacher now gives out words for the children to spell, and they write them on the board. Each word is a pitched ball, and each correctly spelled word is a base hit. The children move around the room from base to base as their teammates spell the words correctly.
>
> The outs seem to increase in frequency as each side gets near the children chosen last. The children have great difficulty spelling "August." As they make mistakes, those in the seats say, "No!" The teacher says, "Man on third." As a child at the board stops and thinks, the teacher says, "There's a time limit; you can't take too long, honey." At last, after many children fail on "August," one child gets it right and returns, grinning with pleasure, to her seat. . . . The motivation level in this game seems terrific. All the children seem to watch the board, to know what's right and wrong, and seem quite keyed up. There is no lagging in moving from base to base. The child who is now writing "Thursday" stops to think after the first letter, and the children snicker. He stops after another letter. More snickers. He gets the word wrong. There are frequent signs of joy from the children when their side is right. (Henry 1963:297–298)

According to Henry, learning to spell is not the most important lesson that students learn from this exercise. They are also learning important cultural values from the way in which spelling is being taught: they are learning to fear failure and to envy success. In exercises like spelling baseball, "failure is paraded before the class minute upon minute" (p. 300), and success is achieved after others fail. And, "since all but the brightest children have the constant experience that others succeed at their expense they cannot but develop an inherent tendency to hate—to hate the success of others" (p. 296).

In an exercise such as spelling baseball, students also learn to be absurd. *To be absurd*, as Henry defines it, means to make connections between unrelated things or events and not to care whether the connections are appropriate or inappropriate. From Henry's point of view, spelling baseball teaches students to be absurd because there is no logical connection between learning to spell and baseball. "If we reflect that one could not settle a baseball game by converting it into a spelling lesson, we see that baseball is bizarrely irrelevant to spelling" (p. 300). Yet, most students participate in classroom exercises like spelling baseball without questioning their purpose. Although some children may ask, "Why are we doing this? What is the point?" and may be told, "So you can learn to spell" or "To make spelling fun," few children challenge further the purpose of this activity. Students go along with the teacher's request and play the game as if spelling is related to baseball because, according to Henry, they are terrified of failure and want so badly to succeed.

Henry argues further that classroom activities such as spelling baseball prepare students to fit into a competitive and consumption-oriented culture. Because the U.S. economy depends on consumption, the country benefits if its citizens purchase nonessential goods and services. The assignments that children do in school do not prepare them to question false or ambiguous statements made by advertisers; schools do not properly prepare demanding individuals to "insist that the world stand up and prove that it is real" (p. 49). Henry argues that this sort of training — this hidden curriculum — makes possible an enormous amount of selling that otherwise could not take place:

> [I]n order for our economy to continue in its present form people must learn to be fuzzy-minded and impulsive, for if they were clear-headed and deliberate, they would rarely put their hands in their pockets. . . . If we were all logicians the economy [as we know it] could not survive, and herein lies a terrify-

Jules Henry maintains that when American children are sent to the blackboard, especially to do math problems, they learn to "fear failure" and "envy success." If they get the problem wrong, a classmate will be called on to correct the mistake; hence, success is achieved at the expense of another's failure.

> ing paradox, for in order to exist economically as we are we must . . . remain stupid. (p. 48)

Henry believes that students who have the intellectual strength to see through absurd assignments such as spelling baseball and who find it impossible to learn to accept such assignments as important may rebel against the system, refuse to do the work, drop out, or come to think of themselves as stupid. We cannot know how many students do poorly in school because they cannot accept the manner in which subjects are taught. The United States has diverse public school systems; different schools teach reading in different ways. Yet, the vast majority of Americans, especially middle- and lower-class Americans, typically learn to carry out these kinds of assignments in the manner described here. In addition, students, especially young students, cannot articulate what they do not like about school; many come to believe that school is not for them and think they have failed rather than believing that school has failed them.

Classroom Morale

A steady diet of absurd assignments and questions also can affect classroom morale. Bracha Alpert (1991), an Israeli social scientist, observed three high school classrooms of upper-middle-class, primarily white students in the United States and was struck by the way students responded to the teachers' questions. Alpert gave the following description of a typical exchange between teacher and students in one classroom:

> Students in this class did not respond to the teacher's questions that attempted to stimulate discussions. On occasions where a student did react, the reaction was short and uttered in a quiet tone of voice that may be described as "mumbling." Since the response rarely reached the whole class, the teacher often had to repeat it aloud. A combination of silence and mumbling is exemplified in the following excerpt of a typical transaction in this classroom.
>
> TEACHER: . . . [N]ow, ah, the first four stanzas certainly do create a mood for the reader. Now, what adjectives would you use to describe that mood?
>
> STUDENTS: (Silence)
>
> TEACHER: Think about it a little bit and then try to run it through your mind. How do we describe moods? Cheerful? Light headed? Sympathetic?
>
> STUDENTS: (Silence)
>
> TEACHER: What adjectives would you use to describe this one?
>
> STUDENTS: (Silence, then a student mumbles) Mellow.
>
> TEACHER: Mellow? OK I can buy that to a certain extent, but what [else]?

To teachers, students' boredom may seem to be a problem of individual motivation or even laziness. But another possibility is that the social environment of the classroom, including some teaching styles, promotes a lack of responsiveness on the part of students.

©Bob Daemmrich/The Image Works

A STUDENT: (Mumbles) Solemn.

TEACHER: OK. Sarah suggests the word "solemn." Does that sound good?

SOME STUDENTS: (Mumble) No. (p. 354)

Alpert maintains that this lack of response from students is likely to occur when teachers emphasize facts and clear-cut answers. When Alpert questioned students about their "reluctant participation," they said that it was a "mode of behavior purposely chosen" as a reaction to the teacher's instructional style.

Tracking

Most schools in the United States arrange students in instructional groups according to similarities in past academic performance and/or on standardized test scores. In elementary school, the practice is often known as ability grouping; in middle school and high school, it is known as streaming or tracking. Under this sorting and allocation system, students may be assigned to separate instructional groups within a single classroom; they may be sorted with regard to selected subjects such as mathematics, science, and English; or they may be separated across the entire array of subjects.

The rationales that underlie ability grouping, streaming, or tracking (hereafter referred to as "tracking") are as follows:

1. Students learn better when they are grouped with those who learn at the same rate: the brighter students are not held back by the slower learners and the slower learners receive the extra time and special attention needed to correct academic deficiencies.
2. Slow learners develop more positive attitudes when they do not have to compete with the more academically capable.
3. Groups of students with similar abilities are easier to teach.

In spite of these rationales, the evidence suggests that tracking does not lead to these benefits.

The Effects of Tracking Sociologist Jeannie Oakes (1985) investigated how tracking affected the academic experiences of 13,719 middle school and high school students in 297 classrooms and 25 schools across the United States.

> The schools themselves were different: some were large, some very small; some in the middle of cities;

> some in nearly uninhabited farm country; some in the far West, the South, the urban North, and the Midwest. But the differences in what students experienced each day in these schools stemmed not so much from where they happened to live and which of the schools they happened to attend but, rather, from differences within each of the schools. (p. 2)

Oakes's findings were consistent with the findings of hundreds of other studies of tracking in terms of how students were assigned to groups, how they were treated, how they viewed themselves, and how well they did:

> *Placement:* Poor and minority students are placed disproportionately in the lower tracks.
>
> *Treatment:* The different tracks are not treated as equally valued instructional groups. There are clear differences with regard to the quality, content, and quantity of instruction and to classroom climate as reflected in the teachers' attitude and in student-student and teacher-student relationships. Low-track students consistently are exposed to inferior instruction — watered-down curriculum and endless repetition — and to a more rigid, more emotionally strained classroom climate.
>
> *Self-image:* Low-track students do not develop positive images of themselves because they are identified publicly and are treated as educational discards, damaged merchandise, or unteachable. Overall, among the average and the low-track groups, tracking seems to foster lower self-esteem and promote misbehavior, higher dropout rates, and lower academic aspirations. Placement in a college preparatory track has positive effects on academic achievement, grades, standardized test scores, motivation, educational aspirations, and attainment. "And this positive relationship persists even after family background and ability differences are controlled" (Hallinan 1988:260).
>
> *Achievement:* The brighter students tend to do well regardless of the academic achievements of the students with whom they learn.

These findings are reflected in the written answers that teachers and students gave to various questions asked by Oakes and her colleagues. For example, when teachers were asked about the classroom climate, high-track teachers tended to reply in positive terms:

> There is a tremendous rapport between myself and the students. The class is designed to help the students in college freshman English composition. This makes them receptive. It's a very warm atmosphere. I think they have confidence in my ability to teach them well, yet because of the class size — 32 — there are times they feel they are not getting enough individualized attention. (Oakes 1985:122)

Low-track teachers replied in less positive terms:

> This is my worst class. Kids [are] very slow — underachievers and they don't care. I have no discipline cases because I'm very strict with them and they are scared to cross me. They couldn't be called enthusiastic about math — or anything, for that matter. (p. 123)

There also were clear differences in the high-track and the low-track students' responses to the question "What is the most important thing you have learned or done so far in this class?" The replies of high-track students centered around themes of critical thinking, self-direction, and independent thought:

> The most important thing I have learned in this [English] class is to loosen up my mind when it comes to writing. I have learned to be more imaginative. (p. 87)
>
> The most important thing I have learned in this [math] class is the benefit of logical and organized thinking, learning is made much easier when the simple processes of organizing thoughts have been grasped. (p. 88)

Low-track students were more likely to give answers that centered around themes of boredom and conformity:

> I think the most important is coming into [math] class and getting out folders and going to work. (p. 89)
>
> To be honest, nothing. (p. 71)
>
> Nothing I'd use in my later life; it will take a better man than I to comprehend our world. (p. 71)

Although many educators have recognized the problems associated with tracking, efforts to detrack have run up against demands for curriculum differentiation from politically powerful parents of high-achieving or identified "gifted" students who insist their children must get something more than the "other" students (Wells and Oakes 1996). As a result, tracking persists "even though many educators and policymakers acknowledge that students in the low and middle tracks are not held to high enough standards and thus are not adequately prepared for either college or the transition to work" (Wells and Oakes 1996:137). In addition to these efforts, tracking can create self-fulfilling prophecies by affecting teachers' expectations of the academic potential and abilities of students placed in each track.

Teachers' Expectations and Self-Fulfilling Prophecies Tracking can become a self-fulfilling prophecy, a deceptively simple yet powerful concept that originated from an insight by William I. and Dorothy Swain Thomas: "If [people] define situations as real, they are real in their consequences" ([1928] 1970:572). A **self-fulfilling prophecy** begins with a false definition of a situation. The false definition, however, is assumed to be accurate, and people behave as if the definition were true. In the end, the misguided behavior produces responses that confirm the false definition (Merton 1957).

A self-fulfilling prophecy can occur if teachers and administrators assume that some children are "fast," "average," or "slow" and expose them to "fast," "average," and "slow" learning environments. Over time, real differences in quantity, quality, and content of instruction cause many students to actually become (and believe that they are) "slow," "average," or "fast." In other words, the prediction or prophecy of academic ability becomes an important factor in determining academic achievement:

> The tragic, often vicious, cycle of self-fulfilling prophecies can be broken. The initial definition of the situation which has set the circle in motion must be abandoned. Only when the original assumption is questioned and a new definition of the situation is introduced, does the consequent flow of events [show the original assumption to be false]. (Merton 1957:424)

In the tradition of symbolic interactionism, Robert Rosenthal and Lenore Jacobson designed an experiment to test the hypothesis that teachers' positive expectations about students' intellectual growth can become a self-fulfilling prophecy and lead to increases in students' intellectual competence.[5]

Rosenthal and Jacobson's experiment took place in an elementary school called Oak School, a name given the school to protect its identity. The student body was largely from lower income families and predominantly white (84 percent); 16 percent of the students were Mexican Americans. Oak School sorted students into ability groups based on reading achievement and teachers' judgments.

At the end of a school year, Rosenthal and Jacobson gave a test, purported to be a predictor of academic "blooming," to those students who were expected to return in the fall. Just before classes began in the fall, all full-time teachers were given the names of the white and Hispanic students from all three ability groups who had supposedly scored in the top 20 percent. The teachers were told that these students "will show a more significant inflection or spurt in their learning within the next year or less than will the remaining 80 percent of the children" (Rosenthal and Jacobson 1968:66). Teachers were also told not to discuss the scores with the students or the students' parents. Actually, the names given to teachers were chosen randomly: the differences between the children earmarked for intellectual growth and the other children were in the teachers' minds. The students were retested after one semester, at the end of the academic year, and after a second academic year.

Overall, intellectual gains, as measured by the difference between successive test scores, were greater for those students who had been identified as "bloomers" than they were for those not identified. Although "bloomers" benefited in general, some bloomers benefited more than others: first- and second-graders, Hispanic children, and children in the middle track showed the largest increases in test scores. It is important to note that the "bloomers" received no special instruction or extra attention from teachers — the only difference between them and the unidentified students was the belief that the "bloomers" bore watching. Rosenthal and Jacobson speculated that this belief was communicated to "bloomers" in very subtle and complex ways, which they could not readily identify:

> To summarize our speculations, we may say that by what she said, by how and when she said it, by her facial expressions, postures, and perhaps by her touch,

[5]Rosenthal and Jacobson were influenced by animal experiments in which trainers' beliefs about the genetic quality of the animals affected the animals' performances. When trainers were told that an animal was genetically inferior, the animal performed poorly; when trainers were told that an animal was genetically superior, the animal's performance was superior. This happened despite the fact that there were no such genetic differences between the animals defined as dull or bright.

Self-fulfilling prophecy A concept that begins with a false definition of a situation, which people assume to be accurate and thus behave as if it were true. The misguided behavior produces responses that confirm the false definition.

> the teacher may have communicated to the ["bloomers"] that she expected improved intellectual performance.
>
> It is self-evident that further research is needed to narrow down the range of possible mechanisms whereby a teacher's expectations become translated into a pupil's intellectual growth. (p. 180)

The research on tracking and teachers' expectations shows that the learning environment affects academic achievement and that tracking and expectations are two mechanisms that contribute to the unequal distribution of knowledge. Because teachers draw on test results to form their expectations about students' academic potential, and because academic personnel place students in different ability groups on the basis of test results (along with teacher evaluations), tests represent another mechanism that contributes to the unequal distribution of knowledge and skills.

We have examined various practices within the schools that contribute to the unequal distribution of knowledge and skills. We cannot blame teachers and other school personnel entirely for this outcome, however. Teachers do not have exclusive control over the classroom environment and cannot single-handedly create students who are interested in learning. For many teachers, the environment in which they work can make teaching problematic.

The Problems That Teachers Face

Teachers' jobs are complex; teachers are expected to undo learning disadvantages generated by larger inequalities in the society and to handle an array of discipline problems. Over 50 percent of elementary and high school teachers surveyed in a recent Gallup poll responded that discipline is a very serious or fairly serious problem in their school, as are uncompleted homework assignments, cheating, stealing, drugs and alcohol, truancy, and absenteeism. Such widespread problems explain why 40 percent of beginning teachers drop out of teaching within five years (Elam 1989). When compared with their counterparts in Germany and Japan, U.S. teachers name "uninterested students," "uninterested parents," "low student morale," "tardiness," and "intimidation or verbal abuse of teachers/staff" as problems that limit teaching effectiveness and/or disrupt the learning environment (see Figure 13.3).

In addition to these problems, U.S. teachers work in environments that discourage systematic learning outside the classroom and collaboration with other teachers. Psychologist Harold Stevenson (1992) finds that Asian schools plan extracurricular activities after school hours. During this time, they teach students computer skills and do not have to use classroom time to do so. In addition, Stevenson finds that Asian teachers work together very closely in preparing lesson plans. The level of collaboration is equivalent to that needed to produce a theatrical production. In contrast, teachers in the United States prepare lesson plans on their own. Asian teachers have more time to collaborate because they teach about 60 percent of the school day; in the remaining time, they discuss ideas with other teachers. American teachers, on the other hand, are in the classroom at least 85 percent of the time.

The job of teaching is further complicated by the social context of education. Teachers in the United States must deal with students from diverse family and ethnic backgrounds and with a student subculture that values and rewards athletic achievement, popularity, social activities, jobs, cars, and appearance at the expense of academic achievement. We examine these aspects of social context in the next section.

The Social Context of Education

We turn to the work of sociologist James S. Coleman (the 1991 president of the American Sociological Association), who has studied both family background and the adolescent student subculture, two factors that affect the classroom atmosphere and the learning experience.

Family Background

James S. Coleman (1966) was the principal investigator of *Equality of Educational Opportunity,* popularly known as the Coleman Report. The project was supported by the U.S. government under the directive of the 1964 Civil Rights Act, which prohibited discrimination for reasons of color, race, religion, or national origin in public places (restaurants, hotels, motels, and theaters); mandated that the desegregation of public schools be addressed; and forbade discrimination in employment (racial segregation in public schools was ruled unconstitutional by the Supreme Court in 1954). Coleman's intent was to examine the degree to which public education is segregated and to explore inequalities of educational opportunity in the United States. Coleman and his six colleagues surveyed 570,000 students and 60,000 teachers, principals, and school su-

Figure 13.3 Student Problems as Seen by Teachers

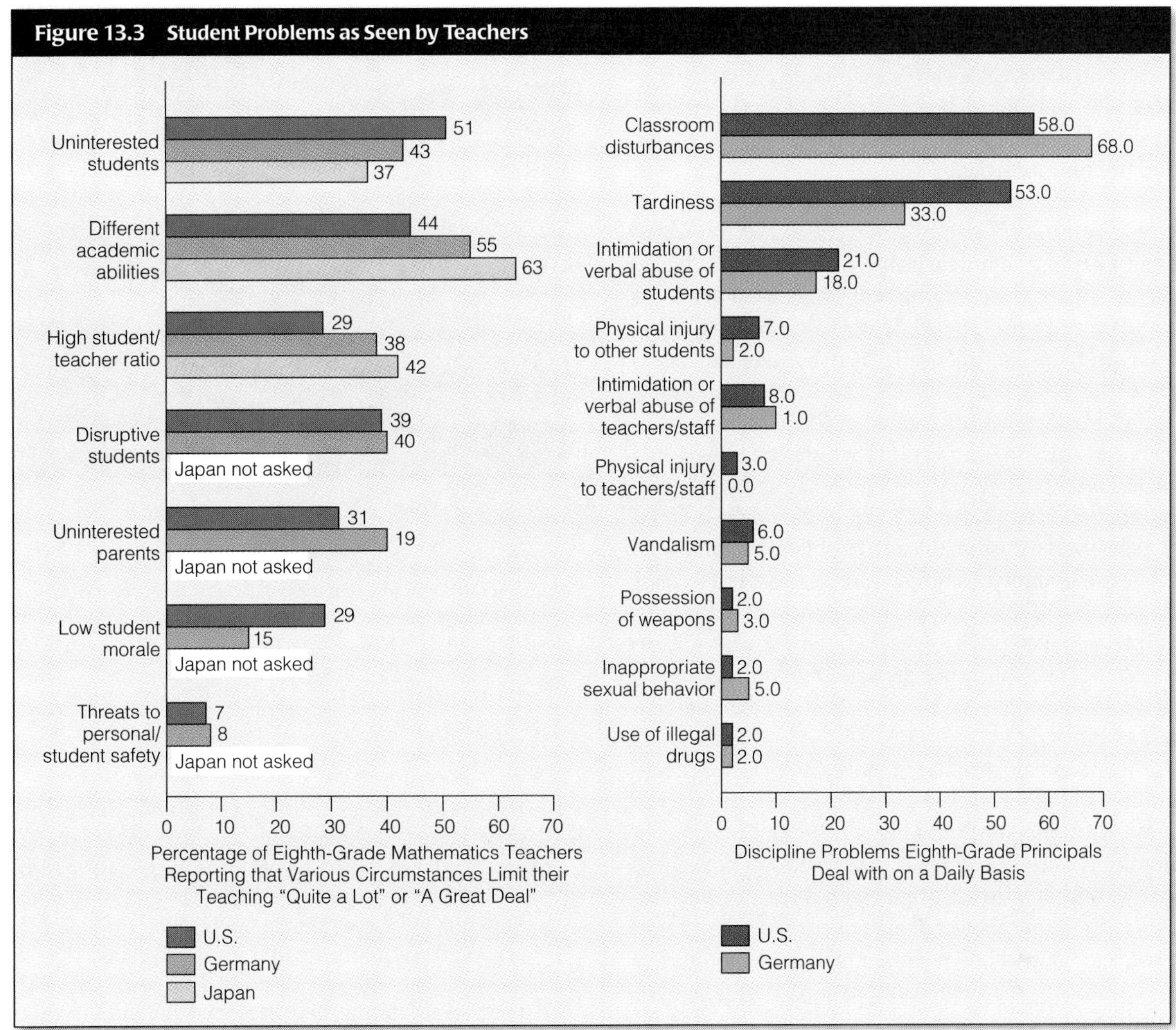

Primary source: Third International Mathematics and Sciences Study, unpublished tabulations, U.S., German, and Japanese teacher surveys, Westat, 1996; secondary source, National Center for Education Statistics (1996).

perintendents in 4,000 schools across the United States. Students filled out questionnaires about their home background and educational aspirations and took standardized achievement tests for verbal ability, nonverbal ability, reading comprehension, mathematical ability, and general knowledge. Teachers, principals, and superintendents answered questionnaires about their backgrounds, training, attitudes, school facilities, and curricula.

Coleman found that a decade after the Supreme Court's famous desegregation decision in 1954 — *Brown v. Board of Education* — the schools were still largely segregated: 80 percent of white children attended schools that were 90 to 100 percent white, and 65 percent of black students attended schools that were more than 90 percent black. Almost all students in the South and the Southwest attended schools that were 100 percent segregated. Although Mexican Americans, Native Americans, Puerto Ricans, and Asian Americans also attended primarily segregated schools, they were not segregated from whites to the same degree as blacks. The Coleman Report also found that white teachers taught black children but that black teachers did not teach whites: approximately 60 percent of the teachers who taught black students were black, whereas 97 percent of the teachers who taught white students were white. When the characteristics of teachers of the average white student were compared with those of teachers of the average African American student, the study found no significant differences in professional qualifications (as measured by degree, major, and teaching experience).

Coleman found sharp differences across racial and ethnic groups with regard to verbal ability, nonverbal ability, reading comprehension, mathematical achievement, and general information as measured by the standardized tests. The white students scored highest, followed by Asian Americans, Native Americans, Mexican Americans, Puerto Ricans, and African Americans.

Contrary to what Coleman expected to find, there were on the average no significant differences in quality between schools attended predominantly by the various ethnic groups and schools attended by whites. (Quality was measured by age of buildings, library facilities, laboratory facilities, number of books, class size, expenditures per pupil, extracurricular programs, and the characteristics of teachers, principals, and superintendents.) Surprisingly, variations in the quality of a school did not have much effect on the students' test scores.

Test scores were affected, however, by family background and other students' attributes. The average minority group member was likely to come from an economically and educationally disadvantaged household and was likely to attend school with students from similar backgrounds. Fewer of his or her classmates would complete high school, maintain high grade point averages, enroll in college preparatory curricula, or be optimistic about their future. Coleman also found some support for the idea "The higher achievement of all racial and ethnic groups in schools with greater proportions of white students is largely, perhaps wholly, related to effects associated with the student body's educational background and aspirations" (1966:307, 310). This finding does not mean that there is something magical about a white environment. The Coleman Report examined the progress of blacks who had participated in school integration programs and found that their scores were higher than those of their counterparts who attended schools with members of the same social class. The important variable is the social class of one's classmates and not ethnicity:

> Taking all these results together, one implication stands out above all: That schools bring little influence to bear on a child's achievement that is independent of his background and general social context; and that this very lack of an independent effect means that the inequalities imposed on children by their home, neighborhood, and peer environment are carried along to become the inequalities with which they confront adult life at the end of school. For equality of educational opportunity through the schools must imply a strong effect of schools that is independent of the child's immediate social environment, and that strong independent effect is not present in American schools. (1966:325)

Once the fond symbol of rural schooling, in recent years the school bus has become instead a symbol of unsuccessful court-ordered integration. Behind the failure of busing to achieve integration is the larger fact that the middle class finds one means or another to segregate its children educationally from the lower class.

Coleman's findings that school expenditures are not an accurate predictor of educational achievement (as measured by standardized tests) was used to support arguments against allocating additional funds to the public school system. Yet, the finding that schools do not make a difference does not mean that schools cannot make a difference. A more accurate interpretation of this finding is that schools, as currently structured, have no significant effect on test scores; this conclusion implies that the educational system needs restructuring.

Coleman's findings about the composition of the student body and about the higher test scores earned by economically disadvantaged black students in predominantly middle-class schools were used to support busing as a means of achieving educational equality. Although Coleman initially supported this policy, he later retracted his endorsement because busing hastened "white flight," or the migration of middle-class white Americans from the cities to the suburbs. This migration only intensified the racial segregation in city and suburban schools. As the ratio of white to black students dropped sharply, the positive effects of desegregation proved to be short-lived. As a result, economically and educationally disadvantaged blacks were sent from their deficient schools into equally deficient lower-class and lower middle-class white neighborhoods. Coleman adamantly maintained that court-ordered busing alone could not achieve integration:

> With families sorting themselves out residentially along economic and racial lines, and with schools tied to residence, the end result is the demise of the common school attended by children from all economic levels. In its place is the elite suburban school . . . the middle-income suburban school, the low-income suburban school, and the central-city schools of several types — low-income white schools, middle-income white schools, and low or middle-income black schools. (1977:3–4)

The findings of this Coleman study do not imply that a person is trapped by family background. Coleman never claimed that family background explains all of the variation in test scores. He did claim, however, that it was the single most important factor in his study.

Coleman's findings about school segregation have changed very little over the past three decades. In 1968, the federal government reported that 76 percent of black students and 55 percent of Hispanic students attended predominantly minority schools (schools in which 50 percent or more of the students are black, Asian, Native American, and/or Hispanic). In 1991, the Harvard Project on School Desegregation found that figure to be 66 percent for blacks and 74.3 percent for Hispanics. In some states, such as Illinois, Michigan, New York, and New Jersey, more than 50 percent of the schools are 90 to 100 percent minority. As in the 1960s, black and other minority students are significantly more likely to find themselves in schools in which overall academic achievement is undervalued and low. The Harvard Project recommended that busing, the most widely used strategy for integrating schools, be supplemented by other strategies such as finding ways to integrate neighborhoods and enforcing desegregation laws (Celis 1993b).

Many subsequent studies support the importance of family background (Hallinan 1988). For example, the International Association for the Evaluation of Educational Achievement tested students in twenty-two countries on six subjects. The association found that the "home environment is a most powerful factor in determining the level of school achievement of students, student interest in school learning, and the number of years of schooling the children will receive" (Bloom 1981:89; Ramirez and Meyer 1980). Yet, in this international study, in the Coleman study, and in other studies, home background (as measured by parents' ethnicity, income, education, and occupation) explains only about 30 percent of the variation in students' achievement. This finding suggests that factors other than socioeconomic status affect academic performance:

A number of studies indicate that the home environment is an important variable—though not the only one—in explaining academic success.

©George Goodwin/Monkmeyer Press

> In most if not all societies, children and youth learn more of the behavior important for constructive participation in the society outside of school than within. This fact does not diminish the importance of school but underlines the nation's dependence on the home, the working place, the community institutions, the peer group and other informal experiences to furnish a major part of the education required for a child to be successfully inducted into society. Only by clear recognition of the school's special responsibilities can it be highly effective in educating its students. (Tyler, quoted in Purves 1974:74c)

In view of these findings, the special responsibility of the schools is not to duplicate the inequalities outside the school. Over the past three decades, researchers have found that "schools exert some influence on an individual's chances of success, depending on the extent to which they provide equal access to learning" (Hallinan 1988:257–258). Unfortunately, as we have learned, several characteristics of U.S. education and practices within the schools — hidden curriculum, test biases, self-fulfilling prophecies, and tracking — work to perpetuate social and economic inequalities. Now we turn to another problematic phenomenon that teachers confront daily — a student value system that deemphasizes academic achievements.

Adolescent Subcultures

Around the turn of the century—the early decades of late industrialization — less than 10 percent of teenagers fourteen to eighteen years of age attended high school in the United States. Young people attended elementary school to learn the three Rs, after which they learned from their parents or from their neighbors the skills needed to make a living. As the pace of industrialization increased, jobs moved away from the home and out of the neighborhood into factories and office buildings. Parents no longer trained their children because the skills they knew were becoming outdated and obsolete, so children came to expect that they would not make their livings as their parents did. In short, as the economic focus in the United States shifted from predominantly farm and small-town work environments to the factory and office, the family became less involved in the training of its children and, by extension, less involved in children's lives. The transfer of work away from the home and neighborhood removed opportunities for parents and children to work together. Under this new arrangement, family occasions became events that were consciously arranged to fit everyone's work schedule.

Coleman argued that this shift in training from the family to the school cut adolescents off from the rest of society and forced them to spend most of the day with their own age group. Adolescents came "to constitute a small society, one that has most of its important interactions within itself, and maintains only a few threads of connection with the outside adult society" (Coleman, Johnstone, and Jonassohn 1961:3).

Coleman surveyed students from 10 high schools in the Midwest to learn about adolescent society. He selected schools representative of a wide range of environments: five schools were located in small towns, one in a working-class suburb, one in a well-to-do suburb, and three in cities of varying sizes; one of the schools was an all-male Catholic school. Coleman was interested in the adolescent status system, a classification of achievements resulting in popularity, respect, acceptance into the crowd, praise, awe, and support, as opposed to isolation, ridicule, exclusion from the crowd, disdain, discouragement, and disrespect. To learn about this system, Coleman asked students questions similar to the following:

- How would you like to be remembered — as an athlete, as a brilliant student, as a leader in extracurricular activities, or as the most popular student?
- Who is the best athlete? The best student? The most popular? The boy the girls go for most? The girl the boys most go for?
- What person in the school would you like most to date? To have as a friend? What does it take to get in with the leading crowd in this school?

Based on the answers to these and other questions, Coleman was able to identify a clear pattern common to all ten schools. "Athletics was extremely important for the boys, and social success with boys [accomplished through being a cheerleader or being good-looking] was extremely important for girls" (1960: 314). Coleman found that girls in particular do not want to be considered as good students, "for the girl in each grade in each of the schools who was most often named as best student has fewer friends and is less often in the leading crowd than is the boy most often named as best student" (Coleman 1960:338). A boy most often could be a good student or dress well or have enough money to meet social expenses, but to really be admired he must also be a good athlete. Coleman also found that the peer group had more influence over and exerted more pressure on adolescents than did their teachers, and he learned that a significant number of adolescents were influenced more by the peer group than by their parents.

Why does the adolescent society penalize academic achievement in comparison with athletic and other achievements? Coleman maintained that the manner in which students are taught contributes to their lack of academic interest: "They are prescribed 'exercises,' 'assignments,' 'tests,' to be done and handed in at a teacher's command" (1960:315). The academic work they do does not require creativity but conformity. Students show their discontent by choosing to become involved in and acquiring things they can call their own — athletics, dating, clothes, cars, and ex-

Coleman's research on adolescent subcultures, published in 1961, is considered a classic study. Do his conclusions about the importance for boys of sports and the importance for girls of success with boys still hold true in U.S. high schools? Has the emergence of female sports or other factors changed things for a significant number of either sex? How could you go about researching this issue? What hypotheses might you explore?

tracurricular activities. Coleman, Johnstone, and Jonassohn (1961) noted that this reaction is inevitable given the passive roles that students are asked to play in the classroom.

> [One] consequence of the passive, reactive role into which adolescents are cast is its encouragement of irresponsibility. If a group is given no authority to make decisions and take action on its own, the leaders need show no responsibility to the larger institution. Lack of authority carries with it lack of responsibility; demands for obedience generate disobedience as well. But when a person or group carries the authority for his own action, he carries responsibility for it. In politics, splinter parties which are never in power often show little responsibility to the political system; a party in power cannot show such irresponsibility. . . . An adolescent society is no different from these. (p. 316)

Athletics is one of the major avenues open to adolescents, especially males, in which they can act "as a representative of others who surround [them]" (1961: 319). Others support this effort, identify with the athletes' successes, and console athletes when they fail. Athletic competition between schools generates an internal cohesion among students that no other event can. "It is as a consequence of this that the athlete gains so much status: he is doing something for the school and the community" (p. 260).

Coleman argues that, because athletic achievement is widely admired, everyone with some ability will try to develop this talent. With regard to the relatively unrewarded arena of academic life, "those who have most ability may not be motivated to compete" (p. 260). This reward structure suggests that the United States does not draw into the competition everyone who has academic potential (see Figure 13.4).

Coleman's findings should deliver the message that the peer group is a powerful influence on learning, but they should not leave the impression that the peer group's world does not overlap with the family or the classroom. In fact, it seems more appropriate to consider how the multiple contexts of students' lives are interrelated. From data collected during interviews and observations with fifty-four ethnically and academically diverse youth in four urban desegregated high schools in California, educators Patricia Phelan, Ann Locke Davidson, and Hanh Cao Yu (1991, 1993; Phelan and Davidson 1994) generated a model of the interrelationships between students' family, peer, and school worlds. The Students' Multiple Worlds Model describes the ways in which sociocultural aspects of students' worlds (for example, norms, values, beliefs, expectations, actions) combine to affect their thoughts and actions with respect to school and learning. These researchers are particularly concerned with understanding students' perceptions of boundaries and borders between worlds and adaptation strategies that students employ as they move from one context to another. Although they report that they found a good deal of variety in students' descriptions of their worlds and in their perceptions of boundaries they also uncovered four distinctive patterns that students utilize as they move between and adapt to different contexts and settings (see "Students' Multiple Worlds Model and Typology").

As Phelan and her colleagues point out, the patterns that students describe "are not necessarily stable for individual students but rather can be affected by external conditions such as classroom or school climate conditions, family circumstances, or changes in peer group affiliation" (1991:228). Also, the typology does not divide students along ethnic, achievement, or gender lines but rather focuses on the congruency of students' worlds and the borders that they face. In other words, youth of the same ethnicity or students who achieve at the same level can be found within any of the four types.

Figure 13.4 Percentage of Students from Selected Nations Scoring Among the Top 10 Percent of Eighth Graders in the Forty-one Countries of the Third International Mathematics and Science Study (TIMSS)

Sociologist James Coleman believes that the reward structure characteristic of U.S. education does not motivate those who have the academic potential to develop it. This may explain why so few students from the United States are among the top 10 percent of students in forty-one countries.

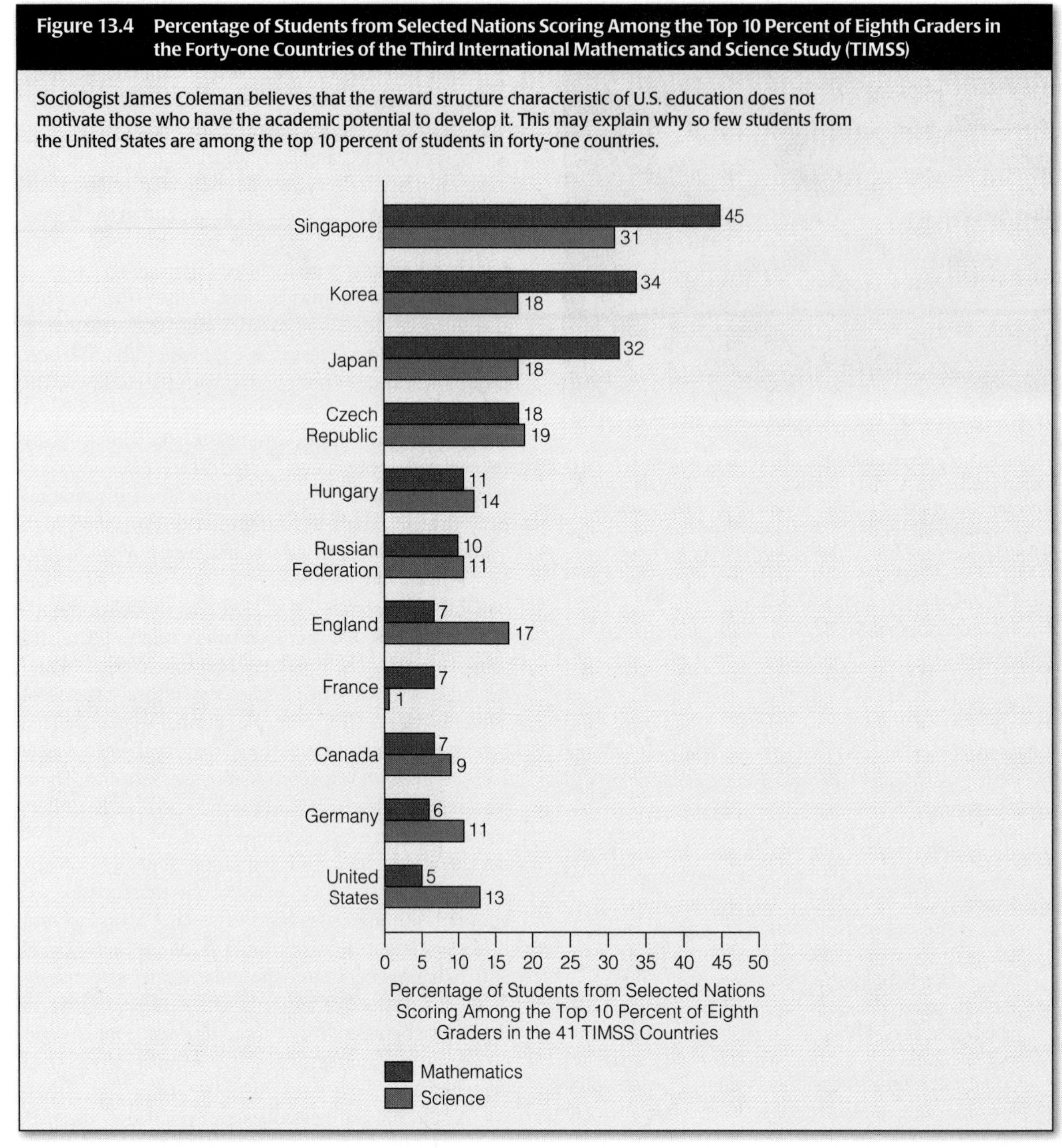

Source: U.S. Department of Education (1996).

Summary and Implications

The material in this chapter leads us to ask, If the United States studies the successes of public education in foreign countries, can it "borrow" those strategies to improve its own system of education? It depends. First, Americans must define concisely the nature and origins of the problems they wish to solve. If the problem is defined superficially as their rank in relation to other countries, without reference to measurable differences, we will have few clues as to what it is about the U.S. system that is in need of reform. Even when there are

Students' Multiple Worlds Model and Typology

Many factors affect students' academic performances, including peer groups, the classroom environment, and family background. Education researchers Patricia Phelan, Ann Locke Davidson, and Hanh Cao Yu generate an important attempt to understand the interrelatedness of students' worlds. Rather than compartmentalizing aspects of students' lives, it gives us a way of looking more holistically at the processes young people use to manage, more or less successfully, the transitions between their various contexts.

CONGRUENT WORLDS/SMOOTH TRANSITIONS

These students describe values, beliefs, expectations, and normative ways of behaving as similar across their worlds. Moving from one setting to another is harmonious and uncomplicated, and boundaries are easily managed. This does not mean that students act exactly the same way or discuss the same things with teachers, friends, and family members but rather that commonalties among worlds override differences. Students who exhibit this pattern say that their worlds are merged by their common sociocultural components rather than bounded by conspicuous differences. Although many of these youths are white, upper-middle-class, and high achieving, this is not always the case. Some minority students describe little difference across their worlds and experience transitions as smooth. Likewise, academically average students can also exhibit patterns that fit this type.

DIFFERENT WORLDS/BORDER CROSSINGS MANAGED

For some adolescents, differences in family, peer, and/or school worlds (with respect to culture, ethnicity, socioeconomic status, and/or religion) require students to adjust and reorient as movement among contexts occurs. For example, a student's family world may be dominated by an all-encompassing religious doctrine in which values and beliefs are contrary to those found in school and peer worlds. For other students, home and neighborhood are viewed as starkly different than school, particularly for students of color who are transported. And for still other students, differences between peers and family are dominant themes. However, regardless of differences, students in this category are able to utilize strategies that enable them to manage crossings successfully (in terms of what is valued in each setting). However, this does not mean that crossings are always easy, that they are made in the same way, or that they cannot result in personal and psychic consequences. It is not uncommon for high-achieving minority youth to exhibit patterns common to this type.

DIFFERENT WORLDS/BORDER CROSSINGS DIFFICULT

In this category, like the former, students define their family, peer, and/or school worlds as distinct. They say they must adjust and reorient as they move across worlds and among contexts. However, unlike students who manage to make adjustments successfully, these students have not learned, mastered, or been willing to adopt all of the strategies necessary for successful transitions. For example, a student may do poorly in a class in which the teacher's interaction style, the student's role, or the learning activity are oppositional to what takes place within the student's peer or family worlds. Likewise, some students in this category describe their comfort and ease at school and with peers but are essentially estranged from their parents. In these cases, parents' values and beliefs are frequently more traditional, more religious, or more constrained than those of their children, making adaptation to their home world difficult and conflictual. Other youth say that the socioeconomic circumstances of their families work against their full engagement in school. For youth who exhibit patterns common to this type, border crossing involves friction and discomfort and in some cases is possible only under particular conditions. This pattern often includes adolescents on the brink between success and failure, involvement and disengagement, commitment and apathy. These are some of the students for whom classroom and school climate conditions can mean the difference between staying in school or dropping out.

DIFFERENT WORLDS/BORDER CROSSINGS RESISTED

In this type, the values, beliefs, and expectations across students' worlds are so discordant that students perceive borders as insurmountable and actively or passively resist transitions. When border crossing is attempted, it is frequently so painful that, over time, these students develop reasons and rationales to protect themselves against further distress. In such cases, borders are viewed as insurmountable, and students actively or passively resist attempts to embrace other worlds. For example, some students say that school is irrelevant to their lives, whereas others immerse themselves fully in the world of their peers. Rather than moving from one setting to another, blending elements of all, these students remain constrained by borders they perceive as rigid and impenetrable. Although low-achieving students (seemingly unable to profit from school and classroom settings) are typical of this type, high-achieving students who do not connect with peers or family also exhibit Type IV patterns.

Source: Phelan, Davidson, and Yu (1991, 1993, 1994); synthesized by Patricia Phelan, University of Washington, Bothell.

measurable differences between the performances of U.S. students and students in particular foreign countries, we have to look beyond the most obvious differences.

For example, Japanese and Taiwanese students outperform U.S. students in science and math. Some researchers believe that these performance differences are related to differences in the amount of time U.S., Japanese, and Taiwanese students spend in school—American students attend school 178 days of the year, whereas Japanese and Taiwanese students attend school 240 days a year. Although U.S. students spend 30.4 hours a week in the classroom, according to one study they spent only 64.5 percent of this time working on academic activities. The remainder of the time is spent maintaining discipline, attending assemblies, going to the bathroom, and so on. In contrast, Taiwanese students spend 40.4 hours in the classroom, 90 percent of which is devoted to academic activities; Japanese students spend 37.3 hours in the classroom, 80 percent of which is devoted to academic activities. On the basis of these findings, Americans could lengthen the school year to 240 days and increase the number of classroom hours to 44 a week. However, such changes would not address the real problem if the content and quality of the curriculum and the issues related to discipline are not addressed as well.

The point is, the United States cannot expect that simply by increasing the length of the school year alone test scores will improve. If we agree that the way the curriculum is designed and delivered is a major factor, then we need to examine how the curriculum is designed and delivered in countries that score measurably higher. In addition, we need to examine how curriculum is designed and delivered in countries where students score measurably lower. If we decide that a foreign technique might be useful in solving U.S. problems, we also have to ask what it is about the country (its history, its family structure, economic incentives, cultural values, the way in which teenagers spend free time) that supports the practice and permits it to work well.

In "Japanese Education: If They Can Do It, Should We?" Japanese studies professor Thomas Rohlen (1986) offers some suggestions for how to use other countries' models of education to improve our model. In the case of Japan, Rohlen argues:

> We would be foolish to see Japanese education as a model for our own efforts; but as a mirror showing us our own weaknesses and as a yardstick against which to measure our efforts, it has great value for us. We cannot allow ourselves to ignore or to imitate its approach. Rather, it is possible to look periodically into the "Japanese mirror" while we quite independently set out to strengthen our schools and our system within our own cultural and social context. (pp. 42–43)

On the other hand, it is important not to overemphasize contextual differences between educational systems and come to believe that no feature can be borrowed without major adjustments. Such a perspective would lead us to exaggerate differences and reject any foreign ideas as inappropriate (Cummings 1989). Likewise, we should not idealize the successes of other systems of education and assume everything is perfect there and assume that the United States has no features that other countries would like to "borrow." As former Secretary of Education William Bennett (1987) concludes in a U.S. Department of Education study, "Much of what seems to work well for Japan in the field of education closely resembles what works best in the United States—and most likely elsewhere. Good education is good education" (p. 71; see "Some Highlights of Japanese Education").

Key Concepts

Use this outline to organize your review of the key chapter ideas.

- **Education**
  - **Informal education**
  - **Formal education**
    - **Schooling**
- **Illiteracy**
- **Formal curriculum**
- **Hidden curriculum**
  - **To be absurd**
- **Tracking/ability grouping**
- **Self-fulfilling prophecy**
- **Status system**

Some Highlights of Japanese Education

Here are some highlights from a study sponsored by the U.S. Department of Education (1987), *Japanese Education Today*. These highlights cover the strengths and weaknesses of the Japanese education system, the reforms underway in Japan, and implications for education in the United States.

STRENGTHS AND WEAKNESSES

- *Japanese society is education-minded to an extraordinary degree: success in formal education is considered largely synonymous with success in life and is, for most students, almost the only path to social and economic status.*
- *Formidable education accomplishments result from the collective efforts of parents, students, and teachers; these efforts are undergirded by the historical and cultural heritage, a close relationship between employers and education, and much informal and supplementary education at preschool, elementary, and secondary levels.*
- *During nine years of compulsory schooling, all children receive a high quality, well-balanced basic education in the 3 Rs, science, music, and art.*
- *Both the average level of student achievement and the student retention rate through high school graduation are very high.*
- *Japanese education also distinguishes itself in motivating students to succeed in school, teaching effective study habits, using instructional time productively, maintaining an effective learning environment, sustaining serious attention to character development, and providing effective employment services for secondary school graduates.*
- *Japanese education is not perfect. Problems include rigidity, excessive uniformity, and lack of choice; individual needs differences that receive little attention in school; and signs of student alienation. A related problem in employment is over-emphasis on the formal education background of individuals.*

EDUCATION REFORM IN JAPAN

- *Japan is concerned about its education system and is making extensive efforts to improve it. The reform movement includes vigorous public debate between proponents of change and defenders of the status quo.*
- *Reformers take a farsighted view of the national interest: they are coming to terms with societal needs in the 21st century while grappling with such complex issues as finding a new balance between group harmony and individual creativity in Japanese education.*

IMPLICATIONS FOR AMERICAN EDUCATION

Some American education ideals may be better realized in Japan than in the United States. A close look at Japanese education provides a potent stimulus for Americans to reexamine the standards, performance, and potential of their own system. Some lessons worth considering:

- *The value of parental involvement from the preschool years on;*
- *The necessity of clear purpose, strong motivation, and high standards and of focusing resources on education priorities;*
- *The importance of maximizing learning time and making effective use thereof;*
- *The value of a competent and committed professional teaching force, and*
- *The centrality of holding high expectations for all children and a firm commitment to developing a strong work ethic and good study habits–recognizing that hard work and perseverance are essential elements in a good education.*

Source: U.S. Department of Education (1987:vi).

internet assignment

The first issue of the innovative journal *CITYSCHOOLS*, a research magazine about urban schools and communities, can be accessed at this site http://www.ncrel.org/ncrel/sdrs/cityschl.htm. *CITYSCHOOLS* rejects the "deficit model" as an approach to solving problems related to urban and inner-city schools and advocates a "resilience model" that emphasizes strengths.

Compare the deficit model with the resilience model and speculate on how this change in perspective might improve the system of education in the United States.

14

Religion

With Emphasis on the Islamic State of Afghanistan

Mountainous terrain in Afghanistan. ©Roger Lemoyne/Gamma-Liaison

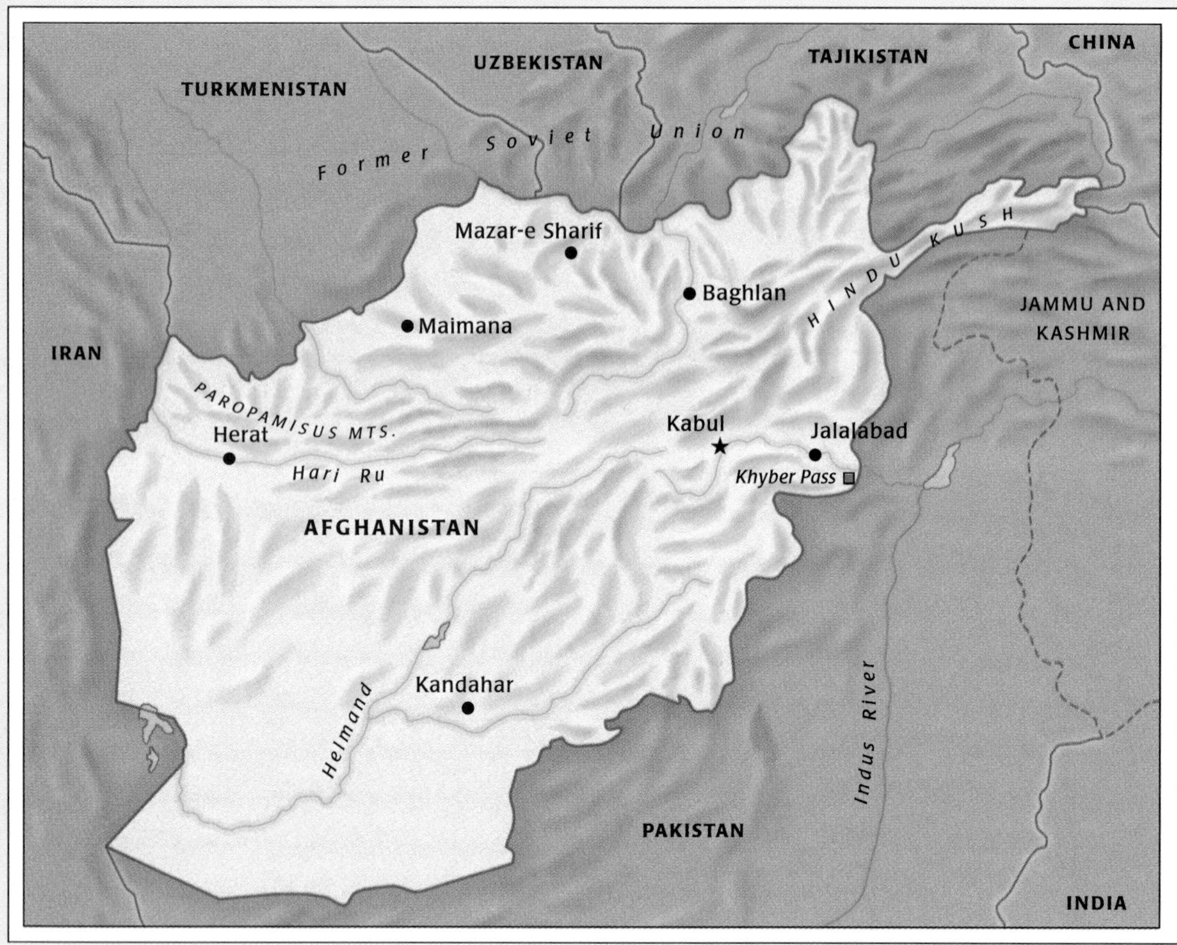

The Historical Context

Before the nineteenth century, mountainous Afghanistan lay in the path of invaders from China, Persia (ancient Iran), and the Indian subcontinent. In the nineteenth and twentieth centuries, Afghanistan became a battleground for the British and Russian empires and, after World War II, for the United States and the Soviet Union. In 1979, the Soviet Union invaded Afghanistan to support a secular government. U.S. observers called Afghanistan "the Soviet Union's Vietnam," a title that history has borne out.

Supported materially by the United States, Afghanistan's military resistance to the Soviets was mobilized in large part through religious institutions proclaiming "holy war." The Soviet exit in 1989 left a ravaged country full of religiously charged armed rival factions. When the Taliban government took power in 1996, it justified many new policies on religious Islamic grounds. Westerners tend to see such policies as simply fanatical and irrational and to overlook the histories that led up to them. They also tend to disregard the parallels that can be drawn between the role of religion in Afghan society and its significant role in our own.

What is religion? What is its connection to society? How have outspoken religious leaders recently become so influential in government in many parts of the world, including the United States?

In explaining how sociologists view and study religion as an aspect of social life, this chapter helps us make sense of events in Afghanistan.

In the early morning hours of September 27, 1996, the Taliban captured Kabul, the capital city of Afghanistan. Here is how the event was reported by Mark Austin of Independent Television News (ITN):

> MARK AUSTIN, ITN: On a hill overlooking Kabul, these are Afghanistan's new soldiers of God, praying they say for peace and stability in a country that's known only conflict for nearly two decades. But below them is a battle-torn city where the fear of war is fast being replaced by a fear of repression. It's symbolized by the white flag of the Taliban militia, heavily armed religious students who patrol the streets, enforcing their vision of Islamic law. The penalties for disobedience: flogging or even death. Their first edict, women must not work, must not be seen uncovered on the streets. Men must grow beards and pray five times a day. The only sounds from the radio, Islamic prayer and poetry. All music and entertainment is banned here. Television shops are being closed down, TV's and video recorders destroyed, tapes hung from trees.
>
> [Taliban] SPOKESMAN: We will confiscate it and destroy it stage by stage.
>
> MARK AUSTIN: At the gates of the presidential palace, we took tea with one group of militia men who told us their goal was a pure Islamic society, free of crime and corruption. But when we toured the palace itself, they proudly showed us works of art they destroyed.
>
> [Taliban] SPOKESMAN: The painting is against Islam.
>
> MARK AUSTIN: After seventeen years of war and suffering, what this city is now experiencing is the most extreme brand of Islam anywhere in the world. The Taliban takeover may have brought temporary peace of a kind, but for the people of Kabul, it's peace at a price. These are the child victims of the Taliban assault on Kabul, appalling injuries caused by shelling and rocket fire. But their tragedy is compounded by the imposition of strict Islamic laws. Eighty percent of the nurses and 40 percent of the doctors here are women, and now most are too frightened even to leave their homes. These are the hands of one of the city's top surgeons. She won't risk being identified but says it's almost as if women no longer exist.
>
> SURGEON: I can't go to my job. I can't help my people because they said the women must sit in the houses, and they can't go outside. It's really bad for us. I'm very sorry, and I want to leave this country.
>
> MARK AUSTIN: Many are already leaving Kabul. Aid workers say more than 100,000 have fled in the last few days. Reports of arrests and beatings abound in this city, and for the women here, the veil conceals the fear that children do not hide. The Taliban are urging people to stay, but there's a sense of panic, and the exodus continues, leaving those who remain to come to terms with life under new rulers with new rules and an existence that many women here say is taking them back to the dark ages. **(Online NewsHour 1996)**

Why Focus on Afghanistan?

In this chapter, we focus on Afghanistan. At the time of this writing, two-thirds of the country was under the control of an Islamic group known as the Taliban. The Taliban rebels received headline news coverage beginning in 1994 (but especially in 1996 after gaining control of the Afghan capital, Kabul). Much of the coverage focused on the Taliban's brutal execution of the ex-Afghan President Najibullal and his brother and their strict and harsh version of Islamic law imposed on the local populations. The focus on this Islamic group, to the neglect of larger historical context, is consistent with how the media cover Islam in general.

The images of repression and violence are consistent with the "snapshot" images many Americans associate with the Muslim religion, or Islam. For many, Islam evokes images of the November 1995 explosion of a U.S. military training and communication center in

Riyadh, the capital of Saudi Arabia; the 1993 World Trade Center bombings in New York City; the 1988 bombing of Pan Am Flight 103 over Lockerbee, Scotland; and the stepped-up airport security in the wake of the Gulf War with Iraq in 1991.

Whatever the event, one idea seems to dominate the coverage: some Islamic group—the Movement of Islamic Amal, the Islamic Jehad organization, the Hezbollah (Party of God), or now the Taliban—is responsible. In other words, each event is reduced to the actions of religious fanatics acting solely from "primitive and irrational" religious conviction. This focus masks more important political, geographic, and economic factors and causes viewers to lose sight of larger questions such as these: Why has Afghanistan experienced civil war for the past twenty years? Is the Taliban's version of Islamic law consistent with "Islamic principles"? How did the Taliban rise to power? In an October 9, 1996, U.S. Department of State press statement, spokesman Nicholas Burns suggested that the situation in Afghanistan somehow goes beyond religion:

> There's a great interest in what happens in Afghanistan, because it is a pivotal country in terms of where it is situated in that part of the world, and I've just listed some of the issues that are of concern to us [such as terrorism, narcotics, human rights, and the treatment of women]. So, we're going to maintain a close view of the events. We'll have contact with the Taliban and others, and we'll just have to proceed on a step-by-step basis to see if we can establish better and closer contacts. (Burns 1996)

In this chapter, we examine religion from a sociological perspective. Such a perspective is useful because it allows us to step back and view in a detached way a subject that is often charged with emotion. Detachment and objectivity are necessary if we wish to avoid making sweeping generalizations about the nature of religions such as Islam that are unfamiliar to many of us.

When sociologists study religion, they do not investigate whether God or some other supernatural force exists, whether certain religious beliefs are valid, or whether one religion is better than another. Sociologists cannot study such questions because they adhere to the scientific method, which requires them to study only observable and verifiable phenomena. Rather, sociologists investigate the social aspects of religion, focusing on the characteristics common to all religions, the types of religious organizations, the functions and dysfunctions of religion, the conflicts within and between religious groups, the way in which religion shapes people's behavior and their understanding of the world, and the way in which religion is intertwined with social, economic, and political issues. We begin with a definition of religion. Defining religion is a surprisingly difficult task and one with which sociologists have been greatly preoccupied.

What Is Religion? Weber's and Durkheim's Views

Figure 14.1 shows what could be called the "major religions" of the world. But what makes something a religion? In the opening sentences of *The Sociology of Religion*, Max Weber (1922) states, "To define 'religion,' to say what it is, is not possible at the start of a presentation such as this. Definition can be attempted if at all, only at the conclusion of the study" (p. 1). Despite Weber's keen interest and his extensive writings about religious activity, he could offer only the broadest of definitions: religion encompasses those human responses that give meaning to the ultimate and inescapable problems of existence—birth, death, illness, aging, injustice, tragedy, and suffering (Abercrombie and Turner 1978). To Weber, the hundreds of thousands of religions, past and present, represented a rich and seemingly endless variety of responses to these problems. In view of this variety, he believed it was virtually impossible to capture the essence of religion in a single definition.

Like Max Weber, Emile Durkheim believed that nothing is as vague and diffused as religion. In the first chapter of his book *The Elementary Forms of the Religious Life*, Durkheim ([1915] 1964) cautions that when

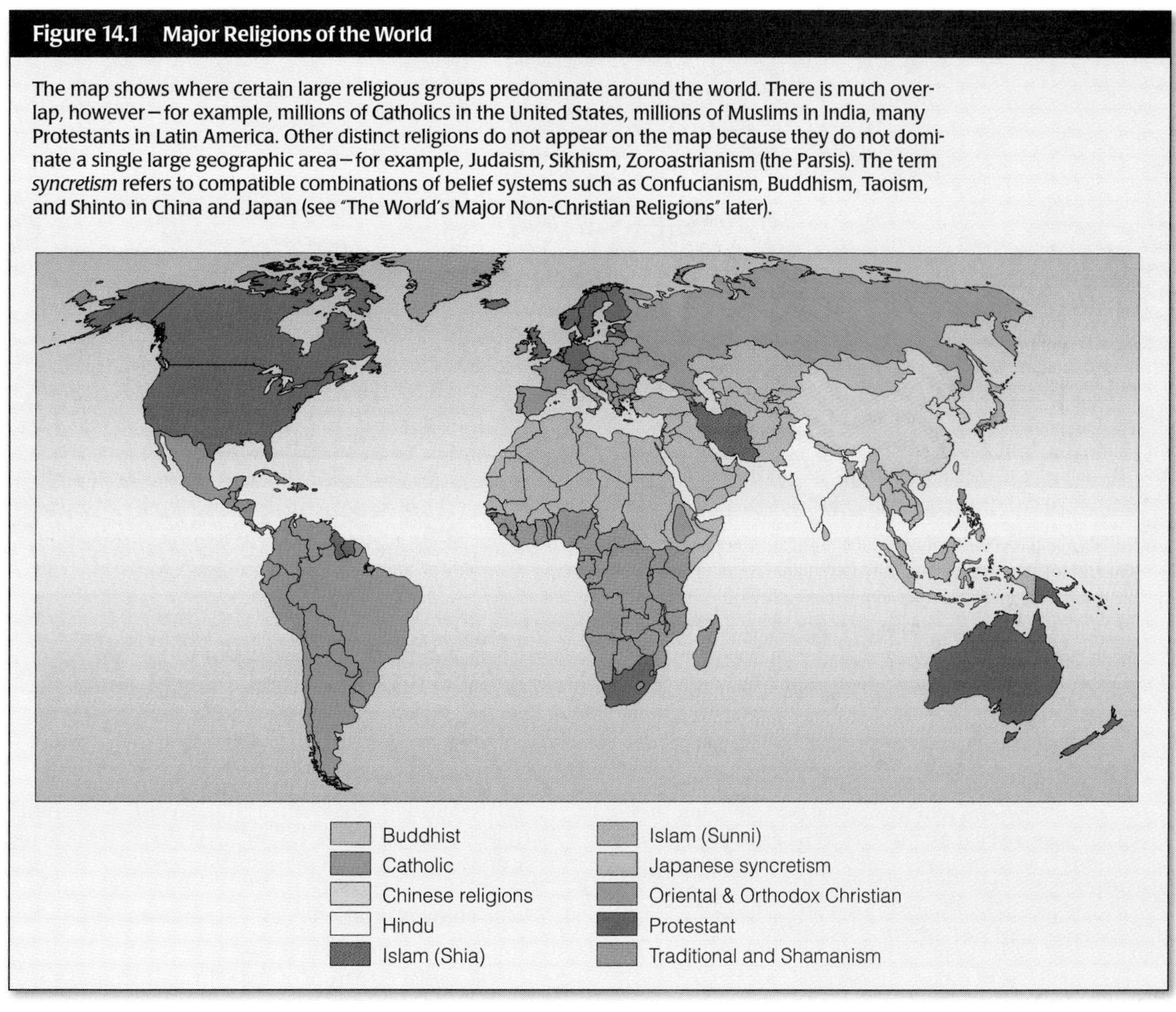

Figure 14.1 Major Religions of the World

The map shows where certain large religious groups predominate around the world. There is much overlap, however—for example, millions of Catholics in the United States, millions of Muslims in India, many Protestants in Latin America. Other distinct religions do not appear on the map because they do not dominate a single large geographic area—for example, Judaism, Sikhism, Zoroastrianism (the Parsis). The term *syncretism* refers to compatible combinations of belief systems such as Confucianism, Buddhism, Taoism, and Shinto in China and Japan (see "The World's Major Non-Christian Religions" later).

studying religions, sociologists must assume that "there are no religions which are false" (p. 3). Like Weber, Durkheim believed that all religions are true in their own fashion — all address in different ways the problems of human existence. Consequently, Durkheim said, those who study religion must first rid themselves of all preconceived notions of what religion should be. We cannot attribute to religion the characteristics that reflect only our own personal experiences and preferences.

In *The Spiritual Life of Children,* psychiatrist Robert Coles (1990) recounts his conversation with a ten-year-old Hopi girl, which illustrates Durkheim's point. The conversation reminds us that if we approach the study of religion with preconceived notions, we will lose many insights about the nature of religion in general:

> "The sky watches us and listens to us. It talks to us, and it hopes we are ready to talk back. The sky is where the God of the Anglos lives, a teacher told us. She [the teacher] asked where our God lives. I said, I don't know. I was telling the truth! Our God is the sky, and lives wherever the sky is. Our God is the sun and the moon, too; and our God is our [the Hopi] people, if we remember to stay here [on the consecrated land]. This is where we're supposed to be, and if we leave, we lose God." [The interviewer then asked the child whether she had explained all of this to the teacher.]
>
> "No."
>
> "Why?"
>
> "Because—she thinks God is a person. If I'd told her, she'd give us that smile."
>
> "What smile?"
>
> "The smile that says to us, You kids are cute, but you're dumb; you're different — and you're all wrong!"
>
> "Perhaps you could have explained to her what you've just tried to explain to me."

It is a challenge to see through one's own preconceptions about what is "right" in everyday life. A Western woman may look on the cover-up Muslim hijab as a sign of sexual oppression. The Muslim woman may look on the Western woman's garb as the result of media pressures on her to display herself as a sex object for men.

©Evan Agostini/Gamma-Liaison (left); ©Anne Dowie (right)

> "We tried that a long time ago; our people spoke to the Anglos and told them what we think, but they don't listen to hear us; they listen to hear themselves." (p. 25)

Consider as a second example that many critics view the *hijab,* or modest dress of Muslim women, as evidence that the women are severely oppressed. Although women in the Middle East certainly do not have the same rights as men, critics should not be so quick to assume that the *hijab* is *the* source of oppression, especially when we consider the view that some Muslim women hold toward American dress customs:

> If women living in western societies took an honest look at themselves, such a question [as to why Muslim women are covered] would not arise. They are the slaves of appearance and the puppets of male chauvinistic society. Every magazine and news medium (such as television and radio) tells them how they should look and behave. They should wear glamorous clothes and make themselves beautiful for strange men to gaze and gloat over them.
>
> So the question is not why Muslim women wear *hijab,* but why the women in the West, who think they are so liberated, do not wear *hijab*? (*Mahjubah* 1984)

The conversation between Cole and the Hopi child and the discussion of *hijab* shows that preconceived notions of what constitutes religion and uninformed opinions about the meaning of religious symbols and practices can close people off to a wide range of religious beliefs and experiences.

In formulating his ideas about religion, Durkheim remained open to the many varieties of religious experiences throughout the world. He identified three essential features that he believed were common to all religions, past and present: (1) beliefs about the sacred and the profane, (2) rituals, and (3) a community of worshipers. Thus, Durkheim defined religion as a system of shared rituals and beliefs about the sacred that bind together a community of worshipers.

Sacred A term describing everything that is regarded as extraordinary and that inspires in believers deep and absorbing sentiments of awe, respect, mystery, and reverence.

Beliefs About the Sacred and the Profane

At the heart of all religious belief and activity stands a distinction between two separate and opposing domains: the sacred and the profane. The **sacred** includes everything that is regarded as extraordinary and that inspires in believers deep and absorbing sentiments of awe, respect, mystery, and reverence. The deep and ab-

"Here's the church, here's the steeple, . . . " the child's rhyme says. Is building a church simply a practical necessity? Or is there some underlying idea in Christianity that tends to make Christian communities feel the need to build large structures?

©Michael Dwyer/Stock, Boston

sorbing sentiments motivate people to safeguard what is sacred from contamination or defilement. To find, preserve, or guard that which they consider sacred, people have gone to war, sacrificed their lives, traveled thousands of miles, and performed other life-endangering acts (Turner 1978).

Definitions of what is sacred vary according to time and place. Sacred things include objects (chalices, sacred documents, and books), living creatures (cows, ants, birds), elements of nature (rocks, mountains, trees, the sea, sun, moon, or sky), places (churches, mosques, synagogues, birthplaces of religious founders), days that commemorate holy events, abstract forces (spirits, good, evil), persons (Christ, Buddha, Moses, Mohammed, Zarathustra, Nanak), states of consciousness (wisdom, oneness with nature), past events (the crucifixion, the resurrection, the escape of the Jews from Egypt, the birth of Buddha), ceremonies (baptism, marriage, burial), and other activities (holy wars, just wars, confession, fasting, pilgrimages).

Durkheim ([1915] 1964) maintains that the sacredness springs not from the item, ritual, or event itself but from its symbolic power and from the emotions that people experience when they think about the sacred thing or when they are in its presence. The emotions are so strong that believers feel part of something larger than themselves and are outraged when others behave inappropriately in the presence of the sacred.

Ideas about what is sacred are such an important element of religious activity that many researchers classify religions according to the type of phenomenon that their followers consider sacred. One such typology consists of three categories: sacramental, prophetic, and mystical religions (Alston 1972).

In **sacramental religions**, the sacred is sought in places, objects, and actions believed to house a god or a spirit. These may include inanimate objects (relics, statues, crosses), animals, trees, plants, foods, drink (wine, water), places, and certain processes (such as the way in which people prepare for a hunt or perform a dance). Examples of a sacramental religion are the various forms of Native Spirituality. An excerpt from the "Statement of Walter Echo-Hawk before the United States Commission on Civil Rights" illustrates the sacramental qualities:

> First, I should tell you something about traditional or tribal religion as native religion is vastly different from the Judeo-Christian religions most of us are familiar with. Because native religions are so different, the religion of the "redman" has never been under-

Sacramental religions Religions in which the sacred is sought in places, objects, and actions believed to house a god or a spirit.

> stood by the non-Indian soldiers, missionaries, or government officials. 1) It is important to note that there are probably as many native religions as there are Indian tribes in this country. 2) None of these religions or religious tenets have been reduced to writing in a holy document such as the Bible or Koran.[1] 3) None of these religions have man-made churches in the Judeo-Christian sense, rather the native religions are practiced in nature, at sacred sites, or in temporary religious structures — such as a tepee or sweat lodge. 4) The religious beliefs are tied to nature, the spiritual forces of nature, the natural elements, and the plants and creatures which make up the environment. . . . Natives are dependent upon all these things in order to practice their many religious ceremonies, rituals and religious observances. (Echo-Hawk 1979:280)

In **prophetic religions**, the sacred revolves around items that symbolize significant historical events or around the lives, teachings, and writings of great people. Sacred books, such as the Christian Bible, the Muslim Koran, and the Jewish Torah hold the records of these events and revelations. In the case of historical events, God or some other higher being is believed to be directly involved in the course and the outcome of the event (a flood, the parting of the Red Sea, the rise and fall of an empire). In the case of great people, the lives and inspired words of prophets or messengers reveal a higher state of being, "the way," a set of ethical principles, or a code of conduct. Followers seek to imitate this life. Some of the best-known prophetic religions include Judaism as revealed to Abraham in Canaan and to Moses at Mount Sinai, Confucianism (founded by Confucius), Christianity (founded by Jesus Christ), and Islam (founded by Mohammed).

In **mystical religions**, the sacred is sought in states of being that, at their peak, can exclude all awareness of one's existence, sensations, thoughts, and surroundings. In such states, the mystic is caught up so fully in the transcendental experience that all earthly concerns seem to vanish. Direct union with the divine forces of the universe is of utmost importance. Not surprisingly, mystics tend to become involved in practices such as fasting or celibacy to separate themselves from worldly attachments. In addition, mystics meditate to clear their minds of worldly concern, "leaving the soul empty and receptive to influences from the divine" (Alston 1972:144). Buddhism and philosophical Hinduism are two examples of religions that emphasize physical and spiritual discipline as a means of transcending the self and earthly concerns.

Keep in mind that the distinctions among sacramental, prophetic, and mystical religions are not clear-cut ones. Most religions incorporate or combine elements of other religions. Consequently, most religions cannot be placed in a single category, although one category often predominates.

According to Durkheim ([1915] 1964), the sacred encompasses more than the forces of good: "There are gods [that is, satans] of theft and trickery, of lust and war, of sickness and of death" (p. 420). Evil and its various representations, however, are almost always portrayed as inferior and subordinate to the forces of good: "in the majority of cases we see the good victorious over evil, life over death, the powers of light over the powers of darkness" (p. 421). Even so, Durkheim considers superordinary evil phenomena to fall under the category "sacred" (as he defines it) because they are endowed with special powers and are the object of rituals (confessions, baptisms, penance, fasting, exorcism) designed to overcome or resist their negative influences.

Religious beliefs, doctrines, legends, and myths detail the origins, virtues, and powers of sacred things and describe the consequences of mixing the sacred with the profane. The **profane** is everything that is not sacred, including things opposed to the sacred (the unholy, the irreverent, the contemptuous, the blasphemous) and things that stand apart from the sacred, although not in opposition to it (the ordinary, the commonplace, the unconsecrated, the temporal, the bodily) (Ebersole 1967). Believers often view contact between the sacred and the profane as being dangerous and sacrilegious, as threatening the very existence of the sacred, and as endangering the fate of the person

Prophetic religions Religions in which the sacred revolves around items that symbolize significant historical events or around the lives, teachings, and writings of great people.

Mystical religions Religions in which the sacred is sought in states of being that, at their peak, can exclude all awareness of one's existence, sensations, thoughts, and surroundings.

Profane A term describing everything that is not sacred, including things opposed to the sacred and things that stand apart from the sacred, although not in opposition to it.

[1]To understand the nature of Indian religion, one must necessarily look to the native practitioners themselves for information.

who made or allowed the contact. Consequently, people take action to safeguard those things they regard as sacred by separating them in some way from the profane. For example, some people refrain from speaking the name of God in frustration, and some believe that a woman must cover her hair or her face during worship and that a man must remove his hat during worship.

The distinctions between the sacred and the profane do not mean that a person, object, or idea cannot pass from one domain to another or that something profane cannot ever come into contact with the sacred. Such transformations and contacts are authorized through rituals — the active and most observable side of religion.

Rituals

In the religious sense, **rituals** are rules that govern how people must behave in the presence of the sacred to achieve an acceptable state of being. These rules may take the form of instructions detailing the appropriate context; the roles of various participants; acceptable dress; and the precise wording of chants, songs, and prayers. Participants must follow instructions closely if they want to achieve a specific goal, whether the goal is to purify the participant's body or soul (confession, immersion, fasting, seclusion), to commemorate an important person or event (pilgrimage to Mecca, the Passover, the Last Supper), or to transform profane items into sacred items (water to holy water, bones to sacred relics). During rituals, behavior is "coordinated to an inner intention to make contact with, or to participate in, the invisible world or to achieve a desired state" (Smart 1976:6).

Rituals can be as simple as closing the eyes to pray or having the forehead marked with ashes; they can be an elaborate combination and sequence of activities such as fasting for three days before entering a sacred place to chant, with head bowed, a particular prayer for forgiveness. Although rituals are often enacted in sacred places, some rituals are codes of conduct aimed at governing the performance of everyday activities — sleeping, walking, eating, defecating, washing, dealing with the opposite sex.

The Taliban have received worldwide attention for the code of conduct it has imposed on those under its rule and for the severe punishments issued to those who violate these codes. For example, women who appear in public must be covered from head to toe; they are not permitted to leave their homes without an acceptable reason or to work outside the home even as health care workers or as distributors of international aid. The Taliban have tried to find and destroy things they consider un-Islamic, including tape recorders, cassettes, toys, TVs, and car radio antennas (Amnesty International 1996a, 1996b).

Durkheim maintains that the nature of the ritual is relatively insignificant. The important element is that the ritual be shared by a community or worshipers and evoke certain ideas and sentiments that help individuals feel part of something bigger than themselves.

Community of Worshipers

Durkheim uses the word **church** to designate a group whose members hold the same beliefs with regard to the sacred and the profane, who behave in the same way in the presence of the sacred, and who gather in body or spirit at agreed-on times to reaffirm their commitment to those beliefs and practices. Obviously, religious beliefs and practices cannot be unique to an individual; they must be shared by a group of people. If this were not the case, the beliefs and practices would cease to exist when the individual who held them died or if he or she chose to abandon them. In the social sense, religion is inseparable from the idea of church. The gathering and the sharing create a moral community and give worshipers a common identity. The gathering, however, need not take place in a common setting. When people perform a ritual on a given day or at given times of day, the gathering is spiritual rather than physical.

Durkheim ([1915] 1964) uses the term *church* loosely, acknowledging that it can assume many forms: "Sometimes it embraces an entire people . . . sometimes it embraces only a part of them . . . sometimes it is directed by a corps of priests, sometimes it is almost completely devoid of any official directing body" (p. 44). Sociologists have identified at least five broad types of religious organizations (communities of worshipers): ecclesiae, denominations, sects, established

Rituals Rules that govern how people must behave in the presence of the sacred to achieve an acceptable state of being.

Church A group whose members hold the same beliefs with regard to the sacred and the profane, who behave in the same way in the presence of the sacred, and who gather in body or spirit at agreed-on times to reaffirm their commitment to those beliefs and practices.

sects, and cults. As is the case with most classification schemes, the categories overlap on some characteristics because the criteria by which religions are classified are not always clear.

Ecclesiae An **ecclesiae** is a professionally trained religious organization governed by a hierarchy of leaders, which claims as its members everyone in a society. Membership is not voluntary; it is the law. Consequently, considerable political alignment exists between church and state officials, which makes the ecclesiae the official church of the state. Ecclesiae formerly existed in England (the Anglican church), France (the Roman Catholic church), and Sweden (the Lutheran church). The 1987 Afghanistan constitution stated that the religion of Afghanistan is Islam. Today Islam is the official religion of Bangladesh and Malaysia, Iran has been an Islamic republic since the Ayatollah Khomeini took power in 1979, and Saudi Arabia is a monarchy based on Islamic law (*The World Almanac and Book of Facts 1991* 1990).

Individuals are born into ecclesiae; newcomers to the society are converted; dissenters often are persecuted. Those who do not accept the official view emigrate or occupy the most marginal status in the society. The ecclesiae claims to be the one true faith and often does not recognize other religions as valid. In its most extreme form, it directly controls all facets of life. In Afghanistan, for example, non-Muslims must conform to Islamic legal and unwritten restrictions and are prohibited from proselytizing (Kurian 1992).

Denominations A **denomination** is a hierarchical organization in a society in which church and state are usually separate; it is led by a professionally trained clergy. In contrast to an ecclesiae, a denomination is one of many religious organizations in the society. For the most part, denominations are tolerant of other religious organizations; they may even collaborate to address and solve some problems in the society. Membership is considered to be voluntary, but most people who belong to denominations did not choose to do so. Rather, they were born to parents who are members. Denominational leaders generally make few demands on the laity, and most members participate in limited and specialized ways: they choose to send their children to church-operated schools, attend church on Sundays and religious holidays, donate money, or attend church-sponsored functions. The leaders of a denomination do not oversee all aspects of the members' lives. Yet, even though laypeople vary widely in lifestyle, denominations frequently attract people of particular races and social classes: that is, their members are drawn disproportionately from specific social and ethnic groups.

Eight major religious denominations exist in the world — Buddhism, Christianity, Confucianism, Hinduism, Islam, Judaism, Taoism, and Shinto — each dominant in different areas of the globe (see "The World's Major Non-Christian Religions"). For example, Christianity predominates in Europe, North and South America, New Zealand, Australia, and the Pacific Islands; Islam in the Middle East and North Africa; Hinduism in India.

Sects A **sect** is a small community of believers led by a lay ministry, with no formal hierarchy or official governing body to oversee the various religious gatherings and activities. Sects typically are composed of people who broke away from a denomination because they came to view it as corrupt. Therefore, they created the offshoot to reform the religion from which they separated.

All of the major religions encompass splinter groups that have sought, at one time or another, to preserve the integrity of their religion. In Islam, for example, the most pronounced split occurred 1,300 years ago, approximately thirty years after the death of the Prophet Mohammed, over the issue of Mohammed's successor. The Shia maintained that the successor should be a blood relative of Mohammed; the Sunni believed that the successor should be selected by the community of believers and need not be related by blood. When Mohammed died, the Sunni (the great majority of Muslims) accepted Abu-Bakr as the caliph (successor). The Shia supported Ali, Mohammad's first cousin and son-in-law, and they called for the overthrow of the existing order and a return to the pure form of Islam. Today Shiaism is the dominant religion in the Islam Republic of Iran (95 percent), and Sunni

Ecclesiae A professionally trained religious organization governed by a hierarchy of leaders, which claims as its members everyone in a society.

Denomination A hierarchical organization in a society in which church and state are usually separate, led by a professionally trained clergy.

Sect A small community of believers led by a lay ministry, with no formal hierarchy or official governing body to oversee the various religious gatherings and activities.

Muslim dominates in the Islamic Republics of Pakistan (77 percent) and Afghanistan (84 percent). These divisions within Islam have existed for a long enough time that Sunni and Shia are regarded as **established sects**, renegades from denominations or ecclesiae that have existed long enough to acquire a large following and widespread respectability. In some ways, established sects resemble both denominations and sects. As you might expect, several divisions exist within each of these established sects, all of which formed in an attempt to reform some policy, practice, or position held by the religious organization from which they separated.

Similarly, several splits have occurred within the Christian churches. In 1054, for example, the Eastern Orthodox churches rejected the pope as the earthly deputy of Christ and questioned the papal claim of authority over all Catholic churches in the world. The Protestant religions owe their origins to Martin Luther (1483–1546), who also challenged the papal authority and protested many of the practices of the Roman Catholic church.[2] Divisions also exist within various Protestant sects and between Catholics. Offshoots of the Roman Catholic church, for example, include Maronites, Greek Catholics, Greek Orthodox, Jacobites, and Gregorians.

People are not born into sects, as they are with denominations; theoretically, they convert. Consequently, newborns are not baptized; they choose membership later in life, when they are considered able to decide for themselves. Sects vary on many levels, including the degree to which they view society as religiously bankrupt or corrupt and the extent to which they take action to change people in society.

Cults Generally, **cults** are very small, loosely organized groups, usually founded by a charismatic leader who attracts people by virtue of his or her personal qualities. Because the charismatic leader plays such a central role in attracting members, cults often dissolve after the leader dies. For this reason, few cults last long enough to become established religions. Even so, a few manage to survive, as evidenced by the fact that the major world religions began as cults. Because cults are formed around new and unconventional religious practices, outsiders view them with considerable suspicion.

Cults vary according to purpose and to the level of commitment that the cult leaders demand of converts. Cults may draw members on the basis of highly specific but eccentric interests such as astrology, UFOs, or transcendental meditation. Members may be attracted by the promise of companionship, a cure from illness, relief from suffering, or enlightenment. A cult may meet infrequently and strictly voluntarily (such as at conventions or monthly meetings), or the cult leaders may require members to break all ties with family, friends, and jobs and thus to come to rely exclusively on the cult to meet all their needs.

A Critique of Durkheim's Definition of Religion

Durkheim's definition of religion revolves around the most outward, most visible characteristics of religion. Critics argue, however, that the three essential characteristics—beliefs about the sacred and the profane, rituals, and a community of worshipers—are not unique to religious activity. This combination of characteristics, they say, can be found at many gatherings—sporting events, graduation ceremonies, reunions, and political rallies — and in many political systems (for example, Marxism, Maoism, and fascism). On the basis of these characteristics alone, it is difficult to distinguish between an assembly of Christians celebrating Christmas, a patriotic group supporting the initiation of a war against another country, and a group of fans eulogizing James Dean or Elvis Presley. In other words, religion is not the only unifying force in society to make use of these three elements that Durkheim defines as characteristic of religion. Civil religion represents another such force that resembles religion as Durkheim defined it.

Established sects Religious organizations, resembling both denominations and sects, composed of renegades from denominations or ecclesiae that have existed long enough to acquire a large following and widespread respectability.

Cults Very small, loosely organized groups, usually founded by a charismatic leader who attracts people by virtue of his or her personal qualities.

[2]This protest is known as the Reformation because it involved efforts to reform the Catholic church and to cleanse it of corruption, especially with regard to paying for indulgences (the forgiveness of sins upon saying specific prayers or performing specific good deeds at the order of a priest). Luther believed that a person is saved not by the intercession of priests or bishops but by private and individual faith (Van Doren 1991).

The tokens of devotion left by "pilgrims" to the grave of James Dean illustrate the difficulty of making absolute distinctions between the sacred and the profane.

©James F. Hopgood

Civil Religion

Civil religion is "any set of beliefs and rituals, related to the past, present and/or future of a people (nation), which are understood in some transcendental fashion" (Hammond 1976:171). A nation's beliefs (such as individual freedom or equal opportunity) and rituals (parades, fireworks, singing of the national anthem, twenty-one-gun salutes, and so on) often assume a sacred quality. Even in the face of internal divisions based on race, ethnicity, region, religion, or gender, national beliefs and rituals can inspire awe, respect, and reverence for the country. These sentiments are most notable on national holidays that celebrate important events or people (such as Washington's Birthday, Martin Luther King, Jr., Day, or Independence Day), in the presence of national monuments or symbols (the flag, the Capitol, the Lincoln Memorial, the Vietnam Memorial), and at times of war or other national crises.

Often political leaders appeal to these sentiments to win an election, to legitimate their policies, to rally a nation around a cause that requires sacrifice, or to motivate a people to defend their country, as President George Bush attempted to do in his State of the Union Address on January 7, 1991:

> I come to this house of the people to speak to you and all Americans, certain that we stand at a defining hour.
>
> Halfway around the world, we are engaged in a great struggle in the skies and on the seas and sands. We know why we're there. We are Americans — part of something larger than ourselves.
>
> For two centuries, we've done the hard work of freedom. And tonight we lead the world in facing down a threat to decency and humanity.
>
> For two centuries, America has served the world as an inspiring example of freedom and democracy. For generations, America has led the struggle to preserve and extend the blessings of liberty. And today, in a rapidly changing world, American leadership is indispensable. Americans know that leadership brings burdens, and requires sacrifice.
>
> But we also know why the hopes of humanity turn to us. We are Americans; we have a unique responsibility to do the hard work of freedom. And when we do, freedom works.

One can argue that the Cold War[3] between the United States and the former Soviet Union elevated

[3]The Cold War is the name given to the political tension and military rivalry that existed between the United States and the former Soviet Union from approximately 1945 until its symbolic end on November 9, 1989 — the date on which the Berlin Wall "fell." The Cold War included an arms race, in which each side competed to match and then surpass any advances made by the other in the number and technological quality of weapons.

Civil religion "Any set of beliefs and rituals, related to the past, present and/or future of a people (nation), which are understood in some transcendental fashion" (Hammond 1976:171).

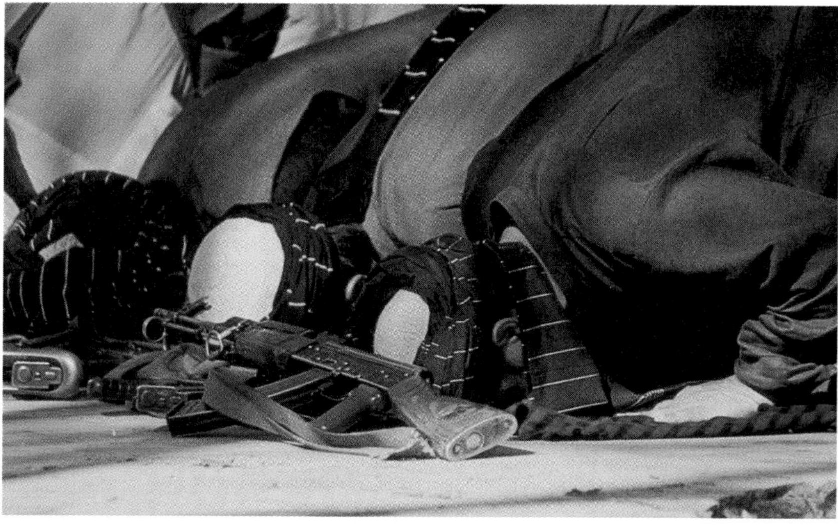

Although many Westerners become uncomfortable hearing Muslims or others describe armed struggle in terms of "holy war," U.S. government officials and politicians have a long tradition of claiming that fighting wars on behalf of the United States involves some "sacred duty." The similar language reflects a similar underlying frame of mind.

each country's economic and political systems to the level of a "religion." During the Cold War, relations between the United States and the Soviet Union fell short of direct, full-scale military engagement; even so, as many as 120 wars were fought in so-called Third World countries. In many of these wars, the United States and the Soviet Union supported opposing factions with weapons and other military equipment, combat training, medical supplies, economic aid, and food. Three of the best-known proxy wars were fought in Korea, Vietnam, and Afghanistan.

Soviet and American leaders justified their direct or indirect intervention in these proxy wars on the grounds that it was necessary to contain the spread of the other side's economic and political system, to protect national and global security, and to prevent the other side from shifting the balance of power in favor of its system.

In the mid-1960s, President Lyndon Johnson justified U.S. involvement in Vietnam in this way:

> If we allow the Communists to win in Vietnam, it will become easier and more appetizing for them to take over other countries in other parts of the world. We will have to fight again some place else—at what cost no one knows. That is why it is vitally important to every American family that we stop the Communists in South Vietnam. (Johnson 1987:907)

The Soviets justified their involvement on the grounds that they were blocking "the export of 'counterrevolution,' or actions by Western powers that would hinder the historic progress of socialism" (Zickel 1991:681). The Soviets defined it as their "internationalist duty" to defend their socialist allies and to give military and economic support to wars of national liberation in Third World countries. They rationalized such actions according to "the Marxist belief that workers around the world are linked by a bond that transcends nationalism" (pp. 999–1000).

Since the end of World War II, the foreign and domestic policies of the United States and the former Soviet Union have been shaped by Cold War dynamics. In the United States, for example, virtually every policy, from the 1949 Marshall Plan to covert aid to the Contras during the Reagan administration, was influenced in some way by America's professed desire to protect the world from the Soviet influence and the spread of Communism. Robert S. McNamara (1989), U.S. secretary of defense under Presidents Kennedy and Johnson, remarks that "on occasion after occasion, when con-

The World's Major Non-Christian Religions

BUDDHISM

*Buddhism has 307 million followers. It was founded by Siddhartha Gautama, known as the Buddha (Enlightened One), in southern Nepal in the sixth and fifth centuries B.C. The Buddha achieved enlightenment through meditation and gathered a community of monks to carry on his teachings. Buddhism teaches that meditation and the practice of good religious and moral behavior can lead to Nirvana, the state of enlightenment, although before achieving Nirvana one is subject to repeated lifetimes that are good or bad depending on one's actions (*karma*). The doctrines of the Buddha describe temporal life as featuring "four noble truths": existence is a realm of suffering; desire, along with the belief in the importance of one's self, causes suffering; achievement of Nirvana ends suffering; and Nirvana is attained only by meditation and by following the path of righteousness in action, thought, and attitude.*

The famous Daibutsu in Kamakura, Japan.

CONFUCIANISM

A faith with 5.6 million followers, Confucianism was founded by Confucius, a Chinese philosopher, in the sixth and fifth centuries B.C. Confucius's sayings and dialogues, known collectively as the Analects*, were written down by his followers. Confucianism, which grew out of a strife-ridden time in Chinese history, stresses the relationship between individuals, their families, and society, based on* li *(proper behavior) and* jen *(sympathetic attitude). Its practical, socially oriented philosophy was challenged by the more mystical precepts of Taoism and Buddhism, which were partially incorporated to create neo-Confucianism during the Sung dynasty (A.D. 960–1279). The overthrow of the Chinese monarchy and the Communist revolution during the twentieth century have severely lessened the influence of Confucianism on modern Chinese culture.*

HINDUISM

A religion with 648 million followers, Hinduism developed from indigenous religions of India in combination with Aryan religions brought to India in c. 1500 B.C. and codified in the Veda and the Upanishads, the sacred scriptures of Hinduism. Hinduism *is a term used to broadly describe a vast array of sects to which most Indians belong. Although many Hindu reject the caste system—in which people are born into a particular subgroup that determines their religious, social, and work-related duties—it is widely accepted and classifies society at large into four groups: the Brahmins or priests, the rulers and warriors, the farmers and merchants, and the peasants and laborers. The goals of Hinduism are release from repeated reincarnation through the practice of yoga, adherence to Vedic scriptures, and devotion to a personal guru. Various deities are worshiped at*

fronted with a choice between support of democratic governments and support of anti-Soviet dictatorships, we have turned our backs on our traditional values and have supported the antidemocratic" and brutally repressive and totalitarian regimes (p. 96).

In light of the brief overview of the Cold War's history, we can see that each government's economic and political beliefs assumed a sacred quality that unified and motivated each side to sacrifice millions of human lives at home and abroad in the name of those principles. The larger point is that the traits that Durkheim cites as characteristics of religion apply to other events, relationships, and forces within society that many people would not define as religious.

Narrower, less inclusive definitions of religion than Durkheim's are problematic as well. Suppose that religion were defined as the belief in an ever-living god. This definition would exclude polytheistic religions

shrines; the divine trinity, representing the cyclical nature of the universe, are Brahma the creator, Vishnu the preserver, and Shiva the destroyer.

ISLAM

Islam has 840 million followers. It was founded by the prophet Muhammad, who received the holy scriptures of Islam, the Koran, from Allah (God) c. A.D. 610. Islam (Arabic for "submission to God") maintains that Muhammad is the last in a long line of holy prophets, preceded by Adam, Abraham, Moses, and Jesus. In addition to being devoted to the Koran, followers of Islam (Muslims) are devoted to the worship of Allah through the Five Pillars: the statement "There is no god but God, and Muhammad is his prophet"; prayer, conducted five times a day while facing Mecca; the giving of alms; the keeping of the fast of Ramadan during the ninth month of the Muslim year; and the making of a pilgrimage at least once to Mecca, if possible. The two main divisions of Islam are the Sunni and the Shiite; the Wahabis are the most important Sunni sect, while the Shiite sects include the Assassins, the Druses, and the Fatimids, among countless others.

JUDAISM

Stemming from the descendants of Judah in Judea, Judaism was founded c. 2000 B.C. by Abraham, Isaac, and Jacob and has 18 million followers. Judaism espouses belief in a monotheistic God, who is creator of the universe and who leads His people, the Jews, by speaking through prophets. His word is revealed in the Hebrew Bible (or Old Testament), especially in that part known as the Torah. The Torah also contains, according to Rabbinic tradition, a total of 613 biblical commandments, including the Ten Commandments, which are explicated in the Talmud. Jews believe that the human condition can be improved, that the letter and the spirit of the Torah must be followed, and that a Messiah will eventually bring the world to a state of paradise. Judaism promotes community among all people of Jewish faith, dedication to a synagogue or temple (the basic social unit of a group of Jews, led by a rabbi), and the importance of family life. Religious observance takes place both at home and in temple. Judaism is divided into three main groups who vary in their interpretation of those parts of the Torah that deal with personal, communal, international, and religious activities: the Orthodox community, which views the Torah as derived from God, and therefore absolutely binding; the Reform movement, which follows primarily its ethical content; and the Conservative Jews, who follow most of the observances set out in the Torah but allow for change in the face of modern life. A fourth group, reconstructionist Jews, rejects the concept of the Jews as God's chosen people, yet maintains rituals as part of the Judaic cultural heritage.

SHINTO

Shinto, with 3.5 million followers, is the ancient native religion of Japan, established long before the introduction of writing to Japan in the fifth century A.D. The origins of its beliefs and rituals are unknown. Shinto stresses belief in a great many spiritual beings and gods, known as kami, *who are paid tribute at shrines and honored by festivals, and reverence for ancestors. While there is no overall dogma, adherents of Shinto are expected to remember and celebrate the* kami, *support the societies of which the* kami *are patrons, remain pure and sincere, and enjoy life.*

TAOISM

Both a philosophy and a religion, Taoism was founded in China by Lao-tzu, who is traditionally said to have been born in 604 B.C. Its number of followers is uncertain. It derives primarily from the Tao-te-ching, which claims that an ever-changing universe follows the Tao, or path. The Tao can be known only by emulating its quietude and effortless simplicity; Taoism prescribes that people live simply, spontaneously, and in close touch with nature and that they meditate to achieve contact with the Tao. Temples and monasteries, maintained by Taoist priests, are important in some Taoist sects. Since the Communist revolution, Taoism has been actively discouraged in the People's Republic of China, although it continues to flourish in Taiwan.

Look back at several of these descriptions. What are some *sacramental, prophetic,* and *mystical* characteristics they mention? Suppose you were asked to write a brief piece describing Christianity in the same ways that the other major religions are described here. What would you say?

Source: Reprinted with permission of Macmillan General Reference USA, a Simon & Schuster Macmillan Company, from *The New York Public Library Desk Reference,* Second Edn. A Stonesong Press book. Copyright © 1989, 1993 by The New York Public Library and The Stonesong Press, Inc. The name, "The New York Public Library," is a trademark and the property of The New York Public Library, Astor, Lenox, and Tilden Foundations.

such as Hinduism, which has more than 640 million adherents. It would also exclude religions in which a deity plays little or no role, such as Buddhism, which has more than 300 million adherents. So, narrow definitions of religion are no improvement over broad ones.

Despite its shortcomings, Durkheim's definition of religion is one of the best and most widely used definitions. No sociologist with any standing in the discipline can study religion without encountering and addressing Durkheim's definition. The question "What is religion?" is not just a sociological question; it is also a question asked by those governments that guarantee its residents freedom of religion (see "Freedom of Religious Expression in U.S. Prisons"). Besides proposing a definition of religion, Durkheim also wrote extensively about the functions of religion. This work laid the foundation for the functionalist perspective of religion.

U.S. In Perspective

Freedom of Religious Expression in U.S. Prisons

The following selection is an excerpt of a statement by Larry F. Taylor, warden at the Federal Correctional Institution in Lompoc, California, speaking before the U.S. Commission on Civil Rights.

> *Thank you, Commissioner Freeman. Mr. Chairman, first of all, let me state that Lompoc is a maximum security institution designed to accommodate about 1,200 offenders [and] . . . a minimum security camp for 420 inmates serving relatively short sentences.*
>
> *Persons confined in the major institutions are serving relatively long sentences for drug offenses, trafficking in drugs, bank robbery, homicides, hijacking, kidnapping, and all the kinds of offenses that you would expect to find at a major institution.*
>
> *We have a mixed racial group: about one-third of our population black, one-third white, one-third Chicano, and about 100 Indians at any one time, Native American Indians.*
>
> *I think it's important to say that the exercise of religion in an institution of this type cannot be absolute and I think it's subject to reasonable regulations designed to protect the welfare of the staff and the inmates, the welfare of the community, control and discipline of the inmates, proper exercise of institutional authority in scheduling activities, etc., and reasonable economic considerations.*
>
> *I don't think, for example, we can provide clergy for every kind of religion represented in our facility; we can't provide facilities for every type of religion represented in our facility. There are some difficult religious issues which a prison administrator has to face each day: first of all, what is a religion? Thank goodness I don't have to decide what a religion is. You've heard the list of religions that are represented in the Bureau of Prisons. That's decided for me by general counsel and by several court cases. But another difficult issue is what are ceremonies, artifacts, and symbols that are mandated by certain recognized religions? And can we make those available to the inmate population?*
>
> *Areas of concern for us have to do primarily with security. Can we operate a safe humane institution and yet allow a number of these kinds of things into the institution?*
>
> *Of course, our goal is to try to be as flexible as possible and allow the greatest amount of flexibility in exercising religion wherever we can. We are concerned about diets and, you know, it is easy to prepare 1,600 meals three times a day if we're preparing the same thing for everybody; but if you have seven or eight different groups who need, who have different kinds of diet requirements, then that task becomes much more difficult. When certain requirements are accepted or recognized, the question of having the necessary experts there is still to advise us on proper preparation. This has been particularly troublesome with the Jewish inmates and also troublesome with Muslim inmates who have special diet requirements.*
>
> *We have a similar problem with what is considered proper religious wearing apparel. Special clothing also becomes a security problem when going in and out of our visiting room. Head gear and items of this sort need to be carefully searched when a man enters or leaves the visiting room.*
>
> *Mr. Cripe indicated one example of Rastafarians, where we've had*

The Functionalist Perspective

As far as we know, some form of religion has existed as long as humans have lived (at least 2 million years). In view of this fact, functionalists maintain that religion must serve some vital social functions for the individual and for the group. On the individual level, people embrace religion in the face of uncertainty: they draw on religious doctrine and ritual to comprehend the meaning of life and death and to cope with misfortunes and injustices (war, drought, illness). Life would be intolerable without reasons for existing or without a higher purpose to justify the trials of existence (Durkheim 1951). Try to imagine, for example, how people cope with the immense devastation and destruction resulting from decades of war. In Afghanistan, after Soviet troops invaded the country in 1979, troops attacked civilian populations, burned village crops, killed livestock, used lethal and nonlethal chemical weapons, planted an estimated 10 million mines, and engaged in large-scale, high-altitude carpet bombing.

> In the countryside it was standard Soviet practice to bombard or even level whole villages suspected of harboring resistance fighters. Sometimes women, children, and old men were rounded up and shot. This devastation of towns and villages forced many civilians to seek refuge in Kabul, whose prewar population of less than one million swelled to nearly two million. (Kurian 1992:5)

Even after the Soviets withdrew from Afghanistan in 1989, civil war continued as various political parties competed to fill the power vacuum. The tragic results of twenty years of war are summarized in Table 14.1. In light of this situation, is it any wonder that the Afghan people turn to religion?

> A crisis like [war] makes a Muslim search within himself for values to counter the bestiality of man. We

some difficulty searching their hair for drugs and things of that nature. Symbols often cause racial tension within our institutions for various reasons. The recognition of holy days and when they should be celebrated is another sensitive issue. Are the inmates allowed off work during those holy days? If one group has the day off from work, then all groups would like their days recognized, and be off from work on that day.

Scheduling of prayer hours so they do not interfere with counts and other institutional functions is also important. Chanting, for example, isn't very popular early in the morning, yet, some of our inmates believe that they must chant at sunrise. In one case at Lompoc, one man assaulted another man because of the noise that was being made and the invasion of privacy.

Special instruments used in the practice of religion is another area that we have to be concerned about. For Native American Indians we allow drums into the institution and although we haven't in the past, we are now looking at allowing a peace pipe into the institution.

Incidentally, in the past when the peace pipe was allowed, there was a conflict in which the pipe disappeared. About forty or fifty inmates decided that that pipe was not going to be returned to its proper place. As a result, there was a great potential for violence and disruption of prison routine because of that incident. Many times, it's a question of what is required of a man's religion as opposed to what he prefers. A question of individual preference over mandated requirements. Clergy don't always agree, courts don't always agree, and certainly prison administrators don't agree on what's required.

One thing that's important to remember: whatever we do for one religious group, we must be willing to do for all religious groups, and so wherever we give a little in one area, we also have to be willing to give in another area. I think the primary barriers to the exercise of religion as far as the Native Americans are concerned is the lack of knowledge, lack of knowledge on my part, lack of knowledge of many administrators who are charged with the responsibility of running the prisons. What is the Indian religion? The same lack of understanding was true of Muslims in the sixties when we first started running into those kind of issues. The Indian religious issue is further complicated by the lack of written definition. I think most of the Indian religion is passed down by word of mouth and only after making some effort and doing some research have I become a little more comfortable with some of the requirements of the Indian religion within our institutions. I think there is a lack of documented history. The Indian religion is a combination of cultural, medical, recreation, and social needs. It serves all those purposes as far as I can understand and so, sometimes, what may seem to be a social event may have, indeed, some very serious religious significance for Indians.

Warden Taylor is happy not to be the one who must "decide what religion is." What is it that he must decide? How is that difficult? What parallels do you see between what he must do and what governments must do to maintain civility or law when a society admits the possibility of more than one religion?

Source: U.S. Commission on Civil Rights (1979).

> look around at what they have done, and the only thing you can say is: Thank God, who is, in a situation like this, the only refuge for those who cannot understand why man would do this to his fellow man. (Ibrahim 1991:A7)

In addition to turning to religion in the face of intolerable circumstances, people also turn to religious beliefs and rituals to help them achieve a successful outcome (the birth of a healthy child, a job promotion) and to gain answers to questions of meaning: How did we get here? Why are we here? What happens to us when we die? According to Durkheim, people who have communicated with their god or with other supernatural forces (however conceived) report that they gain the inner strength and the physical strength to endure and to conquer the trials of existence.

> It is as though [they] were raised above the miseries of the world. . . . Whoever has really practiced a religion knows very well . . . these impressions of joy, of interior peace, of serenity, of enthusiasm which are, for the believer, an experimental proof of his beliefs. (Durkheim [1915] 1964:416–417)

Religion functions in several ways to promote group unity and solidarity. First, the shared doctrine and rituals help create emotional bonds among those who believe. Second, all religions strive to raise individuals above themselves—to help them live a life better than they would lead if they were left to their own impulses. In this sense, religion offers ideas of proper conduct that carry over into everyday life. When believers violate this code of conduct, they feel guilt and remorse. Such feelings, in turn, motivate them to make amends. Third, although observances of many religious rituals function to alleviate individual anxieties, uncertainties, and fears, they also establish, reinforce, or renew social relationships, binding individuals to a

Table 14.1 Profile of the Islamic State of Afghanistan

Population	16 million
World Food Program	Feeds, aids 1.8 million people
Refugees	6.0 million fled the country; 2.0 million displaced internally; 3.0 repatriated
Deaths as a result of war (since 1979)	500,000 military; 1.5 million civilian
Life expectancy	42 years
Malnutrition	15 to 20 percent of children under the age of 7
Damage from war	19,000/22,000 villages bombarded by Soviet air force
Mines	10 million implanted mines
Access to drinking water	5 percent rural; 40 percent urban
Maternal mortality	11,700/100,000
Infant mortality	182/1,000; 257/1,000 die before age 5

Source: U.S. Central Intelligence Agency (1995), Carter Center (1996), and United Nations (1996).

group. Finally, religion functions as a stabilizing force in times of severe social disturbances and abrupt change. During such times, many regulative forces in society break down. When such regulative forces are absent, people are more likely to turn to religion in search of a force that will bind them to a group. This tie helps people think less about themselves and more about some common goal (Durkheim 1951), whether the goal is to work for peace or to participate more fervently in armed conflicts. One can argue that the Taliban hoped to restore order.

The fact that religion functions to meet individual and societal needs, in combination with the fact that people create sacred objects and rituals, led Durkheim to reach a controversial but thought-provoking conclusion: the something out there that people worship is actually society.

Society as the Object of Worship

If we operate under the assumption that all religions are true in their own fashion and that the variety of religious responses is virtually endless, we find support for Durkheim's conclusion that everything encompassed by religion—gods, rites, sacred objects—is created by people. That is, people play a fundamental role in determining what is sacred and how people should act in the presence of the sacred. Consequently, at some level, people worship what they (or those before them) have created. This point led Durkheim to conclude that the real object of worship is society itself—a conclusion that many critics cannot accept (Nottingham 1971).

Let us give Durkheim the benefit of the doubt, however, and ask, Is there anything about the nature of society that makes it deserving of such worship? In reply to this question, Durkheim gave what sociologist W. S. F. Pickering (1984) calls a "virtual hymn to society, a social Gloria in Excelsis" (p. 252). Durkheim maintains that society transcends the individual life because it frees people from the bondage of nature (as in nature and nurture). How is this accomplished? Chapter 5 presents cases that show the consequences of extreme isolation, neglect, and limited social contact. Such cases make it clear that "it is impossible for a person to develop without social interaction" (Mead 1940:135). In addition, studies of mature and even otherwise psychologically and socially sound persons who experience profound isolation—astronauts orbiting alone in space, prisoners of war placed in solitary confinement, individuals who volunteer to participate in scientific experiments in which they are placed in deprivation tanks — show that when people are deprived of contact with others, they lose a sense of reality and personal identity (Zangwill 1987). The fact that we depend so strongly on society supports Durkheim's view that for the individual "it is a reality from which everything that matters to us flows" (Durkheim, cited in Pickering 1984:252).

Durkheim, however, does not claim that society provides us with perfect social experiences:

> Society has its pettiness and it has its grandeur. In order for us to love and respect it, it is not necessary to present it other than it is. If we were only able to love and respect that which is ideally perfect, . . . God Himself could not be the object of such a feeling,

> since the world derives from Him and the world is full of imperfection and ugliness. (quoted in Pickering 1984:253)

Durkheim observes that whenever a group of people has strong conviction (no matter what kind of group it is), that conviction almost always takes on a religious character. Religious gatherings and affiliations become ways of affirming convictions and mobilizing the group to uphold them, especially when they are threatened. This mobilizing function is especially evident when we consider that after the Soviet Union invaded Afghanistan in December 1979, Afghan freedom fighters known as the *mujahidin* resisted, making it impossible for the Soviet Union to establish control outside the major urban centers.

For example, Afghanistan's mountainous terrain made the Khyber Pass a key position in Britain's nineteenth-century colonial strategies. It also made it extremely difficult for massive, highly mechanized Soviet forces to find small bands of *mujahidin* moving on foot (see the chapter opening photo). Analogous problems worked against the United States in Vietnam.

A Critique of the Functionalist View of Religion

To claim that religion functions as a strictly integrative force is to ignore the long history of wars between different religious groups and the many internal struggles among factions within the same religious group. For example, although the Afghan *mujahidin* opposed the Soviet occupation and its secular government, many competing factions existed within the *mujahidin*. After the Soviets withdrew, the former *mujahidin* resistance commanders became the major power brokers, and each took control of different cities outside Kabul. At this point, as has happened in the past, the same rugged terrain that made it impossible for the Soviets to control the country made it difficult for any one internal group to consolidate power. Tribal elders and religious students, in turn, have tried to take control from those in power (U.S. Central Intelligence Agency 1995). The following is a partial list of various political parties: Islamic Society, Islamic Party (which has three factions), Islamic Union for the Liberation of Afghanistan, Islamic Revolutionary Movement, Afghanistan National Liberation Front, National Islamic Front (which has two major factions), Islamic Unity Party, Islamic Movement, and the Taliban (Religious Student Movement).

If religion were entirely an integrative force, religious beliefs and sacred symbols would never be important to ingroup-outgroup distinctions. That is, religious symbols that unite a community of worshipers would not unite them so strongly that they would be willing to destroy persons who did not belong to or support their religion. The Amnesty International (1996c) document, *Afghanistan: Grave Abuses in the Name of Religion,* outlines human rights violations by the Taliban committed in the name of religion. Those abuses include "indiscriminate killings, arbitrary and unacknowledged detention of civilians, physical restrictions on women for reasons of their gender, the beating and ill-treatment of women, children and detainees, deliberate and arbitrary killings, amputations, stoning and executions."

The functionalist perspective, then, tends to overemphasize the constructive consequences associated with religions' unifying, bonding, and comforting functions. Strict functionalists who focus only on the consequences that lead to order and stability tend to overlook the fact that religion can also unify, bond, and comfort believers in such a way that it supports war and other forms of conflict between ingroups and outgroups. The conflict perspective, on the other hand, acknowledges the unifying, comforting functions of religion but, as we shall see, views such functions as ultimately problematic.

The Conflict Perspective

Scholars who view religion from the conflict perspective focus on how religion turns people's attention away from social and economic inequality. This view stems from the work of Karl Marx, who believed that religion was the most humane feature of an inhumane world and that it arose from the tragedies and injustices of human experience. He described religion as the "sigh of the oppressed creature, the sentiment of a heartless world, and the soul of soulless conditions. It is the opium of the people" (*The World Treasury of Modern Religious Thought* 1990:80). People need the comfort of religion to make the world bearable and to justify their existence. In this sense, Marx said, religion is analogous to a sedative.

Even though Marx acknowledged the comforting role of religion, he focused on its repressive, constraining, and exploitative qualities. In particular, he conceptualized religion as an ideology that justifies the status quo, rationalizing existing inequities or downplaying their importance. This aspect of religion is especially relevant with regard to the politically and economically disadvantaged. For them, said Marx, religion is a source of false consciousness. That is, religious teachings encourage the oppressed to accept the economic, political, and social arrangements that constrain their chances in this life because their suffering will be compensated in the next.

Consider how material published in sixty-two languages by the Watchtower Bible and Tract Society (1987) and distributed worldwide describes life in God's Kingdom:

> God's Kingdom will bring earthly benefits beyond compare, accomplishing everything good that God originally purposed for his people to enjoy on earth. Hatreds and prejudices will cease to exist. . . . The whole earth will eventually be brought to a garden-like [paradise]. . . . No longer will people be crammed into huge apartment buildings or run-down slums. . . . People will have productive, satisfying work. Life will not be boring. (pp. 3–4)

This kind of ideology led Marx to conclude that religion justifies social and economic inequities and that religious teachings inhibit protest and revolutionary change. He went so far as to claim that religion would not be needed in a classless society—a propertyless society providing equal access to the means of production. In the absence of material inequality, there would be no exploitation and no injustice—experiences that cause people to turn to religion. In sum, Marx believed that religious doctrines turn people's attention away from unjust political and economic arrangements and that they rationalize and defend the political and economic interests of the dominant social classes. For some contemporary scholars, this legitimating function is reflected by the fact that most religions allow only a specific category of people—men—to be leaders and to handle sacred items. A letter written to the editor of *Christianity Today* in reaction to the article "Women in Seminary: Preparing for What?" shows how the "facts" of the Bible can be used to explain and justify such inequalities:

> If the Lord meant for a woman to lead the church in such roles as preacher, elder, pastor, minister, prophet, priest, et cetera, why didn't he provide early Christians with a scriptural prototype? Where in Scripture can a woman priest be found? A woman (literary) prophet? A woman apostle? A woman elder or pastor? Could it be the Lord didn't intend for a woman to serve in any of these positions? It makes me wonder. (*Christianity Today* 1986:6)

As a more extreme example, in the name of Islam, Taliban leaders placed severe restrictions on female dress, employment, and access to education. The restrictions were so severe that the Save the Children Fund, a humanitarian organization that has been working in Afghanistan and with Afghan refugees in Pakistan since the mid-1970s, suspended operations in western Afghanistan (Save the Children Fund 1996).

Sometimes religion can be twisted in ways that serve the interest of dominant groups. During the days of slavery, for example, some Christians prepared special catechisms for slaves to study. The following questions and answers were included in such catechisms:

> Q: What did God make you for?
>
> A: To make a crop.
>
> Q: What is the meaning of Thou shalt not commit adultery?
>
> A: To serve our heavenly Father, and our earthly Master, obey our overseer, and not steal anything. (Wilmore 1972:34)

Often political leaders use religion to unite their country in war against another. Both Iraqi President Saddham Hussein and U.S. President George Bush invoked the name of God to rally people behind their causes in the Persian Gulf War of 1991:

> In the name of God, the merciful, the compassionate: Our armed forces have performed their holy war duty of refusing to comply with the logic of evil, imposition and aggression. They have been engaged in an epic, valiant battle that will be recorded by history in letters of light. (spokesperson for the Iraqi Government, quoted in "Iraqi Message" 1991:Y1)

> This we do know: Our cause is just. Our cause is moral. Our cause is right. May God bless the United States of America. (Bush 1991:A8)

A Critique of the Conflict Perspective of Religion

The major criticism of Marx and the conflict perspective on religion is that religion is not always the sign of the oppressed creature. On the contrary, the oppressed have often used religion as a vehicle for protesting or working to change social and economic inequities. **Liberation theology** represents such an approach. Libera-

Liberation theology A doctrine that maintains that organized religions have a responsibility to demand social justice for the marginalized peoples of the world, especially landless peasants and the urban poor, and to take an active role at the grassroots level to bring about political and economic justice.

The Nation of Islam was organized in the United States in response to widespread exclusion of African Americans from mainstream society. Despite its representation in the media as a violent organization, the Nation of Islam focuses on empowering members through religious training that defines moral behavior and through small-business enterprise.

©Tannenbaum/Sygma

tion theologians maintain that they have a responsibility to demand social justice for the marginalized peoples of the world, especially landless peasants and the urban poor, and to take an active role at the grassroots level to bring about political and economic justice. Ironically, this doctrine is inspired by Marxist thought in that it advocates raising the consciousness of the poor and teaching them to work together to obtain land and employment and to preserve their cultural identity.

Sociologist J. Milton Yinger (1971) identifies at least two interrelated conditions under which religion can become a vehicle of protest or change. In the first condition, a government or other organization fails to achieve clearly articulated ideals (such as equal opportunity, justice for all, or the right to bear arms). In the second condition, a society is polarized along class, ethnic, or sectarian lines. In such cases, disenfranchised or disadvantaged groups may form sects or cults and may use seemingly eccentric features of the new religion to symbolize their sense of separation (p. 111) and to rally their followers to fight against the establishment or the dominant group. One religion that emerged in reaction to this failure of society to ensure equal opportunity is the Nation of Islam.

In the 1930s, black nationalist W. D. Farad, who went by a variety of names including Farad Mohammed, founded the Nation of Islam and began preaching in the Temple of Islam in Detroit. (When Farad disappeared in 1934, his chosen successor, Elijah Mohammed, took his place.) Farad taught that the white man was the personification of evil and that black people were Muslim but that their religion had been taken from them after they came to America as slaves. Farad also taught that the way out was not through gaining the "devil's" (the white man's) approval but through self-help, discipline, and education. Members received an X to replace their slave names (hence Malcolm X). In the social context of the 1930s, this message was very attractive:

> You're talking about Negroes. You're talking about niggers, who are the rejected and the despised, meeting in some little, filthy, dingy little [room] upstairs over some beer hall or something, some joint that nobody cares about. Nobody cares about these people. . . . You can pass them on the street and in 1930, if they don't get off the sidewalk, you could have them arrested. That's the level of what was going on. (National Public Radio 1984a)

The Nation of Islam is only one example of a religious organization working to improve life for African Americans. Historically, African American churches have reached out to millions who have felt excluded from the system (Lincoln and Mamiya 1990).

> They have demonstrated their capability for empowerment, emancipating black people from the ravages of generations of slavery and equipping them to recover a sense of identity, to forge ties of communal loyalty, to help themselves and others and to create cultural expressions that embody their hopes and aspirations. (Forbes 1990:2)

For example, African American churches were instrumental in the overall successes of the civil rights movement. In fact, some observers argue that the movement would have been impossible if the churches had not been involved (Lincoln and Mamiya 1990).

Religion in a Polarized Society

Although the story of Afghanistan is a very complicated one and beyond the scope of this chapter, we can identify at least one factor that contributes to major polarizations among the Afghan peoples: geographic location, which has made Afghanistan of strategic importance to several foreign powers. And as Durkheim observes, whenever a group of people has strong convictions, those convictions almost always take on a religious character. Religion serves as a vehicle around which convictions can be affirmed and around which a group can be mobilized to uphold them.

As the chapter opening map shows, Afghanistan is landlocked, and two-thirds of its terrain is mountainous. China, Iran, Pakistan, and three former Soviet republics (Tajikistan, Turkmenistan, and Uzbekistan) are on its borders. Given its location, it is not surprising that many foreign governments had national interest in Afghanistan. Consequently, the factional fighting in Afghanistan has been backed by foreign powers.

Afghanistan was of strategic importance to both the expanding British and Russian empires during the nineteenth century. In fact, Britain fought three wars— the Anglo-Afghan wars — in an attempt to control Afghanistan and, by extension, Russian influence in the area. Afghanistan's relationship with Pakistan is influenced by a dispute over its borders stemming from a time when Pakistan was part of India. Pakistan's boundaries with Afghanistan were set in 1893 after the second Anglo-Afghan war and has been a point of dispute ever since. The boundary known as the Durand Line divided the Pashtun and Baluch tribes living in Afghanistan from those living in India and what later became Pakistan.

The Cold War between the United States and the Soviet Union made Afghanistan a focus of that conflict. When the Soviet Union invaded Afghanistan in 1979 and put its Afghan supporters in charge, the United States supported the Afghan freedom fighters known as the *mujahidin,* funneling money through Pakistan. Barnett Rubin (1996) argues that no one paid more for the U.S. Cold War victory than Afghanistan and its people. The political maneuverings of Afghan and foreign leaders "may inspire cynicism or repulsion, but millions of unknown people scarified their homes, their land, their cattle, their health, their families, with barely hope of success or reward, at least in this world" (p. 21).

The case of the Nation of Islam and Afghanistan shows that the larger social, economic, and political context must be incorporated into any discussion of the role of religion in shaping human affairs. This line of investigation was particularly interesting to Max Weber, who examined how religious beliefs direct and legitimate economic activity.

Max Weber: The Interplay Between Economics and Religion

Max Weber was interested in understanding the role of religious beliefs in the origins and development of **modern capitalism**,

> a form of economic life which involved the careful calculation of costs and profits, the borrowing and lending of money, the accumulation of capital in the form of money and material assets, investments, private property, and the employment of laborers and employees in a more or less unrestricted labor market. (Robertson 1987:6)

In his book *The Protestant Ethic and the Spirit of Capitalism,* Weber (1958) asks why modern capitalism emerged and flourished in Europe rather than in China or India (the two dominant world civilizations at the end of the sixteenth century). He also asked why business leaders and capitalists in Europe and the United States were overwhelmingly Protestant. To answer these questions, Weber studied the major world religions and some of the societies in which these religions were practiced. He focused on understanding how

Modern capitalism "A form of economic life which involved the careful calculation of costs and profits, the borrowing and lending of money, the accumulation of capital in the form of money and material assets, investments, private property, and the employment of laborers and employees in a more or less unrestricted labor market" (Robertson 1987:6).

norms generated by different religious traditions influenced the adherents' economic orientations and motivations.

On the basis of his comparisons, Weber concluded that a branch of Protestant tradition — Calvinism — supplied a "spirit" or ethic that supported the motivations and orientations that capitalism required. Unlike other religions that Weber studied, Calvinism emphasized **this-worldly asceticism** — a belief that people are instruments of divine will and that their activities are determined and directed by God. Consequently, people glorify God when they accept the task assigned to them and carry them out in exemplary and disciplined fashion and when they do not indulge in the fruits of their labor (that is, when they do not use money to eat or drink or otherwise relax to excess). In contrast, Buddhism, a religion that Weber defined as the Eastern parallel and opposite of Calvinism, "emphasized the basically illusory character of worldly life and regarded release from the contingencies of the everyday world as the highest religious aspiration" (Robertson 1987:7).

The Calvinists conceptualized God as all-powerful and all-knowing; also they emphasized **predestination**, the belief that God has foreordained all things, including the salvation or damnation of individual souls. According to this doctrine, people could do nothing to change their fate. To compound matters, only relatively few people were destined to attain salvation.

Weber maintained that such beliefs created a crisis of meaning among adherents as they tried to determine how they were to behave in the face of their predetermined fate. Such pressures led them to look for concrete signs that they were among God's chosen people, destined for salvation. Consequently, accumulated wealth became an important indicator of whether one was among the chosen. At the same time, this-worldly asceticism "acted powerfully against the spontaneous enjoyment of possessions; it restricted consumption, especially of luxuries" (Weber 1958:171). Frugal behavior encouraged people to accumulate wealth and make investments, important actions for the success of capitalism.

This calculating orientation was not an official part of Calvinist doctrine per se. Rather, this orientation grew out of and was supported by asceticism and predestination. In view of this distinction, it is important that we do not misread the role that Weber attributed to the Protestant ethic in supporting the rise of a capitalistic economy. According to Weber, the ethic was a significant ideological force; it was not the sole cause of capitalism but "*one* of the causes of *certain aspects* of capitalism" (Aron 1969:204). Unfortunately, many people who encounter Weber's ideas overestimate the importance he assigned to the Protestant ethic for achieving economic success, and they draw a conclusion that Weber himself never reached: the reason that some groups and societies are disadvantaged is simply that they lack this ethic.

Finally, let us remember that Weber was writing about the origins of industrial capitalism, not about the form of capitalism that exists today, which heavily emphasizes consumption and self-indulgence. Weber maintained that once capitalism was established, it would generate its own norms and would become a self-sustaining force. In fact, Weber argued that "[c]apitalism produces a society run along machine-like, rational procedures without inner meaning or value and in which men operate almost as mindless cogs" (Turner 1974:155). In such circumstances, religion becomes increasingly insignificant in maintaining the capitalist system. Some sociologists believe that industrialization and scientific advances cause society to undergo unrelenting secularization, a process in which religious influences become increasingly irrelevant not only to economic life but also to most aspects of social life. On the other hand, some argue that as religion becomes less relevant to economic and social life in general, a significant number of people become fundamentalist —that is, they seek to reexamine their religious principles in an effort to identify and return to the most fundamental and basic principles (from which believers have departed) and to hold those principles up as the definitive and guiding blueprint for life.

Two Opposing Trends: Secularization and Fundamentalism

Secularization and fundamentalism are processes that have become increasingly popular in the recent past. Each has grown in spite of the other's growth, or possi-

This-worldly asceticism A belief that people are instruments of divine will and that their activities are determined and directed by God.

Predestination The belief that God has foreordained all things, including the salvation or damnation of individual souls.

bly in opposition to it. This section examines these two opposing trends.

Secularization

In the most general sense, **secularization** is a process by which religious influences on thought and behavior are reduced. It is difficult to generalize about the causes and consequences of secularization because they vary across contexts. Americans and Europeans associate secularization with an increase in scientific understanding and technological solutions to everyday problems of living. In effect, science and technology assume roles once filled by religious belief and practice. Most Muslims, on the other hand, do not attribute secularization to science or to modernization; there are many devout Muslims who are physical scientists. In a National Public Radio interview, a Muslim geologist explained that Islam was compatible with science:

> I realized that there is something — there's a glue, there's a creator—behind all those marvelous facts of life in the field of geology, of chemistry. [I had] to think and rethink about my attitude towards life, and that there should be a creator and a very wise creator who has already given all of us laws, and produced all these facts of life. (National Public Radio 1984c)

From a Muslim perspective, secularization is a Western-imposed phenomenon — specifically, a result of exposure to what many people in the Middle East consider the most negative of Western values. This point is illustrated by the observations of a Muslim who attended college in Britain and by a Muslim artist:

> If I did not watch out [while I was in college], I knew that I would be washed away in that culture. In one particular area, of course, was exposure to a society where free sexual relations prevailed. There you are not subject to any control, and you are faced with a very serious challenge, and you have to rely upon your own strength, spiritual strength to stabilize your character and hold fast to your beliefs. (National Public Radio 1984b)

> You can't be against foreign influence because art and culture and science are things that you learn from people and [that] we exchange. It depends on what kind of foreign influence is being exerted. For example, if we speak of the United States, the U.S. has a tradition of science and of art and culture . . . which can be extremely useful and can enrich our own culture if there is an exchange. But the type of culture that we are getting from the U.S. at the moment is the *Dallas* series, the cowboy serials, crimes, commodity values, big cars, luxury, things like that. This kind of culture can't help. (National Public Radio 1984b)

Secularization A process by which religious influences on thought and behavior are reduced.

Subjective secularization A decrease in the number of people who view the world and their place in it from a religious perspective.

We can speak of two kinds of secularization: subjective and objective (Berger 1967). **Subjective secularization** is a decrease in the number of people who view the world and their place in it from a religious perspective. In other words, there is a shift in paradigms from a religious understanding of the world, grounded in faith, to an understanding grounded in observable evidence and the scientific method. In the face of uncertainty, then, secular thinkers do not turn to religion or to a supernatural[4] power that they have come to view as a distant, impersonal, and even inactive phenomenon. Consequently, they come to believe less in direct intervention by the supernatural and to rely more strongly on human intervention or scientific explanation:

> Consider the case of the lightning rod. For centuries, the Christian church held that lightning was the palpable manifestation of divine wrath and that safety against lightning could only be conforming to divine will. Because the bell towers of churches and cathedrals tended to be the only tall structures, they were the most common targets of lightning. Following damage or destruction of a bell tower by lightning, campaigns were launched to stamp out local wickedness and to raise funds to repair the tower. Ben Franklin's invention of the lightning rod caused a crisis for the church. The rod demonstrably worked. The laity began to demand its installation on church towers — backing their demands with a threat to withhold funds to restore the tower should lightning strike it. The church had to admit either that Ben Franklin had the power to thwart divine retribution or that lightning was merely a natural phenomenon. (Stark and Bainbridge 1985:432–433)

Considerable debate exists over the extent to which subjective secularization is taking place. Data collected by the Gallup Organization over the past twenty years show little change in the degree of importance that Americans say they assign to religion. More

[4]The supernatural are forces that can defy the laws of nature and affect the outcome of an event.

When Ben Franklin flew his kite in a lightning storm, was he intervening in divine events? Some people think so.
©North Wind Pictures

than 90 percent have a religious preference, and almost 70 percent are members of one of the more than 250,000 places of worship within the United States (White House Press Release 1995); approximately 40 percent attend church weekly; and almost 60 percent state that religion is very important in their lives (Gallup and Castelli 1989). Polls also show that almost 80 percent of Americans are "sometimes very conscious of God's presence" and that more than 80 percent agree that "even today, miracles are performed by the power of God" (Gallup and Castelli 1989:58).

These findings, however, do not necessarily prove that religion influences behavior and thought as strongly today as in the past. To reach such a conclusion, we would have to determine whether people are as willing now as they were in the past to leave matters in God's hands. We might speculate, for example, that people were more willing to put their trust solely in God before the advent of sophisticated medical technology. Today, people who take such a position may be viewed as ignorant, stubborn, or uncaring. Consider the publicity surrounding cases of children who have died of diabetes, meningitis, tumors, and obstructed bowels because their parents believed that physical ailments could be cured by spiritual means and did not seek medical treatment for the children. The publicity and the accompanying outrage suggest that although an overwhelming number of Americans believe in miracles, their belief is not strong enough to leave medical matters in the hands of supernatural forces. On the other hand, although science may have the edge over religion in the harder aspects of medicine, religion still seems to have the advantage in areas of emotional healing and long-term support.

Objective secularization is the decline in the control of religion over education, medicine, law, and politics and the emergence of an environment in which people are free to choose from many equally valid religions the one to which they wish to belong. For several reasons, it is difficult to make generalizations about this type of secularization and the extent to which it exists in a society. First, objective secularization is not an even, inevitable process that eventually results in the dominance of science and rationality and in the end of religion. No matter how strongly science comes to dominate human life, it cannot provide solace for the inescapable problems of existence — birth, death, illness, aging, injustice, tragedy, and suffering — nor can it formulate a coherent plan for life (Stark and Bainbridge 1985:431).

Another reason that it is difficult to generalize about objective secularization is that no aspect of political life seems to be fully secularized. For example, the American currency includes the slogan "In God We Trust"; it does not say "In Federal Reserve We Trust" (Haddad 1991). Additionally, in the United States, public schools are not fully secularized because religious expression is permitted in the schools.

Sometimes religious affiliations mask more important factors such as economic and political inequalities between conflicting parties. On the other hand, because religious differences are used to distinguish the opposing groups and to mobilize people to fight each other, we also must investigate the qualities of religion that make it a factor in such violent conflicts (Stavenhagen 1991). It is especially important to examine how religion influences people's behavior in a specific con-

Objective secularization The decline in the control of religion over education, medicine, law, and politics and the emergence of an environment in which people are free to choose from many equally valid religions the one to which they wish to belong.

It was only in the 1950s that the phrase "under God" was inserted into the U.S. Pledge of Allegiance that most schoolchildren were required to recite each day. At the same time, the United States started printing "In God We Trust" on its coins. American demonization of Communism in the Cold War played a role in these decisions—over the objection of people who saw them as violations of the principle of separation between church and state.

©R. W. Carter/SuperStock

text and how people use it to legitimize actions. In other words, the religion itself is usually not the problem; more often than not, the problem lies with the way people choose to use religion to justify actions.

In this regard, one of the major challenges for the courts that must interpret and enforce the intent of these legal documents is distinguishing situations in which people use religion to justify actions that interfere with others' constitutional rights from situations in which people's freedom to worship and practice their religion are violated. The line is not always clear.

Fundamentalism

In "Popular Conceptions of Fundamentalism," anthropologist Lionel Caplan (1987) offers his readers one of the clearest overviews of a complex religious phenomenon — **fundamentalism**, a belief in the timeless nature of sacred writings and a belief that such writings apply to all kinds of environments. This term is applied popularly to a wide array of religious groups in the United States and around the world, including the Moral Majority in the United States, Orthodox Jews in Israel, and various Islamic groups in the Middle East.

Fundamentalism A belief in the timeless nature of sacred writings and a belief that such writings apply to all kinds of environments.

Religious groups labeled as fundamentalist are usually portrayed as "fossilized relics . . . living perpetually in a bygone age" (Caplan 1987:5). Americans frequently employ this simplistic analysis to explain events in the Middle East, especially the cause of the political turmoil that threatens the interests of the United States (including its need for oil). But such oversimplification misrepresents fundamentalism and cannot explain the widespread appeal of contemporary fundamentalist movements within several of the world's religions.

The Complexity of Fundamentalism Fundamentalism is a more complex phenomenon than popular conceptions lead us to believe. First, it is impossible to define a fundamentalist in terms of age, ethnicity, social class, or political ideology because fundamentalism appeals to a wide range of people. Moreover, fundamentalist groups do not always position themselves against those in power; they are just as likely to be neutral or to support existing regimes fervently. Perhaps the most important characteristic of fundamentalists is the belief that relationship with God, Allah, or some other supernatural force provides answers to personal and social problems. In addition, fundamentalists often wish to "bring the wider culture back to its religious roots" (Lechner 1989: 51).

Caplan (1987) suggests a number of other traits that seem to characterize fundamentalists. First, fundamentalists emphasize the authority, infallibility, and timeless truth of sacred writings as a "definitive blueprint" for life (p. 19). This characteristic should not be taken to mean that a definitive interpretation of sacred writings actually exists. Any sacred text has as many interpretations as there are groups that claim it as their blueprint. For example, even members of the same fundamentalist organization disagree about the true meaning of the texts they follow.

Second, fundamentalists usually conceive of history as a "process of decline from an original ideal state, [and] hardly more than a catalog of the betrayal of fundamental principles" (p. 18). They conceptualize human history as a "cosmic struggle between good and evil": the good is a result of dedication to principles outlined in sacred scriptures, and the evil is an outcome of countless digressions from sacred principles. To fundamentalists, truth is not a relative phenomenon: it does not vary across time and place. Truth is unchanging and is knowable through the sacred texts.

Third, fundamentalists do not distinguish between the sacred and the profane in their day-to-day lives. All areas of life, including family, business, and leisure, are governed by religious principles. Religious behavior, in their view, does not take place only in a church, a mosque, or a temple.

Fourth, fundamentalist religious groups emerge for a reason, usually in reaction to a perceived threat or crisis, real or imagined. Consequently, any discussion of a particular fundamentalist group must include some reference to an adversary.

Fifth, one obvious concern of fundamentalists is to reverse the trend toward gender equality, which they believe is symptomatic of a declining moral order. In fundamentalist religions, women's rights often are subordinated to ideals that the group considers more important to the well-being of the society, such as the traditional family or the right to life. Such a priority of ideals is regarded as the correct order of things.

Islamic Fundamentalism In *The Islamic Threat: Myth or Reality?* professor of religious studies John L. Esposito maintains that most Americans' understanding of fundamentalism does not apply very well to contemporary Islam. The term *fundamentalism* has its roots in American Protestantism and the twentieth-century movement that emphasizes the literal interpretation of the Bible. Fundamentalists are portrayed as static, literalist, retrogressive, and extremist. Just as we cannot apply the term *fundamentalism* to all Protestants in the United States, we cannot apply it to the entire Muslim world, especially when we consider that Muslims make up the majority of the population in at least forty-five countries.

Esposito believes that a more fitting term is **Islamic revitalism** or *Islamic activism*. The form of Islamic revitalism varies from one country to another but seems to be characterized by the following themes:

> A sense that existing political, economic, and social systems had failed; a disenchantment with and at times a rejection of the West; a quest for identity and greater authenticity; and the conviction that Islam provides a self sufficient ideology for state and society, a valid alternative to secular nationalism, socialism, and capitalism. (Esposito 1992:14)

In "Islam in the Politics of the Middle East," Esposito (1986) asks, "Why has religion [specifically Islam] become such a visible force in Middle East politics?" He believes that Islamic revitalism is a "response to the failures and crises of authority and legitimacy that have plagued most modern Muslim states" (p. 53). Recall that after World War I, France and Britain carved up the Middle East into nation-states, with the boundaries drawn to meet the economic and political needs of Western powers. Lebanon, for example, was created in part to establish a Christian tie to the West; Israel was created as a refuge for persecuted Jews when no country seemed to want them; the Kurds received no state; Iraq was virtually landlocked; and resource-rich territories were incorporated into states with very sparse populations (Kuwait, Saudi Arabia, the Emirates). Many of the leaders who took control of these foreign creations were viewed by the citizens "as autocratic heads of corrupt, authoritarian regimes that [were] propped up by Western governments and multinational corporations" (p. 54).

When Arab armies from six states lost "so quickly, completely, and publicly" to Israel in 1967, Arabs were forced to question the political and moral structure of their societies (Hourani 1991:442). Had the leaders and the people abandoned Islamic principles or deviated too far? Could a return to an Islamic way of life restore confidence to the Middle East and give it an identity independent of the West? Questions of social justice also arose. Oil wealth and modernization policies had led to rapid increases in population and urbanization and to a vast chasm between the oil-rich countries such as Kuwait and Saudi Arabia and the poor, densely populated countries such as Egypt, Pakistan, and Bangladesh. Western capitalism, which was believed to be one of the primary forces behind these trends, seemed to be blind to social justice, promoting unbridled consumption and widespread poverty. Marxist socialism (a godless alternative) likewise had failed to produce social justice. It is no wonder that the Taliban and other Muslim groups in Afghanistan reject Western capitalism and Marxist socialism because the disintegration of Afghanistan is a product of the Cold War between the United States and the former Soviet Union.

Islamic revitalism "A sense that existing political, economic, and social systems had failed; a disenchantment with and at times a rejection of the West; a quest for identity and greater authenticity; and the conviction that Islam provides a self sufficient ideology for state and society, a valid alternative to secular nationalism, socialism, and capitalism" (Esposito 1992:14).

For many people, Islam offers an alternative vision for society. According to Esposito (1986), Islamic activists (who are of many political persuasions, from conservative to militant) are guided by five beliefs:

1. Islam is a comprehensive way of life relevant to politics, state, law, and society.
2. Muslim societies fail when they depart from Islamic ways and follow the secular and materialistic ways of the West.
3. An Islamic social and political revolution is necessary for renewal.
4. Islamic law must replace Western-inspired or Western-imposed laws.
5. Science and technology must be used in ways that reflect Islamic values to guard against the infiltration of Western values.

Muslim groups differ dramatically as to how quickly and by what methods these principles should be implemented. Most Muslims, however, are willing to work within existing political arrangements; they condemn violence as a method of bringing about political and social change.

The information presented in this section points to the interplay of religion with political, economic, historical, and other social forces. Fundamentalism cannot be viewed in simple terms. Any analysis must consider the broader context. A focus on context allows us to see that fundamentalism can be a reaction to many events and processes, including secularization, foreign influence, failure or crisis in authority, the loss of a homeland, and rapid change.

Islamic fundamentalism includes the idea that it is wrong to lend money and require interest to be paid. Recently, Islamic banks have found ways to attract money that they can then make available for investment without breaking Islamic law. For example, depositors to the bank can be regarded as owning shares in the bank and receive a legitimate dividend on their ownership. To accommodate Muslims, some globally conscious non-Islamic banks now offer similar arrangements.

©V. Riviere/Sygma

Summary and Implications

In this chapter, sociological concepts and perspectives have helped us see that the civil wars in Afghanistan and the recent initiative by the Taliban cannot be reduced to the actions of fanatics acting from primitive and irrational religious convictions. Religion is important in understanding the nature of these civil wars, but its explanatory importance depends on understanding how religion is used rather than on knowing the religious groups to which the majority in each faction belongs.

Durkheim's ideas about the nature of religion, despite their shortcomings, allow us to consider religion's functions for the individual and the group. Most important, religion is a rich and seemingly endless variety of responses to problems of human existence. People embrace religion in the face of uncertainty to cope with misfortunes and injustices, when they have a great emotional investment in securing a successful outcome, and to find answers to agonizing questions about the meaning of life and death.

In addition, religion contributes to group unity and solidarity. Whenever a group of people has a strong conviction (no matter what kind of group and no matter what the conviction), the conviction almost always takes on a religious character. Religious gatherings and affiliations become ways of affirming beliefs and mobilizing the group members to uphold their beliefs, especially if they are threatened. When religion is used in this way, it can unite a community of worshipers so strongly that they would be willing to destroy those who do not share their views. The convictions that religion is called on to legitimate may be honorable or unprincipled.

Karl Marx was concerned about religion's repressive, constraining, and exploitative qualities. He believed that religious doctrines can be used to turn peo-

Depicted here are American Muslims at prayer. How does this picture compare with the images of Islam that you are likely to see in the mass media?

©John Van Hasselt/Sygma

ple's attention away from unjust political and economic arrangements and that they rationalize and defend the political and economic interests of the dominant classes. For example, those in political power may use religion to unite the society in war against another society or to dominate in some other way. Marx ignored the possibility that the oppressed have used religion as a vehicle by which to protest or to work to change the existing social and economic inequities.

Max Weber alerts us to yet another way in which religion affects economic life by outlining the role that religious beliefs played in the origins and development of modern capitalism. Yet, at the same time Weber argued that once capitalism was established, it would generate its own value-rational logic, which in turn would make religion less relevant to economic activity. (See Chapter 7 for a discussion of value-rational action.) In fact, the advances in science and technology that accompany modern capitalism cause society to become secularized. Ironically, the same forces that support secularization processes also support the rise of fundamentalism.

The uses to which religion are turned show that religion is a multifaceted and complex phenomenon that cannot be discussed in isolation. To understand the wars in Afghanistan, we must understand the larger social, economic, and political context. With this information in hand, we can examine how religion shapes people's behavior and how it legitimates their actions. We can also see how "civil religion" motivated the United States and the Soviet Union to fight for their respective convictions in the same way Islamic principles motivated the *mujahidin*.

Key Concepts

Use this outline to organize your review of the key chapter ideas.

Religion
- **Church**
- **Ecclesiae**
- **Denominations**
- **Sects**
  - **Established sects**
- **Cults**

Civil religion

Rituals

Sacred
- **Sacramental religions**
- **Prophetic religions**
- **Mystical religions**

Profane

Fundamentalism
Secularization
Objective secularization
Subjective secularization
Modern capitalism
Predestination
This-worldly asceticism
Class society

internet assignment

The International Religious Foundation (http://www.rain.org/~origin/wscon.html) is an organization dedicated to promoting world peace through religious dialogue and cooperation. Select "Introduction" for more information on the project's purpose. The foundation has posted the scriptures of several of the major world religions according to how each religion views (1) ultimate reality; (2) divine law, truth, and cosmic principles; (3) the purpose of human life; (4) life beyond death; and (5) the human condition – in addition to sixteen other topics.

Also browse the Web site "Description of Fifty-seven Religions, Faith Groups, and Ethical Systems" maintained by the Ontario Consultants on Religious Tolerance (http://web.canlink.com/ocrt/var_rel.htm), an organization whose aims are to promote tolerance of minority religions, offer useful information on controversial religious topics, and expose hatred and misinformation about any religion.

Compare and contrast the views of Islam with those of one or two other religions. Are there any differences in views or principles that would represent grounds for violent conflict? Explain. What does this say about the role of religion in conflicts?

15

Social Change

With Emphasis on the Internet

Woman with laptop computer. ©David Lissy/The Picture Cube, Inc.

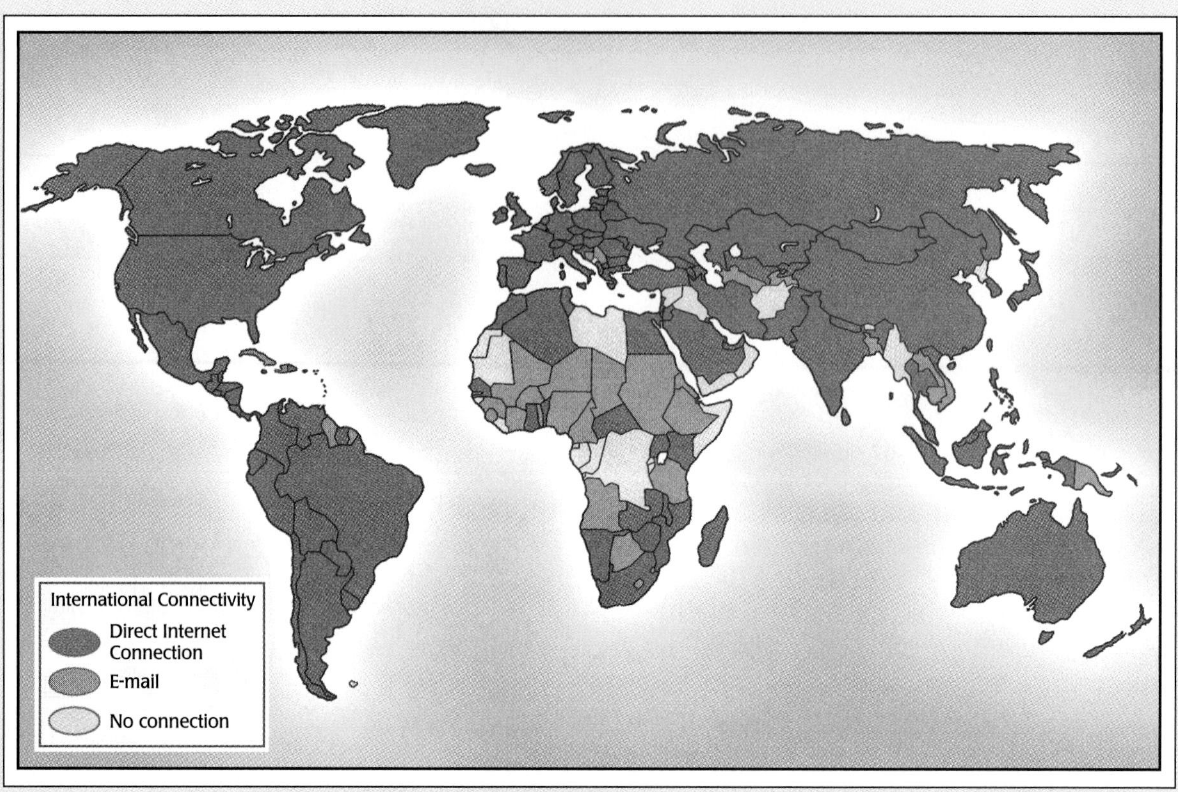

Adapted from a map Copyright ©1995 by Larry Landweber and the Internet Society. Unlimited permission to copy or use is granted subject to inclusion of the copyright notice.

How Worldwide Is the Web?

This map gives an idea of how "worldwide" the World Wide Web actually was as of June 1996, according to a copyrighted report on the Internet from Larry Landweber and the Internet Society. Countries shown in purple had Internet access via telephone and other media such as the "backbone" data transmission lines and relay systems that now connect major computing centers around the world. Green countries had E-mail only. Yellow countries had neither telephone E-mail nor access to the Internet. Of course, within each country that had connections, some regions and people had more and better connections than others.

This map is about power and opportunity in a time of rapid worldwide social change. This chapter is about the sociological study of social change in the context of technological innovation. It is about vast changes at work today whose ultimate effects on people and the organization of social life we must work hard to foresee and understand.

One advantage of the Internet over the print world is that it allows authors to update information whenever they see fit. For example, you may be able to get an update to the data shown on the map at http://www.isoc.org/images/mapv15.gif.

Consider the following newspaper and magazine headlines:

> *"Third-World Pioneers Use Net to Promote Business"*
>
> *"Internet Is Becoming an Essential Tool"*
>
> *"Internet Snares First Criminal as FBI Traps Bank Robber in the Web"*
>
> *"Teachers Explore Online Options"*
>
> *"More Michigan Colleges Bring the Classroom to Cyberspace"*
>
> *"Parishioners Flock Back to Fold as Some Churches Evangelize On-Line"*
>
> *"Judge Turns Back Laws to Regulate Internet Decency"*
>
> *"Are We Creating Internet Introverts?"*

Most of us have seen or heard these kinds of headlines, and even if we have not accessed the Internet, we have heard many things about its potential and peril. Taken together, such headlines suggest that the Internet "is coming straight at us with the speed and momentum of a locomotive or at full steam. We can either prepare to jump on board or be mowed down by it" (Crawford 1995). Sociology offers a framework that can help us conceptualize these changes and place them in a larger social and historical context.

Why Focus on the Internet?

This chapter is about social change and will focus on the Internet—a technology that is expected to set into motion social, economic, and political changes equivalent to those revolutionary changes triggered by the printing press. The **Internet** is a vast network of computer networks connected to one another via special software and phone, fiber-optic, or other type of lines that has the potential to connect millions, even billions, of computers and their users on a global scale. It also has the potential to provide users access to every word, image, or sound that has ever been written, produced, filmed, photographed, painted, or otherwise recorded (Berners-Lee 1996).

Although it is virtually impossible to list all the changes associated with (and expected to be associated with) the Internet, we can say that, for those who can afford it, the Internet will (1) speed up old ways of doing things, (2) give individuals access to the equivalent of a printing press, (3) allow people to bypass the formalized hierarchy devoted to controlling the flow of information, (4) change how students learn, and (5) provide a vehicle by which the lives of people around the world will become more intertwined as they seek to exchange information and mobilize to define and solve a wide range of social problems. In this regard, the Internet facilitates "globalization-from-above" as well as "globalization-from-below."

Globalization-from-above connects those people around the world with educational, economic, and political advantages, excluding those who are not so advantaged. **Globalization-from-below**, however, involves interdependence at the grassroots level that aims to protect, restore, and nurture the environment; to enhance ordinary people's access to the basic resources they need to live a dignified existence; to democratize local, national, and transnational political institutions; and to ease tensions and prevent violent conflict between power centers and authority structures (Brecher, Childs, and Cutler 1993).

Sociologists define **social change** as any significant alteration, modification, or transformation in the organization and operation of social life. Social change is an important topic in the discipline of sociology. In fact, it is fair to say that sociology emerged as a discipline attempting to understand social change. Recall that the early sociologists were obsessed with understanding the nature and consequences of the Industrial Revolution—an event that triggered dramatic and seemingly endless changes in every area of social life. As a result, the discipline of sociology has evolved to encompass a wide range of rich concepts and theories that can be applied to the study of social change.

When sociologists study change, they first must identify the aspect of social life they wish to study that has changed or is undergoing change. The list of possible topics is virtually endless. Some examples include changes in the division of labor; changes in the meaning of what constitutes deviance; changes in how people communicate with each other; changes in the amount of goods and services that people produce, sell, or buy from others; and changes in the size of world population and in the average life span (Martel 1986). Upon identifying a topic, sociologists ask at least two key questions: What factor(s) are causing that change? What are the consequences of that change for social life?

In this chapter, we look at the concepts and theories that sociologists use to answer these complex questions. We give as an example the Internet—specifically, the historical events that led to its invention and the consequences of that invention for social life.

Internet A vast network of computer networks connected to one another via special software and phone, fiber-optic, or other type of lines with the potential to connect millions, even billions, of computers and their users on a global scale.

Globalization-from-above A term describing the Internet's ability to connect those people around the world with educational, economic, and political advantages, excluding those who are not so advantaged.

Globalization-from-below Interdependence at the grassroots level that aims to protect the environment, enhance ordinary people's access to basic resources, democratize political institutions, and ease tensions and prevent violent conflict between power centers and authority structures.

Social change Any significant alteration, modification, or transformation in the organization and operation of social life.

Social Change: Causes and Consequences

When we think about the factors that cause a specific social change, we usually cannot name a single factor as the cause of an identified change. More often than not, change results from a sequence of events. An analogy may help clarify this point. Suppose that a wide receiver, after catching the football and running fifty yards, is tackled at the five-yard line by a cornerback. One could argue that the cornerback caused the receiver to fall. Such an account, however, would not fully explain what actually happened. For one thing, a tackle is not the act of one person; "there is present a simultaneous conflict of forces" (Mandelbaum 1977:54) between the tackler (who is attempting to seize and throw his or her weight onto the person with the ball) and the wide receiver (who is doing everything possible to elude the tackler's grasp). To complicate matters, the wide receiver and the tackler are each members of a team, and their teammates' actions help determine how the play develops and ends. Similarly, we can argue that the Internet is a product of the Cold War and a seemingly endless sequence of interrelated events.

The Cold War is the name given to the political tension and military rivalry that existed between the United States and the former Soviet Union from approximately 1945 until its symbolic end on November 9, 1989, the day the Berlin Wall fell. The Cold War included an arms race, an intense and ongoing buildup by the Soviet Union and the United States in which each side competed to match and surpass any advances made by the other in the quantity and technological quality of weapons on the grounds that it was necessary to contain the spread of the other side's economic and political system. Both countries created a vast, entrenched armaments industry; both made the research and development of weapons (especially nuclear weapons)[1] their highest priorities. U.S. government leaders had pulled together scientists from three sectors—military, industrial, and academic—to coordinate their research and expertise in the Cold War effort. Because these scientists worked in offices and labs located all around the United States, Defense Department

[1]During the Cold War, the number of nuclear weapons grew to 50,000.

Historical factors can spur or retard the pace of innovation. The Cold War stimulated enormous U.S. and Soviet governmental support for science and technological research, including the "conquest of space." When the statues of Lenin came down in Europe (this one was in East Germany), one driving motivation for innovation also came down. But other factors, such as growing global competition in trade, keeps the pressure on.

officials worried about what could happen if a military lab, defense contractor, or university site were attacked. Those in power realized they needed a computer network that would allow information stored at one site to be transferred to another site in the event of an attack, especially a nuclear attack. At the same time, the computer network had to be designed such that if one or more parts of the network failed, the other parts could continue to operate. Such a design meant that there could be no central control over the network. After all, if central control is knocked out, the entire network crashes. Thus, the Internet began in the late 1960s as ARPANET (Advanced Research Projects Agency), linking four universities: the University of California–Los Angeles (UCLA), University of California–Santa Barbara, the University of Utah, and Stanford University.

This brief overview of the history of the Internet suggests that it was designed to (1) transfer information from one site to another quickly and efficiently in the event of war and (2) create an information-sharing system absent central control. Even though the story of the Internet is embedded within a virtually endless stream of interrelated events, we can identify some key types of factors that resulted in the Internet's invention and the subsequent changes this information technology triggered and has the potential to trigger:

- Innovations—the development of something new, whether an idea, a practice, or a tool
- The actions of leaders or of people in positions of authority
- Conflict, or the competition over shares of wealth, power, prestige, and other valued resources
- Capitalism — a continually evolving profit-driven economic system

Sociologists define innovations, the power elite, conflict, and capitalism as some of the most important agents of change in general. Thus, their importance for understanding social change goes beyond the Internet, and we will use them as the conceptual framework for this chapter.

Innovations

Innovation is the invention or discovery[2] of something new—an idea, a process, a practice, a device, or a tool. Innovations can be classified broadly in one of two categories—basic or improving. The distinction between the two is not always clear-cut. **Basic innovations** are

Innovation The invention or discovery of something new—an idea, a process, a practice, a device, or a tool.

Discovery The uncovering of something that had existed before but had remained hidden, unnoticed, or undescribed.

Basic innovations Revolutionary, unprecedented, or ground-breaking inventions that are the cornerstones for a wide range of applications.

[2]A **discovery** is the uncovering of something that had existed before but had remained hidden, unnoticed, or undescribed.

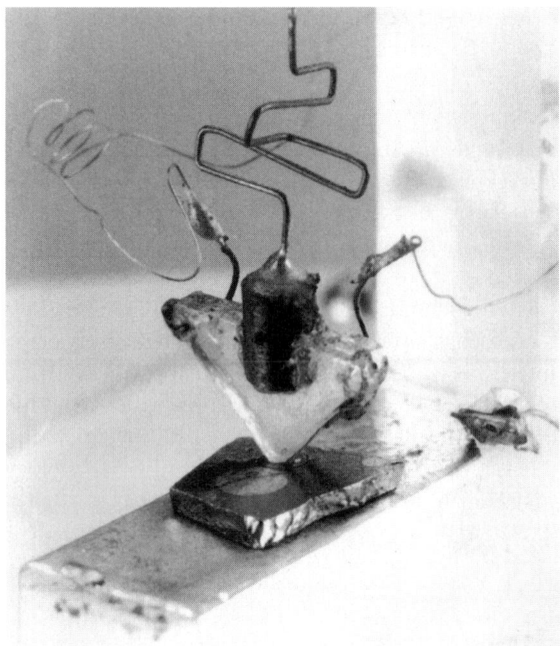

This first semiconductor (transistor) is an example of a basic innovation in that it became the cornerstone for a wide range of applications.

revolutionary, unprecedented, or ground-breaking inventions that are the cornerstones for a wide range of applications. Examples of basic innovations include the first mainframe computer, the first PC (personal computer),[3] computer languages, modems, disks, printers, and scanners.

Improving innovations, on the other hand, represent modifications of basic inventions to improve on them—that is, to make them smaller, faster, less complicated (that is, user-friendly) or more efficient, attractive, or profitable. Each upgrade of a personal computer's CPU (central processing unit) is an improving innovation that increases the power and speed of the machine to handle various tasks such as calculations per second, software applications, and information.

Innovations are sociologically significant because they change the ways in which people think, solve problems, and relate to one another. For example, the Internet has the potential to affect almost every area of life including friendship, job hunting, finding an address, reading the news, adopting, and finding an apartment. The following descriptions of Web sites illustrate:

- Résumé-O-Matic

  http://www.ypn.com/jobs/resumes/index.html

  Résumé-O-Matic makes it easy to post your résumé on the Internet. Simply answer the questions, and Résumé-o-Matic will turn your answers into a résumé that will be posted on the Web for a full year free of charge.

- Interactive Atlas

  http://www.pluto.esc.k12.in.us/home/caview/htm

  Mapquest posts the Interactive Atlas, an on-line service that allows users access to county- and city-level maps from over six continents. This atlas allows you to enter a street address along with its city, state, and ZIP code and to access a map of that address and surrounding streets and landmarks. A one-time free registration is required, which allows you to save and store maps for later use. A graphical browser is required to access this site.

- CReAte Your Own Newspaper (CRAYON)

  http://crayon.net/

  CRAYON is a service maintained by Pressence Incorporated for managing news. After completing a free registration, users select the news sources from which they want to draw information. News sources are available at the international (such as *This Week in Germany*), national (such as *USA Today Nationwide*), and local (such as the *Detroit News*) levels. Users may select information to be taken from specific sections, such as Sports or Weather, of the papers they choose. Users assign a name and motto to their paper and are given a URL address where they can read their paper daily. A graphical browser is required to access this site.

- CyberFriends

  http://dare.com/fr_main

  The purpose of CyberFriends is to provide an easy way for people sharing similar interests and professions to meet on the Internet. You simply fill out an application, which asks approximately eighteen optional and required questions (age, sex, nation-

Improving innovations Modifications of basic inventions to improve on them—that is, to make them smaller, faster, less complicated or more efficient, attractive, or profitable.

[3]The name of the first mainframe was ENIAC (Electronic Numerical Integrator and Computer). The name of the first PC was Altair (Noack 1997).

ality, hobbies, personal philosophy, and so on). In two or three weeks, your application will be posted free of charge. In the meantime you can search CyberFriends' listings for profiles and E-mail addresses for the names of people with whom you might like to correspond.

- Faces of Adoption

  http://www.inetcom.net/adopt

  The National Adoption Center (NAP) and Children Awaiting Parents (CAP) post photographs and descriptions of children awaiting adoption. The Internet serves as one of many avenues both organizations use to feature children. Other avenues include radio, TV (especially talk shows), and magazines.

- Rent•net (Online Apartment Guide)

  http://www.rent•net

  This Web site allows apartment hunters to shop for apartments (unfurnished and furnished) in fifty states, Canada, and five other foreign countries. Its database includes 1,157,000 apartment units. This site also offers a cost-of-living comparison between the city of current location and the city of destination. There is also information on self-storage facilities and other relocation services.

The Cultural Base and the Rate of Change

Anthropologist Leslie White (1949) maintains that once a basic or an improving innovation has been invented, it becomes part of the **cultural base**, which he defines as the number of existing inventions. The size of the cultural base determines the rate of change.

White defines an **invention** as the synthesis of existing inventions. For example, the first airplane was a synthesis of many preexisting inventions, including the gasoline engine, the rudder, the glider, and the wheel. PocketNet, a product of AT&T's Wireless Data Division, combines cell-phone, E-mail, and fax technology to create a device that allows owners to send and receive E-mail and redirect messages to a fax machine (Pulver 1997).

White suggested that the number of inventions in the cultural base increases geometrically — 1, 2, 4, 8, 16, 32, 64, and so on. (Geometric growth is equivalent to a state of runaway expansion.) He argues that if a new invention is to come into being, the cultural base must be large enough to support it. If the Wright brothers had lived in the fourteenth century, for example, they never could have invented the airplane because the cultural base did not contain the ideas, materials, and inventions to support its creation.

The seemingly runaway expansion, or increases in the volume, of new inventions prompted White (1949) to ask, Are people in control of their inventions, or do inventions control people? For all practical purposes, he believed that inventions control us. He supported this conclusion with two arguments. First, he suggested that the old adage "Necessity is the mother of invention" is naive: in too many cases, the opposite idea — that invention is the mother of necessity — is true. That is, an invention becomes a necessity because we find uses for that invention after it comes into being:

> We invent the automobile to get us between two points faster, and suddenly we find we have to build new roads. And that means we have to invent traffic regulations and put in stop lights [and build garages]. And then we have to create a whole new organization called the Highway Patrol — and all we thought we were doing was inventing cars. (Norman 1988:483)

Second, White (1949) argued that when the cultural base is capable of supporting an invention, that invention will come into being whether people want it or not.[4] White supported this conclusion by pointing to **simultaneous-independent inventions**, situations in which the same invention is created by two or more persons working independently of one another at about the same time (sometimes within a few days or months). He cited some 148 such inventions—including the telegraph, the electric motor, the microphone, the telephone, the microscope, the steamboat, and the airplane — as proof that someone will come along to make the necessary synthesis if the cultural base is ready to support a particular invention. In other words, inventions such as the light bulb and the airplane

[4]Physicist Robert Oppenheimer, who led the research and development of the atomic bomb in the 1940s but opposed the building of the hydrogen bomb in the 1950s, would have agreed with White on this idea. Oppenheimer said, "It is a profound and necessary truth that the deep things in science are not found because they are useful; they are found because it was possible to find them" (1986:11).

Cultural base The number of existing inventions.

Invention A synthesis of existing inventions.

Simultaneous-independent inventions Situations in which the same invention is created by two or more persons working independently of one another at about the same time.

Computers started out as room-sized machines invented to perform mathematical calculations for the Department of Defense and the Census Bureau. Today the processing power needed to perform a variety of tasks can be packed into spaces that can fit on a person's lap or the dashboard of a car.

Bettmann Archive (left); ©Mark Richards/PhotoEdit (right)

would have come into being whether or not Thomas Edison and the Wright brothers (the people we traditionally associate with these inventions) had ever been born. According to White's conception, inventors may be geniuses, but they also have to be born at the right place and the right time — that is, in a society with a cultural base sufficiently developed to support their invention.

White's theory suggests that if the parts are present, someone eventually will come along and put them together. The implications are that people have little control over whether an invention should come into being and that they adapt to inventions after the fact. Sociologist William F. Ogburn (1968) calls this failure to adapt to a new invention cultural lag.

Cultural Lag

In his theory of cultural lag, Ogburn (1968) distinguishes between material and nonmaterial culture. Recall from Chapter 4 that material culture includes tangible creations or objects — including resources (oil, trees, land), inventions (paper, guns), and systems (factories, sanitation facilities) — that people have created or, in the case of resources such as oil, have identified as having the properties to serve a particular purpose. Nonmaterial culture, on the other hand, includes intangible creations such as beliefs, norms, values, roles, and language.

Although Ogburn maintains that both material and nonmaterial culture are important agents of social change, his theory of cultural lag emphasizes the material component, which he suggests is the more important of the two. Ogburn believes that one of the most urgent challenges facing people today is adapting to material innovations in thoughtful and constructive ways. He uses the term **adaptive culture** for the portion of the nonmaterial culture (norms, values, and beliefs) that adjusts to material innovations. Such adjustments are not always immediate. Sometimes they take decades; sometimes they are never made. Ogburn uses the term **cultural lag** to refer to a situation in which adaptive culture fails to adjust in necessary ways to a material innovation.

Ogburn, however, is not a **technological determinist** — someone who believes that human beings have no free will and are controlled entirely by their material innovations. For one thing, he notes people do not adjust to new material innovations in predictable and unthinking ways; rather, they choose to create them and after they create them they choose how to use them (see "Pooling Resources and Talent to Create the Internet"). For example, according to Nobel Prize

Adaptive culture The portion of the nonmaterial culture (norms, values, and beliefs) that adjusts to material innovations.

Cultural lag A situation in which adaptive culture fails to adjust in necessary ways to a material innovation.

Technological determinist Someone who believes that human beings have no free will and are controlled entirely by their material innovations.

U.S. In Perspective

Pooling Resources and Talent to Create the Internet

Today, the United States leads the world in many of the technologies that make the information superhighway possible—semiconductors, fiber optics, high-speed switches, supercomputing, advanced software, etc. One reason for this is the billions of dollars that the Federal government, especially the Defense Department and NASA, have invested in information technology over the last forty years. Many of the most important technological breakthroughs in computing and communications are due in part to a Federal research grant or a Federal contract.

The very first electronic computer, the ENIAC, was the product of Federal funding. There are dozens of similar examples. A recent study by the National Research Council on the High-Performance Computing and Communications Program documented how Federal research and development funding played a key role in the development of timesharing, computer networking, workstations, computer graphics, the mouse, the windows interface, VLSI circuit design, RISC computing, parallel computing, and digital libraries.

There has been a very effective partnership between government, academia, and industry. By providing funding, the Federal government enables academic researchers to do the kind of long-term research that businesses can't afford to do. Even more important, that funding makes possible the training of the next generation of scientists and engineers. The research results and the talent from our universities are then applied by the private sector to develop the new products and new processes that they need to stay competitive in world markets.

The Internet provides an excellent example of how the Federal government can work with universities and industry to develop new technologies that can create whole new industries—and thousands of new, high-paying jobs. The ARPANET, the precursor of the Internet, began as a research project funded by the Defense Department's Advance Research Project Agency in 1969. . . . In 1986, the National Science Foundation started the NSFNET program in order to expand the ARPANET and connect more university researchers to the network. The network grew in both size and speed. By the early 1990s, the NSFNET backbone network was operating at 45 megabits per second, almost 1,000 times faster than the original ARPANET.

In the past five years, the NSFNET has evolved into the Internet. . . . There is a similar story for the World Wide Web. . . . According to the latest figures, there is a new Web page being created every four seconds. The Web was developed at CERN, the European Laboratory for Particle Physics, in Switzerland. In 1993, researchers at the National Center for Supercomputing Applications at the University of Illinois, one of the supercomputer centers funded by the National Science Foundation, developed an easy-to-use software package called Mosaic, which made it possible to use the World Wide Web to view not only text, but images, video, and audio as well. In less than a year, there were more than a million copies of Mosaic in use and Internet usage was sky-rocketing. This new technology has made possible companies like Netscape and Spyglass and fundamentally changed the way the Internet is used.

Funding for the development of Mosaic came from the High-Performance Computing and Communications Initiative, a Federal R&D program authorized by the High-Performance Computing Act of 1991, which then-Senator Al Gore first introduced in 1988. That program provides over 1 billion dollars a year to ten participating agencies that fund research in universities, Federal laboratories, and industry. The HPCC Program has pooled the talent and resources available throughout the Federal government to ensure effective coordination between agencies and ensure the efficient use of tax payers' dollars. In many ways, the HPCC Program is a model of how government, industry, and academia can work together to ensure that the U.S. stays at the leading edge of technology.

As the source indicates, this was a White House press release. Naturally, the U.S. federal government argues that its programs benefit the United States. How does it describe its role here in relation to technological innovation and related social change? Is the argument convincing? Who, does it claim, are some specific beneficiaries of its actions? Do its actions favor some groups over others? Which?

Source: White House (1996).

physicist Isidor I. Rabi (1969), considerable choice, leadership, ingenuity, and planning went into the creation of the first and subsequent atomic bombs.

> With very few exceptions, we went around to the various laboratories and took every productive scientist from his laboratory; we took scientists from their teaching laboratories, and we put them into large laboratories for the war effort. The Radiation Laboratory at Cambridge was one of them; another was at Los Alamos, and so on. There they were, in very large numbers, devoted to making weapons, to the problems of war. This was to my knowledge the first great attempt at marrying the military and the scientific. (Rabi 1969:37–38)

Ogburn argues that if people have the power to create material innovations, they also have the power to destroy, ban, regulate, or modify their use. The challenge lies, however, in convincing people that they need to address an invention's potential disruptive consequences before they have a chance to materialize. In this regard, the U.S. government created the Internet as a way of adapting to the potentially devastating consequences of nuclear attacks. On the other hand, it has yet to adapt to the problem of nuclear waste (see "The Problem of Nuclear Waste as an Example of Cultural Lag"). As one critic of the nuclear weapons program commented as early as 1950,"Nuclear waste is like getting on a plane, and in mid-air you ask the pilot, how are we going to land? He says, we don't know — but we'll figure it out by the time we get there" (Bauerlein 1992:34).

In our discussion about innovations as a trigger of social change, we have emphasized material inventions (devices, tools, and equipment). Innovations can also be nonmaterial inventions such as a revolutionary idea.

Revolutionary Ideas

In *The Structure of Scientific Revolutions,* Thomas Kuhn (1975) maintains that most people perceive science as an evolutionary enterprise: that is, over time scientists move closer to finding the solution to problems by building on their predecessors' achievements.

Kuhn takes issue with this evolutionary view; he argues that some of the most significant scientific advances have been made when someone breaks away from or challenges the prevailing paradigms. According to Kuhn, **paradigms** are the dominant and widely accepted theories and concepts in a particular field of study. Paradigms gain their status not because they explain everything but because they offer the "best way" of looking at the world for the time being. The dominant paradigms of an academic discipline generally are recorded in its textbooks, which summarize the body of accepted theory and illustrate it with the most successful applications, the most exemplary observations, and the most supportive experiments. On the one hand, paradigms are important thinking tools; they bind a group of people with common interests into a scientific or national community. Such a community cannot exist without agreed-on paradigms. On the other hand, paradigms can be blinders, limiting the kinds of questions that people ask and the observations they make.

Paradigms The dominant and widely accepted theories and concepts in a particular field of study.

Anomaly An observation or observations that a paradigm cannot explain.

The explanatory value, and hence the status, of a paradigm is threatened by **anomaly**, an observation or observations that it cannot explain. The existence of an anomaly alone, however, is usually not enough to cause people to abandon a particular paradigm. According to Kuhn, before people abandon old paradigms, someone must articulate an alternative paradigm that accounts convincingly for the anomaly. Kuhn hypothesizes that the people most likely to put forth new paradigms are those who are least committed to the old paradigms — the young and those new to a field of study.

A scientific revolution occurs when enough people in the community break with the old paradigm and change the nature of their research or thinking in favor of the incompatible new paradigm. Kuhn (1975) considers a new paradigm incompatible with the one it replaces because it "changes some of the field's most elementary theoretical generalizations" (p. 85). The new paradigm causes converts to see the world in an entirely new light and to wonder how they could possibly have taken the old paradigm seriously. "[W]hen paradigms change, the world itself changes with them. Led by a new paradigm, scientists adopt new instruments and look in new places" (p. 111). To illustrate Kuhn's theory, one can argue that Internet technology is changing the paradigms governing education, specifically the paradigms governing our conceptions of literacy.

The Internet and Changing Paradigms of Literacy

The paradox of the Internet is that, on the one hand, it has the power to put the most meaningful, thought-provoking, and influential of human creations at one's fingertips. On the other hand, the Internet also gives easy and equal access to the most trivial, inane, superficial, and inconsequential of human creations.

One reason that people have access to so much information via the Internet is that the Internet's strongest and most valuable quality is that most of the millions of documents available for access are formatted according to standard specification known as HTML (hypertext makeup language). (Stay with me — do not tune out because the concept hypertext seems too technical.) Hypertext simply refers to a programming feature that connects documents to one another and allows readers to move quickly via linked keywords within and across documents located anywhere on the Internet.

U.S. In Perspective

The Problem of Nuclear Waste as an Example of Cultural Lag

In 1991 a thirteen-member panel made up of an archaeologist, three anthropologists, two astronomers, a geologist, two materials scientists, a linguist, an artist, an architect, and a cognitive psychologist met to discuss the best design for a marker that would warn people to stay away from the Waste Isolation Pilot Plant (WIPP). WIPP is a vast nuclear bedded-salt waste depository 2,150 feet below the earth's surface about a forty-five-minute drive from Carlsbad, New Mexico. Congress authorized the plant's construction in 1979 and the Environmental Protection Agency stipulated that it be constructed in such a way that the estimated 900,000 drums of plutonium-tainted waste from nuclear weapons facilities to be stored there would not leak out for 10,000 years (Burdick 1992; Idelson 1991, 1992; Seltzer 1991). In addition, the agency mandated that a warning marker be constructed that can remain intact for 10,000 years and be understandable to anyone who might approach it. The panel divided into two teams—A and B—which met separately and then reconvened two months later to present their ideas. Representatives from five other countries (England, Belgium, Sweden, France, and Spain) attended to learn how they might mark their own sites. Some of the suggestions they heard are listed below:

Because nuclear waste remains radioactive for 240,000 years, the EPA has mandated that a warning marker be designed to inform future generations about where waste has been stored or buried. What challenges are associated with creating such a marker? Do the markers shown in the photograph represent an acceptable solution?

> *[A]rchaeologist Maureen Kaplan, leader of Team A, flashed transparencies on the overhead projector. To reach all those who might approach WIPP, she explained, the message must come in multiple forms, from the simple ("Danger: Poisonous Radioactive Waste Buried Here. Do Not Dig Until A.D. 12,000") to the complex (a detailed listing of the repository's contents). And in multiple languages: English, Spanish, Chinese, Russian, Arabic, and French; and perhaps, too, in the tongue of the local Mescalero Apache. Of course, as the panelists readily conceded, no modern language will be widely understood in 10,000 years. Who speaks Etruscan anymore?...*
>
> *Team B opted for a series of pictographic panels, a sort of stick-figure version of "See Dick Run (From Radioactive Death)." Team A has disavowed all symbols but two: a Munch-style expression of facial horror; and the radioactive trefoil, which, though conceived only four decades ago, probably will be all-too-well understood as the centuries progress.*
>
> *The conversation turned next to what the marker itself should be made of—or, more precisely, the marker system, since it was becoming evident that any monument agreed upon would have to be composed of numerous elements. . . .*
>
> *And so there was proposed by Team A the Landscape of Thorns: one square mile of randomly spaced basalt spikes, 80 feet high, erupting from the ground at all angles. Or its kin, the Spike Field, with thorns evenly spaced and erectile. Or the Black Hole: a boundless pad of black concrete that would absorb so much heat it would be impossible to approach. . . . Or the team favorite, Menacing Earthworks: an expansive empty square surrounded by 50-foot-high earthen berms jolting outward like jagged bolts of lightning. And at the center of the square, a 2,000-foot-long, walk-on global map displaying all the world's nuclear dumps, including this one. Rudimentary message kiosks would dot the periphery. A sealed room just below-ground would harbor the most intimate details of the repository's contents.*

At one point the Spanish representative sheepishly raised her hand. In halting English she recognized that the discussion of the last two days had addressed the fate of the repository's contents in the coming ten millennia. But, she pointed out, won't the waste be radioactive for some twenty times longer, closer to 240,000 years? Why, please, this interest only in 10,000?

Rip Anderson handled the question confidently. "Because the regulation says so."

Source: Burdick (1992:63–65).

The Internet puts the individual at the center of information in that he or she can access information quickly and directly, bypassing hierarchies devoted to managing and granting access to information (Crawford 1995). As one example, consider that we rely on the media for most of our information about what is going on in the world. Under the current arrangements, reporters go to the source of the news, decide what is important, and then relay it to their viewers and readers for a price. Within the Internet, users can go to the source of the news and read the press releases organizations put out to the media, thus bypassing the hierarchy that brings us an added layer of "filtered" news.

These characteristics of the Internet bring us to the question of the kinds of skills one needs to manage such a powerful information retrieval and distribution tool. In "Redefining Success: Public Education in the 21st Century," Yvonne Katz and Gay Chedester (1992) point out that as more and more students come to school with notebook-size computers tapped into all the world's knowledge, teachers need to adopt a new literacy paradigm in which they cease being the brokers of information and begin teaching the literacies of choice, connectedness, and creativity.

What is *literacy of choice?* The most obvious characteristic of the Internet is the shear glut of useful and trivial information it can deliver at the touch of a few keys. And because of its size, the documents posted on the Internet defy traditional approaches to indexing or categorization. If we do not give people the emotional and conceptual skills to sift through and choose among the resources available, they will simply be carried along in an avalanche of documents, forever surfing from site to site.[5] Two Northern Kentucky University students write about the emotional and conceptual challenges they face doing research on the Internet:

> The amount of information on the Internet is massive. You can find something on almost every topic. However, you must be mentally and physically tough when searching for information. Sometimes you can feel totally overwhelmed to the point of panic. But don't worry, most people feel this way at one time or another. My advice is to try to relax and not feel that you have to look over or reach everything on a topic.
>
> Be patient with the Internet, don't spend too much time at any one site unless it is directly relevant to your task at hand. Look through the sites quickly and if the information isn't useful, keep going and don't look back. Above all, take an "I am in charge attitude." Don't let the Internet overwhelm you. Maneuvering around the Internet is an acquired skill. The more time you spend at it, the more efficient you will become.

A person who possesses choice literacy, then, is someone who is not overwhelmed by information and has the emotional and conceptual ability to choose among the resources available.

The second type of literacy is the *literacy of connectedness.* The fact of the Internet is this: millions of computers across the world are connected to one another. The Internet is not run by some person or a group; it is not a prearranged library of information. Thus, the literacy of connectedness means to understand that we are part of a community of information.

A third type of literacy is *creative literacy.* Creative literacy involves the ability to use others' information in creative ways and to contribute information that others can use. The success of the Internet depends on people being more than passive consumers of information; it has thrived because millions of people were willing to post useful information for others to use. In other words, good citizens of the Internet do not just take information from it; they must give something back. Creative literacy also involves the belief that ideas are works in progress and that authors are free to recreate them at any time. One of my students put it like this:

> The Internet presents a whole new way of thinking about information. My advice is to expect change and learn to be comfortable with it. People who post information on the Internet can choose to take if off or revise it whenever they see fit. In the print world, authors might want to change their ideas but they have to wait until the next edition (if there is a next edition) to get it out there for the world to see. With the Internet, authors can revise their thinking and make changes on the spot. So don't be surprised if you find a document is gone or has changed in dramatic or subtle ways from one visit to the next.

The Actions of Leaders

The actions of leaders represent a second major trigger of social change. In the most general sense, a **leader** is someone who has the power to influence others or who is in charge or in command of a social situation. Recall

Leader Someone who has the power to influence others or who is in charge or in command of a social situation.

[5]As of January 3, 1997, Lycos (a search engine) claimed to have indexed 60,434,860 Web pages.

from Chapter 9 that **power** is the probability that an individual can realize his or her will even against the resistance of others (Weber 1947). The probability increases if that individual can force people to obey his or her commands or if the individual has authority over others. **Authority** is legitimate power in which people believe that the differences in power are just and proper —that is, people see a leader as entitled to give orders. Max Weber identified two types of authority—charismatic and legal-rational[6]—that have important implications with regard to social change.

Charismatic Leaders as Agents of Change

Charismatic authority rests on the exceptional and exemplary qualities of the person issuing the commands. Charismatic leaders are obeyed because their followers believe in and are attracted irresistibly to the vision that they articulate. Because the source of charismatic authority resides in the leader's exceptional qualities and not in tradition or established rules, the charismatic leader's actions and visions are not bound by rules or traditions. Consequently, these leaders, by virtue of their special qualities, have the ability to unleash revolutionary changes; they can ask their followers to behave in ways that depart from rules and traditions.

Charismatic leaders often appear during times of profound crisis (such as economic depressions or wars), when people are most likely to be drawn to someone with exceptional personal qualities who offers them a vision of a new order different from the current, seriously flawed situation. A charismatic leader is more than popular, attractive, likable, or pleasant; a merely popular person, "even one who is continually in our thoughts" (Boudon and Bourricaud 1989:70), is not someone for whom we would break all previous ties and give up our possessions. Charismatic leaders are so demanding as to insist that their followers make extraordinary personal sacrifices, cut themselves off from ordinary worldly connections, or devote their lives to achieving a vision that leaders have outlined.

The source of the charismatic leader's authority, however, does not rest with the ethical quality of the command or vision. Adolf Hitler, Franklin D. Roosevelt, Mao Zedong, and Winston Churchill all were charismatic leaders. All assumed leadership of a country during turbulent times. All conveyed a powerful vision (right or wrong) of their countries' destinies. A description such as the following is typical of charismatic leaders:

> He had a powerful sense of both his nation's destiny and his own. . . . [H]e saw no difference between the two. Difficult, egocentric, vainglorious, he demanded complete loyalty from those beneath him but did not always bestow comparable loyalty on those above him. . . . His belief in his own vision and fate was so strong that few other men dared challenge it. (Halberstam 1986:111)

Charismatic authority is a product of the intense relationships between leaders and followers. From a relational point of view, then, charisma is a "highly asymmetric power-relationship between an inspired guide and a cohort of followers" (Boudon and Bourricaud 1989:70) who believe in the promises and visions offered by the person with special qualities. Many Soviets, for example, believed that Joseph Stalin, the general secretary of the Soviet Union from 1922 to 1953 and the Soviet premier from 1941 to 1953, was laying the foundation for a new kind of society in which people would eventually live in harmony and experience absolute economic security. Similarly, many Germans believed that Adolf Hitler's vision could help them recover from the humiliation and the massive destruction they suffered at the hands of the Allies (Britain, France, Russia, Italy, and the United States) in World War I.

Charismatic leaders and their followers come to constitute an "emotional community" devoted to achieving a goal and sustained by a belief in the leader's special qualities. Weber argues, however, that at some point the followers must be able to return to a normal life and to develop relationships with one another on a basis other than their connections to the leader. Attraction and devotion cannot sustain a community indefinitely, if only because the object of these emotions—the charismatic leader—is mortal. Unless the charisma that bonds a community is routinized, the community

[6]Weber also identified a third type of authority, traditional authority. It rests on the sanctity of time-honored norms that govern the selection of someone to a powerful position (chief, king, queen) and specify responsibilities and appropriate conduct for those selected. People comply because they believe they are accountable to the past and have an obligation to perpetuate it. (They are apt to reason, "It has always been like that.") To give up past ways of doing things is to renounce a heritage and an identity (Boudon and Bourricaud 1989).

Power The probability that an individual can realize his or her will even against the resistance of others.

Authority Legitimate power in which people believe that the differences in power are just and proper.

Charismatic authority Authority that rests on the exceptional and exemplary qualities of the person issuing the commands.

Malcolm X and President Franklin D. Roosevelt illustrate Weber's theory of charismatic leaders whose authority depends on the emotional bonds they forge with their followers. Both were powerful speakers and magnetic personalities. More than thirty years after his death, Malcolm X's words and persona continue to inspire great interest. Within the Nation of Islam, he was an influential leader. He was assassinated soon after leaving the Nation of Islam, and so his leadership had little chance to become "routinized" concretely in a broader context. In contrast, Roosevelt's leadership gave rise to such institutional legacies as Social Security that have endured long after his death in office in 1946.

may disintegrate from exhaustion or from a void in leadership. **Routinized charisma** develops as the community establishes procedures, rules, and traditions to regulate the members' conduct, recruit new members, and ensure the orderly transfer of power. Authority must come to rest on legal-rational grounds. That is, it must be grounded in the position, not in personal qualities.

Routinized charisma A situation in which the community establishes procedures, rules, and traditions to regulate the members' conduct, recruit new members, and ensure the orderly transfer of power.

Legal-rational authority Power that rests on a system of impersonal rules that formally specifies the qualifications for occupying a powerful position.

Great changes Events whose causes lie outside ordinary people's characters or their immediate environments but profoundly affect their life chances.

Power elite Those few people positioned so high in the social structure of leading institutions that their decisions have consequences affecting millions of people worldwide.

The Power Elite: Legal-Rational Authority and Change

Legal-rational authority rests on a system of impersonal rules that formally specifies the qualifications for occupying a powerful position. The rules also regulate the scope of power and the conduct appropriate to someone holding a particular position. In cases of legal-rational authority, people comply with commands, decisions, and directives because they believe that those who have issued them have earned the right to rule.

Sociologist C. Wright Mills (1959, 1963, 1973) argues that the causes of **great changes** — events whose causes lie outside ordinary people's characters or their immediate environments but profoundly affect their life chances — can be traced to the decisions made by the **power elite**, those few people positioned so high in the social structure of leading institutions that their decisions have consequences affecting millions of people worldwide. For the most part, the source of this power

How does this 1945 advertisement support C. Wright Mills's ideas about the power elite?

is legal-rational and resides not in the personal qualities of those in power but in the positions that the power elite have come to occupy. "Were the person occupying the position the most important factor, the stock market would pay close attention to retirements, deaths, and replacements in the executive ranks" (Galbraith 1958:146).

The amount of power wielded by the elite over the lives of others is related to the nature and the quality of instruments that they can use, by virtue of their position, to rule, control, and influence others. Instruments might include weapons, surveillance equipment, and specialized modes of communication. Mills argues that since World War II, rapid advances in technology have allowed power to become concentrated in the hands of a few; those with access to such power can exercise an extraordinary influence not only over their immediate environment but over millions of people, tens of thousands of communities, entire countries, and the globe.

In writing about the power elite, Mills does not focus on any one individual but almost exclusively on the power elite who occupy the highest positions in the leading institutions in the United States. According to Mills, the leading institutions are the military, corporations (especially the 200 or so largest American corporations), and government. "The power to make decisions of national and international consequence is now so clearly seated in political, military, and economic institutions that other areas of society seem off to the side and, on occasion, readily subordinated to these" (Mills 1963:27).

The origins of these institutions' power can be traced to World War II, when the political elite mobilized corporations to produce the supplies, weapons, and equipment needed to fight the war. For one measure of the extent to which government, military, and corporations worked together, we can look at advertisements published in *Life* magazine during the war years. As one example, consider the Bell Telephone System advertisement in the July 30, 1945, issue of *Life* shown at the right.

After the war, as Stalin moved to consolidate his power in Eastern Europe, Japan and the countries of Western Europe — their populations demoralized, their economies in ruins, their infrastructures devastated — had little choice but to accept help from the United States in the form of the Marshall Plan. U.S. corporations, unscathed by the war, were virtually the only companies able to offer the services and products that the war-torn countries needed for rebuilding. The interests of the government, the military, and corporations became further entangled when the political elite decided that a permanent war industry was needed to contain the spread of Communism.

Thus, over the past forty-five or fifty years, these three institutions have become deeply and intricately interrelated in hundreds of ways, as shown in the following examples found on the Web site Project Censored.

- The 1995 Telecommunications Deregulation Bill eased, and in some situations, eliminated antitrust regulation of the U.S. communication industry. Under this bill a single company can control television stations that serve up to 50 percent of the country. In addition, this bill essentially dismantles limits on the number of AM and FM radio stations a single company can own (Nader, Love, Saindon 1995).
- The U.S. Department of Agriculture gives direct subsidies for overseas product advertising for

Joseph Stalin's reign of terror is an extreme case of concentration of power in a single person. But a society does not have to be an autocratic dictatorship for power to be distributed unevenly. According to Mills, in our own society power elites make decisions that have enormous influence on the lives of ordinary citizens.

brand names such as Campbell's soup and for beer, wine, and liquor.

- The U.S. Forest Service spends $100 million annually to build and maintain access roads to assist timber companies' logging operations in the National Forest (Canham-Clyne 1995).
- The Department of Defense (DOD) subsidizes defense contractors post-merger and -acquisitions costs arguing that this plan will save taxpayers money. As one example, under this program the DOD paid Martin Marietta $60 million to purchase a General Dynamics subsidiary (Sloyan 1994).[7]

Because the three realms of institutions (military, government, and corporations) are interdependent and because decisions made by the elite of one realm affect the other two, Mills believes that it is in everyone's interest to cooperate. Shared interests cause those who occupy the highest positions in each realm to interact with one another. Out of necessity, then, a triangle of power has emerged. This is not to say that the alliance among the three is untroubled, that the powerful in each realm are of one mind, that they know the consequences of their decisions, or that they are joined in a conspiracy to shape the fate of a country or the globe:

> At the same time it is clear that they know what is on one another's minds. Whether they come together casually at their clubs or hunting lodges, or slightly more formally at the Business Advisory Council or the Committee for Economic Development or the Foreign Policy Association, they are definitely not isolated from each other. Informal conversation elicits plans, hopes, and expectations. There is a community of interest and sentiment among the elite. (Hacker 1971:141)

Mills gives no detailed examples of the actual decision-making process at the power elite level. He is more concerned about understanding the consequences of this alliance than about understanding the decision-making process or assessing motives. Mills acknowledges that the power elite are not totally free agents, subject to no controls (see Figure 15.1). A chief executive officer of a major corporation is answerable to unions, the Occupational Safety and Health Administration, the Food and Drug Administration, or other regulatory bodies. Pentagon officials are subject to congressional investigations and budget constraints. Defense contractors are liable to the Federal False Claims Act, which gives a share in the settlement to any employee who can prove that the contractor has defrauded the government (Stevenson 1991). The president of the United States is constrained by bureaucratic

[7]Martin Marietta executives persuaded Defense Secretary William Perry and Deputy Defense Secretary John Deutch, both on the company's payroll before assuming their DOD positions, to approve the money-for-merger plan (Sloyan 1994).

Figure 15.1 Public Citizen's Greatest Hits

Public Citizen, founded by Ralph Nader in 1971, is a consumer advocacy organization with 150,000 members. It stands up against "thousands of special interest lobbyists in Washington," including auto makers, drug companies, energy companies, and so on. It accepts no government or corporate support. The following is a select list of its most important victories.

1994 Public Citizen successfully challenges the FDA's failure to restrict use of the drug Parlodel.
1993 Public Citizen lobbies successfully to end funding for the Advanced Breeder Reactor.
1992 Public Citizen helps defeat an industry-backed bill that would restrict the use of the courts by consumers injured by defective products.
1991 Public Citizen publishes *They Love to Fly and It Shows,* a study of lobbyist-funded House travel during 1989–90.
1990 Public Citizen publishes an account of the savings and loan scandal, entitled *Who Robbed America?*
1989 A regulation requiring air bags or passive seat belts in all cars takes effect after a twenty-year battle.
1988 Public Citizen publishes *Worst Pills, Best Pills,* a guide to dangerous drugs for the elderly. By 1995, almost 2 million copies are sold.
1987 Public Citizen forces the chemical companies who produced Agent Orange to make all relevant documents public.
1986 Congress passes a law requiring health warnings on the labels of snuff and chewing tobacco, capping Public Citizen's two-year campaign.
1985 After a four-year battle by Public Citizen, the FDA requires aspirin labels to warn of the risk of Reye's Syndrome.
1984 Public Citizen helps block legislation that would make it easier for manufacturers to use cancer-causing food additives.
1983 *Over the Counter Pills That Don't Work* is published.
1982 Oraflex, an arthritis drug that caused dozens of deaths and hundreds of injuries, is withdrawn from the market after Public Citizen's campaign.
1981 Public Citizen helps thwart President Reagan's attempt to bury the Consumer Product Safety Commission in the Commerce Department.
1980 Public Citizen plays a critical role in passage of the Superfund law to require cleanup of toxic waste dumps.
1979 In response to a 1977 petition by Public Citizen, the EPA bans DBCP, a pesticide that causes sterility in human males and cancer in laboratory animals.
1978 Congress passes the National Consumer Cooperative Bank bill, drafted by Public Citizen, providing $300 million "seed money" for consumer cooperatives.
1977 Public Citizen mobilizes citizens, persuading President Carter to stop construction of the Clinch River Breeder Reactor.
1976 The FDA bans Red Dye No. 2, after Public Citizen's four-year campaign against the carcinogenic food dye.
1975 Public Citizen obtains a court order requiring the Labor Department to reconsider its inadequate rules to protect workers from fourteen carcinogenic chemicals.
1974 Critical Mass Energy Project is founded to mobilize opposition to nuclear power.
1973 Congress Watch, the legislative lobbying arm of Public Citizen, is formed.
1972 The Litigation Group and the Tax Reform Research Group are founded.
1971 Public Citizen is founded by Ralph Nader.

Source: Public Citizen (1996).

red tape and by a sometimes slow-moving, politically oriented Congress. Mills questions, however, whether these constraints on the power elite have "much significance when weighed against the areas of unrestricted action open to [them]" (Hacker 1971:136).

To this point, we have examined two important agents of change: innovations and the actions of people in positions of authority (charismatic leaders and the power elite). We turn now to a third agent of change, conflict, which is intertwined with these two agents. In its most basic form, conflict involves clashes between groups over their shares of wealth, prestige, and other valued resources. Recall that the introduction of innovations (discoveries, inventions, and paradigms) disrupts the balance of power, causing conflict between those who stand to benefit and those who stand to lose from the widespread acceptance of an innovation. Recall also that charismatic leaders and the power elite possess the authority to impose their will (for better or worse), despite resistance by others.

Conflict

Sociologist Lewis Coser (1973) points out in his essay "Social Conflict and the Theory of Social Change" that conflict will always exist, if only because there is never a perfect "concordance between what individuals and groups within a system consider their just due and the system of allocation" (p. 120). Conflict occurs whenever a group takes action to increase its share of or control over wealth, power, prestige, or some other valued resource and when these demands are resisted by those who benefit from the current distribution system. In other words, those who gain control of valued resources strive to protect their own interests against efforts by those without who hope to gain a share.

The Conflict over Scarce and Valued Resources

This fact of life has motivated Michael and Rhonda Hauben (1996) to draft a Declaration of the Rights of Netizens,[8] which argues that access to the Internet should be considered a human right and not something reserved for a privileged elite. Among the rights (and obligations) listed are the right of equal times and quality of connection and the obligation to consider ideas posted according to their merit (see "Who Controls the Internet?").

Conflict, whether it involves violent clashes or public debate, is both a consequence and a cause of change. In general, we can say that any kind of change has the potential to trigger conflict between those who benefit from the change and those who stand to lose because of it. When the bicycle was invented in the 1840s, for example, horse dealers organized against it because it threatened their livelihoods. Some physicians declared that people who rode bicycles risked getting "cyclist sore throat" and "bicycle stoop." Church groups protested that bicycles would swell the ranks of "reckless" women (because bicycles could not be ridden sidesaddle).[9]

Conflict can lead to change as well. It can be a constructive and invigorating force that prevents a social system from becoming stagnant, unresponsive, or inefficient. Conflict such as that resulting from the antinuclear, civil rights, and women's movements can create new norms, relationships, and ways of thinking. Conflict also can generate new and efficient technologies. The Internet is one example of a technology whose origins are rooted in a most destructive form of conflict—war.

Ironically, war has generated advances in lifesaving medical technologies. During World War I, many soldiers fighting on manure-covered farmlands contracted tetanus.[10] In addition, large numbers of soldiers were injured by machine gun shrapnel and bombshells. Physicians experimenting with antitoxins eventually found a cure for tetanus and also made considerable advances in reconstructive surgery. Similarly, the kinds of injuries incurred during World War II motivated doctors to create a system of collecting and preserving blood plasma and to mass-produce an effective drug, penicillin, to treat wound infections (Colihan and Joy 1984). More recently, the Pentagon's Advanced Research Project Agency (ARPA), the same agency out of which the Internet emerged, funds civilian projects that it believes will have military applications. For example, the agency funds a project for electric-vehicle systems that would allow military vehicles to travel silently over enemy-controlled terrain (Wald 1994).

Whether conflict will lead to reform (improvements or alterations in current practices) or to revolution (complete and drastic change) depends on a broad range of factors and contingencies. Sociologist Ralf Dahrendorf has identified and described some of these factors in his essay "Toward a Theory of Social Conflict" (1973).

Structural Origins of Conflict

In his essay, Dahrendorf asks two questions: What is the structural source of conflict, and what forms can conflict take? Dahrendorf's answers rest on the following assumptions. First, in every organization that has a formal authority structure (a state, a corporation, the military, the judicial system, a school system), clear dichotomies exist between those who control the formal system of rewards and punishment (and thus have the authority to issue commands) and those who must obey those commands or face the consequences (loss of job, jail, low grades, and so on). Second, a distinction between "us" and "them" arises naturally from the unequal distribution of power. In view of these assumptions, the structural origins of conflict can be traced to

[8]A *netizen* is a citizen of the Internet.

[9]I read about this resistance to the bicycle many years ago while teaching a Sociology of Sport course. Although the example is very memorable, I cannot remember the source.

[10]Tetanus is an infectious disease characterized by muscular spasms and difficulty in opening the mouth (lockjaw). The tetanus bacterium is dangerous because it produces a toxin that affects the heart and the breathing muscles.

Who Controls the Internet?

On one level, the Internet does put the individual (who can afford it) at the center of information in that he or she can access information quickly and directly, bypassing hierarchies devoted to managing and granting access to information. On another level, however, we cannot assume that individual users entirely control their access to information. For one thing, it would be naive to believe that the Internet provides access to information governments, corporations, and other organizations and agencies wish to keep confidential or secret. Cyberspace law professor Stuart Biegel (1996) discounts the widely held belief that no one is in charge of the Internet. In fact, many persons and groups are "in charge" in some capacity. They include

1. the government, which provided funds to "invent" the Internet and sets policies that affect funding, growth, and access;
2. Internet access providers, who act as the gatekeepers in that users pay them a fee to access the Internet;
3. communication companies investing in the Internet infrastructure such as the high-speed lines that route information;
4. companies that create the hardware and software used to access the Internet and that will shape the future direction and applications;
5. colleges and universities whose faculty develop programs and post large quantities of noncommercial information; and
6. the Internet Society chartered by the U.S. government "for the purpose of commenting on policies and overseeing other boards and task forces dealing with network policy" (Biegel 1996).

the nature of authority relations. That conflict can assume many forms. It can be mild or severe; "it can even disappear for limited periods from the field of vision of a superficial observer" (p. 111). As long as an authority structure exists, however, conflict cannot be abolished.

Dahrendorf outlines a three-stage model of conflict in which progression from one stage to another depends on many things. He does not claim to give an exhaustive list with regard to the possible course of a conflict. In fact, he reminds readers that the conflicts he names are some of the most obvious. The point is that conflict — its course and its resolution — is a complicated phenomenon in which many elements must be considered.

First, every authority structure contains at least two groups with opposite interests. Those with power have an interest in preserving the system; those without power have an interest in changing it. These opposing interests, however, remain below the surface until the groups (especially those without power) organize. "It is immeasurably difficult to trace the path on which a person . . . encounters other people just like himself, and at a certain point . . . [says] 'Let us join hands, friends, so that they will not pick us off one by one'" (Dahrendorf 1973:240).

Often a significant event makes seemingly powerless people aware that they share an interest in seeing the system changed. Vaclav Havel, the president of the Czech Republic, believes that Chernobyl may have played an important role in bringing about the revolutions in central Europe. Chernobyl is located in Ukraine and was the site of a nuclear power plant meltdown in 1986, the most serious kind of accident that can occur at a power plant. Although Ukraine, Russia, and Belarus suffered the most radioactive contamination, areas as far away as Sweden were also affected. Havel maintains that after Chernobyl, people in what was then Czechoslovakia dared to complain openly and loudly to one another (Ash 1989).

Sometimes, too, people organize because they have nothing left to lose. As one East German scientist explained:

> You don't need courage to speak out against a regime. You just need not to care anymore—not to care about being punished or beaten. I don't know why it all happened this year [in 1989]. We finally reached the point where enough people didn't care anymore what would happen if they spoke out. (Reich 1989:20)

In the second stage, if those without authority have opportunities to communicate with one another, the freedom to organize, the necessary resources, and a leader, then they will organize. At the same time, those in positions of authority often use the power of their positions to censor information, restrict resources, and undermine leaders.

One technology that will certainly enhance potential opportunities for people to communicate with one another at the grassroots level is the Internet. I use the word *potential* because, although Internet technology has helped activists organize and pressure authority structures, activists still have not tapped its full potential with regard to "delivering the message" and causing structural impact (Afonso 1997). Activist organizations with Web pages on the Internet range from the American Civil Liberties Union to Voters Telecommunications Watch. One example of activist organizations on the Internet is the Association for Progressive Communication, a host organization that coordinates the operation and development of an evolving global network of individuals and organizations working to achieve social or environmental change (see "What Makes the Association for Progressive Communication [APC] Unique?").

In the third stage, once organized, those without power enter a state of conflict with those in power. The speed and the depth of change depend on the capacity of those who rule to stay in power and on the kind and degree of pressure exerted from below. The intensity of the conflict can range from heated debate to violent civil war, but it is always contingent on many factors, including opportunities for mobility within the organization and the ability of those in power to control the conflict. If those who do not have authority are confident that eventually they will achieve such a position, the conflict is unlikely to become violent or revolutionary. If those in power decide that they cannot afford to compromise and mobilize all their resources to thwart protests, two results are possible. First, the protesters may believe the sacrifices are too great and may then withdraw. Alternatively, the protesters may decide to meet the "enemy" directly, in which case the conflict becomes bloody.

To this point, we have discussed conflict in very general terms. The fourth important agent of change represents a specific kind of conflict, motivated by the pursuit of profit. This agent is capitalism, an economic system whose modern origins can be traced back 500 years.

Capitalism

Karl Marx believed that an economic system—capitalism—ultimately caused the explosion of technological innovation and the enormous and unprecedented increase in the amount of goods and services produced during the Industrial Revolution. In a capitalist system, profit is the most important measure of success. To maximize profit, the successful entrepreneur reinvests profits in order to expand consumer markets and to obtain technologies that allow products and services to be produced at the highest quality and the greatest cost-effectiveness.[11]

The capitalist system is a vehicle of change in that it requires the instruments of production to be revolutionized constantly. Marx believed that capitalism was the first economic system capable of maximizing the immense productive potential of human labor and ingenuity. He also believed, however, that capitalism ignored too many human needs, and that too many people could not afford to buy the products of their labor. Capitalism survived and flourished by sucking the blood of living labor. The drive for profit (which Marx maintained is derived from the labor of those directly involved in the production process) is a "boundless thirst . . . [a] werewolflike hunger . . . [that] takes no account of the health and the length of life of the worker unless society forces it to do so" (Marx, cited in Carver 1987:142). The thirst for profit "chases the bourgeoisie over the whole surface of the globe" ([1881] 1965:531). Marx's theories influenced a group of contemporary sociologists—world system theorists—to write about capitalism as the agent of change underlying global interdependence.

World System Theory

Immanuel Wallerstein (1984) is the sociologist most frequently associated with world system theory, a modern theory concerned with capitalism. Since the early 1970s, he has been writing a four-volume work (three volumes of which have been published) about the ceaseless expansion, over the past 500 years, of a single market force — capitalism. According to Wallerstein, although stagnant periods have occurred and some countries (the Communist countries, for example) have tried to withdraw from the capitalist economy, no real contraction has occurred.

> Hence, by the late nineteenth century, the capitalist world-economy included virtually the whole inhabited earth and it is presently striving to overcome the

[11]Although pursuit of profit is a largely self-centered activity, advocates of capitalism consider it socially legitimate because they assume that what is profitable for entrepreneurs and corporations and their stockholders also benefits society as a whole. Quality and cost-effectiveness are enhanced when competition exists among those who produce or distribute the same goods and services. Ideally, well-informed consumers "vote" for products with their purchases. Manufacturers and providers who cannot sustain consumers' interest and who cannot match or surpass their competitors in cost and quality will be forced to improve, or eventually they will go out of business.

What makes the APC unique?

The APC Networks are distinguished by the depth and quality of their information resources, by the global reach of their communication services, and by the diversity of their membership. APC developed from a dedicated grass-roots constituency and has become a true global community.

The Network provides information in several languages, including:

- Bosnian
- Croation
- English
- Esperanto
- Estonian
- French
- German
- Macedonian
- Polish
- Portuguese
- Russian
- Serbian
- Slovenian
- Spanish
- Swedish

In addition, APC provides extensive user support and training to help people access the system and use it easily and efficiently.

How Are People Using the APC?

On a "typical day" users of the APC networks use the system in a variety of important ways:

- At a meeting in Australia, the trustees of an organization need a 12-page report to proceed with an agenda item, but the only copy is in England. Within half an hour they have the entire report in hand for less than the cost of postage.
- The editor of a newsletter in Brazil receives article submissions on-line, eliminating the need to key in the desired articles herself.
- A solidarity group in England updates itself on events of the last few days in Central America (received in both English and Spanish) and sends a message to the project it supports in El Salvador.

The APC Networks are designed for use by NGOs, other types of not-for-profit citizen-based organizations, and individuals who communicate with NGOs. These include United Nations agencies, research organizations, government departments, educational institutions, multi-stakeholder groups and activists.

People use the APC as an information and organizing tool for a wide range of issues. These include:

- Peace
- Human Rights
- Environmental Issues
- Social & Economic Justice
- Education
- Indigenous Rights
- Women's Empowerment
- Labour Movement
- Public Health
- Population
- Development
- Poverty
- Nonviolence
- Demilitarization

Source: http://www.apc.org/unique.html (1997).

> technological limits to cultivating the remaining corners; the deserts, the jungles, the seas, and indeed the other planets of the solar system. (p. 165)

Wallerstein distinguishes between the terms *world economy* and *world-economy.* People who use the term **world economy** (without the hyphen) envision the world as consisting of 160 or so national economies that have established trade relationships with one another. In this vision, globalization is portrayed as a relatively new phenomenon and as a process in which the countries of the world have moved from relatively isolated, self-sufficient economies to economies that trade with one another to varying degrees. Although popular, this conception of global interdependence is not very accurate. The more accurate term (and conception) is *world-economy.* The **world-economy** (with a hyphen) is not recent; it has been evolving for at least 500 years and is still evolving. People who use the hyphenated term envision a world (encompassing hundreds of countries and thousands of cultures) interconnected by a single division of labor. In the world-economy, economic transactions transcend national boundaries. Although the world-economy is not new, the giant multinationals of today "could not exist at their present scale if there were no computers. Computers are their global nervous systems; their way of keeping track of their billions of moving parts, keeping them synchronized and moving in the same direction for control purposes" (Mander 1997). Although each government seeks to shape the global market in ways that benefit its "own" corporations and national interests, no single political structure (world government) or national government has authority over the system of production and distribution.

Wallerstein (1984) argues that the world-economy is capitalist because "its economy has been dominated by those who operate on the primacy of endless accumulation, . . . driving from the arena those who seek to operate on other premises" (p. 15). Critics counter that this is an exaggeration, that there are many countries with economies that are not capitalist and that no country in the world has an economy that runs on purely capitalist principles.[12]

World economy An economy consisting of 160 or so national economies that have established trade relationships with one another.

World-economy An economy encompassing hundreds of countries and thousands of cultures interconnected by a single division of labor.

Wallerstein counters this criticism in part with the argument that the Communist countries are the equivalent of huge state-owned capitalist corporations. All depend on the world-economy, and all trade on some levels with countries that are bitterly opposed to their political and economic system. Even before the collapse of the Soviet Union, the United States, for example, exported corn and wheat there, and the Soviets exported chemicals, fuels, and minerals to the United States. Likewise, even before the Soviet Union disbanded, the Soviets were exporting natural and enriched uranium to Western countries (Broad 1991). Another feature that makes the world-economy capitalist is the fact that profits from goods and services are distributed unevenly through the global market to a network of beneficiaries, most of whom live in the mechanized rich countries.

The Role of Capitalism in the Global Economy

How has capitalism come to dominate the global network of economic relationships? One answer lies in the ways in which capitalists respond to changes in the economy, especially to economic stagnation. Historically, there have been five important responses, all designed to create economic growth. We will consider how the computer and the Internet are figuring into those responses.

1. *Lowering production costs by hiring employees who will work for lower wages* (for example, by busting unions, buying out workers' contracts, or offering early retirement plans), by introducing labor-saving technologies (such as computerizing the production process), by moving production facilities out of high-wage zones and into lower-wage zones inside or outside the country, or contracting services to be delivered via the Internet. For example, some large U.S. corporations contract computer programmers in India to write software and send it to them via the Internet (Associated Press 1996).
2. *Creating a new product that consumers "need" to buy* (such as the videocassette recorder, the computer,

[12]For example, although most Americans consider the United States to be a capitalist country, capitalist principles do not apply to the way the defense industry operates. In the defense market, a single customer (the Pentagon) describes the weapons or other technologies it desires and contracts with a company to make them. If the industry operated on capitalist principles, the supplier would invest its own money and resources to create and develop the product and then would compete with other companies to sell the product to the buyer (Lambert 1992). Because capitalist principles do not apply to defense contractors, they do not have to control costs, understand customer needs, or be familiar with marketing or sales strategies.

or the fax machine — products introduced in the late 1970s and early 1980s). In 1996, Sony and Fujitsu introduced "Web TV" with the aim of allowing Internet users to surf/browse the Web from an easy chair.

3. *Improving on an existing product and thus making previous versions obsolete.* Every improvement in a personal computer's CPU makes the previous CPU obsolete. As one indicator of the speed at which a "new" CPU becomes obsolete, consider that I used a 1992 reference book to inform myself about CPUs before writing the previous sentences. At the time of the book's printing (Bear and Pozerycki 1992) the highest-grade CPU was 386. Five years later, four versions of a 486 and the pentium have appeared on the market, with an upgrade of the pentium due out in 1997 (Schwartz 1996).
4. *Expanding the outer boundaries of the world-economy and creating new markets.* Since the fall of the Berlin Wall in 1989, American, Western European, and Japanese corporations have been expanding their markets into Eastern Europe, Russia, and the new states of Eurasia. Procter & Gamble, for example, already produces and markets detergent, toothpaste, shampoo, and diapers in the Czech Republic, Hungary, and Poland (Rawe 1991). The Coca-Cola Company moved particularly quickly into Eastern Europe. Almost immediately after the Berlin Wall fell and East Germans started to visit West Germany and West Berlin, Coca-Cola was there, handing out free Cokes.[13] The commercialization of the Internet, beginning in the mid-1990s, represents the attempt to expand the boundaries of the world-economy and/or establish new markets. As of mid-1996, an estimated 12.8 million hosts were on the Internet; 25.8 percent were commercial hosts (Kantor and Neubarth 1996). The commercial Web sites describe a range of products, services, and entertainment activities from antihistamines to zoos.
5. *Redistributing wealth to enable more people to purchase products and services.* Henry Ford was the first to do this on a large scale; in 1908, he came up with the revolutionary concept of paying workers a wage (five dollars per day) large enough to allow them to purchase the products of their labor (Halberstam 1986). Because of the shortage of hard currency in Russia, the new states of Eurasia, and Eastern Europe, American, Western European, and Japanese executives have set up barter systems there (Holusha 1989). "Pepsico, for example, exports wooden chairs from Poland to its Pizza Hut franchises in the United States, and sells its soft drink to the Soviet Union in exchange for old submarines" (O'Sullivan 1990:22). A final example of this strategy is IBM's Global Campus designed to help colleges and universities put their "campuses" on-line. As of early 1997, IBM had enrolled the twenty-three campus California State University systems, twenty-one other U.S. colleges and universities, and a number of other schools outside the United States. The purpose of the program is to help schools with start-up costs and/or enhance already existing distance learning programs (and, not incidentally, to compete with its rival, Apple). The universities who have signed this pilot project have purchased or are considering future purchases of elements of the IBM Global Campus Package (see Figure 15.2).

As a result of these five responses to economic stagnation, capitalism has spread steadily to encompass the globe and facilitate globalization-from-above. In addition, every country of the world has come to play one of three different and unequal roles in the global economy: core, peripheral, and semiperipheral. Their unequal roles in the global economy are also reflected in Internet access (see opening map).

The Roles of Core, Peripheral, and Semiperipheral Economies

Core economies include those of the mechanized rich nations — nations characterized by strong stable governments. Core economies tend to be highly diversi-

Core economies Highly diversified economies of the mechanized rich nations characterized by strong stable governments.

[13]This event was very popular in West Berlin; people sought out the Coca-Cola vendor for samples. Within weeks, the U.S. corporation was in East Germany discussing the distribution of Coca-Cola there:

> Almost at the same time we reorganized: As soon as we saw the changes, we moved East Germany into the West German and E. C. group, and transferred the infrastructure, talent and technology of our West German operations into East Germany. Within weeks, we were shipping Coca-Cola in cans into East Germany to distributors with whom we had made agreements. Within a month or two, we were selling a million cases a month in East Germany. . . .
>
> By July, we had put our first East German production facility in place. Now we're no longer solely importing into East Germany, we're producing and distributing in the country. Next year we expect to sell 30 million cases and, by 1995, 100 million cases a year in East Germany. (Guttman 1990:16)

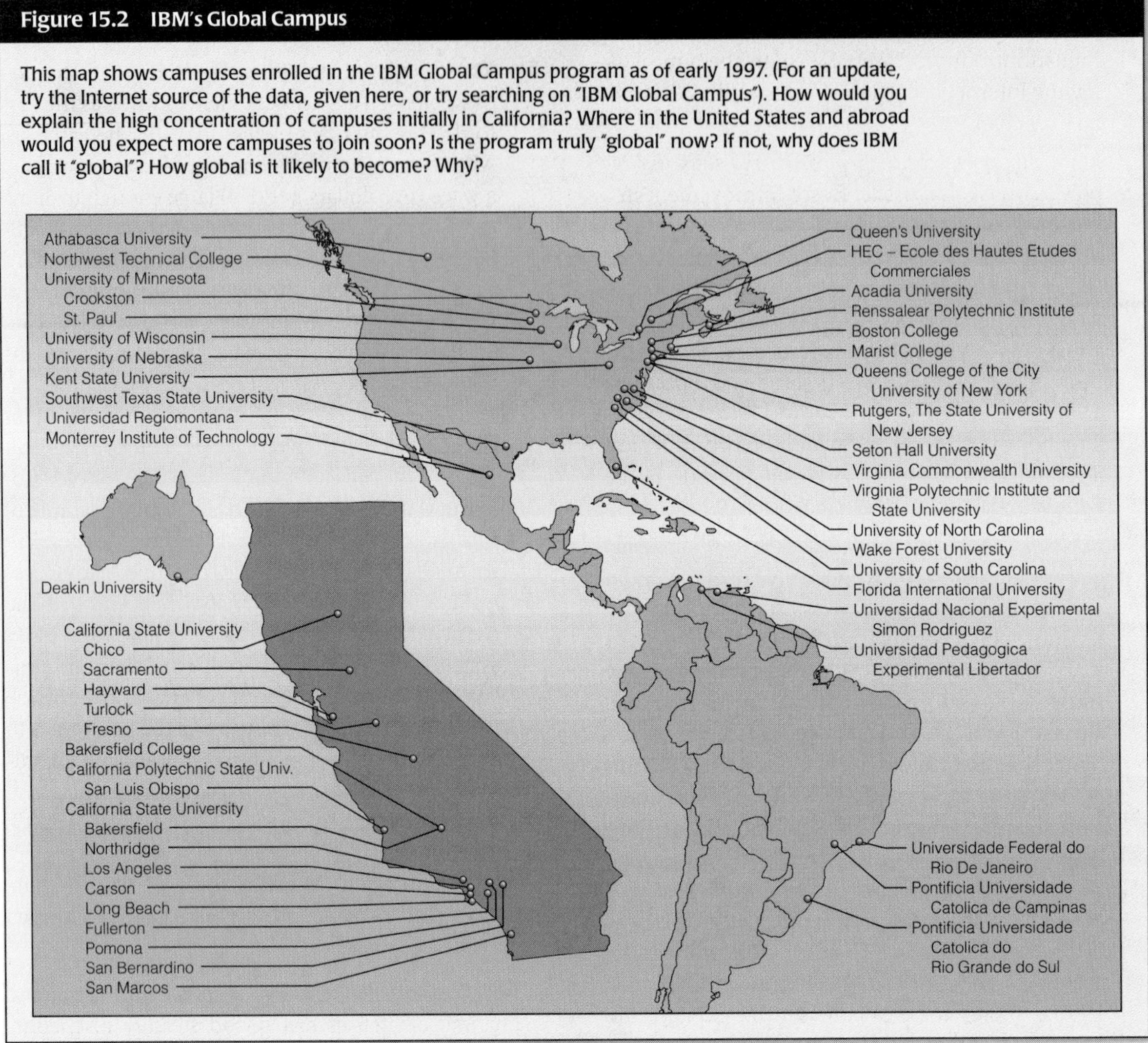

Figure 15.2 IBM's Global Campus

This map shows campuses enrolled in the IBM Global Campus program as of early 1997. (For an update, try the Internet source of the data, given here, or try searching on "IBM Global Campus"). How would you explain the high concentration of campuses initially in California? Where in the United States and abroad would you expect more campuses to join soon? Is the program truly "global" now? If not, why does IBM call it "global"? How global is it likely to become? Why?

Source: http://ike.engr.washington.edu/igc/affiliate/html (1997).

fied. The G-7 countries (Japan, Germany, France, the United States, Canada, Great Britain, and Italy) are examples of core economies absorbing nearly two-thirds of developing countries' exports. They import raw materials from labor-intensive poor countries and make use of free trade zones around the world. The overwhelming majority of "the great global enterprises that make the key decisions — about what people eat and drink, what they read and hear, what sort of air they breathe and water they drink, and, ultimately, which societies will flourish and which city blocks will decay" (Barnet 1990:59) — have their headquarters in countries with core economies. The sales of these corporations exceed the gross national products (GNPs) of many countries (Currie and Skolnick 1988). When economic activity weakens in the industrial world, the labor-intensive poor countries suffer because the amount of exports declines and price levels fall.

Peripheral economies Economies of the labor-intensive poor countries that are not highly diversified, with most of the jobs low-paying and requiring few skills.

Labor-intensive poor countries have **peripheral economies**, which are not highly diversified; most of the jobs are low-paying and require few skills. Peripheral societies depend disproportionately on a single commodity such as coffee, peanuts, or tobacco or a single mineral resource such as tin, copper, or zinc. The aggregate GNP of the peripheral economies is less than that of the European Economic Community (Van Evera 1990). Peripheral economies have a dependent rela-

Only a small portion of the world's population has access to a computer. An even smaller share has access to the Internet.

©Chris Sattlberger/Panos Pictures

tionship with core economies that is rooted in colonialism. Peripheral economies operate on the fringes of the world-economy. In the midst of widespread and chronic poverty are islands of economic activity, including off-shore manufacturing zones, highly vulnerable extractive and single-commodity economies, and tourist zones.

Between the core and the periphery are the **semiperipheral economies**, characterized by moderate wealth (but extreme inequality) and moderate diversification. Semiperipheral economies exploit peripheral economies and are exploited by core economies. By this definition, the Gulf Countries of Bahrain, Kuwait, Oman, Qatar, Saudia Arabia, and United Arab Emirates are semiperipheral economies. The core economies rely on them for cheap oil, and they depend on peripheral economies and other semiperipheral for cheap labor. The extent of that reliance is clear when we consider that 68 percent of the total labor force in these six countries are foreign workers. Egypt, Jordan, Yemen, Sudan, Bangladesh, India, Pakistan, and Sri Lanka are the major countries from which labor flows to these Gulf countries (Omran and Roudi 1993). According to Wallerstein, semiperipheral economies play an important role in the world-economy because they are politically stable enough to provide useful places for capitalist investment if wage and benefit demands become too great in core economies. As a case in point, the U.S. Department of Commerce has identified ten emerging markets, all ten of which can be classified as semiperipheral economies: Chinese Economic Area (which included Hong Kong and Taiwan), India, Indonesia, South Korea, Argentina, Brazil, Mexico, Poland, Turkey, and South Africa.

The chapter opening map identified the "have" and "have not" countries of the world in terms of Internet access. If we use the number of telephones per 100 people as a rough indicator of the potential of a country's population to connect to the Internet, we can see that access reflects that country's role in the global economy (see "How Connected Are We?"). Obviously, the number of telephones only speaks to potential access as only about 30 percent of people in the United States have access to a computer through home, work, or school.[14] In peripheral and semiperipheral economies, 85 percent or more of the people do not have a telephone. Thus, only a privileged elite have access to computers, and even for this elite poor and intermittent phone service constrains Internet use. The commercial uses of the Internet by peripheral and semiperipheral economies reinforce and perpetuate unequal roles in the global economy. For example, the Internet is used to promote the tourist industry, attract jobs, sell the few products the country exports, and solicit contributions and other forms of aid.

[14]Not everyone, however, with access to a computer has access to a computer with a modem. A U.S. Census Bureau survey found that in 1993 (the last year for which data are available) that only 35.3 percent of households with a computer had a telephone modem.

Semiperipheral economies Economies characterized by moderate wealth (but extreme inequality) and moderate diversification.

How Connected Are We?

The map shows about how many telephones there were per 1,000 people in each country, a rough indicator of the potential for a country's general population to connect with the Internet. If and when the Internet can be accessed via cellular phones, poor countries may be able to reduce the infrastructure costs of connection that comes with traditional wire connections – one less barrier to access.

In the United States, economic class is reflected in unequal access: 24 percent of households with income less that $5,000 do not have access to telephone service, whereas almost all households with incomes over $20,000 have access. The data also show large differences in access to phones when households are classified as black, Hispanic, or white.

Suppose you were the secretary of communications development for one of the poor, largely agricultural countries on the map on page 371. What sort of communications technology would be your first priority for the use of larger numbers of people? Would it be the telephone? Something else? If you wanted to dramatically increase the number of people with ready access to a telephone line, how might you try to make that happen? Are there certain groups of people more than others you would want to assist with telephone access? Who and why?

Sources: United Nations Development Programme (1993); U.S. telephone access data from *Statistical Abstract of the United States* (1995).

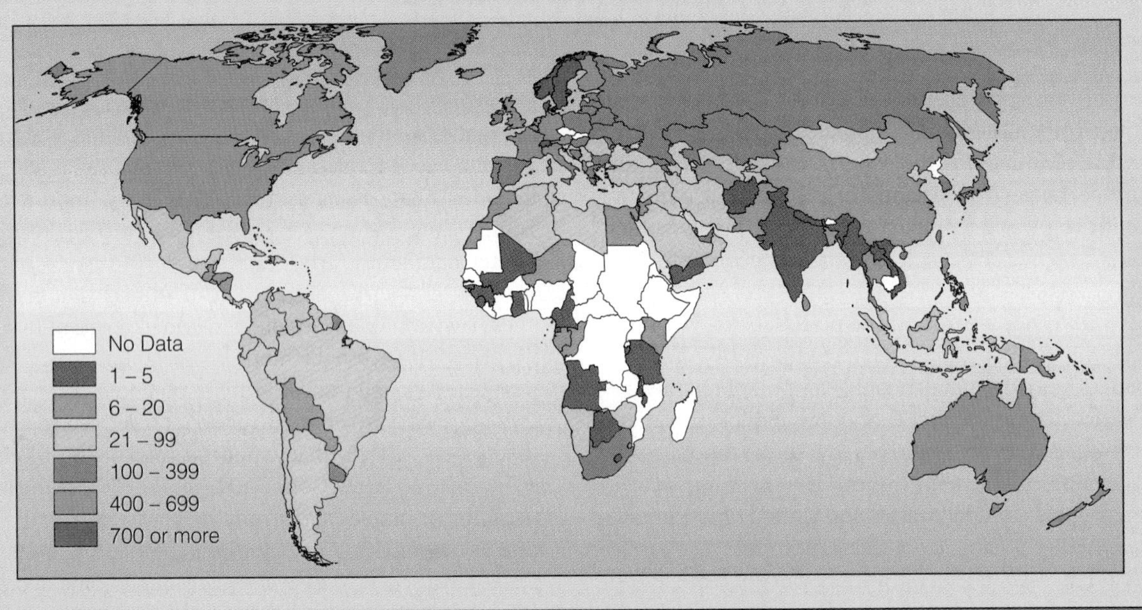

Summary and Implications

Sociologists have identified at least four main agents of change. These agents are innovations, the actions of leaders, conflict over scarce and valued resources, and capitalism. In this chapter, we have considered a major innovation — the Internet — billed by some as "a second printing press." We have studied the role each of the four agents of change have played in creating this technology and the changes (real and potential) associated with the Internet. We have learned that the Internet is a synthesis of a long trail of "existing inventions." It is a product of conflict (the Cold War) and deliberate planning by government leaders, defense contractors, and academic researchers. Contrary to popular opinion, the Internet does not operate independently of controlling organizations. From a sociological point of view, the access to the Internet is a scarce and valued resource that various organizations are attempting to control. These organizations include governments, Internet access providers, communication companies, hardware and software companies, colleges and universities, Internet societies, and grassroots organizations. From a global perspective, a person's access to the Internet is shaped to a large extent by their position in the capitalist-driven economic order. People who live in core economies, but especially the United States, are the most likely to have Internet access. But even within core economies, access to the Internet depends on one's ability to purchase access. From the perspective of a student whose school offers Internet access, it might seem that the Internet is "free." Consider, however, that Internet access is tied to the cost of tuition or a school district's ability to afford it.

By studying these four major agents of change, we can see the ways in which the character of the Internet has been shaped by the past and reflects current economic and political arrangements. The phenomenon globalization-from-above shows the power of the Internet to perpetuate and intensify current arrangements. On the other hand, the Internet has the potential to change the way people think and relate to one another that breaks from past arrangements. One of the best examples is the Internet's effect on conceptions of literacy. Other examples include its potential for invigorating globalization-from-below.

In studying social change, sociologists ask two key questions: What factor(s) are causing that change? What are the consequences of that change for social life? With regard to the first question, we have considered how the four major triggers of change contributed to the Internet's emergence and ongoing development and how the Internet has become a trigger of change as well. Keep in mind that we have covered concepts and theories in Chapters 1 through 14 that can help us conceptualize the consequences of the Internet. A sample of applications follow.

We can use the concept **global interdependence** to think about the Internet as an invention capable of connecting the lives of people around the world and facilitating action directed at solving social problems on a global scale (see Chapter 1). We can use the questions and vocabulary of the **three theoretical perspectives** to frame an analysis of the Internet and its consequences (see Chapter 2). The **methods of social research** can help us evaluate research and conduct studies related to the Internet and its consequences (Chapter 3). We can use the concept **cultural diffusion** to think about the Internet as a tool promoting cultural exchange (Chapter 4). The concepts associated with **socialization** and **resocialization** help us consider the process by which "newcomers" learn about the Internet and come to accept (or reject it) as a tool for solving the problems of living (Chapter 5). We can draw on the concept **solidarity** to frame an analysis of the Internet as a tool that binds (or fails to bind) people to one another (Chapter 6). The concept of **informal and formal aspects of organization** can help us think about the official and unofficial uses of the Internet in organizational settings (Chapter 7). With regard to **deviance, conformity, and social control**, we can think about the Internet as a mechanism of social control (for example, crimestopper Web sites) or a mechanism that needs to be controlled (Chapter 8). The sociological framework for analyzing **social stratification** as it relates to **social class, race, and gender** alerts us to the structural dynamics supporting inequality that affect people's access to the Internet (Chapters 9, 10, 11). We can consider the relationship between a country's role in the worldwide **division of labor** and its citizens' access to the Internet (Chapters 12 and 15). The concepts of **literacy** can help us understand the Internet's role in changing definitions of literacy (Chapter 13). Finally, with regard to religion, we can consider the Internet's role in promoting **secularization** and **fundamentalism** (Chapter 14).

The larger point behind listing possible approaches to studying the Internet is to show the power, vitality, relevance, and versatility of sociological concepts and theories for framing and explaining a variety of social phenomena and issues.

Key Concepts

Use this outline to organize your review of key chapter concepts.

Social change
- **Globalization-from-above**
- **Globalization-from-below**

Innovations
- **Discoveries**
- **Basic innovations**
- **Improving innovations**
- **Inventions**
- **Simultaneous-independent innovations**

Cultural lag
- **Adaptive culture**
- **Technological determinist**

Paradigms

Leader
- **Power**
- **Authority**
  - **Charismatic**
  - **Legal-rational**
    - **Great changes**
    - **Power elite**

World-economy

World economy

Core economies

Peripheral economies

Semiperipheral economies

internet assignment

Test your Internet searching and browsing skills. Find examples of how the Internet is changing the way people solve everyday problems other than those described in this chapter (for example, finding an apartment, determining a location, obtaining news). Find examples of Web sites that illustrate the phenomenon of globalization-from-above. Finally, find Web sites that exemplify globalization-from-below.

Key Concepts

Ability grouping The arranging of elementary school students into instructional groups according to similarities in past academic performance and/or on standardized test scores.

Achieved characteristics Attributes acquired through some combination of choice, effort, and ability. In other words, people must act in some way to acquire the attribute. Examples include occupation, marital status, level of education, and income.

Achieved status Positions earned or lost through effort and ability.

Active adaptation A biologically based tendency to adjust to and resolve environmental challenges.

Adaptive culture The portion of nonmaterial culture (norms, values, and beliefs) that adjusts to material innovations.

Advanced market economics An economic arrangement that offers widespread employment opportunities to women as well as to men.

Affectional A goal pursued in response to an emotion such as revenge, love, or loyalty.

Alienation A state in which human life is dominated by the forces of human inventions.

Altruistic A state of existence such that individuals have no life of their own and, in fact, strive to blend in with the group to have a sense of being.

Annual per capita consumption of energy The average amount of energy each person in a nation consumes over a year.

Anomaly An observation or observations that a paradigm cannot explain and that threaten the paradigm's explanatory value and hence its status. (See also *Paradigms.*)

Anomic A state of existence brought on by dramatic changes in economic circumstances—a recession, a depression, or economic boon.

Anomie See *Structural strain.*

Apartheid A system of laws in which everyone in South Africa was put into a racial category and issued an identity card denoting his or her race.

Archival data Data that have been collected by other researchers for some other purpose.

Ascribed characteristics Attributes that people (1) have at birth (such as skin color, gender, or hair color), (2) develop over time (such as baldness, gray hair, wrinkles, retirement, or reproductive capacity), or (3) possess through no effort or fault of their own (national origin or religious affiliation that was "inherited" from parents).

Ascribed status Positions that people acquire through no fault or virtue of their own.

Aspirational doll A doll that is a role model for a child.

Assimilation A process by which ethnic and racial distinctions between groups disappear.

Authority Legitimate power in which people believe that the differences in power are just and proper and that a leader is entitled to give orders.

Automate, to To use the computer to increase workers' speed and consistency, as a source of surveillance, and to maintain divisions of knowledge and thus a hierarchical arrangement between management and workers.

Back stage The region out of sight where individuals can do things that would be inappropriate or unexpected on the front stage.

Basic innovations Revolutionary, unprecedented, or ground-breaking ideas, practices, and tools that are the cornerstones for a wide range of applications.

Be absurd, to To make connections between unrelated things or events and not to care whether the connections are appropriate or inappropriate.

Beliefs Conceptions that people accept as true about how the world operates and about the place of the individual in the world.

Boundaries The characteristics determining where a culture begins and leaves off, or the qualities marking some people off from others as a unified and distinctive group.

Bourgeoisie The owners of the means of production. (See *Means of production.*)

Bureaucracy In theory, a completely rational organization—one that uses the most efficient means to achieve a valued goal.

Buyers Persons in a targeted territory to which commodities are marketed and sold.

Capitalism An economic system in which natural resources and the means of producing and distributing goods and services are privately owned.

Caste system Any scheme of social stratification in which people are ranked on the basis of traits over which they have no control.

Catechisms Short books covering religious principles written in question-and-answer format.

Charismatic authority Authority that rests on the exceptional and exemplary qualities of the person issuing the commands.

Church According to Durkheim, a group whose members hold the same beliefs with regard to the sacred and the profane, behave in the same way in the presence of the sacred, and gather together in body and spirit at agreed-on times to reaffirm their commitment to those beliefs and practices.

Civil religion "Any set of beliefs and rituals, related to the past, present and/or future of a people (nation), which are understood in some transcendental fashion" (Hammond 1976:171).

Claims makers People who articulate and promote claims and who tend to gain if the targeted audience accepts their claims as true.

Class A term sociologists use to denote a category that designates a person's overall status in society.

Classless society A propertyless society providing equal access to the means of production.

Class system Any scheme of social stratification in which people are ranked on the basis of merit, talent, ability, or past performance.

Cohort A group of people sharing a common characteristic or life event.

Collective memory Experiences shared and recalled by significant numbers of people.

Concepts Powerful thinking and communication tools that enable us to give and receive complex information efficiently and to frame and explain observations.

Confederate Someone who works in cooperation with the experimenter.

Conformists People who have not violated the rules of a group and are treated accordingly.

Conformity (1) Behavior and appearance that follow and maintain standards set by a group; (2) the acceptance of the cultural goals and the pursuit of these goals through legitimate means.

Connotation The set of associations that a word evokes.

Constrictive pyramids Population pyramids that are characteristic of some European societies, most notably Switzerland and western Germany, and that are narrower at the base than in the middle. This shape shows that the population is composed disproportionately of middle-aged and older people. (See also *Population pyramid.*)

Constructionist approach A sociological approach that focuses on the process by which specific groups, activities, conditions, or artifacts become defined as social problems.

Content (of interaction) The cultural factors (norms, values, beliefs, material culture) that guide interpretations, behavior, and dialogue during interaction.

Context (of interaction) The larger historical circumstances that bring people together.

Control variables Variables suspected of causing spurious correlations.

Core economies The highly diversified economies of mechanized rich nations characterized by strong, stable governments.

Corporate crime Crime committed by a corporation as it competes with other companies for market share and profits.

Correlation A relationship between two variables such that a change in one variable is associated with a change in another.

Correlation coefficient A mathematical representative of the extent to which a change in one variable is associated with a change in another.

Counterculture A subculture that conspicuously challenges, rejects, or clashes with the central norms and values of the dominant culture.

Country A political entity, recognized by foreign governments, with a civilian and military bureaucracy to enforce its rules.

Crime Deviance that breaks the laws of society and is punished by formal sanctions.

Cults Generally very small, loosely organized groups, usually founded by a charismatic leader who attracts people by virtue of his or her personal qualities.

Cultural base The number of existing inventions.

Cultural genocide The outcome of a situation in which people of one society define the culture of another society not only as offensive but as so intolerable that they attempt to destroy it.

Cultural lag The failure to adapt to a new invention; a situation in which adaptive culture fails to adjust in necessary ways to a material innovation. (See *Adaptive culture*.)

Cultural markers Distinctive characteristics that are used to clearly classify people into distinct cultural units.

Cultural relativism A perspective in which (1) a foreign culture is not judged by the standards of the home culture and (2) a behavior or way of thinking must be examined in its cultural context.

Culture A way of life, especially general customs and beliefs of a particular group of people at a particular time.

Culture gap Ways of thinking and behaving that vary according to culture.

Culture shock The strain that people from one culture experience when they must reorient themselves to the ways of a new culture.

Data Printed, visual, and spoken materials. Data become information after someone reads, listens to, or views the data.

Dearth of feedback A factor in creating poor-quality data. Much of the data that are televised and published are not subject to honest, constructive feedback because there are too many messages and not enough critical readers and listeners to evaluate the data before the material is released or picked up by the popular media. Without feedback, the creators cannot correct their mistakes; thus, the data they produce diminish in quality.

Demographers Sociologists who study population size, distribution, and age-sex structure.

Demographic gap The difference between birthrates and death rates.

Demographic trap The point at which population growth overwhelms the environment's carrying capacity.

Demography A subdiscipline within sociology that studies population trends.

Denomination A formal, hierarchical, well-integrated organization in a society in which church and state are usually separate.

Denotation The literal definition of a word.

Dependent variable The behavior to be explained or predicted by a change in the independent variable. (See also *Variable*.)

Deviance Any behavior or physical appearance that is socially challenged and condemned because it departs from the norms and expectations of a group.

Deviant subcultures Groups that are part of the larger society but whose members adhere to norms and values that favor violation of the larger society's laws.

Differential association A theory of socialization that explains the origins of delinquent behavior. It refers to the idea that "when persons become criminal, they do so because of contacts with criminal patterns and also because of isolation from anti-criminal patterns" (Sutherland and Cressey 1978:78).

Diffusion The process by which an idea, an invention, or some other cultural item is borrowed from a foreign source.

Discoveries The uncovering of something that had existed before but had remained hidden, unnoticed, or undescribed.

Discrimination The intentionally or unintentionally unequal treatment of individuals or groups on the basis of attributes unrelated to merit, ability, or past performance. The treatment may be based on such attributes as skin color, weight, religion, ethnicity, or social class. Discrimination is behavior aimed at denying members of minority groups equal opportunities to achieve valued social goals (education, health care, long life) and/or blocking their access to valued goods and services.

Disenchantment A great spiritual void accompanied by a crisis of meaning. It occurs when people focus so uncritically on the ways they go about achieving a valued goal that they lose sight of that goal.

Disenchantment of the world Max Weber's phrase for a great spiritual void accompanied by a crisis of meaning.

Dispositional traits Personal or group traits such as motivation level, mood, and inherent ability.

Division of labor Work broken down into specialized tasks, with each task performed by a different set of persons.

Documents Written or printed material, such as magazines, books, calendars, graffiti, birth certificates, and traffic tickets.

Dominant group The ethnic and racial group at the top of the hierarchy.

Doubling time The estimated number of years required for a country's population to double in size.

Downward mobility A change in social class that corresponds to a loss in rank or prestige

Dramaturgical model A model in which interaction is viewed as though it were theater, people as though they were actors, and roles as though they were performances presented before an audience in a particular setting.

Dysfunctions Parts that have disruptive consequences to the system or to some segments of society.

Ecclesiae A formal and well-integrated religious organization led by a hierarchy of leaders that claims as its members everyone in a society.

Education Those experiences that train, discipline, and develop the mental and physical potentials of the maturing person.

Egoistic A state in which the ties attaching the individual to others in the society are weak.

Emigration The departure of individuals from a country.

Engram Physical traces formed by chemicals produced in the brain that store the recollections of experiences.

Established sects Religious organizations that share characteristics of denominations and sects. They are renegades from denominations or ecclesiae but have existed long enough to acquire a significantly large membership and to achieve respectability.

Ethgender A term describing people who share (or are believed by themselves or others to share) the same sex and race and ethnicity.

Ethnic blending "Inter-ethnic unions (interbreeding) and shifts in ethnic affiliation" (Hirschman 1993:549).

Ethnic group A shared national origin, cultural traits, or distinctive physical features.

Ethnicity A term that is used to classify people according to any number of attributes, including national origin, ancestry, distinctive and visible cultural traits (religious practice, dietary habits, style of dress, body ornaments, or languages) and/or socially important physical characteristics.

Ethnocentrism Using one culture as the standard for judging the worth of foreign ways.

Everyday mingling People's routine talking, looking, and/or listening as they live their lives. The settings in which people mingle are endless; they include the workplace, the home, the neighborhood, and the school.

Expansive pyramids Population pyramids that are characteristic of labor-intensive poor nations. They are triangular in shape, broadest at the base, and each successive bar is smaller than the one below it. The relative sizes of the age cohorts in expansive pyramids show that the population is increasing in size and that it is composed disproportionately of young people. (See also *Population pyramid.*)

Externality costs Costs that are not figured into the price of a product but that are nevertheless a price we pay for using or creating a product. An example of externality costs is the cost of restoring contaminated and barren environments and of assisting people to cope.

Facade of legitimacy An explanation that members in dominant groups give to justify their actions; a justifying ideology.

Falsely accused People who have not broken the rules but who are treated as if they have done so.

Family Two or more people related to one another by blood, marriage, adoption, or some other socially recognized criteria.

Fatalistic The opposite of anomic; describes a state in which there is no hope of change and thus oppressive discipline against which there is no chance of appeal or release.

Feeling rules Forms specifying appropriate ways to express the internal sensations that one should feel toward another person.

Femininity (or feminine characteristics) A term that signifies the physical, behavioral, and mental or emotional traits believed to be characteristics of females.

Feminist A man or woman who actively opposes gender scripts and believes that men's and women's self-image, aspirations, and life chances should not be constrained by those scripts.

Folkways Norms that apply to the mundane aspects or details of daily life.

Formal curriculum The various subjects such as mathematics, science, English, reading, physical education, and so on. (See *Hidden curriculum.*)

Formal dimension (of an organization) The official, written guidelines, rules, regulations, and policies that define the goals of the organization and its relationship to other organizations and integral parties. This term also applies to the roles, the nature of the relationships among roles, and the way in which tasks should be carried out to realize the goals.

Formal education A purposeful, planned effort intended to impart specific skills and modes of thought. (See *Informal education* and *Schooling* also.)

Formal sanctions Definite and systematic laws, rules, regulations, and policies that specify (usually in writing) the conditions under which people should be rewarded or punished, and that define the procedures for allocating rewards and imposing punishments. Examples of formal sanctions include medals, cash bonuses, diplomas, fines, prison sentences, and the death penalty. (See *Informal sanctions.*)

Fortified households Preindustrial arrangements in which there is no police force, militia, national guard, or other peacekeeping organization. The household is an armed unit, and the head of the household is its military commander.

Front stage The region where people take care to create and maintain expected images and behavior.

Function The contribution of a part to the larger system and its effect on other parts in the system.

Functionally illiterate A significant percentage of the population in every country that do not possess the level of reading, writing, and calculating skills needed to adapt to the society in which they live.

Fundamentalism A complex religious phenomenon that involves a belief in the timeless nature of sacred writings, and a belief that such writings apply to all kinds of environments.

Games Structured and organized activities that almost always involve more than one person.

Gender Social distinctions based on culturally conceived and learned ideas about appropriate behavior, appearance, and mental or emotional characteristics for males and females.

Gender nonconformists (1) Persons whose primary characteristics are not clear-cut (the intersexed); (2) those whose secondary characteristics depart from the ideal conceptions of masculinity and femininity; (3) those whose interests, feelings, sexual orientation, choice of occupation, or academic major do not match gender-polarized scripts; and (4) those "who actively oppose the gender scripts of the culture" (Bem 1993:167).

Gender polarization "The organizing of social life around the male-female distinction, so that people's sex is connected to virtually every other aspect of human experience, including modes of dress, social roles, and even ways of expressing emotion and experiencing sexual desire" (Bem 1993:192).

Gender schematic decisions Choices related to any aspect of life that are influenced by society's polarized definitions of masculinity and femininity rather than on the basis of other criteria such as self-fulfillment, interest, ability, or personal comfort.

Generalizability The extent to which the findings of a research project can be applied to the larger population from which the sample is drawn.

Generalized other A system of expected behaviors, meanings, and points of view.

Global interdependence A state in which the lives of people around the world are intertwined closely and in which each country's problems—unemployment, employment, drug abuse, pollution, inequality, disease—are part of a larger global situation.

Globalization-from-above Global interdependence characterized by connections among people from around the world with educational, economic, and political advantages, excluding those who are not so advantaged.

Globalization-from-below Global interdependence at the grassroots level that aims to protect, restore, and nurture the environment; enhance ordinary people's access to the basic resources they need to live a dignified existence; democratize local, national, and transnational political institutions; and ease tensions and prevent violent conflict between power centers and authority structures.

Great changes Events whose causes lie outside ordinary people's characters or their immediate environments but profoundly affect their life chances.

Group Two or more people who share a distinct identity, feel a sense of belonging, and interact with one another in direct or indirect, but broadly predictable, ways.

Hate crimes Actions aimed at humiliating members of a minority group and destroying their property or lives.

Hawthorne effect A phenomenon whereby observed persons alter their behavior when they learn they are being observed.

Hidden curriculum All the things that students learn along with the subject matter.

Hidden rape Rape that goes unreported.

Household All related and unrelated persons who share the same dwelling.

Hypertext One software tool that has helped increase the amount of data available at our fingertips.

Hypothesis A trial explanation put forward as the focus of research that predicts how the independent and dependent variables are related. (See also *Independent variable* and *Dependent variable.*)

Ideal type A standard against which real cases can be compared.

Ideologies Fundamental ideas that support the interests of dominant groups.

Ideology A set of ideas that do not hold up under the rigors of scientific investigation and that support the interests of dominant groups.

Idiom A group of words that when taken together have a meaning different from the internal meaning of each word understood on its own.

Illiteracy The inability to understand and use a symbol system, whether it is based on sounds, letters, numbers, pictographs, or other type of character.

Immigration The entrance of individuals into a new country.

Impression management The process by which people in social situations manage the setting, their dress, their words, and their gestures to correspond to the impressions they are trying to make or the image they are trying to project.

Improving innovations Modifications of basic inventions to improve on them—that is, to make them smaller, faster, less complicated, or more efficient, attractive, durable, or profitable. (See also *Basic innovations.*)

Independent variable The variable that explains or predicts the dependent variable; a change in this variable brings about a change in the dependent variable. (See also *Variable.*)

Individual discrimination Any overt action on the part of an individual that depreciates someone from the outgroup, denies the outgroup opportunities to participate, or does violence to lives and property.

Infant mortality The number of deaths in the first year of life for every 1,000 live births.

Informal dimensions (of an organization) Those dimensions of organizational life that include worker-generated norms that evade, bypass, do not correspond with, or are not systematically stated in official policies, rules, and regulations.

Informal education Education that occurs in a spontaneous, unplanned way.

Informal sanctions Spontaneous and unofficial expressions of approval or disapproval; they are not backed by the force of law.

Informate, to To use the computer to empower workers with knowledge of the overall production process, with the expectation that they will make critical and collaborative judgments about production tasks.

Information Data that someone has read, listened to, or viewed.

Information explosion An unprecedented increase in the volume of data due to the development of the computer and telecommunications.

Ingroups Those groups with which people identify and to which they feel closely attached, particularly when that attachment is founded on hatred for another group.

In-migration The movement of people into a designated area. (See *Out-migration.*)

Innovation (as a response to structural strain) The acceptance of the cultural goals but the rejection of legitimate means to obtain these goals. For the innovator, success means winning the game rather than playing by the rules of the game.

Innovation The development of something new—an idea, a practice, or a tool.

Institutionalized discrimination The established and customary ways of doing things in society; the unchallenged rules, policies, and day-to-day practices, which impede or limit minority members' achievements and keep them in a subordinate and disadvantaged position.

Institutionally complete A term that describes a subculture whose members do not interact with anyone outside the subculture.

Instrumental A goal pursued after it has been evaluated in relation to other goals and after thorough consideration of the various means to achieve it.

Intergenerational mobility A change in social class over two or more generations.

Internalization To take as one's own and accept as binding the norms, values, beliefs, and language needed to participate in the larger community.

Internal migration Movement within the boundaries of a single nation from one state, region, or city to another.

International migration Movement of people between countries.

Intersexed A term used to classify people with some mixture of male and female biological characteristics.

Interviews Face-to-face sessions or telephone conversations between an interviewer and a respondent in which the interviewer asks questions and records respondents' answers.

Intragenerational mobility A change in social class during an individual's lifetime.

Involuntary minorities Ethnic and racial groups that do not choose to be a part of a country.

Islamic activism See *Islamic revitalism.*

Islamic revitalism "A sense that existing political, economic, and social systems had failed; a disenchantment with, and at times a rejection of the West; a quest for identity and greater authenticity; and the conviction that Islam provides a self-sufficient ideology for state and society, a valid alternative to secular nationalism, socialism, and capitalism" (Esposito 1992:14).

Issues Matters that can be explained by factors outside an individual's control and immediate environment.

Labor-intensive poor A country differs markedly on doubling time, infant mortality, total fertility, per capita income, annual per capita consumption of energy, and other indicators from countries considered to be industrialized.

Latent dysfunctions The unintended, unanticipated negative consequences that a part causes in some segment of society.

Latent functions The unintended, unrecognized, and unanticipated or unpredicted consequences that contribute to the smooth operation of the system.

Legal-rational authority Power that rests on a system of impersonal rules that formally specifies the qualifications for occupying a powerful position.

Liberation theology An approach to the role of religion in society maintaining that organized religions have a responsibility to demand social justice for the marginalized peoples of the world, especially landless peasants and the urban poor, and to take an active role at the grassroots level to bring about political and economic justice.

Life chances Opportunities that include "everything from the chance to stay alive during the first year after birth to the chance to view fine art, the chance to remain healthy and grow tall, and if sick to get well again quickly, the chance to avoid becoming a juvenile delinquent—and very crucially, the chance to complete an intermediary or higher educational grade" (Gerth and Mills 1954:313).

Looking-glass self A term coined by Charles Horton Cooley to describe the way in which a sense of self develops: we visualize how we appear to others, we imagine a judgment of that appearance, and we develop a feeling somewhere between pride and shame.

Low-technology tribal societies Hunting-and-gathering societies with technologies that do not permit the creation of surplus wealth, or wealth beyond what is needed to meet the basic needs (food and shelter).

Manifest dysfunctions The expected or anticipated disruptions to order and stability that a part causes in some segment of the system.

Manifest functions The intended, recognized, expected, or predictable consequences of a given part of the social system for the whole.

Market A situation that involves transactions between buyers and sellers. (See *Buyers* and *Sellers.*)

Masculinity (or masculine characteristics) A term that signifies physical, behavioral, and mental or emotional traits believed to be characteristics of males.

Master status of deviant An identification that "proves to be more important than most others. One will identified as a deviant first, before other identifications are made" (Becker 1963:33).

Material culture All of the physical objects people have borrowed, discovered, or invented and to which they have attached meaning. It includes natural resources such as plants, trees, and minerals or ores, as well as items people have converted from natural resources into other forms for a purpose.

Means of production The resources—land, tools, equipment, factories, transportation, and labor—essential to the production and distribution of goods and services.

Mechanical solidarity Social order and cohesion based on a common conscience or uniform thinking and behavior.

Mechanisms of social control All of the methods that people employ to teach, persuade, or force others to conform.

Mechanization The addition of external sources of power such as oil or steam to hand tools and to modes of transportation.

Melting pot assimilation A process of cultural blending in which the groups involved accept many new behaviors and values from one another.

Methods of data collection The procedures used to gather relevant data.

Migration The movement of people from one area to another.

Minorities The ethnic and racial groups at the bottom of the hierarchy.

Minority groups Subgroups within a society that can be distinguished from members of the dominant groups by visible and identifying characteristics, including physical and cultural attributes.

Mixed contacts Social situations in which the stigmatized and normals are in each other's company. (See also *Normals.*)

Modern capitalism "A form of economic life which involved the careful calculation of costs and profits, the borrowing and lending of money, the accumulation of capital in the form of money and material assets, investment, private property, and the employment of laborers and employees in a more or less unrestricted labor market" (Robertson 1987:6).

Mores Norms that people define as essential to the well-being of the group.

Mortality crises Frequent and violent fluctuations in the death rate caused by war, famine, and epidemics, during which time the death rate has no limit.

Multinational corporation An enterprise that owns or controls production or service facilities in countries outside the one in which it is headquartered.

Mystical religions Religions in which the sacred is sought in states of being that, at their peaks, can exclude all awareness of one's existence, sensations, thoughts, and surroundings.

Nation A geographic area occupied by the world economy.

Nature Human genetic makeup or biological inheritance.

Negative correlation When one variable increases and the other variable decreases.

Negative sanction An expression of disapproval for noncompliance; the punishment may be withdrawal of affection, ridicule, ostracism, banishment, physical harm, imprisonment, solitary confinement, or even death. (See *Positive sanction.*)

"Negatively privileged" property class Persons completely unskilled, lacking property, and dependent on seasonal or sporadic employment that constitute the very bottom of the class system.

Nonhouseholder class Propertyless laborers and servants usually residing within fortified households. (See *Fortified households.*)

Nonmaterial culture Intangible creations or things that we cannot identify directly through the senses.

Nonparticipant observation A research procedure that involves detached watching and listening; the researcher only observes and does not interact or become involved in the daily life of those being studied.

Nonprejudiced nondiscriminators (all-weather liberals) Persons who accept the creed of equal opportunity and whose conduct conforms to that creed.

Normals Those people who are in the majority or those who possess no discrediting attributes.

Norms The written and unwritten rules that specify behaviors appropriate and inappropriate to particular social situations.

Nurture The environment or the interaction experiences that make up every individual's life.

Objective secularization The decline in the control of religion over education, medicine, law, and politics, and the emergence of an environment in which people are free to choose from many equally valid religions the one to which they wish to belong.

Objectivity A position taken by researchers in which they do not let personal and subjective views about the topic influence the outcome of the research.

Obligations The relationship and behavior that a person enacting a role must assume toward others in a particular status.

Observation A method of data gathering involving watching, listening to, and recording behavior and conversations as they happen.

Oligarchy Rule by the few, or the concentration of decision-making power in the hands of a few persons who hold the top positions in an organization's hierarchy.

Operational definitions Clear and precise definitions and instructions about how to observe and measure the variables being studied.

Organic solidarity Social order based on interdependence and cooperation among people performing a wide range of diverse and specialized tasks.

Organization A coordinating mechanism created by people to achieve stated objectives—whether to maintain order; challenge an established order; keep track of people; grow, harvest, or process food; to produce goods; or provide a service.

Outgroup A group of individuals toward which members of an ingroup feel a sense of separateness, opposition, or even hatred. (See also *Ingroup*.)

Out-migration Movement out of a designated area. (See *In-migration*.)

Paradigms The dominant and widely accepted theories and concepts in a particular field of study.

Participant observation A research procedure in which a researcher joins a group and assumes the role of a group member, interacts directly with those who are being studied, assumes a position critical to the outcome of the study, or lives in a community under study.

Per capita income The average share of income that each person in a country would receive if the country's gross national product were divided evenly.

Peripheral economies The economies of labor-intensive poor countries, in which most of the jobs are low-paying and require few skills. Peripheral economies are not highly diversified.

Play A voluntary and often spontaneous activity with few or no formal rules, which is not subject to constraints of time.

Political parties Organizations "oriented toward the planned acquisition of social power [and] toward influencing social action no matter what its content may be" (Weber 1982:68).

Population The total number of individuals, traces, documents, territories, households, or groups that could be studied.

Population (study of) A specialty within sociology that focuses on the number of people in and composition of various social groupings that live within specified boundaries and the factors that lead to changes in that social grouping's size and composition.

Population pyramid A series of horizontal bar graphs each of which represents a different five-year age cohort. (See *Cohort*.) Two bar graphs are constructed for each cohort, one for males and another for females; the bars are placed end to end, separated by a line that represents zero. Usually, the left-hand side of the pyramid depicts the number or percentage of males that make up each age cohort, and the right-hand side depicts the number or percentage of females. The graphs are stacked according to age; the age 0–4 cohort forms the base of the pyramid, and the 80+ cohort is at the apex of the pyramid. The population pyramid allows us to view the relative sizes of the age cohorts and compare the relative number of males and females.

Positive checks Events that increase mortality, including epidemics of infectious and parasitic disease, war, and famine.

Positive correlation When the value of one variable increases and there is a corresponding increase in the other variable.

Positive sanction An expression of approval and a reward for compliance. Such a sanction may take the form of applause, an approving smile, or a pat on the back.

"Positively privileged" social class Those at the very top of the social class who monopolize the purchase of the highest-priced consumer goods, have access to the most socially advantageous kinds of education, control the highest executive positions, own the means of production, and live on income from property and other investments.

Power elite Those few people positioned so high in the social structure of leading institutions that their decisions have consequences that affect millions of people worldwide.

Predestination The belief that God has foreordained all things, including the salvation or damnation of individual souls.

Prejudice A rigid and usually unfavorable judgment about an outgroup that does not change in the face of contradictory evidence and that applies to anyone who shares the distinguishing characteristics of that group.

Prejudiced discriminators (active bigots) Persons who reject the notion of equal opportunity and profess a right, even a duty, to discriminate.

Prejudiced nondiscriminators (timid bigots) Persons who do not accept the notion of equal opportunity but who refrain from discriminatory actions primarily because they fear the sanctions they may encounter if they are caught.

Primary groups Major socializing agents, especially in the early years, because they give newcomers their first exposure to the "rules of life." These groups are characterized by face-to-face contact and strong ties among members.

Primary sex characteristics The anatomical traits essential to reproduction. Most cultures divide the population into two categories—male and female—largely on the basis of what most people consider to be clear anatomical distinctions.

Private households An arrangement that exists when the workplace is separate from the home, men are heads of households and assume a breadwinner role, and women remain responsible for housekeeping and childrearing.

Probabilistic model A model in which the hypothesized effect does not always result from a hypothesized cause.

Profane Everything that is not sacred, including those things opposed to the sacred (the unholy, the irreverent, the contemptuous, the blasphemous) and those things that, although not opposed to the sacred, stand apart from it (the ordinary, the commonplace, the unconsecrated, the temporal, the bodily).

Professionalization (within organizations) A hiring trend in organizations in which experts are hired who have formal training in a particular subject or activity that is essential to achieving organizational goals.

Proletariat Those who must sell their labor to the bourgeoisie.

Prophetic religions Religions in which conceptions of the sacred revolve around items that symbolize significant historical events or around the lives, teachings, and writings of great people.

Pull factors The conditions that encourage people to move into a particular area. Some of the most common pull factors are employment opportunities, favorable climate, and tolerance.

Pure deviants People who have broken the rules and are caught, punished, and labeled as outsiders.

Push factors The conditions that encourage people to move out of an area. Some of the most common push factors include religious or political persecution, discrimination, depletion of natural resources, lack of employment opportunities, and natural disasters.

Race A term that refers to a group of people who possess certain distinctive physical characteristics.

Racism An ideology that maintains that something in the biological makeup of a specific racial or ethnic group explains its subordinate or superior status.

Random sample A sample drawn in such a way that every case in the population has an equal chance of being selected.

Rationalization As defined by Max Weber, a process in which thought and action rooted in emotion (love, hatred, revenge, joy), superstition, respect for mysterious forces, and tradition are replaced by thought and action grounded in the logical assessment of cause and effect or means and ends.

Rebellion The full or partial rejection of both goals and means and the introduction of a new set of goals and means.

Reentry shock Culture shock in reverse; experienced upon returning home after living in another culture.

Reflexive thinking Stepping outside the self to observe and evaluate it from another's viewpoint.

Refugee "Someone who has a well-founded fear of persecution on the basis of his or her race, religion, nationality, political opinion or membership in a particular social group" (Walsh 1993:A6).

Reliability The extent to which the operational definition gives consistent results.

Religion According to Emile Durkheim, a system of shared beliefs and rituals about the sacred that bind together a community of worshipers. (See also *Sacred.*)

Representative sample A sample with the same distribution of characteristics (such as age, gender, and ethnic composition) as the population from which it was selected.

Research A fact-gathering and fact-explaining enterprise governed by strict rules.

Research design A plan for gathering data to test hypotheses.

Research methods The various techniques that sociologists and other investigators use to formulate meaningful research questions and to collect, analyze, and interpret facts in ways that allow other researchers to check the results.

Research methods literate The ability to know how to collect data that are worth putting into the computer and how to interpret the data that come out of it.

Resocialization The process of discarding values and behaviors unsuited to new circumstances and replacing them with new, more appropriate values and standards of behavior.

Retreatism The rejection of both cultural goals and the means of achieving these goals.

Reverse ethnocentrism The tendency to see the home culture as inferior to a foreign culture.

Rights The behaviors that a person assuming a role can demand or expect from others depending on their status.

Ritualism The rejection of cultural goals but a rigid adherence to the legitimate means of those goals. It is the opposite of innovation (as a response to structural strain); the game is played according to the rules despite defeat.

Rituals Rules that govern how people must behave when in the presence of the sacred to achieve an acceptable state of being. These rules may take the form of instructions detailing the appropriate context, the roles of various participants, acceptable attire, and the precise wording of chants, songs, and prayers.

Role The behavior expected of a status in relation to another status.

Role conflict A predicament in which the expectations associated with two or more roles in a role set are contradictory.

Role set The array of roles associated with every status.

Role strain A predicament in which contradictory or conflicting expectations are associated with the role that a person is occupying.

Role taking Stepping outside the self and viewing its appearance and behavior imaginatively from an outsider's perspective.

Routinized charisma A situation in which the community must establish procedures, rules, and traditions to regulate the members' conduct, to recruit new members, and to ensure the orderly transfer of power.

Sacramental religions Religions in which the sacred is sought in places, objects, and actions believed to house a god or a spirit.

Sacred All phenomena that are regarded as extraordinary and that inspire in believers deep and absorbing sentiments of awe, respect, mystery, and reverence.

Sample A portion of cases from a larger population.

Sampling frame A complete list of every case in the population.

Sanctions Reactions of approval and disapproval to behavior and appearances. Sanctions can be positive or negative, formal or informal.

Scapegoat A person or a group that is assigned blame for conditions that cannot be controlled, threaten a community's sense of well-being, or shake the foundations of a trusted institution.

Schooling A program of formal and systematic instruction that takes place primarily in a classroom but also includes extracurricular activities and out-of-classroom assignments.

Scientific method An approach to data collection guided by two assumptions: (1) knowledge about the world is acquired through observation, and (2) the truth of the knowledge is confirmed by verification by other persons making the same observations.

Scientific revolution A condition that occurs when enough people in the community break with an old paradigm and change the nature of their research in favor of the incompatible new paradigm.

Secondary sex characteristics Physical traits not essential to reproduction (breast development, quality of voice, distribution of facial and body hair, and skeletal form) that result from the action of so-called male hormones (androgen) and female hormones (estrogen). Although testes produce androgen and ovaries estrogen, the adrenal cortex produces androgen and estrogen in both sexes.

Secondary sources Data that have been collected by other researchers for some other purpose.

Secret deviants People who have broken the rules but whose violation goes unnoticed, or, if it is noticed, no one reacts to enforce the law.

Sect A small community of believers led by a lay ministry, with no formal hierarchy or official governing body to oversee the various religious gatherings and activities.

Sectarian community A geographically distinct group of people who profess allegiance to a particular religion, such as Islam or Christianity, and who also have strong ties to a powerful family, clan, or ethnic group.

Secularization A process by which religious influences on thought and behavior are reduced.

Selective perception The process by which prejudiced persons notice only those behaviors or events that support their stereotypes about an outgroup.

Self-administered questionnaire A set of questions given or mailed to respondents, who read the instructions and fill in the answers themselves.

Self-fulfilling prophecy A concept that begins with a false definition of a situation. The false definition is assumed to be accurate, and people behave as if the definition were true. In the end, the misguided behavior produces responses that confirm the false definition.

Sellers Persons who conduct one or more of the extensive transactions (producing, advertising, shipping, storing, selling) needed to sell a commodity or service.

Semiperipheral economies The moderately diversified economies of moderately wealthy countries.

Sexist ideologies The ideologies that justify one sex's social, economic, and political dominance over the other.

Sexual property "The relatively permanent claim to exclusive sexual rights over a particular person" (Collins 1971:7).

Sick role A term coined by sociologist Talcott Parsons to represent the rights and obligations accorded people when they are sick.

Significant others People or characters who are important in a child's life in that they have considerable influence on a child's self-evaluation and encourage the child to behave in a certain manner.

Significant symbols Words, gestures, and other learned signs that are used to convey a meaning from one person to another and are the same for both the communicator and the recipient. Particularly important significant symbols are language and symbolic gestures.

Simultaneous-independent inventions Situations in which the same invention is created by two or more persons working independently of one another at about the same time (sometimes within a few days or months).

Situational factors Forces outside an individual's control, such as environmental conditions or bad luck.

Small groups Groups of two to about twenty people who interact with one another in meaningful ways.

Social action Behavior or actions that people take in response to others with emphasis on the meanings that the involved parties attach to their behavior.

Social change Any alteration, modification, or transformation of social phenomena over a specified period.

Social control The methods used to teach, persuade, or force their members, and even nonmembers, to comply and not to deviate from norms and expectations.

Social emotions Internal bodily sensations that people experience in relationships with others.

Social identity The category to which a person belongs and the qualities that others believe, rightly or wrongly, to be "ordinary and natural" (Goffman 1963:2) for a member of that category.

Social interactions Everyday events in which at least two people communicate, interpret, and respond to affect one another's behavior and thinking.

Social mobility Movement from one class to another.

Social movements Organized, deliberate efforts by believers to transform, reform, or replace some element of culture, and in the process to convert nonbelievers to their position.

Social promotion Passing students from one grade to another on the basis of age rather than academic competency.

Social relativity The view that ideas, beliefs, and behavior vary according to time and place.

Social status A position in a social structure.

Social stratification A systematic process by which people are divided into categories that are ranked according to a scale of social worth.

Social structure Two or more people interacting and interrelating in specific expected ways, regardless of the unique personalities involved.

Socialization A complex, lifelong process of learning about the social world.

Society A group of people living in a given territory who share a culture and interact with people of that territory more than with people of another territory.

Sociological imagination The ability to connect seemingly impersonal and remote historical forces to the most basic incidents of an individual's life. The sociological imagination enables people to distinguish between personal troubles and public issues.

Sociological theory A set of principles and definitions that tell how societies operate and how people relate to one another.

Sociology The systematic study of social interaction.

Solidarity The ties that bind people to one another in a society.

Spurious correlation A correlation that is coincidental or accidental.

State (1) A political entity recognized by foreign governments, with a civilian and military bureaucracy to carry out its policies, enforce its rules, and regulate other activities within its borders; (2) a governing body organized to manage and control specified activities of people living in a given territory.

Stationary pyramids Population pyramids that are characteristic of most developed nations and that are similar to constrictive pyramids except that all of the age cohorts in the population are roughly the same size, and fertility is at replacement level. (See also *Population pyramid* and *Constrictive pyramids.*)

Status group A plurality of persons held together by virtue of a common lifestyle, formal education, family background, or occupation and "by level of social esteem and honor accorded them by others" (Coser 1977:229).

Status system A classification of achievements resulting in popularity, respect, and acceptance into the crowd versus disdain, discouragement, and disrespect.

Status value A situation in which persons who possess one feature of a characteristic (white skin versus brown skin, blond hair versus dark hair) are regarded and treated as more valuable or worthy than persons who possess other categories (Ridgeway 1991).

Stereotypes Exaggerated and inaccurate generalizations about people who are members of an outgroup.

Stigmas Statuses that are deeply discrediting in the sense that they overshadow all other statuses that a person occupies.

Streaming The arranging of middle school and high school students into instructional groups according to similarities in past academic performance and/or on standardized test scores.

Structural constraints The established and customary rules, policies, and day-to-day practices that affect a person's life chances.

Structural strain (or anomie) A situation in which (1) the valued goals have no clear limits, (2) people are unsure whether the legitimate means that society provides will lead to the valued goals, and (3) legitimate opportunities for meeting the goals are closed to a significant portion of the population.

Structure of opportunities The chances available in a society to achieve a valued goal.

Structured interview An interview in which the wording and the sequence of questions are set in advance and cannot be altered during the course of the interview.

Subcultures Groups that share in some parts of the dominant culture but have their own distinctive values, norms, language, or material culture.

Subjective secularization A decrease in the number of people who view the world and their place in it from a religious perspective.

Symbol Any kind of physical phenomenon—a word, an object, a color, a sound, a feeling, an odor, a movement, a taste—to which people assign a meaning or value.

Symbolic gestures Nonverbal cues that include tone of voice, inflection, facial expression, posture, and other body movements or positions that convey meaning from one person to another.

Technological determinist Someone who believes that human beings have no free will and are controlled entirely by their material innovations.

Technology The knowledge, skills, and tools used to transform resources into forms with specific purposes, and the skills and knowledge required to use them.

Territories Settings that have borders or that are set aside for particular activities.

Theory A framework that can be used to comprehend and explain events.

Theory of the demographic transition A model that outlines historical changes in birth- and death rates among the mechanized rich countries and the factors underlying those changes. Some demographers have theorized that a country's birth- and death rates are linked to its level of industrial or economic development.

This-worldly asceticism A belief that people are instruments of divine will and that their activities are determined and directed by God.

Total fertility The average number of children women bear over their lifetime.

Total institutions Settings in which people surrender control of their lives, voluntarily or involuntarily, to an administrative staff and in which they (as inmates) carry out daily activities in the presence of other inmates.

Traces Materials or other evidence that yield information about human activity, such as the items that people throw away, the number of lights on in a house, or changes in water pressure.

Tracking The arranging of middle school and high school students into instructional groups according to similarities in past academic performance and/or on standardized test scores.

Traditional A goal pursued because it was pursued in the past.

Trained incapacity The inability to respond to new and unusual circumstances or recognize when official rules and procedures are outmoded or no longer applicable.

Transformative powers of history The dramatic consequences of important historical events on people's opportunity.

Troubles Personal problems and difficulties that can be explained in terms of individual characteristics such as motivation level, mood, personality, or ability.

Unit of analysis Who or what is to be studied in a research project.

Unprejudiced discriminators (fair-weather liberals) Persons who believe in equal opportunity but engage in discriminatory behaviors because it is to their advantage to do so or because they fail to consider the discriminatory consequences of some of their actions.

Unstructured interview An interview that is flexible and open-ended; the question-answer sequence is spontaneous and like a conversation in that the questions are not worded in advance and are not asked in a set order.

Upward mobility A change in social class that corresponds to a gain in rank or prestige.

Urbanization An increase in the number of cities and in the proportion of the population living in cities.

Urban underclass The ghetto poor, or a "heterogeneous grouping of families and individuals in the inner city that are outside the mainstream of the American occupational system and that consequently represent the very bottom of the economic hierarchy" (Wilson 1983:80).

Validity The degree to which an operational definition measures what it claims to measure.

Value rationale A goal pursued because it is valued, and it is pursued with no thought of foreseeable consequences and often without consideration of the appropriateness of the means chosen to achieve it.

Values General, shared conceptions about what is good, right, appropriate, worthwhile, and important with regard to conduct, appearance, and states of existence.

Variable Any trait or characteristic that can change under different conditions or that consists of more than one category. (For example, gender is a variable with two categories: male and female.)

Vertical mobility A change in class status that corresponds to a gain or loss in rank or prestige.

Voluntary minorities Racial and ethnic groups that come to a country expecting to improve their way of life.

White-collar crime "Crimes committed by persons of respectability and high social status in the course of their occupations" (Sutherland and Cressey 1978:44).

Witch-hunt A campaign against subversive elements with the purpose of investigating and correcting behavior that undermine a group or a country.

World-economy An economy in which economic transactions transcend national boundaries.

References

CHAPTER 1

Abercrombie, Nicholas, Stephen Hill, and Bryan S. Turner. 1988. *The Penguin Dictionary of Sociology.* New York: Penguin.

Bacon, Kevin, and Bill Davis. 1995. "Interview with Peter Carey." *Glimmer Train Stories* (Spring):123–233.

Barnet, Richard J., and John Cavanagh. 1994. *Global Dreams.* New York: Simon & Schuster.

Bates College. 1996. "Sociology." http://www.bates.edu/pubs/Dept.Letters/sociology.html

Berger, Peter L. 1963. *Invitation to Sociology: A Humanistic Perspective.* New York: Anchor.

Boden, Deirdre, Anthony Giddens, and Harvey L. Molotch. 1990. "Sociology's Role in Addressing Society's Problems Is Undervalued and Misunderstood in Academe." *The Chronicle of Higher Education* (February 21):B1, B3.

Carver, Terrell. 1987. *A Marx Dictionary.* Totowa, NJ: Barnes & Noble.

Center for International Financial Analysis and Research, Inc. 1992. *CIFAR's Company Handbook.* Princeton, NJ: The Center.

Chaliand, Gerard, and Jean-Pierre Rageau. 1995. *The Penguin Atlas of Diasporas.* New York: Penguin Group.

Charyn, Jerome. 1978. "Black Diamond." *The New York Review of Books* (August 17):41.

Conrad, Robert Edgar. 1996. "Slave Trade." Pages 127–128 in *Encyclopedia of Latin American History and Culture,* ed. B. A. Tenenbaum. New York: Scribner's.

Coser, Lewis A. 1977. *Masters of Sociological Thought,* ed. R. K. Merton. New York: Harcourt Brace Jovanovich.

Du Bois, W. E. B. [1899] 1996. *The Philadelphia Negro: A Social Study.* Philadelphia: University of Pennsylvania Press.

———. [1903] 1996. *The Souls of Black Folk.* http://www.cc.columbia.edu/acis/bartleby/dubois/

Durkheim, Emile. [1888] 1978. *On Institutional Analysis,* ed. and trans. M. Traugott. Chicago: University of Chicago Press.

———. 1951. *Suicide.* New York: Free Press.

Engels, Frederick. [1883] 1993. "Frederick Engels' Speech at the Grave of Karl Marx." http://csf.Colorado.EDU/psn/marx/Archive/1883-Death/burial.htm Transcribed by Mike Lepore.

Erlewine, Michael. 1994. *All American Music Guide: The Best CD's, Albums, and Tapes.* San Francisco: Miller Freeman.

Fletcher, Max E. 1974. "Harriet Martineau and Ayn Rand: Economics in the Guise of Fiction." *American Journal of Economics and Sociology* 33(4):367–379.

Freund, Julien. 1968. *The Sociology of Max Weber.* New York: Random House.

Gates, Henry Louis. 1995. "The Political Scene: Powell and the Black Elite." *The New Yorker* (September 25):64–80.

Gordon, John Steele. 1989. "When Our Ancestors Became Us." *American Heritage* (December):106–221.

Gould, Stephen Jay. 1981. *The Mismeasure of Man.* New York: Norton.

H. [1879] 1996. "Interview with Karl Marx." http://csf.Colorado. EDU/psn/marx/Archive/Interviews/1879int1.htm

Henslin, James M. 1993. *Down to Earth Sociology: Introductory Readings.* New York: Free Press.

Hilburn, Robert, and Chuck Philips. 1992. "Rock's New World Order." *Los Angeles Times* (November 29):7+.

Holloway, Thomas H. 1996. "Immigration." Pages 239–242 in *Encyclopedia of Latin American History and Culture,* ed. B. A. Tenenbaum. New York: Scribner's.

Hoover's Guide to Media Companies. 1996. Austin, TX: Hoover's Business Press.

Lengermann, Patricia M. 1974. *Definitions of Sociology: A Historical Approach.* Columbus, OH: Merrill.

Lewis, David Levering. 1993. *Biography of a Race, 1868–1919.* New York: A John Macrae Book/Holt.

Lohr, Steve. 1993. "New Appeals to Pocketbook Patriots." *The New York Times* (January 22):23+.

Martin, Waldo E., Jr. 1993. "A Great and Difficult Man." *The New York Times Book Review* (December 12):1.

Martineau, Harriet. [1837] 1968. *Society in America,* ed. and abridged by S. M. Lipset. Gloucester, MA: Peter Smith.

Marx, Karl. [1881] 1965. "The Class Struggle." Pages 529–535 in *Theories of Society,* ed. T. Parsons, E. Shils, K. D. Naegele, and T. R. Pitts. New York: Free Press.

Miller, S. M. 1963. *Max Weber: Selections from His Work.* New York: Crowell.

Mills, C. Wright. 1959. *The Sociological Imagination.* New York: Oxford University Press.

The New Columbia Encyclopedia. 1975. "Steamship." New York: Columbia University Press.

Ornstein, Robert, and Paul Ehrlich. 1989. *New World, New Mind.* New York: Touchstone.

Page, Clarence. 1996. *Showing My Color.* New York: HarperCollins.

Pennsylvania State University. 1996. "What Is Sociology?" http://rudolph.la/psu.edu/soc/welcome.htm

Princeton University. 1996. "Sociology." http://www.princeton.edu/~sociolog/ugrad_program.html

Quest Interactive Media. 1996. Tour Schedules.

Random House Encyclopedia. 1990. "European Imperialism in the 19th Century." New York: Random House.

Reich, Robert B. 1988. "Corporation and Nation." *The Atlantic Monthly* (May):76–81.

Terry, James L. 1983. "Bringing Women . . . In: A Modest Proposal." *Teaching Sociology* 10(2):251–261.

University of Waterloo. 1996. "Sociology." http://watarts.uwaterloo.ca/SOC/overview.html#subject

U.S. Centers for Disease Control. 1994. *State Injury Mortality Statistics.* http://www.cdc.gov/ncipc.osp.states/0004.htm

U.S. Central Intelligence Agency. 1995. *World Factbook 1995.* http://www.odci.gov/cia/publications/95fact/index.html

U.S. Department of Commerce. 1993. *Globalization of the Mass Media.* Washington, D.C.: U.S. Government Printing Office.

U.S. Department of Defense. 1996. "Active Duty Military Personnel Strengths by Regional Area and by Country." http://web1.whs.osd.mil/mmid/military/309ab966.htm (June 30).

U.S. Department of the Interior, Bureau of Mines. *Minerals Yearbook. Vol. 3: Area Reports: International 1994.* Washington, D.C.: U.S. Government Printing Office.

Variety. 1995. "Worldwide Rentals Beat Domestic Take." February 13–19:280.

Webb, R. K. 1960. *Harriet Martineau, A Radical Victorian.* New York: Columbia University Press.

White House press release. 1995. "Miami International to Be Designated 'Reinvention Lab.'" Office of the Vice President, August 30.

Zuboff, Shoshana. 1988. *In the Age of the Smart Machine.* New York: Basic Books.

CHAPTER 2

Bearak, Barry. 1996. "Stitching Together a Crusade." *Los Angeles Times* (July 25): A1+.

Bearden, Tom. 1990. "Finally—Bus Stop (Report on Greyhound Bus Strike That Began Two Months Ago)." *MacNeil/Lehrer Newshour* (transcript #3723). New York: WNET.

Blumer, Herbert. 1962. "Society as Symbolic Interaction." In *Human Behavior and Social Processes,* ed. A. Rose. Boston: Houghton Mifflin.

Carver, Terrell. 1987. *A Marx Dictionary.* Totowa, NJ: Barnes & Noble.

Case, Brendan M. 1996. "Cashing in on Immigration." *Los Angeles Times* (September 14):17.

Castillo, Arture. 1996. Personal correspondence (December 6).

Cimons, Marlene. 1994. "Stricken Veterans Blame the Military for Health Hazards." *Los Angeles Times* (May 7):A1.

Cornejo, José A. Perez. 1988. "The Implications for the U.S. Economy of Tariff Schedule Item 807 and Mexico's Maquila Program." *Maquiladora Newsletter* 15(5):2–5.

Darling, Juanita, and Sebastian Rotella. 1992. "Free Trade Could Worsen Clogged U.S.–Mexican Border." *Los Angeles Times* (July 13):A1.

Dodge, David. 1988. "Insights into the Mexicans." Pages 46–55 in *Fodor's Mexico,* ed. A. Beresky. New York: Fodor's Travel Publications.

Ehrenreich, Barbara, and Annette Fuentes. 1985. "Life on the Global Assembly Line." Pages 373–388 in *Crisis in American Social Institutions,* 6th ed., ed. J. H. Skolnick and E. Currie. Boston: Little, Brown.

Engels, Friedrich. [1881] 1996. "A Fair Day's Wage for a Fair Day's Work." http://www.idbsu.edu/surveyrc/Staff/jaynes/marxism/fairwage.htm

Essential Organization. 1996. "Executive Summary." *Monitoring Border and Environmental and Health Conditions Two Years After NAFTA.* http://www.essential.org/orgs/public_citizen/pctrade/borderexec

Excelsior. 1994. Advertisement circular (February 25).

Gans, Herbert. 1972. "The Positive Functions of Poverty." *American Journal of Sociology* 78:275–289.

Garcia, Juan Ramon. 1980. *Operation Wetback: The Mass Deportation of Mexican Undocumented Workers in 1954.* Westwood, CT: Greenwood.

Glionna, John M. 1992. "The Paper Chase." *Los Angeles Times* (November 8):E1.

Goldberg, Raymond. 1995. "We're Creating Jobs in the Third World." *New York Times* (August 1):A10.

Golden, Tim. 1995. "Mexicans Find Dreams Devalued." *New York Times* (January 8):E5.

Halberstam, David. 1986. *The Reckoning.* New York: Morrow.

Hamashige, Hope. 1995. "Wiring for Dollars." *Los Angeles Times* (February 9):D1.

Herzog, Lawrence A. 1985. "The Cross-Cultural Dimensions of Urban Land Use Policy on the U.S.–Mexico Border: A San Diego–Tijuana Case Study." *Social Science Journal* (July):29–46.

Holland, Gale. 1996. "A Losing Battle Against Illegal Immigration." *USA Today.*

Jacobson, Gary. 1988. "The Boom on Mexico's Border." *Management Review* (July):21–24.

Kilborn, Peter T. 1990. "When Plant Shuts Down, Retraining Laid-Off Workers Is Toughest Job Yet." *New York Times* (April 23):A12.

Koenenn, Connie. 1992. "The Power of Pulling Purse Strings." *Los Angeles Times* (December 1):E1.

Kraul, Chris. 1996. "State and Baja Team Up to Promote Business Deal." *Los Angeles Times* (July 25):A1.

Krause, Charles. 1996. "A Tale of Two Cities." *Online NewsHour: Across the Border.* http://www.pbs.org/newshour/bb/election/august96/sandiego_8-9.html

Langley, Lester D. 1988. *MexAmerica: Two Countries, One Future.* New York: Crown.

Lee, Patrick, and Chris Kraul. 1993. "Uniqueness of Maquiladora Could Fade." *Los Angeles Times* (November 19):D1.

Lekachman, Robert. 1985. "The Specter of Full Employment." Pages 74–80 in *Crisis in American Institutions,* ed. J. H. Skolnick and E. Currie. Boston: Little, Brown.

Levering, Robert, and Milton Moskowitz. 1993. *The 100 Best Companies to Work for in America.* New York: Currency & Doubleday.

Levi, Issaac A. 1995. "Peso Crisis Haunts Mexico." *USA Today* (December 20):4B.

Los Angeles Times. 1996. Advertisements (October 31).

Magaziner, Ira C., and Mark Patinkin. 1989. *The Silent War: Inside the Global Business Battles Shaping America's Future.* New York: Random House.

Marx, Karl. [1888] 1961. "The Class Struggle." Pages 529–535 in *Theories of Society,* ed. T. Parsons, E. Shils, K. D. Naegele, and J. R. Pitts. New York: Free Press.

McAllen Economic Development Corporation. 1996. http://www.medc.org/wage.html

Mead, George H. 1934. *Mind, Self and Society.* Chicago: University of Chicago Press.

Merton, Robert K. 1967. "Manifest and Latent Functions." Pages 73–137 in *On Theoretical Sociology: Five Essays, Old and New.* New York: Free Press.

Moyers, Bill. 1986. "One River, One Country: The U.S.–Mexican Border" (tape of documentary). New York: CBS.

Myerson, Allen R. 1995. "Greyhound: The Airline of the Road." *New York Times* (January 18):C1.

The Nafta. 1993. Washington, D.C.: U.S. Government Printing Office.

New York Times. 1995. "We're Creating Jobs in the Third World." (August 1):A10.

Noll, Cheryl L. 1992. "Mexican Maquiladora Workers: An Attitude Toward Working." *Southwest Journal of Business and Economics* (Spring):1–7.

O'Hare, William P. 1988. "The Rise of Poverty in Rural America." *Population Trends and Public Policy* (July). Washington, D.C.: Population Reference Bureau.

Parra, Angelo, Edmundo Elias Fernandez, and Carol S. Osmond. 1996. "Maquila Program: Key Features of the Maquiladora Program." *Twin Plant News* (July):21–24.

Partida, Gilbert A., and Cesar Ochoa. 1990. "Border Waste Program." *Twin Plant News* (April):29–31.

Patten, Mike. 1996. "Border University Offers Engineering Help to Maquilas." *Twin Plant News* (June):26–32.

Pear, Robert. 1996. "U.S. Strengthens Patrols Along the Mexican Border." *New York Times* (July 13):G7.

Pearce, Jean. 1987. "Mexico Holds Special Place in the Heart of Japan: An Interview with Sergio Gonzalez Galvez, Ambassador of Mexico." *Business Japan* (November/December):47–49.

Pina, Rudy R. 1996. "Verifying Origin." *Twin Plant News* (April):39–40.

Risen, James. 1992. "U.S., Mexico, Canada Agree to Form Huge Common Market." *Los Angeles Times* (August 13):A1+.

Roderick, Larry M., and J. Rene Villalobos. 1992. "Pollution at the Border." *Twin Plant News* (November):64–67.

Rodriguez, Richard. 1992. *Days of Obligation: An Argument with My Mexican Father.* New York: Viking.

Rose, Kenneth J. 1988. *The Body in Time.* New York: Wiley.

Sanders, Thomas G. 1987. "Tijuana, Mexico's Pacific Coast Metropolis." *UFSI Reports* 38:1–8.

Sears. 1996. Phone calls to housewares and electronics (December 4).

Sexton, Joe. 1995. "A Factory Reinvents the Sweatshop." *New York Times* (May 29):Y18.

Suro, Roberto. 1991. "Border Boom's Dirty Residue Imperils U.S.–Mexico Trade." *New York Times* (March 31):Y1.

Tuleja, Tad. 1987. *Curious Customs: The Stories Behind 296 Popular American Rituals.* New York: Harmony.

Tumin, Melvin. 1964. "The Functionalist Approach to Social Problems." *Social Problems* 12:379–388.

Twin Plant News. 1990a. Advertisement: "Maquila's Multi-Billion Dollar Market: At Your Fingertips!" (July):53.

———. 1990b. Advertisement: "Shopping the Interior? Picture This. . . . " (February):7.

———. 1992. Advertisement: "What Industrial City Is 3 Times the Size of Dallas and 45 Times Closer to You Than Taiwan?" (July):2.

———. 1996a. Advertisement: "TPN at 11: A Look Ahead." (August):36–37.

———. 1996b. Advertisement: "Yucatan." (August):34.

———. 1996c. "Maquila Scoreboard." (July):51.

Uchitelle, Louis. 1993. "Those High-Tech Jobs Can Cross the Border, Too." *New York Times* (March 28):E4.

U.S. Central Intelligence Agency. 1995. *World Factbook 1995.* http://www.odci.gov/cia/publications/95fact/index.html

U.S. Department of Labor, Bureau of Labor Statistics. 1996. "Occupational Compensation Survey: Pay Only, Dallas, Texas Metropolitan Area." February 1995, Bulletin 308-4. Washington, D.C.: U.S. Government Printing Office.

U.S. Department of State. 1996. *Background Notes: Mexico, May 1996.* gopher://dosfan.lib.uic.edu/1D-%3A22525%ABackground%20Ser

Weisman, Alan. 1986. *La Frontera: The United States Border with Mexico.* New York: Harcourt Brace Jovanovich.

White, George, and Andrea Maier. 1992. "A Closer Look at the Trade Agreement." *Los Angeles Times* (August 13):A7.

White, Leslie A. 1949. *The Science of Culture: A Study of Man and Civilization.* New York: Farrar, Straus.

Wilson, Patricia A. 1992. *Exports and Local Development Mexico's New* Maquiladoras. Austin: University of Texas Press.

Yim, Yong Soon. 1989. "American Perceptions of Korean-Americans: An Analytical Study of a 1988 Survey." *Korean and World Affairs* 13 (Fall):519–542.

CHAPTER 3

Allison, Anne. 1991. "Japanese Mothers and Obentōs: The Lunch-Box as Ideological State Apparatus." *Anthropological Quarterly* 64:195–208.

Bailey, William T., and Wade C. Mackey. 1989. "Observations of Japanese Men and Children in Public Places: A Comparative Study." *Psychological Reports* 65:731–734.

Cameron, William B. 1963. *Informal Sociology.* New York: Random House.

Collins, Randall. 1988. "Theoretical Continuities in Goffman's Work." Pages 41–63 in *Erving Goffman: Exploring the Interaction the Interaction Order,* ed. P. Drew and A. Wootton. Boston: Northeaster University Press.

Deemer, Charles. 1994. "What Is Hypertext?" http://www.teleport.com/~cdeemer/essay.html

Drew, Paul, and Anthony Wooton. 1988. "Introduction." Pages 1–13 in *Erving Goffman: Exploring the Interaction Order.* Boston: Northeastern University Press.

Durkheim, Emile. 1951. *Suicide: A Study in Sociology,* trans. J. A. Spaulding and G. Simpson. New York: Free Press.

Dye, Lee. 1995. "Duplication of Research Isn't as Bad as It Sounds." *Los Angeles Times* (April 26):D5+.

Fallows, James. 1989. "The Real Japan." *New York Review of Books* (July 20): 23–28.

Giddens, Anthony. 1988. "Goffman as a Systematic Social Theorist." Pages 250–279 in *Erving Goffman: Exploring the Interaction Order,* ed. P. Drew and A. Wooton. Boston: Northeastern University Press.

Goffman, Erving. 1959. *The Presentation of Self in Everyday Life.* New York: Anchor.

———. 1961. *Asylums: Essays on the Social Situation of Mental Patients and Other Inmates.* New York: Anchor.

———. 1981. *Forms of Talk.* Oxford: Blackwell.

———. 1983. "The Interaction Order." *American Sociological Review* (February):1–17.

Goodman, Roger. 1993. *Japan's International Youth: The Emergence of a New Class of School Children.* Oxford: Clarendon.

Hagan, Frank E. 1989. *Research Methods in Criminal Justice and Criminology.* New York: Macmillan.

Hilts, Philip J. 1991. "U.S. Abandons Idea of Carrying Out Household Survey on Cases of AIDS." *New York Times* (January 11):A10.

Horan, Patrick M. 1995. "Review of 'Working with Archival Data: Studying Lives.'" *Contemporary Sociology* (May): 423–424.

Institute of International Education. 1993/94. *Academic Year Abroad.* New York: Institute of International Education.

———. 1994/95. *Open Doors.* New York: Institute of International Education.

Internal Revenue Service. 1995. *Statistics of Income Bulletin* (Fall). Washington, D.C.: U.S. Government Printing Office.

Ishii-Kuntz, Masako. 1992. "Are Japanese Families 'Fatherless'?" *Sociology and Social Research* 76(3):105–110.

Joseph, Michael. 1982. *The Timetable of Technology.* London: Marshal Editions.

Jussaume, Ramond A., Jr., and Yoshiharu Yamada. 1990. "A Comparison of the Viability of Mail Surveys in Japan and the United States." *Public Opinion Quarterly* 54:219–228.

Katzer, Jeffrey, Kenneth H. Cook, and Wayne W. Crouch. 1991. *Evaluating Information: A Guide for Users of Social Science Research,* 3d ed. New York: McGraw-Hill.

Klapp, Orrin E. 1986. *Overload and Boredom: Essays on the Quality of Life in the Information Society.* New York: Greenwood.

Kōji, Kata. 1983. "Pachinko." Page 143 in *Kodansha Encyclopedia of Japan.* New York: Kodansha International.

Lester, David, Yutaka Motohashi, and Bijou Yang. 1992. "The Impact of the Economy on Suicide and Homicide Rates in Japan and the United States." *International Journal of Social Psychiatry* (Winter):314–317.

Lewis, Catherine C. 1988. "Japanese First-Grade Classrooms: Implications for U.S. Theory and Research." *Comparative Education Review* 32(2):159–172.

Lucky, Robert W. 1985. "Message by Light Wave." *Science* (November):112–113.

Lynd, Robert S., and Helen M. Lynd. [1929] 1956. *Middletown: A Study in Modern American Culture.* New York: Harcourt, Brace & World.

Macer, Darryl. 1995. "Editorial: Why a New Journal?" *Eubois Journal of Asian and International Bioethics.* http://20bell.biol.tsukuba.ac.jp/~macer/EJ51A.html

Marsa, Linda. 1992. "Scientific Fraud." *Omni* (June):39+.

Michael, Donald. 1984. "Too Much of a Good Thing? Dilemmas of an Information Society." *Technological Forecasting and Social Change* 25(4):347–354.

Mouer, Ross, and Yoshio Sugimoto. 1990. *Images of Japanese Society.* London: Kegan Paul International.

Ogawa, Noahiro, and Robert D. Retherford. 1993. "Care of the Elderly in Japan: Changing Norms and Expectations." *Journal of Marriage and the Family* (August):585–597.

Paulos, John A. 1988. *Innumeracy: Mathematical Illiteracy and Its Consequences.* New York: Hill & Wang.

Rathje, William L., and Cullen Murphy. 1992. *Rubbish: The Archaeology of Garbage.* New York: HarperCollins.

Robinson, Richard D. 1985. "Another Look at Japanese–United States Trade Relations." *UFSI Reports* 19:1–7.

Roethlisberger, F. J., and William J. Dickson. 1939. *Management and the Worker.* Cambridge, MA: Harvard University Press.

Rossi, Peter H. 1988. "On Sociological Data." Pages 131–154 in *Handbook of Sociology,* ed. N. Smelser. Newberry Park, CA: Sage.

Ryūzo, Satō. 1991. "Maturing from We-ism to Global You-ism." *Japan Quarterly* 38:273–282.

Sanger, David E. 1992. "A Defiant Detroit Still Depends on Japan." *New York Times* (February 27):A1+.

Schonberg, Harold C. 1981. "Sumō Embodies Ancient Rituals." *New York Times*:B9.

Sengupta, Subir. 1995. "The Influence of Culture on Portrayals of Women in Television Commercials: A Comparison Between the United States and Japan." *International Journal of Advertising* 14:314–333.

Shotola, Robert W. 1992. "Small Groups." Pages 1796–1806 in *Encyclopedia of Sociology,* vol. 4, ed. E. F. Borgatta and M. L. Borgatta. New York: Macmillan.

Singleton, Royce A., Jr., Bruce C. Straits, and Margaret Miller Straits. 1993. *Approaches to Social Research,* 2d ed. New York: Oxford University Press.

Smith, Joel. 1991. "A Methodology for the Twenty-First Century Sociology." *Social Forces* 70(1):1–17.

Swazey, Judith P., Melissa S. Anderson, and Karen Seashore Lewis. 1993. "Ethical Problems in Academic Research." *American Scientist* (November–December):542–553.

Thayer, John E. III. 1983. "Sumo." Pages 270–274 in *Kodansha Encyclopedia of Japan,* vol. 7. Tokyo: Kodansha.

Thomas, Bill. 1992. "King Stacks." *Los Angeles Times Magazine* (November 15):31+.

Totten, Bill. 1990. "Japan's Mythical Trade Surplus." *New York Times* (December 9):F13.

Tuss, Paul, Jules Zimmer, and Hsiu-Zu Ho. 1995. "Causal Attributions of Underachieving Fourth-Grade Students in China, Japan, and the United States." *Journal of Cross-Cultural Psychology* (July):408–425.

U.S. Central Intelligence Agency. 1995. *World Factbook 1995.* http://www.odci.gov/cia/publications/95fact/index.html

U.S. Department of Commerce. 1996. *U.S. Foreign Trade Highlights 1995.* Washington, D.C.: U.S. Government Printing Office.

U.S. Department of State. 1996. *Background Notes: Japan, August 1996.* gopher://dosfan.lib.uic.edu/1D-%3A22525%ABackground%20Ser

White, R. D., ed. 1989. *Science and the Human Spirit.* Belmont, CA: Wadsworth.

Whyte, William H. 1988. *City: Rediscovering the Center.* New York: Doubleday.

Winkin, Yves. 1989. "Erving Goffman (1922–1982)." Pages 223–225 in *International Encyclopedia of Communications,* ed. E. Barnouw, G. Berbner, W. Schramm, T. L. Worth, and L. Gross. New York: Oxford University Press.

CHAPTER 4

An, Heejung. 1997. "Sports Categories in Korea." http://www.itp.tsoa.nyu.edu/~student/heejung/sports5.htn

Behrangi, Samad. 1994. Quoted in "International Rural Education Teacher and Literacy Critic: Samad Behrangi's Life and Thought." *Journal of Global Awareness* 21(1):27–35.

Berkhofer, Robert F., Jr. 1978. *The White Man's Indian: Images of the American Indian from Columbus to the Present.* New York: Knopf.

Berreby, David. 1995. "Unabsolute Truths: Clifford Geertz." *New York Times Magazine* (April 9):44–47.

Bestani, William. 1994. "Essays on Korea." *Korean Culture* (Winter):37–40.

Bok, Lee Suk. 1987. *The Impact of U.S. Forces in Korea.* Washington, D.C.: National Defense University Press.

Breton, Raymond. 1967. "Institutional Completeness of Ethnic Communities and the Personal Relations of Immigrants." *American Journal of Sociology* 70:193–205.

Brown, Rita Mae. 1988. *Rubyfruit Jungle.* New York: Bantam.

Cambridge International Dictionary of English. 1995. "Culture." Cambridge: Cambridge University Press.

Chambers, Kevin. 1988. *The Travelers' Guide to Asian Customs & Manners.* New York: Meadowbrook.

The Chronicle of Higher Education. 1996. "Foreign Students' Countries of Origin, 1994–95" (September 2):20.

Clifton, James A. 1989. *Being and Becoming Indian: Biographical Studies of North American Frontiers.* Chicago: Dorsey.

The Comenius Group. 1996. *The Weekly Idiom.* http://www.comenius.com/idiom/index.html

Doi, Takeo, M.D. 1986. *The Anatomy of Dependence,* trans. J. Bester. Tokyo: Kodansha International.

Fallows, James. 1988. "Trade: Korea Is Not Japan." *Atlantic Monthly* (October): 22–33.

———. 1990. "Wake Up, America!" *New York Review of Books* (March 1):14–19.

Frontline. 1988. "American Game, Japanese Rules" (transcript #611). Boston: WGBH Educational Foundation.

Geertz, Clifford. 1995. *After the Fact: Two Countries, Four Decades, One Anthropologist.* Cambridge, MA: Harvard University Press.

Gordon, Emily Fox. 1995. "Faculty Brat: A Memoir." *Boulevard* 10(1–2):1–17.

Gordon, Steven L. 1981. "The Sociology of Sentiments and Emotion." Pages 562–592 in *Social Psychology Sociological Perspectives,* ed. M. Rosenberg and R. H. Turner. New York: Basic Books.

Grove, Cornelius, and Constance Franklin. 1990. "Using the Right Fork Is Just the Beginning: Intercultural Training in the Global Era." *International Public Relations Review* 13 (1):13–21.

Halberstam, David. 1986. *The Reckoning.* New York: Morrow.

Hannerz, Ulf. 1992. *Cultural Complexity: Studies in the Social Organization of Meaning.* New York: Columbia University Press.

———. 1993. "When Culture is Everywhere: Reflections on a Favorite Concept." *Ethnos* 58 (1/2):95–111.

Henry, William A. 1988. "No Time for the Poetry: NBC's Cool Coverage Stints on the Drama." *Time* (October 3):80.

Herskovits, Melville J. 1948. *Man and His Works: The Science of Cultural Anthropology.* New York: Knopf.

Hochschild, Arlie R. 1976. "The Sociology of Feeling and Emotion: Selected Possibilities." Pages 280–307 in *Another Voice,* ed. M. Millman and R. Kanter. New York: Octagon.

———. 1979. "Emotion Work, Feeling Rules, and Social Structure." *American Journal of Sociology* 85:551–575.

Hughes, Everett C. 1984. *The Sociological Eye: Selected Papers.* New Brunswick, NJ: Transaction.

Hurst, G. Cameron. 1984. "Getting a Piece of the R.O.K.: American Problems of Doing Business in Korea." *UFSI Reports* 19.

Ingram, Erik. 1992. "Water Use Continues to Decline: Bay Area Districts Report Record Savings." *San Francisco Chronicle* (July):A15.

Institute of International Education. 1996. "Foreign Student's Countries of Origin, 1994–95," cited in *Chronicle of Higher Education* (September):20.

International Baseball Association. 1993. *International Baseball Association Membership List.* (July). Indianapolis: International Baseball Association.

Iyer, Pico. 1988. "The Yin and Yang of Paradoxical, Prosperous Korea." *Smithsonian* (August):47–58.

Jun, Suk-ho, and Daniel Dayan. 1986. "An Interactive Media Event: South Korea's Televised 'Family Reunion.'" *Journal of Communication* (Spring):73–82.

Kephart, William M. 1987. *Extraordinary Groups.* New York: St. Martin's.

Kim, Bo-Kyung, and Kevin Kirby. 1996. Personal correspondence (April 25).

Kim, Choong Soon. 1989. "Attribute of 'Asexuality' in Korean Kinship and Sundered Koreans during the Korean War." *Journal of Comparative Family Studies* 20(3):309–325.

Kim, Eun Mee. 1993. "Contradictions and Limits of a Developmental State: With Illustrations from the South Korean Case." *Social Problems* 40(2):228–249.

Kluckhohn, Clyde. 1949. *Mirror for Man: Anthropology and Modern Life.* New York: McGraw-Hill.

Korean Overseas Information Service. 1995–1996. *Tourism.* http://korea.emb.washington.dc.us/Kois/explore/Facts/tourism.html

Koehler, Nancy. 1986. "Re-Entry Shock." Pages 89–94 in *Cross-Cultural Reentry: A Book of Readings.* Abilene, TX: Abilene Christian University Press.

Kristof, Nicholas D. 1995. "Where a Culture Clash Lurks Even in the Noodles." *New York Times* (September 4):Y4.

Lamb, David. 1987. *The Arabs: Journeys Beyond the Mirage.* New York: Random House.

Larson, James F., and Nancy K. Rivenburgh. 1991. "A Comparative Analysis of Australian, U.S., and British Telecasts of the Seoul Olympic Opening Ceremony." *Journal of Broadcasting and Electronic Media* 35(1):75-94.

Lee, Jennifer. 1994. "The Invisible Nation of Korean Emigrants." *Korean Culture* (Winter):39–40.

"Letters." 1988. "Reading for Meaning, and Not Just for Sounds." *New York Times* (November 12):Y4.

Lidz, Theodore. 1976. *The Person: His and Her Development Throughout the Life Cycle.* New York: Basic Books.

Lie, John. 1995. "Sea Change: Pacific Asia as the New World Industrial Center." *Contemporary Sociology* (March):207.

Linton, Ralph. 1936. *The Study of Man: An Introduction.* New York: Appleton-Century-Crofts.

Liu, Hsein-Tung. 1994. "Intercultural Relations in an Emerging World Civilization." *Journal of Global Awareness* 2(1):48–53.

Mahmood, Cynthia K., and Sharon Armstrong. 1992. "Do Ethnic Groups Exist? A Cognitive Perspective on the Concept of Cultures." *Ethnology* 31(1):1–14.

Magaziner, Ira C., and Mark Patinkin. 1989. *The Silent War: Inside the Global Business Battles Shaping America's Future.* New York: Random House.

Mendelsohn, Harold. 1964. "Listening to Radio." Pages 239–249 in *People, Society, and Mass Communications,* ed. Lewis Anthony Dexter and David Manning White. London: Collier-Macmillan.

Moran, Robert T. 1987. "Cross-Cultural Contact: What's Funny to You May Not Be Funny to Other Cultures." *International Management* 42 (July/August):74.

Murphy, Dean. 1994. "New East Europe Retailers Told to Put on a Happy Face." *Los Angeles Times* (November 26):A1.

Nasaw, David. 1991. "The Stuff That Made Us What We Are." *New York Times Book Review* (December 15):10.

Park, Myung-Seok. 1979. *Communication Styles in Two Different Cultures: Korean and American.* Seoul: Han Shin.

Perry, Charles. 1993. "The American Grain." *Los Angeles Times* (July 1):H12.

Peterson, Mark. 1977. "Some Korean Attitudes Toward Adoption." *Korea Journal* 17(12):28–31.

Protzman, Ferdinand. 1991. "As Marriage Nears, Germans in the Wealthy West Fear Cost in Billions." *New York Times* (September 24):A6.

Reader, John. 1988. *Man on Earth.* Austin: University of Texas Press.

Redfield, Robert. 1962. "The Universally Human and the Culturally Variable." Pages 439–453 in *Human Nature and the Study of Society: The Papers of Robert Redfield,* vol. 1, ed. M. P. Redfield. Chicago: University of Chicago Press.

Reid, Daniel P. 1988. *Korea: The Land of the Morning Calm.* Lincolnwood, IL: Passport.

Reinhold, Robert. 1989. "The Koreans' Big Entry into Business." *New York Times* (September 24):E4.

Rohner, Ronald P. 1984. "Toward a Conception of Culture for Cross-Cultural Psychology." *Journal of Cross-Cultural Psychology* 15(2):111–138.

Rokeach, Milton. 1973. *The Nature of Human Values.* New York: Free Press.

Rosenfeld, Jeffrey P. 1987. "Barking up the Right Tree." *American Demographics* (May):40–43.

Sanchez, Jesus. 1987. "L.A. Top Choice for Koreans in Business in U.S." *Los Angeles Times* (June 6):IV, 1.

Sapir, Edward. 1949. "Selected Writings of Edward Sapir," in *Language, Culture and Personality,* ed. D. G. Mandelbaum. Berkeley: University of California Press.

Schoenberger, Karl. 1992. "Moving Between Two Worlds." *Los Angeles Times* (July 12):A1+.

Shapiro, Laura. 1992. "In the American Grain." *New York Times Book Review* (August 2):9.

Sobie, Jane Hipkins. 1986. "The Cultural Shock of Coming Home Again." Pages 95–102 in *The Cultural Transition: Human Experience and Social Transformation in the Third World and Japan,* ed. M. I. White and S. Pollack. Boston: Routledge & Kegan.

Sterngold, James. 1991. "New Doubts on Uniting Two Koreas." *New York Times* (May 30):C1.

Sumner, William Graham. 1907. *Folkways.* Boston: Ginn.

Tuleja, Tad. 1987. *Curious Customs: The Stories Behind 296 Popular American Rituals.* New York: Harmony.

UNESCO. 1995. *UNESCO Statistical Yearbook 1995.* Paris: UNESCO Publishing and Bernan.

U.S. Department of Agriculture. 1995–96. *Agricultural Statistics 1995–96.* Washington, D.C.: U.S. Government Printing Office.

U.S. Department of Commerce. 1995. "Selected Data for Nonbank Foreign Affiliates, by Country and by Major Industry, 1992 and 1993, Table 10," in *Survey of Current Business,* vol. 75(6). Washington, D.C.: U.S. Government Printing Office.

U.S. Department of Defense. 1995. *Worldwide Military Strength as of December 31, 1994.* Washington, D.C.: U.S. Government Printing Office.

U.S. Department of Justice. 1995. "Immigrants Admitted in Fiscal Years 1993 and 1994 for Regions and Top 15 Countries of Birth, Table 4," in *INS Fact Book: Summary of Recent Immigration Data.* Washington, D.C.: U.S. Government Printing Office.

U.S. Department of State. 1996. "Korea International Adoption." http://travel.state.gov/adoption_korea.html

U.S. Department of Transportation. 1993. "Passenger Travel Between U.S. and Foreign Countries Distribution by U.S. and Foreign Flag Carriers Commercial Traffic Only, Table IIa," in *U.S. International Air Travel Statistics.* Washington, D.C.: U.S. Government Printing Office.

U.S. Federal Communications Commission. 1993–94. *Statistics of Communications Common Carriers,* Table 4.1. Washington, D.C.: U.S. Government Printing Office.

Visser, Margaret. 1988. *Much Depends on Dinner.* New York: Grove.

———. 1989. "A Meditation on the Microwave." *Psychology Today* (December):38–42.

Wallace, Charles P. 1994. "Singapore Affirms Flogging of American." *Los Angeles Times* (April 1):A5.

Wallace, F. C. 1952. "Notes on Research and Teaching." *American Sociological Review* (December):747–751.

Wallerstein, Immanuel. 1990. "Culture as the Ideological Battleground of the Modern World-System." *Theory, Culture, and Society* 7:31–55.

Welch, Wilford H. 1991. "Regional Powerhouse Reshapes World." *World Paper* (November):3.

Werkman, Sidney L. 1986. "Coming Home: Adjustment of Americans to the United States After Living Abroad." Pages 5–18 in *Cross-Cultural Reentry: A Book of Readings.* Abilene, TX: Abilene Christian University Press.

Winchester, Simon. 1988. *Korea: A Walk Through the Land of Miracles.* New York: Prentice Hall.

World Monitor. 1992. "The Map: Batters Up!" (April):11.

———. 1993. "The Map: Hoop-la" (February):10–11.

Yeh, May. 1991. "A Letter." *Amerasia* 17(2):1-7.

Yoo, Yushin. 1987. *Korea the Beautiful: Treasures of the Hermit Kingdom.* Los Angeles: Golden Pond.

CHAPTER 5

Al-Batrawi, Khaled, and Mouin Rabbani. 1991. "Break Up of Families: A Case Study in Creeping Transfer." *Race and Class* 32(4):35–44.

Atta, Nasser. 1989. "Letters to the Editor: Israel a Military Power, Violates Human Rights." *News Record* (February 8):5.

Ben-David, Amith, and Yoav Lavee. 1992. "Families in the Sealed Room: Interaction Patterns of Israeli Families During SCUD Missile Attacks." *Family Process* 31(1):35–44.

Bourne, Jenny. 1990. "The Rending Pain of Reenactment." *Race and Class* 32(2): 67–72.

Broder, John M., and Norman Kempster. 1993. "'Enough of Blood and Tears': Israel and P.L.O. Adopt Framework for Peace." *Los Angeles Times* (September 14):A1+.

Chaliand, Gerard, and Jean-Pierre Rageau. 1995. *The Penguin Atlas of Diasporas.* New York: Viking.

Clinton, William Jefferson. 1996. "Remarks by the President on Announcing Middle East Peace Summit Meeting." Office of the Press Secretary (September 29).

Cooley, Charles Horton. 1909. *Social Organization.* New York: Scribner's.

———. 1961. "The Social Self." Pages 822–828 in *Theories of Society: Foundations of Modern Sociological Theory,* ed. T. Parsons, E. Shils, K. D. Naegele, and J. R. Pitts. New York: Free Press.

———. 1964. *Human Nature and the Social Order.* New York: Schocken.

Corsaro, William A. 1985. *Friendship and Peer Culture in the Early Years.* Norwood, NJ: Ablex.

Coser, Lewis A. 1992. "The Revival of the Sociology of Culture: The Case of Collective Memory." *Sociological Forum* 7(2):365–373.

Davis, Kingsley. 1940. "Extreme Isolation of a Child." *American Journal of Sociology* 45:554–565.

———. 1947. "Final Note on a Case of Extreme Isolation." *American Journal of Sociology* 3(5):432–437.

Dyer, Gwynne. 1985. *War.* New York: Crown.

Elon, Amos. 1993. "The Jews' Jews." *New York Review of Books* (June 10):14–18.

Encyclopedia Judaica Yearbook. 1987. Jerusalem: Keter.

Facts About Israel. 1985. *Ministry of Foreign Affairs, Information Division.* Jerusalem: Israel Information Centre.

Faris, Ellsworth. 1964. "The Primary Group: Essence and Accident." Pages 314–319 in *Sociological Theory: A Book of Readings,* ed. L. A. Coser and B. Rosenberg. New York: Macmillan.

Fields, Rona M. 1979. "Child Terror Victims and Adult Terrorists." *Journal of Psychohistory* 7(1):71–75.

Figler, Stephen K., and Gail Whitaker. 1991. *Sport and Play in American Life.* Dubuque, IA: Brown.

Fontana, Andrea. 1962. *The Last Frontier: The Social Meaning of Growing Old.* Beverly Hills, CA: Sage.

Freeman, Norman H. 1987. "Children's Drawings of Human Figures." Pages 135–139 in *The Oxford Companion to the Mind,* ed. R. L. Gregory. Oxford: Oxford University Press.

Freud, Anna, and Dorothy T. Burlingham. 1943. *War and Children.* New York: Willard.

Freud, Anna, and Sophie Dann. 1958. "An Experiment in Group Upbringing." Pages 127–168 in *The Psychoanalytic Study of the Child,* vol. 6, ed. R. S. Eissler, A. Freud, H. Hartmann, and E. Kris. New York: Quadrangle.

Friedman, Robert I. 1992. *Zealots for Zion: Inside Israel's West Bank Settlement Movement.* New York: Random House.

Friedman, Thomas L. 1989. *From Beirut to Jerusalem.* New York: Farrar, Straus & Giroux.

Goffman, Erving. 1961. *Asylums: Essays on the Social Situation of Mental Patients and Other Inmates.* New York: Anchor.

Goldman, Ari L. 1989. "Mementos to Preserve the Record of Anguish." *New York Times* (February 29):Y14.

Gorkin, Michael. 1986. "Countertransference in Cross-Cultural Psychotherapy: The Example of Jewish Therapist and Arab Patient." *Psychiatry* 49:69–79.

Griffith, Marlin S. 1977. "The Influences of Race on the Psychotherapeutic Relationship." *Psychiatry* 40:27–40.

Grossman, David. 1988. *The Yellow Wind,* trans. H. Watzman. New York: Farrar, Straus & Giroux.

Halbwachs, Maurice. 1980. *The Collective Memory,* trans. F. J. Ditter, Jr., and V. Y. Ditter. New York: Harper & Row.

Hannerz, Ulf. 1992. *Cultural Complexity: Studies in the Social Organization of Meaning.* New York: Columbia University Press.

Hellerstein, David. 1988. "Plotting a Theory of the Brain." *New York Times Magazine* (May 22):17+.

Kagan, Jerome. 1988. Interview on "The Mind," Public Broadcast Service (transcript). Boston: WGBH Educational Foundation.

———. 1989. *Unstable Ideas: Temperament, Cognition, and Self.* Cambridge: Harvard University Press.

Mannheim, Karl. 1952. "The Problem of Generations." Pages 276–322 in *Essays on the Sociology of Knowledge,* ed. P. Kecskemeti. New York: Oxford University Press.

Matar, Ibrahim. 1983. "Israeli Settlements and Palestinian Rights." Pages 117–141 in *Occupation: Israel over Palestine,* ed. N. H. Aruri. Belmont, MA: AAUG.

McNeill, William H. 1995. *Keeping Together in Time.* Cambridge, MA: Harvard University.

Mead, George Herbert. 1934. *Mind, Self and Society.* Chicago: University of Chicago Press.

Merton, Robert K. 1976. *Sociological Ambivalence and Other Essays.* New York: Free Press.

Montgomery, Geoffrey. 1989. "Molecules of Memory." *Discover* (December):46–55.

Nova. 1986. "Life's First Feelings" (February 11).

Ornstein, Robert, and Richard F. Thompson. 1984. *The Amazing Brain.* Boston: Houghton Mifflin.

Patai, Raphael. 1971. "Zionism." Page 1262 in *Encyclopedia of Zionism and Israel.* New York: McGraw-Hill.

Pawel, Ernst. 1989. *The Labyrinth of Exile: A Life of Theodor Herzl.* New York: Farrar, Straus & Giroux.

Penfield, Wilder, and P. Perot. 1963. "The Brain's Record of Auditory and Visual Experience: A Final Summary and Discussion." *Brain* 86:595–696.

Piaget, Jean. 1923. *The Language and Thought of the Child,* trans. M. Worden. New York: Harcourt, Brace & World.

———. 1929. *The Child's Conception of the World,* trans. J. Tomlinson and A. Tomlinson. Savage, MD: Rowan & Littlefield.

———. 1932. *The Moral Judgement of the Child,* trans. M. Worden. New York: Harcourt, Brace & World.

———. 1946. *The Child's Conception of Time,* trans. A. J. Pomerans. London: Routledge & Kegan Paul.

———. 1967. *On the Development of Memory and Identity.* Worchester, MA: Clark University Press.

Pickering, W. S. F. 1984. *Durkheim's Sociology of Religion.* London: Routledge & Kegan Paul.

Population Reference Bureau. 1996. *World Population Data Sheet.* Washington, D.C.: Population Reference Bureau.

Rabin, Yitzhak. 1993. "Making a New Middle East: 'Shalom, Salaam, Peace': Views of Three Leaders." *Los Angeles Times* (September 14):A7.

Restak, Richard M. 1988. *The Mind.* New York: Bantam.

Rose, Peter I., Myron Glazer, and Penina M. Glazer. 1979. "In Controlled Environments: Four Cases of Intensive Resocialization." Pages 320–338 in *Socialization and the Life Cycle,* ed. P. I. Rose. New York: St. Martin's.

Rosenthal, Elisabeth. 1989. "Mystery on Arrival." *Discover* (December):78–82.

Rubinstein, Danny. 1988. "The Uprising: Reporter's Notebook." *Present Tense* 15:22–25.

———. 1991. *The People of Nowhere: The Palestinian Vision of Home,* trans. T. Friedman. New York: Random House.

Sacks, Oliver. 1985. *The Man Who Mistook His Wife for a Hat and Other Clinical Tales.* New York: Summit.

———. 1989. *Seeing Voices: A Journey into the World of the Deaf.* Los Angeles: University of California Press.

Satterly, D. J. 1987. "Jean Piaget (1896–1980)." Pages 621–622 in *The Oxford Companion to the Mind,* ed. R. I. Gregory. Oxford: Oxford University Press.

Shadid, Mohammed, and Rick Seltzer. 1988. "Political Attitudes of Palestinians in the West Bank and Gaza Strip." *Middle East Journal* 42:16–32.

Shipler, David. 1986. *Arab and Jew: Wounded Spirits in a Promised Land.* New York: Times Books.

Sichrovsky, Peter. 1991. *Abraham's Children: Israel's Young Generation.* New York: Pantheon.

Smooha, Sammy. 1980. "Control of Minorities in Israel and Northern Ireland." *Society for Comparative Study of Society and History* 10:256–280.

Spitz, Rene A. 1951. "The Psychogenic Diseases in Infancy: An Attempt at Their Etiological Classification." Pages 255–278 in *The Psychoanalytic Study of the Child,* vol. 27, ed. R. S. Eissler and A. Freud. New York: Quadrangle.

Steiner, George. 1967. *Language and Silence: Essays on Language, Literature, and the Inhuman.* New York: Atheneum.

Theodorson, George A., and Achilles G. Theodorson. 1979. *A Modern Dictionary of Sociology.* New York: Barnes & Noble.

U.S. Central Intelligence Agency. 1996. *World Factbook 1995.* http://www.odci.gov/cia/publications/95fact/index.html

Usher, Graham. 1991. "Children of Palestine." *Race and Class* 32(4):1–18.

Yishai, Yael. 1985. "Hawkish Proletariat: The Case of Israel." *Journal of Political and Military Sociology* 13:53–73.

CHAPTER 6

American National Red Cross. 1996. "Biomedical Services 1995–96." http://www.crossnet.org/biomed/bio-fact.html

Altman, Lawrence K. 1986. "Anxiety on Transfusions." *New York Times* (July 18):A1, B4.

———. 1995. "Long-Term Survivors May Hold Key Clues to Puzzle of AIDs." *New York Times* (January 24):B1.

Anderson, Elijah. 1990. "The Police and the Black Male." Pages 190–206 in *Streetwise: Race, Class, and Change in an Urban Community.* Chicago: University of Chicago Press.

Barr, David. 1990. "What Is AIDS? Think Again." *New York Times* (December 1): Y15.

Bloor, Michael, David Goldberg, and John Emslie. 1991. "Research Note: Ethnostatics [sic] and the AIDs Epidemic." *British Journal of Sociology* 42(1): 131–138.

Brooke, James. 1987. "In Cradle of AIDs Theory, a Defensive Africa Sees a Disguise for Racism." *New York Times* (November 19):B13.

———. 1988a. "In Africa, Tribal Hatreds Defy the Borders of State." *New York Times* (August 28):E1.

———. 1988b. "Mobutu's Village Basks in His Glory." *New York Times* (September 9):Y4.

Clark, Matt, with Stryker McGuire. 1980. "Blood Across the Border." *Newsweek* (December 29):61.

Clarke, Thurston. 1988. *Equator: A Journey.* New York: Morrow.

Colby, Ron. 1986. Quoted in "Did Media Sensationalize Student AIDS Case?" by John McGauley. *Editor and Publisher* 119:19.

Conrad, Joseph. 1971. *Heart of Darkness,* rev. and ed. R. Kimhough. New York: Norton.

De Cock, Kevin M., and Joseph B. McCormick. 1988. "Correspondence: Reply to HIV Infection in Zaire." *New England Journal of Medicine* 319(5):309.

Doyal, Lesley, with Imogen Pennell. 1981. *The Political Economy of Health.* Boston: South End Press.

Durkheim, Emile. [1933] 1964. *The Division of Labor in Society,* trans. G. Simpson. New York: Free Press.

The Economist. 1981. "America the Blood Bank" (October 17):87.

———. 1983. "Vein Hopes, Mainline Profits" (January 22):63–64.

Fox, Renée. 1988. *Essays in Medical Sociology: Journeys into the Field.* New Brunswick, NJ: Transaction.

Frontline. 1993. "AIDS, Blood, and Politics." Boston: WGBH Educational Foundation and Health Quarterly.

Giese, Jo. 1987. "Sexual Landscape: On the Difficulty of Asking a Man to Wear a Condom." *Vogue* 177 (June):227+.

Global Childnet. 1995. "Seattle and Minneapolis Named Alternative Medicine Centers." http://edie.cprost.sfu.ca/gcnet/ISS4-05f.html

Goffman, Erving. 1959. *The Presentation of Self in Everyday Life.* New York: Anchor.

———. 1963. *Stigma: Notes on the Management of Spoiled Identity.* Upper Saddle River, NJ: Prentice Hall.

Grmek, Mirkod. 1990. *History of AIDS: Emergence and Origin of a Modern Pandemic,* trans. R. C. Maulitz and J. Duffin. Princeton, NJ: Princeton University Press.

Grover, Jan Zita. 1987. "AIDS: Keywords." *October* 43:17–30.

Halberstam, David. 1986. *The Reckoning.* New York: Morrow.

Harris, Robert, and Jenny Paxman. 1982. *A Higher Form of Killing.* New York: Hill & Wang.

Hiatt, Fred. 1988. "Tainted U.S. Blood Blamed for AIDS' Spread in Japan." *Washington Post* (June 23):A29.

Hilts, Philip J. 1988. "Dispelling Myths About AIDS in Africa." *Africa Report* 33:27–31.

Hunt, Charles W. 1989. "Migrant Labor and Sexually Transmitted Disease: AIDS in Africa." *Journal of Health and Social Behavior* 30:353–373.

Hurley, Peter, and Glenn Pinder. 1992. "Ethics, Social Forces, and Politics in AIDS-Related Research: Experience in Planning and Implementing a Household HIV Seroprevalence Survey." *Milbank Quarterly* 70(4):605–628.

International Federation of Pharmaceutical Manufacturers Associations. 1981. Personal correspondence.

Irwin, Kathleen. 1991. "Knowledge, Attitudes and Beliefs About HIV Infection and AIDS Among Healthy Factory Workers and Their Wives, Kinshasa, Zaire." *Social Science and Medicine* 32(8):917–930.

Johnson, Diane, and John F. Murray, M.D. 1988. "AIDS Without End." *New York Review of Books* (August 18):57–63.

Kaptchuk, Ted, and Michael Croucher, with the BBC. 1986. *The Healing Arts: Exploring the Medical Ways of the World.* New York: Summit.

Kerr, Dianne L. 1990. "AIDS Update: Ryan White's Death." *Journal of School Health* 60(5):237–238.

Kolata, Gina. 1989. "AIDS Test May Fail to Detect Virus for Years, Study Finds." *New York Times* (June 1):Y1.

Kornfield, Ruth. 1986. "Dr., Teacher, or Comforter? Medical Consultation in a Zairian Pediatrics Clinic." *Culture, Medicine and Psychiatry* 10:367–387.

Kramer, Reed. 1993. "Ties That Bind: Pressure Points Considered to Back Change in Zaire." *Africa News* (March 8–21):2.

Kramer, Staci D. 1988. "The Media and AIDS." *Editor and Publisher* 121:10–11, 43.

Krause, Richard. 1993. Quoted in *A Dancing Matrix: Voyage Along the Viral Frontier,* by Robin Marantz Henig. New York: Knopf.

Kraut, Alan M. 1994. *Silent Travelers: Germs, Genes, and the "Immigrant Menace."* New York: Basic Books.

Kurian, George. 1992. *Atlas of the Third World,* 2d ed. New York: Facts on File.

Lamb, David. 1987. *The Africans.* New York: Random House.

Lasker, Judith N. 1977. "The Role of Health Services in Colonial Rule: The Case of the Ivory Coast." *Culture, Medicine and Psychiatry* 1:277–297.

Lippmann, Walter. 1976. "The World Outside and the Pictures in Our Heads." Pages 174–181 in *Drama in Life: The Uses of Communication in Society,* ed. J. E. Combs and M. W. Mansfield. New York: Hastings House.

Liversidge, Anthony. 1993. "Heresy! 3 Modern Galileos." *Omni* (June):43–51.

Mahler, Halfdan. 1989. Quoted in *AIDS and Its Metaphors,* by Susan Sontag. New York: Farrar, Straus & Giroux.

Mannheim, Karl. 1952. "The Problem of Generations." Pages 276–322 in *Essays on the Sociology of Knowledge,* ed. P. Kecskemeti. New York: Oxford University Press.

McNeill, William H. 1976. *Plagues and People.* New York: Anchor.

Meltzer, Milton. 1960. *Mark Twain: A Pictorial Biography.* New York: Bonanza.

Merton, Robert K. 1957. *Social Theory and Social Structure.* Glencoe, IL: Free Press.

Naisbitt, John. 1984. *Megatrends: Ten New Directions Transforming Our Lives.* New York: Warner.

Noble, Kenneth B. 1989. "More Zaire AIDS Cases Show Less Underreporting." *New York Times* (December 26):J4.

———. 1992. "As the Nation's Economy Collapses, Zairians Squirm Under Mobutu's Heel." *New York Times* (August 30):Y4.

The Panos Institute. 1989. *AIDS and the Third World.* Philadelphia: New Society.

Parsons, Talcott. 1975. "The Sick Role and the Role of the Physician Reconsidered." *Milbank Memorial Fund Quarterly: Health and Society* 53(1):257–278.

Peretz, S. Michael. 1984. "Providing Drugs to the Third World: An Industry View." *Multinational Business* 84 (Spring): 20–30.

Postman, Neil. 1985. *Amusing Ourselves to Death.* New York: Penguin.

Rozenbaum, Willy. 1989. Quoted in *AIDS and Its Metaphors,* by Susan Sontag. New York: Farrar, Straus & Giroux.

Sanchez, Jesus. 1994. "Flying in Formation." *Los Angeles Times* (March 27): D1+.

Shilts, Randy. 1987. *And the Band Played On: Politics, People, and the AIDS Epidemic.* New York: St. Martin's.

Sontag, Susan. 1989. *AIDS and Its Metaphors.* New York: Farrar, Straus & Giroux.

Stolberg, Sheryl. 1996. "Officials Find Rare HIV Strain in L.A. Woman." *Los Angeles Times* (July 5):A1+.

Swenson, Robert M. 1988. "Plagues, History, and AIDS." *The American Scholar* 57:183–200.

Thomas, William I., and Dorothy Swain Thomas. [1928] 1970. *The Child in America.* New York: Johnson Reprint.

Times Atlas of World History. 1984. Maplewood, NJ: Hammond.

Turnbull, Colin M. 1961. *The Forest People.* New York: Simon & Schuster.

———. 1962. *The Lonely African.* New York: Simon & Schuster.

———. 1965. *Wayward Servants.* New York: Doubleday.

———. 1983. *The Human Cycle.* New York: Simon & Schuster.

Twain, Mark. 1973. "King Leopold's Soliloquy on the Belgian Congo." Pages 41–60 in *Mark Twain and the Three R's,* ed. Maxwell Geismar. New York: International Publishers.

Urban Institute. 1996. "Why Teenagers Do Not Use Condoms." http://www.urban.org/periodc/prr25_2c.htm

U.S. Bureau of the Census. 1992. *U.S. Exports and General Imports by Hormonized Commodity by Country,* Report FT947/91-A. Washington, D.C.: U.S. Government Printing Office.

———. 1996a. "Merchandise Trade—Exports by Commodity." *National Trade Data Bank* (January 2). Washington, D.C.: U.S. Government Printing Office.

———. 1996b. "Merchandise Trade—Imports by Commodity." *National Trade Data Bank* (January 2). Washington, D.C.: U.S. Government Printing Office.

U.S. Bureau for Refugee Programs. 1988. *World Refugee Report.* Washington, D.C.: U.S. Government Printing Office.

U.S. Central Intelligence Agency. 1995. *World Factbook 1995.* http://www.odci.gov/cia/publications/95fact/cg.html

U.S. Department of Commerce. 1994. *Statistical Abstract of the United States,* 114th ed. Washington, D.C.: Government Printing Office.

U.S. Department of Health and Human Services. 1990. *HIV/AIDS Surveillance Report.* Washington, D.C.: U.S. Government Printing Office.

———. 1992. *AIDS Knowledge and the Attitudes for January–March 1991: Provisional Data from the National Health Interview Survey.* No. 216. Washington, D.C.: U.S. Government Printing Office.

U.S. General Accounting Office. 1987. *AIDS: Information of Global Dimensions and Possible Impacts.* Washington, D.C.: U.S. Government Printing Office.

Watson, William. 1970. "Migrant Labor and Detribalization." Pages 38–48 in *Black Africa: Its Peoples and Their Cultures Today,* ed. J. Middleton. London: Collier-Macmillan.

Whitaker, Jennifer Seymour. 1988. *How Can Africa Survive?* New York: Harper & Row.

Wilson, Mary E. 1996. "Travel and the Emergence of Infectious Diseases." *Emerging Infectious Disease* (April–June 1995). http://www.cdc.gov/ncidod/EID/vol1no2/wilson.htm

World Health Organization. 1988. "A Global Response to AIDS." *Africa Report* (November/December):13–16.

Yasuda, Yukuo. 1994. "Japanese Hemophiliacs Suffering from HIV Infection." http://www.nmia.com/~mdibble/japan2.html

Zuck, Thomas F. 1988. "Transfusion-Transmitted AIDS Reassessed." *New England Journal of Medicine* 318: 511–512.

CHAPTER 7

Abercrombie, Nicholas, Stephen Hill, and Bryan S. Turner. 1988. *The Penguin Dictionary of Sociology.* New York: Penguin.

Aldrich, Howard E., and Peter V. Marsden. 1988. "Environments and Organizations." Pages 361–392 in *Handbook of Sociology,* ed. N. J. Smelser. Newbury Park, CA: Sage.

Barnet, Richard J., and John Cavanagh. 1994. *Global Dreams: Imperial Corporations and the New World Order.* New York: Simon & Schuster.

Barnet, Richard J., and Ronald E. Müller. 1974. *Global Reach: The Power of the Multinational Corporations.* New York: Simon & Schuster.

Basu, Amrita. 1994. "On the Tenth Anniversary of the Bhopal Disaster." *Bulletin of Concerned Asian Scholars* 26(1):3–14.

Blau, Peter M. 1974. *On the Nature of Organizations.* New York: Wiley.

Blau, Peter M., and Richard A. Schoenherr. 1973. *The Structure of Organizations.* White Plains, NY: Longman.

Bleiberg, Robert M. 1987. "Thought for Labor Day: In the Bhopal Disaster, There's Plenty of Blame to Go 'Round." *Barron's* 67 (September 7):9.

Bonine, John E. 1994. "A Descent into Toxic Hell in India." http://www.igc.apc.org/elaw/asia/india/descent.html

Burns, John F. 1996. "India Now Winning U.S. Investment." *New York Times* (February 1):C1+.

Chenevière, Alain. 1987. *Vanishing Tribes.* Garden City, NY: Doubleday.

Clark, Andrew. 1993. "Learning the Rules of Global Citizenship: Transnationals Need to Meet Their Challenges, or Be Overwhelmed by Them." *The World Paper* (February):5.

Crossette, Barbara. 1989. "New Delhi Prepares Attempt to Control Pervasive Pollution." *New York Times* (July 4):Y24.

———. 1996. "Report Blames Poor Farmers for Much of Deforestation." *New York Times* (August 4):8.

Dahlberg, John-Thor. 1994. "Bhopal Marks 10th Anniversary of Gas Disaster." *Los Angeles Times* (December 4):A1+.

———. 1996. "A United Front: India's New Ruling Party Eager for More Foreign Capital." *Los Angeles Times* (June 6):D1+.

de la Torre, Jose. 1995. "Multinational Expansion Underscores Management Challenges." *Los Angeles Times* (January 29):D2.

Derdak, Thomas. 1988. *International Directory of Company Histories,* vol. 1. Chicago: St. James.

Diamond, Stuart. 1985a. "The Bhopal Disaster: How It Happened." *New York Times* (January 28):A1+.

———. 1985b. "The Disaster in Bhopal: Workers Recall Horror." *New York Times* (January 30):A1+.

———. 1985c. "Problems at Chemical Plants Raise Broad Safety Concerns." *New York Times* (November 25):A1, D11.

Directory of American Firms Operating in Foreign Countries. 1987. 11th ed., vol. 3. New York: World Trade Academy Press.

———. 1996. 14th ed., vol. 3. New York: World Trade Academy Press.

The Economist. 1994. "The Lesson From Bhopal" (December 3–9):78.

Engler, Robert. 1985. "Many Bhopals: Technology Out of Control." *The Nation* 240 (April 27):488–500.

Environmental Law Alliance Worldwide. 1996. "India: Goldman Prize Winner M.C. Mehta, Indian Council for Enviro-Legal Action." http://www.igc.apc.org/elaw/asia/india/icela.html

Essential Organization. 1996. http://www.essential.org/EI.html

Everest, Larry. 1986. *Behind the Poison Cloud: Union Carbide's Bhopal Massacre.* Chicago: Banner.

Fortune. 1996. "The World's Largest Corporations" (August 5):F-1.

Franklin, Ben A. 1985. "Toxic Cloud Leaks at Carbide Plant in West Virginia." *New York Times* (August 12):1+.

Freund, Julien. 1968. *The Sociology of Max Weber.* New York: Random House.

Gross, Edward, and Amitai Etzioni. 1985. *Organizations in Society.* Upper Saddle River, NJ: Prentice Hall.

Hazarika, Sanjoy. 1984. "Stores Reopen in Indian City as Poison Disposal Proceeds." *New York Times* (December 19): A10.

———. 1994. "In India's City of Death, Time Has Healed Little." *New York Times* (December 2):A7.

Holmes, Bob. 1992. "The Joy Ride Is Over: Farmers Are Discovering That Pesticides Increasingly Don't Kill Pests." *U.S. News & World Report* (September 14):72–73.

Hutchison, Robert A. 1989. "A Tree Hugger Stirs Villagers in India to Save Their Forests." *Smithsonian* (May):185–196.

Internal Revenue Service. 1995. *Statistics of Income Bulletin* (Fall). Washington, D.C.: U.S. Government Printing Office.

Iritani, Evelyn. 1996. "Primed for Power." *Los Angeles Times* (January 18):D1+.

Jacobs, Paul. 1996. "UC Relishes Power of the Patent." *Los Angeles Times* (February 14):A1+.

Jasanoff, Sheila. 1988. "The Bhopal Disaster and the Right to Know." *Social Science and Medicine* 27(10):1113–1123.

Keller, George M. 1986. "International Business and the National Interest." *Vital Speeches of the Day* (December 1): 124–128.

Kennedy, Paul. 1993. *Preparing for the Twenty-First Century.* New York: Random House.

Khan, Rahat Nabi. 1986. "Multinational Companies and the World Economy: Economic and Technological Impact." *Impact of Science on Society* 36(141): 15–25.

Kurzman, Dan. 1987. *A Killing Wind: Inside Union Carbide and the Bhopal Catastrophe.* New York: McGraw-Hill.

Lengermann, Patricia M. 1974. *Definitions of Sociology: A Historical Approach.* Columbus, OH: Merrill.

Lepkowski, Wil. 1985. "Chemical Safety in Developing Countries: The Lessons of Bhopal." *Chemical and Engineering News* 63:9–14.

———. 1994. "Union Carbide out of India with Bhopal Sale." *Chemical and Engineering News* (September):9.

McNeely, C. L. 1996. "Review of *Global Dreams: Imperial Corporations and the New World Order.*" *Contemporary Sociology* 25(5):610–611.

Michels, Robert. 1962. *Political Parties,* trans. E. Paul and C. Paul. New York: Dover.

Moskowitz, Milton. 1987. *The Global Marketplace.* New York: Macmillan.

Morehouse, Ward. 1993. "The Ethics of Industrial Disasters in a Transnational World: The Elusive Quest for Justice and Accountability in Bhopal." *Alternatives* (Fall):475–504.

Mosher, Steven W. 1993. "No Refuge." *Reason* (December 1):54–55.

Multinational Monitor. 1994. "The Corporate Hall of Shame" (December). http://www.essential.org/monitor/hyper/

———. 1995. "Harnessing the Law to Clean Up India" (July/August). http://www.essential.org/monitor/hyper/mm0795.09.html

The New York Times. 1985a. "Most at Plant Thought Poison Was Chiefly Skin-Eye Irritant" (January 1):A6.

———. 1985b. "Strong Odor Alerted Residents to Chemical Plant Problem" (August 12):A12.

Nikore, Monika, and Marian Leahy. 1993. *World Monitor* (February):46–52.

Organization for Economic Co-operation and Development. 1987. *OECD Environmental Data Compendium.* Paris: OECD.

Prewitt, Kenneth. 1983. "Scientific Illiteracy and Democratic Theory." *Daedalus* 112 (Spring):49–64.

Reich, Robert B. 1988. "Corporation and Nation." *The Atlantic Monthly* (May): 76–81.

Reinhold, Robert. 1985. "Pesticide Plant Started as a Showpiece but Ran into Troubles." *New York Times* (February 3):A8.

Sadri. Mahmoud. 1996. "Book Review of Occidentalism: Images of the West." *Contemporary Sociology* (September): 612.

Schumacher, E. F. 1985. Quoted in "Technology Out of Control," by Robert Engler. *Nation* 240 (April 27):490.

Scientific American. 1995. "Persistently Toxic" (June):16–18.

Sekulic, Dusko. 1978. "Approaches to the Study of Informal Organization." *Sociologija* 20(1):27–43.

Shabecoff, Philip. 1988a. "A Guide to Some of the Scariest Things on Earth." *New York Times* (December 25):E3.

———. 1988b. "Military Is Accused of Ignoring Rules on Hazardous Waste." *New York Times* (June 14):C4.

———. 1989. "U.S. Only Narrowly Avoided 17 Bhopal-Like Disasters, Study Says." *New York Times* (April 30):A16.

Slavin, Barbara. 1995. "Taking Stock of Their Beliefs." *Los Angeles Times* (June 1):E1+.

Snow, Charles P. 1961. *Science and Government.* Cambridge, MA: Harvard University Press.

Standke, Klaus-Heinrich. 1986. "Technology Assessment: An Essentially Political Process." *Impact of Science on Society* 36(141):65–76.

Stevens, William K. 1984. "Workers at Site of Leak Described as Unskilled." *New York Times* (December 6):A10.

Trade Environment Project. 1996. "Bhopal, Trade and Environment (Bhopal Case)." http://gurukul.ucc.american.edu/TED/BHOPAL.HTM

Union Carbide Annual Report. 1984. "After Bhopal." Danbury, CT: Union Carbide.

U.S. Central Intelligence Agency. 1995. *World Factbook 1995.* http://www.odci.gov/cia/publications/95fact/index.html

U.S. Environmental Protection Agency. 1996. "Chapter 3 Year-to-Year Comparison of TRI Data." *Toxic Release Inventory 1992.* http://www.epa.gov/docs/TRI-Chem/

U.S. General Accounting Office. 1978. *U.S. Foreign Relations and Multinational Corporations: What's the Connection?* Washington, D.C.: U.S. Government Printing Office.

Veblen, Thorstein. 1933. *The Engineers and the Price System.* New York: Viking.

Wald, Matthew L. 1990. "Where All That Gas Goes: Drivers' Thirst for Power." *New York Times* (November 21):A1, C17.

Weber, Max. 1947. *The Theory of Social and Economic Organization*, ed. and trans. A. M. Henderson and T. Parsons. New York: Macmillan.

Weir, David. 1987. *The Bhopal Syndrome: Pesticides, Environment, and Health.* San Francisco: Sierra Club Books.

Wexler, Mark N. 1989. "Learning from Bhopal." *The Midwest Quarterly* 31(1): 106–129.

Wilkins, Lee. 1986. "Media Coverage of the Bhopal Disaster: A Cultural Myth in the Making." *International Journal of Mass Emergencies and Disasters* (March):16–33.

Williams, Malcolm T. 1995. Personal correspondence. December 21.

Young, T. R. 1975. "Karl Marx and Alienation: The Contributions of Karl Marx to Social Psychology." *Humboldt Journal of Social Relations* 2(2):26–33.

Zuboff, Shoshana. 1988. *In the Age of the Smart Machine: The Future of Work and Power.* New York: Basic Books.

CHAPTER 8

Author X. 1992. "Mao Fever—Why Now?" trans. and adapted from the Chinese by R. Terrill. *World Monitor* (December):22–25.

Becker, Howard S. 1963. *Outsiders: Studies in the Sociology of Deviance.* New York: Free Press.

———. 1973. "Labelling Theory Reconsidered" in *Outsiders: Studies in the Sociology of Deviance.* New York: Free Press.

Belkin, Lisa. 1990. "Airport Anti-Drug Nets Snare Many People Fitting 'Profiles.'" *New York Times* (March 20):A1, A11.

Bem, Sandra. 1990. Quoted in "Blurring the Line: Androgyny on Trial," by Don Monkerud. *Omni* (October):81–86+.

Bernstein, Richard. 1982. *From the Center of the Earth: The Search for the Truth About China.* Boston: Little, Brown.

Bernstein, Thomas P. 1983. "Starving to Death in China." *The New York Review of Books* (June 16):36–38.

Best, Joel. 1989. *Images of Issues: Typifying Contemporary Social Problems.* New York: Aldine de Gruyter.

Bonavia, David. 1989. *The Chinese.* London: Penguin.

Bracey, Dorothy H. 1985. "The System of Justice and the Concept of Human Nature in the People's Republic of China." *Justice Quarterly* 2(1):139–144.

Broaded, C. Montgomery. 1991. "China's Lost Generation." *Journal of Contemporary Ethnography* (October):352–379.

Broadfoot, Robert. 1993. Quoted in "Ancient Power Steps into Asian Spotlight," by D. Holley (a special Pacific Rim edition of *World Report*). *Los Angeles Times* (June 15):H15.

Butterfield, Fox. 1976. "Mao Tse-Tung: Father of Chinese Revolution." *New York Times* (September 10):A13+.

———. 1980. "The Pragmatists Take China's Helm." *The New York Times Magazine* (December 28):22–35.

———. 1982. *China: Alive in the Bitter Sea.* New York: Times Books.

Calhoun, Craig. 1989. "Revolution and Repression in Tiananmen Square." *Society* (September/October):21–38.

Carrel, Todd, and Richard Hornik. 1994. "A Chinese Gold Rush? Don't Hold Your Breath." *New York Times* (September 14):A15.

Chambliss, William. 1974. "The State, the Law, and the Definition of Behavior as Criminal or Delinquent." Pages 7–44 in *Handbook of Criminology*, ed. D. Glaser. Chicago: Rand McNally.

Chang Jung. 1991. *Wild Swans: Three Daughters of China.* New York: Simon & Schuster.

———. 1992. Quoted in "Literature of the Wounded," by Jonathan Mirsky. *The New York Review of Books* (March 5):6.

Chinese Embassy. 1996. "Response of Chinese Embassy in UK to the Film *Return to the Dying Rooms.*" http://oneworld.org/news/partner_news/china_top.html

Chiu Hungdah. 1988. "China's Changing Criminal Justice System." *Current History* (September):265–272.

Clark, John P., and Shirley M. Clark. 1985. "Crime in China—As We Saw It." *Justice Quarterly* 2(1):103–110.

Collins, Randall. 1982. *Sociological Insight: An Introduction to Nonobvious Sociology.* New York: Oxford University Press.

Cowell, Alan. 1992. "Strike Hits Tobacco Industry, and 13 Million Italians Suffer." *New York Times* (November 19):A5.

Deng Xiaoping. 1995. Quoted in "Tide Turning Against China's 'Special Zones,'" by Rone Tempest. *Los Angeles Times* (November 3):A5.

Durkheim, Emile. [1901] 1982. *The Rules of Sociological Method and Selected Texts on Sociology and Its Method*, ed. S. Lukes and trans. W. D. Halls. New York: Free Press.

Erikson, Kai T. 1966. *Wayward Puritans.* New York: Wiley.

Fairbank, John King. 1987. *The Great Chinese Revolution 1800–1985.* New York: Harper & Row.

———. 1989. "Why China's Rulers Fear Democracy." *The New York Review of Books* (September 28):32–33.

Farley, Maggie. 1995. "China's Neighborly Capitalists." *Los Angeles Times* (December 7):A17.

Feng Jicai. 1991. *Voices from the Whirlwind: An Oral History of the Chinese Cultural Revolution.* New York: Pantheon.

Goldman, Merle. 1989. "Vengeance in China." *The New York Review of Books* (November 9):5–9.

Gould, Stephen Jay. 1990. "Taxonomy as Politics: The Harm of False Classification." *Dissent* (Winter):73–78.

Han Xu. 1989. "The Chinese Ambassador's Version. . . . " *New York Times* (August 21):Y19.

Hareven, Tamara K. 1987. "Divorce, Chinese Style." *The Atlantic Monthly* (April):70–76.

Henriques, Diana B. 1993. "Great Men and Tiny Bubbles: For God, Country and Coca-Cola." *The New York Times Book Review* (May 23):13.

Holley, David. 1993. "Ancient Power Steps into Asian Spotlight" (a special Pacific Rim edition of *World Report*). *Los Angeles Times* (June 15):H1+.

Hong Kong Trade Development Council. 1996. "Will the U.S. Focus on Trade Deficit with China in 1996?" *Business Alert* (January 4). http://www.tdc.org.hk/alert/us9611.htm

Ignatius, Adi. 1988. "China's Birthrate Is Out of Control Again as One-Child Policy Fails in Rural Areas." *Asian Wall Street Journal Weekly* (July 18):18.

Jerome, Richard. 1995. "Suspect Confessions." *The New York Times Magazine* (August 13):28–31.

Kitsuse, John I. 1962. "Societal Reaction to Deviant Behavior: Problems of Theory and Method." *Social Problems* 9 (Winter):247–256.

Kometani, Foumiko. 1987. "Pictures from Their Nightmare." *The New York Times Book Review* (July 19):9–10.

Kristof, Nicholas D. 1989. "China Is Planning 2 Years of Labor for Its Graduates." *New York Times* (August 13):Y1.

Kwong, Julia. 1988. "The 1986 Student Demonstrations in China." *Asian Survey* 28(9):970–985.

Lemert, Edwin M. 1951. *Social Pathology.* New York: McGraw-Hill.

Leys, Simon. 1989. "After the Massacres." *The New York Review of Books* (October 12):17–19.

———. 1990. "The Art of Interpreting Nonexistent Inscriptions Written in Invisible Ink on a Blank Page." *The New York Review of Books* (October 11):8–13.

Lilly, J. Robert. 1991. "Prisons: How to Help Eastern Europe." *The Angolite* 16(1):1.

Link, Perry. 1989. "The Chinese Intellectuals and the Revolt." *The New York Review of Books* (June 29):38–41.

Liu Binyan. 1993. "An Unnatural Disaster," trans. P. Link. *The New York Review of Books* (April 8):3–6.

Liu Zaifu. 1989. Quoted in "The Chinese Intellectuals and the Revolt," by Perry Link. *The New York Review of Books* (June 29):40.

Lubman, Stanley. 1983. "Comparative Criminal Law and Enforcement: China." Pages 182–193 in *Encyclopedia of Crime and Justice,* ed. S. H. Kadish. New York: Free Press.

Mao Zedong (Mao Tse-tung). 1965. "Report on an Investigation of the Peasant Movement in Hunan (March 1927)." In *Selected Works of Mao Tse-tung.* Peking: Foreign Language Press.

Mathews, Jay, and Linda Mathews. 1983. *One Billion: A China Chronicle.* New York: Random House.

Merton, Robert K. 1957. *Social Theory and Social Structure.* Glencoe, IL: Free Press.

Milgram, Stanley. 1974. *Obedience to Authority: An Experimental View.* New York: Harper & Row.

———. 1987. "Obedience." Pages 566–568 in *The Oxford Companion to the Mind,* ed. R. L. Gregory. Oxford: Oxford University Press.

Monkerud, Don. 1990. "Blurring the Lines: Androgyny on Trial." *Omni* (October):81–86+.

Montalbano, William D. 1993. "Lifetime of Change in a Dozen Years" (a special Pacific Rim edition of *World Report*). *Los Angeles Times* (June 15):H2.

Mosher, Steven W. 1991. "Chinese Prison Labor." *Society* (November/December): 49–59.

National Council for Crime Prevention in Sweden. 1985. *Crime and Criminal Policy in Sweden.* Report no. 19. Stockholm: Liber Distribution.

The New York Times. 1996. "Clearing the Air: Significant Signposts." (June 30) 12xx.

Oxman, Robert. 1993a. "China in Transition." Interview on *MacNeil/Lehrer Newshour* (December 27), transcript #4828. New York: WNET.

———. 1993b. "China in Transition: Mao to Markets." Interview on *MacNeil/Lehrer Newshour* (December 28), transcript #4829. New York: WNET.

———. 1993c. "China in Transition: Status Report (Chinese Women)." Interview on *MacNeil/Lehrer Newshour* (December 29), transcript #4830. New York: WNET.

———. 1993d. "Focus, Olympic Hurdle." Interview on *MacNeil/Lehrer Newshour* (September 21), transcript #4759. New York: WNET.

———. 1994a. "China in Transition: Any Progress? (Human Rights)." Interview on *MacNeil/Lehrer Newshour* (January 31), transcript #4853. New York: WNET.

———. 1994b. "China in Transition: Taking the Plunge (Education in China; Going into Business in China)." Interview on *MacNeil/Lehrer Newshour* (January 4), transcript #4834. New York: WNET.

Personal correspondence. 1993. Comments by an anonymous reviewer.

Piazza, Alan. 1996. Quoted in "In China's Outlands, Poorest Grow Poorer," by P .E. Tyler. *New York Times* (October 26):A1, A4.

Ramos, Francisco Martins. 1993. "My American Glasses." Pages 1–10 in *Distant Mirrors: America as a Foreign Culture,* ed. Philip R. DeVita and James D. Armstrong. Belmont, CA: Wadsworth.

Reinarman, Craig, and Harry G. Levine. 1989. "The Crack Attack: Politics and Media in America's Latest Drug Scare." Pages 115–138 in *Images of Issues: Typifying Contemporary Social Problems,* ed. J. Best. New York: Aldine de Gruyter.

Robert Wood Johnson Foundation. 1992. *1992 Annual Report: Substance Abuse.* http://sparc5.ep.cursci.com/library/pbar3.htm

Rojek, Dean G. 1985. "The Criminal Process in the People's Republic of China." *Justice Quarterly* 2(1):117–125.

Rorty, Amelie Oksenberg. 1982. "Western Philosophy in China." *Yale Review* 72(1):141–160.

Schell, Orville. 1996. "China's 'Model' State Orphanages Serve as Warehouses for Death." *Los Angeles Times* (January 7):M2+.

Shipp, E. R., Dean Baquet, and Martin Gottlieb. 1991. "Slaying Casts a New Glare on Law's Uncertain Path." *New York Times* (June 23):A1+.

Simmons, J. L., with Hazel Chambers. 1965. "Public Stereotypes of Deviants." *Social Problems* 3(2):223–232.

Skoloss, Gary N. 1996. "Divorce Laws: Some Grounds for Absolute Divorce." P. 729 in *The World Almanac and Book of Facts 1996.* Mahwak, NJ: Funk and Wagnalls.

Spector, Malcolm, and J. I. Kitsuse. 1977. *Constructing Social Problems.* Menlo Park, CA: Cummings.

Stevenson, Richard W. 1996. "Tread Lightly with China, Business Leaders Urge U.S." *New York Times* (May 11):Y7.

Strebeigh, Fred. 1989. "Training China's New Elite." *The Atlantic Monthly* (April):72–80.

Sturm, Roland, and Junsen Zhang. 1994. "When Do Couples Sign the One-Child Certificate in Urban China?" *Population Research and Policy Review* (March):69–82.

Sumner, William Graham. 1907. *Folkways.* Boston: Ginn.

Sutherland, Edwin H., and Donald R. Cressey. 1978. *Principles of Criminology,* 10th ed. Philadelphia: Lippincott.

Tannenbaum, Frank. 1938. *Crime and the Community.* New York: Ginn.

Theroux, Paul. 1989. "Travel Writing: Why I Bother." *The New York Times Book Review* (July 30):7–8.

Tien H. Yuan. 1990. "Demographer's Page: China's Population Planning After Tiananmen." *Population Today* 18(9): 6–8.

Tien H. Yuan, Zhang Tianlu, Ping Yu, Li Jingneng, and Liang Zhongtang. 1992. "China's Demographic Dilemmas." *Population Bulletin* 47(1):1–44.

Tobin, Joseph J., David Y. H. Wu, and Dana H. Davidson. 1989. *Preschool in Three Cultures: Japan, China and the United States.* New Haven, CT: Yale University Press.

Tyler, Patrick E. 1996a. "In China's Outlands, Poorest Grow Poorer." *New York Times* (October 26):1+.

———. 1996b. "Chinese Maltreatment at Orphanage." *New York Times* (January 9):A4.

———. 1996c. "U.S. Rights Group Asserts China Lets Thousands of Orphans Die." *New York Times* (January 6):1+.

U.S. Bureau of Justice Statistics. 1996a. *Criminal Victimization 1994: National Crime Victimization Survey.* http://www.ndjrs.org/txtfiles/cv94.txt

———. 1996b. *Statistics About Crime and Victims.* http://www.ojp.usdoj.gov/bjs/cvict.htm

U.S. Central Intelligence Agency. 1995. *World Factbook 1995: China.* http://www.odci.gov/cia/publications/95fact/ch.html

U.S. Department of State. 1996. *China Country Commercial Guide 1995–1996.* http://www.usia.gov/abtusia/posts/HK1/wwheh10html

Wang Ruowang. 1989. Quoted in "The Chinese Intellectuals and the Revolt," by Perry Link. *The New York Review of Books* (June 29):40.

Williams, Terry. 1989. *The Cocaine Kids: The Inside Story of a Teenage Drug Ring.* Reading, MA: Addison-Wesley.

Wu Han. 1981. Quoted in *Coming Alive: China After Mao,* by Roger Garside. New York: McGraw-Hill.

WuDunn, Sheryl. 1993. "Booming China Is a Dream Market for West." *New York Times* (February 15):A1+.

Xu, Xinyi. 1995. "The Impact of Western Forms of Social Control on China." *Crime, Law, and Social Change* 67–87.

Zhou, Joseph. 1995. Quoted in "The Impact of Western Forms of Social Control on China: A Preliminary Evaluation," by Xinyi Xu. *Crime, Law, and Social Change* 23:67–87.

CHAPTER 9

African National Congress. 1996. "The African National Congress on the Working Class Struggle for National Liberation." gopher://gopher.anc.org nc/history/bababenz.pak

Angelou, Maya. 1987. "Intra-Racism." Interview on *The Oprah Winfrey Show,* Journal Graphics transcript #W172:2.

Berreman, Gerald D. 1972. "Race, Caste, and Other Invidious Distinctions in Social Stratification." *Race* 13(4): 385–414.

Bodley, Hal. 1996. "Baseball Payroll Poised to Top $1 Billion." *USA Today* (November 14):C1.

Boudon, Raymond, and François Bourricaud. 1989. *A Critical Dictionary of Sociology,* selected and trans. P. Hamilton. Chicago: University of Chicago Press.

Chapkis, Wendy. 1986. *Beauty Secrets: Women and the Politics of Appearance.* Boston: South End Press.

Chass, Murray. 1992. "A Zillionaire at the Bat." *International Herald Tribune* (February 28):16.

———. 1993. "25 Men on a Team and 7 Figures Per Man." *New York Times* (April 11):S4.

Constitution of South Africa. 1996. "Preamble." gopher://gopher.anc.org.za/00/anc/misc/sacon96l.txt

Coser, Lewis A. 1977. *Masters of Sociological Thought,* 2d ed., ed. R. K. Merton. New York: Harcourt Brace Jovanovich.

Crapanzano, Vincent. 1985. *Waiting: The Whites of South Africa.* New York: Random House.

Crystal, Graef. 1995. "Growing the Pay Gap." *Los Angeles Times* (July 23):D2.

Davis, Kingsley, and Wilbert E. Moore. 1945. "Some Principles of Stratification." Pages 413–445 in *Ideological Theory: A Book of Readings,* ed. L. A. Coser and B. Rosenberg. New York: Macmillan.

Drogin, Bob. 1995. "Apartheid Brutality on Trial." *Los Angeles Times* (February 20):A6+.

———. 1996. "South Africa Bringing Power to the People." *Los Angeles Times* (January 31):A1+.

Eiseley, Loren. 1990. "Man: Prejudice and Personal Choice." Pages 640–943 in *The Random House Encyclopedia,* 3d ed. New York: Random House.

Federal Election Commission. 1996. "Top 50 PAC's—Contributions to Candidates January 1, 1995–June 30, 1996." http://www.fec.gov/press/pac18ctr.htm

Finnegan, William. 1986. *Crossing the Line: A Year in the Land of Apartheid.* New York: Harper & Row.

Franklin, John Hope. 1990. Quoted in "That's History, Not Black History," by Mark Megurl. *The New York Times Book Review* (June 3):13.

Frontline. 1985. "A Class Divided," transcript #309. Boston: WGBH Educational Foundation.

Gerth, Hans, and C. Wright Mills. 1954. *Character and Social Structure: The Psychology of Social Institutions.* London: Routledge & Kegan Paul.

Graham, Lawrence Otis. 1995. *Member of the Club: Reflections on Life in a Racially Polarized World.* New York: HarperCollins.

Haney López, Ian F. 1994. "The Social Construction of Race: Some Observations on Illusion, Fabrication, and Choice." *Harvard Civil Rights–Civil Liberties Law Review* 29(1):1–62.

Jencks, Christopher. 1990. Quoted in "The Rise of the 'HyperPoor,'" by David Whitman. *U.S. News & World Report* (October 15):40–42.

Jones, Arthur. 1994. "Reality Check." *World Trade* (February):34–38.

Keller, Bill. 1993. "South Africa's Wealth Is Luring Black Talent." *New York Times* (February 12):A1+.

———. 1994. "Mandela's Party Publishes Plan to Redistribute Wealth." *New York Times* (January 15):Y3.

Lamb, David. 1987. *The Africans.* New York: Vintage.

Lambert, Father Rollins. 1988. "A Day in the Life of Apartheid: The Editor's Interview with Father Rollins Lambert." *U.S. Catholic* 53:26–32.

Liebenow, J. Gus. 1986. "South Africa: Home, 'Not-So-Sweet,' Homelands." *UFSI Reports* no. 23.

Lock, Margaret. 1993. "The Concept of Race: An Ideological Construct." *Transcultural Psychiatric Research Review* 30:203–227.

Loy, John W., and Joseph F. Elvogue. 1971. "Racial Segregation in American Sport." *International Review of Sport Sociology* 5:5–24.

Mabuza, Lindiwe. 1990. "Apartheid: Far from Over." *New York Times* (June 20): A15.

Mackay, C. J. 1996. "The Development of Housing Policy in South Africa in the Post Apartheid Period." *Housing Studies* (January 1):133–146.

Mandela, Nelson. 1990. "I Am the First Accused" (Rivonia Trial Statement 1964). *One Nation, One Country: The Phelps-Stokes Fund* 4 (May):17–45.

Marx, Karl. 1909. *Capital: A Critique of Political Economy,* vol. III, ed. F. Engles, trans. E. Untermann. Chicago: Kerr.

———. [1895] 1976. *The Class Struggles in France 1848–1850.* New York: International.

Medoff, Marshall H. 1977. "Positional Segregation and Professional Baseball." *International Review of Sport Sociology* 12:49–56.

Menaker, Drusilla. 1995. "Doing Worse by Doing Good?" *Business Week* (June 12):50–51.

Myrdal, Gunnar. 1944. *An American Dilemma: The Negro Problem and Modern Democracy.* New York: Harper & Brothers.

Office of the Press Secretary. 1994. "Press Conference by President Clinton and President Mandela of South Africa." Washington, D.C.: White House.

O'Hare, William P. 1996. "A New Look at Poverty in America." *Populations Bulletin* (September):2–48.

O'Hare, William P., and Brenda Curry-White. 1992. "Demographer's Page: Is There a Rural Underclass?" *Population Today* 20(3):6–8.

Page, Clarence. 1996. *Showing My Color: Impolite Essays on Race and Identity.* New York: HarperCollins.

Passell, Peter. 1994. "South Africa's Huge Challenge: Bringing the Good Life to Blacks." *New York Times* (June 9):C2.
Plaschke, Bill. 1995. "Campanis Embarrassment: A Call to Sports Diversity." *Los Angeles Times* (July 23):A18+.
Poussaint, Alvin. 1987. "Intra-Racism." Interview on *The Oprah Winfrey Show,* Journal Graphics transcript #W172:7.
Ridgeway, Cecilia. 1991. "The Social Construction of Status Value: Gender and Other Nominal Characteristics." *Social Forces* 70(2):367–386.
Rockefeller Foundation. 1996a. "Where the American Public Would Set the Poverty Line." *Poverty Research Brief* no. 23. http://www.cdinet.com/Rockefeller/Briefs/brief23.html
———. 1996b. "White Poverty in America." *Poverty Research Brief* no. 28. http://www.cdinet.com/Rockefeller/Briefs/brief28.html
———. 1996c. "Working Their Way Out of Poverty?" *Poverty Research Brief* no. 11. http://www.cdinet.com/Rockefeller/Briefs/brief11.html
Roberts, Margaret. 1994. "The Ending of Apartheid: Shifting Inequalities in South Africa." *Journal of the Geographical Association* (January 1):53–64.
Ross, Edward Alsworth. [1908] 1929. *Social Psychology: An Outline and Source Book.* New York: Macmillan.
Russell, Diana E. H. 1989. *Lives of Courage: Women for a New South Africa.* New York: Basic Books.
Seidman, Judy. 1978. *Ba Ye Zwa: The People Live.* Boston: South End Press.
Simon, David. 1994. "The Apartheid City." Pages 312–314 in *Atlas of World Development,* ed. Tim Unwin. New York: Wiley.
Simpson, Richard L. 1956. "A Modification of the Functional Theory of Social Stratification." *Social Forces* 35:132–137.
South Africa Labour Development Research Unit. 1994. *South Africans Rich and Poor: Baseline Household Statistics.* Cape Town: University of Cape Town.
Sparks, Allister. 1990. *The Mind of South Africa.* New York: Knopf.
Thomas, Isaiah. 1987. Quoted in "The Coloring of Bird," by Ira Berkow. *New York Times* (June 2):D27.
Tucher, A., ed. 1990. *Bill Moyers, A World of Ideas II: Public Opinion from Private Citizens.* New York: Doubleday.
Tumin, Melvin M. 1953. "Some Principles of Stratification: A Critical Analysis." *American Sociological Review* 18:387–394.
U.S. Bureau of the Census. 1992. *Money Income of Households, Families, and Persons in the United States: 1991.* Washington D.C.: U.S. Government Printing Office.
———. 1995. *Statistical Abstract of the United States 1995.* Washington, D.C.: U.S. Government Printing Office.
———. 1996. *1990 Census Lookup.* http://census.census.gov/cdrom/lookup
U.S. Central Intelligence Agency. 1995. *World Factbook 1995.* http://www.odci.gov/cia/publications/95fact/index.html
U.S. Department of Labor. 1995. *Employment, Hours, and Earnings, United States, 1990–1995.* Washington, D.C.: U.S. Government Printing Office.
U.S. Department of State. *1996 Background Notes: South Africa, November 1994.* gopher://dosfan.lib.uic.edu/1D-%3A22525%ABackground%20Ser
U.S. Postal Service. 1996. *City State/ZIP Code Associations.* http://www.usps.gov/cgi-bin/zip4/ctystzip
Wacquant, Loic J. D. 1989. "The Ghetto, the State, and the New Capitalist Economy." *Dissent* (Fall):508–520.
Wacquant, Loic J. D., and William Julius Wilson. 1989. "The Cost of Racial and Class Exclusion in the Inner City." *The Annals of the American Academy* (January):8–25.
Weber, Max. 1982. "Status Groups and Classes." Pages 69–73 in *Classes, Power, and Conflict: Classical and Contemporary Debates,* ed. A. Giddens and D. Held. Los Angeles: University of California.
———. [1947] 1985. "Social Stratification and Class Structure." Pages 573–576 in *Theories of Society: Foundations of Modern Sociological Theory,* ed. T. Parsons, E. Shils, K. D. Naegele, and J. R. Pitts. New York: Free Press.
Weekend World: Johannesburg. 1977. "Lesson 8: Basic Economics at People's College" (April 24).
Wilson, Francis, and Mamphela Ramphele. 1989. *Uprooting Poverty: The South African Challenge.* New York: Norton.
Wilson, William Julius. 1983. "The Urban Underclass: Inner-City Dislocations." *Society* 21:80–86.
———. 1987. *The Truly Disadvantaged: The Inner City, the Underclass, and Public Policy.* Chicago: University of Chicago Press.
———. 1991. "Studying Inner-City Social Dislocations: The Challenge of Public Agenda Research" (1990 presidential address). *American Sociological Review* (February):1–14.
———. 1994. "Another Look at the Truly Disadvantaged." *Political Science Quarterly* 106(4):639–656.
Wirth, Louis. [1945] 1985. "The Problem of Minority Groups." Pages 309–315 in *Theories of Society: Foundations of Modern Sociological Theory,* ed. T. Parsons, E. Shils, K. D. Naegele, and J. R. Pitts. New York: Free Press.
The World Almanac and Book of Facts 1994. 1993. Mahwah, NJ: World Almanac.
World Development Report. 1990. "Poverty." New York: Oxford University Press.
Wren, Christopher S. 1990a. "De Klerk Hopes to Show Bush the Change Is Real." *New York Times* (September 23):Y8.
———. 1991. "South Africans Desegregate Some White Public Schools." *New York Times* (January 10):A1.
Yeutter, Clayton. 1992. "When 'Fairness' Isn't Fair." *New York Times* (March 24):A13.

CHAPTER 10

Adams, Anne V. 1992. "Translator's Afterword." Pages 234–237 in *Showing Our Colors: Afro-German Women Speak Out,* ed. M. Opitz, K. Oguntoye, and D. Schultz. Amherst: University of Massachusetts Press.
Adomako, Abena. 1992. "Mother: Afro-German Father: Ghanian." Pages 199–203 in *Showing Our Colors: Afro-German Women Speak Out,* ed. M. Opitz, K. Oguntoye, and D. Schultz. Amherst: University of Massachusetts Press.
Alba, Richard D. 1992. "Ethnicity." Pages 575–584 in *Encyclopedia of Sociology,* vol. 2, ed. E. F. Borgatta and M. L. Borgatta. New York: Macmillan.
Anson, Robert Sam. 1987. *Best Intentions: The Education and Killing of Edmund Perry.* New York: Random House.
Atkins, Elizabeth. 1991. "For Many Mixed-Race Americans, Life Isn't Simply Black or White." *New York Times* (June 5):B8.
Barrins, Adeline. 1992. Quoted in "The Tallest Fence: Feelings on Race in a White Neighborhood." *New York Times* (June 21):Y12.
Breton, Raymond, Wsevolod W. Isajiw, Warren E. Kalbach, and Jeffrey G. Reitz. 1990. *Ethnic Identity and Equality: Varieties of Experience in a Canadian City.* Toronto: University of Toronto.
Buckley, Jerry. 1991. "Mt. Airy, Philadelphia." *U.S. News and World Report* (July 22):22–28.
Carver, Terrell. 1987. *A Marx Dictionary.* Totowa, NJ: Barnes & Noble.

Castles, Stephen. 1986. "The Guest-Worker in Western Europe—An Obituary." *International Migration Review* 20(4):761–778.

Cheliand, Gerard, and Jean-Pierre Rageau. 1995. *The Penguin Atlas of Diasporas.* New York: Penguin.

Cornell, Stephen. 1990. "Land, Labour and Group Formation: Blacks and Indians in the United States." *Ethnic and Racial Studies* 13(3):368–388.

Crapanzano, Vincent. 1985. *Waiting: The Whites of South Africa.* New York: Random House.

Davis, F. James. 1978. *Minority–Dominant Relations: A Sociological Analysis.* Arlington Heights, IL: AHM.

Dunne, John Gregory. 1991. "Law and Disorder in Los Angeles." *New York Review of Books* (October 10):26.

Emde, Helga. 1992. "An 'Occupation Baby' in Postwar Germany." Pages 101–111 in *Showing Our Colors: Afro-German Women Speak Out,* ed. M. Opitz, K. Oguntoye, and D. Schultz. Amherst: University of Massachusetts Press.

Encyclopedia of Latin American History and Culture. 1996. New York: Scribner's.

Faist, Thomas, and Hartmut Haubermann. 1996. "Immigration, Social Citizenship and Housing in Germany." *International Journal of Urban and Regional Geography* (March):83–98.

Federal Bureau of Investigation. 1996. *Hate Crime—1995 Uniform Crime Reports.* http://www.fbi.gov/ucr/hatecm.htm

Fein, Helen. 1978. "A Formula for Genocide: Comparison of the Turkish Genocide (1915) and the German Holocaust (1939–1945)." *Comparative Studies in Sociology* 1:271–294.

Glassner, Martin Ira. 1996. *Political Geography.* New York: Wiley.

Goffman, Erving. 1963. *Stigma: Notes on the Management of Spoiled Identity.* Upper Saddle River, NJ: Prentice Hall.

Gordon, Milton M. 1978. *Human Nature, Class, and Ethnicity.* New York: Oxford University Press.

Hacker, Andrew. 1992. *Two Nations: Black and White, Separate, Hostile, Unequal.* New York: Scribner's.

Haney López, Ian F. 1994. "The Social Construction of Race: Some Observations on Illusion, Fabrication, and Choice." *Harvard Civil Rights—Civil Liberties Law Review* 29(1):1–62.

Herbert, Ulrich. 1995. "Immigration, Integration, Foreignness: Foreign Workers in Germany Since the Turn of the Century." *International Labor and Working-Class History* (Fall):91–93.

Hirschman, Charles. 1993. "How to Measure Ethnicity: An Immodest Proposal." Pages 547–560 in *Challenges of Measuring an Ethnic World: Science, Politics and Reality,* ed. Statistics Canada and U.S. Bureau of the Census. Washington D.C.: U.S. Government Printing Office.

Holzner, Lutz. 1982. "The Myth of Turkish Ghettos: A Geographic Case Study of West German Responses Towards a Foreign Minority." *Journal of Ethnic Studies* 9(4):65–85.

Houston, Velin Hasu. 1991. "The Past Meets the Future: A Cultural Essay." *Amerasia Journal* 17(1):53–56.

Jones, Tamara, and Hugh Pope. 1993. "Kurds Raid Turk Offices in Europe." *New York Times* (June 25):A1+.

Jopke, Christian. 1996. "Multiculturalism and Immigration: A Comparison of the United States, Germany, and Great Britain." *Theory and Society* 25(4): 449–500.

Kaw, Eugenia. 1993. "Medicalization of Racial Features: Asian American Women and Cosmetic Surgery." *Medical Anthropology Quarterly,* pp. 74–89.

King, Lloyd. 1992. "Lloyd King." Pages 397–401 in *Race: How Blacks and Whites Think and Feel About the American Obsession,* ed. Studs Terkel. New York: New Press.

Kramer, Jane. 1993. "Letter from Europe: Neo-Nazis: A Chaos in the Head." *The New Yorker* (June 14):52–70.

Krell, Gert, Hans Nicklas, and Anne Ostermann. 1996. "Immigration, Asylum, and Anti-Foreigner Violence in Germany." *Journal of Peace Research,* pp. 153–170.

Kurthen, Hermann. 1995. "Germany at the Crossroads: National Identity and the Challenges of Immigration." *International Migration Review* (Winter): 914–937.

Lawrence, Cecile Ann. 1996. "Racelessness." *Interracial Voice.* http://www.webcom.com/%7Eintvoice/

Lieberman, Leonard. 1968. "The Debate over Race: A Study in the Sociology of Knowledge." *Phylon* 39 (Summer): 127–141.

Los Angeles Times. 1992. "Probe Finds Pattern of Excessive Force, Brutality by Deputies" (July 21):A18.

Mandel, Ruth. 1989. "Turkish Headscarves and the 'Foreigner Problem': Constructing Difference Through Emblems of Identity." *New German Critique* 46(Winter):27–46.

Marshall, Tyler. 1992. "Saying 'No' to Nazis in Germany." *Los Angeles Times* (December 5):A1, 13.

———. 1993. "Arson Attacks on Foreigners No Longer Big News in Germany." *Los Angeles Times* (July 17):A8.

Martin, Philip L., and Mark J. Miller. 1990. "Guests or Immigrants? Contradiction and Change in the German Immigration Policy Debate Since the Recruitment Stop." *Migration World* 15(1):8–13.

McClain, Leanita. 1986. *A Foot in Each World: Essays and Articles,* ed. C. Page. Evanston, IL: Northwestern University Press.

McDowell, Jeanne. 1989. "He's Got to Have It His Way." *Time* (July 17):92–94.

McIntosh, Peggy. 1992. "White Privilege and Male Privilege: A Personal Account of Coming to See Correspondences Through Work in Women's Studies." Pages 70–81 in *Race, Class, and Gender: An Anthology,* ed. M. L. Andersen and P. H. Collins. Belmont, CA: Wadsworth.

Merton, Robert K. 1957. *Social Theory and Social Structure.* New York: Free Press.

———. 1976. "Discrimination and the American Creed." Pages 189–216 in *Sociological Ambivalence and Other Essays.* New York: Free Press.

The New York Times. 1990. "Advertising: New Group Makes the Case for Black Agencies and Media" (October 26): C18.

Northern Kentucky University student. 1996. Response paper (September).

O'Connor, Peggy. 1992. Quoted in "The Tallest Fence: Feelings on Race in a White Neighborhood." *New York Times* (June 21):Y12.

Ogbu, John U. 1990. "Minority Status and Literacy in Comparative Perspective." *Daedalus* 119(2):141–168.

Opitz, May. 1992a. "Recapitulation and Outlook." Pages 228–233 in *Showing Our Colors: Afro-German Women Speak Out,* ed. M. Opitz, K. Oguntoye, and D. Schultz. Amherst: University of Massachusetts Press.

———. 1992b. "Three Afro-German Women in Conversation with Dagmar Schultz: the First Exchange for This Book." Pages 145–164 in *Showing Our Colors: Afro-German Women Speak Out,* ed. M. Opitz, K. Oguntoye, and D. Schultz. Amherst: University of Massachusetts Press.

Page, Clarence. 1990. "Black Youth Need More Help, Not More Scorn." *Cincinnati Post* (March 9):A10.

Poston, Dudley L., and Mei-Yu Yu. 1990. "The Distribution of the Overseas Chinese in the Contemporary World." *International Migration Review* (Fall): 480–508.

Raspberry, William. 1991. *MacNeil/Lehrer Newshour* (June). New York: WNET.

Rawley, James A. 1981. *The Transatlantic Slave Trade: A History.* New York: Norton.

Reynolds, Larry T. 1992. "A Retrospective on 'Race': The Career of a Concept." *Sociological Focus* 25(1):1–14.

Safran, William. 1986. "Islamization in Western Europe: Political Consequences and Historical Parallels." *Annals* 485 (May):98–112.

Sayari, Sabri. 1986. "Migration Policies of Sending Countries: Perspectives on the Turkish Experience." *Annals* 485 (May):87–97.

Scales-Trent, Judy. 1995. *Notes of a White Black Woman: Race, Color, Community.* University Park: Pennsylvania State University Press.

Schneider, Peter. 1989. "If the Wall Came Tumbling Down." *The New York Times Magazine* (June 25):22+.

Segal, Aaron. 1993. *An Atlas of International Migration.* London: Zell.

Smokes, Saundra. 1992. "A Lifetime of Racial Rage Control Snaps with a Telephone Call." *Cincinnati Post* (May 13):14A.

Starr, Paul D. 1978. "Ethnic Categories and Identification in Lebanon." *Urban Life* 7(1):111–142.

Steele, Shelby. 1990. "A Negative Vote on Affirmative Action." *The New York Times Magazine* (May 13):46–49+.

Strasser, Hermann. 1993. "The German Debate over Multicultural Society: Climax or Test of Organized Capitalism?" Paper presented at American Sociological Association Annual Meeting, Miami, FL.

Teraoka, Arlene Akiko. 1989. "Talking 'Turk': On Narrative Strategies and Cultural Stereotypes." *New German Critique* 46:104–128.

Terkel, Studs. 1992. *Race: How Blacks and Whites Think and Feel About the American Obsession.* New York: New Press.

Thränhardt, Dietrich. 1989. "Patterns of Organization Among Different Ethnic Minorities." *New German Critique* 46: 10–26.

The Times Atlas of World History. 1984. Maplewood, NJ: Hammond.

Toro, Luis Angel. 1995. "'A People Distinct from Others': Race and Identity in Federal Indian Law and the Hispanic Classification in OMB Directive No. 15." *Texas Tech Law Review* 26:1219–1274.

U.S. Bureau for Refugee Programs. 1992. *World Refugee Report: A Report Submitted to the Congress as Part of the Consultations on FY1993 Refugee Admission to the United States.* Washington, D.C.: U.S. Government Printing Office.

U.S. Central Intelligence Agency. 1995. *World Factbook 1995.* http://www.odci.gov/cia/publications/95fact/index.html

U.S. Commission on Civil Rights. 1981. *Affirmative Action in the 1980s: Dismantling the Process of Discrimination (A Proposed Statement).* Clearinghouse Publication 65. Washington, D.C.: U.S. Government Printing Office.

U.S. Department of Justice. 1996a. *Hate Crime Report 1995.* http://www.fbi.gov/ucr

———. 1996b. Press release (October 31). http://www.usdoj.gov/gopherdata/press_releases/previous/1996/Oct96/532cr.htm

U.S. Department of State. 1996. *Background Notes: Germany, July 1995.* gopher://dosfan.lib.uic.edu/1D-%3A22525%ABackground%20Ser

Walton, Anthony. 1989. "Willie Horton and Me." *The New York Times Magazine* (August 20):52+.

Wiedenroth, Ellen. 1992. "What Makes Me So Different in the Eyes of Others?" Pages 165–177 in *Showing Our Colors: Afro-German Women Speak Out,* ed. M. Opitz, K. Oguntoye, and D. Schultz. Amherst: University of Massachusetts Press.

Wilpert, Czarina. 1991. "Migration and Ethnicity in a Non-Immigration Country: Foreigners in a United Germany." *New Community* 18(1):49–62.

Wirth, Louis. 1945. "The Problem of Minority Groups." Pages 347–372 in *The Science of Man,* ed. R. Linton. New York: Columbia University Press.

CHAPTER 11

Alderman, Craig, Jr. 1990. "10 February 1989 Memo for Mr. Peter Nelson." Page 108 in *Gays in Uniform: The Pentagon's Secret Reports,* ed. K. Dyer. Boston: Alyson.

Almquist, Elizabeth M. 1992. Review of "Gender, Family, and Economy: The Triple Overlap." *Contemporary Sociology* 21(3):331–332.

American Medical Association Bureau of Investigation. 1929. "The Tricho System: Albert C. Gryser X-Ray Method of Depilation." *Journal of the American Medical Association* 92:252.

Anspach, Renee R. 1987. "Prognostic Conflict in Life-and-Death Decisions: The Organization as an Ecology of Knowledge." *Journal of Health and Social Behavior* 28(3):215–231.

Anthias, Floya, and Nira Yuval-Davis. 1989. "Introduction." Pages 1–15 in *Woman-Nation-State,* ed. N. Yuval-Davis and F. Anthias. New York: St. Martin's.

Baumgartner-Papageorgiou, Alice. 1982. *My Daddy Might Have Loved Me: Student Perceptions of Differences Between Being Male and Being Female.* Denver: Institute for Equality in Education.

Bem, Sandra Lipsitz. 1993. *The Lenses of Gender: Transforming the Debate on Sexual Inequality.* Binghamton, NY: Vail-Ballou.

Borden, Anthony. 1992. "The Yugoslav Conflict." *The European Security Network* 1:1–8.

Boroughs, Don L. 1990. "Valley of the Doll?" *U.S. News & World Report* (December 3):56–59.

Brew, Jo. 1994. "European Feminists Meet." *Off Our Backs* 24(1):1.

Brownmiller, Susan. 1975. *Against Our Will: Men, Women and Rape.* New York: Simon & Schuster.

Canadian Broadcasting Corporation. 1994. *Sunday Morning* (January 9).

Collins, Randall. 1971. "A Conflict Theory of Sexual Stratification." *Social Problems* 19(1):3–21.

Cordes, Helen. 1992. "What a Doll! Barbie: Materialistic Bimbo or Feminist Trailblazer?" *Utne Reader* (March/April):46, 50.

Curtis, Glenn E. 1992. *Yugoslavia: A Country Study,* 3d ed. Washington, D.C.: U.S. Government Printing Office.

Darville, Ray L., and Joy B. Reeves. 1992. "Social Inequality Among Yugoslav Women in Directoral Positions." *Sociological Spectrum* 12(3):279–292.

Dewhurst, Christopher J., and Ronald R. Gordon. 1993. Quoted in "How Many Sexes Are There?" *New York Times* (March 12):A15.

Doherty, Jake. 1993. "Conference to Focus on Plight of Wartime 'Comfort Women.'" *Los Angeles Times* (February 20):B3.

Drakulić, Slavenka. 1990. "Women of Eastern Europe." *Ms.* (July/August):36–47.

———. 1992. *How We Survived Communism and Even Laughed.* New York: Norton.

———. 1993a. *The Balkan Express: Fragments from the Other Side of War.* New York: Norton.

———. 1993b. "Women and the New Democracy in the Former Yugoslavia." Pages 123–130 in *Gender Politics and Post-Communism: Reflections from Eastern Europe and the Former Soviet Union,* ed. N. Funk and M. Mueller. New York: Routledge.

Enloe, Cynthia. 1993. *The Morning After: Sexual Politics at the End of the Cold War.* Los Angeles: University of California Press.

Fagot, Beverly, Richard Hagan, Mary Driver Leinbach, and Sandra Kronsberg. 1985. "Differential Reactions to Assertive and Communicative Acts of Toddler Boys and Girls." *Child Development* 56(6):1499–1505.

Fausto-Sterling, Anne. 1993. "How Many Sexes Are There?" *New York Times* (March 12):A15.

Ferrante, Joan. 1988. "Biomedical Versus Cultural Constructions of Abnormality: The Case of Idiopathic Hirsutism in the United States." *Culture, Medicine and Psychiatry* 12:219–238.

Garb, Frances. 1991. "Secondary Sex Characteristics." Pages 326–327 in *Women's Studies Encyclopedia. Vol. 1: Views from the Sciences,* ed. H. Tierney. New York: Bedrick.

Gauguin, Paul. [1919] 1985. *Noa Noa: The Tahitian Journal,* trans. O. F. Theis. New York: Dover.

Geschwender, James A. 1992. "Ethgender, Women's Waged Labor, and Economic Mobility." *Social Problems* 39(1):1–16.

Glenny, Mirsha. 1992. "Yugoslavia: The Revenger's Tragedy." *The New York Review of Books* (August 13):37–43.

Grady, Denise. 1992. "Sex Test of Champions." *Discover* (June):78–82.

Gross, Jane. 1993. "Big Grocery Chain Reaches Landmark Sex-Bias Accord." *New York Times* (December 17):A1, B11.

Hall, Edward T. 1959. *The Silent Language.* New York: Doubleday.

Harper's. 1990. "Employment History: Zeitgeist Barbie" (August):20.

Henry, Jules. 1963. *Culture Against Men.* New York: Vintage.

Hoon, Shim Jae. 1992. "Haunted by the Past." *Far Eastern Economic Review* (February 6):20.

Ignatieff, Michael. 1993. *Blood and Belonging: Journeys into the New Nationalism.* Toronto: Viking.

Johnson, G. David, Gloria J. Palileo, and Norma B. Gray. 1992. " 'Date Rape' on a Southern Campus: Reports from 1991." *SSR* 76(2):37–44.

Katzarova, Mariana. 1993. "Opening the Door." *The Nation* (July 26):148–150.

Kifner, John. 1994. "Bosnian Serbs Order General Mobilization for 'Conclusion of War.'" *New York Times* (February 1):A4.

Kinzer, Stephen. 1993. "Feminist Gadfly Unappreciated in Her Own Land." *New York Times* (December 11):Y4.

Kirka, Danica. 1993. "Slavenka Drakulić: Out of Sync in a Country Unaccustomed to Democracy." *Los Angeles Times* (December 19):M3.

Kolata, Gina. 1992. "Track Federation Urges End to Gene Test for Femaleness." *New York Times* (February 12): A1, B11.

Komarovsky, Mirna. 1991. "Some Reflections on the Feminist Scholarship in Sociology." Pages 1–25 in *Annual Review of Sociology,* vol. 17, ed. W. R. Scott and J. Blake. Palo Alto, CA: Annual Reviews.

Koss, Mary P., Christine A. Gidycz, and Nadine Wisniewski. 1987. "The Scope of Rape: Incidence and Prevalence of Sexual Aggression and Victimization in a National Sample of Higher Education Students." *Journal of Consulting and Clinical Psychology* 55(2):162–170.

Laber, Jeri. 1993. "Bosnia: Questions About Rape." *The New York Review of Books* (March 25):3–6.

Lambert, Bruce. 1993. "Abandoned Filipinas Sue U.S. over Child Support." *New York Times* (June 21):A3.

Lemonick, Michael D. 1992. "Genetic Tests Under Fire." *Time* (February 24):65.

Lewin, Tamar. 1993. "At Bases, Debate Rages over Impact of New Gay Policy." *New York Times* (December 24):A1+.

Milic, Andgelka. 1993. "Women and Nationalism in the Former Yugoslavia." Pages 109–122 in *Gender Politics and Post-Communism: Reflections from Eastern Europe and the Former Soviet Union,* ed. N. Funk and M. Mueller. New York: Routledge.

Mills, Janet Lee. 1985. "Body Language Speaks Louder Than Words." *Horizons* (February):8–12.

Mirsada. 1993. "Testimony." *Los Angeles Times Magazine* (January 31):28–32.

Mladjenovic, Lepa. 1993. "Universal Soldier: Rape Is War by a Feminist in Serbia." *Off Our Backs* (March):14–15.

Morawski, Jill G. 1991. "Femininity." Pages 136–139 in *Women's Studies Encyclopedia. Vol. 1: Views from the Sciences,* ed. H. Tierney. New York: Bedrick.

Morgenson, Gretchen. 1991. "Barbie Does Budapest." *Forbes* (January 7):66–69.

Pion, Alison. 1993. "Accessorizing Ken." *Origins* (November):8.

Ramet, Sabrina P. 1991. *Social Currents in Eastern Europe.* Durham, NC: Duke University Press.

Rank, Mark R. 1989. "Fertility Among Women on Welfare: Incidence and Determinants." *American Sociological Review* 54(4):296–304.

Rieff, David. 1992. "Letter from Bosnia: Original Virtue, Original Sin." *The New Yorker* (November 23):82–95.

Sarbin, Theodore R., and Kenneth E. Karols. 1990. "Nonconforming Sexual Orientations and Military Suitability." Pages 6–49 in *Gays in Uniform: The Pentagon's Secret Reports,* ed. K. Dyer. Boston: Alyson.

Schaller, Jane Green, and Elena O. Nightingale. 1992. "Children and Childhoods: Hidden Casualties of War and Civil Unrest." *Journal of the American Medical Association* 268(5):642–644.

Schmalz, Jeffrey. 1993. "From Midshipman to Gay-Rights Advocate." *New York Times* (February 4):B1+.

Segal, Lynne. 1990. *Slow Motion: Changing Masculinities, Changing Men.* London: Virago.

Shweder, Richard A. 1994. "What Do Men Want? A Reading List for the Male Identity Crisis." *The New York Times Book Review* (January 9):3, 24.

Solinger, Rickie. 1992. *Wake Up Little Susie: Single Pregnancy and Race Before Roe v. Wade.* New York: Routledge.

Stevenson, Samantha. 1994. "Sister Strikeout Does the Job Well." *New York Times* (February 23):B10.

Sturdevant, Saundra Pollock, and Brenda Stoltzfus. 1992. *Let the Good Times Roll: Prostitution and the U.S. Military in Asia.* New York: New Press.

Swiss, Shana, and Joan E. Giller. 1993. "Rape as a Crime of War." *Journal of the American Medical Association* 270(5): 612–615.

Tattersall, Ian. 1993. "Focus—All in the Family" (*Homo sapiens* exhibit at New York's American Museum of Natural History). Interview on *MacNeil/Lehrer Newshour,* July 12, transcript #4708. New York: WNET.

Tierney, Helen. 1991. "Gender/Sex." Page 153 in *Women's Studies Encyclopedia. Vol. 1: Views from the Sciences,* ed. H. Tierney. New York: Bedrick.

United Nations. 1991. *The World's Women 1970–1990: Trends and Statistics.* New York: United Nations.

U.S. Bureau of the Census. 1991. *World Population Profile: 1991.* Washington, D.C.: U.S. Government Printing Office.

U.S. Department of Defense. 1990. "DOD Directive 1332.14." Page 19 in *Gays in Uniform: The Pentagon's Secret Reports,* ed. K. Dyer. Boston: Alyson.

U.S. Department of Education. 1995. *Digest of Education Statistics,* by Thomas D. Snyder and Charlene Hoffman. NCES-95-029. Washington, D.C.: U.S. Government Printing Office.

Williams, Carol J. 1993. "Postscript: 'Ethnic Cleansing' Threatens to Wipe Away Memory of Tito." *Los Angeles Times* (October 26):H1+.

Williams, Lena. 1993. "Pregnant Teenagers Are Outcasts No Longer." *New York Times* (December 2):B1+.

CHAPTER 12

American Demographics. 1996. "Estimating Child-Raising Costs" (October 1996). http://www.marketingpower.com/Publications/FC/sample/sampFQ8.HTM

Atlas of World Development. 1994. New York: Wiley.

Behnam, Djamshid. 1990. "An International Inquiry into the Future of the Family: A UNESCO Project." *International Social Science Journal* 42: 547–552.

Berelson, Bernard. 1978. "Prospects and Programs for Fertility Reduction: What? Where?" *Population and Development Review* 4:579–616.

Beresky, Andrew E., ed. 1991. *Fodor's Brazil: Including Bahia and Adventures in the Amazon.* New York: Fodor's Travel Publications.

Boccaccio, Giovanni. [1353] 1984. "The Black Death." Pages 728–740 in *The Norton Reader: An Anthology of Expository Prose,* 6th ed., ed. A. M. Eastman. New York: Norton.

Brooke, James. 1989. "Decline in Births in Brazil Lessens Population Fears." *New York Times* (August 8):Y1+.

Brown, Christy. 1992. "The Letter 'A.'" Pages 85–90 in *One World, Many Cultures,* ed. S. Hirschberg. New York: Macmillan.

Brown, Lester R. 1987. "Analyzing the Demographic Trap." *State of the World 1987: A Worldwatch Institute Report on Progress Toward a Sustainable Society.* New York: Norton.

Burke, B. Meredith. 1989. "Ceausescu's Main Victims: Women and Children." *New York Times* (January 16):Y15.

Butts, Yolanda, and Donald J. Bogue. 1989. *International Amazonia: Its Human Side.* Chicago: Social Development Center.

Calsing, Elizeu Francisco. 1985. "Extent and Characteristics of Poverty in Brazil. Estimation of Social Inequalities." *Revista Paraguaya de Sociologia* 22:29–53.

Caufield, Catherine. 1985. "A Reporter at Large: The Rain Forests." *The New Yorker* (January 14):41+.

Cowell, Adrian. 1990. *The Decade of Destruction: The Crusade to Save the Amazon Rain Forest.* New York: Holt.

Davis, Kingsley. 1984. "Wives and Work: The Sex Role Revolution and Its Consequences." *Population and Development Review* 10(3):397–417.

Dean, Warren. 1991. *Review of Coffee, Contention, and Change: In the Making of Modern Brazil,* ed. Mauricio A. Font and Charles Tilly. *Journal of Latin American Studies* 23(3):649–650.

Dickenson, John. 1994. "Company Towns: The Brazilian Experience." Pp. 186–187 in *Atlas of World Development,* edited by T. J. Unwin. New York: Wiley.

Durning, Alan B. 1990. "Ending Poverty." Pages 135–153 in *State of the World 1990: A Worldwatch Institute Report on Progress Toward a Sustainable Society,* ed. L. Starke. New York: Norton.

Dychtwald, Ken, and Joe Flower. 1989. *Age Wave: The Challenges and Opportunities of an Aging America.* Los Angeles: Tarcher.

Eckholm, Erik. 1990. "An Aging Nation Grapples with Caring for the Frail." *New York Times* (March 27):A1+.

Feder, Ernest. 1971. *The Rape of the Peasantry: Latin America's Landholding System.* Garden City, NY: Anchor.

Fonseca, Claudia. 1986. "Orphanages, Foundlings, and Foster Mothers: The System of Child Circulation in a Brazilian Squatter Settlement." *Anthropological Quarterly* 59:15–27.

Glascock, Anthony P. 1982. "Decrepitude and Death Hastening: The Nature of Old Age in Third World Societies (Part I)." *Studies in Third World Societies* 22: 43–66.

Goldani, Ana Maria. 1990. "Changing Brazilian Families and the Consequent Need for Public Policy." *International Social Science Journal* 42(4):523–538.

Goldenberg, Sheldon. 1987. *Thinking Sociologically.* Belmont, CA: Wadsworth.

Goode, Judith. 1987. "Gaminismo: The Changing Nature of the Street Child Phenomenon in Colombia." *UFSI Reports,* no. 28.

Gutis, Philip S. 1989a. "Family Redefines Itself, and Now the Law Follows." *New York Times* (May 28):B1.

———. 1989b. "What Makes a Family? Traditional Limits Are Challenged." *New York Times* (August 31):Y15+.

Harrison, Paul. 1987. *Inside the Third World: The Anatomy of Poverty,* 2d. ed. New York: Viking Penguin.

Johansson, S. Ryan. 1987. "Status Anxiety and Demographic Contraction of Privileged Populations." *Population and Development Review* 13(3):439–470.

Lewin, Tamar. 1990. "Strategies to Let Elderly Keep Some Control." *New York Times* (March 28):A1, A11.

Light, Ivan. 1983. *Cities in World Perspective.* New York: Macmillan.

Malthus, Thomas R. [1798] 1965. *First Essay on Population.* New York: Kelley.

Molano, Alfredo. 1993. Quoted in "Colombia's Vanishing Forests," *World Press Review* (June):43.

Nations, Marilyn K., and Mara Lucia Amaral. 1991. "Flesh, Blood, Souls, and Households: Cultural Validity in Morality Inquiry." *Medical Anthropology Quarterly* 5(3):204–220.

Nolty, Denise, ed. 1990. *Fodor's '90 Brazil: Including the Amazon and Bahia.* New York: Fodor's Travel Publications.

Olshansky, S. Jay, and A. Brian Ault. 1986. "The Fourth Stage of the Epidemiologic Transition: The Age of Delayed Degenerative Diseases." *The Milbank Quarterly* 64(3):355–391.

Omran, Abdel R. 1971. "The Epidemiologic Transition: A Theory of the Epidemiology of Population Change." *The Milbank Quarterly* 49(4):509–538.

Perlman, Janice. 1967. *The Myth of Marginality.* Berkeley: University of California Press.

Revkin, Andrew. 1990. *The Burning Season: The Murder of Chico Mendes and the Fight for the Amazon Rain Forest.* Boston: Houghton Mifflin.

Rock, Andrea. 1990. "Can You Afford Your Kids?" *Money* (July):88–99.

Rusinow, Dennison. 1986. "Mega-Cities Today and Tomorrow: Is the Cup Half Full or Half Empty?" *UFSI Reports,* no. 12.

Sanders, Thomas G. 1986. "The Politics of Agrarian Reform in Brazil." *UFSI Reports,* no. 32.

———. 1987a. "Brazilian Street Children. Pt. I: Who They Are." *UFSI Reports,* no. 17.

———. 1987b. "Brazilian Street Children. Pt. II: The Public and Political Response." *UFSI Reports,* no. 18.

———. 1988. "Happiness Also Rises Up There: The Favelas of Rio." *UFSI Reports,* no. 2.

Sayre, Robert F. 1983. "The Parents' Last Lessons." Pages 124–142 in *Life Studies: A Thematic Reader,* ed. D. Cavitch. New York: St. Martin's.

SEADE Foundation. 1994. *Survey of Living Conditions in the Metropolitan Area of Sao Paulo,* Research Series 101. Geneva: International Institute for Labour Studies.

Sehoen, R., and R. M. Weinck. 1993. "The Slowing Metabolism of Marriage: Figures from 1988 U.S. Marital Status Life Tables." *Demography* 30(4):737–746.

Semana. 1993. "Colombia's Vanishing Forests." *World Press Review* (June):43.

Simons, Marlise. 1988. "Man-Made Amazon Fires Tied to Global Warming." *New York Times* (August 12):Y1+.

Skidmore, Thomas E. 1993. "Bi-racial U.S.A. vs. Multi-racial Brazil: Is the Contrast Still Valid?" *Journal of Latin American Studies* 25:373–386.

Soldo, Beth J., and Emily M. Agree. 1988. "America's Elderly." *Population Bulletin* 43(3):5+.

Sorel, Nancy Caldwell. 1984. *Ever Since Eve: Personal Reflections on Childbirth.* New York: Oxford.

Statistical Abstract of Latin America. 1989. Vol. 27, ed. James W. Wilkie and Enrique Ochoa. Los Angeles: University of California Press.

Stockwell, Edward G., and H. Theodore Groat. 1984. *World Population: An Introduction to Demography.* New York: Watts.

Stockwell, Edward G., and Karen A. Laidlaw. 1981. *Third World Development: Problems and Prospects.* Chicago: Nelson-Hall.

Stone, Robyn, Gail Lee Cafferata, and Judith Sangl. 1987. "Caregivers of the Frail Elderly: A National Profile." *The Gerontologist* 27(5):616–626.

Stub, Holger R. 1982. *The Social Consequences of Long Life.* Springfield, IL: Thomas.

Targ, Dena B. 1989. "Feminist Family Sociology: Some Reflections." *Sociological Focus* 22(3):151–160.

Tremblay, Hélène. 1988. *Families of the World: Family Life at the Close of the Twentieth Century. Vol. 1: The Americas and the Caribbean.* New York: Farrar, Straus & Giroux.

United Nations. 1983. *World Population Trends and Policies: 1983 Monitoring Report,* vol. 1. New York: United Nations.

U.S. Bureau of the Census. 1947. *Statistical Abstract of the United States, 1947.* Washington, D.C.: U.S. Government Printing Office.

———. 1961. *Statistical Abstract of the United States, 1961.* Washington, D.C.: U.S. Government Printing Office.

———. 1991. *World Population Profile: 1991.* Washington, D.C.: U.S. Government Printing Office.

———. 1993. *Statistical Abstract of the United States, 1993.* Washington, D.C.: U.S. Government Printing Office.

U.S. Central Intelligence Agency. 1992. *World Factbook 1992.* Washington, D.C.: U.S. Government Printing Office.

———. 1995. *World Factbook 1995: Brazil.* http://www.odci.gov/cia/publications/95fact/br.html

U.S. Department of the Army. 1983. *Brazil, A Country Study,* 4th ed., ed. Richard F. Nyrop. Washington, D.C.: U.S. Government Printing Office.

U.S. Department of State. 1996. *Background Notes: Brazil, March 1995.* gopher://dosfan.lib.uic.edu/1D-%3A22525%ABackground%20Ser

van de Kaa, Dirk J. 1987. "Europe's Second Demographic Transition." *Population Bulletin* 42(1):1–59.

Wark, John T. 1995. "Raising Child to 22 Can Cost $265,000." *Detroit News* (October 19). http://detnews.com/menu/stories/20850.htm

Watkins, Susan C., and Jane Menken. 1985. "Famines in Historical Perspective." *Population and Development Review* 11(4):647–675.

The World Bank. 1990. *World Development Report, 1990.* New York: Oxford University Press.

———. 1994. *World Development Report, 1994: Infrastructure for Development.* New York: Oxford University Press.

CHAPTER 13

Alpert, Bracha. 1991. "Students' Resistance in the Classroom." *Anthropology and Education Quarterly* 22(4):350–366.

Bennett, William J. 1987. "Epilogue: Implications for American Education." Pages 69–71 in *Japanese Education Today.* Washington, D.C.: U.S. Government Printing Office.

Blakeslee, Sandra. 1989. "Crack's Toll Among Babies: A Joyless View of Even Toys." *New York Times* (September 17): Y1+.

Bloom, Benjamin S. 1981. *All Our Children Learning: A Primer for Parents, Teachers and Other Educators.* New York: McGraw-Hill.

Botstein, Leon. 1990. "Damaged Literacy: Illiteracies and American Democracy." *Daedalus* 119(2):55–84.

Boyer, Ernest. 1986. "Forum: How Not to Fix the Schools." *Harper's* (February): 39–51.

Celis, William III. 1992. "A Texas-Size Battle to Teach Rich and Poor Alike." *New York Times* (February 12):B6.

———. 1993a. "International Report Card Shows U.S. Schools Work." *New York Times* (December 9):A1+.

———. 1993b. "Study Finds Rising Concentration of Black and Hispanic Students." *New York Times* (December 14):A1+.

Chira, Susan. 1991. "Student Tests in Other Nations Offer U.S. Hints, Study Says." *New York Times* (May 20):A1+.

Cohen, David K., and Barbara Neufeld. 1981. "The Failure of High Schools and the Progress of Education." *Daedalus* (Summer):69–89.

Coleman, James S. 1960. "The Adolescent Subculture and Academic Achievement." *American Journal of Sociology* 65:337–347.

———. 1966. *Equality of Educational Opportunity.* Washington, D.C.: U.S. Government Printing Office.

———. 1977. "Choice in American Education." Pages 1–12 in *Parents, Teachers, and Children: Prospects for Choice in American Education.* San Francisco: Institute for Contemporary Studies.

Coleman, James S., John W. C. Johnstone, and Kurt Jonassohn. 1961. *The Adolescent Society.* New York: Free Press.

Dorfman, Andrea. 1989. "Alcohol's Youngest Victims." *Time* (August 28):60.

Dorris, Michael. 1989. *The Broken Cord.* New York: Harper & Row.

Durkheim, Emile. 1961. "On the Learning of Discipline." Pages 860–865 in *Theories of Society: Foundations of Modern Sociological Theory,* vol. 2, ed. T. Parsons, E. Shils, K. D. Naegele, and J. R. Pitts. New York: Free Press.

———. 1968. *Education and Sociology,* trans. S. D. Fox. New York: Free Press.

Elam, Stanley M. 1989. "The Second Gallup/Phi Delta Kappa Poll of Teachers' Attitudes Toward the Public Schools." *Phi Delta Kappan* (June): 785–798.

Foster, Jack D. 1991. "The Role of Accountability in Kentucky's Education Reform Act of 1990." *Education Leadership,* pp. 34–36.

Gardner, John W. 1984. *Excellence: Can We Be Equal and Excellent Too?* New York: Norton.

Guzzardi, Walter, Jr. 1976. "The Uncertain Passage from College to Job." *Fortune* (January):126–129, 168–172.

Hallinan, M. T. 1988. "Equality of Educational Opportunity." Pages 249–268 in *Annual Review of Sociology,* vol. 14, ed. W. R. Scott and J. Blake. Palo Alto, CA: Annual Reviews.

Henry, Jules. 1965. *Culture Against Man.* New York: Random House.

Hirsch, E. D., Jr. 1989. "The Primal Scene of Education." *The New York Review of Books* (March 2):29–35.

———. *Cultural Literacy: What Every American Needs to Know.* New York: Houghton Mifflin.

Hirsch, E. D., Jr., Joseph F. Kett, and James Trefil. 1993. *The Dictionary of Cultural Literacy.* New York: Houghton Mifflin.

Lynn, Richard. 1988. *Educational Achievement in Japan: Lessons for the West.* London: Macmillan.

Merton, Robert K. 1957. *Social Theory and Social Structure.* Glencoe, IL: Free Press.

Meyer, John W., and David P. Baker. 1996. "Forming American Educational Policy with International Data: Lessons from the Sociology of Education." *Sociology of Education* (Extra Issue):123–130.

Oakes, Jeannie. 1985. *Keeping Track: How Schools Structure Inequality.* Binghamton, NY: Vail-Ballou.

———. 1986a. "Keeping Track. Pt. 1: The Policy and Practice of Curriculum Inequality." *Phi Delta Kappan* 67(September):12–17.

———. 1986b. "Keeping Track. Pt. 2: Curriculum Inequality and School Reform." *Phi Delta Kappan* 67(October):148–154.

Online NewsHour. 1996a. "Ready for Work" (March 27). http://www1.pbs.org/newshour

———. 1996b. "Education Report Card" (November 21). http://www1.pbs.org/newshour

Organization for Economic Co-operation and Development and Statistics Canada. 1995. *Literacy, Economy, and Society: Results of the First International Adult Literacy Survey.* Paris: OECD.

Ouane, Adama. 1990. "National Languages and Mother Tongues." *UNESCO Courier* (July):27–29.

Phelan, Patricia, and Ann Locke Davidson. 1994. "Looking Across Borders: Students' Investigations of Family, Peer, and School Worlds as Cultural Therapy." Pages 35–59 in *Pathways to Cultural Awareness: Cultural Therapy with Teachers and Students,* ed. George and Louise Spindler. Thousand Oaks, CA: Corwin.

Phelan, Patricia, Ann Locke Davidson, and Hanh Cao Yu. 1993. "Students' Multiple Worlds: Navigating the Borders of Family, Peer, and School Cultures." Pages 89–107 in *Renegotiating Cultural Diversity in American Schools,* ed. Patricia Phelan and Ann Locke Davidson. New York: Teachers College Press.

Phelan, Patricia, Ann Locke Davidson, and Hanh Thanh Cao. 1991. "Students' Multiple Worlds: Negotiating the Boundaries of Family, Peer, and School Cultures." *Anthropology and Education Quarterly* 22(3):224–250.

Potter, J. Hasloch, and A. E. W. Sheard. 1918. *Catechizings for the Church and Sunday Schools,* 2d series. London: Skeffington.

Purves, Alan C. 1974. "Divergent Views on the Schools: Some Optimism Justified." *New York Times* (January 16):C74.

Ramirez, Francisco, and John W. Meyer. 1980. "Comparative Education: The Social Construction of the Modern World System." Pages 369–399 in *Annual Review of Sociology,* vol. 6, ed. A. Inkeles, N. J. Smelser, and R. H. Turner. Palo Alto, CA: Annual Reviews.

Resnick, Daniel P. 1990. "Historical Perspectives on Literacy and Schooling." *Daedalus* 119(2):15–32.

Richardson, Lynda. 1994. "More Schools Are Trying to Write Textbooks Out of the Curriculum." *New York Times* (January 31):A1+.

Rohlen, Thomas P. 1986. "Japanese Education: If They Can Do It, Should We?" *The American Scholar* 55:29–43.

Rosenthal, Robert, and Lenore Jacobson. 1968. *Pygmalion in the Classroom.* New York: Holt, Rinehart & Winston.

Sanchez, Claudio. 1993. Interview on National Public Radio, *Morning Edition* (December 8):11–13 (transcript). New York: WNET.

Shelley, Kristina J. 1992. "The Future of Jobs for College Graduates." *Monthly Labor Review* (July):13–21.

Sowell, Thomas. 1981. *Ethnic America: A History.* New York: Basic Books.

Stevenson, Harold. 1992. Interview on National Public Radio, *Morning Edition* (December 10):8–10 (transcript). New York: WNET.

Stevenson, Harold W., Shin-ying Lee, and James W. Stigler. 1986. "Mathematics Achievement of Chinese, Japanese, and American Children." *Science* 231: 693–699.

Third International Mathematics and Science Study. 1996. http://www.ed.gov/NCES/times/index.html

Thomas, William I., and Dorothy Swain Thomas. [1928] 1970. *The Child in America.* New York: Johnson Reprint.

Tyack, David. "Forming the National Character." *Harvard Educational Review* 36:29–41.

Tyack, David, and Elisabeth Hansot. 1981. "Conflict and Consensus in American Public Education." *Daedalus* (Summer):1–43.

U.S. Bureau of the Census. 1993. *Statistical Abstract of the United States,* 113th ed. Washington, D.C.: U.S. Government Printing Office.

U.S. Department of Education. 1987. *Japanese Education Today.* Washington, D.C.: U.S. Government Printing Office.

———. 1993a. *Adult Literacy in America: A First Look at the Results of the National Literacy Survey.* Washington, D.C.: U.S. Government Printing Office.

———. 1993b. *Occupational and Educational Outcomes of Recent College Graduates 1 Year After Graduation: 1991.* Washington, D.C.: U.S. Government Printing Office.

———. 1995. *Digest of Education Statistics 1995,* by Thomas D. Snyder and Charlene M. Hoffman. NCES 95-029. Washington, D.C.: U.S. Government Printing Office.

———. 1996. *The Condition of Education, 1996,* by Thomas Smith. NCES 96-304. Washington D.C.: U.S. Government Printing Office.

Wells, Amy Stuart, and Jeannie Oakes. 1996. "Potential Pitfalls of Systematic Reform: Early Lessons from Research on Detracking." *Sociology of Education* (Extra Issue).

CHAPTER 14

Abercrombie, Nicholas, and Bryan S. Turner. 1978. "The Dominant Ideology Thesis." *British Journal of Sociology* 29(2):149–170.

Alston, William P. 1972. "Religion." Pages 140–145 in *The Encyclopedia of Philosophy,* vol. 7, ed. P. Edwards. New York: Macmillan.

Amnesty International. 1996a. "Afghanistan: AI Appeals to International Community to Take the Initiative to Promote and Protect Human Rights" (January 16). http://www.oneworld.org/amnesty/ai_afghan_jan18.html

———. 1996b. "Afghanistan: International Community Should Act Now to Prevent Possible Bloodbath" (September 26). http://www.oneworld.org/amnesty/press/afghanistan_sept26.html

———. 1996c. "Afghanistan: Taleban Take Hundreds of Civilians Prisoner" (October 2). http://www.oneworld.org/amnesty/press/afghanistan_oct12.html

Aron, R. 1969. Quoted in *The Sociology of Max Weber* by Julien Freund. New York: Random House.

Berger, Peter L. 1967. *The Sacred Canopy: Elements of a Sociological Theory of Religion.* New York: Doubleday.

Burns, Nicholas. 1996. "U.S. Department of State Daily Press Briefing" (September 30) http://www.state.gov/www/briefs/index/%20 Briefings%3A9609 %20Press%20Briefings%3A9960930% 20Daily%20Briefing

Bush, George H. 1991. *State of the Union Address by the President of the United States* (January 29). Washington, D.C.: U.S. Government Printing Office.

Caplan, Lionel. 1987. "Introduction: Popular Conceptions of Fundamentalism." Pages 1–24 in Studies in *Religious Fundamentalism,* ed. L. Caplan. Albany: State University of New York Press.

Carter Center. 1996. "State of World Conflict Report: Afghanistan, 1994–1995." http://www.emory.edu/CARTER_CENTER/PUBS/SWCR9495/afghan.htm#summary

Christianity Today. 1986. "Letters" (October 17):6.

Coles, Robert. 1990. *The Spiritual Life of Children.* Boston: Houghton Mifflin.

Durkheim, Emile. [1915] 1964. *The Elementary Forms of the Religious Life,* 5th ed., trans. J. W. Swain. New York: Macmillan.

———. 1951. *Suicide: A Study in Sociology,* trans. J. A. Spaulding and G. Simpson. New York: Free Press.

Ebersole, Luke. 1967. "Sacred." Page 613 in *A Dictionary of the Social Sciences,* ed. J. Gould and W. L. Kolb. New York: UNESCO.

Echo-Hawk, Walter. 1979. "A Prepared Statement from Walter Echo-Hawk." Pages 280–287 in *Religious Discrimination: A Neglected Issue by U.S. Commission on Civil Rights.* Washington, D.C.: U.S. Government Printing Office.

Esposito, John L. 1986. "Islam in the Politics of the Middle East." *Current History* (February):53–57, 81.

———. 1992. *The Islamic Threat: Myth or Reality?* New York: Oxford University Press.

Forbes, James. 1990. "Up from Invisibility: Review of *The Black Church in the African American Experience* by C. Eric Lincoln and Lawrence H. Mamiya." *The New York Times Book Review* (December 23):1–2.

Gallup, George, Jr., and Jim Castelli. 1989. *The People's Religion: American Faith in the '90s.* New York: Macmillan.

Haddad, Yvonne. 1991. Interview with Bill Moyers on "Images of God in the Arab World," Public Broadcast Service (transcript). Boston: WGBH Educational Foundation.

Hammond, Phillip E. 1976. "The Sociology of American Civil Religion: A Bibliographic Essay." *Sociological Analysis* 37(2):169–182.

Hourani, Albert. 1991. *A History of the Arab Peoples.* Cambridge, MA: Belknap.

Ibrahim, Youssef M. 1991. "In Kuwait, Ramadan Has a Bitter Taste." *New York Times* (March 19):A1, A7.

Johnson, Lyndon B. 1987. Quoted in *America's History Since 1865* by James A. Henretta, W. Elliott Brownlee, David Brody, and Susan Ware. Chicago: Dorsey.

Kurian, George Thomas. 1992. *Encyclopedia of the Third World.* New York: Facts on File.

Lechner, Frank J. 1989. "Fundamentalism Revisited." *Society* (January/February): 51–59.

Lincoln, C. Eric, and Lawrence H. Mamiya. 1990. *The Black Church in the African American Experience.* Durham, NC: Duke University Press.

Mahjubah: The Magazine for Moslem Women. 1984 (July).

McNamara, Robert S. 1989. *Out of the Cold: New Thinking for American Foreign and Defense Policy in the 21st Century.* New York: Simon & Schuster.

Mead, George Herbert. 1940. *Mind, Self and Society,* 3d ed. Chicago: University of Chicago Press.

National Public Radio. 1984a. "Black Islam." *The World of Islam* (tape). Washington, D.C.: NPR.

———. 1984b. "Decay or Rebirth: The Plight of Islamic Art." *The World of Islam* (tape). Washington, D.C.: NPR.

———. 1984c. "Voices of Resurgence." *The World of Islam* (tape). Washington, D.C.: NPR.

The New York Public Library Desk Reference. 1989. New York: Simon & Schuster.

Nottingham, Elizabeth K. 1971. *Religion: A Sociological View.* New York: Random House.

Online NewsHour. 1996. "Veiled in Fear" (October 9). . . . pbs.org/newshour/bb/asia/july-dec96/afghan_background_10-9.html

Pickering, W. S. F. 1984. *Durkheim's Sociology of Religion.* London: Routledge & Kegan Paul.

Robertson, Roland. 1987. "Economics and Religion." Pages 1–11 in *The Encyclopedia of Religion.* New York: Macmillan.

Rubin, Barnett R. 1996. Quoted in *A Nation in Arms* by Karl E. Meyer. *The New York Times Book Review* (August 11):21.

Save the Children Fund. 1996. "Restrictions on Women Lead to Withdrawal from Western Afghanistan." http://www.oneworld.org/scf/press_mar7.html

Smart, Ninian. 1976. *The Religious Experience of Mankind.* New York: Scribner's.

Spokesperson for the Iraqi government. 1991. Quoted in "Iraqi Message: 'Duty' Fulfilled." *New York Times* (February 26):Y1.

Stark, Rodney, and William S. Bainbridge. 1985. *The Future of Religion: Secularization, Revival and Cult Formation.* Berkeley: University of California Press.

Stavenhagen, Rodolfo. 1991. "Ethnic Conflicts and Their Impact on International Society." *International Social Science Journal* (February):117–132.

Turner, Bryan S. 1974. *Weber and Islam: A Critical Study.* Boston: Routledge & Kegan Paul.

Turner, Jonathan H. 1978. *Sociology: Studying the Human System.* Santa Monica, CA: Goodyear.

United Nations. 1996. *General Assembly Fifty-First Session Agenda Item 21* (December 3). gopher://gopher.un.org/00/ga/docs/51/plenary/A51-704.EN

U.S. Central Intelligence Agency. 1995. *World Factbook 1995.* http://www.odci.gov/cia/publications/95fact/af.html

U.S. Commission on Civil Rights. 1979. *Religious Discrimination: A Neglected Issue.* Washington, D.C.: U.S. Government Printing Office.

Van Doren, Charles L. 1991. *A History of Knowledge: Past, Present, and Future.* New York: Carol.

Watchtower Bible and Tract Society. 1987. *Life in a Peaceful New World.* Brooklyn: Watchtower.

Weber, Max. 1922. *The Sociology of Religion,* trans. E. Fischoff. Boston: Beacon.

———. 1958. *The Protestant Ethic and the Spirit of Capitalism,* 5th ed., trans. T. Parsons. New York: Scribner's.

White House press release. 1995. "Remarks by the President on Religious Liberty in America" (July 12). http://www.whitehouse.gov/WH/EOP/OP/html/book3-plain.html

Wilmore, Gayraud S. 1972. *Black Religion and Black Radicalism.* Garden City, NY: Doubleday.

The World Almanac and Book of Facts 1991. 1990. New York: Pharos.

The World Treasury of Modern Religious Thought. 1990. Ed. Jaroslav Pelikan and Clifton Fadiman. Boston: Little, Brown.

Yinger, J. Milton. 1971. *The Scientific Study of Religion.* New York: Macmillan.

Zangwill, O. L. 1987. "Isolation Experiments." Pages 393–394 in *The Oxford Companion to the Mind,* ed. R. L. Gregory. New York: Oxford University Press.

Zickel, Raymond E. 1991. *Soviet Union: A Country Study,* 2d ed. Washington, D.C.: U.S. Government Printing Office.

CHAPTER 15

Afonso, Carlos A. 1997. "The Internet and Social Strategies." *Corporate Watch.* http://www.corpwatch.org/trac/feature/feature1/Afonso.html

Ash, Timothy Garton. 1989. *The Uses of Adversity: Essays on the Fate of Central Europe.* New York: Random House.

Associated Press. 1996. "Third World Pioneers Use Net to Promote Business." *USA Today* (December 30). http://usatoday.com/80+USATODAY_ONLINE+ USATODAY_ONLINE+NEWS+NEWS++internet

Association for Progressive Communication. 1997. "What Makes the APC Unique?" http://www.apc.org/unique.html

Barnet, Richard J. 1990. "Reflections: Defining the Moment." *The New Yorker* (July 16):45–60.

Bauerlein, Monika. 1992. "Plutonium Is Forever: Is There Any Sane Place to Put Nuclear Waste?" *Utne Reader* (July/August):34–37.

Bear, John, and David M. Pozerycki. 1992. *Computer Wimp No More: The Intelligent Beginner's Guide to Computers.* Berkeley: Ten Speed Press.

Bell Telephone System. 1945. "Advertisement: Millions of Military Telephones." *Life* (August 10):3.

Berners-Lee, Wright Tim. 1996. Quoted in "Seek and You Shall Find (Maybe)" by Steve G. Steinberg. *Wired* (May):111.

Biegel, Stuart. 1996. "Does Anyone Control the Internet?" *The UCLA Online Institute for Cyberspace Law and Policy.* http://www.gse.ucla.edu/iclp/control.html

Boudon, Raymond, and François Bourricaud. 1989. *A Critical Dictionary of Sociology,* selected and trans. P. Hamilton. Chicago: University of Chicago Press.

Brecher, Jeremy, John Brown Childs, and Jill Cutler. 1993. *Global Visions: Beyond the New World Order.* Boston: South End Press.

Burdick, Alan. 1992. "The Last Cold-War Monument: Designing the 'Keep Out' Sign for a Nuclear Waste Site." *Harper's* (August):62–67.

Carver, Terrell. 1987. *A Marx Dictionary.* Totowa, NJ: Barnes & Noble.

Colihan, Jane, and Robert J. T. Joy. 1984. "Military Medicine." *American Heritage* (October/November):65.

Coser, Lewis A. 1973. "Social Conflict and the Theory of Social Change." Pages 114–122 in *Social Change: Sources, Patterns, and Consequences,* ed. E. Etzioni-Halevy and A. Etzioni. New York: Basic Books.

Crawford, Jack. 1995. "Renaissance Two: Second Coming of the Printing Press?" http://www.lincoln.ac.nz/reg/futures/renaiss2.htm

Currie, Elliott, and Jerome H. Skolnick. 1988. *America's Problems: Social Issues and Public Policy,* 2d ed. Boston: Little, Brown.

Dahrendorf, Ralf. 1973. "Toward a Theory of Social Conflict." Pages 100–113 in *Social Change: Sources, Patterns, and Consequences,* 2d ed., ed. E. Etzioni-Halevy and A. Etzioni. New York: Basic Books.

Galbraith, John K. 1958. *The Affluent Society.* Boston: Houghton Mifflin.

Guttman, Robert J. 1990. "Interview with John Georgas: The Coca-Cola Company." *Europe* (July/August):16.

Hacker, Andrew. 1971. "Power to Do What?" Pages 134–146 in *The New Sociology: Essays in Social Science and Social Theory in Honor of C. Mills,* ed. I. L. Horowitz. New York: Oxford University Press.

Halberstam, David. 1986. *The Reckoning.* New York: Morrow.

Hauben, Michael, and Ronda Hauben. 1996. *Proposed Declaration of the Rights of Netizens.* http://www.columbia.edu/~rh120/netizen-rights.txt

Holusha, John. 1989. "Eastern Europe: Its Lure and Hurdles." *New York Times* (December 18):Y25+.

Kantor, Andrew, and Michael Newbarth. 1996. "Off the Charts: The Internet 1996." *Internet World* (December): 46–57.

Katz, Yvonne, and Gay Chedester. 1992. "Redefining Success: Public Education in the 21st Century." *Catalyst.* gopher://borg.lib.vt.edu/00/catalyst/v22n3/katz.v22n3

Kuhn, Thomas S. 1975. *The Structure of Scientific Revolutions.* Chicago: University of Chicago Press.

Lambert, Michael. 1992. "Defense Contractors Use Down-Sizing, Not Diversification, to Maintain Profits." *Aviation Week and Space Technology* (March 9): 61–63.

Mandelbaum, Maurice. 1977. *The Anatomy of Historical Knowledge.* Baltimore: Johns Hopkins University Press.

Mander, Jerry. 1997. "The Net Loss of the Computer Revolution." *Corporate Watch.* http://www.corpwatch.org/trac/feature/feature1/mander.htm

Martel, Leon. 1986. *Mastering Change: The Key to Business Success.* New York: Simon & Schuster.

Marx, Karl. [1881] 1965. "The Class Struggle." Pages 529–535 in *Theories of Society,* ed. T. Parsons, E. Shils, K. D. Naegele, and J. R. Pitts. New York: Free Press.

Mills, C. Wright. 1959. *The Sociological Imagination.* New York: Oxford University Press.

———. 1963. "The Structure of Power in American Society." Pages 23–38 in *Power, Politics and People: The Collected Essays of C. Wright Mills,* ed. I. L. Horowitz. New York: Oxford University Press.

———. 1973. "The Sources of Societal Power." Pages 123–130 in *Social Change: Sources, Patterns, and Consequences,* 2d ed., ed. E. Etzioni-Halevy and A. Etzioni. New York: Basic Books.

Nader, Ralph, James Love, and Andrew Saindon. 1995. "Project Censored's Top Ten Censored Stories of 1995." *Consumer Project on Technology* (July 14). http://censored.sonoma.edu/ProjectCensored/Stories1995.html

Noack, David. 1997. "The Origin of PCs." *Internet World* (February):49–54.

Norman, Donald A. 1988. Quoted in "Management's High-Tech Challenge." *Editorial Research Report* (September 30):482–491.

Ogburn, William F. 1968. "Cultural Lag as Theory." Pages 86–95 in *Culture and Social Change,* 2d ed., ed. O. D. Duncan. Chicago: University of Chicago Press.

Omran, Abdel R., and Farzaneh Roudi. 1993. "The Middle East Population Puzzle." *Population Bulletin* 48(1): 1–40.

Oppenheimer, Robert. 1986. Quoted in *The Making of the Bomb* by Richard Rhodes. New York: Touchstone.

O'Sullivan, Anthony. 1990. "Eastern Europe." *Europe* (September):21–22.

Public Citizen. 1996. http://www.citizen.org/

Pulver, Jeff. 1997. "Unwired." *Internet World* (February):100–102.

Rabi, Isidor I. 1969. "The Revolution in Science." Pages 28–66 in *The Environment of Change,* ed. A. W. Warner, D. Morse, and T. E. Cooney. New York: Columbia University Press.

Rawe, Dick. 1991. "P&G Expands Entry in Eastern Europe." *Cincinnati Post* (June 20):10B.

Reich, Jens. 1989. Quoted in "People of the Year." *Newsweek* (December 25): 18–25.

Schwartz, Bruce. 1996. "Weigh Pros and Cons Before Beefing Up the PC." *USA Today* (November 18):15E.

Sloyan, Patrick J. 1994. "Flak for Defense Merger." *Top Censored News Stories of 1994.* http://censored.sonoma.edu/ProjectCensored/Stories1994.html

Stevenson, Richard W. 1991. "Northrop Settles Workers' Suit on False Missile Tests for $8 Million." *New York Times* (June 25):A7.

United Nations Development Programme. 1993. *Human Development Report.* New York: Oxford University Press.

Van Evera, Stephen. 1990. "The Case Against Intervention." *The Atlantic Monthly* (July):72–80.

Wald, Matthew L. 1994. "A Military-Industrial Alliance Turns Plowshares to Swords." *New York Times* (March 16): A1+.

Wallerstein, Immanuel. 1984. *The Politics of the World-Economy: The States, the Movements and the Civilizations.* New York: Cambridge University Press.

Weber, Max. 1947. *The Theory of Social and Economic Organization,* trans. A. M. Henderson and T. Parsons. Glencoe, IL: Free Press.

White, Leslie A. 1949. *The Science of Culture: A Study of Man and Civilization.* New York: Grove.

White House. 1996. "The Internet." http://www.whitehouse.gov/WH/EOP/OVP/24hours/internet.html

Index